GaAs
DEVICES AND CIRCUITS

MICRODEVICES
Physics and Fabrication Technologies

Series Editors: Ivor Brodie and Julius J. Muray
SRI International
Menlo Park, California

GaAs DEVICES AND CIRCUITS
Michael Shur

SEMICONDUCTOR LITHOGRAPHY
Principles, Practices, and Materials
Wayne M. Moreau

GaAs
DEVICES AND CIRCUITS

Michael Shur
University of Minnesota
Minneapolis, Minnesota

PLENUM PRESS • NEW YORK AND LONDON

Library of Congress Cataloging in Publication Data

Shur, Michael.
  GaAs devices and circuits.

  (Microdevices: physics and fabrication technologies)
  Includes bibliographical references and index.
  1. Gallium arsenide semiconductors. I. Title. II. Series: Microdevices.
TK7871.15.G3S55   1986   621.3815′2   86-25323
ISBN 0-306-42192-5

© 1987 Plenum Press, New York
A Division of Plenum Publishing Corporation
233 Spring Street, New York, N.Y. 10013

Printed in the United States of America

To the memory of my father,
Saul Shur

Preface

GaAs devices and integrated circuits have emerged as leading contenders for ultra-high-speed applications. This book is intended to be a reference for a rapidly growing GaAs community of researchers and graduate students. It was written over several years and parts of it were used for courses on GaAs devices and integrated circuits and on heterojunction GaAs devices developed and taught at the University of Minnesota.

Many people helped me in writing this book. I would like to express my deep gratitude to Professor Lester Eastman of Cornell University, whose ideas and thoughts inspired me and helped to determine the direction of my research work for many years. I also benefited from numerous discussions with his students and associates and from the very atmosphere of the pursuit of excellence which exists in his group.

I would like to thank my former and present co-workers and colleagues—Drs. Levinstein and Gelmont of the A. F. Ioffe Institute of Physics and Technology, Professor Melvin Shaw of Wayne State University, Dr. Kastalsky of Bell Communications, Professor Gary Robinson of Colorado State University, Professor Tony Valois, and Dr. Tim Drummond of Sandia Labs—for their contributions to our joint research and for valuable discussions. My special thanks to Professor Morkoç, for his help, his ideas, and the example set by his pioneering work.

Since 1978 I have been working with engineers from Honeywell, Inc.—Drs. Nick Cirillo, Max Helix, Steve Jamison, Andy Peczalski, T. C. Lee, Chente Chao, Jon Abrokwah, Dave Arch, Obert Tufte, Bob Daniels, Peter Roberts, David Lamb, and Don Long, Mr. Tho Vu, and others—and have spent with them countless hours discussing GaAs-related issues and working on GaAs devices and integrated circuits. I would like to thank them for their help and for creating an environment of intellectual challenge.

Over the years I have presented my work at different institutions. I would like to thank Drs. Hans Rupprecht, Paul Solomon, Norm Braslau, Marshal Nathan, and Bob Rosenberg of IBM, Dr. Dick Eden of Gigabit Logic, Dr. Zucca of Rockwell International, Dr. Zuleeg of McDonnell Douglas, Dr. Kim of Ford Microelectronics, Dr. Jim Oakes of Westinghouse, Drs. S. S. Pei and S. Luryi of AT&T Bell Labs, and Dr. Conwell of Xerox for suggestions and comments.

This book would not have been possible without my former and present graduate students—Drs. Kwyro Lee, Tzu-Hung Chen, Kang Lee, Chong Hyun, Chung-Hsu Chen, and many others. I would also like to express my thanks to the students who took my courses on GaAs devices and circuits at the University of Minnesota, for their enthusiasm and encouragement, and to Messrs. Pailu Wang, Jingming Xu, Jun-ho Baek, Phil Jenkins, and Young Byun for reading parts of the manuscript.

I would like to thank my colleagues—faculty members at the Department of Electrical Engineering of the University of Minnesota.

I am thankful to the Microelectronics and Information Sciences Center, and the Supercomputer Institute at the University of Minnesota, for their continuous support of our research.

My wife, Paulina Shur, provided me with indispensable support over the years, helping me enormously with this project in many different ways. And I would like to finish this Preface with my wife's favorite quote from Thomas Mann:

> Again and further are the right words, for the unresearchable plays a kind of mocking game with our research ardours; it offers apparent holds and goals, behind which, when we have gained them, new reaches of the past still open out—as happens to the coastwise voyager, who finds no end to his journey, for behind each headland of clayey dune he conquers, fresh headlands and new distances lure him on.

Michael Shur

Minneapolis, Minnesota

Contents

CHAPTER 10. Modulation Doped Field Effect Transistors

CHAPTER 11. Novel GaAs Devices

1

Chemical Bonds and Crystal Structure

1-1. ATOMIC STATES

The energy levels and wave functions for an electron in a hydrogen atom are found from the solution of the time-independent Schrödinger equation:

$$\left[-\frac{\hbar^2\nabla^2}{2m_e} + U(r) \right]\psi = E\psi \tag{1-1-1}$$

where $\hbar$ is the reduced Planck constant, ψ is the wave function, m_e is the electron mass, U is the potential energy,

$$U = -q^2/4\pi\varepsilon_0 r \tag{1-1-2}$$

r is the distance from the nucleus, q is the electronic charge, and ε_0 is the permittivity of vacuum.

The time dependence of the wave functions for the stationary states is given by $\exp(-iEt/\hbar)$, where E is the energy and t is time.

The solutions of Eq.(1-1-1), which approach zero at large r, are found to exist only when the energy E is given by

$$E_n = -E_B/n^2 \tag{1-1-3}$$

where

$$n = 1, 2, 3, 4, \ldots \tag{1-1-4}$$

is the principal quantum number,

$$E_B = q^2/8\pi\varepsilon_0 a_B \tag{1-1-5}$$

is the Bohr energy, and

$$a_B = \frac{4\pi\varepsilon_0\hbar^2}{m_e q^2} \tag{1-1-6}$$

is the Bohr radius.

The angular dependence of ψ is determined by the orbital quantum number l and the magnetic quantum number m:

$$l = 0, 1, 2, \ldots, n - 1 \tag{1-1-7}$$

$$m = -l, -l + 1, \ldots, l - 1, l \tag{1-1-8}$$

The simplest possible approximation in the treatment of many-electron atoms is to assume that the wave functions are the same as those for the hydrogen atom, with the modification that the nuclear charge is not q but Zq, where Z is the atomic number equal to the number of protons in the nucleus. According to the Pauli exclusion principle, no two electrons can be in states in which all four quantum numbers (including spin) are the same. Hence the electronic structure of complex atoms may be understood in terms of filling up higher and higher energy levels, with the assumption that the chemical properties are determined mainly by the outermost (valence) electrons.

The average distance of an electron from the nucleus depends on n. The innermost "shell" of electrons consists of those in the $n = 1$ states, the next shell corresponds to the $n = 2$ states, etc. These shells are sometimes identified by capital letters starting with K:

$$n = 1, 2, 3, 4, \ldots$$

$$\text{shell} = K, L, M, N, \ldots$$

The energy levels in many-electron atoms depend not only on n but also on l (owing to the screening of the nucleus potential by the electrons). Therefore, electrons are subdivided into "subshells" according to the value of l. These subshells are often labeled by lower case letters:

$$l = 0, 1, 2, 3, 4, 5, \ldots$$

$$\text{subshell} = s, p, d, f, g, h, \ldots$$

For a given l there are $2l + 1$ values of m, each corresponding to a distinct state. For each of these there are two possible values of the spin quantum number. The resulting number of levels in different subshells is given in Table 1-1-1.

TABLE 1-1-1. Shells and Subshells

Shell	n	l	m	Spectroscopic notation	Number of levels
K	1	0	0	$1s$	2
L	2	0	0	$2s$	2
	2	1	0, ± 1	$2p$	6
M	3	0	0	$3s$	2
	3	1	0, ± 1	$3p$	6
	3	2	0, ± 1, ± 2	$3d$	10
N	4	0	0	$4s$	2
	4	1	0, ± 1	$4p$	6
	4	2	0, ± 1, ± 2	$4d$	10
	4	3	0, ± 1, ± 2, ± 3	$4f$	14

1-2. CHEMICAL BONDS

Bonds in a crystal can be presented as a linear combination of atomic wave functions (atomic orbitals). The atomic orbitals for the valence electrons are determined by the valence configurations. For Si, Ge, Ga, and As the valence configurations are given by

$$
\begin{array}{ll}
\text{Si} & \text{core} + 3s^2 3p^2 \\
\text{Ge} & \text{core} + 4s^2 4p^2 \\
\text{Ga} & \text{core} + 4s^2 4p^1 \\
\text{As} & \text{core} + 4s^2 4p^3
\end{array}
$$

where superscripts denote the number of electrons in the subshells.

However, the atomic orbitals that form the bonding orbitals are not the same as in the ground state of the atoms. In semiconductors with four valence electrons per atom each atom is tetrahedrally coordinated (see Fig. 1-2-1) in order to share eight electrons (two electrons per bond) corresponding to a complete shell. Such a configuration can be formed using hybridized orbitals ψ_i for each atom:

$$\psi_1 = \tfrac{1}{2}(|s\rangle + |P_x\rangle + |P_y\rangle + |P_z\rangle) \tag{1-2-1}$$

$$\psi_2 = \tfrac{1}{2}(|s\rangle + |P_x\rangle - |P_y\rangle - |P_z\rangle) \tag{1-2-2}$$

$$\psi_3 = \tfrac{1}{2}(|s\rangle - |P_x\rangle + |P_y\rangle + |P_z\rangle) \tag{1-2-3}$$

$$\psi_4 = \tfrac{1}{2}(|s\rangle - |P_x\rangle - |P_y\rangle + |P_z\rangle) \tag{1-2-4}$$

Here $|s\rangle$, $|P_x\rangle$, $|P_y\rangle$, and $|P_z\rangle$ are the atomic orbitals. Orbital $s\rangle$ is isotropic and the angular distributions of $|P_x\rangle$, $|P_y\rangle$, and $|P_z\rangle$ in the spherical coordinate system are given by

$$|P_x\rangle = (3/4\pi)^{1/2} \sin \theta \cos \phi \tag{1-2-5}$$

$$|P_y\rangle = (3/4\pi)^{1/2} \sin \theta \sin \phi \tag{1-2-6}$$

$$|P_z\rangle = (3/4\pi)^{1/2} \cos \theta \tag{1-2-7}$$

Using Eqs. (1-2-1)–(1-2-7) one can check that the hybridized orbitals ψ_1, ψ_2, ψ_3, and ψ_4 have maximum values in the directions shown in Fig. 1-2-2. Comparison

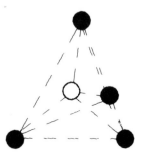

FIGURE 1-2-1. Tetrahedral atomic configuration.

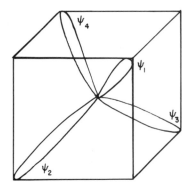

FIGURE 1-2-2. Hybridized orbitals ψ_1, ψ_2, ψ_3, and ψ_4.

between Fig. 1-2-1 and Fig. 1-2-2 clearly illustrates the significance of the hybridized orbitals in the bond formation.

Hybridized orbitals do not correspond to the ground state of the atom (the atomic orbitals do). It usually requires on the order of 5–10 eV to promote an electron into the hybridized state. This energy is recovered in the crystal due to the interaction between the atoms. The bonding energy level is lower than the ground atomic state by an amount of the order of 1 eV/valence electron (cohesion energy).

In a solid, hybridized (or directed) orbitals are combined into bonding, anti-bonding, and nonbonding orbitals forming covalent bonds.

A bonding orbital ψ_b consists of two directed orbitals associated with the nearest-neighbor atoms and combined in phase in such a way that ψ_b is large in the bonding region between the atoms, as shown in Fig. 1-2-3 [1].

An antibonding orbital ψ_a is similar to the bonding orbital except that the phase between the directed orbitals is reversed and ψ_a has a node in the bonding region.

A nonbonding orbital is centered on one atom and does not have a directional character.

The bonding state has the lowest energy due to the overlap of the Coulomb potential between the nearest ion cores (see Fig. 1-2-4). The antibonding state has the largest energy and the nonbonding state is in the middle. Usually the bonding and nonbonding states are occupied and the antibonding states are empty.

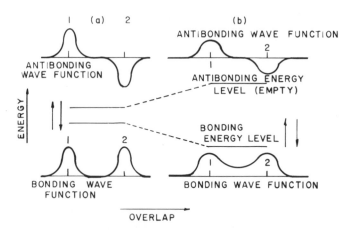

FIGURE 1-2-3. Formation of bonding and antibonding states [1]. (a) Atom; (b) crystal.

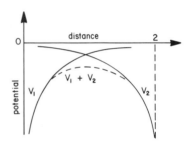

FIGURE 1-2-4. Overlap of the ion core potential between the
nearest neighbors.

If all atoms in a crystal are identical (as in Si or Ge, for example) the bonding
orbital between the nearest neighbors 1 and 2 has the following form:

$$\psi_b \sim \psi_1 + \psi_2 \qquad (1\text{-}2\text{-}8)$$

This corresponds to a purely covalent bond. In a compound semiductor such as
GaAs, the bonding orbital includes the weighting factor λ:

$$\psi_b \sim \psi_1 + \lambda\psi_2 \qquad (1\text{-}2\text{-}9)$$

which may be interpreted in terms of fractions of time f_1 and f_2 the bonding electron
spends on the first and second atom, respectively:

$$f_1 = \frac{1}{1+\lambda^2}; \qquad f_2 = \frac{\lambda^2}{1+\lambda^2} \qquad (1\text{-}2\text{-}10)$$

Electrons in the bonding states spend a greater fraction of time on the anions;
electrons in the antibonding states spend a greater fraction of time on the cations.
    In covalent crystals the difference in the energies of the bonding and antibonding
states depends on the bond length. In ionic compounds there is also a contribution
which depends on the difference of the atomic potentials. According to Phillips [1]:

$$E_{ba}^2 = E_h^2 + C^2 \qquad (1\text{-}2\text{-}11)$$

Here E_{ba} is the energy gap between the bonding and antibonding state, E_h is the
energy gap produced by the covalent part of the potential, and C is the magnitude
of the energy gap produced by the ionic potential. The fractional ionicity and
covalency of the bond are then defined as

$$f_i = C^2/E_{ba}^2 \qquad (1\text{-}2\text{-}12)$$

$$f_c = E_h^2/E_{ba}^2 \qquad (1\text{-}2\text{-}13)$$

Values of E_h, C, f_i, and E_{ba} for binary tetrahedrally coordinated crystals are given
in Table 1-2-1 [1].
    Many properties of semiconductors such as their crystalline structure, cohesive
energy, elastic constants, vibrational spectra, etc. are strongly dependent on the
degree of the ionicity of the bonds. As can be seen from Fig. 1-2-5 [1], heteropolar
compounds are crystallized in the rock salt structure and more covalent crystals
have diamond, zinc-blende, or wurtzite structure.

TABLE 1-2-1. Energies E_h, C, E_{ba} and Ionicity f_i
in Binary Tetrahedrally Coordinated Crystals[a]

Crystal	E_h (eV)	C (eV)	E_{ba} (eV)	f_i
C	13.5	0	13.5	0
Si	4.77	0	4.77	0
Ge	4.31	0	4.31	0
Sn	3.06	0	3.06	0
BAs	6.55	0.38	6.56	0.002
BP	7.44	0.68	7.47	0.006
BeTe	4.54	2.05	4.98	0.169
SiC	8.27	3.85	9.12	0.177
Alsb	3.53	2.07	4.14	0.250
BN	13.1	7.71	15.2	0.256
GaSb	3.55	2.10	4.12	0.261
BeSe	5.65	3.36	6.57	0.261
AlAs	4.38	2.67	5.14	0.274
BeS	6.31	3.99	7.47	0.286
AlP	4.72	3.14	5.67	0.307
GaAs	4.32	2.90	5.20	0.310
InSb	3.08	2.10	3.73	0.321
GaP	4.73	3.30	5.75	0.327
InAs	3.67	2.74	4.58	0.357
InP	3.93	3.34	5.16	0.421
AlN	8.17	7.30	11.0	0.449
GaN	7.64	7.64	10.8	0.500
MgTe	3.20	3.58	4.80	0.554
InN	5.93	6.78	8.99	0.578
BeO	11.5	13.9	18.0	0.602
ZnTe	3.59	4.48	5.74	0.609
ZnO	7.33	9.30	11.8	0.616
ZnS	4.82	6.20	7.85	0.623
ZnSe	4.29	5.60	7.05	0.630
HgTe	2.92	4.0	5.0	0.65
HgSe	3.43	5.0	6.1	0.68
CdS	3.97	5.90	7.11	0.685
CuI	3.66	5.50	6.61	0.692
CdSe	3.61	5.50	6.58	0.699
CdTe	3.08	4.90	5.79	0.717
CuBr	4.14	6.90	8.05	0.735
CuCl	4.83	8.30	9.60	0.746
CuF	8.73	15.8	18.1	0.766
AgI	3.09	5.70	6.48	0.770
MgS	3.71	7.10	8.01	0.786
MgSe	3.31	6.41	7.22	0.790
HgS	3.76	7.3	8.3	0.79

[a] Reference 1.

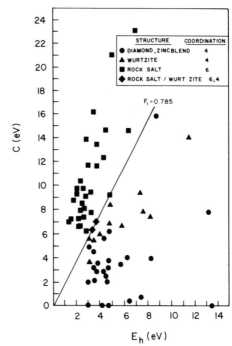

FIGURE 1-2-5. Values of E and C for A_3B_5 crystals
[1]. Note that the line $f = 0.785$ separates all fourfold from all sixfold coordinated crystals.

Another measure of the bond ionicity is the effective charge which characterizes
the interaction of the lattice transverse optic mode with the transverse electromagnetic field (see Table 1-2-2 [1]).

1-3. CRYSTAL STRUCTURE

Most A_3B_5 compounds including GaAs crystallize in the zinc blende structure.
The primitive cell of the structure contains two atoms, A and B, which are repeated
in space, with each species forming a face-centered cubic lattice. It can be described

TABLE 1-2-2. Values of the Effective Charge for
Tetrahedrally Coordinated Crystals[a]

Crystal	$|e^*|$	Crystal	$|e^*|$
BN	0.55	GaN	0.42
SiC	0.41	AlN	0.40
BP	0.10	BeO	0.62
AlP	0.28	ZnO	0.53
AlSb	0.19	ZnS	0.41
GaP	0.24	ZnSe	0.34
GaAs	0.20	ZnTe	0.27
GaSb	0.15	CdS	0.40
InP	0.27	CdSe	0.41
InAs	0.22	CdTe	0.34
InSb	0.21	CuCl	0.27

[a] Reference 1.

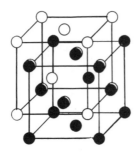

FIGURE 1-3-1. GaAs structure. ○, Ga; ●, As.

as mutually penetrating fcc lattices of Ga and As (in the case of GaAs) shifted relative to each other by a quarter of the body diagonal (see Fig. 1-3-1).

If atoms A and B are identical this crystal structure becomes a diamond structure which has an inversion symmetry.

The space group of the zinc blende structure is T^2d. It is a symmetry group with three twofold axes, eight threefold axes, four fourfold axes, and six symmetry planes. It lacks the inversion symmetry.

Directions in a crystal are specified by a set of three integers u, v, and w defining a vector $u\hat{a} + v\hat{b} + w\hat{c}$, which points along the given direction. Here $\hat{a}$, $\hat{b}$, and $\hat{c}$ are primitive base vectors. The integers u, v, and w are enclosed in square brackets and have no common integral divisor (Miller notation). Examples are [100], [111], [110], etc. (see Fig. 1-3-2 [3]). A set of all equivalent directions is denoted as $\langle uvw \rangle$. Symbol $\langle 110 \rangle$ for a cubic lattice, for instance, stands for all 12 directions along the face diagonals. A negative integer is represented by placing a bar above the integer. A symbol (uvw) denotes a crystal plane. A set of equivalent planes is denoted as

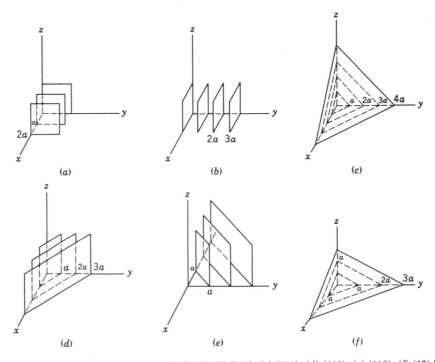

FIGURE 1-3-2. Important cubic planes: (a) (100); (b) (010); (c) (111); (d) (110); (e) (110); (f) (12) [3].

FIGURE 1-3-3. Schematic drawing of the orientations of the FETs with respect to the GaAs substrate. A in [011] direction, B in [011] direction, C [010] direction, and D in [001] direction. The shapes of the etched grooves are in two different [011] directions as shown by the dashed lines. The difference in the etched pattern enables us to identify these two different [011] directions [4].

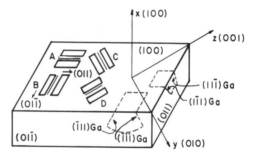

$\{uvw\}$. Some important planes and directions for a cubic crystal are shown in Fig. 1-3-2.

As can be seen from Fig. 1-3-1, the bonds between the nearest Ga and As atoms are in the $\langle 111 \rangle$ directions. The closest equivalent atoms are in the $\langle 110 \rangle$ directions.

Owing to the lack of the inversion symmetry in GaAs, the directions [111] and [$\overline{111}$] are not equivalent. The direction from Ga to the nearest As is usually denoted as [111], whereas [$\overline{111}$] corresponds to the opposite direction. The surface of the crystal cleaved along the (111) plane consists either of Ga atoms, which have three bonds with the crystal, or of As atoms, which have just one bond with the crystal. The opposite is true for the ($\overline{111}$) plane. The (111) plane is called the Ga plane and the ($\overline{111}$) plane is called the As plane. The difference between these two planes becomes apparent when the crystal is chemically etched (see Fig. 1-3-3), subjected to the ion implantation, or covered by a passivating dielectric layer. As a result of

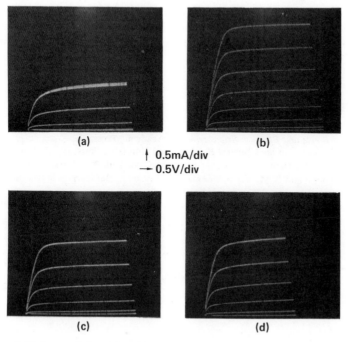

FIGURE 1-3-4. $I-V$ characteristics of four identical FETs oriented in four directions. Pictures (a), (b), (c), and (d) correspond to FETs oriented in the A, B, C, and D directions as shown in Fig. 1-3-3. The vertical scale is 0.5 mA/div, the horizontal scale is 0.5 V/div, and ΔV_{os} is 0.2 V/step [4].

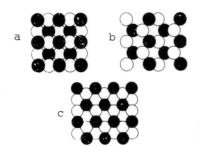

FIGURE 1-3-5. GaAs crystal planes. (a) {100}; (b) {110}; (c) {111}; ○, Ga; ●, As [5].

this difference, the properties of GaAs field effect transistors, for example, may strongly depend on the gate orientation on the substrate (see Fig. 1-3-4) [4].

Three basic crystal planes of the GaAs lattice are shown in Fig. 1-3-5. Each As atom on the (100) surface has two bonds with Ga atoms from the layer below. Two other bonds are free. The (110) plane contains the same number of Ga and As atoms. Each atom has one bond with the layer below. Atoms on the (111) surface have three bonds with the Ga atoms from the layer below. The fourth bond is free.

The distance between the nearest neighbors is 2.44 Å. It is equal to the sum of the atomic radii of As (1.18 Å) and Ga (1.26 Å). The lattice constant is equal to 5.65 Å.

Partially heteropolar bonds in GaAs are stronger than homopolar bonds in Si or Ge. It leads to a smaller amplitude of the lattice vibrations (and as a consequence to a higher mobility), higher melting point, and wider energy gap.

This difference in the chemical bonds becomes apparent when the crystals are cleaved. In diamond, for example, the crystals are cleaved primarily along (111) planes. As shown above, in GaAs adjacent {111} planes are formed by planes of Ga and As atoms. Electrostatic interaction between the planes makes the cleavage difficult. GaAs crystals are easily cleaved along {110} planes which contain equal numbers of Ga and As atoms.

REFERENCES

1. J. C. Phillips, *Bonds and Bands in Semiconductors*, Academic Press, New York, 1973.
2. L. Pauling, *The Nature of the Chemical Bond*, Cornell University Press, Ithaca, New York, 1960.
3. C. A. Wert and R. W. Thomson, *Physics of Solids*. McGraw-Hill, New York, 1964.
4. C. P. Lee, R. Zucca, and B. M. Welch, Orientation effect on planar GaAs Schottky barrier field effect transistors, *Appl. Phys. Lett.* **37**(3), 311–314 (1980).
5. D. N. Nasleolov and F. P. Kesamanly, *Gallium Arsenide. Fabrication, Properties and Application*, Nauka, Moscow, 1973 (in Russian).

2

Band Structure and
Transport Properties

2-1. BAND STRUCTURE

According to the Bloch theorem a one-electron wave function in a crystal is of the form

$$\psi_k(r) = e^{ik \cdot r} u_k(r) \tag{2-1-1}$$

where $u_k(r)$ is a function with the same spatial periodicity as the crystal lattice. The wave functions ψ_k are found from the solution of the Schrödinger equation:

$$-\frac{\hbar^2}{2m}\nabla^2\psi_k + V(r)\psi_k = E\psi_k \tag{2-1-2}$$

where $V(r)$ is the crystal potential, k is the wave vector, and E is the electron energy.

Physically significant and unique solutions may be assigned to the wave vectors k in the first Brillouin zone which is defined using the reciprocal lattice in the k space. The primitive vectors of the real and reciprocal lattices a_j and K_i satisfy the conditions

$$K_i \cdot a_j = 2\pi\delta_{ij} \tag{2-1-3}$$

Based on Eq. (2-1-3) one can show that

$$K_1 = 2\pi\frac{a_2 \times a_3}{a_1 \cdot (a_2 \times a_3)}$$

$$K_2 = 2\pi\frac{a_3 \times a_1}{a_1 \cdot (a_2 \times a_3)} \tag{2-1-4}$$

$$K_3 = 2\pi\frac{a_1 \times a_2}{a_1 \cdot (a_2 \times a_3)}$$

The first Brillouin zone is defined as the region in the k space within which two points differ no more than one of the reciprocal lattice vectors. Thus we can obtain

11

the first Brillouin zone by constructing planes which normally bisect the reciprocal lattice vectors emanating from the origin in the k space. All values of k in the first Brillouin zone are inequivalent in the sense that subtracting any vector of the reciprocal lattice from k inside the zone does not give another vector also inside this zone.

The boundaries of the first, second, third, and higher-order Brillouin zones may be defined using the Bragg equations:

$$2k \cdot K - K^2 = 0 \qquad\qquad (2\text{-}1\text{-}5)$$

where K is a vector of the reciprocal lattice. However, the geometry of the Brillouin zones beyond the first is rather complicated. For this reason the energy states in a crystal are labeled using the wave vectors in the first Brillouin zone.

The first Brillouin zone for the zinc blende structure is the same as for the face-centered cubic lattice (see Fig. 2-1-1).

The symmetry points and lines of the Brillouin zone are marked in Fig. 2-1-1 (see also Table 2-1-1). A detailed study of the symmetry of electronic states in the zinc blende structure can be found in Refs. 1 and 2.

A simple approach to the calculation of the electronic spectra is to present an electron wave function as a linear combination of the atomic orbitals and consider the crystal potential as perturbation. The band structure is then determined using the perturbation theory [3]. This method is called the linear combination of atomic orbitals (LCAO) method.

A simplified description of the band structure can be obtained in the frame of the nearly free electron model (see Fig. 2-1-2). In this model the unperturbed electron wave functions are assumed to be the plane waves (just as in free space):

$$\psi_k(r) = c\, e^{ik \cdot r} \qquad\qquad (2\text{-}1\text{-}6)$$

The periodic crystal potential is considered as a small perturbation. In this model a nearly parabolic E vs. k curve has the energy gaps at the boundaries of the Brillouin zones due to the Bragg reflections of the electron wave functions from the crystal planes. The resulting E vs. k curve can be folded back into the first Brillouin zone. In the scheme based on the extended bands at zero temperature the states with $k < k_F$ are filled and the states with $k > k_F$ are empty with the energy gap in

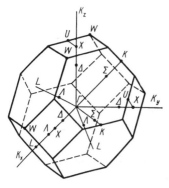

FIGURE 2-1-1. The first Brillouin zone of the face-centered cubic lattice.

TABLE 2-1-1. Symmetry Points of the Brillouin Zone of the Zinc Blende Structure
(see Fig. 2-1-1)[a]

No. of equivalent K vectors	Point	Symmetry operations	Point groups
1	$\Gamma(000)$	$E, 8C_3, 3C_2, 6S_4, 6\sigma$	T_d
3	$X\frac{2\pi}{a}(100)$	$E, 3C_2, 2S_4, 2\sigma$	D_{2d}
4	$L\frac{2\pi}{a}(\frac{1}{2}\frac{1}{2}\frac{1}{2})$	$E, 2C_3, 3\sigma$	C_{3v}
6	$W\frac{2\pi}{a}(10\frac{1}{2})$	$E, 2S_4, C_2$	S_4
6	$\Delta(K00)$	$E, 2\sigma, C_2$	C_{2v}
4	$\Lambda(KKK)$	$E, 2C_3, 3\sigma$	C_{3v}
12	$\Sigma(K0K)$	E, σ	C_s
12	$K\frac{2\pi}{a}(\frac{3}{4}0\frac{3}{4})$	E, C_2	C_s

[a] For GaAs $2\pi/a = 1.111 \cdot 10^8$ cm^{-1}.

between. Here k_F is the Fermi wave vector of the valence electrons, i.e.,

$$k_F = (3\pi^2 n)^{1/3} \qquad (2\text{-}1\text{-}7)$$

where n is the concentration of the valence electrons (four per atom).

This simple picture represents a good qualitative description of the band structure. In quantitative analysis, however, there are problems related to very big values of the crystal Coulomb potential near the centers of the ion cores and its rapid variation in this region. This leads to the necessity to take into account a very large number of the plane waves with large values of k to approach the correct solution. To some extent the method can be improved if the combinations of the plane waves, which have proper symmetry properties, are used as the basic functions.

More accurate calculations of the band structure, however, are based on the pseudopotential method (see, for example, Refs. 4 and 5). The idea of this approach is to use a model potential which leads to the same energy levels as the real potential but not to the same wave functions. The pseudopotential has a form that makes it possible to improve the convergence of the series of the plane waves which represent

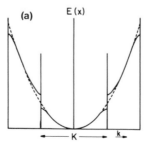
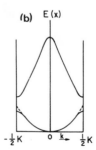

FIGURE 2-1-2. Electron energy in one dimension in the frame of the nearly free electron model. (a) Extended zone scheme. Notice the energy gaps due to the Bragg reflection from the crystal planes. (b) Band structure "folded back" into the first Brillouin zone (reduced zone scheme).

the electron pseudowave functions. In many cases it is convenient to choose the pseudopotential to be a constant within the ion core (see Fig. 2-1-3). The parameters of the pseudopotential can be determined from the spectroscopic data for the individual atoms.

The results of the pseudopotential calculations of energy bands of GaAs are shown in Fig. 2-1-4 [5].

The part of the band structure near the bottom of the lowest conduction band and near the top of the highest valence band is depicted in Fig. 2-1-4b [6]. When the spin–orbit splitting is neglected, the energy levels corresponding to the top of the valence band become triply degenerate. The wave functions corresponding to these states have the same symmetry properties as p functions in an atom. The wave function corresponding to the lowest state in the conduction band has the same symmetry as an s function in an atom. The bottom "split-off" valence band is lower by about 0.34 eV owing to the spin–orbit splitting.

The model which summarizes the most important features of the band structure of the cubic semiconductors is shown in Fig. 2-1-5 [7]. The bands shown in Fig. 2-1-5a are typical for Si, Ge, and AB semiconductors. The band structure of Fig. 2-1-5b occurs in gapless semiconductors such as grey tin and HgTe.

The band structure shown in Fig. 2-1-5a has three sets of conduction band minima located at points $\Gamma(0, 0, 0)$, $L(1/2, 1/2, 1/2)$ and at a point along the $(k, 0, 0)$ lines of the Brillouin zone (Fig. 2-1-1). The latter point may be close to point $X(1, 0, 0)$. The tops of the valence bands are located at Γ; two of these bands (light and heavy holes) are degenerate at this point, while the third one is split by the spin–orbit interactions. Near the minima of the conduction band the energy versus wave vector is given by one of the following functions:

$$E(k) = \hbar^2 k^2/(2m) \qquad \text{(spherical)} \qquad\qquad (2\text{-}1\text{-}8)$$

$$E(k) = (\hbar^2/2)(k_x^2/m_x + k_y^2/m_y + k_z^2/m_z) \qquad \text{(ellipsoidal)} \qquad (2\text{-}1\text{-}9)$$

A wave vector k and energy E in Eqs. (2-1-8)–(2-1-9) are measured from the corresponding minimum of the energy band.

Equation (2-1-8) represents a band with a single scalar effective mass and spherical surfaces of equal energy. It is an appropriate model for the Γ minimum of the conduction band and for the split-off valence band. It is also the simplest possible model of the band structure and is generally adopted for rough estimates of transport properties.

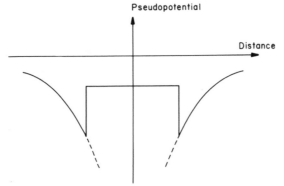

FIGURE 2-1-3. Model ion pseudo-potential.

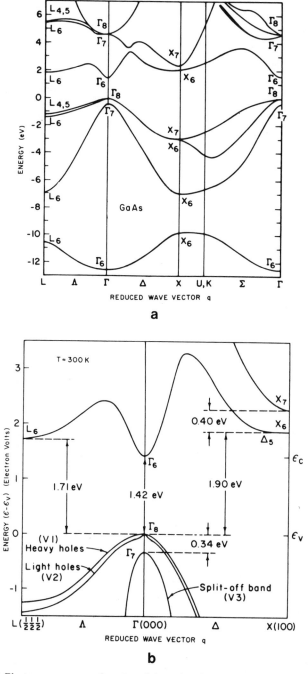

FIGURE 2-1-4. (a) Electron energy as a function of the reduced wave vector [5]. (b) Electron energy as a function of the reduced wave vector for energies close to the top of the valence band and the bottom of the conduction band [6]. The values of energies shown are for room temperature.

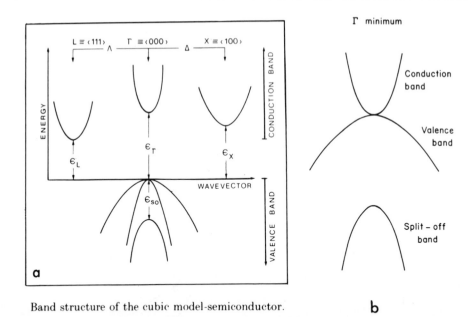

Band structure of the cubic model-semiconductor.   **b**

FIGURE 2-1-5. Lowest minima of the conduction band and highest maxima of the valence band for the cubic semiconductors [7]. (a) Si, Ge, and A_3B_5 bands. (b) Γ minimum for the gray Sn band structure.

Equation (2-1-9) represents a band with ellipsoidal equienergy surfaces. It can be used to describe the L and X minima of the conduction band. Because of the rotational symmetry of the ellipsoids of equal energy (see Fig. 2-1-6), Eq. (2-1-9) can be rewritten in a simpler form:

$$E(k) = \frac{\hbar^2}{2}\left(\frac{k_l^2}{m_l} + \frac{k_t^2}{m_t}\right)$$

where $1/m_l$ and $1/m_t$ are the longitudinal transverse components of the inverse effective mass tensor and k_l and k_t are the longitudinal and transverse components of the wave vector.

Energy versus wave vector near the top of the valence band is given by [8]

$$E(k) = E_V - \hbar^2 k^2 g(k)/2m \qquad (2\text{-}1\text{-}10)$$

where

$$g = A \pm [B^2 + C^2(k_x^2 k_y^2/k^4 + k_x^2 k_z^2/k^4 + k_y^2 k_z^2/k^4)]^{1/2} \qquad (2\text{-}1\text{-}11)$$

(warped bands, see Fig. 2-1-6). Here A, B, and C are constants. Plus and minus signs in Eq. (2-1-11) correspond to the heavy and light holes, respectively.

More realistic models should take into account the deviation of the E vs. k relationship from parabolicity. In the simplest way it can be done substituting E in Eqs. (2-1-8)–(2-1-10) by

$$\gamma(E) = E(1 + \alpha E) \qquad (2\text{-}1\text{-}12)$$

For example, for the minimum of the conduction band in the direct gap semiconduc-

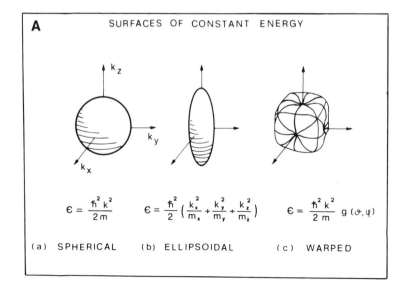

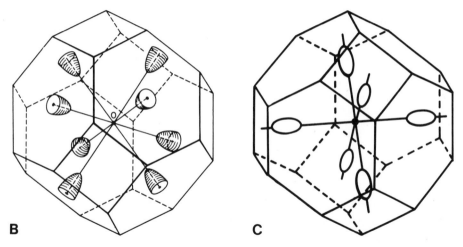

B

C

FIGURE 2-1-6. (A) Surfaces of equal energy for electrons and holes in cubic semiconductors [7]. (B) Constant energy surfaces for the lowest minima of the conduction band in Ge. There are eight ellipsoids of revolution along ⟨111⟩ axes, and the Brillouin zone boundaries are in the middle of the ellipsoids [16]. (C) Constant energy surfaces for the lowest minima of the conduction band in Si. There are six ellipsoids along ⟨100⟩ axes with the centers of the ellipsoids located at about three quarters of the distance from the Brillouin zone center to the Brillouin zone boundary. For comparison, in GaAs the constant energy surfaces for the lowest minima of the conduction band are spheres with the center coinciding with the center of the Brillouin zone.

tors such as GaAs we have

$$E(1 + \alpha E) = \frac{\hbar^2 k^2}{2m} \tag{2-1-13}$$

where

$$\alpha = \frac{1}{E_g}\left(1 - \frac{m}{m_e}\right)^2 \tag{2-1-14}$$

Here E_g is the energy gap, m is the effective mass, and m_e is the free electron mass [9]. A more accurate expression for the nonparabolicity constant may be found in Ref. 6.

Equations (2-1-13), (2-1-14) were derived based on the so-called k-p method [9]. In this method Eq. (2-1-1) is substituted into the Schrödinger equation for the Bloch amplitude $u(r)$ near the symmetry points of the Brillouin zone. The resulting equation contains a term $k \cdot p$, where p is the momentum operator which is considered as perturbation. (For this reason the approach is called the k-p method.) The solution of the equation for $u(r)$ is then found based on the assumption that $u(r)$ can be presented as a linear combination of the wave functions corresponding to the top valence bands and the bottom conduction band.

Energy versus wave vector for the lowest minimum of the conduction band in GaAs is shown in Fig. 2-1-7 [10]. As can be seen from this figure, besides the nonparabolicity, there is also a slight anisotropy of the conduction band. The nonparabolicity becomes quite important at energies ~ 0.1 eV above the bottom of the conduction band.

Band gaps and effective masses of some cubic semiconductors are given in Table 2-1-2 [7].

All parameters characterizing the band structure are temperature dependent. According to Ref. [6] for GaAs

$$E(T) = E_{\Gamma 0} - aT^2/(T + \beta) \tag{2-1-15}$$

where $E_{\Gamma 0} = 1.519$ eV, $a = 5.405 \times 10^{-4}$ eV/K, and $\beta = 204$ K. Equation (2-1-15) is valid in the temperature range $0 < T < 1000$ K,

$$E_L = 1.815 - 6.05 \times 10^{-4} T^2/(T + 204) \text{ (eV)} \tag{2-1-16}$$

and

$$E_X = 1.981 - 4.60 \times 10^{-4} T^2(T + 204) \text{ (eV)} \tag{2-1-17}$$

The calculation based on the k-p method yields the temperature dependence of the electron effective mass in the Γ minimum shown in Fig. 2-1-8 [6]. In particular $m = 0.067 m_e$ at $T \to 0$ and $m = 0.0632 m_e$ at 300 K.

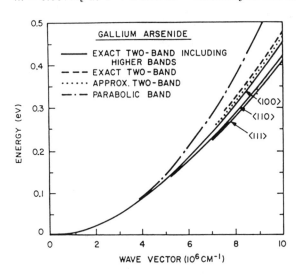

FIGURE 2-1-7. Energy vs. wave vector for the central minimum of the conduction band in GaAs [10, 6]. The parabolic approximation is for $m = 0.063 m_e$.

TABLE 2-1-2. Band Gaps and Effective Masses of Some Cubic Semiconductors[a]

	E_Γ (eV)	E_L (eV)	E_Δ (eV)	E_{s0} (eV)	Electron effective masses			Hole effective masses[b]	
					m_1	m	m_t	m_{hh}	m_{h1}
Si	4.08	1.87	1.13	0.04	0.98	—	0.19	0.53	0.16
Ge	0.89	0.76	0.96	0.29	1.64	—	0.082	0.35	0.043
AlP	3.3	3.0	2.1	0.05	—	—	—	0.63	0.2
AlAs	2.95	2.67	2.16	0.28	2.0	—	—	0.76	0.15
AlSb	2.5	2.39	1.6	0.75	1.64	—	0.23	0.94	0.14
GaP	2.24	2.75	2.38	0.08	1.12	—	0.22	0.79	0.14
GaAs	1.42	1.71	1.90	0.34	—	0.067	—	0.62	0.074
GaSb	0.67	1.07	1.30	0.77	—	0.045	—	0.49	0.046
InP	1.26	2.0	2.3	0.13	—	0.080	—	0.85	0.089
InAs	0.35	1.45	2.14	0.38	—	0.023	—	0.6	0.027
InSb	0.17	1.5	2.0	0.81	—	0.014	—	0.47	0.015
ZnS	3.8	5.3	5.2	0.07	—	0.28	—	—	—
ZnSe	2.9	4.5	4.5	0.43	—	0.14	—	—	—
ZnTe	2.56	3.64	4.26	0.92	—	0.18	—	—	—
CdTe	1.80	3.40	4.32	0.91	—	0.096	—	—	—

[a] Reference [7].
[b] Reference [15].

The energies of the conduction band minima and effective masses also depend on pressure [6, 11–14]:

$$E_\Gamma = E_\Gamma(0) + 0.0126P - 3.77 \times 10^{-5}P^2 \text{ eV} \qquad (2\text{-}1\text{-}18)$$

where P is the hydrostatic pressure in kilobars,

$$E_L = E_L(0) + 0.0055P \text{ (eV)} \qquad (2\text{-}1\text{-}19)$$

and

$$E_X = E_X(0) - 0.0015P \text{ (eV)} \qquad (2\text{-}1\text{-}20)$$

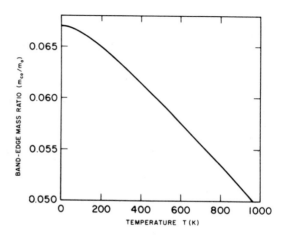

FIGURE 2-1-8. Conduction band effective mass in GaAs as a function of temperature as predicted by the k–p model [6].

At pressures larger than 35 kbar GaAs becomes an indirect gap semiconductor. Because of the nonparabolicity, the effective mass and the density of states also depend on the electron concentration in the conduction band. For the parabolic band the equilibrium concentration n_0 is given by

$$n_0 = N_{c0} \frac{2}{\sqrt{\pi}} \int_0^\infty \frac{x^{1/2}\,dx}{1 + \exp(x - y)} \equiv N_{c0} F_{1/2}(y) \qquad (2\text{-}1\text{-}21)$$

where $y = (E_F - E_c)/k_B T$, E_F is the Fermi level, E_c is the bottom of the conduction band, $F_{1/2}$ is the Fermi integral, and

$$N_{c0} \equiv 2(2\pi m_{c0} k_B T/h^2)^{3/2} \qquad (2\text{-}1\text{-}22)$$

is the effective density of states in the conduction band for the parabolic band. At room temperature for GaAs $N_{c0} = 3.99 \times 10^{17}$ cm^{-3}. When the nonparabolicity is taken into account, density of states N_{c0} in Eq. (2-1-21) should be replaced by the nonparabolic effective density of states N_c^1 [6]:

$$N_c^1 = 8.63 \times 10^{13} T^{3/2}(1 - 1.93 \times 10^{-4} T - 4.19 \times 10^{-8} T^2),\ \text{cm}^{-3}$$

N_c^1 and N_{c0} are shown in Fig. 2-1-9 as a function of temperature [6].

Equation (2-1-21), even with N_c^1 instead of N_{c0}, takes into account only the lowest minimum of the conduction band. At high temperatures upper X and L minima may also become populated:

$$
\begin{aligned}
n_X &= 2(2\pi m_X k_B T/h^2)^{3/2} \exp[(\eta - \Delta_{\Gamma X})/k_B T] \\
n_L &= 2(2\pi m_L k_B T/h^2)^{3/2} \exp[(\eta - \Delta_{\Gamma L})/k_B T]
\end{aligned}
\qquad (2\text{-}1\text{-}23)
$$

where $\eta = E_F - E_c$, $\Delta_{\Gamma X} = E_X - E_\Gamma$, $\Delta_{\Gamma L} = E_L - E_\Gamma$. The value of the electron density n in the central valley is still given by Eq. (2-1-21) and

$$n_0 = n_\Gamma + n_X + n_L \qquad (2\text{-}1\text{-}24)$$

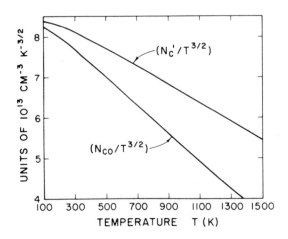

FIGURE 2-1-9. Density of states in the central minimum in GaAs for the parabolic model (N_{c0}) and nonparabolic model (N_c^1) [6].

The values of the effective masses for the X and L minima are given by [12]

L minimum:
$$m_l \simeq 1.9 m_e$$
$$m_t \simeq 0.075 m_e$$
$$m_L = (16 m_l m_t^2)^{1/3} = 0.56 m_e$$

X minimum:
$$m_l \simeq 1.9 m_e$$
$$m_t \simeq 0.19 m_e$$
$$m_X = (9 m_l m_t^2)^{1/3} \simeq 0.85 m_e$$

Here m_X and m_L are density-of-states effective masses. The fraction of the electron concentration in the three lowest conduction band minima of GaAs is shown in Fig. 2-1-10 [6]. As can be seen from the figure, the electron concentration in upper valleys may actually exceed n_Γ at elevated temperatures. When all three sets of the conduction minima are taken into account the electron concentration as a function of temperature and position of the Fermi level is given by [6]

$$n_0 = N_c^* \exp[(E_F - E_c)/k_B T] \quad (\text{for } n_0 < N_c^*) \tag{2-1-25}$$

where

$$N_c^* = 8.63 \times 10^{13} T^{3/2}[(1 - 1.93 \times 10^{-4} T - 4.19 \times 10^{-8} T^2)$$
$$+ 21 \exp(-\Delta_{\Gamma L}/k_B T) + 44 \exp(-\Delta_{\Gamma X}/k_B T)] \tag{2-1-26}$$

The equilibrium concentration of holes in the valence band is given by

$$P_0 = N_v F_{1/2}(y_1) \tag{2-1-27}$$

where $y_1 = (E_V - E_F)/k_B T$, and

$$N_v = 2(2\pi m_v k_B T/h^2)^{3/2}$$

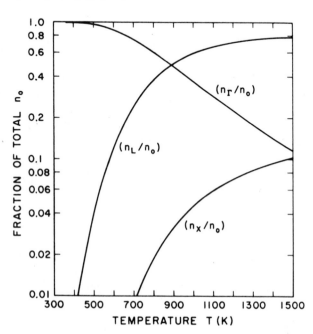

FIGURE 2-1-10. Fraction of the electron concentration in three lowest minima of the conduction band in GaAs vs. temperature [6].

The hole density of states effective mass m_v is given by

$$m_v = (m_{hh}^{3/2} + m_{lh}^{3/2})^{2/3} \qquad (2\text{-}1\text{-}28)$$

As suggested in Ref. 6, $m_{hh}/m_e \approx 0.51$, $m_{lh}/m_e \simeq 0.082$, leading to $m_v/m_e \simeq 0.53$.

2-2. THE BOLTZMANN TRANSPORT EQUATION

Electrons in a crystal with a given band structure may, in many cases, be regarded as a collection of fictitious particles with the given relationship between the energy E and the wave vector k.

In a six-dimensional phase space (r, k) the motion of each particle can be represented by a moving point. The equations of motion are

$$\dot{r} = v = \frac{1}{\hbar} \nabla_k E(k) \qquad (2\text{-}2\text{-}1)$$

$$\hbar \dot{k} = q(F + v \times B) \qquad (2\text{-}2\text{-}2)$$

where v is the velocity, F is the electric field, q is the electronic charge, and B is the magnetic field. However, this ballistic motion is interrupted by collisions. If the collisions are instantaneous the particle disappears owing to the collision and instantaneously reappears in a different point of the phase space.

The average occupancy $f(k, r, t)$ of a point in the phase space is called the distribution function. The distribution function may change owing to the scattering, the flow of electrons in real space determined by their velocity v, and the flow of electrons in the k space, which is determined by the time derivative $\dot{k}$. The equation for the distribution function f is called the Boltzmann transport equation:

$$\frac{\partial f}{\partial t} + v \cdot \nabla_r f + \dot{k} \cdot \nabla_k f = \left(\frac{\partial f}{\partial t}\right)_{\text{coll}} \qquad (2\text{-}2\text{-}3)$$

As noticed above, this equation is based on the assumption that the collisions are instantaneous. Also the external fields should be sufficiently small and the band-to-band transition should be absent [17].

The term in the right-hand part of Eq. (2-2-3) represents the collisions (the collision integral):

$$\left(\frac{\partial f}{\partial t}\right)_{\text{coll}} = \int [W(k', k)f_{k'}(1 - f_k) - W(k, k')f_k(1 - f_{k'})] \, dV_{k'} \qquad (2\text{-}2\text{-}4)$$

where the integration is over the first Brillouin zone. Here $W(k, k') \, dV_{k'} \, dt$ is the conditional probability of the transition from the state k to a state k' in $dV_{k'}$ in time dt given that there is initially an electron in state k and the state k' is empty.

The principle of the detailed balance requires that the transition probability be symmetric for scattering processes which conserve electron energy, i.e.,

$$W(k, k') = W(k', k) \qquad (2\text{-}2\text{-}5)$$

For such processes the collision integral can be simplified:

$$\left(\frac{\partial f}{\partial t}\right)_{coll} = \int W(k, k')\,(f_{k'} - f_k)\,dV_{k'} \tag{2-2-6}$$

In uniform equilibrium the Boltzmann equation reduces to

$$\left(\frac{\partial f}{\partial t}\right)_{coll} = 0 \tag{2-2-7}$$

The solution of this equation is

$$f_0 = \frac{1}{e^{(E_k - E_F)/k_B T} + 1} \tag{2-2-8}$$

i.e., the Fermi–Dirac function. Here E_F is the Fermi level and T is temperature.

If the transition probabilities are independent of r and t, Eq. (2-2-7) is also valid for the non-uniform equilibrium state. An example of such a system would be a p–n junction or nonuniformly doped semiconductor with zero electric current (no external voltage applied).

The general non-uniform equilibrium solution of the Boltzmann equation is given by

$$f = \frac{1}{\exp[E_k - E_F + q\phi)/k_B T] + 1} \tag{2-2-9}$$

Here ϕ is the electric potential.

In an external electric field the probabilities $W(k, k')$ have to be calculated in order to evaluate the collision integral and to solve the Boltzmann equation. By the Golden Rule of quantum mechanics

$$W(k, k') = \frac{2\pi}{\hbar}|H|^2\delta(E' - E) \tag{2-2-10}$$

where E' and E are the initial and final energies of the crystal, $|H|^2$ is the squared matrix element. For the non-umklapp processes (i.e., for the processes which do not involve the change in k by a vector of a reciprocal lattice)

$$|H|^2 = AG(k, k') \tag{2-2-11}$$

where A depends on a particular scattering mechanism and $G(k, k')$ is the overlap factor:

$$G(k, k') = \left|\int_{cell} u_k(r)u_{k'}(r)\,dr\right|^2 \tag{2-2-12}$$

between the periodic parts $u_k(r)$ and $u_{k'}(r)$ of the Bloch wave functions of the initial and final states.

For the electrons in the central valley the difference of G from unity is related to the nonparabolicity [18]

$$G(k, k') = \frac{\{[1 + \alpha E(k)]^{1/2}[1 + \alpha E(k')]^{1/2} + \alpha[E(k)E(k')]^{1/2} \cos \theta\}^2}{[1 + 2\alpha E(k)][1 + 2\alpha E(k')]} \qquad (2\text{-}2\text{-}13)$$

Here θ is the angle between k and k'. For the intraband transitions within heavy or light hole bands we have [19]

$$G_{h,1}(k, k') = \tfrac{3}{4}(1 + 3 \cos^2 \theta) \qquad (2\text{-}2\text{-}14)$$

while for interband hole transitions

$$G_{h,1}(k, k') = \tfrac{3}{4}\sin^2 \theta \qquad (2\text{-}2\text{-}15)$$

The expressions for the squared matrix element A for different scattering mechanisms may be found, for example, in Ref. 7.

2-3. SCATTERING RATES

The most important scattering mechanisms for GaAs are polar optical phonon scattering, ionized impurity scattering, piezoelectric scattering, and acoustic scattering (with polar optical scattering and ionized impurity scattering being dominant) and intervalley scattering in a high electric field.

A convenient way to characterize the relative strengths of the different scattering mechanisms is to compute the total scattering rate

$$\lambda(k) = \int W(k, k') \, dV_{k'} \qquad (2\text{-}3\text{-}1)$$

which is the probability per unit time that an electron with the wave vector k will be scattered. Here the integral is over the Brillouin zone.

The following expressions for the electron scattering rates are obtained using Eqs. (2-2-10)–(2-2-13) and Eq. (2-3-1) [18].

2-3-1. Polar Optical Scattering

$$\lambda_0(E) = 5.61 \times 10^{15} \left(\frac{m}{m_e}\right)^{1/2} E_0 \left(\frac{1}{K_\infty} - \frac{1}{K_0}\right) \frac{1 + 2\alpha E'}{\gamma^{1/2}(E)} F_0(E, E')$$

$$\times \begin{cases} N_0 & \text{(absorption)} \\ (N_0 + 1) & \text{(emission)} \end{cases} \quad (\text{sec}^{-1}) \qquad (2\text{-}3\text{-}2)$$

where $E_0 = \hbar\omega_0/q$, i.e., the energy of the optical phonon in eV,

$$E' = \begin{cases} E + E_0 & \text{(absorption)} \\ E - E_0 & \text{(emission)} \end{cases}$$

$$\gamma(E) = E(1 + \alpha E) \qquad (2\text{-}3\text{-}3)$$

$$F_0(E, E') = C^{-1}\left[A \ln\left|\frac{\gamma^{1/2}(E) + \gamma^{1/2}(E')}{\gamma^{1/2}(E) - \gamma^{1/2}(E')}\right| + B\right] \qquad (2\text{-}3\text{-}4)$$

$$N_0 = \frac{1}{\exp(\hbar\omega_0/k_B T) - 1} \qquad (2\text{-}3\text{-}5)$$

$$A = \{2(1 + \alpha E)(1 + \alpha E') + \alpha[\gamma(E) + \gamma(E')]\}^2 \qquad (2\text{-}3\text{-}6)$$

$$B = -2\alpha\gamma^{1/2}(E)\gamma^{1/2}(E')\{4(1 + \alpha E)(1 + \alpha E') + \alpha[\gamma(E) + \gamma(E')]\} \qquad (2\text{-}3\text{-}7)$$

$$C = 4(1 + \alpha E)(1 + \alpha E')(1 + 2\alpha E)(1 + 2\alpha E') \qquad (2\text{-}3\text{-}8)$$

Here E is the electron energy (in eV), (m/m_e) is the relative electron effective mass, and K_∞ and K_0 are high- and low-frequency dielectric constants (10.92 and 12.9 for GaAs).

For the polar optical scattering with the emission of a phonon Eq. (2-3-2) is valid only for $E' = E - \hbar\omega_0 > 0$. For $E < \hbar\omega_0$ the scattering rate λ corresponding to the phonon emission is zero.

For the parabolic bands $\alpha = 0$ and

$$A = C = 4 \qquad (2\text{-}3\text{-}9)$$

$$B = 0 \qquad (2\text{-}3\text{-}10)$$

so that Eq. (2-3-2) can be simplified.

The energy dependence of the polar scattering rate in GaAs computed from Eq. (2-3-2) is shown in Fig. 2-3-1. The scattering rates for processes with absorption and emission of an optical polar phonon are shown in Figs. 2-3-2 and 2-3-3 for 300 K and 77 K, respectively [18]. (The values of the scattering parameters for GaAs are given in Table 2-3-1 [20].)

The energy of the final state in the polar optical scattering is equal to $E - \hbar\omega_0$ for the emission and $E + \hbar\omega_0$ for the absorption of a phonon. The angular probability distribution of wave vectors of the final states can be found from Eqs. (2-2-11)–(2-2-13) [18]:

$$P(\theta)\, d\theta \sim \frac{[\gamma^{1/2}(E)\gamma^{1/2}(E') + \alpha EE' \cos\theta]^2 \sin\theta\, d\theta}{[\gamma(E) + \gamma(E') - 2\gamma^{1/2}(E)\gamma^{1/2}(E') \cos\theta]} \qquad (2\text{-}3\text{-}11)$$

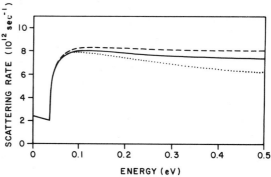

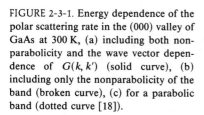

FIGURE 2-3-1. Energy dependence of the polar scattering rate in the (000) valley of GaAs at 300 K, (a) including both non-parabolicity and the wave vector dependence of $G(k, k')$ (solid curve), (b) including only the nonparabolicity of the band (broken curve), (c) for a parabolic band (dotted curve) [18].

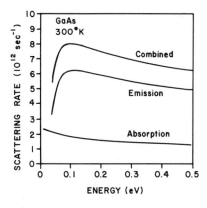

FIGURE 2-3-2. Energy dependence of the polar scattering rate (emission, absorption, and combined) in the (000) minimum of GaAs at 300 K (for a parabolic band). Parameters used in the calculation are the same as in Ref. 18.

(see Fig. 2-3-4). Here θ is the angle between k and k'. As can be seen from Fig. 2-3-4 at high electron energies, the scattering in the same direction is more probable.

2-3-2. Acoustic Scattering

The scattering rate is given by

$$\lambda_a(k) = \frac{0.449 \times 10^{18}(m/m_e)^{3/2} TE_1^2}{\rho u^2}\gamma^{1/2}(E)\,(1 + 2\alpha E)F_a(E)\,(\text{s}^{-1}) \qquad (2\text{-}3\text{-}12)$$

where

$$F_a(E) = \frac{(1 + \alpha E)^2 + 1/3(\alpha E)^2}{(1 + 2\alpha E)^2} \qquad (2\text{-}3\text{-}13)$$

Here ρ is the crystal density in g/cm^3, E_1 is the acoustic deformation potential (in eV), T is the lattice temperature (K), and u is the sound velocity (cm/s). It is not a very important scattering mechanism in the (000) valley. The acoustic scattering is practically elastic because the energy of the acoustic phonons involved is small compared to $k_B T$.

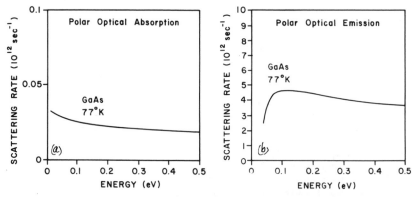

FIGURE 2-3-3. Energy dependence of the polar scattering rate in the (000) minimum of GaAs at 77 K (for a parabolic band). (a) Absorption; (b) emission. Parameters used in the calculation are the same as in Ref. 18.

TABLE 2-3-1. Parameters for Electron Scattering
in GaAs[a]

Energy separation	$\Gamma - L$	0.33
between valleys (eV)	$\Gamma - X$	0.52
Effective mass	Γ	0.063
(m/m_e)	L	0.17
	X	0.58
Intervalley coupling	$\Gamma - L$	0.18
constant (10^9 eV/cm)	$\Gamma - X$	1
	$L - X$	0.1
	$L - L$	0.5
	$X - X$	1
Nonparabolicity	Γ	0.62
(eV^{-1})	L	0.5
	X	0.3
Intervalley phonon energy (eV)		0.0299
Acoustic deformation potential (eV)		7
LO phonon energy (eV)		0.0362
Static dielectric constant		12.9
Optical dielectric constant		10.92
Sound velocity (cm/s)		5.2×10^5
Density (g/cm^3)		5.37

[a] Reference 20.

The angular probability distribution for the acoustic scattering is given by

$$P(\theta)\, d\theta \sim [1 + \alpha E(1 + \cos \theta)^2] \sin \theta\, d\theta \qquad (2\text{-}3\text{-}14)$$

which reduces to

$$P(\theta)\, d\theta \sim \sin \theta\, d\theta \qquad (2\text{-}3\text{-}15)$$

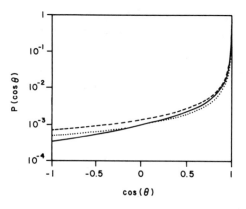

FIGURE 2-3-4. Angular probability $P(\cos \theta)$ for polar optical scattering in the (000) valley of GaAs at 300 K for the electron energy 0.4 eV. The labeling of the curves is the same as in Fig. 2-3-1 [18].

For $\alpha = 0$ (i.e., for the parabolic band) Eq. (2-3-15) is equivalent to the statement that all final states are equally probable and, hence, the probability of the angle between k' and k being between θ and $\theta + d\theta$ is proportional to the number of states on a circle with the radius $|k| \sin \theta$. In other words the acoustic scattering processes are randomizing.

2-3-3. Impurity Scattering

The ionized impurity scattering rate is given by [19]

$$\lambda_i(k) = 4.84 \times 10^{11} \left(\frac{m}{m_e}\right)^{1/2} \frac{T}{K_0} \frac{1 + 2\alpha E}{[E(1 + \alpha E)]^{1/2}} \, (\text{s}^{-1}) \qquad (2\text{-}3\text{-}16)$$

Here K_0 is the static dielectric constant ($K_0 = 12.9$ for GaAs).

The scattering rate given by Eq. (2-3-16) is independent of N_D. This result is the consequence of the long-range Coulomb potential leading to the divergence of the scattering cross section for small scattering angles. The scattering cross section is limited only by the average distance between impurities. The magnitude of the electron concentration affects the transport properties by changing the angular distribution of scattering, since, as N_D increases, the electrons are on average scattered through to a larger angle θ. From Eqs. (2-2-11)–(2-2-13) $P(\theta) \, d\theta$ can be found to be

$$P(\theta) \, d\theta \sim \left[\frac{1 + \alpha E(1 + \cos \theta)}{2k^2(1 - \cos \theta) + \beta^2}\right]^2 \sin \theta \, d\theta \qquad (2\text{-}3\text{-}17)$$

where

$$\beta = \left(\frac{q^2 N_D}{\varepsilon_0 k_B T}\right)^{1/2} \qquad (2\text{-}3\text{-}18)$$

is the inverse screening length. When $N_D \to 0$, $\beta \to 0$ and $P(\theta)$ diverges as $\theta \to 0$, so that only zero-angle scattering remains and impurity scattering has no effect. We should notice that $\lambda_i(k)$ is independent of N_D only when $(\beta/2k)^2 \ll 1$ [19].

2-3-4. Intervalley Scattering between Nonequivalent Valleys

In high electric fields the electron energy in GaAs may become sufficiently high to enable electrons to scatter into L and X valleys. Then, in addition to polar optical scattering, the intervalley scattering becomes important. The intervalley scattering rate for the scattering between the nonequivalent 000 and upper valleys is given by [18]

$$\lambda_{ij}(E) = 1.129 \times 10^{-5} Z_j \left(\frac{m_j}{m_e}\right)^{3/2} \frac{D_{ij}^2}{\rho E_{ij}} \gamma_j^{1/2}(1 + 2\alpha_j E_j') F_{ij}(E_i, E_j')$$

$$\times \begin{cases} N_{ij} & \text{(absorption)} \\ (N_{ij} + 1) & \text{(emission)} \end{cases} (\text{s}^{-1}) \qquad (2\text{-}3\text{-}19)$$

Here

$$N_{ij} = \frac{1}{\exp(\hbar\omega_{ij}/k_BT) - 1} \qquad (2\text{-}3\text{-}20)$$

$$F_{ij}(E, E') = \frac{(1 + \alpha_i E_i)(1 + \alpha_j E'_j)}{(1 + 2\alpha_i E_i)(1 + 2\alpha_j E'_j)} \qquad (2\text{-}3\text{-}21)$$

The energies are measured in eV and are counted from the bottoms of the minima,

$$E'_j = E_i \pm E_{ij} \qquad (2\text{-}3\text{-}22)$$

$E_{ij} = \hbar\omega_{ij}/q$ is the energy of the intervalley scattering phonon in eV, Z_j is the number of the equivalent valleys, and D_{ij} is the deformation potential constant in eV/cm. This scattering is caused by the nonpolar optical phonons. It is randomizing and the angular probability distribution is given by Eq. (2-3-14).

2-3-5. Intervalley Scattering between Equivalent Valleys

This scattering is also caused by the nonpolar optical phonons. The scattering rate is given by

$$\lambda_e(k) = 1.129 \times 10^{-5}(Z_e - 1)\left(\frac{m}{m_e}\right)^{3/2} \frac{D_e^2}{\rho E_e}(E')^{1/2}$$

$$\times \begin{cases} N_e & \text{(absorption)} \\ (N_e + 1) & \text{(emission)} \end{cases} \quad (\text{s}^{-1})$$

where Z_e is the number of the equivalent valleys,

$$N_e = \frac{1}{\exp(\hbar\omega_e/k_BT) - 1} \qquad (2\text{-}3\text{-}23)$$

E_e is the energy of the equivalent intervalley scattering phonon (in eV) ($E_e = \hbar\omega_e/q$), D_e is the equivalent intervalley scattering constant (in eV/cm), and nonparabolicity is neglected.

This scattering is also randomizing.

When the scattering mechanisms are established, the Boltzmann transport equation has to be solved to yield the distribution function. The average drift velocity $\bar{v}_d$, the average electron energy $\bar{E}$, etc. can then be found.

2-4. MONTE CARLO SIMULATION

The Monte Carlo method has become a standard numerical method of solving the Boltzmann equation. It is based on the approach developed by Kurosawa [21]. The idea of this technique is to simulate the electron motion in the k space. We consider a free electron flight which is interrupted by the scattering processes. If we observe a single electron for a sufficiently long time, the distribution of times the electron spends in the vicinity of the points in the k space will reproduce the shape of the distribution function $f(k)$.

Between the scattering events the electron wave vector changes with the rate determined by the electric field F:

$$k(t) = k_0 + \frac{qF}{\hbar}t \tag{2-4-1}$$

where k_0 is the initial value of the wave vector. The moments of time when the scattering events occur can be determined using the computer-generated random number r.

These moments of time depend on the total scattering rate

$$\lambda(k) = \sum_{i=1}^{n} \lambda_i(k) \tag{2-4-2}$$

Here $\lambda_i(k)$ are scattering rates for different scattering mechanisms (see Section 2-3).

The total scattering rate $\lambda(k)$ is a complicated function of $k(t)$ which makes it difficult to reproduce the probability distribution of the scattering events, generating random numbers. This difficulty can be avoided by the introduction of an additional fictitious (so-called "self-scattering") process, which does not change the wave vector. The scattering probability of the self-scattering process is defined as

$$W_0(k, k') = \lambda_0(k)\delta(k - k') \tag{2-4-3}$$

The function $\lambda_0(k) + \lambda(k)$ is chosen in such a way that

$$\lambda_0(k) + \lambda(k) = \Gamma \tag{2-4-4}$$

where Γ is a constant and $\lambda_0(k)$ is positive. Now the total probability for an electron to scatter during the time interval between t and $t + dt$ is given by

$$P(t)\,dt = \Gamma\,e^{-\Gamma t}\,dt \tag{2-4-5}$$

Here t is the time passed from the previous scattering event. The time of the free flight t_s can now be related to the random number r uniformly distributed with equal probability between 0 and 1:

$$r = \int_0^{t_s} P(t)\,dt \tag{2-4-6}$$

or

$$r = 1 - e^{-\Gamma t_s} \tag{2-4-7}$$

$$t_s = -\frac{1}{\Gamma}\ln(1 - r) \tag{2-4-8}$$

After the time t_s of the free flight has been determined, it is necessary to establish the scattering process responsible for terminating the flight. Since the probability

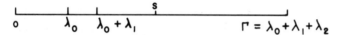

FIGURE 2-4-1. Selection of the scattering mechanism terminating the free flight. S is a random number uniformly distributed between 0 and Γ. When many scattering events are simulated the numbers of the different scattering mechanisms selected are proportional to the corresponding scattering rates λ_i.

of the free flight termination by the scattering process i is proportional to $\lambda_i(k)$ and since $\sum_{i=0}^{n} \lambda_i = \Gamma$, it is only necessary to generate a random number s between 0 and Γ and test the inequalities

$$s < \sum_{i=0}^{m} \lambda_i(k) \qquad (2\text{-}4\text{-}9)$$

for $m = 0, 1, 2, \ldots, n$. Here n is the number of the real scattering mechanisms. If the inequality is satisfied for $m = 0$, the self-scattering process (in the increasing m order) is selected. If it is satisfied for $m = 1$, then the first scattering mechanism is selected, and so on. This process is illustrated by Fig. 2-4-1.

The next step is to determine the final state after the scattering. As mentioned in Section 2-2, we assume that, as a result of scattering, the electron disappears from point k of the phase space it occupies and instantaneously reappears in point k' corresponding to the final state. This new position k' is a starting point for the next segment of the free flight, i.e., it replaces k_0 in Eq. (2-4-1).

The final state is known, of course, for the self-scattering process—because it does not change the wave vector k [see Eq. (2-4-3)]. [It is clearly an advantage to choose $\lambda_0(k)$ and hence Γ as small as possible (keeping, however, $\lambda_0(k) > 0$) to minimize the probability of the self-scattering process being selected.]

For all real scattering processes additional random numbers have to be generated in order to simulate the probability distribution of the final states. For randomizing scattering events, such as acoustic scattering in a parabolic valley and intervalley scattering, all states on the energy surface corresponding to the final state are equally probable (see Section 2-3). Hence, the probability that the angle between k' and the direction of the electric field F is θ is proportional to the number of states in a circle with radius $|k'| \sin \theta$ (see Fig. 2-4-2):

$$P(\theta)\, d\theta = \tfrac{1}{2}\sin \theta\, d\theta \qquad (2\text{-}4\text{-}10)$$

Here θ varies between 0 and π and $1/2$ is a normalizing constant. A random number S_1 uniformly distributed between 0 and 1

$$S_1 = \int_0^\theta P(\theta)\, d\theta \qquad (2\text{-}4\text{-}11)$$

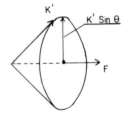

FIGURE 2-4-2. Selection of the final state. The number of final states with the same energy ($k' = \text{const.}$) and with the same angle θ between k' and the electric field F is proportional to the length of the circle ($2\pi k' \sin \theta$).

is selected to determine θ. It produces the probability distribution of the final states given by Eq. (2-4-10). Substituting Eq. (2-4-10) into Eq. (2-4-11), we find

$$S_1 = \tfrac{1}{2}(1 - \cos \theta) \tag{2-4-12}$$

For the acoustic scattering in a nonparabolic valley (such as Γ minimum in GaAs) and for the polar optical scattering a more complicated procedure has to be adopted [18]. First, the probability distribution $P(\cos \theta) = P(\theta)/\sin \theta$ is computed, where θ is the angle between k and k' [see Eqs. (2-3-11) and (2-3-14)]. Then two random numbers S_2 and S_3 are generated with S_2 uniformly distributed between -1 and 1 and S_3 uniformly distributed between 0 and $P(1)$. If S_3 is less than $P(S_2)$, then S_2 is taken as the value of $\cos \theta$. If S_3 is larger than $P(S_2)$, a second pair of random numbers S_2 and S_3 are generated and the process is repeated until S_3 is less than $P(S_2)$. As shown in Ref. 18, the accepted values of S_2 follow the required probability distribution.

In a particular case of GaAs three sets of valleys have to be taken into account, Γ, L, and X minima (see Fig. 2-1-5). In order to determine the distribution functions in the Γ, L, and X valleys histograms are set up in (k_z, k_ρ) space, where k_z is parallel and k_ρ is perpendicular to the electric field. Counts proportional to the time the electron spends in each cell of the histograms are recorded. Some quantities such as the drift velocity, for example, are calculated directly from the initial and final values of k for each free flight. The average drift velocity v_d in valley j, for example, is given by [18]

$$v_j = \frac{1}{\hbar k_j} \sum (E_f - E_i) \tag{2-4-13}$$

where E_f and E_i are the energies of the final and initial state for each free flight, and k_j is the total length of the electron trajectory in k space in the valley:

$$k_j = \frac{qFT_j}{\hbar} \tag{2-4-14}$$

where T_j is the total time the electron spends in the valley. The summation in Eq. (2-4-13) is over all free flights in valley j.

The Monte Carlo algorithm described above is summarized in Table 2-4-1.

TABLE 2-4-1. Monte Carlo Algorithm

Generate random number r and determine the time of free flight.

Record time the electron spends in each cell of the k-space during the free flight, drift velocity, mean energy, etc.

Generate random numbers, determine which scattering process has occurred and the final state.

Repeat until the desired number of scattering events is reached.

Calculate the distribution function, drift velocity, mean energy, etc.

In order to determine when the number of simulated scattering events is sufficiently large, the simulation is split into a number of successive time intervals of equal duration. The electron drift velocity (or other quantity of interest) is computed for each of them. From these partial "measurements" the mean value and standard deviations are found.

The Monte Carlo technique is also used to determine the diffusion coefficient tensor D_{xy}. If x and y are the distances traveled along the x and y directions, respectively, during the interval t the diffusion coefficient tensor is given by

$$D_{xy} = \frac{1}{2}\frac{d}{dt}\langle(x - \langle x\rangle)(y - \langle y\rangle)\rangle$$

$$= \frac{1}{2}\frac{d}{dt}(\langle xy\rangle - \langle x\rangle\langle y\rangle)$$

where $\langle\ \rangle$ indicates the average over all time intervals.

A typical electron trajectory in the k space simulated by the Monte Carlo method is shown in Fig. 2-4-3 [22]. In this particular case only two conduction band minima (the low and upper valleys) are taken into account.

Figure 2-4-4 shows the average drift velocity in GaAs as a function of the simulation time [22]. It can be seen from this figure how the fluctuations near the stationary value of the drift velocity decrease with time. The average time between collisions in this case is of the order of 10^{-13} s so that it takes roughly 15,000 collisions for the drift velocity to reach the stationary value. This is not of course a real transient process but just a consequence of the Monte Carlo approach.

The Monte Carlo technique can be used to describe the transient effects and to investigate the electron transport in small semiconductor devices (see, for example, [23–38]). For such applications the Monte Carlo simulation is performed not for one but for thousands of electrons with the initial wave vectors k_0 distributed according to the equilibrium initial distribution. The number of collisions for each electron in such a simulation can be relatively small.

As an example, the carrier distribution under the gate of a GaAs MESFET computed by this technique is shown in Fig. 2-4-5 [31]. In this simulation 8192 particles were moving in a 256×8 mesh cell epilayer with squared mesh cells of width 0.02 μm. The epilayer doping density was 10^{17} cm^{-3}.

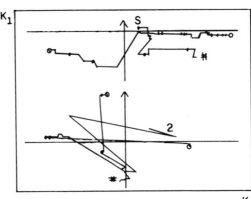

FIGURE 2-4-3. Electron trajectory in k-space simulated by Monte Carlo technique [22]. S, starting point; 1, low valley; 2, upper valley. Circles and stars denote intervalley transitions; points denote the self-scattering events.

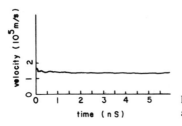

FIGURE 2-4-4. Average drift velocity vs. simulation time for GaAs at 300 K [22]. Electric field 1.5 kV/cm was turned on at $t = 0$.

The results of the Monte Carlo simulation for long samples [18–22, 35–41] are used as an input for the phenomenological device modeling (see Section 2-9).

The symmetrical part of the distribution function in Γ and L valleys of GaAs is shown in Fig. 2-4-6 [20]. The distribution function in the Γ valley is drastically different from the Maxwell–Boltzmann distribution with the effective electron temperature T_e (given by a straight line in a semilog scale). Moreover, the distribution function actually becomes inverted at $F > 10$ kV/cm. The peak in the distribution function corresponds to the energy close to X-Γ separation.

The scattering between L and Γ valleys is only essential for 2.5 kV/cm $< F <$ 7kV/cm because of the relatively small coupling constant (see Table 2-3-1). As a result, the distribution function in the L valleys is not drastically different from the Maxwell–Boltzmann distribution up to $F \simeq 20$ kV/cm.

The electron drift velocity as a function of electric field is shown in Fig. 2-4-7 for $N_d = 0$. The fractions of electrons in L and X valleys and the mean energies of electrons in Γ, L, and X valleys are shown in Fig. 2-4-8.

The role of impurity scattering is illustrated by Fig. 2-4-9, where the drift velocity and the longitudinal diffusion coefficient at 300 and 77 K are presented. As shown in Ref. 20, an increase in the impurity concentration N_d causes the decrease of drift velocity v_d and the increase of the field strength F_p at which a negative differential mobility appears. The decrease of v_d with the increase of N_d is due to the decrease of mean velocity in Γ valley as well as in L minima. At the field strengths near F_p a decisive role in the reduction of v_d is played by the decrease of mean velocity in Γ valley, v_Γ. For example, at $F = 3.5$ kV/cm and $T_1 = 300$ K the increase of N_d from 0 to 10^{17} cm^{-3} leads to the reduction of v_d by 20%, of which 17% is due to the

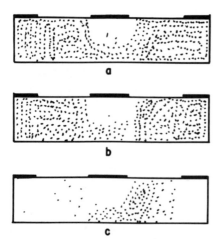

FIGURE 2-4-5. Distribution of carriers under the gate of a GaAs FET computed by the Monte Carlo method [31]. (a) All carriers; (b) central valley only; (c) upper valley only. Gate voltage, 1.02 V; drain voltage, 3 V.

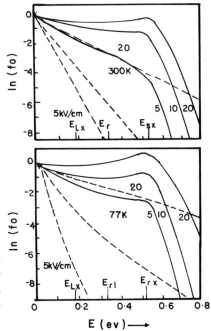

FIGURE 2-4-6. Energy dependence of the spherically symmetric part of the electron distribution function in Γ (solid curves) and L (dashed curves) valleys at 300 and 77 K. Electric field strengths are indicated in the figure. $N_d = 0$ [20].

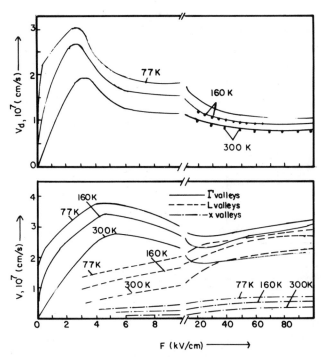

FIGURE 2-4-7. The average drift velocity of electrons v_d and the mean velocity in Γ, L, and X valleys vs. electric field at 77, 160, and 300 K [20]. Circles correspond to the experimental values of the drift velocity [42].

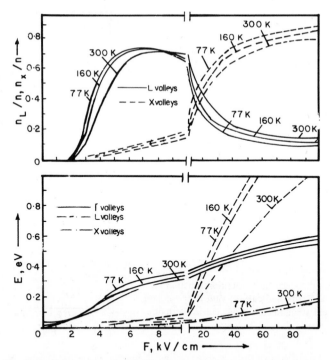

FIGURE 2-4-8. Fraction of electrons in L and X valleys, n_L and n_X (a) and mean energy ε in Γ, L, and X valleys (b) as a function of electric field strength at 77, 160, and 300 K, $N_d = 0$ [20].

decrease in v_Γ. At stronger fields the influence of impurity scattering on v_Γ weakens because of a significant electron heating in Γ valley. At $F > 7\,\text{kV/cm}$ and $N_d \leqslant 10^{17}\,\text{cm}^{-3}$ this influence vanishes. At these fields the reduction of v_d with the increase of N_d is caused mainly by the decrease of mean velocity in L valleys, v_L, where electron heating is significantly lower.

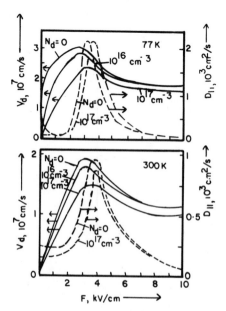

FIGURE 2-4-9. Dependence of the average drift velocity (solid curves) and of the longitudinal diffusion coefficient (dashes) on the electric field at 77 and 300 K taking into account ionized impurity scattering. Concentration indicated in the figure [20].

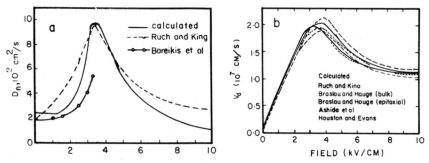

FIGURE 2-4-10. Comparison of computed and measured curves $D(F)$ (a) and $v(F)$ (b) [20]. $N_D = 0, T = 300$ K used in the calculation. Experimental data of Houston and Evans [42], Braslau and Hauge [43], Ashide *et al.* [44], Ruch and Kino [45], and Bareikis *et al.* [46].

Impurity scattering leads also to the decrease of electron fraction in the upper valleys, especially in L minima. At lower lattice temperatures the decrease of n_L is more pronounced, e.g., if at $T = 300$ K and $F = 3.5$ kV/cm the increase of N_d from 0 to 10^{17} cm^{-3} reduces n_L from 0.22 to 0.20; then at $T = 77$ K and at the same field strength, n_L reduces from 0.41 to 0.28. At larger electric fields the influence of impurity scattering is reduced. At $F > 7$ kV/cm it becomes negligibly small.

The calculated [20] and experimental [42–46] curves are compared in Fig. 2-4-10. Some quantitative disagreement in the experimental diffusion curves may be attributed to the impurity scattering, which could influence the results reported in Ref. 46. Besides, the measurements reported in Ref. 46 were performed at 10 GHz and the effective diffusion coefficient at this frequency should be somewhat smaller [20].

2-5. ELECTRON AND HOLE MOBILITIES AND DRIFT VELOCITIES

In low electric fields a carrier drift velocity is proportional to the electric field:

$$v_d = \mu F \qquad (2\text{-}5\text{-}1)$$

The low field mobility μ is determined by different scattering mechanisms. The polar optical phonon scattering, the impurity scattering, and deformation potential acoustic phonon scattering are dominant for electrons in low valleys in GaAs at low energies, i.e., at low electric fields. The piezoelectric scattering may also be important at low temperatures and in pure samples.

The relative contributions of three major scattering mechanisms are illustrated by Fig. 2-5-1 [47, 6], where the electron Hall mobility is shown as a function of temperature for three samples with different doping densities.

For elastic scattering (such as impurity scattering, acoustic, or piezoelectric scattering) one may use a relaxation time approximation in order to solve the Boltzmann equation in low electric fields. According to this approximation the collision integral in the Boltzmann equation is presented as $-f^1/\tau$, where τ is the relaxation time, which is assumed to be a function of the wave vector k, or in the isotropic media, a function of the electron energy E, and f^1 is the correction to the equilibrium distribution function caused by the electric field. For a long

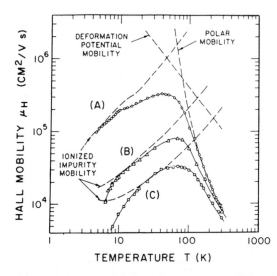

FIGURE 2-5-1. Temperature dependence of the electron Hall mobility μ_H (for $B = 5\,kG$) for three N-type GaAs samples, after Stillman $et\ al.$ [47]. They estimated donor densities of (A) 5×10^{13} cm^{-3}, (B) 10^{15} cm^{-3}, and (C) 5×10^{15} cm^{-3} for the three samples so identified, with $(N_0/N_d) \simeq$ 0.3–0.4 in each case. Expected contributions of three major processes toward the scattering are shown. Reproduced from the review paper by J. S. Blakemore [6]. Here $N_0 = N_d - N_a$, N_d is the donor density, and N_a is the acceptor density.

uniform sample kept at constant temperature the Boltzmann equation (2-2-3) becomes

$$q \frac{df_0}{dE} \boldsymbol{v}_k \cdot \boldsymbol{F} = \frac{f^1}{\tau} \qquad (2\text{-}5\text{-}2)$$

so that the distribution function in a low electric field may be found. Here $\boldsymbol{v}_k$ is the electron velocity in state k:

$$\boldsymbol{v}_k = \hbar k / m$$

Once f^1 is known, the current density j may be calculated as

$$\boldsymbol{j} = -\frac{q}{4\pi^3} \int \boldsymbol{v}_k f^1(\boldsymbol{k})\, dV_k \qquad (2\text{-}5\text{-}3)$$

Here dV_k is the infinitesimal volume in the k space and the integration is over the Brillouin zone. For the simple case of nondegenerate statistics and parabolic spherical bands

$$f_0 = \frac{n}{N_c} \exp\left(-\frac{E}{k_B T}\right) \qquad (2\text{-}5\text{-}4)$$

where energy E is measured from the bottom of the conduction band, n is the electron concentration, k_B is the Boltzmann constant, T is the lattice temperature, and N_c is the density of states in the conduction band. From Eqs. (2-5-3)–(2-5-4) we find

$$\mu = \frac{q\langle\tau\rangle}{m} \qquad (2\text{-}5\text{-}5)$$

where

$$\langle\tau\rangle = \frac{\int_0^\infty \tau E^{3/2} \exp(-E/k_B T)\, dE}{\int_0^\infty E^{3/2} \exp(-E/k_B T)\, dE} \qquad (2\text{-}5\text{-}6)$$

For the ionized impurity scattering the relaxation time τ_{ii} is given by the Brooks–Herring formula [48]:

$$1/\tau_{ii} = 2.4 - \frac{n}{K_0^2}\left(\frac{m}{m_e}\right)^{1/2} T^{-1.5} X^{-1.5}[\ln(1+\beta) - \beta/(1+\beta)] \qquad (2\text{-}5\text{-}7)$$

wkere K_0 is the static dielectric constant (12.9 for GaAs), n is the electron concentration in cm^{-3}, $X = E/k_B T$, and

$$\beta = 4.31 \times 10^{13} \frac{K_0}{n}\frac{m}{m_e} T^2 X \qquad (2\text{-}5\text{-}8)$$

In most cases β is large so that Eq. (2-5-7) may be simplified:

$$1/\tau_{ii} = 2.41 - \frac{n}{K_0^2}\left(\frac{m}{m_e}\right)^{1/2} T^{-1.5} X^{-1.5} \ln(\beta) \qquad (2\text{-}5\text{-}9)$$

For the acoustic deformation potential scattering we have [49]

$$1/\tau_{ac} = 4.15 \times 10^{19} \frac{(m/m_e)^{3/2} T^{3/2}(E_1)^2 X^{1/2}}{\rho u^2} \qquad (2\text{-}5\text{-}10)$$

Here ρ is the density in g/cm^3 ($\rho = 5.37\ g/cm^3$ for GaAs), E_1 is the acoustic deformation potential in eV ($E_1 = 7$ eV for GaAs), and u is the sound velocity in cm/s ($u = 5.2 \times 10^5$ cm/s for GaAs) (see Table 2-3-1).

For the piezoelectric scattering [50]

$$1/\tau_{pe} = 1.05 \times 10^7 (h_{14})^2 (4/C_t + 3/C_l)(m/m_e)^{1/2} T^{1/2} X^{-1/2} \qquad (2\text{-}5\text{-}11)$$

where h_{14} is the piezoelectric constant (1.41×10^7 V/cm for GaAs) and

$$C_l = (3C_{11} + 2C_{12} + 4C_{44})/5$$
$$C_t = (C_{11} - C_{12} + C_{44})/5$$

C_{11}, C_{12} and C_{44} are elastic constants in dyn/cm^2. For GaAs $C_{11} = 1.221 \times 10^{12}$ dyn/cm^2, $C_{12} = 0.566 \times 10^{12}$ dyn/cm^2, and $C_{44} = 0.599 \times 10^{12}$ dyn/cm^2.

Polar optical phonon scattering is a dominant scattering mechanism in lightly or moderately doped GaAs samples at room temperature (see Fig. 2-5-1). It is inelastic scattering because the optical phonon energy $\hbar\omega_0/q$ is comparable to the thermal energy at room temperature ($\hbar\omega_0/q = 0.0362$ eV for GaAs; see Table 2-3-1). Hence, the correction to the distribution function depends not only on the energy of the electron prior to the scattering but also on the energy of the final state and, strictly speaking, the relaxation time approximation cannot be used. The Boltzmann equation may be solved by the variation method [51] or by the iteration technique as proposed by Rode [51]. However, a simple and useful semiempirical relaxation time has been introduced by Harrison and Hauser [53]:

$$1/\tau_{op} = [1/X^{1/2} + \exp(Z)/(X-Z)^{1/2}]/\tau_0 \qquad (2\text{-}5\text{-}12)$$

where

$$\tau_0 = 1.93 \times 10^{-14} \frac{T^{1/2}}{T_{po}(m/m_e)^{1/2}(1/K_\infty - 1/K_0)} [\exp(Z) - 1] \quad (2\text{-}5\text{-}13)$$

Here T_{po} is the Einstein temperature ($T_{po} = \hbar\omega_0/k_B$, 420 K for GaAs), $Z = T_{po}/T$, K_∞ is the high-frequency dielectric constant (10.92 for GaAs).

The accuracy of this approximation is illustrated by Fig. 2-5-2, where the polar optical low field mobility for GaAs, calculated using the relaxation time approximation, is compared with the exact solution and with the Hall mobility. As can be seen from the figure, the agreement may be considered as adequate for most applications.

The parameters needed for a mobility calculation are given in Table 2-3-1.

The Hall mobility μ_H is related to the conductivity mobility μ defined above:

$$\mu_H = \mu r_H \quad (2\text{-}5\text{-}14)$$

where r_H is the Hall factor which depends on temperature, doping, magnetic field, and other factors. In very large magnetic fields r_H tends to unity. Hall measurements are performed under the conditions which in most cases correspond to a low magnetic field limit. When the relaxation time approximation is valid the Hall factor in the

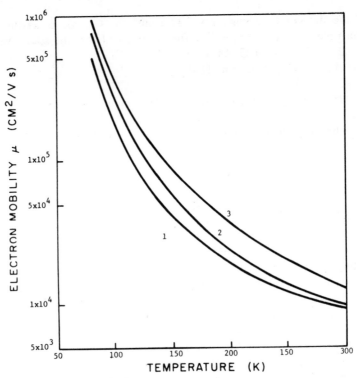

FIGURE 2-5-2. Comparison of mobility vs. temperature using the variation method [52] for conductivity mobility, curve 1, and the relaxation time-approximation formula [Eq. (2-5-12)] to calculate conductivity mobility, curve 2, and to calculate Hall mobility, curve 3 [53]. (Only optical polar scattering is taken into account.)

weak magnetic field is given by

$$r_H = \langle \tau^2 \rangle / \langle \tau \rangle^2 \tag{2-5-15}$$

(see, for example, Ref. 54). The relaxation time τ (also called momentum relaxation time) is found as

$$1/\tau = 1/\tau_{ii} + 1/\tau_{op} + 1/\tau_{ac} + 1/\tau_{pe} \tag{2-5-16}$$

The average relaxation time $\langle \tau \rangle$ is then calculated using Eq. (2-5-6). The average squared relaxation time $\langle \tau^2 \rangle$ is found as

$$\langle \tau^2 \rangle = \frac{\int_0^\infty \tau^2 E^{3/2} \exp(-E/k_B T)\, dE}{\int_0^\infty E^{3/2} \exp(-E/k_B T)\, dE} \tag{2-5-17}$$

In the case when

$$\tau = A/E^s \tag{2-5-18}$$

where A and s are constants, integrals in expressions for $\langle \tau \rangle$ and $\langle \tau^2 \rangle$ may be evaluated analytically. (For example, for the acoustic scattering and for piezoelectric scattering $s = -1/2$, for impurity scattering $s \approx 3/2$, etc.). In this case we have

$$\langle \tau \rangle^2 = A^2(k_B T)^{-2s}\Gamma(5/2 - 2s)/\Gamma(5/2) \tag{2-5-19}$$

$$\langle \tau^2 \rangle = [A(k_B T)^{-s}\Gamma(5/2 - s)/\Gamma(5/2)]^2 \tag{2-5-20}$$

(see, for example, Ref. 54). Here $\Gamma(N)$ is the gamma function. Some of the values of the gamma function are given in Table 2-5-1.

Using Eqs. (2-5-19) and (2-5-20) and Table 2-5-1 we find that $r_H = 1.93$ for the impurity scattering and $r_H = 1.18$ for the acoustic or piezoelectric scattering. The

TABLE 2-5-1. Gamma Function

x	$\Gamma(x)$
1	1.00
1.25	0.906
1.5	0.886
1.75	0.919
2	1.00

For positive integer values of x

$\Gamma(x) = (x - 1)!$

The values of $\Gamma(x)$ for $x < 1$ and for $x > 2$ may be evaluated using the following formulas:

$\Gamma(x) = \Gamma(x + 1)/x$

$\Gamma(x) = (x - 1)\Gamma(x - 1)$

Example: $\Gamma(2.5) = 1.5\Gamma(1.5) = 1.33$

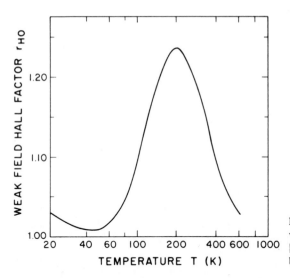

FIGURE 2-5-3. Hall factor for GaAs for a weak magnetic field for low doped GaAs [10]. Reproduced from the review paper by J. S. Blakemore [6].

Hall factor for the polar optical phonon scattering may be estimated from Fig. 2-5-2. The weak field Hall factor which includes all important scattering mechanisms was calculated for GaAs by Rode [10] (see Fig. 2-5-3).

In many papers the low field mobility is estimated as

$$1/\mu = 1/\mu_{\rm op} + 1/\mu_{\rm ii} + 1/\mu_{\rm ac} + 1/\mu_{\rm pe} \qquad (2\text{-}5\text{-}21)$$

where the terms on the right-hand side are calculated using the corresponding average relaxation times. Although this approach may introduce considerable error,

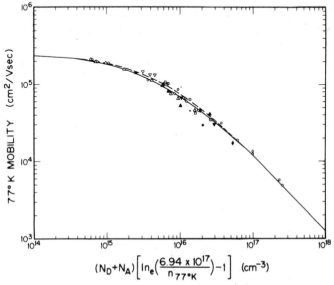

FIGURE 2-5-4. Summary of determined values for N_D and N_A in GaAs presented as the 77 K mobility as a function of $(N_D + N_A)$ times the Brooks–Herring screening factor to normalize the data [55]. $\Box$, Hall data [56]; ∇, Hall data [57]; $\bullet$, Hall data [58]; $\blacktriangle$, Hall data [59]; $\blacksquare$, Hall data [60]; $\blacktriangledown$, calculated [61]; $\blacklozenge$, calculated [62].

it may be adequate for a crude estimate of the low field mobility. A useful estimate for the electron mobility in low doped samples at temperature close to room temperature was proposed by Blakemore [6]:

$$\mu = 8000(300/T)^{2.3} \quad (\text{cm}^2/\text{V s}) \tag{2-5-22}$$

$$\mu_H = 9400(300/T)^{2.3} \quad (\text{cm}^2/\text{V s}) \tag{2-5-23}$$

Here μ and μ_H are electron drift and Hall mobility respectively.

The electron mobility at 77 K gives a good indication of the concentration of ionized impurities and is frequently considered as a merit figure characterizing the material quality.

Figure 2-5-4 from the paper of Wolfe and Stillman [55] shows the measured and calculated values of the mobility at 77 K. As can be seen from the figure, the expected values of the 77 K mobility may be close to 250,000 cm^2/V s in pure GaAs samples. Owing to the compensation the ionized impurity density may be quite a bit larger than the concentration of shallow donors. The results presented in Fig. 2-5-4 combined with the Hall data may be used for a crude estimate of the compensation ratio N_D/N_A.

As can be seen from Fig. 2-5-1, the mobility has a broad maximum roughly between 60 and 90 K. The maximum shifts toward low temperatures with the decrease in the impurity concentration. In the electron inversion layers in modulation doped structures (see Chapter 11) the ionized donors are removed from the electron gas and the impurity scattering is very much weakened. As a result the mobility continuously increases as temperature decreases and may reach several million at liquid helium temperature in a high quality material.

The temperature dependence of the low field mobility at temperatures above 300 K is shown in Fig. 2-5-5 [6]. The mobility decrease at elevated temperatures (which are typical for high-power GaAs devices) is quite substantial. Most of this decrease is caused by the intervalley transitions into higher minima of the conduction

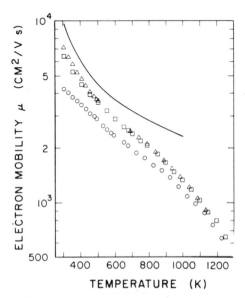

FIGURE 2-5-5. Data of Nichols et al. for Hall mobility (at 3.5 kG) vs. T for N-type GaAs epitaxial layers, above room temperature. The solid line shows the curve, calculated by Rode for high-purity N-type GaAs [10]. Doping of the three samples here is as follows: $N_d \simeq 4N_a \simeq 1.2 \times 10^{17}$ cm^{-3}; $N_d \simeq 4N_a \simeq 10^{16}$ cm^{-3}; $N_d \simeq 3N_a \simeq 2 \times 10^{15}$ cm^{-3}. Reproduced from the review paper by J. Blakemore [6].

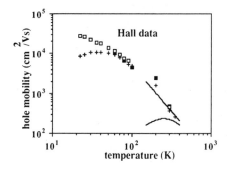

FIGURE 2-5-6. Temperature dependence of the hole mobility in GaAs. Experimental data from Hill [64]; Mears and Stradling [65]; Zshauer [66] (in increasing order). Solid lines – calculated using Eq. (2-5-22) for $N_I = 10^{14}$ cm^{-3} and $N_I = 10^{17}$ cm^{-3}.

band (mostly into the L minima), where the mobility is smaller than in the central Γ minimum and decreases faster with temperature.

The hole mobility in GaAs should be primarily determined by the heavy holes due to the larger density of states. The polar optical phonon scattering and impurity scattering are the dominant scattering mechanisms. The formulas for the electron scattering times given above may be used for crude estimates of the hole mobility and Hall factor if the electron effective mass is replaced by the effective mass of heavy holes ($m_{hh}/m_e \approx 0.5$). A useful empirical equation for temperatures close to 300 K was proposed by Blakemore [6]

$$\mu_p = [0.0025(T/300)^{2.3} + 4 \times 10^{-21} N_I (300/T)^{1.5}]^{-1} \qquad (2\text{-}5\text{-}22)$$

Here N_I is the concentration of ionized impurities in cm^{-3} and the hole mobility μ_p is in cm^2/V s.

The temperature dependence of the hole mobility is shown in Fig. 2-5-6 [15]. The dependence of the hole mobility on the hole concentration is reproduced in Fig. 2-5-7 [15].

The electron drift velocity in high electric fields has been discussed in Section 2-4 in regard to the results of the Monte Carlo simulation (see Figs. 2-4-7 and

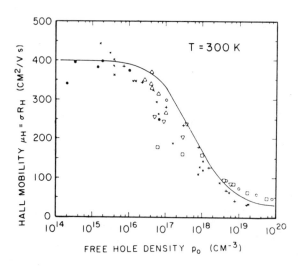

FIGURE 2-5-7. Variation of 300 K Hall mobility with hole concentration [15]. $\triangle$, Rosi et al. [67]; $\square$, Hill [68]; $\bullet$, Hill [64]; $\times$, Vilms and Garrett [69]; +, Rosztoczy et al. [70]; $\bigcirc$, Emeliyanenko et al. [71]; ∇, Gasanli et al. [72].

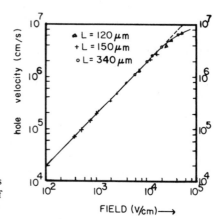

FIGURE 2-5-8. Hole velocity vs. electric field in GaAs after Dalal [74]. The dotted line is the extrapolation of the linear dependence from the low field region.

2-4-10). As can be seen from these figures, the most interesting feature of the drift velocity vs. electric field curve is the region with a negative differential mobility. Blakemore [6] suggested that the temperature dependence of the threshold field F_{th} of the negative differential mobility region, the maximum drift velocity v_p, and the electron drift velocity v_d at $F = 50\,\text{kV/cm}$ may be approximated by the following expressions:

$$F_{th} = 4.7 - T/215 \quad (\text{kV/cm}) \tag{2-5-23}$$

$$v_p = (3.3 - 0.004 \times T) \times 10^5 \quad (\text{m/s}) \tag{2-5-24}$$

$$v_d = (1.28 - 0.0015 \times T) \times 10^5 \quad (\text{m/s}) \tag{2-5-25}$$

According to the experimental data of Houston and Evans [42] the electron velocity monotonically decreases with the increase in the electric field for $F > F_{th}$. For $50\,\text{kV/cm} < F < 220\,\text{kV/cm}$ their results may be approximated as

$$v(F) = (0.928 - 1.35 \times 10^{-3}F) \times 10^5 \quad (\text{m/s}) \tag{2-5-26}$$

where F is in kV/cm. The Monte Carlo calculation by Shichijo and Hess [73] seems to indicate that at still larger electric fields (up to $500\,\text{kV/cm}$) the electron velocity is nearly constant (about $0.6 \times 10^5\,\text{m/s}$ at 300 K).

The drift velocity also depends on the doping density and the sample dimensions. In short samples the effective carrier velocity may be substantially higher due to overshoot or ballistic effects (see Section 2-10).

The hole drift velocity at 300 K is shown in Fig. 2-5-8 as a function of the electric field [74]. As can be seen from the figure, the hole saturation velocity is close to $1 \times 10^5\,\text{m/s}$ and the velocity saturation field is considerably larger than the threshold field F_{th}.

2-6. MAXWELLIAN AND DISPLACED MAXWELLIAN DISTRIBUTION FUNCTIONS AND PHENOMENOLOGICAL TRANSPORT EQUATIONS

Monte Carlo simulation technique described in Section 2-2 has become an important tool for semiconductor device modeling. This technique is equivalent to the exact solution of the Boltzmann equation. However, the parameters characterizing different scattering mechanisms, especially at high electron energies, and detailed structure of the energy bands are not yet established with the desired accuracy. Moreover, the validity of the Boltzmann equation for very large electric fields is being debated (see, for example, Refs. 75–79). Therefore less accurate but simpler analytical models describing the electron transport are quite useful. Such theories may make it possible to gain a better insight into the device physics. We will describe three simplified models which are frequently used in the literature: phenomenological equations based on the Maxwellian or displaced Maxwellian distribution function (see, for example, Refs. 80–83), phenomenological equations based on the field-dependent velocity and diffusion (see, for example, Ref. 84), and phenomenological equations based on energy-dependent relaxation times [85].

In this section we discuss the equations based on the Maxwellian and displaced Maxwellian distribution functions.

When the electrons exchange energy through the electron–electron collisions at a faster rate than they lose it through electron–phonon scattering, the symmetrical part of the distribution function becomes Maxwellian with the effective electron temperature T_e:

$$f = c \exp(-E/k_B T_e) \tag{2-6-1}$$

At even higher electron concentrations the electron–electron collisions redistribute both electron energy and momentum leading to the so-called displaced Maxwellian distribution function

$$f(\boldsymbol{p}) = c \exp\left[-\frac{(\boldsymbol{p} - \boldsymbol{p}_0)^2}{2mk_B T_e} \right] \tag{2-6-2}$$

where $\boldsymbol{p}_0$ is the drift momentum.

For this approximation to be valid the average time between the electron–electron collisions should be much smaller than the momentum relaxation time. If the number of electrons in the conduction band is determined by the concentration of the ionized donors this can never be true and therefore the displaced Maxwell distribution function can be considered only as a crude approximation.

If, however, this distribution function is assumed, the phenomenological transport equations can be derived from the Boltzmann equation. These equations are derived by substitution of Eq. (2-6-2) into the Boltzmann equation [80]. Integration of this equation then yields the particle conservation equation for each valley:

$$\frac{\partial n_i}{\partial t} + \nabla(\boldsymbol{v}_i \cdot n_i) = \left(\frac{\partial n_i}{\partial t}\right)_c \tag{2-6-3}$$

where n_i is the concentration of electrons in the ith valley, $\boldsymbol{v}_i$ is the average electron drift velocity in the ith valley, and $(\partial n_i/\partial t)_c$ is the collision term describing the intervalley transitions.

Multiplying the Boltzmann equation by momentum before the integration yields the continuity equation

$$\frac{\partial \boldsymbol{p}_i}{\partial t} + \nabla(\boldsymbol{v}_i \cdot \boldsymbol{p}_i) = qn_i \boldsymbol{F} - \nabla(n_i k_B T_i) + \left(\frac{\partial \boldsymbol{p}_i}{\partial t}\right)_c \tag{2-6-4}$$

where p_i is the momentum density in the ith valley, F is the electric field, and T_i is the electron temperature in the ith valley.

Finally multiplying the Boltzmann equation by the squared velocity before the integration we find the equation for the kinetic energy density in the ith valley W_i

$$\frac{\partial W_i}{\partial t} + \nabla \cdot (v_i W_i) = q n_i v_i \cdot F - \nabla \cdot (v_i n_i k_B T_i) - \nabla \cdot q_i + \left(\frac{\partial W_i}{\partial t}\right)_c \qquad (2\text{-}6\text{-}5)$$

where

$$W_i = \tfrac{3}{2} k_B T_i n_i + \frac{m_i n_i v_i^2}{2} \qquad (2\text{-}6\text{-}6)$$

The heat flow vector q_i appears as a third moment of the distribution function. With the assumption of a displaced Maxwellian distribution (or any symmetric distribution) these moments vanish. In spite of this, we will allow for the electron "heat" conduction by assuming [80]

$$q_i = -K_i \nabla T_i \qquad (2\text{-}6\text{-}7)$$

where K_i is the heat conductivity of the electron gas in the ith valley. This term will be included because they are believed to represent some of the most important effects of a non-Maxwellian distribution function. K_i could be estimated using the Wiedemann–Franz law [86]

$$\frac{K_i}{\sigma_i T_i} = L \qquad (2\text{-}6\text{-}8)$$

where σ_i is the conductivity of electrons in the ith valley and L is the Lorentz number. The calculation of L yields [86]

$$L = \frac{\pi^2}{3} \frac{k_B^2}{q^2} \qquad (2\text{-}6\text{-}9)$$

for metals,

$$L = \frac{3}{2} \frac{k_B^2}{q^2} \qquad (2\text{-}6\text{-}10)$$

for the classical electron gas, and

$$L \geqslant \frac{k_B^2}{q^2} \qquad (2\text{-}6\text{-}11)$$

for semiconductors.

The collision terms can be defined using the relaxation time approximation. In case of just two valleys we find

$$\left(\frac{\partial n_i}{\partial t}\right)_c = -\frac{n_i}{\tau_{nij}(W_i)} + \frac{n_j}{\tau_{nji}(W_j)} \qquad (2\text{-}6\text{-}12)$$

$$\left(\frac{\partial \boldsymbol{p}_i}{\partial t}\right)_c = -\frac{m_i n_i \boldsymbol{v}_i}{\tau_{pi}(W_i)} \tag{2-6-13}$$

$$\left(\frac{\partial W_i}{\partial t}\right)_c = -\frac{n_i(W_i - \frac{3}{2}k_B T_0)}{\tau_{wii}(W_i)} - \frac{n_i W_i}{\tau_{wij}(W_i)} + \frac{n_j W_j}{\tau_{wji}(W_j)} \tag{2-6-14}$$

Here τ_{nij} is the particle relaxation time for scattering from the ith to the jth valley. It depends on the average kinetic energy of electrons W_i in the ith valley.

The relaxation time τ_{pi} for momentum includes both intravalley and intervalley scattering:

$$\frac{1}{\tau_{pi}} = \frac{1}{\tau_{pii}} + \frac{1}{\tau_{pij}} \tag{2-6-15}$$

where the two terms represent intravalley and intervalley scattering, respectively. Equivalent valley scattering is included in the intravalley term.

The energy relaxation is described by three terms: the first describes the relaxation toward the lattice temperature T_0 in intravalley scattering, including equivalent valley scattering; the second describes the rate of energy leaving the ith valley in intervalley scattering; and the third describes the rate of energy supplied to the ith valley by intervalley scattering from the jth valley.

The detailed analysis of the transport equations discussed above is given in Ref. 80. The expressions for the relaxation times can be found in Ref. 81.

GaAs device simulation based on this set of transport equation is described, for example, in Ref. 91. Recent experimental data [157, 158] and theoretical studies [159] clearly show the inadequacy of the effective electron temperature concept in very short devices.

2-7. PHENOMENOLOGICAL EQUATIONS BASED ON THE FIELD-DEPENDENT VELOCITY AND DIFFUSION

In a low electric field the drift and diffusion current densities for electrons and holes are given by

$$\boldsymbol{j}_n = q(n\mu_n \boldsymbol{F} + D_n \nabla n) \tag{2-7-1}$$

$$\boldsymbol{j}_p = q(p\mu_p \boldsymbol{F} - D_p \nabla p) \tag{2-7-2}$$

where the low field mobilities and diffusion coefficients are given by

$$\mu_n = \frac{q\langle \tau_n \rangle}{m_n}, \qquad D_n = \frac{\mu_n k_B T}{q} \tag{2-7-3}$$

$$\mu_p = \frac{q\langle \tau_p \rangle}{m_p}, \qquad D_p = \frac{\mu_p k_B T}{q} \tag{2-7-4}$$

Here $\langle \tau_n \rangle$, $\langle \tau_p \rangle$, m_n, and m_p are momentum relaxation times and effective masses of electrons and holes, respectively. In a high electric field when carriers are heated by the electric field the drift velocity is no longer proportional to the electric field. In this case the following phenomenological equations are frequently used in the device modeling:

$$j_n = q[-nv_n(F) + D_n(F)\nabla n] \qquad (2\text{-}7\text{-}5)$$

$$j_p = q[pv_p(F) - D_p(F)\nabla p] \qquad (2\text{-}7\text{-}6)$$

Here $v_n(F)$, $v_p(F)$, $D_n(F)$, $D_p(F)$ are assumed to be the same functions of electric field as computed or measured for the uniform sample under the steady state conditions. In a low electric field

$$v_n = -\mu_n F$$

$$v_p = \mu_p F$$

and Eqs. (2-7-5) and (2-7-6) reduce to Eqs. (2-7-1) and (2-7-2). The critical analysis of the applicability of equations (2-7-5) and (2-7-6) in a high electric field may be found in Ref. 80. These equations are not derived from the Boltzmann equation. They may lead to qualitative errors in describing the hot electron behavior, for example, in magnetic field [88].

At high frequencies comparable to the inverse energy relaxation time (which is of the order of 2 ps for electrons in the central minimum of the conduction band in GaAs), the velocity and diffusion do not instantaneously follow the variations of the electric field. Therefore, the effective differential mobility, for example, becomes frequency dependent (see Fig. 2-7-1) and Eqs. (2-7-5) and (2-7-6) do not apply [89].

The advantage of using Eqs. (2-7-5) and (2-7-6) or even similar equations with field-independent diffusion coefficient is a relative simplicity of the analysis which makes it possible to achieve some insight into the device physics.

Equations (2-7-5) and (2-7-6) [or Eqs. (2-7-1) and (2-7-2) in a low electric field] should be solved together with the Poisson equation

$$\nabla \cdot F = \rho/\varepsilon \qquad (2\text{-}7\text{-}7)$$

and continuity equations

$$\frac{\partial n}{\partial t} - \frac{1}{q}\nabla \cdot j_n = G - R \qquad (2\text{-}7\text{-}8)$$

$$\frac{\partial p}{\partial t} + \frac{1}{q}\nabla \cdot j_p = G - R \qquad (2\text{-}7\text{-}9)$$

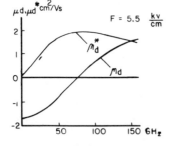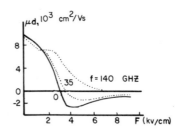

FIGURE 2-7-1. Electron differential mobility in GaAs at different frequencies [89]. μ_d is a real part of the differential mobility; μ_d^* is an imaginary part of the differential mobility.

Here ε is the dielectric permittivity,

$$\rho = q(N_D - N_A - n + p) + \rho_T \qquad (2\text{-}7\text{-}10)$$

is the space charge density, N_D is the concentration of the ionized donors, N_A is the concentration of the ionized acceptors, ρ_T is the charge density of the charged recombination centers and traps, G is the generation rate of electron–hole pairs (due to the light excitation or impact ionization), and R is the recombination rate.

The detailed theory that allows us to calculate ρ_T and R was developed by Sah, Noyce, and Shockley [90]. According to this theory the net rate of recombination per unit volume R for just one recombination level E_t is given by

$$R = \frac{pn - n_i^2}{\tau_{pl}(n + n_t) + \tau_{nl}(p + p_t)} \qquad (2\text{-}7\text{-}11)$$

Here n_i is the intrinsic concentration, τ_{nl} and τ_{pl} are electron and hole lifetimes,

$$n_t = n_i \exp\left(\frac{E_t - E_i}{k_B T}\right) \qquad (2\text{-}7\text{-}12)$$

$$p_t = n_i \exp\left(\frac{E_i - E_t}{k_B T}\right) \qquad (2\text{-}7\text{-}13)$$

Another useful equation may be derived from equations given above when the trapped charge ρ_T in Eq. (2-7-10) may be neglected. Indeed, subtracting Eq. (2-7-8) from Eq. (2-7-9), we obtain

$$q\frac{\partial}{\partial t}(p - n) + \nabla \cdot (j_n + j_p) = 0 \qquad (2\text{-}7\text{-}14)$$

From Eq. (2-7-10), neglecting ρ_T, we find

$$q\frac{\partial}{\partial t}(p - n) = \frac{\partial \rho}{\partial t} \qquad (2\text{-}7\text{-}15)$$

Differentiating Eq. (2-7-7) with respect to time and substituting the resulting expression for $(\partial/\partial t)(p - n)$ into the continuity equation, we obtain

$$\varepsilon\frac{\partial}{\partial t}\nabla F + \nabla \cdot (j_n + j_p) = 0 \qquad (2\text{-}7\text{-}16)$$

Integration of Eq. (2-7-16) over the space coordinates leads to

$$I(t) = j_n + j_p + \varepsilon\frac{\partial F}{\partial t} \qquad (2\text{-}7\text{-}17)$$

where $I(t)$ is the total current density (including the displacement current $\varepsilon\partial F/\partial t$:

$$i = \int_s I \cdot ds \qquad (2\text{-}7\text{-}18)$$

where i is the current in the circuit and s is the sample cross section. In a one-dimensional case

$$I = i/s \qquad (2\text{-}7\text{-}19)$$

for a sample with a constant cross section.

2-8. ENERGY-DEPENDENT RELAXATION TIMES AND PHENOMENOLOGICAL TRANSPORT EQUATIONS

The major drawback of the phenomenological equation for the current density (2-7-5) is that the inertia effects due to the finite electron mass are not taken into account.

In case when the electron diffusion may be neglected, the following set of phenomenological equations has been proposed to substitute for Eq. (2-7-5) [85]:

$$\frac{dm(E)v}{dt} = -qF - \frac{mv}{\tau_p(E)} \qquad (2\text{-}8\text{-}1)$$

$$\frac{dE}{dt} = \frac{j \cdot F}{n} - \frac{E - E_0}{\tau_E(E)} \qquad (2\text{-}8\text{-}2)$$

$$j = -qnv \qquad (2\text{-}8\text{-}3)$$

Here E is the average electronic energy, τ_p is the effective momentum relaxation time, τ_E is the effective energy relaxation time, $E_0 = \frac{3}{2}kT_0$, where T_0 is the lattice temperature, $m(E)$ is the effective mass of electrons. $m(E)$, $\tau_p(E)$, and $\tau_E(E)$ can be determined from the steady state Monte Carlo calculations for the constant electric field [see Sections (2-4) and (2-5)]. For this fit we use the following expressions:

$$\tau_p(E) = \left\{ \frac{m[F(E)]v[F(E)]}{qF(E)} \right\}_{\text{steady state}} \qquad (2\text{-}8\text{-}4)$$

$$\tau_E(E) = \frac{E - E_0}{q\{F(E)v[F(E)]\}_{\text{steady state}}} \qquad (2\text{-}8\text{-}5)$$

where the steady state curves are taken directly from the Monte Carlo calculations. This empirical approach has been shown to be in a very good agreement with the direct Monte Carlo simulation for nonequilibrium conditions [85, 34, 36] (see Fig. 2-8-1). The agreement is very good not only for GaAs but also for InP and $Ga_{0.47}In_{0.53}As$ samples (see Fig. 2-8-2 [36]).

The dependences of the steady state velocity, energy, and effective mass on the electric field deduced from the Monte Carlo simulation are shown in Fig. 2-8-3, 2-8-4, and 2-8-5 for samples with doping densities 10^{17} cm^{-3} and 3×10^{17} cm^{-3} [34]. These curves have been used for the simulation of GaAs field effect transistors [34].

In short semiconductor devices the diffusion effects may be very important. These effects have been incorporated into this model by using the following equation for the electric current density:

$$j = -qnv + qD(E)\nabla n \qquad (2\text{-}8\text{-}6)$$

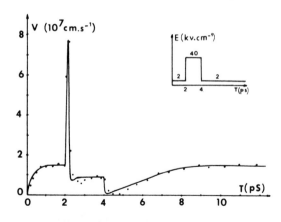

FIGURE 2-8-1. Electron velocity vs. time when the electric field step is applied [34]. Lines, solution of phenomenological equations based on the relaxation time approximation [see Eqs. (2-8-1)–(2-8-3)]; points, Monte Carlo simulation.

where the diffusion coefficient $D(E)$ is determined from the steady state Monte Carlo simulation as

$$D = \text{var}[v(E)\tau_p(E)] \qquad (2\text{-}8\text{-}7)$$

This model has been criticized on the grounds that the results obtained using this model were in disagreement with the results of a simulation in the frame of the temperature model (see Section 2-6) [91]. However, as has been pointed out in Section 2-6, the temperature model is also an empirical model and it cannot be used for checking the validity of this system of equations. As will be shown in Section 2-9, the results obtained in the frame of this model are in good agreement with the nonstationary Monte Carlo simulations of an n–i–n structure [92]. The agreement with the transient Monte Carlo simulations (see, for example, Figs. 2-8-1 and 2-8-2 and Refs. 85, 34, 36) is also quite impressive. Nevertheless, further studies are needed to establish the range of the applicability of this model, which may develop into a useful tool for modeling transient nonstationary effects in short device structures.

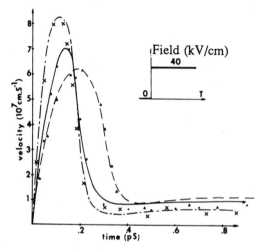

FIGURE 2-8-2. Electron velocity vs. time when the electric field step is applied [36]. GaAs, InP, $Ga_{0.47}In_{0.53}As$, relaxation time equations; GaAs, InP, $Ga_{0.47}In_{0.53}As$, Monte Carlo results.

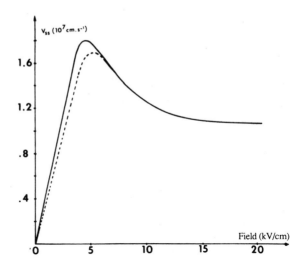

FIGURE 2-8-3. Steady state drift
velocity vs. electric field [34].
——, $N_d = 10^{17}$ cm^{-3}; ---, $N_d = 3 \times 10^{17}$ cm^{-3}.

2-9. ELECTRON TRANSPORT IN SMALL SEMICONDUCTOR DEVICES

Characteristic dimensions of modern semiconductor devices may be well below
1 μm (see, for example, Refs. 93-101). Under such conditions the length of the
device may become comparable with the mean free path of carriers in a semiconductor and the transit time may become comparable to or even smaller than an average
relaxation time. As was pointed by Ruch [24], the equilibrium distribution is not
established under such conditions and the average drift velocity of electrons in the
channel may substantially exceed the value of the saturation or even peak velocity
in long samples. This increase in the velocity due to the nonstationary effects has
been referred to as "overshoot" effects. In the limiting case when the electron
collisions with the lattice phonons and impurities may be totally neglected, the
electron transport was called "ballistic" [102]. Because of the possible improvement
in device performance due to higher velocities in smaller size devices this area of
research has attracted considerable attention [102, 92, 98, 99, 103–139]. The electron
transport in short device structures is quite complicated. More theoretical and

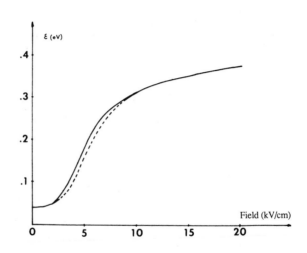

FIGURE   2-8-4.   Steady   state
average electron energy vs. electric
field [34]. ——, $N_d = 10^{17}$ cm^{-3};
..., $N_d = 3 \times 10^{17}$ cm^{-3}.

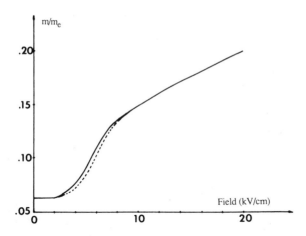

FIGURE 2-8-5. Steady state effective mass ratio vs. electric field [34]. ———, $N_d = 10^{17}$ cm^{-3}; ⋯, $N_d = 3 \times 10^{17}$ cm^{-3}.

experimental work has to be done to establish a better understanding of the electron motion in very small devices. In this section we will present an approximate and crude descriptions of new effects that may occur in submicron structures. Novel device structures which attempt to utilize the ballistic motion and velocity enhancement will be considered in Chapter 12.

We will start from the consideration of the electron transport in a small device in a low electric field [116], even though most, if not all, practical devices operate in the high field region where the electron energy may considerably exceed the thermal energy. However, even in the low field region new and interesting effects occur in submicron structures. Moreover, the analysis of the low field transport will enable us to introduce characteristic scales determining the electron motion in short structures.

When device dimensions are much bigger than the de Broglie wavelength (approximately 260 Å for GaAs devices at 300 K), the electron motion can be described by the semiclassical equation of motion. In this regime the electron transport is determined by three characteristic lengths: device length a, mean free path λ, and Debye radius $L_D = \left(\dfrac{\varepsilon k_B T}{q^2 N_d}\right)^{1/2}$ (see Fig. 2-9-1). When

$$a \gg \lambda \qquad (2\text{-}9\text{-}1)$$

the conventional collision dominated transport takes place. At low voltages, which is the case we consider here, the spatial carrier distribution in the sample is uniform when

$$a \gg L_D \qquad (2\text{-}9\text{-}2)$$

Equation (2-9-2) is equivalent to the requirement

$$V_T \ll U_{\text{po}} \qquad (2\text{-}9\text{-}3)$$

where $V_T = k_B T/q$ is the thermal voltage and

$$U_{\text{po}} = q N_d a^2 / 2\varepsilon_r \varepsilon_0 \qquad (2\text{-}9\text{-}4)$$

is the pinch-off voltage. Here N_d is the doping density assumed to be uniform, and

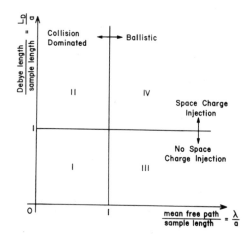

FIGURE 2-9-1. Classification of conduction mechanisms depending on three lengths: device length a, mean free path λ, and Debye length L_D.

$\varepsilon_o \varepsilon_r$ is the dielectric permittivity. The regime defined by inequalities (2-9-1) and (2-9-2) corresponds to region I of Fig. 2-9-1.
When

$$a \ll L_D \quad \text{and} \quad a \gg \lambda \qquad \qquad (2\text{-}9\text{-}5)$$

the distribution of carriers in the sample is not uniform. This regime, which may be defined as the space charge injection in the collision dominated regime, corresponds to region II of Fig. 2-9-1.
When the inequality opposite to inequality (2-9-5) holds,

$$a \ll \lambda \qquad \qquad (2\text{-}9\text{-}6)$$

we have a ballistic transport (regions III and IV of Fig. 2-9-1). Here again we may have a uniform carrier distribution when

$$L_D \ll a \qquad \qquad (2\text{-}9\text{-}7)$$

(region III of Fig. 2-9-1) or a nonuniform carrier distribution (space charge injection) when

$$L_D \gg a \qquad \qquad (2\text{-}9\text{-}8)$$

We will first consider a low field electron transport in a uniform sample (regions I and III of Fig. 2-9-1). This calculation makes it possible to follow the transition from a collision-dominated regime in a long sample (Ohmic conduction) to a collision-free transport in a very short sample. We will then consider the effects in samples with nonuniform carrier distribution (regions II and IV of the diagram in Fig. 2-9-1).

2-9-1. Low-Field Electron Transport in a Uniform Sample [116]

The contacts (i.e., boundary conditions) should play an important role in determining the characteristics of a small semiconductor device. In this section we will limit ourselves to the simplest possible device structure—an n^+-n^--n^+ diode

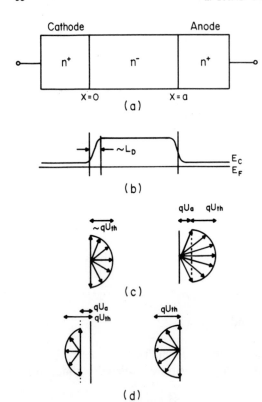

FIGURE 2-9-2. (a) n^+-n^--n^+ diode structure. (b) Energy band diagram of the n^+-n-n^+ structure at thermal equilibrium. (c) Velocity distribution of electrons traveling ballistically from cathode to anode (acceleration). (d) Velocity distribution of electrons traveling ballistically from anode to cathode (deceleration).

(see Fig. 2-9-2). We assume that the length, a, of the n^- region is much bigger than the Debye length L_D in the same region. Hence the electron concentration is uniform in most of the active n^- region.

We start from the Boltzmann transport equation in the relaxation time approximation:

$$\frac{\partial f_k}{\partial t} + v_k \cdot \nabla f_k + k \cdot \nabla_k f_k = -\frac{f_k - f_k^{(0)}}{\tau_k} \tag{2-9-9}$$

If we let

$$f_k = f_k^{(0)} + f_k^{(1)} \tag{2-9-10}$$

and assume that

$$|f_k^{(1)}| \ll f_k^{(0)} \tag{2-9-11}$$

then for a uniform field F, Eq. (2-9-9) reduces to

$$\frac{\partial f_k^{(1)}}{\partial t} + qF \cdot v_k \frac{\partial f_k^{(0)}}{\partial E} = -\frac{f_k^{(1)}}{\tau_k} \tag{2-9-12}$$

To solve Eq. (2-9-9) we used the following approximation:

$$\frac{1}{\tau_{k,\,\text{eff}}} = \frac{1}{\tau_{kb}} + (1 - p)\frac{|v_k \cos \theta|}{a/2} \tag{2-9-13}$$

where $1/\tau_{kb}$ is the bulk scattering rate, $|v_k \cos \theta|/a$ is the inverse of the carrier transit time across the n region, and p is an energy-dependent constant. The second term in the right-hand part of Eq. (2-9-13) may be interpreted as describing the electron scattering by the sample boundaries with the scattering rate inversely proportional to the transit time.

The physical reasons for using Eq. (2-9-13) are as follows. For a long collision-dominated semiconductor, Eq. (2-9-13) reduces to the familiar relaxation time approximation. For a short collision-free transport with $p = 0$, it leads to the following expression for the sample conductance [111, 112, 116]:

$$G = \left(\frac{1}{2\pi}\right)^{1/2} \frac{q^2 n}{(mk_B T)^{1/2}} \qquad (2\text{-}9\text{-}14)$$

Equation (2-9-14) has been derived in Ref. 112 assuming a total loss of the drift velocity when electrons scatter from the contacts. It is interesting to notice that this equation is identical to the equation predicted by the thermionic emission theory for small applied voltage $U \ll V_T$.

The same result [i.e., Eq. (2-9-14)] can be obtained assuming that the finite sample length limits the time the electron spends in the electric field (see Ref. 111). Furthermore we can also derive Eq. (2-9-14) in the following way. The ballistic electron wave packet moves according to the semiclassical equation of motion, and thus Liouville's theorem is satisfied [140]. Hence

$$J(x) = J_+(x) + J_-(x) \qquad (2\text{-}9\text{-}15)$$

where

$$J_+(x) = -qn \int_0^\infty \left[v_{x0}^2 + \frac{2qU(x)}{m}\right]^{1/2} f_0(v_{x0})\, dv_{x0} \qquad (2\text{-}9\text{-}16)$$

$$J_-(x) = -qn \int_{-\{2q[U_a - U(x)]/m\}^{1/2}}^{-\infty} \left[v_{x0}^2 + \frac{2q(U - U_a)}{m}\right]^{1/2}$$
$$\times f_0(v_{x0})\, dv_{x0} \qquad (2\text{-}9\text{-}17)$$

Here $J_+(x)$ is the current density flowing from cathode to anode and $J_-(x)$ vice versa (see Figs. 2-9-2c and 2-9-2d), and $f_0(v_{x0})$ is the thermal equilibrium distribution function at the cathode and/or anode.

First-order expansion of Eqs. (2-9-15)–(2-9-17) with respect to qU/kT leads again to Eq. (2-9-14).

Thus, the limiting solution for a very short sample which follows from Eq. (2-9-13) seems to be correct as it can be derived using several different approaches. Hence, Eq. (2-9-13) describes correctly both limiting cases—a very short and a very long sample—and we use this heuristic equation on this ground.

The energy-dependent constant p is a quantum mechanical energy-dependent reflection coefficient at the n^+-n interfaces. This reflection is negligible in most cases, because the de Broglie wavelength of an electron (approximately 260 Å at 300 K for GaAs) is much less than the Debye length (approximately 1360 Å at 300 K for GaAs, whose electron density is 10^{15} cm^{-3}). This situation is quite analogous to the specular and diffusive scattering at the surface for electron transport parallel to the surface of the film [141].

Equation (2-9-13) with $\cos \theta$ being replaced by $\sin \theta$ was used by Many *et al.* [142] for the carrier transport parallel to the surface of thin film semiconductors at flat band. The agreement between the exact solution of the Boltzmann transport equation and the approximation using this equation was shown to be quite good [143].

Here again it is very interesting to notice that there is indeed some similarity in the mathematics of the electron transport perpendicular and parallel to the surface of a semiconductor film, even though the physical mechanisms are quite different from each other. For example, $2|v_k \cos \theta|/L$ is a measure of the diffusive scattering rate for the electron transport parallel to the surface. But in the perpendicular transport case, the inverse of this parameter is proportional to the energy gained from the electric field, which should be limited by the applied voltage. This also can be interpreted by invoking the concept of the boundary scattering.

For the electric field

$$F = F_0 e^{j\omega t} \tag{2-9-18}$$

the solution of Eqs. (2-9-12) and (2-9-13) is given by

$$f_k^{(1)} = \frac{qF \cdot v_k \, \partial f_k^{(0)}/\partial E}{1/\tau_k + |v_k \cos \theta|/a_{\text{eff}} + j\omega} \tag{2-9-19}$$

where

$$a_{\text{eff}} = \frac{a/2}{1 - p} \tag{2-9-20}$$

For a nondegenerate *n*-type semiconductor with a parabolic spherical conduction band

$$f_k^{(0)} = \frac{n}{N_c} \exp(-E/k_B T) \tag{2-9-21}$$

$$N_c = 2 \left(\frac{m k_B T}{2\pi\hbar^2} \right)^{3/2} \tag{2-9-22}$$

Substituting (2-9-21) and (2-9-22) into Eq. (2-9-19), one can show that Eq. (2-9-11) leads to the requirement

$$U_a \ll k_B T/q \tag{2-9-23}$$

where U_a is the applied voltage.

When this condition is fulfilled, the changes in the distribution function caused by the applied bias are small even in the ballistic mode of operation. This leads to the voltage-independent conductance at low voltages, which is consistent with the results of experiments [98, 99, 110], with those Monte Carlo simulation [92], and with those of analytical derivation [117].

The current density is given by

$$J = -\frac{q}{4\pi^3} \int v_k f_k^{(1)} \, dV_k + \varepsilon_r \varepsilon_0 \frac{\partial F}{\partial t}$$

$$= \frac{q^2}{4\pi^3} \int \frac{v_k F \cdot v_k \, \partial f_k^{(0)}/\partial E}{1/\tau_k + |v_k \cos \theta|/a_{\text{eff}} + j\omega} \, dV_k + \varepsilon_r \varepsilon_0 \frac{\partial F}{\partial t} \tag{2-9-24}$$

We consider the bulk limiting case first (a sample much larger than the mean free path).

When a_{eff} tends to infinity, Eq. (2-9-24) reduces to

$$J = \frac{q^2}{4\pi^3} \int \frac{v_k F \cdot v_k \, \partial f_k^{(0)}/\partial E \, dV_k}{1/\tau_k + j\omega} + \varepsilon_r \varepsilon_0 \frac{\partial F}{\partial t} \qquad (2\text{-}9\text{-}25)$$

In a simple case when $\tau_{\bar{k}} = \tau = $ const Eq. (2-9-25) corresponds to the equivalent circuit shown in Fig. 2-9-3a, which can be transformed into the equivalent circuit shown in Fig. 2-9-3b. The value of components obtained from Eq. (2-9-25) is then given by

$$G(0) = \frac{nq^2\tau}{ma} \qquad (2\text{-}9\text{-}26)$$

$$C = \frac{\varepsilon_r \varepsilon_0}{a} \qquad (2\text{-}9\text{-}27)$$

$$L = \frac{ma}{nq^2} \qquad (2\text{-}9\text{-}28)$$

The resonant frequency is equal to the plasma frequency

$$\omega_0 = \omega_p = \frac{1}{(LC)^{1/2}} = \left(\frac{nq^2}{\varepsilon_r \varepsilon_0 m}\right)^{1/2} \qquad (2\text{-}9\text{-}29)$$

and the Q factor is equal to

$$Q = \frac{\omega_p L}{R} = \omega_p \tau \qquad (2\text{-}9\text{-}30)$$

At low frequencies such that $\omega L \ll 1/G$ or $\omega\tau \ll 1$, $G(\omega) = G(0)$ and $X_L = G^2(0)L$. Thus the negative susceptance due to the inductance becomes inversely proportional to a.

Equation (2-9-24) can be simplified:

$$J = \frac{4}{\sqrt{\pi}} \frac{q^2 nF}{m} \int_0^{\infty} x^{3/2} e^{-x}\tau(x)[R(x) - jI(x)] \, dx + \varepsilon_r \varepsilon_0 j\omega F \qquad (2\text{-}9\text{-}31)$$

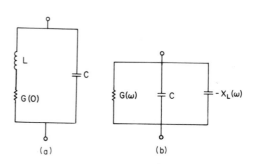

FIGURE 2-9-3. (a) Equivalent circuit for the bulk limited sample. (b) Transformation of Fig. 2-9-3a.

where

$$R(x) = V\left[\frac{1}{2} - V + \frac{V^2 - W^2}{2}\ln\frac{(1+V)^2 + W^2}{V^2 + W^2}\right.$$
$$\left. + 2VW\tan^{-1}\frac{W}{W^2 + V(1+V)}\right] \tag{2-9-32}$$

$$I(x) = VW\left[1 - V\ln\frac{(1+V)^2 + W^2}{V^2 + W^2} + \frac{V^2 - W^2}{W}\tan^{-1}\frac{W}{W^2 + V(1+V)}\right] \tag{2-9-33}$$

and

$$V(x) = \frac{a/2}{\tau v_k} = \frac{\alpha_0}{\tau\sqrt{x}} \tag{2-9-34}$$

is the normalized scattering rate,

$$W(x) = \frac{\omega a/2}{v_k} = \frac{\omega\alpha_0}{\sqrt{x}} \tag{2-9-35}$$

is the normalized frequency,

$$\alpha_0 = \frac{a}{2}\left(\frac{m}{2k_BT}\right)^{1/2} \tag{2-9-36}$$

and $x = E/k_BT$ is the normalized energy. Here we have assumed $p = 0$.

We first consider the current density J in two limited cases at low frequency ($W \ll 1$). When $V \gg 1$ (i.e., a tends to infinity—bulk limit)

$$R(x) = 1/3 \tag{2-9-37}$$

This leads to

$$G = \frac{nq^2\tau}{ma} \tag{2-9-38}$$

as it should.

Furthermore,

$$I(x) = \frac{\omega}{3\tau} \tag{2-9-39}$$

Thus

$$L = X_L/G^2(0) = \frac{ma}{nq^2} \tag{2-9-40}$$

On the other hand, when $V \ll 1$ (which corresponds to the short sample limit, i.e., $a_{\text{eff}} \ll \lambda$)

$$R(x) = V(x)/2 \tag{2-9-41}$$

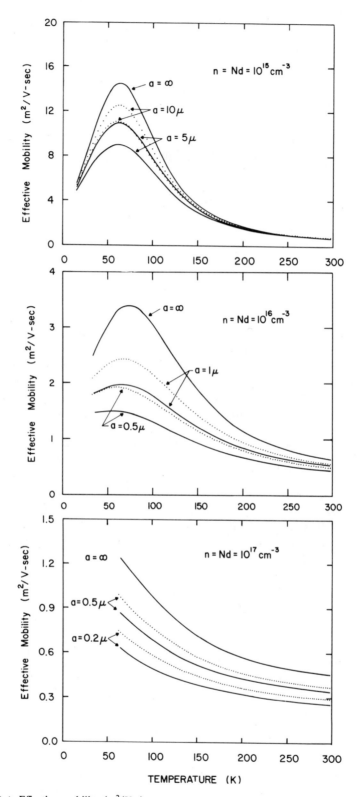

FIGURE 2-9-4. Effective mobility (m²/V s) vs. temperature. Solid line, numerical calculation; dotted line, simple model.

leading to

$$G(0) = \left(\frac{1}{2\pi}\right)^{1/2} \frac{nq^2}{(mk_BT)^{1/2}}$$

(2-9-42)

(see Refs. 111, 112, 116). In other words, G becomes independent of a.
Furthermore, it can be proved using Eqs. (2-9-31)–(2-9-33) that

$$\frac{1}{(LC)^{1/2}} \simeq \omega_p$$

(2-9-43)

and thus

$$X_L(\omega \to 0) = \left(\frac{1}{2\pi}\right) \frac{nq^2}{k_BT} a$$

(2-9-44)

For the bulk limit X_L is inversely proportional to a, while in the short sample limit X_L becomes proportional to a. Thus it can be expected that X_L reaches a maximum value when the sample length is comparable to the mean free path.

The results of the numerical calculation of the effective mobility, admittance, and impedance of a GaAs thin film based on the solution of Eq. (2-9-31) are presented in Figures 2-9-4–2-9-8. The device length was chosen such that the space charge injection can be negligible. Three dominant scattering mechanisms were taken into account in this calculation: a polar optical phonon scattering, an impurity scattering, and piezoelectric scattering. The relaxation time approximation was used (as described in Section 2-5).

The results of the computation can be interpreted in the frame of a simple analytical model which postulates the equivalent circuit of the finite sample shown in Fig. 2-9-9. This equivalent circuit includes the bulk resistance, in series with the resistance due to effective boundary scattering, in series with the inductance L due to the inertia effect of the carrier transport, and all in parallel with the geometric capacitance C. The sample inductance L is equal to $(\omega_p^2 C)^{-1}$, where ω_p is the plasma frequency. This simple model agrees fairly well with the results of the numerical calculation (see Figs. 2-9-5 and 2-9-6).

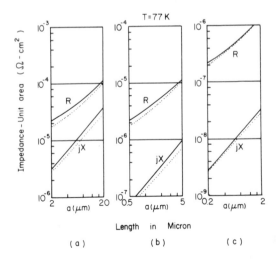

FIGURE 2-9-5. Impedance of unit area ($\Omega \cdot cm^2$) vs. length at 10 GHz at 77 K. R, resistance; X, reactance; solid line, numerical calculation; dotted line, simple model (Fig. 2-9-9).

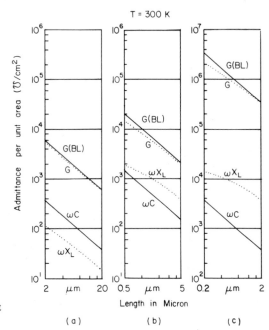

FIGURE 2-9-6. Admittance vs. length at
10 GHz at 300 K. BL, Bulk limit.

We have also checked that at direct current the numerical calculation of the
sample conductance is in a good agreement with the model as well as with the
calculations in Refs. 111 and 112 repeated for the same set of parameters (except
a factor of 2 omitted in Refs. 111 and 112).

As can be seen from Figure 2-9-6, the length dependence of the admittance at
300 K shows the bulk limit characteristics because the mean free path is quite small.
At 77 K, however, the situation is quite different from the bulk limit (see Fig. 2-9-7).
Even though we cannot obtain short sample limit characteristics for the device

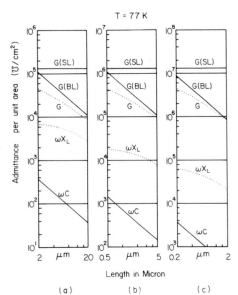

FIGURE 2-9-7. Admittance vs. length at 10 GHz at
77 K. SL, Short sample limit; BL, bulk limit.

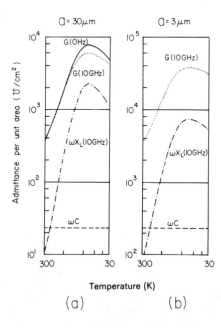

FIGURE 2-9-8. Effect of temperature variation.

length range studied here, there is a tendency toward them as the length becomes smaller. Indeed $G(0)$ becomes nearly independent of a, and X_L has maximum value when a is comparable to λ. (It should be noted that the device length was taken to be longer than ten times the Debye length to have uniformity in carrier distribution.)

Figure 2-9-8 shows the temperature variation of the admittance components. It illustrates the reduction of $G(\omega)$ and $X_L(\omega)$ from the low-frequency values when the frequency ω becomes comparable to $\langle 1/\tau_k \rangle$ and/or $\langle v_k/a \rangle$. This effect is pronounced in a short sample due to the increase of the effective boundary scattering with the sample length decrease.

These results can be checked experimentally using the microwave measurements of a small signal impedance of the thin samples similar to the measurements of the bulk Ge and Si samples reported in Ref. 144 (see also Ref. 125).

2-9-2. Electron Transport in Short n^+–n–n^+ and n^+–p–n^+ Structures [117]

For small applied voltages (say, less than a thermal voltage) the carrier and field distribution are determined by the balance between the diffusion and drift with a characteristic length equal to the Debye length. The low field resistance of the sample is then determined either by diffusion and drift or by thermionic emission

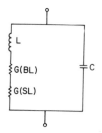

FIGURE 2-9-9. Equivalent circuit including the effective boundary scattering. $G(\text{BL}) = q^2 n\tau/ma$, bulk conductance; $G(\text{SL}) = (1/2\pi)^{1/2} q^2 n/(mk_B T)^{1/2}$, short sample limited conductance; $L = ma/nq^2$, effective inductance; $C = \varepsilon_0 \varepsilon_r/a$, geometric capacitance.

depending on temperature, doping, and sample length. In Section 2-9-1 we analyzed the case when the device length was much larger than the Debye radius and the nonuniformity of the carrier distribution may be neglected.

In this section we consider the effects of the nonuniform carrier distribution in short structures and compare the results of the calculation with the experimental values of low-field resistances for short GaAs n^+-n^--n^+ structures. Our approach is based on solving the equations describing the equilibrium potential distribution in the device structure and then determining a response to a small perturbation from the equilibrium.

We first consider an n^+-n^--n^+ diode (Fig. 2-9-10) in which one of the n^+ regions extends from $-\infty$ to $-L/2$, the other from $L/2$ to ∞, and the n^- region extends from $-L/2$ to $+L/2$. Poisson's equation states

$$\frac{d^2\psi}{dx^2} = -\frac{\rho(x)}{\varepsilon}$$

(2-9-45a)

where $\psi(x)$ is the potential and $\rho(x)$ is the space charge density:

$$\rho(x) = q(N_{d+} - n(x)) \qquad \text{for } L/2 \leq |x|$$

(2-9-45b)

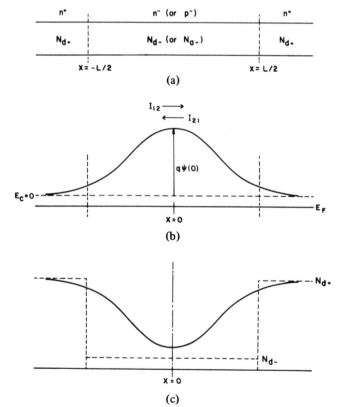

FIGURE 2-9-10. (a) Diode structure. (b) Energy band diagram at equilibrium (E_F is the Fermi level and E_c is the conduction band edge). I_{12} and I_{21} are the two current components at equilibrium. (c) Carrier profile $n(x)$ at equilibrium.

and

$$\rho(x) = q(N_{d-} - n(x)) \qquad \text{for } |x| \leq L/2 \qquad (2\text{-}9\text{-}45c)$$

Here N_{d+} is the ionized donor density in the bulk n^+ region, N_{d-} is the ionized donor density in the n^- region, and $n(x)$ is the carrier density given by

$$n(x) = N_{d+} \exp[q\psi(x)/k_BT] \qquad (2\text{-}9\text{-}45d)$$

The boundary conditions are

$$\psi = 0 \qquad \text{at } x = \pm\infty \qquad (2\text{-}9\text{-}45e)$$

and, owing to the symmetry of the problem,

$$\frac{d\psi}{dx} = 0 \qquad \text{at} \qquad x = 0 \quad \text{and} \quad x = \pm\infty \qquad (2\text{-}9\text{-}45f)$$

To solve Eq. (2-9-45a) we multiply both sides by $2d\psi$ and integrate from 0 to x. This yields

$$\left(\frac{d\psi}{dx}\right)^2 = -\left(\frac{2}{\varepsilon}\right) \int_0^\psi \rho(\psi')\, d\psi' \qquad (2\text{-}9\text{-}46a)$$

Introducing the dimensionless variables

$$u(x) = \frac{q\psi(x)}{k_BT}, \qquad y = \frac{x}{L_{d+}} \qquad (2\text{-}9\text{-}46b)$$

where

$$L_{d+} = \left(\frac{\varepsilon k_BT}{q^2 N_{d+}}\right)^{1/2} \quad \text{and} \quad L_{d-} = \left(\frac{\varepsilon k_BT}{q^2 N_{d-}}\right)^{1/2} \qquad (2\text{-}9\text{-}46c)$$

the Debye lengths in the n^+ and n^- regions, respectively, and integrating Eq. (2-9-46a) from 0 to x we obtain for $|x| < L/2$

$$\frac{du}{dy} = \sqrt{2}\left\{-\left(\frac{N_{d-}}{N_{d+}}\right)[u(x) - u(0)] + \exp[u(x)] - \exp[u(0)]\right\}^{1/2} \qquad (2\text{-}9\text{-}47a)$$

while for $|x| > L/2$

$$\frac{du}{dy} = \sqrt{2}\{\exp[u(x)] - u(x) - 1\}^{1/2} \qquad (2\text{-}9\text{-}47b)$$

so that $du/dy = 0$ at $x = +\infty$. Equating (2-9-47a) and (2-9-47b) for $x = L/2$ gives

$$u\left(\frac{L}{2}\right)\left[1 - \left(\frac{N_{d-}}{N_{d+}}\right)\right] = \exp[u(0)] - 1 - \left(\frac{N_{d-}}{N_{d+}}\right)u(0) \qquad (2\text{-}9\text{-}48)$$

Integration of Eqs. (2-9-47a) and (2-9-47b) leads to

$$\int_{u(0)}^{u(x)} \frac{du}{\{\exp(u) - \exp[u(0)] - (N_{d-}/N_{d+})[u - u(0)]\}^{1/2}}$$

$$= \sqrt{2}\left(\frac{x}{L_{d+}}\right) \qquad \text{for } 0 \leqslant x \leqslant L/2 \qquad (2\text{-}9\text{-}49a)$$

and

$$\int_{u(L/2)}^{u(x)} \frac{du}{[\exp(u) - 1 - u]^{1/2}} = \sqrt{2}\,\frac{(x - L/2)}{L_{d+}} \qquad \text{for } x \geqslant L/2 \qquad (2\text{-}9\text{-}49b)$$

From Eqs. (2-9-49a) and (2-9-49b) we can evaluate u as a function of x if $u(0)$ is known. In order to find $u(0)$ we substitute $x = L/2$ into Eq. (2-9-49a):

$$\int_{u(0)}^{u(L/2)} \frac{du}{\{\exp(u) - \exp[u(0)] - (N_{d-}/N_{d+})[u - u(0)]\}^{1/2}} = \frac{1}{\sqrt{2}}\,\frac{L}{L_{d+}} \qquad (2\text{-}9\text{-}50)$$

from which $u(0)$ can be evaluated.

In order to simplify the numerical evaluation of the integrals in Eqs. (2-9-49a) and (2-9-49b), we introduce a new variable z defined by

$$z = \{\exp[u - u(0)] - 1\}^{1/2}$$

Using z we have instead of Eq. (2-9-50)

$$\sqrt{2}\int_{0}^{z(x)} \frac{dz}{(1 + z^2)\{(N_{d+}/N_{d-})\exp[u(0)] - [\ln(1 + z^2)/z^2]\}^{1/2}} = \frac{x}{L_{d-}} \qquad (2\text{-}9\text{-}51a)$$

and instead of (2-9-50b)

$$\sqrt{2}\int_{0}^{z(L/2)} \frac{dz}{(1 + z^2)\{(N_{d+}/N_{d-})\exp[u(0)] - [\ln(1 + z^2)/z^2]\}^{1/2}} = \frac{L/2}{L_{d-}} \qquad (2\text{-}9\text{-}51b)$$

If the ratio N_{d+}/N_{d-} is large, Eqs. (2-9-51a) and (2-9-51b) are reduced to

$$\left(2\frac{N_{d-}}{N_{d+}}\right)^{1/2}\exp\left[-\frac{u(0)}{2}\right]\tan^{-1} z = \frac{x}{L_{d-}} \qquad (2\text{-}9\text{-}52a)$$

$$\left(2\frac{N_{d-}}{N_{d+}}\right)^{1/2}\exp\left[-\frac{u(0)}{2}\right]\tan^{-1}(\{\exp[-1 - u(0)] - 1\}^{1/2}) = \frac{L/2}{L_{d-}} \qquad (2\text{-}9\text{-}52b)$$

[When the ratio N_{d+}/N_{d-} is large, $\exp[u(0)]$ is also small and from Eq. (2-9-48) $u(L/2) = -1$.]

The results of the numerical integration of Eqs. (2-9-51a) and (2-9-51b) are shown in Figs. 2-9-11 and 2-9-12. The analytical approximations of Eqs. (2-9-52a) and (2-9-52b) are shown in the same figures.

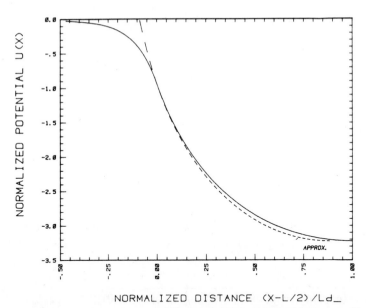

FIGURE 2-9-11. Normalized potential u vs. normalized distance $(x - L/2)/L_{d-}$ for a specific n^+-n^--n^+ diode with $L/(2L_{d-}) = 1$ and $N_{d+}/N_{d-} = 100$. ———, numerical calculation; - - -, analytical approximation for n^- region.

The analytical solutions are obtained by neglecting the factor $\ln(1 + z^2)/z^2$ in Eqs. (2-9-51a) and (2-9-51b). This is equivalent to neglecting the space charge due to ionized impurities. This case was analyzed by Knol and Diemer [145] for symmetrical double-cathode valves where the space charge is due to electrons only. For $N_{d-}/N_{d+} \ll 1$ the two problems become identical and our Eq. (2-9-52a) is consistent with Eq. (16) in the above reference.

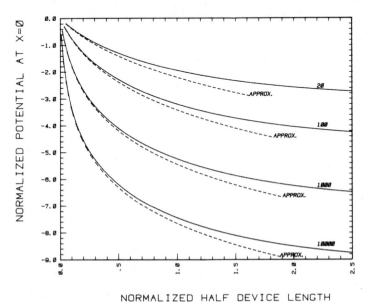

FIGURE 2-9-12. Curves of normalized potential at $x = 0$ vs. normalized half device length $L/(2L_{d-})$ for an n^+-n^--n^+ diode with N_{d+}/N_{d-} as a parameter. Approx., analytical approximation.

An even simpler (but far less accurate) expression may be obtained if we also neglect the space charge of the injected electrons at the center of the structure. In this case the analytical solution reduces to

$$L_d(n) \simeq \frac{L/2 - x}{\sqrt{2}} \tag{2-9-53}$$

where

$$L_d(n) = \left(\frac{\varepsilon k_B T}{q^2 n}\right)^{1/2} \tag{2-9-54}$$

or in a different form

$$n = \frac{2\varepsilon k_B T}{q^2 (L/2 - x)^2} \tag{2-9-55}$$

In this limiting case the minimum concentration at the center of the structure can be calculated exactly [153]

$$n(0) = \frac{2\pi^2 \varepsilon k_B T}{q^2 L^2} \tag{2-9-56a}$$

For GaAs $\varepsilon = 1.14 \times 10^{-10}$ F/m and we find

$$n(0) \sim \frac{3.6 \times 10^{14}}{L^2(\mu m)} \frac{T}{300\ K}\ (cm^{-3}) \tag{2-9-56b}$$

Solution (2-9-53)–(2-9-55) is very similar to the solution obtained for an n^+-n interface by Chandra [146]. The physical meaning of Eq. (2-9-56a) is transparent; $n(0)$ corresponds to the electron concentration injected into the structure by the thermal voltage $k_B T/q$.

The point where carrier spillover becomes important can be determined from Eq. (2-9-56a). Here by spillover we mean the carrier injection at thermal equilibrium. At zero bias it happens if

$$N_{d-} \leqslant n(0) \tag{2-9-57}$$

i.e.,

$$N_{d-} \leqslant \frac{3.6 \times 10^{14}}{L^2(\mu m)} \frac{T}{300\ K} \tag{2-9-58}$$

for GaAs. However, when an external bias V is applied, more space-charge injection takes place, in addition to the spillover. When $V > k_B T/q$ the space-charge injection becomes greater than the spillover electron concentration and thus the spillover effects become masked by the space-charge injection due to the external bias.

We will now consider n^+-p^--n^+ diodes. In this case the space charge densities in the n^+ and p^- regions are given by (see Fig. 2-9-10)

$$\rho(x) = q(N_{d+} - n(x)) \qquad \text{for } L/2 \leqslant |x| \tag{2-9-59a}$$

and

$$\rho(x) = q(-N_{a-} - n(x) + p(x)) \qquad \text{for } |x| \leqslant L/2 \tag{2-9-59b}$$

Carrying out the integration in Eq. (2-9-46a), we have

$$\frac{du}{dy} = \sqrt{2}\left\{\left(\frac{N_{a-}}{N_{d+}}\right)[u - u(0)] + \exp(u) - \exp[u(0)] - \left(\frac{n_i}{N_{d+}}\right)^2\right.$$
$$\left. \times [\exp(-u) - \exp[-u(0)]]\right\}^{1/2} \qquad \text{for } |x| \le L/2 \qquad (2\text{-}9\text{-}60a)$$

and

$$\frac{du}{dy} = \sqrt{2}[\exp(u) - u(x) - 1]^{1/2} \qquad \text{for } |x| \ge \frac{L}{2} \qquad (2\text{-}9\text{-}60b)$$

Integrating (2-9-60a) and introducing again the variable z yields

$$\sqrt{2}\int_0^{z(x)} \frac{dz}{(1+z^2)\left\{\dfrac{N_{d+}}{N_{a-}}\exp[u(0)] + \ln(1+z^2)/z^2 + \dfrac{e^{-u(0)}}{1+z^2}\dfrac{n_i^2}{N_{d+}N_{a-}}\right\}^{1/2}}$$
$$= \frac{x}{L_{d-}} \qquad (2\text{-}9\text{-}61a)$$

and for $x = L/2$

$$\sqrt{2}\int_0^{zL/2} \frac{dz}{(1+z^2)\left\{\dfrac{N_{d+}}{N_{a-}}\exp[u(0)] + \ln(1+z^2)/z^2 + \dfrac{e^{-u(0)}}{1+z^2}\dfrac{n_i^2}{N_{d+}N_{a-}}\right\}^{1/2}}$$
$$= \frac{L}{2L_{d-}} \qquad (2\text{-}9\text{-}61b)$$

Figure 2-9-13 is derived from Eq. (2-9-61b); in this figure the analytical approximation is not included because it becomes poor for long devices where space-charge of ionized donors cannot be neglected.

For a relatively thin p^- region there are very few holes left when the electron spillover occurs. For example, in GaAs for $n > 10^{12}$ cm^{-3} we have $p = n_i^2/n < 10^{13}/10^{12} = 10^1$ cm^{-3}, so that we can neglect $p(x)$ altogether. In this case we can still use Eq. (2-9-46a) with

$$\rho(x) = q(N_{d+} - n(x)) \qquad \text{for } L/2 \le |x| \qquad (2\text{-}9\text{-}59c)$$

and

$$\rho(x) = q(-N_{a-} - n(x)) \qquad \text{for } |x| \le L/2 \qquad (2\text{-}9\text{-}59d)$$

The solution is the same as for an $n^+\text{-}n^-\text{-}n^+$ diode. The only difference is that N_{a-} replaces N_{d-} and the sign of the ln term changes. Thus instead of (2-9-51a) and (2-9-51b) we have

$$\sqrt{2}\int_0^{z(x)} \frac{dz}{(1+z^2)\left\{\dfrac{N_{d+}}{N_{a-}}\exp[u(0)] + \dfrac{\ln(1+z^2)}{z^2}\right\}^{1/2}} = \frac{x}{L_{d-}} \qquad (2\text{-}9\text{-}62a)$$

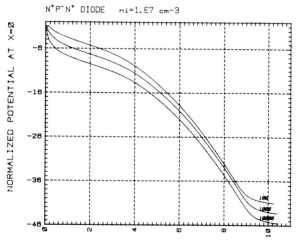

(a)

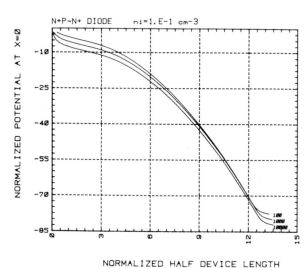

FIGURE 2-9-13. Curves of normalized potential at $x = 0$ [$u(0)$] vs. $L/(2L_{d-})$ for an n^+-p^--n^+ diode with N_{d+}/N_{a-} as a parameter; (a) $n_i = 10^7$ cm^{-3}; (b) $n_i = 10^{-1}$ cm^{-3}.

(b)

and

$$\sqrt{2} \int_0^{z(L/2)} \frac{dz}{(1+z^2)\left\{\dfrac{N_{d+}}{N_{a-}} \exp[u(0)] + \dfrac{\ln(1+z^2)}{z^2}\right\}^{1/2}} = \frac{L}{2L_{d-}} \qquad (2\text{-}9\text{-}62b)$$

For $N_{d+}/N_{a-} \gg 1$ we get equations (2-9-52a) and (2-9-52b) with N_{a-} replacing N_{d-}. Figure 2-9-14 is derived from Eqs. (2-9-52a), (2-9-52b), (2-9-62a), and (2-9-62b).

The spillover electron concentration at the center of a diode is shown in Figs. 2-9-15 and 2-9-16. These results are obtained by solving Eqs. (2-9-51b) and (2-9-62b)

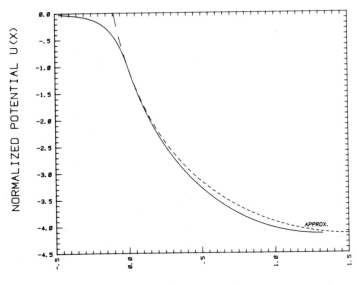

NORMALIZED DISTANCE (X-L/2)/Ld_

FIGURE 2-9-14. Normalized potential u vs. normalized distance for a specific $n^+-p^--n^+$ diode with $L/(2L_{d-}) = 1.32$ and $N_{d+}/N_{a-} = 100$. ——, numerical calculations; - - -, analytical approximation for p^- region.

with the substitution

$$N_{d+} = \frac{n(0)}{\exp[u(0)]} \tag{2-9-63}$$

The corresponding field profiles, for $n^+-n^--n^+$ diodes, derived from Eqs. (2-9-47a) and (2-9-47b) are shown in Figs. 2-9-17 and 2-9-18.

The resistance of the diodes just mentioned above may be limited at low voltages either by thermionic emission over the barrier (see Fig. 2-9-10) or by the processes of the diffusion and drift within the n^- layer. This situation is very similar to the electron transport in metal–semiconductor junctions where two limiting cases corre-

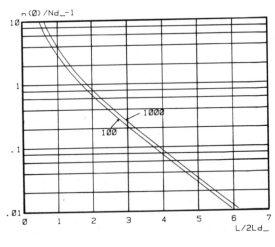

FIGURE 2-9-15. Normalized spillover electron density at $x = 0$ [$(n(0)/N_{d-}) - 1$] vs. $L/(2L_{d-})$ for an $n^+-n^--n^+$ diode for different N_{d+}/N_{d-}.

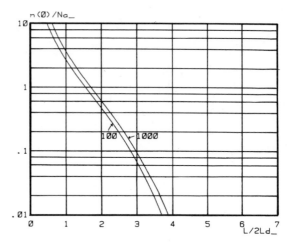

FIGURE 2-9-16. Normalized spillover electron density at $x = 0$ $[n(0)/N_{a-}]$ vs. normalized half device length $L/(L_{d-})$ for an n^+-p^--n^+ diode for different N_{d+}/N_{a-}.

spond to the thermionic and diffusion model [147]. These two limiting cases are considered and compared below.

In the case when the thermionic emission dominates, the current I_{th} through the device at low voltages $V \ll k_B T/q$ is approximately given by [147]

$$I_{th} = qn(0)\frac{v_{th}}{4}\frac{qV}{k_B T} A \qquad (2\text{-}9\text{-}64a)$$

where A is the cross section and

$$v_{th} = \left(\frac{8k_B T}{\pi m}\right)^{1/2} \qquad (2\text{-}9\text{-}64b)$$

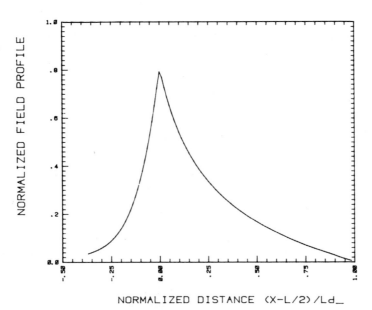

NORMALIZED DISTANCE (X-L/2)/Ld_

FIGURE 2-9-17. Field profile, normalized with respect to $k_B T/qL_{d+}$ vs. normalized distance $L/(L_{d-})$ for a specific n^+-n^--n^+ diode with normalized half device length $L/(2L_{d-}) = 1$ and $N_{d+}/N_{d-} = 100$.

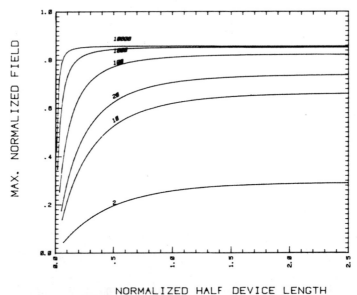

NORMALIZED HALF DEVICE LENGTH

FIGURE 2-9-18. Maximum normalized field strength vs. normalized half device length $L/(2L_{d-})$ for an n^+-n^--n^+ diode for different values of N_{d+}/N_{d-}.

is the thermal velocity. Thus the sample resistance is given by [for a sample with $N_{d-} \ll n(0)$, see Eq. (2-9-56a)]

$$R_{\text{th}} = \frac{2L^2}{\pi^2 \varepsilon v_{\text{th}} A} = \left(\frac{m}{2k_B T}\right)^{1/2} \frac{L^2}{\pi^{3/2} \varepsilon A} \qquad (2\text{-}9\text{-}65)$$

As shown below, when the diffusion-drift model applies [147] and $N_{d-} \ll n(0)$

$$I_D \approx 2qn(0)\mu \frac{V}{L} A \qquad (2\text{-}9\text{-}66)$$

where μ is the low field mobility. This model applies only if $L \gg \lambda$, where λ is the mean free path. Hence for a low doped sample the low field resistance is given by

$$R_D = \frac{qL^3}{4\pi^2 k_B T \varepsilon \mu A} \qquad (2\text{-}9\text{-}67)$$

When the sample size becomes small compared to the mean free path and the spillover is small so that $n(0) \simeq N_{d-}$, the effective low field mobility becomes limited by the boundary scattering [111, 112, 116]. In this case from Eq. (2-9-42) we find

$$\mu_1 = \frac{LG(0)}{qn} = \frac{qL}{mv_{\text{th}}} \frac{2}{\pi} \qquad (2\text{-}9\text{-}68)$$

and Eq. (2-9-67) reduces to Eq. (2-9-65).

Using Eqs. (2-9-67) and (2-9-65) we establish that diffusion and drift dominate when

$$L > \frac{8\mu k_B T}{q v_{\text{th}}} \qquad (2\text{-}9\text{-}69)$$

or

$$L > \frac{\mu}{q}(8\pi k_B T m)^{1/2} \qquad (2\text{-}9\text{-}70)$$

The criterion of Eq. (2-9-70) is similar to $L > \lambda$. For GaAs at 300 K, $\mu \approx 0.8\ \text{m}^2/\text{V s}$ and $m \approx 0.063 m_e$ and this criterion yields

$$L > 0.4\ \mu\text{m}$$

However, at 77 K and $\mu \approx 14\ \text{m}^2/\text{V s}$ we find

$$L > 3.4\ \mu\text{m}$$

Thus we expect that for GaAs samples with $L = 0.4\ \mu\text{m}$ the thermionic emission dominates at 77 K.

More accurate results based on the computer calculation are described below.

A. *Diffusion-Drift Model.* Assuming that diffusion and drift within the n^- layer determine the sample resistance, we find

$$I = -q\mu n(x) A \frac{d\psi_F(x)}{dx} \qquad (2\text{-}9\text{-}71)$$

Here A is the cross-sectional area of the device, $\psi_F(x)$ is the quasi-Fermi potential, and μ is the low field mobility and we assume that the electron mobility is independent of position for $|x| < L/2$. Integration of Eq. (2-9-71) yields

$$\int_{-L/2}^{L/2} \frac{I}{q\mu n(x) A}\, dx = -\int_{-L/2}^{L/2} d\psi_F(x) \equiv V \qquad (2\text{-}9\text{-}72)$$

So that

$$R_D = \frac{V}{I}\bigg|_{I\to 0} = \int_{-L/2}^{L/2} \frac{dx}{q\mu n(x) A} = \frac{2}{q\mu A n(0)}\int_0^{L/2} \frac{dx}{\exp[u(x)-u(0)]}$$

$$= \frac{\sqrt{2}L_{d+}}{q\mu A n(0)}\int_{u(0)}^{u(L/2)} \qquad (2\text{-}9\text{-}73)$$

$$\times \frac{du}{\exp[u(x)-u(0)]\{-(N_{d-}/N_{d+})[u(x)-u(0)]+\exp[u(x)]-\exp[u(0)]\}^{1/2}}$$

where Eq. (2-9-47a) has been used. Equations (2-9-73) can be rewritten as

$$R_D = \frac{L}{q\mu n(0) A}\frac{I_2}{I_1} = \frac{L}{q\mu N_{d-} A}\left[\frac{N_{d-}}{n(0)}\frac{I_2}{I_1}\right] = R_0\left(\frac{N_{d-}}{n(0)}\right)\left(\frac{I_2}{I_1}\right) \qquad (2\text{-}9\text{-}74)$$

where

$$I_1 = \int_0^{z(L/2)} \frac{dz}{(1 + z^2)\{(N_{d+}/N_{d-})\exp[u(0)] - [\ln(1 + z^2)/z^2]\}^{1/2}} \qquad (2\text{-}9\text{-}75)$$

(i.e. $I_1 = L/(2\sqrt{2}L_{d-})$ (see Eq. [2-9-51b]) and

$$I_2 = \int_0^{z(L/2)} \frac{dz}{(1 + z^2)^2\{(N_{d+}/N_{d-})\exp[u(0)] - [\ln(1 + z^2)/z^2]\}^{1/2}} \qquad (2\text{-}9\text{-}76)$$

For $N_{D-}/n(0) \ll 1$ we obtain $I_2/I_1 \simeq \frac{1}{2}$ and Eq. (2-9-74) leads to Eq. (2-9-66). For the n^+-p^--n^+ diode the expression for R_D is the same as in Eq. (2-9-73) after the substitution of N_{d-} with N_{a-} and the change of the sign of the ln term in the denominator of the integrands for I_1 and I_2.

Figures 2-9-19 and 2-9-20 give the results of numerical calculations using Eq. (2-9-74a) for the n^+-n^--n^+ and n^+-p^--n^+ diodes, respectively.

B. *Thermionic Emission Case.* When the device length becomes far less than the mean free path of the carriers, meaning that there are no collisions, then thermionic emission over the energy barrier in the center of the diode will limit the current.

At thermal equilibrium there are two current components flowing in the diode (see Fig. 2-9-10) [147]

$$I_{12} = I_{21} = I_s = A\frac{qn(0)v_{th}}{4} \qquad (2\text{-}9\text{-}77)$$

When a small bias V is applied, the two new currents I'_{12} and I'_{21} no longer cancel each other resulting in a net current I. This situation is the same as the double-cathode valve problem under bias [145]; we find

$$I = I'_{12} - I'_{21} = I_{12}\exp[qV/k_BT] - I_{21} = I_s[\exp(qV/k_BT) - 1] \qquad (2\text{-}9\text{-}78)$$

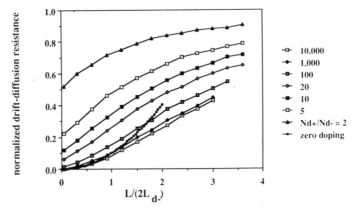

FIGURE 2-9-19. Normalized diffusion drift resistance (R_D/R_0), where $R_0 = L/(q\mu N_{d-}A)$, vs. $L/(L_{d-})$ for an n^+-n^--n^+ diode for different values of N_{d+}/N_{d-}. Also shown is the analytical result for $N_{d-} \to 0$ (see Eq. [2-9-66]).

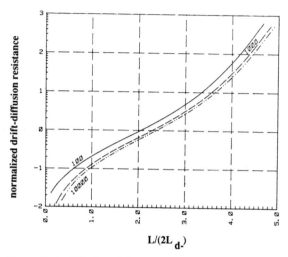

FIGURE 2-9-20. Normalized diffusion drift resistance (R_D/R_0), where $R_0 = L/(q\mu N_{d-}A)$, vs. $L/(L_{d-})$ for an n^+-p^--n^+ diode for different values of N_{d+}/N_{a-}.

Tables 2-9-1 and 2-9-2, which show the diffusion-drift resistance and thermionic current density of a GaAs n^+-n^--n^+ diode with $L = 4000$ Å, were constructed from the diffusion-drift current of Fig. 2-9-19 and the thermionic emission current of Eq. (2-9-76) and Fig. 2-9-21.

The derived curves are quite universal in the sense that all device parameters are included in L/L_D except the doping ratio, which effect is negligible if it is much larger than one.

TABLE 2-9-1a. Diffusion-Drift Resistance and Thermionic Current Density for an n^+-n^--n^+ Diode (V is in units of mV)

$$N_{d+} = 10^{18}\ cm^{-3}, \qquad N_{d-} = 10^{15}\ cm^{-3}, \qquad L = 4000\ Å$$

T (°K)	$\dfrac{L}{2L_{d-}}$	$\dfrac{n(0)}{N_{d-}}$	$\dfrac{R_D}{R_0}$	μ (cm²/Vs)	R_0 (Ωcm²)	R_D (Ωcm²)	J_{th} (A/cm²)
300	1.47	2.65	0.21	7,400	$\dfrac{1}{2.96 \times 10^4}$	$\dfrac{1}{1.41 \times 10^5}$	$4.54 \times 10^3 (e^{V/25.8} - 1)$
77	2.90	1.28	0.47	140,000	$\dfrac{1}{5.6 \times 10^5}$	$\dfrac{1}{1.19 \times 10^6}$	$1.11 \times 10^3 (e^{V/6.62} - 1)$

TABLE 2-9-1b. As in (a) for $N_{d-} = 10^{16}\ cm^{-3}$

T (°K)	$\dfrac{L}{2L_{d-}}$	$\dfrac{n(0)}{N_{d-}}$	$\dfrac{R_D}{R_0}$	μ (cm²/Vs)	R_0 (Ωcm²)	R_D (Ωcm²)	J_{th} (A/cm²)
300	4.65	1.04	0.69	6,400	$\dfrac{1}{2.56 \times 10^5}$	$\dfrac{1}{3.71 \times 10^5}$	$1.78 \times 10^4 (e^{V/25.8} - 1)$
77	9.19	1	0.85	34,000	$\dfrac{1}{1.36 \times 10^6}$	$\dfrac{1}{1.6 \times 10^6}$	$8.67 \times 10^3 (e^{V/6.62} - 1)$

TABLE 2-9-2. Comparison between Theoretical and Measured Values of the Resistance for $n^+-n^--n^+$ and $n^+-p^--n^+$ Diodes
(The Measured Values, R_{meas}, are from Ref. 148)

Device # 1: $N_{d+} = 1 \times 10^{18}$, $N_{d-} = 1 \times 10^{15}$, $L = 0.4 \ \mu m$, $n^+-n^--n^+$
Device # 2: $N_{d+} = 5 \times 10^{17}$, $N_{d-} = 2 \times 10^{15}$, $L = 0.24 \ \mu m$, $n^+-n^--n^+$
Device # 3: $N_{d+} = 1 \times 10^{18}$, $N_{a-} = 6 \times 10^{14}$, $L = 0.47 \ \mu m$, $n^+-p^--n^+$

(all doping concentrations in cm^{-3})

	$T(°K)$	$\dfrac{L}{2L_{d-}}$	$\dfrac{n(0)}{N_{d-}}$	$\dfrac{R_D}{R_0}$	μ (cm²/Vs)	R_0 (Ωcm²)	R_D (Ωcm²)	R_{th} (Ωcm²)	$R_D + R_{th}$ (Ωcm²)	R_{meas} (Ωcm²)
Device # 1	300	2.08	1.72	0.38	7,400	$\dfrac{1}{5.92 \times 10^4}$	$\dfrac{1}{1.55 \times 10^5}$	$\dfrac{1}{2.28 \times 10^5}$	$\dfrac{1}{0.92 \times 10^5}$	$\dfrac{1}{0.8 \times 10^5}$
	77	4.1	1.08	0.62	16,000	$\dfrac{1}{1.28 \times 10^5}$	$\dfrac{1}{2.06 \times 10^5}$	$\dfrac{1}{2.82 \times 10^5}$	$\dfrac{1}{1.2 \times 10^5}$	$\dfrac{1}{0.9 \times 10^5}$
Device # 2	300	1.25	3.2	0.18	7,400	$\dfrac{1}{9.86 \times 10^4}$	$\dfrac{1}{5.44 \times 10^5}$	$\dfrac{1}{4.24 \times 10^5}$	$\dfrac{1}{2.38 \times 10^5}$	$\dfrac{1}{1.1 \times 10^5}$
	77	2.46	1.43	0.42	16,000	$\dfrac{1}{2.13 \times 10^5}$	$\dfrac{1}{5.09 \times 10^5}$	$\dfrac{1}{3.74 \times 10^5}$	$\dfrac{1}{2.15 \times 10^5}$	$\dfrac{1}{0.8 \times 10^5}$
Device # 3	300	1.34	1.95	0.26	7,400	$\dfrac{1}{3.02 \times 10^4}$	$\dfrac{1}{1.16 \times 10^5}$	$\dfrac{1}{7.76 \times 10^4}$	$\dfrac{1}{4.65 \times 10^4}$	$\dfrac{1}{1.3 \times 10^3}$
	77	2.63	0.21	2.0	16,000	$\dfrac{1}{6.54 \times 10^4}$	$\dfrac{1}{3.27 \times 10^4}$	$\dfrac{1}{1.65 \times 10^4}$	$\dfrac{1}{1.1 \times 10^4}$	$\dfrac{1}{1.0 \times 10^3}$

Figures 2-9-11–2-9-16 show that the spillover effect becomes important when the device length is of the order of several extrinsic Debye lengths of the n^-(or p^-) region as expected. In considering the spillover effect, two approximations were introduced. The first approximation made in Eqs. (2-9-52a), (2-9-59a), and (2-9-59b), which were derived by neglecting the space charge due to ionized impurities in the $n^-(p^-)$ region, showed a good agreement with the exact solution [see Eqs. (2-9-51a) and (2-9-51b)], as shown in Figs. 2-9-11, 2-9-12, and 2-9-14. The second approximation made in Eq. (2-9-53), which was derived by neglecting the electron density at the center of the diodes, can be used as a rough estimate of the spillover effects; see, for example, Eqs. (2-9-56a) and (2-9-67).

The spillover electron density at the center of a diode can be found easily from Fig. 2-9-12 or 2-9-15 for $n^+-n^--n^+$ structures, and Fig. 2-9-13 or 2-9-16 for $n^+-p^--n^+$ structures.

Figures 2-9-19 and 2-9-20 were derived for the diffusion-drift dominated case. They show the decrease of resistance due to spillover for short devices. In Fig. 2-9-19 for the $n^+-n^--n^+$ diode R_D/R_0, where $R_0 = L/(q\mu N_{d-} A)$, approaches 1 for long devices as expected; for the $n^+-p^--n^+$ diode the resistance increases rapidly with device length (Fig. 2-9-20) because the mobile carrier density decreases very fast. Comparing Fig. 2-9-19 with Fig. 2 in Ref. 112 we have an exact agreement even though our curves were derived for the equilibrium limit (I tends to 0).

There are two factors in the ratio R_D/R_0 [see Eq. (2-9-74a)]: I_1/I_2 and $N_{d-}/n(0)$. The factor I_1/I_2, which ranges between 1 and 2 in most cases, is due to the nonuniformity of carrier distribution. Thus the dominant contribution to the ratio R_D/R_0 comes from $N_{d-}/n(0)$.

The total resistance of the device can be approximated by

$$R = R_{th} + R_D \tag{2-9-79}$$

In cases when the spillover is not important $N_{d-} \simeq n(0)$ and this equation is equivalent to

$$\mu_{eff} = (\mu_0^{-1} + \mu_1^{-1})^{-1} \tag{2-9-79a}$$

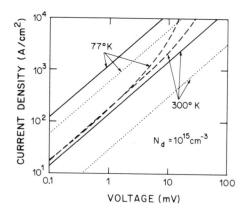

FIGURE 2-9-21. Current density J (A/cm^2) vs. applied voltage V (mV) characteristics for the $n^+-n^--n^+$ diode of Table 2-9-1a. . . . ., Diffusion-drift current without spillover; ———, diffusion-drift current with spillover; - - -, thermionic emission current.

Here μ_0 is the low field bulk mobility and μ_1 is given by Eq. (2-9-68). Equation (2-9-79a) was derived in Ref. 111. The physical meaning of Eq. (2-9-79a) is that in the limiting case of a very short sample ($L \ll \lambda$, where λ is the mean free path), the effective scattering time is determined by the transit time ("collisions" with the boundaries). In the opposite limiting case ($L \gg \lambda$) μ_{eff} is reduced to the low field mobility μ_0.

In other words, when the resistance of the sample is limited by thermionic emission, the electron drift velocity is determined by the transit time of electrons (L/v_{th}), which is the time they spend in being accelerated in the applied electric field. The collisions within the sample are not important in this case ("ballistic" transport).

From Fig. 2-9-21 it can be seen that for $N_{d-} = 10^{15}$ cm^{-3} the thermionic emission current is higher than the diffusion-drift current at 300 K, but the opposite is true at 77 K. This means that at 77 K the current is controlled by thermionic emission while at 300 K it is controlled by diffusion and drift. Similar results for $N_{d-} = 10^{16}$ cm^{-3} are shown in Fig. 2-9-22.

In order to compare the theoretical results with experimental data [148], Table 2-9-2 was constructed using measured values for the mobility rather than theoretical values as in Table 2-9-1. In Table 2-9-2 the total resistance $R_D + R_{\text{th}}$ was calculated by adding the diffusion-drift and thermionic resistances. The agreement between theoretical and experimental results is quite good for device No. 1 (n^+-n^--n^+) and reasonably good for device No. 2 (n^+-n^--n^+); device No. 3 (n^+-p^--n^+) shows an order of magnitude difference. Possibly this is due to the fact that the resistance of the n^+-p^--n^+ diode is a very sensitive function of the p^- region doping level and even a small compensation can increase the resistance dramatically.

Characteristic scales involved in the problem discussed above—the Debye radius and the mean free path—are shown in Fig. 2-9-23 as functions of the low field mobility and concentration for GaAs samples. This figure allows us to establish the relationship between the Debye length, the mean free path, and the device length and thus determine which results should apply to a particular sample.

2-9-3. Overshoot and Ballistic Transport in High Electric Field

Some important features of the electron transport in the high electric field in short semiconductor structures may be understood by studying the electron velocity response to an almost instantaneous electric field increase. Such a dependence is shown in Fig. 2-9-24 for a GaAs sample [34]. As can be seen from the figure there are four distinct regions. At a time scale smaller than the effective momentum relaxation time τ_p the velocity increases linearly with time:

$$v = qFt/m \qquad (2\text{-}9\text{-}80)$$

This corresponds to a "ballistic," i.e., collision-free transport. At a time scale larger than τ_p we have an "overshoot" region where the electron drift velocity may substantially exceed the equilibrium value of the velocity reached when t tends to infinity. The overshoot region is followed by an "undershoot" region where the velocity is slightly smaller than the equilibrium value. A comparison between the electron transit time across the device and the time scale of Fig. 2-9-24 may be used for an approximate qualitative characterization of the electron transport in a short structure.

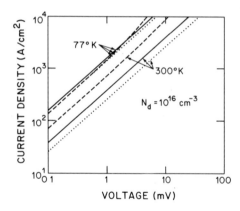

FIGURE 2-9-22. Same as in Fig. 2-9-21 from Table 2-9-1b.

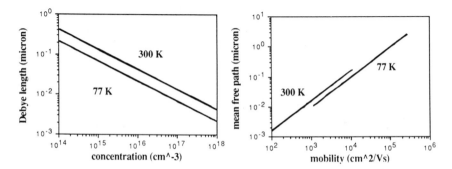

FIGURE 2-9-23. Debye length vs. electron concentration and mean free path vs. low field mobility for GaAs.

Such an analysis applied to GaAs devices indicates that the overshoot effects should be quite noticeable in devices with a characteristic size of an order of a micron. This conclusion is in agreement with numerous studies of these effects; see, for example, Refs. 24, 30-38, 75-77, 83, 85, 98, 99, 103-108, 115, 137, 138, 139, 149. In submicron devices the electron transport may exhibit characteristics typical of a ballistic mode of operation (see, for example, Refs 102-110, 118-139, 92, 151). As in Ref. 129, we define a ballistic transport as the electron transport with no collisions (except at the contacts), a near ballistic transport as a transport with only few collisions, and overshoot transport as a transport with a substantial increase of the electron velocity above its equilibrium value.

Our analysis of the ballistic and near ballistic transport will follow Refs. 102, 105. It is based on the Poisson equation:

$$\frac{dF}{dx} = \frac{q}{\varepsilon}(n - n_0) \qquad (2\text{-}9\text{-}81)$$

the equation for the current density j

$$j = qnv \qquad (2\text{-}9\text{-}82)$$

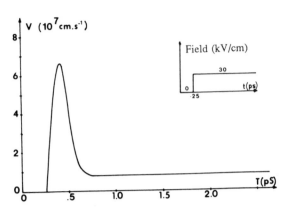

FIGURE 2-9-24. Electron velocity in GaAs as a function of time when an electric field step is applied [34].

and the equation of motion

$$m \, dv/dt = qF - mv/\tau \qquad (2\text{-}9\text{-}83)$$

As is frequently done in the literature, we here treat electrons as positive particles emitted from the cathode in order to have fewer negative signs in the equations. We also assume for simplicity that the electron effective mass m and momentum relaxation time τ are energy-independent. This is not a realistic assumption for a semiconductor device where the relaxation times and effective masses are energy-dependent. Hence, for a quantitative analysis the phenomenological equations based on energy-dependent energy and momentum relaxation times with an energy-dependent effective mass (see Section 2-8) should be used to describe the near-ballistic regime or overshoot effects as has been done in Refs. 103, 34, 36, and others. Many-particle Monte Carlo simulation may be used for a more accurate simulation [92]. However, in the ballistic regime the collisions are not important and the inclusion of the momentum relaxation time in Eq. (2-9-83) allows us to model the transition from the ballistic to the near-ballistic and then to the collision dominated transport elucidating the physics of short semiconductor structures.

Equation (2-9-82) does not include the diffusion current, which may be essential, especially at low voltages. An estimate using the results obtained in Section 2-9-2 indicates that for an n^+-n cathode contact this may be justified at bias voltages much larger than the thermal voltage (around 6.6 mV at 77 K) when the electron spill over is less than the concentration of the space charge electrons injected by the applied field.

The solutions of Eq. (2-9-81)–(2-9-83) are strongly dependent on the boundary conditions [108, 105, 109]. Here we assume

$$v = 0 \qquad \text{at } x = 0 \qquad (2\text{-}9\text{-}84)$$

and

$$dv/dx = 0 \qquad \text{at } x = 0 \qquad (2\text{-}9\text{-}85)$$

or $U(x = 0) = 0$ and $F(x = 0) = 0$, where x is the space coordinate ($x = 0$ corresponds to the cathode contact). These boundary conditions have been used to derive both the Child–Langmuir law describing the space charge injection with no collisions and the Mott–Gurney law [150] for the injection into the collision-dominated semiconductor and thus allow us to follow the transition between these two limited cases. However, these boundary conditions are not unique. More realistic boundary conditions were analyzed in Ref. 151. An interesting idea to utilize a heterojunction cathode contact to inject electrons with high kinetic energy into the sample in order to boost the electron drift velocity was discussed and reviewed, for example, in Refs. 154 and 155. All these boundary conditions underscore a special role played by a cathode contact in n-type devices which was first emphasized in Ref. 152. However, the anode contact may also play an important role [92]. Hence, our simplified treatment cannot cover all complicated aspects of the electron transport in submicron structures. However, it will allow an analytical treatment of many new and interesting physical phenomena which should take place in submicron devices.

We will consider here a stationary case when $dv/dt = v\,dv/dx$. In this case we may derive the following equation for the electron velocity v from Eqs. (2-9-81)–(2-9-83):

$$\frac{\varepsilon_0\varepsilon_r m v}{2q}\frac{d^2 v^2}{dx^2} + \frac{\varepsilon_0\varepsilon_r m}{2q\tau}\frac{dv^2}{dx} + an_0 v - j = 0 \qquad (2\text{-}9\text{-}86)$$

We now introduce dimensionless variables explained in Table 2-9-3. In terms of the dimensionless squared velocity u, dimensionless coordinate w, and dimensionless scattering frequency β, Eq. (2-9-86) may be rewritten as

$$\frac{d^2 u}{dw^2} + \frac{\beta}{\sqrt{u}}\frac{du}{dw} + 1 - \frac{1}{\sqrt{u}} = 0 \qquad (2\text{-}9\text{-}87)$$

with the boundary conditions

$$u = 0 \qquad \text{at } w = 0 \qquad (2\text{-}9\text{-}88)$$

$$du/dw = 0 \qquad \text{at } w = 0 \qquad (2\text{-}9\text{-}89)$$

In a limited case $\beta \to 0$ (i.e., τ tends to infinity)

$$\frac{mv^2}{2} = qU \qquad (2\text{-}9\text{-}90)$$

because of the energy conservation. In this case Eq. (2-9-87) corresponds to the purely ballistic motion [102] and it becomes universal, i.e., it does not contain any parameters. Such parameters as the electron concentration, effective mass, dielectric constant, sample length, and current density only determine the characteristic scales (see Table 2-9-3). In a general case there is only one parameter β.

For a subsequent analysis it is convenient to rewrite Eq. (2-9-87) in a different form introducing variables $t = \beta^4 u$ and $z = \beta^3 w$. (The relationship between real and dimensionless variables is explained in Table 2-9-4.) Using variables t and z we find from Eq. (2-9-87)

$$\frac{d^2 t}{dz^2} + \frac{1}{\sqrt{t}}\frac{dt}{dz} - \frac{1}{\sqrt{t}} + \frac{1}{\beta^2} = 0 \qquad (2\text{-}9\text{-}91)$$

In a particular case when

$$1/\sqrt{t} \gg 1/\beta^2 \qquad (2\text{-}9\text{-}92)$$

we obtain the following universal equation (containing no parameters):

$$\frac{d^2t}{dz^2} + \frac{1}{\sqrt{t}}\frac{dt}{dz} - \frac{1}{\sqrt{t}} = 0 \qquad (2\text{-}9\text{-}93)$$

Using Eq. (2-9-92) we may show that Eq. (2-9-93) is valid when $n \gg n_0$, i.e., in a strong injection case. This equation may be rewritten as

$$\frac{dg}{dz} = \frac{1-g}{\sqrt{t}} \qquad (2\text{-}9\text{-}94)$$

$$\frac{dt}{dz} = g \qquad (2\text{-}9\text{-}95)$$

Dividing Eq. (2-9-94) by Eq. (2-9-95) and integrating the result with the boundary condition $g = 0$ at $t = 0$ we get

$$-\ln(1-g) - g = 2\sqrt{t} \qquad (2\text{-}9\text{-}96)$$

TABLE 2-9-3. Dimensionless Variables and Characteristic Parameters

Parameter	Definition	Comment
u	$u = \dfrac{q^2 n_0^2}{j^2} v^2$	Dimensionless squared velocity
w	x/x_0	Dimensionless coordinate
x_0	$\dfrac{j}{\sqrt{2}qn_0}\dfrac{1}{\omega_p}$	Characteristic distance
ω_p	$\left(\dfrac{q^2 n_0}{\varepsilon_0 \varepsilon_r m}\right)^{1/2}$	Plasma frequency
β	$\dfrac{1}{\sqrt{2}\omega_p \tau}$	Dimensionless scattering frequency
F_{ch}	$\dfrac{j}{\sqrt{2}\varepsilon_0 \varepsilon \omega_p}$	Characteristic electric field
t	$\beta^4 u$	Dimensionless squared velocity
z	$\beta^3 w$	Dimensionless coordinate
g	$\dfrac{dt}{dz}$	$g = \beta\dfrac{du}{dw}$
j_{ch}	$q^2 L\left(\dfrac{2n_0^3}{\varepsilon_0 \varepsilon_r m}\right)^{1/2}$	Characteristic current density
U_{po}	$\dfrac{qn_0 L^2}{2\varepsilon_0 \varepsilon_r}$	Punch-through voltage
cur	j/j_{ch}	Dimensionless current density
vol	U/U_{po}	Dimensionless voltage

This equation has two roots with the only relevant root corresponding to $g = dt/dz > 0$. It can be solved analytically in two limited cases $t \ll 1$ and $t \gg 1$ (see Fig. 2-9-25). The solution for the intermediate values of t is blown up in the insert in Fig. 2-9-25. For $t \ll 1$, g is also small and the expansion of the left-hand side of Eq. (2-9-96) into the Taylor series yields

$$g \approx 2t^{1/4} \tag{2-9-97}$$

and, hence,

$$2t^{3/4}/3 \approx z \tag{2-9-98}$$

assuming the boundary condition $t = 0$ at $z = 0$ discussed above. In terms of real variables Eq. (2-9-98) states

$$j = \frac{4}{9}\left(\frac{2q}{m}\right)^{1/2} \frac{\varepsilon_o \varepsilon_r}{L^2} U^{3/2} \tag{2-9-99}$$

TABLE 2-9-4. Relationships between Real and
Dimensionless Variables

Velocity:

$$v = \frac{j}{qn_0}\sqrt{u} \qquad \text{general case}$$

$$v = \frac{j}{qn_0\beta^2}\sqrt{t} \qquad \text{high injection case}$$

Field:

$$F = F_{ch}\left(\frac{du}{dw} + 2\beta\sqrt{u}\right) \qquad \text{general case}$$

$$F = \frac{F_{ch}}{\beta}(g + 2\sqrt{t}) \qquad \text{high injection case}$$

Voltage:

$$U = F_{ch}x_0\left(u + 2\beta\int_0^w \sqrt{u}\, dw'\right) \qquad \text{general case}$$

$$U = \frac{F_{ch}x_0}{\beta^4}\left(t + 2\int_0^z \sqrt{t}\, dz'\right) \qquad \text{high injection case}$$

Current:

$$j = j_{ch}\frac{1}{w_L} \qquad \text{general case}$$

$$j = j_{ch}\frac{\beta^3}{z_L} \qquad \text{high injection case}$$

Coordinate:

$$x = x_0 w \qquad \text{general case}$$

$$x = x_0\frac{z}{\beta^3} \qquad \text{high injection case}$$

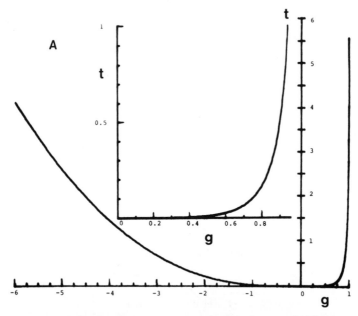

FIGURE 2-9-25. Universal dimensionless solution for a high injection case [105]. t is proportional to the squared velocity, $g = dt/dz$; z is the dimensionless coordinate; (a) t vs. g; (b) t vs. z; (c) t vs. z for $z \ll 1$. Solid line, computer solution; dashed line, analytical solution. (Computer and analytical solutions practically coincide in Fig. 2-9-25a).

which is a well-known expression for the current voltage characteristic of a vacuum diode (the Child–Langmuir Law). Using Eq. (2-9-83), one can express the electric field F in terms of g and t:

$$F = F_{ch}(g + 2\sqrt{t})/\beta \qquad (2\text{-}9\text{-}100)$$

(see also Tables 2-9-3 and 2-9-4). From Eq. (2-9-100) we find that close to the cathode contact the electric field varies as

$$F = F_{ch}(12x/x_0)^{1/3} \qquad (2\text{-}9\text{-}101)$$

Equations (2-9-93)–(2-9-101) are valid if $t \ll \beta^4$. This condition of the high injection regime may be rewritten as

$$U \gg 3.43 \frac{qn_0 L^2}{2\varepsilon_0\varepsilon_r} \qquad (2\text{-}9\text{-}102)$$

This condition ensures that the injected electron concentration is much larger than the background doping density. At the same time we also require $t \ll 1$ or

$$\beta^4 u \ll 1 \qquad (2\text{-}9\text{-}103)$$

This condition is more stringent than $u \ll 1$ only if $\beta > 1$. Converting to real variables we find from Eq. (2-9-103)

$$U \gg \frac{9}{8} \frac{mL^2}{q\tau^2} \qquad (2\text{-}9\text{-}104)$$

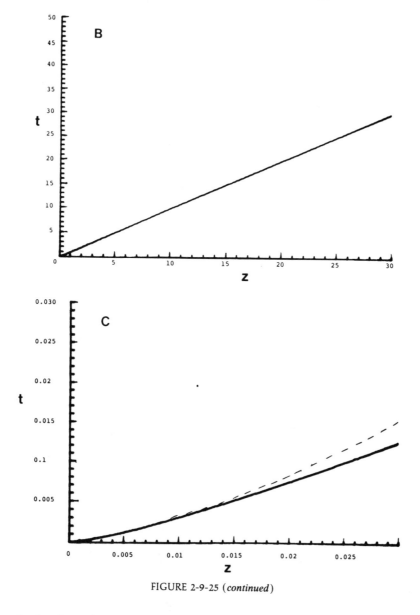

FIGURE 2-9-25 (*continued*)

This criterion is equivalent to the requirement that the transit time should be much shorter than the collision time τ because in the absence of collisions $qU = mv^2(L)/2$, where $v(L)$ is the electron velocity at the anode.

Now we consider the opposite limited case when $t \gg 1$ and collisions dominate. In this case the solution of Eq. (2-9-93) is given by

$$\frac{dt}{dz} \equiv g \simeq 1, \qquad t = z + c \tag{2-9-105}$$

As can be seen from Fig. 2-9-25b, Eq. (2-9-105) with $c = 0$ practically coincides with the computer solution for large values of z. In terms of variables u and w we then have

$$u = w/\beta \qquad (2\text{-}9\text{-}106)$$

Taking into account that Eq. (2-9-106) is valid when $z \gg 1$ and using Eq. (2-9-106) we find from Eq. (2-9-100)

$$F \simeq 2F_{\text{ch}}(\beta w)^{1/2} = 2F_{\text{ch}}\left(\frac{\beta x}{x_0}\right)^{1/2} = \left(\frac{2jm}{q\varepsilon_0\varepsilon_r\tau}\right)^{1/2} x^{1/2} \qquad (2\text{-}9\text{-}107)$$

$$U = \int_0^x F \, dx' = \frac{2\sqrt{2}}{3}\left(\frac{jm}{q\varepsilon_0\varepsilon_r\tau}\right)^{1/2} x^{3/2} \qquad (2\text{-}9\text{-}108)$$

Substituting $x = L$ into Eq. (2-9-108) we find the current–voltage characteristic which is given in this case by the Mott–Gurney law [150]:

$$j = \frac{9}{8}\varepsilon_0\varepsilon_r\mu\frac{U^2}{L^3} \qquad (2\text{-}9\text{-}109)$$

Hence, in the high injection case the current–voltage characteristic changes from the one described by the Child–Langmuir law for a scattering time much larger than the transit time to the one described by the Mott–Gurney law for the scattering time much smaller than the transit time.

For the intermediate values of z (not much smaller or much larger than one) the current–voltage characteristic can be deduced from the solution presented in Fig. 2-9-25 using Tables 2-9-3 and 2-9-4:

$$\text{vol} = \frac{2(\text{cur})^2}{\beta^4}\left(t_L + 2\int_0^{z_L}\sqrt{t}\,dz'\right) \qquad (2\text{-}9\text{-}110)$$

$$\text{cur} = \frac{\beta^3}{z_L} \qquad (2\text{-}9\text{-}111)$$

Here $\text{vol} = U/U_{\text{po}}$ is the dimensionless voltage, $U_{\text{po}} = qn_0L^2/2\varepsilon_r\varepsilon_0$ is the pinch-off voltage, and $\text{cur} = j/j_{\text{ch}}$ is the dimensionless current density and $z_L = \beta^3 L/x_0$, $t_L = t(z_L)$. Field, potential, and concentration profiles and I–V characteristics for the high injection case are presented in Figs. 2-9-26–2-9-30. These curves are universal and the characteristic scales can be calculated and the condition of the high injection case can be checked using Tables 2-9-3 and 2-9-4.

In a more general case of an arbitrary injection level Eq. (2-9-91) should be solved numerically. The solutions of Eq. (2-9-91) for different values of β are presented in Figs. 2-9-31–2-9-35. As β increases these solutions show a transition from a damped mode (see curves with $\beta = 1$) to an oscillatory mode (see curves with $\beta = 0$ and $\beta = 0.1$). Such oscillations are also present in computer solutions based on phenomenological equations of Section 2-8 (see Ref. 103). They may lead to an S-type current–voltage characteristic for small values of voltages. The analytical solution for $\beta = 0$ is given by

$$-2u^{1/4}\left(1 - \frac{u^{1/2}}{2}\right)^{1/2} + 2^{3/2}\sin^{-1}\left(\frac{u^{1/4}}{2^{1/2}}\right) = w \qquad (2\text{-}9\text{-}112)$$

From this equation and Table 2-9-3 we find the oscillation wavelength λ to be equal to

$$\lambda = w_\lambda x_0 = \frac{2\pi j}{q n_0 \omega_p} \tag{2-9-113}$$

or, in a slightly different form,

$$\lambda = \frac{2\pi (\varepsilon_0 \varepsilon_r m)^{1/2} j}{q^2 n_0^{3/2}} \tag{2-9-114}$$

This coincides with the value predicted by the small signal analysis [103] and clearly indicates that the plasma effects are involved.

In Fig. 2-9-35 the current–voltage characteristic for $\beta = 0$ is compared with the Child–Langmuir law (curve 2), which is given by

$$\mathrm{cur} = \frac{\sqrt{2}}{9}(\mathrm{vol})^{3/2} \tag{2-9-115}$$

in terms of the dimensionless variables, and with the electron beam asymptotic solution (curve 1):

$$\mathrm{cur} = \left(\frac{\mathrm{vol}}{8}\right)^{1/2} \tag{2-9-116}$$

The S-type current–voltage characteristic at low voltages may be indicative of instabilities in the sample leading to the current filamentation.

Let us describe the space oscillation cycle following Figs. 2-9-32–2-9-34. At

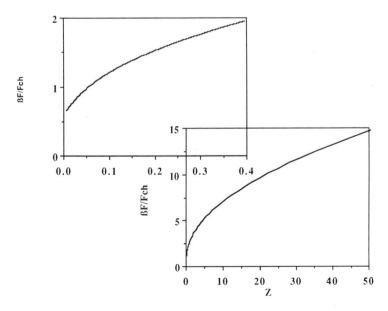

FIGURE 2-9-26. Normalized electric field vs. normalized distance z for the high injection case [105].

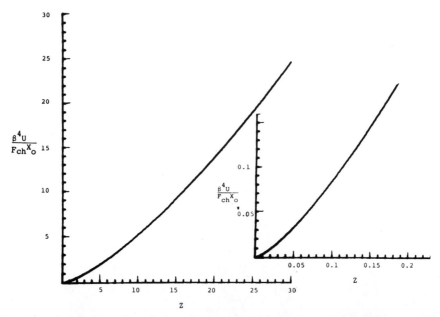

FIGURE 2-9-27. Dimensionless voltage vs. dimensionless distance for the high injection case [105].

$w = 0$ the velocity and field are zero and the concentration is infinitely large. When w increases, n drops and F and v both increase. However, because the electron concentration n drops the space charge density $q(n - n_0)$ also decreases, changing sign when n becomes smaller than n_0. After this point the field starts to decrease, leading to a decrease in the acceleration force qF and, hence, to a decrease in the acceleration. However, v is still going up, though at a diminishing acceleration rate,

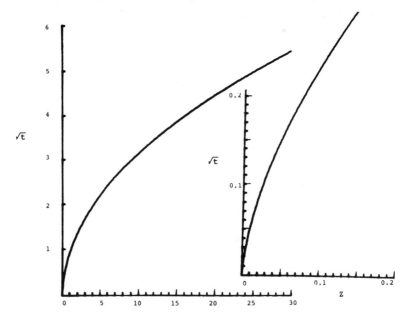

FIGURE 2-9-28. Dimensionless electron velocity vs. dimensionless distance for the high injection case [105].

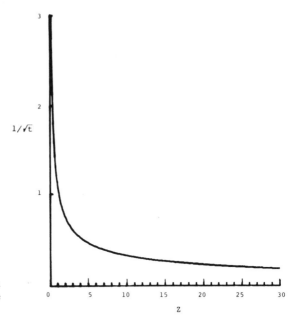

FIGURE 2-9-29. Dimensionless electron concentration vs. dimensionless distance for the high injection case [105].

forcing n to drop even further to maintain the current continuity ($j = qnv$ must be independent of position). This will continue until at some point n reaches the minimum value and the electric field F changes sign. Therefore, at larger distances the velocity v decreases, leading to the increase of n, etc. For $\beta = 0$ such cycles may be repeated indefinitely in an infinite sample. When the collisions are introduced ($\beta > 0$) the number of cycles may be limited to a few and for $\beta > 1$ the oscillations disappear altogether.

Oscillations similar to those shown in Figs. 2-9-31 through 2-9-34 were observed in electron beams [156]. In order to observe these oscillations in a semiconductor such as GaAs several conditions have to be met. In particular the sample length has to be much greater than the Debye length but much smaller than the mean free path. Also, the applied voltage should be much larger than the thermal voltage but much smaller than the pinch-off voltage U_{po}. The increase in scattering rates at higher voltages leads to the increase in damping (corresponding to larger values of β in these curves). At high enough voltages the intervalley transitions may become important increasing an effective mass. Therefore, a situation corresponding to an

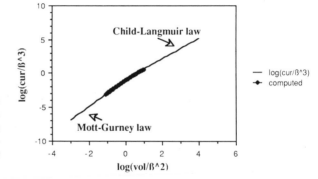

FIGURE 2-9-30. Dimensionless current vs. dimensionless bias voltage for the high injection case [105]. Solid line, computer solution; dotted line, Mott–Gurney law.

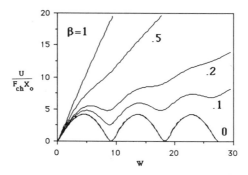

FIGURE 2-9-31. Dimensionless potential $u/F_{ch}x_0$ vs. dimensionless distance w for different values of β [105].

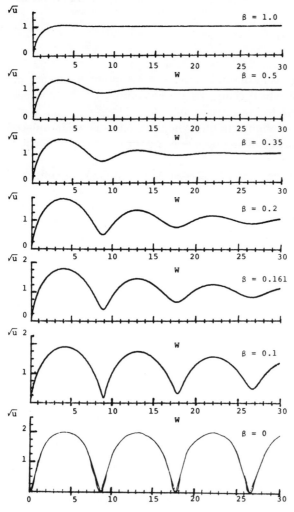

FIGURE 2-9-32. Dimensionless velocity $\sqrt{u}$ vs. dimensionless distance w for different values of β [105].

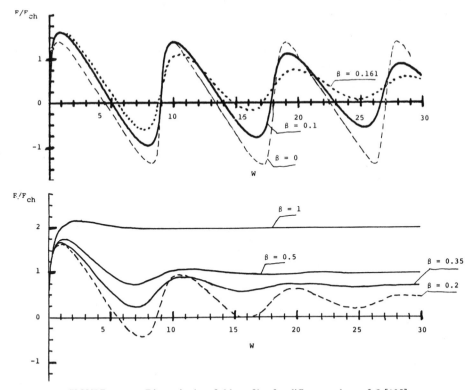

FIGURE 2-9-33. Dimensionless field profiles for different values of β [105].

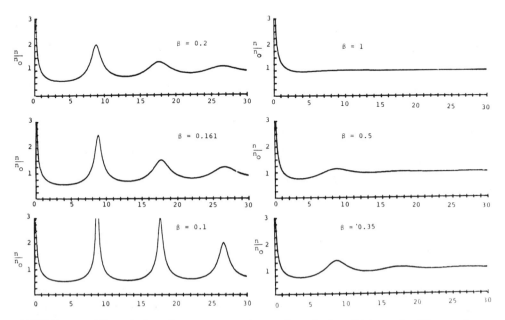

FIGURE 2-9-34. Dimensionless electron concentration vs. dimensionless distance w for different values of β [105].

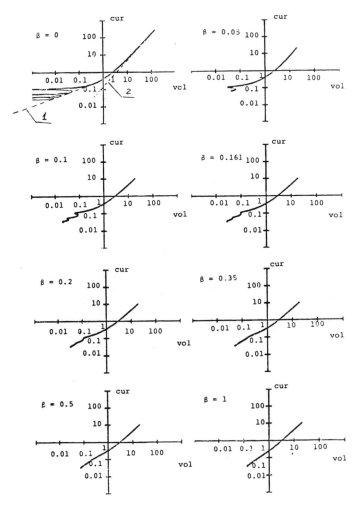

FIGURE 2-9-35. Dimensionless current–voltage characteristics for different values of β [105].

electron spillover at low voltages and to a high injection regime at high voltages is more typical for submicron GaAs n^+-n-n^+ structures.

A more accurate simulation to date is the Monte Carlo simulation of a short GaAs n-i-n structure [92]. The results of this simulation are presented in Figs. 2-9-36 and 2-9-37, where they are compared with the results obtained using the relaxation time equations of Section 2-8 and the collision free model. The Monte Carlo simulation [92] did not take into account the nonparabolicity which overestimated the averaged drift velocity at 0.36 V by about 30%. The resulting current–voltage characteristics are not dramatically different from the collision free model [102–105] and quite close to the relaxation time model [103]. The most important new result of the Monte Carlo simulation is the indication of the back reflection of electrons from the anode contact. This means that the anode is not simply a drain of electrons, as implied by the analytical model, and may play an important role in obtaining a high electron velocity.

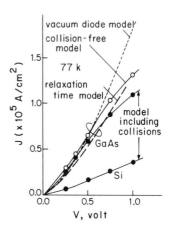

FIGURE 2-9-36. Current-voltage characteristic of 0.25-μm GaAs and Si n^+-n-n^+ diodes at 77 K [92]. Also shown are the results obtained in the frame of the relaxation time model (see Section 2-8) for $v(0) = 1000$ m/s. Part of the difference between the Monte Carlo simulation and the relaxation time model results is due to the different band structure and scattering parameters used in these calculations.

The electron velocity and concentration profiles computed in the frame of the relaxation time model [85] for a 0.2-μm GaAs n^+-n structure are shown in Fig. 2-9-38. As can be seen from the figure, the velocity is smaller than the "ballistic" limit but substantially larger than the saturation drift velocity. Experimental results [110, 126, 130, 148, 99] seem to confirm this conclusion, at least indirectly. As can be seen from Fig. 2-9-39 [104], the increase in the electron velocity follows the predicted ballistic behavior near the injecting contact but then deviates from it owing to the collisions and, at higher voltages, owing to the intervalley transfer to higher valleys of the conduction band. Closer to the anode the velocity actually decreases, which is a reflection of the oscillatory behavior predicted by the analytical model discussed above.

A chart illustrating the velocity enhancement in short GaAs structures was proposed by L. F. Eastman [128–130]; see Fig. 2-9-40. He also presented approximate experimental dependences of the average drift velocity as a function of the effective electrical length (see Fig. 2-9-41). The results presented in Fig. 2-9-41 are in agreement with the theory of the ballistic motion considered in this section.

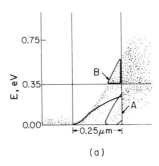

(a)

FIGURE 2-9-37. Energy and velocity distributions for a 0.25-μm GaAs n^+-n-n^+ diode. Points and solid lines from the Monte Carlo simulation [92]; dashed lines, the solution of the relaxation time equations of Section 2-8.

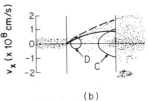

(b)

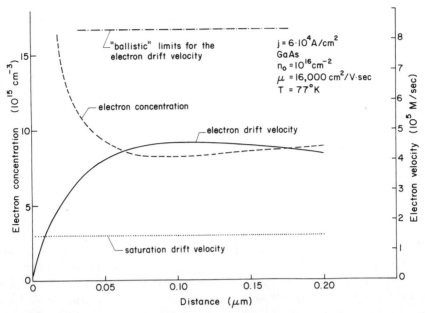

FIGURE 2-9-38. Electron concentration and electron drift velocity vs. distance for a short GaAs diode [103].

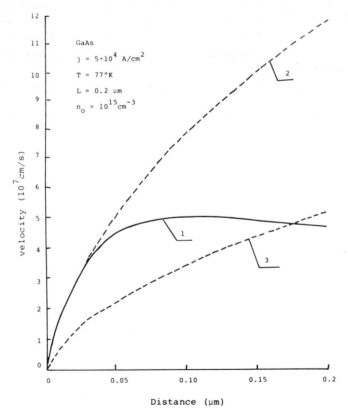

FIGURE 2-9-39. Velocity vs. distance in a short GaAs n^+-n-n^+ diode, comparison with the ballistic model [104]. 1, computer solution of the relaxation time equations; 2, "ballistic limit" [102] ($m = 0.067m_e$); 3, "ballistic limit" [102] ($m = 0.35m_e$).

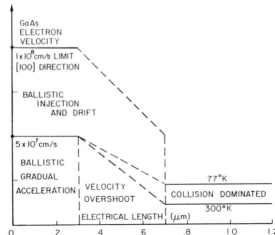

FIGURE 2-9-40. Average electron velocity range vs. effective electrical length for both gradual acceleration and fast electron injection in GaAs, after L. F. Eastman [128].

Recently direct observation of the ballistic transport was reported by M. Heiblum *et al.* [157] and A. F. Levi *et al.* [158]. The device used in [157] is shown in Fig. 2-9-42. At negative collector base voltages only tunneling electrons with energies higher than the collector barrier height are collected. The energy distribution of the collected electrons is proportional to the derivative of the collector current with respect to the collector base voltage $G_c = \partial I_c / \partial V_{cb}$. The main peak in the dependence of G_c on V_{cb} (see Fig. 2-9-43) corresponds to the ballistic electrons. One interesting conclusion reached in [157] was that only about 50% of electrons behave ballistically. The loss mechanism was not identified. In this regard we should mention theoretical work [159] where the evolution of the electron distribution function in short sample was considered based on the simultaneous solution of the Boltzmann transport equation and Poisson equation. The results predict an appearance of the ballistic peak in the distribution function.

In [158] the hot electron distribution was measured using a GaAs structure with two planar barriers and a transit region of 650 Å. The measured hot electron

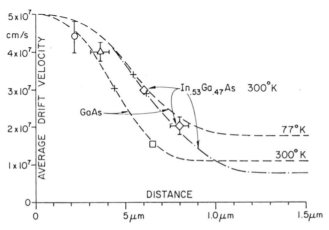

FIGURE 2-9-41. Experimental values of average electron velocity vs. effective electrical length for GaAs and $In_{0.53}Ga_{0.47}As$, after L. F. Eastman [128].

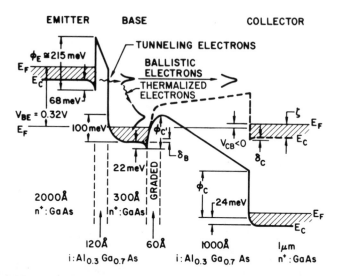

FIGURE 2-9-42. Band diagram of the test ballistic structure [157]. The dashed line corresponds to the negative value of the collector-base voltage.

spectrum showed two pronounced peaks with a sharp peak corresponding to quasi-ballistic electrons.

A considerable increase in the electron drift velocity may be quite important in short semiconductor devices such as field effect transistors (see Chapter 7), modulation doped field effect transistors (see Chapter 10), bipolar junction transistors, Schottky diodes, and avalanche devices. The velocity enhancement has also stimulated some interesting work on novel semiconductor devices, such as permeable base transistors, vertical ballistic transistors, and planar doped barrier transistors (see Chapter 11).

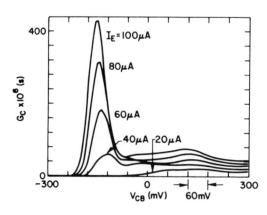

FIGURE 2-9-43. The energy distribution of the collected hot electrons (G_c vs. V_{cb}) for different emitter currents [157].

REFERENCES

1. R. H. Parmenter, *Phys. Rev.* **100**, 573 (1955).
2. G. Dresselhaus, *Phys. Rev.* **100**, 580 (1955).
3. W. A. Harrison, *Electronic Structure and Properties of Solids*, Freeman, San Francisco, 1980.
4. V. Heine, Pseudopotential concept, in *Solid State Physics, Advances in Research and Applications*, Vol. 24, Academic, New York, 1970.
5. J. R. Chelikowsky and M. L. Cohen, Non-local pseudopotential calculations for the electronic structure of eleven diamond and zinc blend semiconductors, *Phys. Rev. B* **14**(2), 556–582 (1976).
6. J. S. Blakemore, Semiconductor and other major properties of GaAs, *J. Appl. Phys.* **53**(10), R123–R181 (1982).
7. C. Jacoboni and L. Reggiani, Bulk hot-electron properties of cubic semiconductors, *Adv. Phys.* **28**(4), 493–553 (1979).
8. G. Dresselhaus, A. F. Kip, and C. Kittel, *Phys. Rev.* **98**, 368 (1955).
9. E. O. Kane, *J. Phys. Chem. Solids* **1**, 249 (1957).
10. D. L. Rode, in *Semiconductor and Semimetals*, Ed. by R. K. Willardson and A. C. Beer, Academic, New York, 1975, Vol. 10, p. 1.
11. B. Weber, M. Cardona, C. K. Kim, and S. Rodriguez, *Phys. Rev. B* **12**, 5729 (1975).
12. D. E. Aspnes and J. Lees, *Phys. Rev. B* **2**, 4144 (1970).
13. G. D. Pitt and J. Lees, *Phys. Rev. B* **2**, 4144 (1970).
14. G. D. Pitt, *J. Phys. C.* **6**, 1586 (1973).
15. J. D. Wiley, Mobility of holes in III–V compounds, in *Semiconductors and Semimetals*, Academic, New York, 1975, Vol. 10, pp. 91–174.
16. A. Nussbaum, *Semiconductor Device Physics*, Prentice-Hall, Englewood Cliffs, New Jersey, 1962.
17. A. C. Smith, J. F. Janak, and R. B. Adler, *Electronic Conduction in Solids*, McGraw-Hill, New York, 1967.
18. W. Fawcett, D. A. Boardman, and S. Swain, Monte Carlo determination of electron transport properties in gallium arsenide, *J. Phys. Chem. Solids* **31**, 1963–1990 (1970).
19. J. G. Ruch and W. Fawcett, Temperature dependence of the transport properties of gallium arsenide determined by a Monte Carlo method, *J. Appl. Phys.* **41**(9), 3843–3849 (1970).
20. Yu. K. Pozhela and A. Reklaitis, Electron transport properties in GaAs at high electric fields, *Solid-State Electron.* **23**, 927–933 (1980).
21. T. Kurosawa, *J. Phys. Soc. Jpn. Suppl.* **21**, 424 (1966).
22. Yu. K. Pozhela, *Plasma and Current Instabilities in Semiconductors*, Nauka, Moscow, 1977 (in Russian).
23. P. A. Lebwohl and P. J. Price, *Appl. Phys. Lett.* **19**, 530 (1971).
24. J. G. Ruch, Electron dynamics in short channel field-effect transistors, *IEEE Trans. Electron Devices* **ED-19**(5), 652 (1972).
25. A. Matulionis, Y. Pozhela, and A. Reklaitis, *Solid State Commun.* **16**, 1133 (1975).
26. A. Matulionis, Y. Pozhela, and A. Reklaitis, *Phys. Stat. Sol.* (a) **31**, 83 (1975); **35**, 43 (1976).
27. M. Tomizawa, A. Yoshii, and K. Yokoyama, *IEEE Electron Device Lett.* **6**(7), 332–334 (1985).
28. G. Baccarani, C. Jacoboni, and A. Mazzone, *Solid State Electron* **20**, 5 (1977).
29. T. J. Maloney and J. Frey, *J. Appl. Phys.* **48**, 781 (1977).
30. N. Rees, G. S. Sanghera, and R. A. Warriner, Low-temperature F.E.T. for low-power high-speed logic, *Electron. Lett.* **13**(6), 156 (1977).
31. A. Warriner, Computer simulation of gallium arsenide field-effect transistors using Monte Carlo methods, *Solid State Electron Devices* **1**(4), 105–108 (1977).
32. J. Zimmermañ, Y. Leroy, and E. Constant, *J. Appl. Phys.* **49**, 3378 (1978).
33. C. Moglestue, Computer simulation of a dual gate GaAs field-effect transistor using the Monte Carlo method, *Solid State Electron Devices* **3**(1), 133 (1979).
34. B. Carnez, A. Cappy, A. Kaszinski, E. Constant, and G. Salmer, Modeling of submicron gate field-effect transistor including effects of non-stationary electron dynamics, *J. Appl. Phys.* **51**(1), 784 (1980).
35. G. S. Sanghera, A. Chryssafis, and C. Moglestue, Monte Carlo particle simulation of *n*-type GaAs field-effect transistors with a *p*-type buffer layer, *IEE Proc.* **127**, Pt. I(4), 203 (1980).
36. A. Cappy, B. Carnez, R. Fauquembergues, G. Salmer, and E. Constant, Comparative potential performance of Si, GaAs, GaInAs, InAs submicrometer-Gate FET's, *IEEE Trans. Electron Devices* **ED-27**(11), 2158 (1980).

37. J. Y. Tang and K. Hess, Investigation of transient and electronic transport in GaAs following high energy injection, *IEEE Trans. Electron Devices* **ED-29**, 1906-1910 (1982).

38. K. Brennan, K. Hess, and G. J. Iafrate, Monte Carlo simulation of reflecting contact behavior on ballistic device speed, *IEEE Electron Device Lett.* **EDL-4**(9), 332-334 (1983).

39. M. A. Littlejohn, J. R. Hauser, and T. H. Glisson, Velocity-field characteristics of GaAs with Γ6-L6-X6 conduction-band ordering, *J. Appl. Phys.* **48**, 4587, 4590 (1977).

40. W. Czubatyj, M. S. Shur, and M. P. Shaw, *Solid State Electron.* **21**, 75 (1978).

41. P. J. Price, Monte Carlo calculation of electron transport in solids, in *Semiconductors and semimetals*, Vol. 14, Lasers, Junctions, Transport, Ed. by R. K. Willardson and A. C. Beer, pp. 249-30, Academic, New York, 1979.

42. P. A. Houston and A. G. R. Evans, *Solid State Electron.* **20**, 197 (1977).

43. N. Braslau and P. S. Hauge, *IEEE Trans. Electron Devices* **ED-17**, 616 (1970).

44. K. Ashida, M. Inoue, J. Shirafuji, and Y. Inuisi, *J. Phys. Soc. Jpn* **37**, 408 (1974).

45. J. G. Ruch and G. S. Kino, *Phys. Rev.* **174**, 921 (1968).

46. V. Bareikis, A. Galdikas, R. Milisyte, and V. Viktoravicius, Program and papers of 5th Int. Conf. on Noise in Physical Systems, p. 212, 13-16 March, Bad Nauheim, West Germany, 1978.

47. G. E. Stillman, C. M. Wolfe, and J. O. Dimmock, *J. Phys. Chem. Solids* **31**, 1199 (1970).

48. H. Brooks, *Adv. Electron. Electron Phys.* **7**, 158 (1955).

49. J. Bardeen and W. Shockley, *Phys. Rev.* **80**, 72 (1950).

50. H. J. G. Meijer and D. Polder, *Physica* **19**, 255 (1953).

51. H. Ehrenreich, *J. Chem. Solids* **12**, 97 (1959).

52. D. L. Rode, Electron mobility in direct gap polar semiconductors, *Phys. Rev. B* **2**(4), 1012-1024 (1970).

53. J. W. Harrison and J. R. Hauser, *J. Appl. Phys.* **47**, 292 (1976).

54. S. M. Sze, *Physics of Semiconductor Devices*, 2nd Edition, Wiley Interscience, New York, 1981.

55. C. M. Wolfe and G. E. Stillman, Self-compensation of centers in high purity GaAs, *Appl. Phys. Lett.* **27**(10), 564, November (1975).

56. D. E. Bolger, J. Franks, J. Gordon, and J. Whitaker, *Proceedings of the First International Conference on GaAs*, Institute of Physics and the Physical Society, London, 1967, p. 16.

57. M. Maruyama, S. Kikuchi, and O. Mizuno, *J. Electrochem. Soc.* **116**, 413 (1969).

58. J. C. Carballes, D. Diguet, and J. Lebailly, *Proceedings of the Second International Conference on GaAs*, Institute of Physics and the Physical Society, London, 1969, p. 28.

59. V. F. Dvoryankin, O. V. Emel'yanenko, D. N. Nasledov, D. D. Nedeoglo, and A. A. Telegin, *Sov. Phys.-Semicond.* **5**, 1936 (1972).

60. I. Akasaki and T. Hara, *Proceedings of the 9th International Conference on the Physics of Semiconductors*, Nauka, Moscow, 1969, p. 787.

61. D. Kranzer and G. Eberharter, *Phys. Status Solidi A* **8**, K89 (1971).

62. H. Ehrenreich, *Phys. Rev.* **120**, 1951 (1960).

63. K. H. Nicholas, C. M. L. Yee, and C. M. Wolfe, *Solid State Electron.* **23**, 109 (1980).

64. D. E. Hill, *J. Appl. Phys.* **41**, 1815 (1970).

65. A. L. Mears and R. L. Stradling, *J. Phys. C.* **4**, L22 (1971).

66. K. H. Zschaues, in GaAs and related compounds, *Inst. Phys. Conf. Ser.* **17**, 3 (1973).

67. F. D. Rosi, D. Meyerhofer, and R. V. Jensen, *J. Appl. Phys.* **31**, 1105 (1960).

68. D. E. Hill, *Phys. Rev.* **133**, A866 (1964).

69. J. Vilms and J. P. Garrett, *Solid State Electron.* **15**, 443 (1972).

70. F. E. Rosztoczy, F. Ermanis, I. Hayashi, and B. Schwartz, *J. Appl. Phys.* **41**, 264 (1970).

71. O. V. Emil'yanenko, T. S. Lagunova, and D. N. Nasledov, *Sov. Phys. Solid State* **2**, 176 (1960).

72. S. M. Gasanli, O. V. Emil'yanenko, V. K. Ergakov, F. P. Kesamanly, T. S. Lagunova, and D. N. Nasledov, *Sov. Phys. Semicond.* **5**, 1641 (1972).

73. H. Shichijo and K. Hess, *Phys. Rev. B* **23**, 4197 (1981).

74. V. L. Dalal, Hole velocity in *p*-GaAs, *Appl. Phys. Lett.* **16**(12), 489-491, 1970.

75. J. R. Barker and D. K. Ferry, On the physics and modeling of small semiconductor devices-I, *Solid State Electron.* **23**, 519-530 (1978).

76. D. K. Ferry and J. R. Barker, On the physics and modeling of small semiconductor devices-II, *Solid State Electron.* **23**, 531-544 (1980).

77. D. K. Ferry and J. R. Barker, Physics and modeling of small semiconductor devices-III. Transient response in the finite collision regime, *Solid State Electron.* **23**, 545-549 (1980).

78. J. R. Barker, Quantum transport theory of high field conduction in semiconductors, *J. Phys. C: Solid State Phys.* **6**, 2663-2684 (1973).

79. J. R. Barker, High-field collision rates in polar semiconductors, *Solid State Electron.* **21**, 267–272 (1978).

80. K. Blotekjaer, Transport equations for two-valley semiconductors, *IEEE Trans. Electron Devices* **ED-17**(1), 38–47 (1970).

81. K. Blotekjar and E. B. Lunde, Collision integrals for displaced Maxwellian distributions, *Phys. Status Solidi* **35**, 581 (1969).

82. R. Bosch and H. W. Thim, *IEEE Trans. Electron. Devices* **ED-21**(1), 16–25 (1974).

83. W. R. Curtice and Y.-H. Yun, A temperature model for the GaAs MESFET, *IEEE Trans. Electron Devices* **ED-28**(8), 954–962 (1981).

84. P. N. Butcher, W. Fawcett, and C. Hilsum, A simple analysis of stable domain propagation in the Gunn effect, *Brit. J. Appl. Phys.* **17**(7), 841–850 (1966).

85. M. S. Shur, Influence of non-uniform field distribution on frequency limits of GaAs field-effect transistors, *Electron Lett.* **12**(23), 615–616 (1976).

86. J. M. Ziman, *Electrons and Phonons*, Oxford University Press, London, 1960.

87. D. E. McCumber and A. G. Chynoweth, Theory of negative conductance amplification and of Gun instabilities in "two-valley" semiconductors, *IEEE Trans. Electron Devices* **ED-13**, 5 (1966).

88. V. B. Gorvinkel, MM. E. Levinstein, and D. V. Mashovets, Influence of strong transverse magnetic field on the Gunn effect, *Sov. Phys. Semicond.* **13**(3), 331 (1979).

89. H. D. Rees, Hot electron effects at microwave frequencies in GaAs, *Solid Commun.* **7**(2), 267–269 (1969).

90. C. T. Sah, R. N. Noyce, and W. Shockley, Carrier generation and recombination in p-n junctions and *p-n* junctions characteristics, *Proc. IRE* **45**, 1228 (1957).

91. R. K. Cook and J. Frey, Diffusion effects and "Ballistic Transport", *IEEE Trans. Electron Devices* **ED-28**(8), 951–953 (1981).

92. Y. Awano, K. Tomizawa, N. Hashizume, and M. Kawashima, Monte Carlo simulation of a submicron sized GaAs n^+-$i(n)$-n^+ diode, *Electron. Lett.* **18**(3), 133–134 (1982).

93. M. T. Elliot, M. R. Splinter, A. B. Jones, and J. P. Reekstin, Size effects in E-beam fabricated MOS devices, *IEEE Trans. Electron Devices* **ED-26**, 469–475 (1979).

94. W. Hunter, T. C. Holloway, P. K. Chatterjee, and A. F. Tasch, Jr., A new edge-defined approach for submicrometer MOSFET fabrication, *IEEE Electron Device Lett.* **EDL-2**, 4–6 (1980).

95. P. C. Chao, W. H. Ku and J. Nulman, A high aspect ratio 0.1 micron gate technique for low-noise MESFET's, *IEEE Electron Device Lett.* **EDL-3**(1), 24–26 (1982).

96. C. Chao, W. H. Ku, and C. Lowe, A pile-up masking technique for the fabrication of sub-half-micron gate length GaAs MEFET's, *IEEE Electron Device Lett.* **EDL-3**(10), 286–288 (1982).

97. P. C. Chao, W. H. Ku, and J. Nulman, Experimental comparisons in the electric performance of long and ultrashort gate length GaAs MESFETs, *IEEE Electron Device Lett.* **EDL-3**(8), 187–190 (1982).

98. M. Hollis, N. Dandekar, L. F. Eastman, M. Shur, D. Woodward, R. Stall, and C. Wood, Transverse magnetoresistance in GaAs two-terminal devices, A characterization of electron transport in the near-ballistic regime, in *IEDM Technical Digest*, pp. 622–625, Dec. 1980, IEEE, Washington DC.

99. M. A. Hollis, L. F. Eastman, and C. E. C. Wood, Measurement of J/V characteristics of a GaAs submicron n^+-n^--n^+ diode, *Electron. Lett.* **18**(13), 570–572 (1982).

100. R. A. Sadler and L. F. Eastman, High speed logic at 300 K with self-aligned submicrometer-gate GaAs MESFETs, *Electron Device Lett.* **EDL-4**(7), 215–217 (1983).

101. H. M. Levy, R. E. Lee, and R. A. Sadler, A submicron self-aligned GaAs MESFET technology for digital integrated circuits, in Proceedings Device Res. Conf., paper IV B-3, Ft. Collins, Colorado, June 1982.

102. M. S. Shur and L. F. Eastman, Ballistic transport in semiconductors at low temperature for low-power high speed logic, *IEEE Trans. Electron Devices* **ED-26**, 1677–1683 (1979).

103. M. S. Shur and L. F. Eastman, Near ballistic electron transport in GaAs devices at 77 K, *Solid State Electron.* **24**, 11–18 (1981). Also in Proc. Biennial Cornell Conf. on Microwave Devices, August, 1979.

104. M. S. Shur and L. F. Eastman, Ballistic and near ballistic transport in GaAs, *IEEE Electron Device Lett.* **EDL-1**(8), 147–148 (1980).

105. M. S. Shur, Ballistic transport in a semiconductor with collisions, *IEEE Trans. Electron Devices*, **ED-28**(10), 1120–1130 (1981).

106. K. Hess, Ballistic electron transport in semiconductors, *IEEE Trans. Electron Devices*, **ED-28**, 937–940 (1981).

107. M. S. Shur, Ballistic and collision dominated transport in a short semiconductor diode, in *IEDM Technical Digiset*, pp. 618–621, Dec. 1980, IEEE.

108. J. R. Barker, D. K. Ferry, and H. L. Grubin, On the nature of ballistic transport in short-channel semiconductor devices, *IEEE Electron. Device Lett.* **EDL-1**, 209-210 (1980).

109. J. J. Rosenberg, E. J. Yoffa, and M. Nathan, Importance of boundary conditions to conduction in short samples, *IEEE Trans. Electron Devices* **ED-28**, 941-944 (1981).

110. L. Eastman, R. Stall, D. Woodward, N. Dandekar, C. Wood, M. Shur, and K. Board, Ballistic electron motion in GaAs at room temperature, *Electron. Lett.* **16**(13), 524 (1980).

111. A. A. Kastalsky and M. S. Shur, Conductance of small semiconductor devices, *Solid-State Commun.* **39(b)**, 715 (1981).

112. A. A. Kastalsky, M. S. Shur, and Kwyro Lee, Conductance of small semiconductor devices, Proceeding 8th Biennial Cornell Electr. Eng. Conf., 1981.

113. M. S. Shur and D. Long, Performance prediction for submicron GaAs SDFL logic, *IEEE Electron Device Lett.* **EDL-2**(4), 124-127 (1982).

114. R. L. Fork, C. V. Shank, B. I. Greene, F. K. Reinhart, and R. A. Logan, Experimental observation of nonequilibrium carrier transport in GaAs, *IEEE Trans. Electron Devices* **ED-27**(11), 2198 (1980).

115. C. V. Shank, R. L. Fork, B. I. Greene, F. K. Reinhart, and R. A. Logan, Picosecond nonequilibrium carrier transport in GaAs, *Appl. Phys. Lett.* **38**, 104 (1981).

116. Kwyro Lee and Michael Shur, Impedance of thin semiconductor films in low electric field, *J. Appl. Phys.* **54**(7), 4028-4034 (1983).

117. Aldert van der Ziel, Michael Shur, Kwyro Lee, Tzu-Hung Chen, and Kostas Amberiadis, Carrier distribution and low-field resistance in short $n^+-n^--n^+$ and $n^+-p^--^+n$ structures, *IEEE Trans. Electron Devices* **ED-30**(2), 128-137 (1983).

118. M. S. Shur, Ballistic regime in semiconductor devices, *Bull. Am. Phys. Soc.* **26**(3), 466 (1981).

119. M. S. Shur and L. F. Eastman, GaAs n^+-p-n^+ ballistic structure, *Electron. Lett.* **16**, 522-523 (1980).

120. M. S. Shur, Ballistic and collision dominated transport in a short semiconductor diode, *IEDM Technical Digest*, pp. 618-621, Dec. 1980, IEEE, Washington DC.

121. B. Abraham-Shrauner, Instabilities of inertial transport in semiconductors, *IEEE Trans. Electron Devices* **ED-28**(8), 945-950 (1981).

122. J. Frey, Ballistic transport in semiconductor devices, in *Tech. Dig. Int. Electron Devices Meet. 1980*, p. 613, IEEE, New York, 1980.

123. M. Muller, The effect of non-parabolic bands, in Semiconductor Millimeter Wavelength Electronics, report No. ONR 80-1, Washington University, October 1980.

124. T. J. Maloney, Polar mode scattering in ballistic transport GaAs devices, *IEEE Electron Device Lett.* **EDL-1**, 54 (1980).

125. W. R. Frensley, High-frequency effects of ballistic electron transport in semiconductors, *IEEE Trans. Electron Devices Lett.* **EDL-1**, 137-139 (1980).

126. R. Zuleeg, Ballistic effects in current limiters, *IEEE Electron Device Lett.* **EDL-1**, 234 (1980).

127. S. Teitel and J. W. Wilkins, Ballistic transport and velocity overshoot in semiconductors: Part I—Uniform field effects, *IEEE Trans. Electron Devices* **ED-30**(2), 150-153 (1983).

128. L. F. Eastman, Very high velocities in short Gallium arsenide structures, *Feskorperprobleme* **XXII**, 173-187 (1982).

129. L. F. Eastman, The limits of electron ballistic motion in compound semiconductor transistors, *Inst. Phys. Conf. Ser.* **63**, 245-250 (1981).

130. L. F. Eastman, Experimental studies of ballistic transport in semiconductors, *J. Phys.* (*Paris*) *Col.* **C7**, suppl. au10, **42**, C7-263 (1981).

131. P. E. Schmidt, M. Octavio, and P. D. Esqueda, Single-carrier space-charge controlled conduction vs. ballistic transport in GaAs devices at 77 K, *IEEE Electron Device Lett.* **EDL-2**, 205 (1981).

132. J. B. Socha and L. F. Eastman, Comments on "Single-carrier space-charge controlled conduction vs. Ballistic transport in GaAs devices at 77 K," *IEEE Electron Device Lett.* **EDL-3**(1), 27 (1982).

133. Johnson Lee, The probability for ballistic electron motion in *n*-GaAs, *IEEE Electron Device Lett.* **EDL-2**, 167-169 (1981).

134. R. O. Grondin, P. Lugli, and D. K. Ferry, Ballistic transport in semiconductors, *IEEE Electron Device Lett.* **EDL-3**, 373-375 (1982).

135. F. Capasso and G. B. Bachelet, Ballistic electron impact ionization in GaAs and in InP avalanche devices, *IEDM Technical Digest*, pp. 633, Dec. 1980, IEEE, Washington, DC.

136. J. B. Socha and G. J. Rees, Ballistic electron transport in a nonparabolic band, *Electronics Lett.* **16**(23), Nov. (1980).

137. G. J. Rees and J. B. Socha, Ballistic electron transport in a transverse magnetic field, *Solid State Electron.* **24**, 695-698 (1981).

138. B. Fauquemberque, M. Pernisek, and E. Constant, Monte Carlo simulation of a space charge Injection FET, *Electron. Lett.* **18**(15), 670–671 (1982).

139. C. K. Williams, T. M. Glisson, M. A. Littlejohn, and J. R. Hauser, *IEEE Electron Device Lett.* **EDL-4**(6), 161–163 (1983).

140. See, for example, N. W. Ashcroft and N. D. Mermin, *Solid State Physics*, Holt, Rinehart, and Winston, New York, 1976.

141. E. H. Sondheimer, *Adv. Phys.* **1**, 1 (1952).

142. A. Many, Y. Goldstein, and N. B. Grover, *Semiconductor Surfaces*, North-Holland, Amsterdam, 1965.

143. C. Anderson, *Adv. Phys.* **19**, 311 (1970).

144. K. S. Champlin, D. B. Armstrong, and P. D. Gunderson, *IEEE Proc.* **52**, 677 (1964).

145. K. S. Knol and G. Diemer, Theory and experiments on electrical fluctuations and damping of double-cathode valves, *Philips Res. Labs., Tech. Rep.* **5**, 131–154 (1950).

146. Amitabh Candra, Calculation of the free carrier density profile in a semiconductor near an ohmic contact, *Solid-State Electron.* **23**, 516–517 (1980).

147. E. H. Rhoderick, *Metal–Semiconductor Contacts*, Clarendon Press, Oxford, 1978.

148. E. H. Hollis, An Investigation of electron transport in two-terminal submicron GaAs devices, M.S. thesis, Cornell University, Ithaca, New York, May 1981.

149. T. J. Maloney and J. Frey, Transient and steady-state electron transport properties of GaAs and InP, *J. Appl. Phys.* **48**, 781 (1977).

150. N. F. Mott and R. W. Gurney, *Electronic Process in Ionic Crystals*, Clarendon Press, Oxford, 1940.

151. A. J. Holden and B. T. Debney, Improved theory of ballistic transport in one dimension, *Electron. Lett.* **18**(13), 558–559 (1982).

152. M. P. Shaw, P. R. Solomon, and H. L. Grubin, Circuit-controlled current instabilities in *n*-GaAs, *Appl. Phys. Lett.* **17**(2), 535–537 (1970).

153. M. Shur and M. Hack, *J. Appl. Phys.* **59**(3), 803 (1986).

154. H. Kroemer, *Proc. IEEE* **70**, 13–25, Jan. (1982).

155. D. Ankri and L. F. Eastman, *Electron Lett.* **18**, 750 (1982).

156. T. G. Mihrah, *J. Appl. Phys.* **33**(4), 1382 (1962).

157. M. Heiblum, M. I. Nathan, D. C. Thomas, and C. M. Knoedler, *Phys. Rev. Lett.* **55**(20), 2200 (1985).

158. A. F. J. Levi, J. R. Hayes, P. M. Platzman, and W. Wiegmann, *Phys. Rev. Lett.* **55**(19), 2071 (1985).

159. H. V. Baranger and J. W. Wilkins, *Phys. Rev. B.* **30**(12), 7349–7351 (1984).

3

GaAs Technology

3-1. GALLIUM AND ARSENIC

Gallium is a rare element. It is produced as a by-product in Al or Zn production. The physical properties of this metal are described in Table 3-1-1. Standard purification processes make it possible to obtain Ga as pure as 99.99999%. Liquid Ga reacts with quartz at high temperatures leading to impurities in GaAs grown in quartz containers. Ga is considered to be toxic.

Arsenic is mainly produced from sulfur ores such as As_2S_4 or As_2S_3. Oxidation is used to obtain As_2O_3, which is then reduced to As in reaction with carbon. Arsenic can be obtained in three different modifications. The most stable structure is a metallic crystalline As, which can be easily oxidized by air. The oxidation is accompanied by the phase transformation of the surface layer into a black amorphous phase. Abrupt cooling of As vapor leads to formation of yellow crystalline As_4. This compound is unstable and rapidly transforms into a metallic As. Amorphous As is transformed into the metal phase when heated in vacuum at 300°C. The physical properties of As are described in Table 3-1-2.

Arsenic is more difficult to purify than Ga. The purest As is obtained by the thermal decomposition of arsin. Arsenic is toxic. Concentrations of As more than 3 mg/m^3 are dangerous. Arsin is especially toxic. The concentrations of arsin in air more than 0.3 mg/m^3 may be toxic. Arsenic vapors are chemically very active. This limits the choice of materials for containers used in GaAs growth.

TABLE 3-1-1. Physical Properties of Ga

Lattice constants	$a = 4.5197$ Å
(rhombic crystal structure)	$b = 7.6601$ Å
	$c = 4.5257$ Å
Melting temperature	29.8°C
Boiling temperature	2230°C
Vapor pressure at 1000°C	0.001 mm Hg
Density (liquid phase)	6.095 g/cm³ near the melting point
Density (solid phase)	5.904 g/cm³

TABLE 3-1-2. Physical Properties of Arsenic

Density (metallic arsenic)	5.73 g/cm^3
(amorphous arsenic)	4.7 g/cm^3
(yellow arsenic)	2.0 g/cm^3
Melting temperature	814°C
(at 36 atm)	

3-2. CRYSTAL GROWTH OF BULK MATERIAL

The basic difficulty in GaAs growth is that GaAs decomposes when heated. Therefore, the As vapor pressure should be chosen in such a way that the stochiometry of the melt is maintained. Another difficulty lies in chemical activity of Ga and As which interact with the material of the container. Finally GaAs expands during the solidification, which may lead to large stresses in the crystal imposed by the container walls.

Techniques for GaAs growth include crystal pulling from the melt (Czochralski method) and zone melting of an ingot held in a container.

The Czochralski method allows the continuously observed growth of mechanically unconstrained GaAs. The higher yield and better crystal quality make it the most widely used technique. An additional advantage is that the crystallographic orientation of the boule is controlled by the crystallographic orientation of the seed.

The schematic diagram of a crystal puller is shown in Fig. 3-2-1. The equipment should be vibration-free. It should allow pull rates as low as 0.1 mm/h. Early techniques for pulling GaAs crystals used "syringe pullers," where a bearing surface of boron nitride provided a mechanical seal capable of working at temperatures higher than 600°C. This helped avoid problems related to the decomposition of GaAs at the melting point. Another approach involves magnetic coupling to a high-Curie-point alloy attached to the pull rod. However, these techniques have been almost entirely superseded by the liquid encapsulation technique (see Fig. 3-2-2).

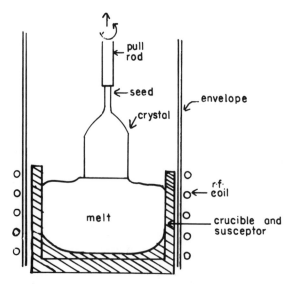

FIGURE 3-2-1. Crystal puller [1].

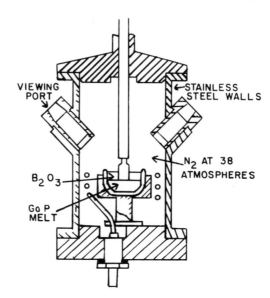

VIEWING PORT

STAINLESS STEEL WALLS

N_2 AT 38 ATMOSPHERES

B_2O_3

GaP MELT

FIGURE 3-2-2. Liquid encapsulation technique [1].

In this technique a layer of boric oxide is used to cover the melt. Ga and As are virtually insoluble in this glassy, viscous substance. The pressure in the growth chamber is raised to a pressure greater than the dissociation pressure. Pulling is performed by lowering the seed through the boric oxide layer with the growth occurring at the interface of the two liquids. Moreover, a thin coating remains on the surface of the grown crystal boule and prevents evaporation from the solid during the cooling.

The description of Bridgman and other techniques of bulk GaAs growth can be found in Refs. 1 and 2.

Unintentionally doped bulk GaAs grown at 1238°C is usually n-type with silicon donors from original chemical or the quartz container. Typical values of electron mobility are of the order of 3500–4000 cm^2/V s. The material contains levels which are identified by DLTS (see Table 3-2-1).

One of the most important attributes of GaAs technology is that the semi-insulating substrates allow us to achieve a low parasitic capacitance to the ground.

Chromium deep acceptor trap density up to a solubility limit of 2–3×10^{17} cm^{-3} can be obtained by adding chromium to GaAs melt. These intentional deep acceptors trap electrons supplied by the unintentional net donors. As a result the resistivity of GaAs could become as high as 10^8 Ω cm. There is, however, a high concentration of ionized impurities in the semi-insulating chromium doped GaAs (up to 5×10^{17} cm^{-3}) and the electron mobility is low (down to 3000 cm^2/V s at room temperature). The dislocation density is high (up to 10^5 cm^{-2} [3]). The trap density is also high, and the traps may rapidly diffuse into the active layer grown on the substrate (with the diffusion constant up to 4×10^{-8} cm^2/s at 800°C [3]).

Baking the substrates at 750°C in hydrogen for about 20 h reduces sharply the diffusion of deep acceptor traps into the active layer [4]. The difficulty with this technique is that the surface of the substrate may become conductive (p-type) after the bake to a depth of 0.3–3 μm. A possible explanation involves the movement of silicon atoms from Ga sites to As sites near the surface where As is lost owing to the baking. Replacement of Si with intentional tellurium donors helps to prevent this problem [3].

TABLE 3-2-1. Impurities in GaAs [3]

	$(E_c - E)$	Comment
Shallow donors	~ 5.8 meV	Requires Zeeman splitting to identify
Oxygen donors (EL2)	0.82 eV	Diffuses rapidly, ~ 10^{-8} cm^2/s at 750°C
Chromium acceptor (EL1)	0.61 eV	Excited state 0.86 eV above ground state, a radiative transition
Deep acceptor (EL3)	0.58 eV	Diffuses rapidly, 4×10^{-8} cm^2/s at 800°C; absent in N$^+$, 2×10^{17}/cm^3 in intrinsic
Electron trap (EB3)	0.90 eV	Both present in electron irradiated GaAs,
Electron trap (EB6)	0.41 eV	or at grain boundaries

	$(E - E_v)$	Comment
Shallow acceptors	10's meV	Identify by photoluminescence
Tin acceptor	0.17 eV	Self compensates n-type GaAs
Copper acceptor	0.42 eV	Diffuses very rapidly, ~ 4×10^{-6} cm^2/s at 800°C
Hole trap (HB2)	0.71 eV	Ambient in LPE GaAs, at grain boundaries
Hole trap (HB6)	0.29 eV	Present in electron irradiated GaAs, grain boundaries
Hole trap (HC1)	~ 0.15 eV	Another copper level, diffuses rapidly

3-3. LIQUID EPITAXY

The word "epitaxy" is derived from Greek and means "arranged upon." Epitaxial growth is the growth of a crystalline film on a crystalline substrate. The grown film reproduces the crystallographic structure of the substrate.

Epitaxial layers may be either "homoepitaxial," i.e., grown onto a crystal of the same composition, or "heteroepitaxial," i.e., grown onto a crystal of different composition or even of different crystal structure.

The phase diagram of the Ga–As system is shown in Fig. 3-3-1. In a liquid epitaxy liquid Ga is saturated with As. The saturation temperature is chosen close

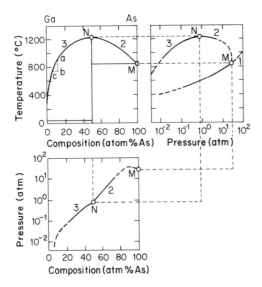

FIGURE 3-3-1. Phase diagram of Ga–As system [2]. Line 1, equilibrium between As liquid and As vapor. Line 2, equilibrium between GaAs solid, melt (with more than 50% As), and vapor. Line 3, equilibrium between GaAs solid, melt (with less than 50% As), and vapor. Point N, congruent melting point, equilibrium between GaAs solid, melt (with 50% As), and vapor.

to 850 K. When the solution is cooled, a crystal GaAs film is grown onto a substrate. Relatively low temperature of the epitaxial growth compared to the congruent melting point leads to a small contamination and good quality of the epitaxially grown films.

A typical temperature cycle for an LPE growth is shown in Fig. 3-3-2. The beginning of the cooling corresponds to point a of curve 3 of the phase diagram shown in Fig. 3-3-1. The solution is cooled to the point b corresponding to the two-phase region, so that a driving force for the precipitation of GaAs exists until point c on the phase diagram is reached, at which point precipitation ceases.

Steps involved in the liquid epitaxial growth (supersaturation of melt, introduction of the substrate, the growth itself, and removal of the substrate with the film from the melt; see Fig. 3-3-2) may be repeated for multiple layer growth.

The so-called "slider technique" [6] and its variations are widely used at the present time. Earlier techniques [7, 8] are similar in principle of operation and cooling cycle but differ in method of controlling contact between substrate and melt.

A diagram of an LPE system is shown in Fig. 3-3-3. In this system the position of the substrate is controlled by sliding the substrate held in a graphite tray under the appropriate melt. This technique is the best for growing multilayer LPE structures. It provides the best uniformity of films and the most complete melt removal.

A typical growth procedure may include the following steps: First Ga is loaded into the bins of the slider boat. It is then baked at 1000°C for 4–10 h to evaporate the impurities. Next Ga is saturated with As by adding polycrystalline GaAs to Ga solution. A subsequent bake at approximately 800°C for 40 h removes volatile impurities such as oxygen and sulfur (see Figs. 3-3-4 and 3-3-5). At longer bakes the impurity concentration rises again [9]. The reasons for that rise are not well understood at the present time.

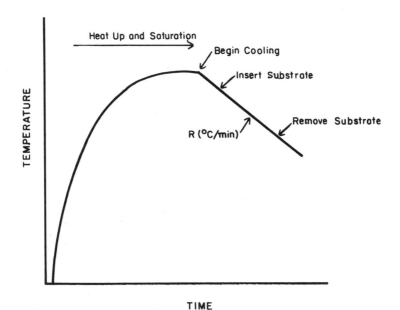

FIGURE 3-3-2. Temperature cycle for a typical LPE growth process [7].

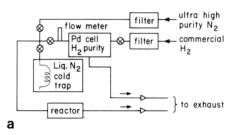

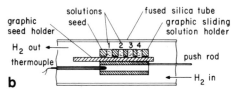

a

b

FIGURE 3-3-3. LPE system. (a) Block diagram; (b) slider boat.

After the bakeout the substrate is loaded and impurities are introduced. Dopants with very low vapor pressure and low distribution coefficient are used to reduce the cross-contamination of different melts.

The next step involves raising the temperature to about 15°C higher than the growth temperature for 2-3 h and cooling down to the growth temperature (700–800°C) (see Fig. 3-3-2).

The slider technique may be further improved by adding the confined melt feature [11], when a GaAs slice is placed on top of the melt. The slice is constrained to remain parallel to the substrate, which is separated from it by a melt only a few millimeters thick. The thin melt with GaAs at the top and bottom leads to a uniform saturation of the melt, better temperature uniformity, and convection suppression, thus facilitating uniformity of composition and thickness in the grown layer.

A typical growth thickness versus growth time dependence for GaAs is shown in Fig. 3-3-6. This particular film was grown in a slider system at 800°C with a cooling rate of 0.6°C/min and a melt thickness of 0.44 cm [9]. The growth rate is limited by the diffusion of As through the liquid phase with the diffusion rate typically of the order of $D \simeq 10^{-5}$ cm^2/s. The width of the melt which can be

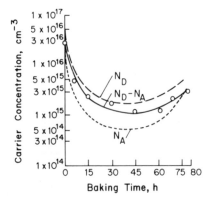

FIGURE 3-3-4. Dependence of the free carrier concentration on accumulated bakeout time at bakeout temperature of 775°C [10].

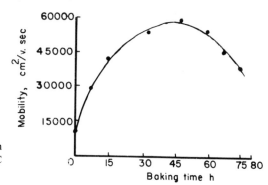

FIGURE 3-3-5. Mobility variation with bakeout time at bakeout temperature of 775°C [10].

equilibrated with the substrate is approximately [9]:

$$d_m \simeq \frac{3}{2\pi^{1/2}}(Dt)^{1/2} \tag{3-3-1}$$

where t is time.

The growth rate W can be found from the thermodynamics consideration:

$$W = \frac{Rd_m}{MC_s} \tag{3-3-2}$$

where R is the cooling rate, °C/s, C_s is the number of As atoms per unit volume in the solid, and

$$M = \frac{dT}{dC_v} \tag{3-3-3}$$

is the slope of liquidus (in °C/number of As atoms per unit volume). Thus, in agreement with the experimental data the layer thickness is proportional to $t^{3/2}$ (see Fig. 3-3-6).

Higher growth rates may be achieved in the LPE systems where convection dominates mass transport [10]. However, thickness control and uniformity are usually poor in such systems.

A big advantage of the LPE technique is that the growth layer may contain fewer impurities than the components used in the growth because most impurities tend to stay in a liquid phase during the crystallization. The distribution coefficient is temperature-dependent and this may lead to a non-uniform distribution of impurities along the layer. However, during the growth the concentration of impurities in the liquid increases. For the optimum growth regime these two effects almost compensate each other, leading to practically uniform layers.

Dopants with low distribution coefficient and low vapor pressure are preferred because the cross-contamination of different melts is reduced. Group II and IV dopants have too high vapor pressures and distribution coefficients for practical use (see Table 3-3-1) [14].

Group IV elements such as Sn or Si can substitute either Ga or As sites and behave as donors or acceptors. In LPE material grown at low temperatures Si behaves as a compensated acceptor. At growth temperatures higher than 800°C Si could change from the Ga site to the As site and behave as a donor. Ge behaves

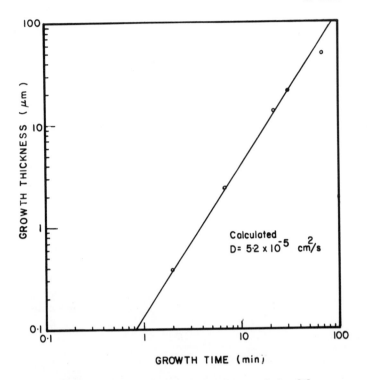

FIGURE 3-3-6. GaAs epilayer thickness vs. growth time [9].

as an uncompensated donor under LPE conditions and Sn as a donor. Sn as a donor and Ge an acceptor are typical choices for LPE GaAs grown for microwave applications.

The layer quality is strongly dependent on the quality of substrates. Substrates should be clean, free of oxide, and have a low dislocation density because the structural imperfections of the substrate tend to appear in the growth of epilayers as well.

One way to obtain quality layers is to grow buffer unintentionally doped epilayers between the substrate and the epilayer. Using LPE one can grow films from 0.1 μm to several hundred micrometers. The difficulty in growing very thick layers is related to the large growth time, which makes it difficult to maintain strictly the same parameters in the system during the growth.

TABLE 3-3-1. Vapor Pressure and Distribution Coefficients for Dopants in GaAs When Grown from Ga Solution [14]

		Vapor pressure at 1000 K (Torr)	Distribution coefficient
Donor	Te	80	3
	Se	1×10^3	10
	Sn	5×10^{-8}	8×10^{-5}
Acceptor	Zn	100	1×10^{-2}
	Cd	500	—
	Ge	6×10^{-10}	8×10^{-3}

LPE GaAs can be repeatedly grown with 77 K mobilities exceeding 150,000 cm^2/V s and with electron concentrations near 10^{14} cm^{-3}. Such material has electron trap densities below 10^{10} cm^{-3} [3].

3-4. VAPOR EPITAXY

The most frequently used process for GaAs vapor epitaxy is the chloride transport technique. It exists in two modifications, liquid and solid source techniques. With the liquid source technique the arsenic is obtained from AsCl$_3$ and the gallium from a liquid gallium source. With the solid source technique the solid GaAs is used as the Ga source and As is derived from both AsCl$_3$ and the GaAs source. The schematic representation of the vapor epigrowth process is shown in Fig. 3-4-1. The Ga or GaAs source is heated to about 850°C. The substrate temperature is lower (750°C). Upon heating AsCl$_3$ decomposes to form As$_4$ and HCl:

$$4AsCl_3 + 6H_2 \rightarrow As_4 + 12HCl$$

The HCl combined with Ga (or GaAs) gives GaCl:

$$2Ga + 2HCl \rightarrow 2GaCl + H_2$$

In the case of the liquid gallium source the arsenic is initially taken by the gallium until it becomes saturated and a thin crust of gallium arsenide forms upon it. This is called the source saturation; GaCl and As pass downstream and over the substrate where the deposition occurs according to the reactions

$$6GaCl + As_4 \rightarrow 4GaAs + 2GaCl_3$$

$$2GaCl + As_2 + H_2 \rightarrow 2GaAs + 2HCl$$

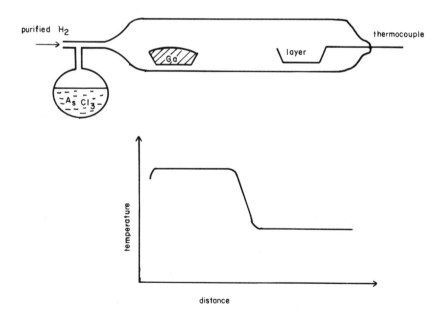

FIGURE 3-4-1. Schematic diagram of Ga–AsCl$_3$–H$_2$ process for vapor epitaxial growth of GaAs.

The purity of the grown layers is strongly dependent on the molar fraction of $AsCl_3$ flowing into the reactor. At low molar fractions (10^{-3}) unintentional doping level can be as high as 10^{16} cm^{-3}. At larger molar fractions (up to 10^{-2}) this number can drop to 10^{13} cm^{-3} or below in a good system. This effect is due to the reduction of the HCl interaction with the SiO_2 reactor tube.

The grown layers can be doped by introducing gases containing the dopants such as H_2S, H_2Se, $Zn + H_2$, etc. The dopants can be also added to the Ga source.

The growth rate can reach tens of micrometers per hour.

A different process known as the hydride process utilizes AsH_3 instead of $AsCl_3$ as a source of As. The HCl gas is added to H_2 in order to transport Ga from the source downstream. This method has a greater flexibility but requires very pure HCl gas.

Similar techniques can be used to grow other A_3B_5 compounds. For instance, indium phosphide can be grown if PCl_3 is used instead of $AsCl_3$ and In source instead of Ga. The exceptions are Al compounds because $AlCl_3$ reacts with the silica reactor walls. The so-called alkyl process is used for the vapor growth of the aluminum compounds.

3-5. METAL-ORGANIC CHEMICAL VAPOR DEPOSITION

A new version of the vapor epitaxial growth is a metal–organic vapor deposition (MO CVD or MO VPE) technique. It was first successfully demonstrated by H. M. Manasevit and W. I. Simpson in 1969 [15].

To a certain extent, this technique has replaced a halide VPE process involving Ga–As Cl_3–H_2 (see Section 3-4) and LPE (see Section 3-3) because of its simplicity, ease of control, and other advantages.

TABLE 3-5-1. III–V Compound Semiconductors Formed on Insulators from Metal-Organics and Hydrides [19]

Compound	Insulating substrate	Reactants	Growth Temperature (°C)
GaAs	Al_2O_3,$MgAl_2O_4$ BeO,ThO$_2$	TMGa–AsH$_3$	650–750
GaP	Al_2O_3,$MgAl_2O_4$	TMGa–PH$_3$	700–800
GaAs$_{1-x}$P$_x$($x = 0.1$–0.6)	Al_2O_3,$MgAl_2O_4$	TMGa–AsH$_3$–PH$_3$	700–725
GaAs$_{1-x}$Sb$_x$($x = 0.1$–0.3)		TMGa–AsH$_3$–TMSb	
GaSb	Al_2O_3	TEGa–TMSb	500–550
AlAs	Al_2O_3	TMAl–AsH$_3$	700
Ga$_{1-x}$Al$_x$As	Al_2O_3	TMGa–TMAl–AsH$_3$	700
AlN	Al_2O_3,α-SiC	TMAl–NH$_3$	1250
GaN	Al_2O_3,α-SiC	TMGa–NH$_3$	925–975
GaN	Al_2O_3	TEGa–NH$_3$(unstable)	800
InAs	Al_2O_3	TEIn–AsH$_3$	650–700
InP	Al_2O_3	TEIn–PH$_3$	725
Ga$_{1-x}$In$_x$As	Al_2O_3	TEIn–TMGa–AsH$_3$	675–725
InSb	Al_2O_3	TEIn–TESb–AsH$_3$	460–475
InAs$_{1-x}$Sb$_x$ ($x = 0.1$–0.7)	Al_2O_3	TEIn–TESb–AsH$_3$	460–500

TABLE 3-5-2. II–VI Compounds Formed on Insulators by MO-CVD [19]

Compound	Substrate	Reactants	Growth temperature (°C)
ZnS	Al_2O_3,BeO,$MgAl_2O_4$	DEZn–H_2S	~ 750
ZnSe	Al_2O_3,BeO,$MgAl_2O_4$	DEZn–H_2Se	725–750
ZnTe	Al_2O_3	DEZn–DMTe	~ 500
CdS	Al_2O_3	DMCd–H_2S	475
CdSe	Al_2O_3	DMCd–H_2Se	600
CdTe	Al_2O_3,BeO,$MgAl_2O_4$	DMCd–DMTe	~ 500

In a typical MO CVD process a metal organic compound [trimethylgallium (TMGa) or triethylgallium (TEGa)] is used as a source of Ga, and arsin (AsH_3) is used as a source of As.

Whereas a standard VPE growth occurs in a hot reactor (see Section 3-4), only the substrate has to be heated in the MO CVD process. This is usually done either inductively [15] or by radiation [16]. The supersaturation of the vapor phase is very large. The high nucleation rate and the cold reactor walls make the MO CVD technique uniquely suited for heteroepitaxy [17, 18].

GaAs and other A_3B_5, A_2B_6, A_4B_6 and ternary compounds could be grown on insulators from metal organics and hydrides (see Tables 3-5-1, 3-5-2, and 3-5-3 [19]).

A schematic of the MO CVD process is shown in Fig. 3-5-1. Fig. 3-5-2 presents a typical MO CVD apparatus [20].

A growth procedure includes substrate preparation (cleaning and etching), system purging by the hydrogen flow, the inductive heating of the substrate, the deposition process itself, and annealing after the deposition. A typical growth rate is of the order of 0.1 μm/min (see Fig. 3-5-3).

Carbon contamination from the organic source is one potential problem of the MO CVD process. However, the material quality is comparable with the quality achieved by other epitaxial techniques (see Fig. 3-5-4 [21]). The biggest advantages of MO CVD growth are relative simplicity and a possibility to grow $Al_xGa_{1-x}As$ compounds.

TABLE 3-5-3. IV–VI Compounds Formed on Insulators by MO-CVD [19]

Compound	Reactants	Growth temperature (°C)
PbTe[a]	TMPb–DMTe,TEPb–DMTe	500–625
$Pb_{1-x}Sn_xTe$	TMPb–TESn–DMTe	550–625
PbS[b]	TMPb–H_2S	~ 550
PbSe[b]	TMPb–H_2Se	~ 550
SnTe	TESn–DMTe	~ 625
(SnS)	TESn–H_2S	~ 550
(SnSe)	TESn–H_2Se	~ 500

[a] $(111)PbTe/\!/(0001)Al_2O_3;[\bar{1}10]PbTe/\!/[\bar{1}2\bar{1}0]Al_2O_3;$ ~ $(111)PbTe/\!/(111)MgAl_2O_4;$ $[\bar{2}11]PbTe/\!/[\bar{1}01]MgAl_2O_4;$ $(111)PbTe/\!/(111)BaF_2;$ $[1\bar{1}0]PbTe/\!/[0\bar{1}1]BaF_2$
[b] IbS and PbSe possessed the same relationships as PbTe on α-Al_2O_3 and $(111)BaF_2$; growth on $(111)CaF_2$ was the same as on $(111)BaF_2$.

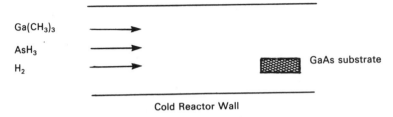

FIGURE 3-5-1. Schematic diagram of MO CVD process.

3-6. MOLECULAR BEAM EPITAXY

Molecular beam epitaxy (MBE) is one of the most promising and modern techniques for growing thin single-crystal semiconductor structures. In MBE technology the controlled evaporation from a thermal source (or coevaporation from several thermal sources) is used to deposit epitaxial films under ultrahigh vacuum conditions.

A typical MBE apparatus is shown in Fig. 3-6-1 [22]. An ion pumped ultrahigh vacuum system contains a substrate holder and molecular or atomic beam sources—effusion cells—small heated containers with openings facing the substrate. The effusion cells are separated by the liquid-nitrogen shrouds in order to minimize thermal cross-talk and prevent intercontamination of the sources. A liquid-nitrogen shroud behind the sample holder helps to reduce the contamination of the background gases. For the same purpose source and substrate holders are manufactured from materials with low vapor pressure (such as alumina, tantalum, graphite). The working vacuum in the growth chamber is close to 10^{-10} torr. In order to preserve the vacuum the samples are inserted into the chamber through the vacuum interlock.

In situ monitoring devices such as a reflection high-energy diffractometer, a mass spectrometer, an Auger spectrometer, and an ion gauge are used to control the molecular or atomic beams and the grown layers. An ion gauge monitors neutral atomic beams. A mass spectrometer is used to analyze both the atomic or molecular beams and background gases. The reflection high- or medium-energy diffractometer yields information about the layer surface during and after growth. This diffractometer consists of the electron gun emitting the electron beam under the glancing

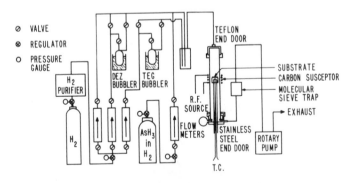

FIGURE 3-5-2. MO CVD apparatus [20].

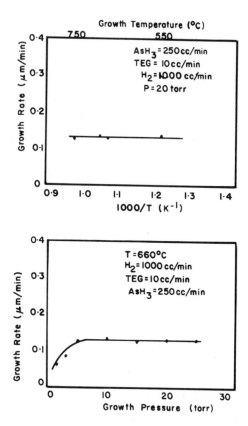

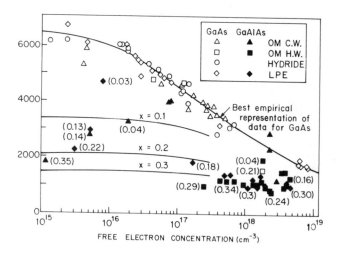

FIGURE 3-5-3. Growth rate of an MO CVD process vs. temperature and pressure [20].

FIGURE 3-5-4. Electron mobility at 300 K vs. free carrier concentration for GaAs and $Al_xGa_{1-x}As$ grown by various epitaxial techniques [21]. Curves for $x = 0.1$, 0.2, and 0.3 are calculated.

angle of about 2° and a fluorescent screen to observe the diffraction patterns. Auger electron spectroscopy (AES) provides semiquantitative information about alloy composition, interface sharpness, and interdiffusion. *In situ* secondary ion mass spectroscopy is a sensitive but destructive technique for analyzing the layer composition and impurity profiles. *In situ* monitoring of the epitaxial growth is one of the advantages of the MBE technology. However, many other characterization techniques such as photoluminescence studies, electron energy loss spectroscopy, Raman scattering, laser scattering, etc. have also been used for studies of MBE-grown layers.

In the MBE growth the substrate temperature may be kept relatively low (500–650°C for GaAs) leading to low growth rates (of the order of 1 Å/s) and a low bulk diffusion rate. The beam flux can be altered very rapidly using the main shutter or effusion cell shutters (see Fig. 3-6-1). It makes it possible to change the composition or doping of the grown structures literally within one atomic distance.

In the MBE of GaAs layers Ga atoms and As_2 and As_4 molecules impinge on the GaAs substrate. Practically all Ga atoms stick to the surface, but only one As atom per each Ga atom remains on the surface forming a stoichiometric composition of the MBE layer for the excess As flux. The intensity of molecular beams and, hence, the deposition rate may be controlled by varying the temperature of the Ga source. Typically the Ga flux is close to 10^{15} atoms per cm^2 per second and arsenic flux is 5–10 times higher [23]. The source of arsenic molecules is usually solid arsenic. Gallium source is used for the gallium beam. An additional aluminum source is needed for growing AlGaAs compounds where the ratio of Al and Ga is proportional to the flux ratio of aluminum and gallium beams. Besides the temperature of the effusion oven the flux depends on the molecular weight of the emitted atoms or molecules, on the orifice area, and the distance to the wafer. For a 2-in. wafer the nonuniformity of the grown film may be reduced to less than 2% if the sample is rotated during the growth. Even a better uniformity may be achieved by using multiple sources.

Nominally undoped GaAs layers grown by MBE are typically *p*-type with the

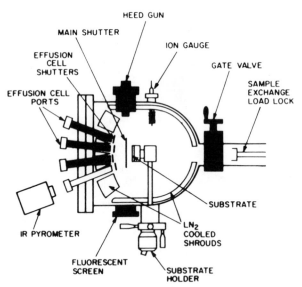

FIGURE 3-6-1. Schematic diagram of an MBE growth chamber [22].

carrier concentration between 10^{13} and 10^{15} cm^{-3}. This leads to the development of the depletion regions at the boundary between the semi-insulating substrate and the unintentionally doped MBE layer and also between this layer and the active device layer (usually n-type).

Group IV elements such as Si, Ge, and Sn are used as n-type donors in MBE growth of GaAs. They can occupy either Ga or As sites in the GaAs crystal lattice with the type of doping dependent on the concentrations of the Ga and As vacancies. Tin is the least sensitive to the arsenic–gallium ratio and yields an n-type material [23]. Si is probably the most common n-type dopant, yielding the highest liquid–nitrogen mobilities, which are frequently considered as a crude figure of merit characterizing GaAs quality. The most promising p-type dopant is beryllium in spite of its extreme toxicity.

In MBE growth of AlGaAs oxygen may be used to produce a semi-insulating material. The material resistivity may be greater than 10^9 Ω cm. The oxygen doped layer of AlGaAs on GaAs may be converted back into the semiconducting state using the diffusion of impurities. Another way to obtain an insulating layer is to grow an AlAs layer, which is then oxidized by reaction with heated water. The thickness of the insulating layer consisting of aluminum and gallium oxides is determined by the thickness of the AlAs layer [23]. This technique may be used to fabricate GaAs MIS field effect transistors.

The electron concentration in GaAs layers produced by MBE may exceed 5×10^{18} cm^{-3}, which is considerably higher than what may be achieved in samples grown by vapor epitaxy. Such high doping levels may be used to produce nonalloyed ohmic contacts due to a very thin depletion region of the Schottky barrier at the interface with the contact metal. The lowest contact resistance of 10^{-7} Ω cm^2 was achieved using this technique [24].

Another advantage of the MBE technology is smoothing of the surface of the GaAs layer during the growth. This is illustrated by Fig. 3-6-2 [23], where the photographs of the surface taken using an electron microscope are presented. This property of the MBE growth makes it particularly attractive for the growth of heterojunctions, superlattices, and multilayered structures.

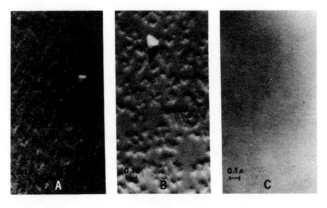

FIGURE 3-6-2. Smoothing of GaAs surface during MBE growth [23]. (A) Highly polished surface of GaAs substrate under electron microscope. (B) Same after MBE growth of 0.015-μm-thick layer of GaAs. (C) Same after MBE growth of a 1-μm-thick GaAs layer.

As in other epitaxial techniques, the substrate preparation is crucial for the quality of the MBE-grown layers. The main problem is to achieve a damage-free, atomically clean, stoichiometric surface. A typical preparation includes chemical treatment by Br_2/methanol solution and various sulfuric acid–hydrogen peroxide-water mixtures (typically 7:1:1) [25]. The substrates are then mounted on molybdenum heater blocks with In or In–Ga alloys and loaded into the system. Subsequent heating to 555°C + 5°C under arsenic flux is used to remove the oxide and traces of carbon. Sputter ion cleaning at elevated temperatures is also used for the removal of carbon.

Significant oxygen concentrations (in excess of 10^{11} cm^{-2}) have been found at the substrate layer interface by SIMS analysis [25]. SIMS also identified an interfacial pile-up of chromium from chromium doped semi-insulating substrates.

Rapid diffusion of accumulated impurities (Cr, O, etc.) during the MBE growth may contaminate the bottom part of the grown layer, creating deep acceptor centers, reducing the carrier mobility, and altering a carrier profile. The growth of the high resistivity buffer layer may alleviate these problems.

A high quality of GaAs MBE layers is illustrated by Fig. 3-6-3 [26]. In this particular work As_2H_3 was used as an As source, and peak mobilities over 130,000 cm^2/V s have been obtained.

A large variety of materials have been grown by MBE including silicon [27], A_3B_5 compounds [26, 28–35], A_2B_6 compounds [36–39], and ternary and quarternary alloys [40–41].

Two new masking methods have been explored for device fabrication based on the MBE technology [23]. In one technique, which has been used to fabricate microwave mixer diodes at Bell Labs, the substrate was covered by a thin mask layer of SiO_2. High-quality epitaxial material was grown on the uncovered areas of the substrate. The layer grown on the masked part of the substrate was polycrystalline and semi-insulating. In the other technique a sheet of silicon with precisely dimensioned openings was placed over the wafer during the MBE process. The

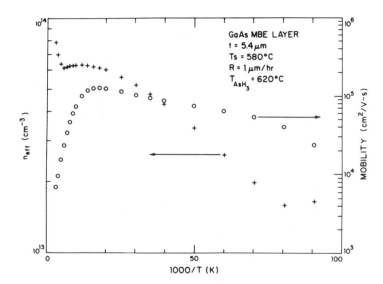

FIGURE 3-6-3. Temperature dependence of Hall mobility and electron concentration of a GaAs MBE layer [26].

material grows only in the unmasked areas of the substrate. Preliminary demonstrations of this technique produce mesas of epitaxial GaAs and AlGaAs only 1 μm wide [23].

Metal grids with very narrow spacings (up to 1600 Å) have been fabricated inside the GaAs epitaxial layer using overgrowth of GaAs on top of the metal grid [42].

These novel approaches clearly demonstrate the unique and exciting potential of MBE technology for fabricating multilevel three-dimensional integrated circuits and devices.

Numerous devices have been fabricated with MBE including mixer and varactor diodes [43], light-emitting diodes [44], semiconductor lasers [45-47], field effect transistors [48-50], camel gate field effect transistors (CAMFETs) [51], modulation doped field effect transistors (also called HEMTs and TEGFETs) [52, 53], heterojunction bipolar junction transistors [54], planar doped barrier devices [55, 56], permeable base transistors [42], Gunn diodes [57], ultra-high-speed integrated circuits [58], and solar cells [59]. However, much more research has to be done before MBE will evolve from an extremely interesting novel laboratory technique into an established production technology.

A comparison of different epitaxial techniques is given in Table 3-6-1 [21].

3-7. CHARACTERIZATION OF EPITAXIAL LAYERS

Characterization of epitaxial layers includes the measurement of the layer thickness, chemical composition, and uniformity; the determination of the carrier mobility, Hall factor, and doping profile; and the study of galvanomagnetic properties and defect centers.

3-7-1. Determination of Layer Thickness

Mechanical, optical, or electrical properties of the layer may be used to determine the layer thickness.

The weight of a film per unit of the surface area is proportional to the thickness. Weights of the order of 10^{-9} g/cm^2 (which corresponds to less than a 1/100th of a monolayer) may be detected by gravimetrical measurements in vacuum [60]. A typical commercial system measures a change of the resonance frequency of a quartz crystal oscillator caused by the added weight and may also be used to monitor the rate of deposition.

In an MBE system a crude measurement of the deposition rate may be performed using an ion gauge. This gives an approximate estimate of a film thickness.

Two optical methods are widely used for measuring the layer thickness [61-63]. The first method is called an angle lapping and staining technique [62]. In this technique the sample is lapped to provide a section inclined at a small angle to the surface. The layer–substrate interface is delineated by a chemical stain. The sample is then illuminated with a monochromatic light. The layer thickness is determined by counting the number of the interference fringes. This technique can be used only for relatively thick layers (several microns thick).

TABLE 3-6-1. Comparison of Epitaxial Growth Techniques [21]

Technique	Characteristic		Inherent advantages	Inherent disadvantages	Status
	Solid composition	Purity			
LPE	Thermodynamics (phase diagram)	III melt container, gettering	Simple, *high purity*	Volume limited, inflexible morphology	Laboratory technique
VPE (hydride)	Thermodynamics	Gases, leaks, reactor materials	Flexible, large scale	No AlGaAs or other Al alloys	Production technique (GaAsP)
OMVPE	Kinetics, arrival at surface	OM sources, AsH$_3$, C contamination, leaks	*Most versatile*, large scale (?), simple	C contamination, problems with In (?)	Potential commercial AlGaAs
MBE	Kinetics, flux, sticking coefficient	Vacuum sources, system (walls)	Most abrupt (2-10 Å), low temperature	Expensive, slow growth rate, problems with phosphorus	Special structures

The second optical technique is based on measurements of the infrared reflectance of the epitaxial layer as a function of a wavelength [63]. It utilizes the change in the refractive index at the layer–substrate interface caused by the change in doping. As a result interference fringes are produced by the superposition of the reflected radiation from the layer surface and the layer–substrate interface. The fringes depend on the wavelength of the incident radiation, and this dependence makes it possible to estimate the layer thickness quite accurately.

3-7-2. Determination of the Doping Profile from C–V Measurements

A standard technique of determining the impurity profile in an epitaxial layer is based on C-V measurements. A Schottky contact or a p-n junction may be used to create a depletion layer with the width controlled by the doping profile. In a typical scheme two Schottky contacts with areas different by orders of magnitude are used to provide two capacitances in series connected by the undepleted portion of the active layer (see Fig. 3-7-1). The capacitance of a bigger area contact is then so much bigger that it does not affect the measurements too much. This scheme makes it possible to avoid the necessity of making an ohmic contact to the bottom of the layer. In a typical commercial setup a mercury probe may be used to provide a Schottky contact.

The relationship between the doping profile and the C-V characteristic of a Schottky contact may be found as follows.

The charge Q in the depletion layer is given by

$$Q = qS \int_0^W N_D(x) \, dx \qquad (3\text{-}7\text{-}1)$$

where S is the Schottky contact area, N_D is the doping density, and x is the coordinate in the direction perpendicular to the layer (see Fig. 3-7-1). The voltage drop between the Schottky contact and the conducting channel in the layer is $-V + V_{bi}$, where V_{bi} is the built-in voltage and V is the applied voltage:

$$V - V_{\text{Bi}} = - \int_0^W F(x) \, dx \qquad (3\text{-}7\text{-}2)$$

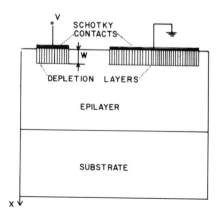

FIGURE 3-7-1. Depletion layers under Schottky contacts. The doping profile may be deduced from the measurements of the capacitance between the contacts as a function of the applied voltage.

where F is the electric field. Assuming the total depletion we have

$$\frac{dF}{dx} = \frac{qN_D(x)}{\varepsilon} \tag{3-7-3}$$

Multiplying Eq. (3-7-3) by x and integrating by parts yields

$$-V + V_{bi} = \frac{q}{\varepsilon} \int_0^W N_D(x)x\,dx \tag{3-7-4}$$

where the condition

$$F(W) = 0 \tag{3-7-5}$$

has been taken into account.

From Eqs. (3-7-1) and (3-7-4) we find the differential capacitance:

$$C = \frac{dQ}{dV} = \frac{dQ/dW}{dV/dW} = \frac{\varepsilon S}{W} \tag{3-7-6}$$

Hence the depletion capacitance is determined by the depletion width independently of the doping profile. Finally differentiating Eq. (3-7-6) with respect to W and using Eq. (3-7-4) we find

$$N_D(W) = \frac{C^3}{q\varepsilon S^2\,dC/dV} \tag{3-7-7}$$

This equation together with Eq. (3-7-6) can be used to determine the doping profile from the C-V measurements.

The results obtained may be corrected to account for the fringing capacitance C_S of the Schottky contact [64]:

$$W = W_m\left(1 - \frac{W_m C_S}{\varepsilon S}\right)^{-1} \tag{3-7-8}$$

$$N_d(W) = N_{dm}(W)\left(1 + \frac{W C_S}{\varepsilon S}\right)^{-3} \tag{3-7-9}$$

Here W_m and N_{dm} are the measured values [obtained from Eqs. (3-7-6) and (3-7-7)] and W and N_d are the values corrected to account for the stray capacitance.

This method is less accurate at the forward bias because of the forward current through the Schottky barrier. It results in a somewhat limited depth range over which the profile can be measured because at zero bias the depletion layer may extend through a considerable portion of the layer.

For thick and highly doped layers the reverse voltage required to deplete the layer may become larger than the breakdown voltage.

Another limitation of this technique is related to the series resistance of the undepleted portion of the layer, which becomes increasingly important when the depletion region reaches the substrate [64].

3-7-3. Resistivity and Hall Measurements

The method developed by van der Pauw [65, 66] has become a standard technique for measuring the sheet resistance and Hall concentration and mobility in epitaxial and thin films.

In this technique four ohmic contacts are placed on the surface of the film and located at the corners of a square as shown in Fig. 3-7-2 [67]. In resistivity measurements contacts A and B serve as a source and sink of an applied current, respectively. Contacts D and C are used as potential probes. In the Hall measurements contacts A and C are used as current electrodes with the Hall voltages measured between B and D.

For a perfectly square electrode arrangement and uniform, homogeneous, and isotropic layer the film resistivity is given by

$$\rho = \pi R d / \ln(2) \tag{3-7-10}$$

where d is the film thickness and

$$R = V/I \tag{3-7-11}$$

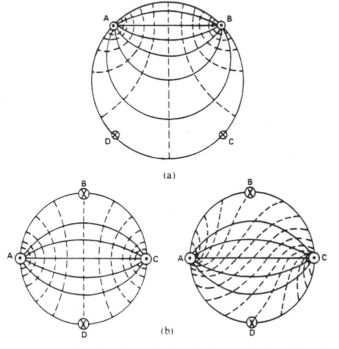

FIGURE 3-7-2. Current streamlines (continuous lines) and equipotentials (dotted lines) for (a) resistivity measurements and Hall measurements for (b) $B = 0$ and (c) $B > 0$ [67].

Here V is the voltage difference between the voltage probes D and C and I is the input current. Equation (3-7-10) is obtained from the solution of the Laplace equation in the film plane with appropriate boundary conditions.

In the Hall measurements (see Fig. 3-7-2b) the Hall voltage V_H is proportional to the magnetic field B and to the input current I:

$$V_H = R_h BI / d \tag{3-7-12}$$

Here R_h is the Hall constant.

For a more accurate measurement, which compensates for thermoelectric effects, and for a misalignment the Hall voltage is taken as an average of four measurements corresponding to two different voltage polarities and two opposite directions of the magnetic field perpendicular to the film plane.

In a practical measurement errors are introduced by a finite size of electrodes, asymmetry of contacts, spatial nonuniformity of the film, etc. A detailed analysis of these factors is given in Ref. 67.

In particular, the film resistivity for four contacts with different distances between the adjacent contacts may be found as

$$\rho = \left(\frac{V_{5,6}}{I_{1,4}}\right) \frac{\pi d}{\ln(r_{1,6}r_{4,5}) - \ln(r_{1,5}r_{4,6})} \tag{3-7-13}$$

where $r_{i,j}$ is the distance between contact i and j (see Fig. 3-7-3).

The Hall constant for this contact configuration can be found from

$$V_{5,6} = \frac{I_{1,4}\rho[\ln(r_{1,6}r_{4,5} - \ln(r_{1,5}r_{4,6})]}{\pi d} + \frac{R_h IB(\theta_1 + \theta_4)}{\pi d} \tag{3-7-14}$$

The notation is explained in Fig. 3-7-3.

The Hall constant is inversely proportional to the electron concentration n

$$R_h = -r_H / qn \tag{3-7-15}$$

for an n-type material and to the hole concentration p

$$R_h = r_H / qp \tag{3-7-16}$$

for a p-type material. Here r_H is the Hall factor which is determined by dominant scattering mechanisms (see Section 2-5). In particular $r_H = 1.93$ for the impurity scattering, and $r_H = 1.18$ for acoustic or piezoelectric scattering (see also Fig. 2-5-3 for GaAs). Once the Hall constant and film resistivity are found, the Hall mobility is determined as

$$\mu_H = |R_h / \rho| \tag{3-7-17}$$

The conductivity mobility is given by

$$\mu = \mu_H / r_H \tag{3-7-18}$$

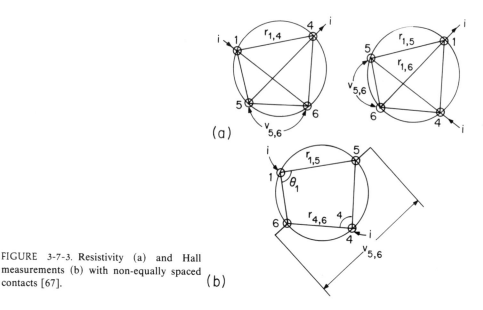

FIGURE 3-7-3. Resistivity (a) and Hall measurements (b) with non-equally spaced contacts [67].

An alternative way to determine the low field mobility involves the measurement of the geometric magnetoresistance. This technique has been used for characterization of transferred electron devices (see Section 4-15).

The film resistivity is often found using a four-probe measurement (see Fig. 3-7-4 [68]). In this technique the current is passed through two outer probes, and the voltage is measured between the inner contacts. The sheet resistance R is then given by

$$R = CFV/I \qquad (3\text{-}7\text{-}19)$$

where the correction factor CF may be found from Fig. 3-7-4.

Once resistivity and Hall constant are established the comparison between the measured and calculated (see Section 2-5) mobilities may be used to estimate the concentration of ionized impurities. A crude estimate may be obtained using Fig. 2-5-4.

3-7-4. Deep Level Transient Spectroscopy

The deep level transient spectroscopy (DLTS) technique provides information about trap concentrations, energy levels, and capture rates [69]. In addition, this technique is able to distinguish between majority- and minority-carrier traps, and between bulk and interface states.

In DLTS the capacitance transient associated with the thermal emission of carriers from deep trap levels in the depletion region is monitored by a fast capacitance or current meter and a correlator. The correlator sets a rate window such that only transients with the desired time constant will produce a maximum output. The spectroscopic nature of this technique is achieved by scanning the temperature which allows traps at different energy levels to be thermally activated [69].

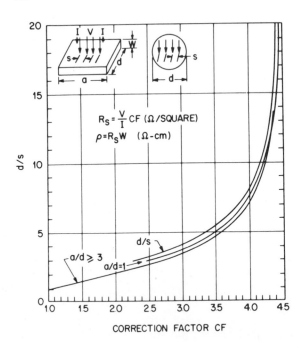

FIGURE 3-7-4. Correction factor for measurements of resistivity using a four-probe technique [68]. The sample geometry and probe location for rectangular and round wafers are shown in the insert.

The Schottky barrier diodes are the most widely used structure for DLTS measurements because they can be easily fabricated. Let us consider a Schottky barrier formed on an n-type semiconductor with only two discrete trapping levels: an electron-trapping level at E_1 and a hole-trapping level at E_2. In addition, as shown in Fig. 3-7-5a, there is a donor level near E_c which is fully ionized. In the absence of bias the bands bend due to the difference in work functions of two materials. The entire system is in thermal equilibrium and the Fermi level E_F is constant everywhere. The occupation of the traps by electrons is given by the Fermi–Dirac distribution.

When the Schottky barrier is reverse biased and a steady state is achieved, the Fermi level "splits" into electron and hole quasi-Fermi levels. As shown in Fig. 3-7-5b, the hole quasi-Fermi level, E_{Fp}, remains flat and coincides with the metal Fermi level, E_{Fm}, while the electron quasi-Fermi level, E_{Fn}, also remains flat but coincides with the bulk Fermi level in the semiconductor. This description of the quasi-Fermi level behavior is a good approximation only in the depletion region away from the edges.

Under nonequilibrium conditions, the Fermi–Dirac distribution is no longer valid. However, under quasiequilibrium, the occupation probability of traps by electrons is determined by Shockley–Read–Hall statistics given by [70]

$$f_T = \frac{\sigma_n \langle v \rangle n + e_p}{\sigma_n \langle v \rangle n + e_n + \sigma_p \langle v \rangle p + e_p} \qquad (3\text{-}7\text{-}20)$$

where σ_n and σ_p are the electron and hole capture cross sections, $\langle v \rangle$ is the mean thermal velocity, n and p are the concentration of electrons and holes, and e_n and e_p are the thermal emission rates for electrons and holes. In addition, from the principle of detailed balance, it can be shown that the thermal emission rates are given by

$$e_n = \frac{1}{g} \sigma_n \langle v \rangle N_c \exp\left[-\frac{(E_c - E_T)}{k_B T} \right] \qquad (3\text{-}7\text{-}21)$$

and

$$e_p = \frac{1}{g} \sigma_p \langle v \rangle N_v \exp\left[-\frac{(E_T - E_v)}{k_B T} \right] \qquad (3\text{-}7\text{-}22)$$

where E_T is the trap energy level, g is the trap level degeneracy and N_c and N_v are effective densities of states in the conduction and valence bands.

A detailed consideration of Eqs. (3-7-20)–(3-7-22) shows that the occupation probability for traps in the upper half of the gap is given by a Fermi–Dirac distribution with E_F equal to E_{Fn}. Then, assuming zero temperature approximation, all traps above E_{Fn} are empty and all traps below are filled. Hence, E_{Fn} determines the occupation of electron traps. Similarly, the occupation of traps in the lower half of the gap is determined by E_{Fp} [71].

If we now reduce the quiescent reverse bias V_R by applying a "filling" pulse $V_P > V_R$ (Fig. 3-7-6a), some of the formerly empty traps in the upper half of the gap are filled by capturing electrons. The occupation of most traps below midgap remains unchanged, unless the flat band voltage is greater than $E_g/2$, during the "filling pulse." Thus voltage pulses are useful only to observe electron traps. The identification of the hole traps, however, can be achieved using optical excitation such as by a laser pulse.

When the reverse bias is restored to the value V_R, a non-quasi-equilibrium condition exists. The additional traps neutralized by electron capture reduce the space charge density and increase the depletion layer beyond the quiescent width.

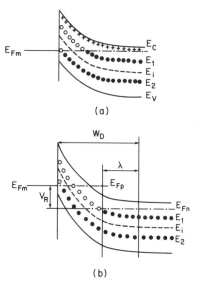

(a)

(b)

FIGURE 3-7-5. Band diagram of a Schottky diode with (a) zero and (b) reverse bias.

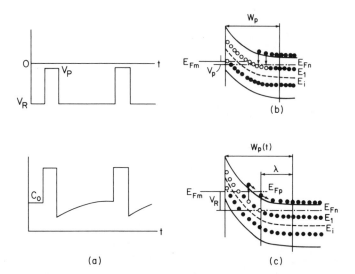

FIGURE 3-7-6. (a) DLTS waveforms and (b, c) related band diagrams.

As electrons are thermally emitted, the depletion layer width decreases back to the quiescent width (see Figs. 3-7-6b and 3-7-6c).

The small signal depletion capacitance of a Schottky barrier with uniform doping density N_d^+ and charged trap density N_T^+ is given by

$$C_0 = \frac{\varepsilon S}{W_0} = S\left[\frac{\varepsilon q (N_d^+ + N_T^+)}{2(V_R + V_{bi})}\right]^{1/2}$$

The transient capacitance is approximately given by

$$C(t) = S\left\{\frac{\varepsilon q [N_d^+ + N_T^+(1 - e^{-e_n t})]}{2(V_R + V_{bi})}\right\}^{1/2}$$

The magnitude of the transient $\Delta C = C(t \to \infty) - C(t = 0)$ is given by

$$\Delta C = C_0 - S\left[\frac{\varepsilon q (N_s - N_T^+)}{2(V_R + V_{bi})}\right]^{1/2} \tag{3-7-23}$$

where $N_s = N_d^+ + N_T^+$. The quantity N_s can be determined from C–V measurements using the relationship derived in Section 3-7-2 [see Eq. (3-7-7)]:

$$N_s(x) = \frac{C^3}{q\varepsilon S^2}\left(\frac{dC}{dV}\right)^{-1} \tag{3-7-24}$$

We can now solve Eq. (3-7-23) for N_T^+, which yields

$$N_T^+ = \left[1 - \frac{(C_0 - \Delta C)^2}{C_0^2}\right] N_s \tag{3-7-25}$$

In most crystalline semiconductors, the condition $N_d^+ \gg N_T^+$ is satisfied. Then Eq. (3-7-25) reduces to

$$N_T^+ = \frac{2\Delta C}{C_0} N_d^+ \qquad\qquad (3\text{-}7\text{-}26)$$

These basic principles of capacitance transients apply not only to DLTS but also to other space change spectroscopy techniques. The distinguishing feature of DLTS is the concept of the "rate window." The rate window sets the total integration period $t_M = t_2 - t_1$, for the cross-correlation between the capacitance signal and a known weighting function $W(t)$[72]. The correlation procedure can be considered as a filtering operation of the capacitance signal with an output

$$F(t_M) = \int_{t_1}^{t_2} C(t)\,W(t)\,dt \qquad\qquad (3\text{-}7\text{-}27)$$

such that, by a proper choice of $W(t)$, only transients with a desired time constant will produce a maximum output. The rate window concept can be implemented by a variety of methods. One method uses the dual-gated signal averager (double boxcar) to sample the transient amplitude at two different times t_1 and t_2. The difference $C(t_1) - C(t_2)$ is the DLTS signal.

As shown in Fig. 3-7-7, if we repetitively apply a filling pulse while slowly scanning the sample temperature, we get a series of transient signals with different time constants. The time constant t of the transients is an exponential function of both the energy depth of the trap and the temperature. A plot of the DLTS signal vs. temperature, called a DLTS spectrum, will exhibit peaks at the temperatures where the emission rate approximately equals to the rate window. The relation between the peak emission rate $\tau_{\max}$ and the rate window is

$$\tau_{\max} = \frac{t_1 - t_2}{\ln(t_1/t_2)} \qquad\qquad (3\text{-}7\text{-}28)$$

The choice of t_1 and t_2 is determined by such considerations as range of temperature scan and estimate of the emission rate. In addition, if the ratio t_1/t_2 is kept constant (Fig. 3-7-6), the shapes of the DLTS spectra for different t_1 and t_2 remains unchanged

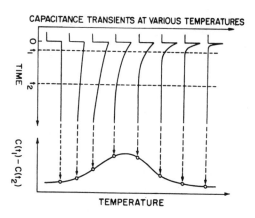

FIGURE 3-7-7. DLTS spectrum [69].

and only shift in temperature. A smaller ratio t_1/t_2 improves the S/N ratio, whereas a larger t_1/t_2 extends the range of time constants that can be interpolated for a maximum time t_2.

For simple exponentials, the DLTS signal can be related to ΔC by

$$\Delta C = \frac{C(t_1) - C(t_2)}{e^{-t_1/\tau} - e^{-t_2/\tau}} \qquad (3\text{-}7\text{-}29)$$

Now Eq. (3-7-25) or (3-7-26) can be used to derive the trap concentration. However, because only part of the depletion layer contributes to the DLTS spectrum (the region between points 1 and 2 in Fig. 3-7-5b) the trap concentration given by Eq. (3-7-25) or by Eq. (3-7-26) is underestimated. It has been found empirically that the real value of N_T^+ is larger by approximately 50% [73].

The activation energy of the trap level is determined from the slope of the $\ln(e_n/T^2)$ vs. $1/kT$ plot. The T^2 factor is included to take into account the temperature dependence of the $\langle v \rangle N_c$ term in the emission rate. The temperature dependence of the capture cross section σ, however, is neglected. Once e_n and $E_c - E_T$ are known, σ can be calculated using the emission rate equation or determined independently by measurements of DLTS peak amplitudes vs. filling pulse widths.

In general, the trap density resolution limit of a DLTS system is determined by the resolution of the $\Delta C/C$ measurement. Typical sensitivities range from $10^{-4} \times N_D$ to $2 \times 10^{-7} \times N_D$. When leakage currents are significant, the sensitivity is further reduced owing to leakage current noise modulating the diode capacitance.

Another limiting feature concerns the long dielectric relaxation times in high resistivity materials. When the condition $\omega \ll \omega_D$ breaks down, where ω_D is the reciprocal of the dielectric relaxation time $\tau_d = \rho\varepsilon$, a high enough series resistance is introduced to affect the capacitance measurements. This applies not only to the ac signal for the capacitance meter but also to the Fourier components of the filling pulse.

The above discussion was limited to DLTS measurements using Schottky barrier diodes. Similar principles apply, however, for measurements using p-n junctions and MOS diodes. For p^+-n junctions, all the basic equations derived for Schottky barriers are valid. The only significant difference is that with p-n junctions, hole traps can also be observed using voltage pulses $V_P > 0$.

The application of DLTS to MOS diodes and MOS structures is more difficult. The oxide layer capacitance, the presence of large interface states, and possible formation of a surface inversion layer must all be taken into consideration. The inversion layer formation can be avoided by operating in the deep depletion mode. For typical values of minority-carrier lifetime and doping density of 10^{15} cm^{-3}, the time to form an inversion layer is about 0.2 s [74]. Hence, this is not a major constraint. Any minority carrier generated will not be accumulating during the temperature scan since they are instantaneously recombined during the filling pulse cycle.

Then the DLTS spectra may be interpreted under the same assumptions as in the Schottky barrier case but the oxide capacitance has to be taken into account. Also, the surface states are of fundamental importance in MOS devices [75].

By using a computer based data acquisition system to measure and record the entire transient at each temperature, it is possible to obtain the entire set of data from only one temperature scan.

The digital DLTS system shown in Fig. 3-7-8 [72, 76] consists of three sections: the sample bias control and capacitance measurement instruments, the temperature measurement and control section, and the system controller for instrumentation control and data storage and analysis.

The sample placed in the cryostat is cooled using liquid nitrogen. Its temperature is measured by a thermocouple and digitized using the HP3455 DVM. The heating rate is controlled by setting the current to the resistive heaters.

The entire measurement process is controlled by the HP9825 desktop computer. The different instruments are interfaced to the computer through the HP-IB (IEEE-488) bus. The computer programs the system voltmeter for the desired delay times and number of readings. Then the programmable power supply is programmed to generate the voltage pulse of desired width. The trailing edge of the pulse triggers the system voltmeter, which reads the capacitance meter output and transfers the

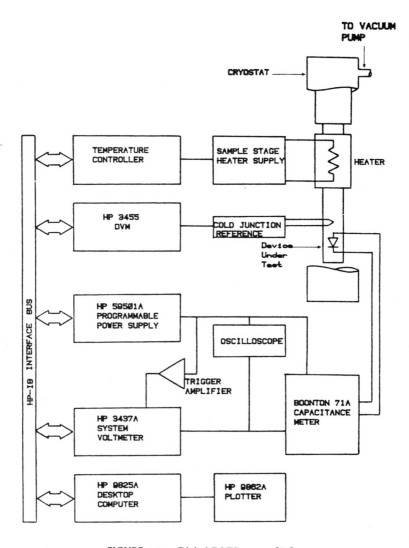

FIGURE 3-7-8. Digital DLTS system [76].

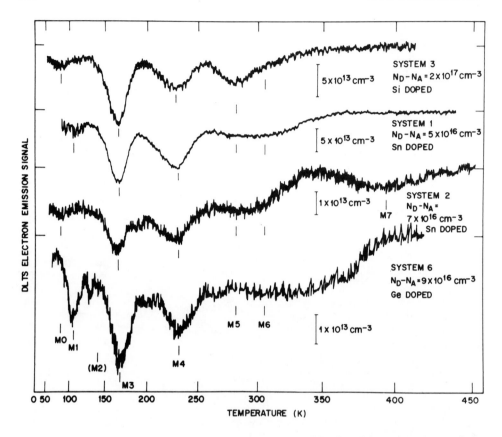

FIGURE 3-7-9. DLTS spectra of electron traps in n-type GaAs grown in four different MBE systems [78].

data to the bus. This cycle is repeated for a number of measurements to be averaged at each temperature. The sequentially averaged transients are stored on the magnetic tape.

The heating rate is kept at about 2 K per minute to avoid temperature tracking errors. Then for a typical temperature scan from 80 to 440 K the total measurement time is about 3 h.

The stored transients can be retrieved and analyzed using a variety of techniques implemented in the software. The newest approach is to use the fast Fourier transform method and the method of moments, which allows a more accurate analysis of nonexponential capacitance transient [77].

DLTS spectra of n-GaAs grown in four different MBE systems with three different dopants are shown in Fig. 3-7-9 as an example [78].

3-7-5. Film Characterization by Beam Interaction Techniques

A variety of characterization techniques utilizes photon, electron, or ion beams. These techniques include photoluminescence and photoconductivity measurements. Photoluminescence spectra help to identify deep acceptor levels in GaAs. Photoconductivity studies provide information about shallow donors.

TABLE 3-7-1. Survey of Major Methods of Materials Characterization (after R. E. Honig [79])

Primary excitation	Detected emission			
	Optical	X-rays	Electrons	Ions (+ and −)
Photons Optical	AA: Atomic absorption IR: Infrared visible } spectroscopy UV: Ultraviolet		ESCA: electron spectroscopy for chemical analysis UPS: VAC. UV photoelectron spectroscopy—outer shell XPS: X-ray photoelectron spectroscopy—inner shell	
X-rays		X-ray fluorescence spectrometry X-ray diffraction		
Electrons		EPM: Electron probe microanalysis	AES: Auger electron spectroscopy SAM: Scanning Auger microanalysis SEM: Scanning electron microscopy TEM: Transmission electron microscopy	
Ions (positive and negative)	[SCANIIR: Surface composition by analysis of neutral and ion impact radiation]	[Ion-induced X-rays]		SIMS: Secondary ion mass spectrometry IPM: Ion-probe microanalysis ISS: Ion scattering spectrometry [RBS: Rutherford back-scattering spectrometry] SSMS: Spark source mass spectrography
Radiation	ES: Emission spectroscopy			

TABLE 3-7-2. Survey of Methods for Surface and Thin Film Analysis (after R. E. Honig [79])

Method	Probe diameter (μm)	Sampling depth		Optimum detection sensitivity (ppm atomic)	Reproducibility (%)	Coverage of elements	Special features
		μm	Atomic layers				
X-ray fluorescence spectrometry	10^4	3–100	10^4–3×10^5	1–100	±1	Nearly complete ($Z \geq 9$)	Quantitive; nondestructive; insulators
Electron-probe microanalysis	1	0.03–1	102–3×10^3	100–1000	±2	Complete ($Z \geq 4$)	Quantitative; "nondestructive"
Solids mass spectrography	10–100	1–10	3×10^3–3×10^4	0.01–10	±20 ±2	Nearly complete	Semi-quantitative ion-sensitive plates electrical readout
Ion scattering spectrometry	10^3		1	0.1–1%	±20	Nearly complete no H, He	Semi-quantitative; in depth concentration profile insulators
Secondary ion mass spectrometry	10^3		3	0.1–100	±2	Nearly complete	Semi-quantitative; in depth concentration profile
Ion-probe microanalysis	1–300		10–1000	0.1–100	±2	Nearly complete	Semi-quantitative; three-dimensional concentration profile
Auger electron spectrometry	25–100		2–10	0.01–0.1%	±20	Nearly complete no H, He	Semi quantitative; three-dimensional concentration profile
SAM: scanning Auger microanalysis	4–15		2–10	0.1–1%	±20	Nearly complete no H, He	Semi quantitative; three-dimensional concentration profile; two-dimensional Auger images
XPS: X-ray photoelectron spectroscopy (ESCA)	10^4		2–10	0.1–1%	±20	Nearly complete no H, He	Semi-quantitative; valence states

TABLE 3-7-3. Surface and Thin Film Methods: Capabilities and Limitations (after R. E. Honig [79])

Method	Detection sensitivity Optimum (ppm)	Detection sensitivity Range factor	Effects Matrix	Effects Geometrical	Charge-up and field problems	Beam-induced chemical changes	Depth resolution raster/gating Without	Depth resolution raster/gating With	Lateral resolution	Elemental identification	Typical analytical time	Capabilities/Limitations
X-ray fluorescence spectrometry	1–100				No				None	Good	15 min	C: Nondestructive; quantitative; fast. L: $Z > 9$
Electron-probe microanalysis	100–1000	10	Some	Yes	Yes	Yes			Exc.	Good	1h	C: Nondestructive; quantitative; area and line scans. L: $Z \geq 4$
Solids mass spectrography	0.01–10	10			Yes				Fair	Exc.	1 h	C: Sensitive survey method
Ion scattering spectrometry	0.1–1%	10		Yes	No	Yes	Poor	Fair	Poor	Fair	3h	C: True "surface" analysis: insulators; depth profile
Secondary ion mass spectrometry	0.1–100	10^4	Severe		Yes	Yes	Poor	Good	Fair	Good	30 min	C: Depth profile. L: Matrix effects
Ion-probe microanalysis	0.1–100	10^4	Severe	Yes	Yes	Yes	Poor	Good	Good	Good	30 min	C: Area and line scans; depth profile. L: Matrix effects
Auger electron spectrometry	0.01–0.1%	20			Yes	Yes		Good	Fair	Good	30 min	C: Depth profile; multiplexing of six elements
SAM: scanning Auger microanalysis	0.1–1%	20			Yes	Yes	Good	Good	Good	Good	1h	C: Area and line scans; depth profiles of six elements
XPS: X-ray photoelectron spectroscopy (ESCA)	0.1–1%				No				None	Fair	3h	C: Molecular information; valence states. L: Slow method

Different methods used for surface and thin film analysis have been surveyed by R. E. Honig [79] (see Tables 3-7-1, 3-7-2, and 3-7-3). In Table 3-7-1 the basic terminology for different methods is introduced. In Tables 3-7-2 and 3-7-3 the capabilities and limitations of each method are compared. Below we briefly discuss several of the most important techniques.

In Auger electron spectroscopy (AES) a beam of electrons strikes the surface and causes the emission of Auger electrons from surface atomic layers. The analysis of the energies of emitted electrons yields information about the chemical composition of the surface layer. When this method [or scanning Auger microanalysis (SAM)] is used in conjunction with ion sputtering, concentration profiles of different elements composing the film are determined as functions of depth with sensitivity between 0.1 and 1 atomic %. The lateral resolution of SAM may reach 4 μm. In certain samples with high vapor pressure, the electron beam may induce chemical changes [79].

Ion probe microanalysis may also be used to establish the chemical composition of the epitaxial films. In this method the sample surface is scanned by a beam of rare gas ions. This leads to the emission of secondary ions from a few atomic layers below the surface. The analysis of the secondary ion emission provides information about the composition of the wafer.

A simpler version of this technique is second ion mass spectrometry (SIMS), where the scanning of the surface is not used. The sensitivity of this technique can be as high as 0.1 parts per million. The limitations of this method include strong influence of the surface and oxygen on ion intensities.

X-ray photoelectron spectroscopy (ESCA) provides valuable information about chemical structure of atoms near the surface. The detection sensitivity is about 1%. More detailed information regarding these techniques may be found in Ref. 79.

3-8. SCHOTTKY CONTACTS

3-8-1. Schottky Barriers

The electrostatic potential barrier (Schottky barrier) exists at the boundary between a semiconductor and a metal. In order to understand this effect, one should first consider the boundary of an n-type semiconductor (see Fig. 3-8-1). The potential energy of electrons inside the crystal is smaller because the electrons are attracted by the positive ions of the crystal lattice. However, owing to the thermal motion some electrons have energy higher than $E_c + X_{so}$ and may leave the crystal.

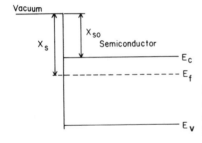

FIGURE 3-8-1. Potential distribution at the semiconductor surface. E_c, the bottom of the conduction band; E_v, the top of the valence band; E_F, the Fermi level; X_s, the work function; X_{so}, the electron affinity.

The electron flux I perpendicular to the surface can be calculated based on the Maxwell–Boltzmann distribution:

$$I_s = \frac{m_n (k_B T)^2}{2\pi^2 \hbar^3} e^{-X_s/k_B T} \qquad (3\text{-}8\text{-}1)$$

where m_n is the electron mass and T is the lattice temperature. This effect is called the thermionic emission. The escaping electrons leave the unbalanced positive charge inside the crystal. The resulting electric field leads to band bending near the surface. This positive charge also attracts electrons back, thus establishing the thermodynamic equilibrium. The simple potential distribution shown in Fig. 3-8-1 does not take the band bending into account and is therefore inaccurate.

When a metal and a semiconductor are suddenly placed close to each other, there are two competing electron fluxes from the semiconductor to the metal [see Eq. (3-8-1)] and from the metal to the semiconductor:

$$I_m = \frac{m_n (k_B T)^2}{2\pi^2 \hbar^3} e^{-X_m/k_B T} \qquad (3\text{-}8\text{-}2)$$

where X_m is the work function of the metal. If $X_m > X_s$, then $I_m < I_s$, the metal will be charged negatively and semiconductor will be charged positively, with the resulting potential difference

$$V_{bi} = X_m - X_s \qquad (3\text{-}8\text{-}3)$$

corresponding to the barrier height

$$\phi_b = X_m - X_{so} \qquad (3\text{-}8\text{-}4)$$

where X_{so} is the electron affinity (see Fig. 3-8-1). The negative charge in the metal is practically localized at the surface atomic layer (due to a very large free electron density). The positive charge density in the semiconductor is limited by the concentration of ionized donors and the space charge region extends into the semiconductor. The resulting potential distribution is shown in Fig. 3-8-2.

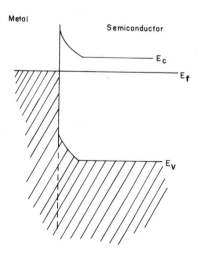

FIGURE 3-8-2. Barrier between metal and semiconductor.

The variation of the space charge, electric field, and potential in the semiconductor space charge region can be found using the depletion approximation:

$$\rho = qN_d \tag{3-8-5}$$

$$F = -\frac{qN_d(A_0 - x)}{\varepsilon_s} \tag{3-8-6}$$

$$V = -\frac{qN_d(A_0 - x)^2}{2\varepsilon_s} \tag{3-8-7}$$

(see Fig. 3-8-3). Here A_0 is the width of the space charge region, and ε_s is the permittivity of the semiconductor.

In the case when Eq. (3-8-4) is valid, the built-in voltage is given by

$$V_{bi} = X_m - X_{so} - E_c + E_F \tag{3-8-8}$$

In practice, however, Schottky barrier heights are quite different from the energies given by Eq. (3-8-4). This difference is caused by the surface states at the boundary between the semiconductor and a thin oxide layer which is always present at the surface. The oxide is so thin that electrons can easily tunnel through; the surface states change the barrier height. These states, which can be thought of as "dangling" bonds at the surface, are continuously distributed in energy within the energy gap. They are characterized by a "neutral" level ϕ_0 such that the states below ϕ_0 are neutral when filled by electrons, and states above ϕ_0 are neutral when empty (see Fig. 3-8-4). It can be then shown that the barrier height [71]

$$\phi_b = \gamma(X_m - X_{so}) + (1 - \gamma)(E_g - \phi_0) - \gamma F_m \delta \varepsilon_s / \varepsilon_i \tag{3-8-9}$$

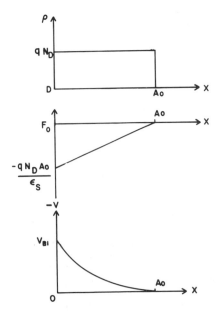

FIGURE 3-8-3. Space charge, electric field, and potential distributions in the depletion region.

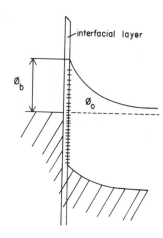

FIGURE 3-8-4. Surface states at metal–semiconductor boundary.

where

$$\gamma = \frac{\varepsilon_i}{\varepsilon_i + qN_s\delta} \tag{3-8-10}$$

E_g is the energy gap, ε_i is the permittivity of the interfacial layer, δ is the thickness of the interfacial layer, N_s is the density of the surface states (in $eV^{-1}\,m^{-2}$)

$$F_m = \left(\frac{2qN_dV_{bi}}{\varepsilon_s}\right)^{1/2} \tag{3-8-11}$$

$$V_{bi} = \phi_b - E_c + E_F \tag{3-8-12}$$

The last term in the right-hand part of Eq. (3-8-9) is a part of the voltage drop across the interfacial layer, which is small in most cases.

The barrier height can be determined experimentally from the I–V characteristics, photoelectric measurements, and capacitance measurements [80]. The results of such measurements are summarized in Fig. 3-8-5, where the barrier heights on chemically etched Si and GaAs surfaces are shown for different methods.

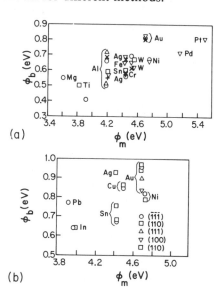

FIGURE 3-8-5. Barrier heights on chemically etched surfaces of (a) Si and (b) GaAs [80].

The barrier heights depend on the method of the surface preparation and often change with time. There is a definite correlation between ϕ_b and ϕ_m, with higher ϕ_m leading to larger ϕ_b.

Schottky barriers on p-type Si were studied by Smith and Rhoderick [81], who found that for a given metal, the sum of barrier heights on n-type and p-type Si was equal to the band gap.

The comparison of experimental data with Eq. (3-8-9) leads to the following parameters for n-type Si [80]:

$$\gamma \simeq 0.5$$
$$N_S \simeq 10^{18} \text{ eV}^{-1} \text{ m}^{-2}$$
$$\phi_0 - E_v \simeq 0.3 \text{ eV}$$

The barrier height in n-type GaAs depends on the crystal orientation. For (111) crystal face the best fit to experimental data is given by [80]

$$N_s \simeq 4 \times 10^{17} \text{ eV}^{-1} \text{ m}^{-2}$$
$$\phi_0 - E_v \simeq 0.52 \text{ eV}$$

For the (110) GaAs surface Seiranyan and Tkhorik [82] found

$$N_s = 3 \times 10^{17} \text{ eV}^{-1} \text{ m}^{-2}$$
$$\phi_0 - E_v \simeq 0.45 \text{ eV}$$

(3-8-13)

3-8-2. Current–Voltage Characteristics (Thermionic Emission Model)

The current through the Schottky barrier may be limited either by the drift and diffusion of carriers in the space charge region (in low mobility samples) or by the thermionic emission over the barrier (in high mobility semiconductors). The thermionic model is valid when the mean free path of electrons exceeds the distance over which the barrier decreases by $k_B T / q$:

$$\lambda > \frac{k_B T}{q F_{max}}$$

(3-8-14)

where F_{max} is the maximum electric field at the metal–semiconductor interface:

$$F_{max} = \left(\frac{2q N_D V_{bi}}{\varepsilon_s}\right)^{1/2}$$

(3-8-15)

and the mean free path

$$\lambda \simeq \frac{\mu}{q}(1.5 k_B T m)^{1/2}$$

(3-8-16)

Inequality (3-8-14) is fulfilled in GaAs at room temperature for $N_D > 10^{14} \text{ cm}^{-3}$, assuming $V_{bi} = 0.7$ V and $\mu = 0.5 \text{ m}^2/\text{V} \cdot \text{s}$.

The band diagrams of a Schottky barrier under forward and reverse bias are shown in Fig. 3-8-6. The quasi-Fermi levels for electrons are practically flat throughout the depletion regions except a narrow region near the interface.

The current–voltage characteristics predicted by the thermionic model are given by [83]

$$j = j_0 \exp\left(\frac{qV}{nk_BT}\right)\left[1 - \exp\left(-\frac{qV}{k_BT}\right)\right] \qquad (3\text{-}8\text{-}17)$$

where

$$j_0 = A^* T^2 \exp\left(-\frac{q\phi_b}{k_BT}\right) \qquad (3\text{-}8\text{-}18)$$

and A^* is the Richardson constant:

$$A^* = f_p f_Q \frac{4\pi mqk_B^2}{h^3} \qquad (3\text{-}8\text{-}19)$$

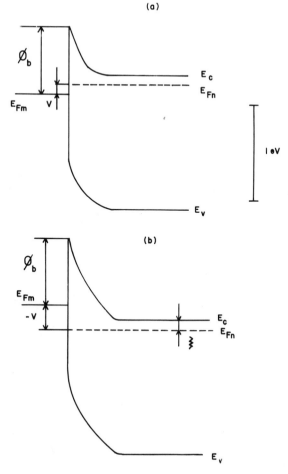

FIGURE 3-8-6. Band diagram of a Schottky barrier under (a) forward and (b) reverse bias.

Here f_p is the probability of an electron reaching the metal without being scattered by an optical phonon after having passed the top of the barrier, f_Q is the transmission coefficient ($f_p f_Q \simeq 0.5$), and m is the effective mass. For a nonparabolic semiconductor with N conduction minima

$$m = N(m_D)^{3/2}/(m_C)^{1/2} \qquad (3\text{-}8\text{-}20)$$

where m_D is the density of states effective mass $m_D = (m_t^2 m_l)^{1/3}$ and m_C is the conduction effective mass. For (111) surfaces of Si and GaAs A^* is equal to 9.6×10^5 A/m^2 K^2 and 4.4×10^4 A/m^2 K^2, respectively.

The factor n in Eq. (3-8-17) is called the "ideality factor." This factor is related to the voltage dependence of the barrier height. The review of different mechanisms leading to this dependence may be found in Refs. 80 and 84.

It is instructive to compare the value of the saturation current j_o for the Schottky barrier with the saturation current in p–n junction J_{opn}. As shown in Ref. 87,

$$\frac{j_0}{J_{opn}} \simeq \left(\frac{\tau_{nl}}{\tau_n}\right)^{1/2} \exp\left(\frac{V_{bi} - \phi_b}{k_B T}\right) \qquad (3\text{-}8\text{-}21)$$

where τ_{nl} is the electron lifetime in the p-type region, τ_n is the average time between collisions, and V_{bi} is the build-in voltage in the p–n junction. The ratio (τ_{nl}/τ_n) is of the order of 10^4 in GaAs and even higher in Si. Also, typically V_{bi} is greater than ϕ_b and hence the saturation current in a Schottky diode is bigger than in a p–n junction by many orders of magnitude and the cut-in voltage is smaller by several tens of volts.

3-8-3. Current–Voltage Characteristics. Thermionic-Field Emission and Field Emission

In highly doped semiconductors the Schottky barrier becomes so thin that electrons near the top of the barrier can tunnel through the barrier (see Fig. 3-8-7) [84]. This process is called thermionic-field emission. In degenerate semiconductors, especially in semiconductors with small electron effective mass, such as GaAs, electrons can tunnel through the barrier near the Fermi level (field emission). The current–voltage characteristic in case of the thermionic-field emission or field emission is determined by the competition between the thermal activation and tunneling and is given by [84]

$$J = J_{stf} \exp\left(\frac{qV}{E_0}\right) \qquad (3\text{-}8\text{-}22)$$

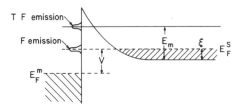

FIGURE 3-8-7. Field and thermionic emission under forward bias [84].

where

$$E_0 = E_{00} \coth(E_{00}/k_B T) \qquad (3\text{-}8\text{-}23)$$

and

$$E_{00(\text{eV})} = \frac{h}{4\pi}\left(\frac{N_D}{m\varepsilon_s}\right)^{1/2} = 1.85 \times 10^{-14}\left[\frac{N_D}{(m/m_e)(\varepsilon_s/\varepsilon_0)}\right]^{1/2} \qquad (3\text{-}8\text{-}24)$$

where SI units are used. The preexponential term was calculated by Crowell and Rideout [85]:

$$J_{\text{stf}} = \frac{A^* T[\pi E_{00} q(\phi_b - V - \xi)]^{1/2}}{k_B \cosh(E_{00}/k_B T)} \exp\left[-\frac{q\xi}{k_B T} - \frac{q}{E_0}(\phi_b - \xi)\right] \qquad (3\text{-}8\text{-}25)$$

Here $\xi = (E_c - E_{\text{Fn}})/q$ (see Fig. 3-8-6) and is negative for a degenerate semiconductor, ϕ_b is the barrier height, and A^* is the Richardson constant.

In GaAs the thermionic field emission occurs roughly for $N_D > 10^{23}$ m^{-3} at 300 K and for $N_D > 10^{22}$ m^{-3} at 77 K. In silicon the corresponding values of N_D are roughly four times bigger.

At very high doping levels the width of the depletion region becomes so narrow that direct tunneling from the semiconductor to the metal may take place, as shown (see Fig. 3-8-7). This happens when E_{00} becomes much greater than $k_B T$. The current–voltage characteristics in this regime are given by

$$J \approx J_{sf} \exp(qV/E_{00}) \qquad (3\text{-}8\text{-}26)$$

where

$$J_{sf} = \frac{\pi A^* T}{k_B C_1 \sin(\pi k_B T C_1)} \exp\left(-\frac{q\phi_b}{E_{00}}\right)$$

and

$$C_1 = (2E_{00})^{-1} \ln[-4(\phi_b - V)/\xi]$$

The effective resistance of the Schottky barrier in the field emission regime is quite low. Therefore the metal–n^+ Schottky barrier is used to make ohmic contacts. To this end a highly doped layer is fabricated on top of the moderately or low doped active layer (see Section 3-9).

3-8-4. Small Signal Circuit of Schottky Diode

The small signal equivalent circuit of a Schottky diode is shown in Fig. 3-8-8. It includes the parallel combination of differential resistance of the Schottky barrier

$$R_d = \frac{dV}{dI} \qquad (3\text{-}8\text{-}27)$$

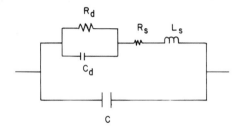

FIGURE 3-8-8. Equivalent circuit of a Schottky diode.

and the differential capacitance of the space charge region (which may be estimated using the depletion approximation):

$$C_d = S\left[\frac{qN_d\varepsilon_s}{2(V_{bi} - V)}\right]^{1/2} \qquad (3\text{-}8\text{-}28)$$

Here S is the device cross section. These circuit elements are in series with series resistance (comprised of the contact resistance and the resistance of the neutral semiconductor region between the ohmic contact and the depletion region) and the equivalent inductance. The device geometric capacitance appears in parallel across the contacts

$$C = \varepsilon S/L \qquad (3\text{-}8\text{-}29)$$

Here L is the device length.

The major difference between the equivalent circuit of Fig. 3-8-8 and an equivalent circuit for a $p\text{-}n$ junction is the absence of the diffusion capacitance in the equivalent circuit of a Schottky diode. This leads to a much faster response under the forward bias conditions and makes it possible to use Schottky diodes as microwave mixers, detectors, etc.

In most cases $R_s \ll R_d$ and $C_d \gg C$ so that the characteristic time constant limiting the frequency response of a Schottky diode is given by

$$\tau_{\text{Schottky}} = R_s C_d \qquad (3\text{-}8\text{-}30)$$

3-8-5. Practical Schottky Contacts

In practical devices the semiconductor surface is chemically etched just prior to the metal deposition to form the Schottky barriers. Under such conditions the formation of the interfacial layer is unavoidable and the surface states determine the barrier height. Though the kind of metal has a noticeable effect on the barrier height, the choice of the metal is determined primarily by such factors as the ease of deposition, reliability, metal resistance, etc. In particular, aluminum has been widely used as a gate metal in GaAs devices. TiW/Au contacts have been also used for applications in GaAs IC's [86]. TiW alloys have good adhesion, are corrosion resistant, and act as an effective diffusion barrier. The gold film is deposited on top to decrease the metal resistance. Tungsten contacts are used when high temperatures must be reached (for example, in annealing after ion implantation with the Schottky contacts in place). However, the metal resistance of such a contact is over twice that of Al.

A good review of early work on Schottky contacts is given in Ref. 87. A more recent bibliography may be found in Ref. 88. A recent thorough review of metal-semiconductor contacts to the III–V semiconductors was published by G. Y. Robinson [89].

3-9. OHMIC CONTACTS

3-9-1. Minimum Specific Contact Resistance

As was mentioned in Section 3-8, a practical way to obtain a low resistance "ohmic" contact is to increase the doping near the metal–semiconductor interface to a high value so that the depletion layer caused by the Schottky barrier becomes very thin and the current transport through the barrier is enhanced by tunneling (field emission regime).

A band diagram of a metal-n^+-GaAs-n-GaAs contact, which illustrates the role played by the n^+ layer, is shown in Fig. 3-9-1 [90].

Based on this diagram we can estimate a low bound for a specific contact resistance r_c. This estimate of a minimum value of r_c is based on the assumption that the specific contact resistance at the metal–semiconductor interface is small compared to the specific resistance of an n^+-n junction. From the theory of an n^+-n junction we then find [see Eq. (2-9-64a)]

$$(r_c)_{min} = \frac{(2\pi m k_B T)^{1/2}}{q^2 n_0} \qquad (3\text{-}9\text{-}1)$$

Here m is the effective mass and n_0 is the doping of the active layer. For GaAs $m = 0.067\, m_e$ and

$$(r_c)_{min}(\Omega\ \text{cm}^2) = 1.55 \times 10^{-5}(T/300)^{1/2}(10^{15}/n_0) \qquad (3\text{-}9\text{-}2)$$

Here sample temperature T is in degrees Kelvin and n_0 is in cm^{-3}.

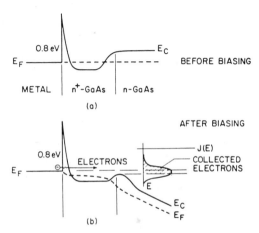

FIGURE 3-9-1. Band diagram of metal-n^+-GaAs-n-GaAs ohmic contact [90] (a) without a bias; (b) under bias.

3-9-2. Alloyed, Implanted, and Heterojunction Ohmic Contacts

A conventional approach is to use alloyed ohmic contacts which makes it possible to form a very highly doped region near the surface. However, as will be discussed below, a nonuniformity of the current flow through the alloyed contact does not allow us to reach a theoretical value of the contact resistance in this case.

Ion implantation or diffusion has been also used to create a highly doped region near the surface to facilitate the formation of ohmic contacts. The doping concentration is limited by the impurity solubility and may reach 5×10^{19} cm^{-3} for n-type GaAs and 10^{20} cm^{-3} for p-type GaAs [91]. Using ion implantation doping concentrations in excess of the solubility limit may be reached near the surface [89].

Still another technique utilizes an epitaxial growth of highly doped layers between the active layer and the metal contact. A promising modification of this method is to use a heterojunction contact [24, 92, 93]. The idea is to reduce the barrier height and hence the contact resistance by introducing an intermediate semiconductor layer between the metal and the active layer of the device. (It has been found empirically that in many cases the barrier height is close to 2/3 of the energy gap). Stall et $al.$ [24, 92] fabricated MBE heterojunction contacts using an n^+-Ge layer grown on n^+-GaAs. The barrier height for Ge is only 0.5 eV (compared to 0.7–0.8 eV for GaAs). Also, a very high doping level (up to 10^{20} cm^{-3}) may be achieved in the Ge layer. Contact resistances as low as 5×10^{-8} Ω cm^2 and 1.5×10^{-7} Ω cm^2 have been obtained for GaAs doped at 1.5×10^{18} cm^{-3} and 1×10^{17} cm^{-3}, respectively [24].

Woodall et $al.$ [93] reported InAs/Ga$_x$In$_{1-x}$As/GaAs heterojunction contacts fabricated by MBE. An intermediate Ga$_x$In$_{1-x}$As layer had a graded composition in order to avoid the band discontinuities which appear at the abrupt InAs/GaAs interface and impede the current flow.

An evaporated Au/Ge eutectic with a Ni overlayer alloyed at a temperature greater than the eutectic melting temperature was introduced by Braslau et $al.$ [94] and has become a very widely used technique for contacting n-GaAs. Other alloyed contacts use AgInGe [95], AuGe/AgAu [96], In/AuGe [97], and AuGe/Pt [98]. During the alloying process Ga vacancies are produced. When Au/Ge contacts are used, these vacancies are populated by diffusing Ge atoms which act as a donor when located on a Ga site. The effective doping density of the alloyed layer could reach 5×10^{19} cm^{-3}.

The contact resistance r_c, however, does not seem to be determined by this high doping concentration $N_{d\,\text{HIGH}}$ of the alloyed region, as could be expected from the field emission theory. This theory predicts (see Section 3-8-3)

$$r_c = r_0 \exp[A_0/(N_{d\,\text{HIGH}})^{1/2}] \tag{3-9-3}$$

where r_0 and A_0 are constants.

Instead r_c seems to vary as $1/(\mu N_d)$, where N_d is the doping density and μ is a low-field mobility of the active layer (see fig. 3-9-2 [90] and Fig. 3-9-3 [89]).

A recent explanation of this effect proposed by Braslau [90] is based on the nonuniformity of the alloyed layer (see Fig. 3-9-4 [90]). An optical microphotograph of GaAs surface from which an AuGeNi contact has been removed corroborates this model (see Fig. 3-9-5 [90]).

In the Braslau model the protruding contact areas are approximated by hemispherical regions of radius $\langle r \rangle$ which are separated by an average distance $\langle A \rangle$. The

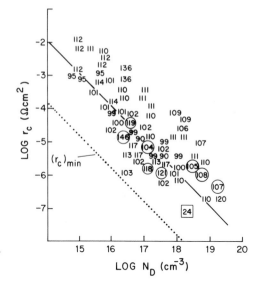

FIGURE 3-9-2. Observed contact resistances vs. doping of an active layer for n-type GaAs [90]. Circles correspond to laser or electron beam annealed contacts. The point in the square is for an MBE-grown heterojunction contact. Numbers on the figure correspond to the reference numbers. Dotted line—predicted minimum value of the specific contact resistance.

resistance of each contact area is

$$r_{c1} = \frac{r_c}{2\pi\langle r\rangle^2 f} \tag{3-9-4}$$

where r_{c1} is the contact resistance predicted by the field emission theory for a perfect planar junction, and the factor f takes into account the field enhancement (and hence the reduction in r_{c1}) due to the nonuniform field distribution. The measured contact resistance of unit area is then given by

$$r_{cmeas} = \langle A\rangle^2 \frac{\rho}{\pi\langle r\rangle} + \frac{r_c}{2f\pi\langle r\rangle^2} \tag{3-9-5}$$

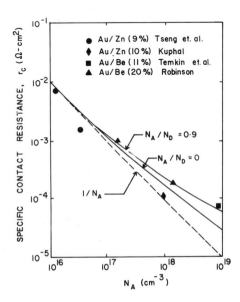

FIGURE 3-9-3. Specific contact resistance of Au-based ohmic contacts on p-type InP, plotted against substrate doping [89]. The two solid lines are calculated for two different levels of compensation by donors taking into account the dependence of the low field mobility on doping. The data points are taken from the following references: 89, 122, 123, 124.

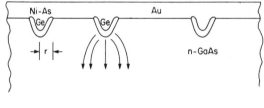

$$r_c)_{MEAS} \simeq \langle A \rangle^2 \left[\rho_n / \pi \langle r \rangle + r_c / 2 \pi f \langle r \rangle^2 \right]$$

$$f \gg 1$$

FIGURE 3-9-4. Suggested model of alloyed ohmic contact to GaAs. Conduction takes place through parallel array of Ge-rich protrusions of negligible contact resistance as compared to spreading resistance in series with them (from Ref. 90).

Here

$$\rho = 1/(q\mu N_d) \tag{3-9-6}$$

is the resistivity of the active layer.

In Eq. (3-9-5) we assume that the protrusions are connected by the metal layer. The first term in the left-hand side of Eq. (3-9-5) is the spreading resistance; the second term may be neglected if $\rho > 10^{-3} \ \Omega$ cm.

The predictions of the model are compared with experimental results in Figs. 3-9-2 and 3-9-3. We also show in Fig. 3-9-2 the minimum value of the specific contact resistance predicted by Eq. (3-9-2). As can be seen from Figs. 3-9-2 and 3-9-3, the model seems to be in good agreement with experimental results. The minimum value of the contact resistance follows the same trend as the measured values of the contact resistance [$(r_c)_{\min}$ is proportional to $1/n_0$].

However, as will be discussed in Section 3-9-4, the measurements of the specific contact resistance on planar structures do not agree with the most important assumption of the Braslau model, i.e., that the boundary between the alloyed and nonalloyed regions solely determines the contact resistance. In order to explain the experimental results we are forced to assume that the metal–semiconductor interface

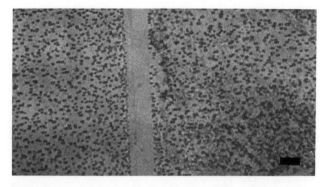

FIGURE 3-9-5. Optical microphotograph of AuGeNi contact after contact metal has been removed, showing irregular pitting of GaAs surface, from Ref. 90. Scale bar: 5 μm.

also plays an important role [130], and hence more complicated and realistic models of an ohmic contact have to be developed.

According to the Braslau model one way to minimize the contact resistance is to decrease the separation between the protrusions. However, their size and spacing are difficult to control during processing. A more practical way to decrease r_c is first to create a uniform heavily doped region using multiple ion implants or melocular beam epitaxy and then make alloyed contact thus reducing the spreading resistance [125, 126, 107].

Another interesting possibility discussed above is to decrease the effective barrier height of the Schottky barrier using a heterojunction contact [24, 92, 93].

3-9-3. Fabrication of Alloyed Contacts

Fabrication of alloyed contacts includes surface preparation prior to deposition, metal deposition, and alloying. First samples are rinsed in organic solvents, then etched and rinsed in DI water. Metal deposition is usually performed with substrate at room temperature. Au–Ge is evaporated from the eutectic (88% Au, 12% Ge by weight). A subsequently deposited Ni (or Pt) layer improves the melting of the eutectic film and increases the solubility of GaAs [90]. Typical thicknesses are 1000–1500 Å for AuGe and 100–500 Å for Ni. Excessive amounts of Ni degrade the contact performance because of Ni diffusion into GaAs [127, 128]. A refracting metal and then a thicker layer of gold may be deposited to reduce the sheet resistance of the metal [129].

A typical alloying cycle is described in Ref. 90. The sample is rapidly (400°C/min) brought to 450°C, and after 30 s rapidly cooled in a hydrogen atmosphere.

A laser or electron beam annealing is also used to decrease interdiffusion and segregation because of the short heating times achieved in these techniques [104, 105].

The "recipe" of the fabrication of ohmic contacts to n-type GaAs has been summarized by N. Braslau [131] (see Table 3-9-1).

TABLE 3-9-1. Alloyed Ohmic Contacts to n-Type GaAs
(after Braslau [131])

1. Surface preparation	
Remove oxide with HF of HCl or dilute etch	
Rinse in DI water	
Rinse in $NH_4CH:H_2O$ (1 : 1)	
Blow dry in N_2 with minimum exposure to air	
2. Evaporation (pressure less than 10^{-6} torr)	
AuGe (88% : 12%)	1000 Å
Ni (E-beam evaporation)	500 Å
Au	300 Å
3. Alloying	
Furnace temperature 450°C, forming gas	
Temperature vs. time as shown in Fig. 3-9-7	
4. Resulting contact parameters (for $N_d = 1 \times 10^{17}$ cm^{-3})	
Featureless	
Resistance per square	1 Ω/square
Resistance per mm of contact width	0.5 Ω mm
Specific contact resistance	5×10^{-6} Ω cm^2

Good alloyed ohmic contacts to n-type GaAs have a low contact resistance (less than $10^{-6}\,\Omega\,cm^2$) and high reliability. Some of the results of a reliability study of the ohmic contacts for GaAs are depicted in Fig. 3-9-6 [98].

Au–Zn [132], Au–Zn–Au [133], and Ag–Zn [134] alloyed contacts (with 5–15 wt % of Zn) are used for p-type GaAs. The results of Sanada and Wada may be approximated by [89]

$$r_c(\Omega\,cm^2) = 1.8 \times 10^{18}/p^{1.3}\,(cm^{-3})$$

in crude agreement with the Braslau model. Here p is the hole concentration.

The discussion of ohmic contacts to InP and GaP may be found in Ref. 89. Ohmic contacts to A_3B_5 alloys were considered in Ref. 135. Nakato *et al.* [136] studied ohmic contacts to p-type InGaAs alloys.

A contact fabrication procedure for ohmic contacts to AlGaAs described in the literature is practically identical to the one used for GaAs.

A high contact resistance is often a factor limiting the performance of high-speed semiconductor devices. More detailed experimental and theoretical studies are necessary in order to learn how to produce low resistance, reliable, and reproducible contacts to GaAs and related compounds.

3-9-4. Resistance of Planar Ohmic Contacts

Schematic diagrams of planar ohmic contacts are shown in Fig. 3-9-7. Figure 3-9-7a corresponds to an alloyed contact in the case when the alloyed low resistivity region goes through the entire active layer. This situation was considered in Refs. 137–139. Figure 3-9-7b illustrates the case when the alloyed region occupies only part of the active layer. Figure 3-9-7c (no low resistivity region under the contact metal) may be considered as an adequate model only for an epitaxially grown (nonalloyed) contact.

The resistance of ideal planar contacts shown in Figs. 3-9-7a and 3-9-7c may be found using a transmission line model first introduced for diffused ohmic contacts in silicon [140–144].

The current distribution under such a contact (see Fig. 3-9-8) is described by the following equation:

$$\frac{dI}{dx} = -J(x)\,W \tag{3-9-7}$$

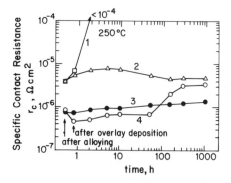

FIGURE 3-9-6. Specific contact resistance as a function of thermal aging at 250°C (from Ref. 98). Curve 1, AuGe/Pt with Ti/Pt/Au overlay; curve 2, AuGe/Pt without overlay; curve 3, AuGe/Ni without overlay; curve 4, AuGe/Ni with Ti/Pt/Au overlay.

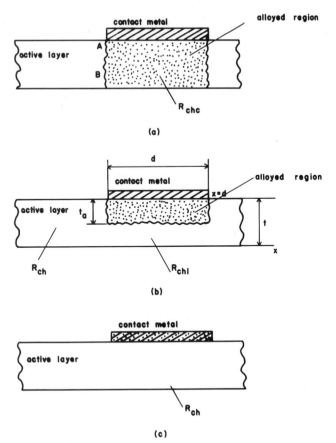

FIGURE 3-9-7. Schematic diagram of alloyed ohmic contacts. (a) Alloyed region goes through the entire active layer; (b) alloyed region does not go through the entire active layer; (c) nonalloyed contact. R_{ch} is the sheet resistance of the active layer outside the contact, R_{ch1} is the sheet resistance of the unalloyed portion of the active layer under the contact; R_{chc} is the sheet resistance of the alloyed region under the contact when it goes through the entire active region.

where I is the current, x is the coordinate (in the direction parallel to the current flow in the active layer), J is the current density:

$$J(x) = V(x)/r_c \qquad (3\text{-}9\text{-}8)$$

Here r_c is the specific contact resistance and $V(x)$ is the channel potential with respect to the potential of the contact metal:

$$\frac{dV}{dx} = -IR_{chc}/W \qquad (3\text{-}9\text{-}9)$$

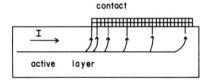

FIGURE 3-9-8. Current distribution under the planar contact.

R_{chc} is the sheet resistance of the semiconductor film under the contact in Ω per square and W is the contact width. (For the contact shown in Fig. 3-9-7c $R_{chc} = R_{ch}$, where R_{ch} is the sheet resistance of the semiconductor layer.) Equations (3-9-7)–(3-9-9) may be reduced to the equation similar to that describing a transmission line

$$\frac{d^2V}{dx^2} = \frac{V}{(L_T)^2} \qquad (3\text{-}9\text{-}10)$$

where

$$L_T = (r_c/R_{chc})^{1/2} \qquad (3\text{-}9\text{-}11)$$

is called a transfer length. Equation (3-9-10) is only valid when the active layer thickness is much smaller than L_T.

The boundary conditions for Eq. (3-9-10) are

$$\frac{dV}{dx}(x = 0) = \frac{R_{chc}}{W} I_0 \qquad (3\text{-}9\text{-}12)$$

$$\frac{dV}{dx}(x = d) = 0 \qquad (3\text{-}9\text{-}13)$$

Here d is the contact length, I_0 is the electric currents in the semiconductor layer.

The solution of Eq. (3-9-10) with boundary conditions (3-9-12) and (3-9-13) is given by

$$V = A \exp(x/L_T) + B \exp(-x/L_T) \qquad (3\text{-}9\text{-}14)$$

where

$$A = \frac{I_0 R_{chc} L_T \exp(-d/L_T)}{W[\exp(d/L_T) - \exp(-d/L_T)]} \qquad (3\text{-}9\text{-}15)$$

$$B = \frac{I_0 R_{chc} L_T \exp(d/L_T)}{W[\exp(d/L_T) - \exp(-d/L_T)]} \qquad (3\text{-}9\text{-}16)$$

From Eq. (3-9-14) we find

$$V(0) = I_0 R_c \qquad (3\text{-}9\text{-}17)$$

where the contact resistance

$$R_c = R_{chc}(d/W) F_{t1m} \qquad (3\text{-}9\text{-}18)$$

Here

$$F_{t1m} = (L_T/d) \coth(d/L_T) \qquad (3\text{-}9\text{-}19)$$

For $d/L_T \ll 1$ $F_{t1m} \approx (L_T/d)^2$. For $d/L_T \gg 1$ $F_{t1m} \approx L_T/d$ (see Fig. 3-9-9).

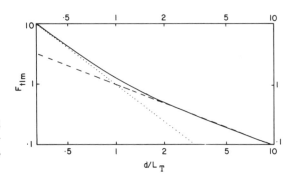

FIGURE 3-9-9. F_{tlm} vs. d/L_T. Dashed and dotted lines show asymptotic approximations for large and small values of d/L_T, respectively.

The value of the potential of the channel at the end of the contact $V(d)$ is also found from Eq. (3-9-14):

$$V(d) = \frac{I_0 R_{chc} L_T}{W \sinh(d/L_T)} \qquad (3\text{-}9\text{-}20)$$

Hence, the so-called "end" resistance defined as

$$R_{end} = V(d)/I_0 \qquad (3\text{-}9\text{-}21)$$

is given by

$$R_{end} = \frac{R_{chc} L_T}{\sinh(d/L_T) W} \qquad (3\text{-}9\text{-}22)$$

From Eqs. (3-9-18) and (3-9-22) we find

$$R_c/R_{end} = \cosh(d/L_T) \qquad (3\text{-}9\text{-}23)$$

Equations (3-9-18), (3-9-22), and (3-9-23) are used for the contact characterization based on the transmission line measurements (see Section 3-10).

The values of R_{ch} and R_{chc} measured using transmission line technique [based on Eqs. (3-9-18) and (3-9-22)] are given in Table 3-9-2 for alloyed contacts to n-GaAs and sintered contacts to p-type Si [139]. As can be seen from the table, the change of the sheet resistance under the contact is quite dramatic in both cases.

It is interesting to discuss the results reported in Ref. 139 in relation with the Braslau model. According to this model discussed in Section 3-9-2, the interface between the alloyed and nonalloyed regions (and not the interface between the

TABLE 3-9-2. Channel and Contact Resistances Determined from TLM Measurements [139]

Material	Contact procedure	Active layer depth (μm)	R_{ch} ($\Omega/\square$)	R_{chc} ($\Omega/\square$)	r_c ($\Omega.cm^2$)
GaAs	Alloyed	3.0	430	22	7×10^{-5}
Si	Sintered	0.03	2100	430	4.9×10^{-3}

metal and the semiconductor) determines r_c, which is essentially the spreading resistance between small protruding regions with highly conductive interfaces. However, as shown in Ref. 130, the results of the TLM measurements [139] indicate that the metal–semiconductor interface must also play an important role in determining the specific contact resistance.

The conventional transmission line model is based on the assumption that r_c is determined by the metal–semiconductor interface and the semiconductor resistivity under the contact is uniform (though it may be different from the resistivity of the semiconductor layer outside the contact) [137–139]. In this case the contact resistance R_c and R_{end} are given by Eqs. (3-9-18) and (3-9-22), respectively.

The solution of the TLM equations for the case when the interface between the alloyed region and the active region determines the specific contact resistance is analyzed in Ref. 130. The results of this derivation may be easily understood by considering two important limiting cases:

1. The alloyed region goes through the whole epitaxial layer (see Fig. 3-9-7a).

2. The alloyed region does not go through the whole epitaxial region (see Fig. 3-9-7b) and the transfer length $L_{T1} = (r_c/R_{ch2})^{1/2}$ is greater than t_a. Here t_a is the depth of the alloyed region, $R_{ch2} = R_{ch}/(1 - t_a/t)$ is the sheet resistance of the unalloyed portion of the channel under the contact, and R_{ch} is the sheet resistance of the active layer outside the contact.

In the first limiting case we find

$$R_c = r_c/(tW) \qquad (3\text{-}9\text{-}24)$$

In this case the "end" resistance is equal to the contact resistance as most of the voltage drop across the contact is at the interface between the alloyed and unalloyed region (interface AB in Fig. 3-9-7a). In fact, the "end" resistance reported in Ref. 139 was several times smaller than R_c for long contacts ($d > L_T$) and approached R_c for short contacts ($d < L_T$).

It may be expected that for the GaAs devices studied in Ref. 139 the low resistivity alloyed region did not go through the entire epitaxial layer. The depth of the alloyed region is expected to be several tenths of a micrometer or less, whereas the active layer thickness in Ref. 139 was 3 micrometers. In this second limiting case the calculation in the frame of the transmission line model shows that Eqs. (3-9-18) and (3-9-23) remain valid if R_{chc} is replaced by the sheet resistance of the unalloyed portion of the channel under the contact R_{ch1}. Hence, the value of R_{ch1} deduced from the TLM measurements should be larger than R_{ch}, contrary to the experimental results given in Ref. 139: the measured values of R_{ch} and R_{ch1} of 430 and 22 Ω per square, respectively.

Based on this discussion and on the general trend of the specific contact resistance decreasing with the doping of the active layer [139], we speculate that both the interface between the contact metal and the semiconductor and the interface between the alloyed and nonalloyed regions may contribute to the specific contact resistance depending on the conditions of the contact fabrication. It also means that the values of r_c deduced from the conventional TLM measurements may be inaccurate. More sophisticated models taking into account two resistive interfaces with different specific contact resistances may be required for the interpretation of the experimental TLM results.

3-10. CHARACTERIZATION OF OHMIC CONTACTS

3-10-1. Cox–Strack Technique [95]

This technique is used for the experimental determination of the contact resistance for thick (bulk) samples when the contacts may be made to both sides of the wafer. For the configuration shown in Fig. 3-10-1, the total resistance R between the top and the bottom contacts is given by [95]

$$R = R_c + R_b \qquad (3\text{-}10\text{-}1)$$

where

$$R_c = r_c/(\pi a^2) \qquad (3\text{-}10\text{-}2)$$

is the contact resistance and

$$R_b = \frac{\rho F(a/t)}{a} \qquad (3\text{-}10\text{-}3)$$

is the bulk resistance of the layer. Here t is the film thickness, a is the radius of the top contact, ρ is the resistivity of the active layer, and the function F takes into account the spreading of the current streamlines in the active layer.

According to [95], F may be approximated as

$$F(a/t) \approx \frac{\tan^{-1}(2t/a)}{\pi} \qquad (3\text{-}10\text{-}4)$$

A more accurate numerical calculation of F was reported in Ref. 145.

In practical measurements the values of R are determined for top contacts of different areas and the values of $R - R_b$ are plotted as a function of $1/a^2$ to determine r_c and to find the correction introduced by the interconnect wires [89]. The values of r_c that are as small as $1 \times 10^{-6}\,\Omega\,\text{cm}^2$ can be measured with approximately 25% error for n-type GaAs epitaxial layers [89].

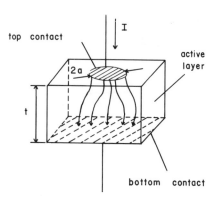

FIGURE 3-10-1. Cox–Strack characterization technique.

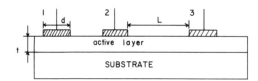

FIGURE 3-10-2. TLM pattern. Three or more ohmic contacts with different distances between adjacent contacts are used in the TLM measurement.

3-10-2. TLM (Transmission Line Model) Measurements

The TLM pattern used for the contact resistance measurements [139–144] is shown in Fig. 3-10-2. The resistance between two adjacent pads of width W separated by distance L is given by

$$R = R_{ch} L / W + 2R_c + R_P \qquad (3\text{-}10\text{-}5)$$

where R_{ch} is the sheet resistance of the active layer between the contacts (i.e., the film resistance per square), R_c is the contact resistance, and R_P is the resistance of the interconnect wires.

The value of resistance R_P of the interconnect wires may be estimated by bonding two interconnect wires to the same contact. The value of R_c is determined from the intercept of the R vs. L curve. The sheet resistance of the channel outside the contact R_{ch} is found from the slope of this line (see Fig. 3-10-3).

In the case when the sheet resistance of the active layer under the contact R_{chc} is different from R_{ch}, an additional "end" resistance measurement is needed to deduce R_{chc} and r_c (see Section 3-9-4). The "end" resistance R_{end} is defined as

$$R_{end} = V_{23} / I_{12} \qquad (3\text{-}10\text{-}6)$$

where I_{12} is the current flowing through contacts 1 and 2 and V_{23} is the voltage difference between contact 2 and floating contact 3 (see Fig. 3-10-4). An alternative but equivalent technique of measuring of R_{end} was proposed in Ref. 139.

As discussed in Section 3-9-4, in the simple case when the specific contact resistance is determined by either the contact metal–semiconductor interface or by the interface between the alloyed and nonalloyed portions of the active layer, the specific contact resistance r_c and the sheet resistance of the channel under the contact R_{chc} may be determined from R_c and R_{end} using equations of the transmission line

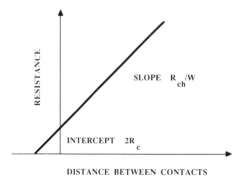

FIGURE 3-10-3. Interpretation of the TLM measurements. R_c and R_{ch} are found from the intercept and slope of the R vs. L curve where L is the distance between TLM contacts. (The resistance of the interconnect wires, R_p, is neglected.)

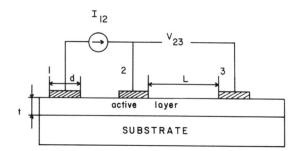

FIGURE 3-10-4. "End" resistance
measurement.

model:

$$R_c = R_{chc}(L_T/W) \coth(d/L_T)$$ (3-9-18')

$$R_{end} = \frac{R_{chc}(L_T/W)}{\sinh(d/L_T)}$$ (3-9-22')

From Eqs. (3-10-2) and (3-10-3) we find

$$L_T = d/\cosh^{-1}(R_c R_{end}/R_c)$$ (3-10-7)

Then R_{chc} and r_c are determined from Eqs. (3-9-11):

$$L_T = (r_c/R_{chc})^{1/2}$$ (3-9-11)

and (3-9-18).

The results of such a measurement for ohmic contacts to n-type GaAs are summarized in Table 3-9-2 [139]. As can be seen from the table, the sheet resistance of the channel under the alloyed contact is much smaller than the sheet resistance of the active layer. As a consequence, the transfer length is much larger than the value which is deduced when R_{chc} is assumed to be equal to R_{ch}.

If both interfaces (the interface between the contact metal and semiconductor and the interface between the alloyed and nonalloyed portions of the active layer) contribute to the contact resistance, more sophisticated models will have to be developed to deduce the specific contact resistance from the TLM measurements.

As discussed in Section 3-9-4 this seems to be the case for alloyed contacts to GaAs.

3-10-3. Four-Point Method

The four-point method of measuring the specific contact resistance is illustrated by Fig. 3-10-5. This technique [146–148] may be used for the characterization of contacts to planar structures. The input current flows between contacts 1 and 4. The voltage is measured between contacts 2 and 3.

As shown in Ref. 148, when the following unequality is valid

$$\rho a^2 < r_c t$$ (3-10-8)

where a is the radius of the contacts, t is the thickness of the layer, ρ is the layer resistivity and r_c is the specific contact resistance, the spreading resistance is much

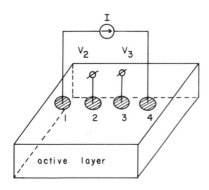

FIGURE 3-10-5. Four-point method of measuring specific contact resistance.

smaller than the contact resistance. In this case the specific contact resistance may be found from [147]

$$r_c \approx \pi a^2 \left\{ \frac{V_2}{I} - \left(\frac{V_3}{I} \right) \frac{\ln[(3d/2a) - 1/2]}{(2 \ln 2)} \right\} \qquad (3\text{-}10\text{-}9)$$

This equation is valid when $a \ll d$ and $t \ll d$. Here d is the distance between the centers of the contacts.

The values of r_c close to 10^{-6} Ω cm^2 may be measured by this technique within approximately 10% error.

3-11. ION IMPLANTATION

Ion implantation has developed into one of the most important techniques for fabricating GaAs microwaves devices and integrated circuits [149–151].

In ion implantation doping or isolation in a semiconductor is achieved by bombarding the semiconductor surface with a high-velocity ion beam. The doping density and the distribution of dopants in a semiconductor are controlled by the ion flux and velocity. The defects created as a result of an ion bombardment are annealed at temperatures about 800°C.

The advantages of this technique include independent control of doping level and doping profile, relatively good reproducibility, and ease of selective doping of the desired areas of the surface using conventional masking techniques. Multiple implants may be used to create complicated doping profiles with a good lateral resolution which is difficult or impossible to achieve by epitaxial techniques. These advantages are especially important in compound semiconductors such as GaAs where diffusion doping is more difficult than in silicon.

The distribution of implanted ions has an approximately Gaussian shape which is characterized by an average projected range R and its standard deviation ΔR (see Fig. 3-11-1). The tables of the average projected range and its standard deviation for ions implanted into GaAs are given in Ref. 152.

The ion-implantation theory [153–154] predicts a Gaussian shape of ion-implanted doping profile for amorphous targets where the deflection of an ion from its trajectory is completely random. In crystalline targets the distribution of implanted ions depends on the crystallographic orientation of the target surface. When the ion beam propagates in the crystallographic direction, where relatively large openings (channels) exist between the semiconductor atoms, some ions propagate relatively deeply into the semiconductor. This effect may lead to a non-Gaussian doping

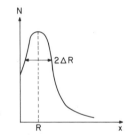

FIGURE 3-11-1. Typical implantation profile. R, projected range; ΔR, standard deviation.

profile and even to the second peak in the ion distribution. This problem can be minimized by a misalignment of the beam with the crystallographic direction.

The implanted ions displace the semiconductor atoms from their sites. When the ion energy is high enough, the areas of amorphization are created. At large implant dose the surface layer of the semiconductor may become amorphous. In GaAs the formation of the amorphous layers is suppressed if the ions are implanted at temperatures higher than 150°C. Temperature annealing after the implantation at temperatures close to 800°C is necessary to reduce the density of the defects created in the crystal.

The surface of GaAs loses As when heated up at temperatures higher than 600°C. Plasma deposited Si_3N_4 is used as an encapsulant which allows the annealing of GaAs at temperatures up to 950°C [155]. At temperatures up to 750°C AlN has been found to be an even better encapsulant because of a smaller difference with GaAs in the thermal expansion coefficients [156–157]. Another technique is annealing in As vapor overpressure (capless annealing) [158–160]. Heat treatment by infrared radiation from halogen lamps [161] and laser annealing of ion implanted GaAs may also be a promising capless annealing technique [162, 163].

The diffusion of the implaned dopants during the annealing modifies the Gaussian distribution of the impurities. Many encapsulant films act as barriers for the out-diffusion leading to the increase in the doping concentration near the surface compared to the original near-Gaussian profile. The diffusion coefficients of different impurities in GaAs are given in Ref. 164.

Depending on the implanted species GaAs may be doped n-type or p-type, or made highly resistive. The n-type GaAs is produced by implanting column 6 (Se, Te, S) or column 4 (Si, Sn) impurities. Column 6 ions occupy As sites when implanted into the samples heated to 150°C or more. Implantation of these ions at such elevated temperatures results in a higher doping efficiency (i.e., in a higher ratio of the doping density to the density of the implanted ions); see Fig. 3-11-2 [165]. Column 4 elements must occupy Ga sites in order to behave like donors. Implantation of Si into the heated samples leads only to a small increase in the doping efficiency [149], but for Sn atoms the increase of the doping efficiency due to the sample heating during the implantation is substantial [166].

Subsequent anneal temperature is another important factor. Typically an anneal temperature is chosen between 850 and 950°C. The doping efficiency also depends on the implantation dose. It is substantially higher at smaller doses owing to less damage to the material (see Fig. 3-11-3 [149]). Pulsed laser beam [163, 168–172] or electron beam [173–174] annealing has been demonstrated to remove the radiation damage from high dose implanted layers. In pulsed laser annealing, short (about 20-ns) laser pulses are used to convert high dose implanted amorphous layer into a crystalline layer. The required energy of the pulse depends on the thickness of

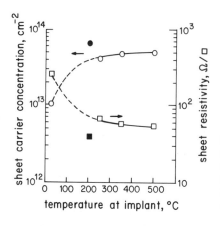

FIGURE 3-11-2. Sheet carrier concentration and sheet resistivity vs. the substrate temperature [165]. Implant does 1×10^{14} cm^{-2}. Se implant, 400 keV implant energy, annealed at 900°C for 15 min, Si$_3$N$_4$ encapsulation. Open squares, implanted through 700 Å of Si$_3$N$_4$; dark squares, implanted directly into GaAs.

the amorphous layer. For a 2300-Å-thick amorphous layer produced in the semiinsulating GaAs by implanting a 10^{15}-cm^{-2} dose of 400 kV Te ions at room temperature the threshold energy density is 1 J/cm^2 [172]. At lower energies the implanted layer remains polycrystalline. At higher energies the recrystallization takes place and the implanted ions move into the substitutional lattice sites. The peak carrier concentration in laser annealed implanted GaAs layers could be as high as 2×10^{19} cm^{-3}, which is substantially higher than what can be obtained with the thermal annealing [163]. Similar peak electron concentrations have been achieved with the electron beam annealing [173].

The electron mobility in the implanted layers depends on the peak carrier concentration and varies from about 3500 cm^2/V s for the doping densities of the order of 10^{17} cm^{-3}, to 2000 cm^2/V s for 10^{18} cm^{-3} and 300 cm^2/V s for 10^{19} cm^{-3}.

The properties of n-type dopants in GaAs are summarized in Table 3-11-1 [149]. The data in the table are for the conventional thermal annealing.

Typical p-type implanted dopants in GaAs are Zn, Be, Cd, Mg, and C [167, 175–177]. For low doses (up to 10^{14} cm^{-2}) the doping efficiency is close to 100%. At larger doses the doping efficiency drops owing to the solubility limit. The effect of the anneal temperature on the doping efficiency is shown in Fig. 3-11-4. Using the pulsed laser beam annealing peak hole concentrations of $1–3 \times 10^{19}$ cm^{-3} have been achieved with hole mobilities from 40 to 80 cm^2/V s. The properties of the p-type dopants in GaAs are summarized in Table 3-11-2 [149].

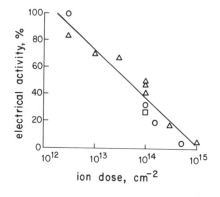

FIGURE 3-11-3. Electric activity vs. dose for 400-keV Se$^+$ implant into Cr doped semi-insulating GaAs [149]. Implant temperature 350°C. Anneal temperature 900°C. Anneal time from 10 to 15 min. △, [155]; □, [188]; ○, [189].

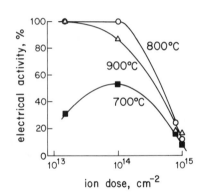

FIGURE 3-11-4. Electrical activity vs. dose for different anneal temperatures [167]. Cd^+ ions implanted at room temperature. Implant energy 60 keV. SiO_2 encapsulation.

The ion implantation changes the material stoichiometry. When implanted electrically active donors occupy As sites the number of vacant As sites is reduced. The product of arsenic and gallium vacancy concentrations is constant under the thermal equilibrium. Therefore, the number of Ga vacancies increases [178]. An additional implantation of Ga atoms has been used to reduce the number of Ga vacancies [179, 180] (dual ion implantation). Different combinations have been tried [149] and some enhancement of the doping efficiency compared to the single implants has been observed.

The quality and reproducibility of GaAs ion-implanted layers strongly depend on the quality of GaAs substrates used for the ion implantation. (A good review of the properties of GaAs semi-insulating substrates is given in Refs. 3 and 181.) The effects of the substrates are especially important in low dose implants used for active layers of GaAs FETs (with a total dose less than 10^{12} cm^{-2}). The possible complications due to poor substrate quality include the formation of a surface conducting layer in the nonimplanted regions of the semi-insulating substrate during the anneal process (this effect is called surface conversion). Also, the ion-implanted profile may vary across the wafer, and extended "tails" in the doping profile (going as far as $1\,\mu m$ from the surface) may appear. An indication of poor substrate quality may be an apparent activation energy of implanted species greater than 100% [150]. Careful substrate qualification tests help to eliminate these problems, which may be related to the diffusion of Cr in Cr-doped semi-insulating substrates. Good results have also been reported for implantation into semi-insulating epitaxial buffer layers and for implantation into undoped semi-insulating substrates grown by the liquid-encapsulated Czochralski (LEC) technique [150].

Semi-insulating GaAs may be produced by ion implantation due to the ion induced damage. Proton bombardment has been used to isolate the IMPATT structures [182] and GaAs-AlGaAs lasers [183], to form simple optical waveguide structures [183, 184], and to make GaAs ICs using epitaxial material [185]. Oxygen implantation (followed by annealing at 650-800°C) has also been used to produce semi-insulating GaAs layers [186-189].

A variety of different devices from GaAs and other compound semiconductors including FETs [190-193], integrated circuits [194-201], varactor diodes [202], IMPATTs [203-204], solar cells [205], and bipolar junction transistors [206] have been produced using ion implantation. Some of the examples of this technology will be considered in the corresponding sections of the book.

TABLE 3-11-1. Characteristics of n-Type Dopants in GaAs [149]

Ion	Encapsulant	Substrate temperature (°C)	Energy (keV)	Dose (cm⁻²)	Anneal temperature (°C)	Activity A (%)	Comments
Tellurium	Si₃N₄	150–350	90–400	$2 \times 10^{12} - 1 \times 10^{13}$	700–950	$50 \geq A \geq 0.3$	Peak of carrier concentration profile in good agreement with theory (LSS) for (i) low-dose implants ($< 1 \times 10^{14}$ cm⁻²), (ii) high anneal (>800°C). Considerable diffusion takes place for high-dose implants. Enhanced electrical activity for hot implants (>200°C).
	AlN	350	220–400	$1 \times 10^{14} - 3 \times 10^{14}$	900–950	$32 \geq A \geq 13$	
	SiO₂	150	100–220	$1 \times 10^{13} - 1 \times 10^{14}$	750	6–7	
	Al	200	90–600	$1 \times 10^{13} - 1 \times 10^{13}$	700	$19 \geq A \geq 0.4$	
Selenium	Si₃N₄	200–400	100–400	$3 \times 10^{12} - 1 \times 10^{15}$	700–900	$100 \geq A \geq 1$	Approximately same properties as tellurium implantation.
	AlN	350	400	$3 \times 10^{13} - 2 \times 10^{15}$	800–1000	$70 \geq A \geq 1$	
	SiO₂	RTᵃ	50	$2 \times 10^{12} - 1 \times 10^{14}$	800–850	$23 \geq A \geq 6$	
	SiO₂	200–400	50	$2 \times 10^{12} - 1 \times 10^{14}$	800–850	$40 \geq A \geq 7$	
	Al	200	210–500	$3 \times 10^{12} - 1 \times 10^{15}$	700–830	$85 \geq A \geq 1$	
	Ga-rich SiO₂	RT	1000	1×10^{14}	825	9	
	Ga-rich SiO₂	280	1000	1×10^{14}	825	29	
	As vapor overpressure	350	400	1.8×10^{12}	850	70	

Element	Target						Comments
Silicon	Si_3N_4	RT	50–400	1×10^{12}–2.5×10^{15}	600–900	$88 \gg A \geq 0.2$	Same properties as for tellurium and selenium except no significant improvement in electrical activation when implanted at elevated temperatures.
		200–340	50–400	1×10^{13}–2.5×10^{15}	600–900	$50 \gg A \geq 0.6$	
	SiO_2	RT	70	1×10^{13}	700	~ 30	
	As-rich SiO_2	RT	1000	1×10^{14}	825	19	
	As-rich SiO_2	235	1000	1×10^{14}	825	22	
	As vapor overpressure	RT	120–130	5.6×10^{12}–2.3×10^{13}	700–850	$89 \gg A \gg 52$	
Sulfur	Si_3N_4	RT	400	2×10^{14}–2.5×10^{13}	700	$1 \gg A \geq 0.4$	Considerable diffusion takes place during annealing. No significant improvement in electrical activity for low-dose implants at elevated temperatures, although enhancement observed at higher doses.
	Si_3N_4	200	400	2×10^{14}–2.5×10^{13}	700–900	$85 \gg A \geq 0.3$	
	SiO_2	RT	70–350	2×10^{12}–1×10^{14}	600–900	$100 \gg A \geq 0.3$	
	Ga-rich SiO_2	RT	1000	1×10^{14}	825	16	
	Ga-rich SiO_2	360	1000	1×10^{14}	825	40	
	As vapor overpressure	RT	110–150	1×10^{13}–1×10^{15}	800–950	$82 \gg A \geq 1$	
Tin	Si_3N_4	200	400	1×10^{14}–2.5×10^{15}	700	$3.5 \gg A \geq 0.6$	For high-dose implants carrier concentration profiles deeper than predicted by theory. Increased electrical activity obtained for hot implants.
	Al	200	90–300	1×10^{13}–1×10^{15}	700–750	$17 \gg A \geq 1.5$	

[a] RT—Room temperature.

TABLE 3-11-2. Characteristics of p-Type Dopants in GaAs [149]

Ion	Encapsulant	Substrate temperature (°C)	Energy (keV)	Dose (cm^{-2})	Anneal temperature (°C)	Electrical activity A (%)	Comments
Zinc	SiO$_2$	RT^a	20–85	1×10^{12}–1×10^{16}	600–800	$100 \geqslant A \geqslant 6$	Low-dose implants ($\leqslant 10^{14}$ cm^{-2}) annealed at ~800°C produce high electrical activities (~100%). No significant improvement in activity by implanting at elevated temperatures for low-medium doses. For high doses considerable diffusion takes place during high-temperature annealing ($\geqslant$800°C).
	SiO$_2$	400	20–60	6×10^{13}–1×10^{16}	800–900	$100 \geqslant A \geqslant 18$	
	Aluminum	RT	60–450	$\leqslant 10^{15}$	700–800	$100 \geqslant A \geqslant 17$	
	none	400	20		600	p-type	
Cadmium	Si$_2$N$_4$	RT	135	1×10^{13}–1×10^{16}	800–900	$100 \geqslant A \geqslant 0.4$	Basically same properties as for zinc.
	SiO$_2$	RT	60	1.5×10^{13}–1×10^{15}	700–900	$100 \geqslant A \geqslant 8$	
	SiO$_2$	400	20	1×10^{16}	900	6	
Beryllium	Si$_3$N$_4$	RT	100–400	1.3×10^{14}–1.3×10^{14}	900	$100 \geqslant A \geqslant 10$	High electrical activation obtained for low-medium doses annealed at $\geqslant$600°C. Considerable diffusion occurs when annealing high-doses implants at temperatures $\geqslant$800°C. No significant improvement in electrical activity by implanting at elevated temperatures.
	SiO$_2$	RT	40–350	1×10^{13}–1×10^{16}	500–900	$100 \geqslant A \geqslant 12$	
Magnesium	Si$_3$N$_4$	RT	120	1×10^{13}–1×10^{13}	700–900	$85 \geqslant A \geqslant 6$	Basically same properties as for zinc and cadmium.
	SiO$_2$	RT	45–60	5×10^{13}–1×10^{15}	800	6	
Carbon	SiO$_2$	RT	10–50	4×10^{13}–4×10^{14}	600	p-type	Very little work done on carbon implantation. For anneal temperatures >600°C p-type activation observed.

^{a} RT—Room temperature.

REFERENCES

1. D. T. J. Hurle, *Current Growth Techniques in Crystal Growth, A Tutorial Approach*, North-Holland, Amsterdam, 1979.
2. F. P. Kesamanly and D. N. Nasledov, Eds., *GaAs Growth, Properties, and Applications*, Nauka, Moscow, 1973 (in Russian).
3. L. F. Eastman, Semi-insulating GaAs substrates for integrated circuit devices: promises and problems, *J. Vac. Sci. Technol*, **16**, (6), 2050 (1979).
4. C. O. Bozler, J. P. Donnelly, W. T. Lindley, and R. A. Reynolds, *Appl. Phys. Lett.* **29**, 698 (1976).
5. J. Van den Boomgaard and K. Schol, *Phil. Res. Rep.* **12**, 127 (1957).
6. M. B. Panish, I. Hayashi, and S. Sumski, A technique for the preparation of low-threshold room-temperature GaAs laser diode structures, *IEEE J. Quantum Electron.* **QE-5**, 210-211 (1969).
7. M. Nelson, Epitaxial growth from the liquid state and its applications to the fabrication of tunnel and laser diodes, *RCA Rev.* **24**, 603-615 (1963).
8. H. Rupprecht, New aspects of solution regrowth in the device technology of gallium arsenide, *Proc. 1966 Symp. on GaAs*, Inst. of Physics and Physics Society of London, 1967, pp. 57-62.
9. G. B. Stringfellow, *LPE of III/V Semiconductors in Crystal Growth, A Tutorial Approach*, North-Holland, Amsterdam, 1979.
10. T. N. Bhar and R. Dat, The effect of baking on the quality of GaAs LPE layers, *Solid State Electron.* **22**, 743-744 (1979).
11. H. F. Lockwood and M. Ettenberg, Thin solution multiple layer epitaxy, *J. Cryst. Growth* **15**, 81-83 (1972).
12. J. J. Hsieh, Thickness and surface morphology of GaAs LPE layers grown by supercooling, step-cooling, equilibrium cooling, and two phase solution technology, *J. Cryst. Growth* **27**, 49-61 (1974).
13. I. Crossley and M. B. Small, Computer simulations of liquid phase epitaxy of GaAs in Ga solution, *J. Crystal Growth* **11**, 157-165 (1971).
14. F. E. Rosztoczy, S. I. Long, and J. Kinoshita, *J. Crystal Growth* **32**, 95-100 (1976).
15. H. M. Manaserit and W. I. Simpson, *J. Electrochem. Soc.* **116**, 1725 (1969).
16. H. Beneking, A. Escobosa, and H. Krautle, presented at the Electron Material Conference, Cornell University, Ithaca, New York, 1980.
17. C. C. Wang and S. M. MacFarlane III, *Thin Solid Films* **31**, 3 (1976).
18. P. Balk and E. Venhoff, Deposition of III-V compounds by MO-CVD and in halogen transport systems, A critical comparison, *J. Crystal Growth* **55**, (1), 35-41 (1981).
19. H. M. Manasevit, Recollections and reflection of MO-CVD, *J. Crystal Growth* **55**, (1), 1-9 (1981).
20. C. Y. Chang, Y. K. Su, M. K. Lee, L. G. Chen, and M. P. Houng, Characterization of GaAs epitaxial layers by low pressure MO VPE using TEG as Ga source, *J. Crystal Growth* **55**, (1), 24-29 (1981).
21. G. B. Stringfellow, *J. Appl. Phys.* **50**, 4178 (1979).
22. T. J. Drummond, H. Morkoc, and A. Y. Cho, Dependence of electron mobility on spatial separation of electrons and donors in $Al_xGa_{1-x}As/GaAs$ heterostructures, *J. Appl. Phys.* **52**(3), 1380-1386 (1981).
23. M. G. Panish and A. Y. Cho, Molecular beam epitaxy, *Spectrum* **17**(4), 18 (1980).
24. R. Stall, C. E. C. Wood, K. Board, and L. F. Eastman, Ultra low resistance ohmic contacts to n-GaAs, *Electron. Lett.* **15**, 800-801 (1979).
25. C. E. C. Wood, Progress, problems, and applications of molecular beam epitaxy, in *Physics of Thin Films*, Ed. by G. Hass and M. Francone, Academic, New York, 1981.
26. A. R. Calawa, On the use of AsH_3 in the molecular beam epitaxial growth of GaAs, *Appl. Phys. Lett.* **38**(9), 701-703 (1981).
27. J. C. Bean, Proc. Int. Electron Device Meeting, Washington, DC, 1981, p. 6.
28. A. Y. Cho and J. R. Arthur, *Progress in Solid State Chemistry*, Ed. by G. Somorjai and J. McCaldin, Pergamon, New York, 1975, Vol. 10, p. 157.
29. A. Y. Cho, Recent developments in molecular beam epitaxy (MBE), *J. Vac. Sci. Technol.* **16**, 275 (1979).
30. J. N. Walpole, A. R. Calawa, S. R. Chinn, S. H. Groves, and T. C. Harman, *Appl. Phys. Lett.* **29**, 307 (1976).
31. K. Ploog, Molecular beam epitaxy of III-V compounds, in *Crystals—Growth, Properties, and Applications*, Ed. by L. F. Boschke, Springer-Verlag, Heidelberg, 1979.
32. C. E. C. Wood, in *Technology and Physics of MBE*, Ed. by E. H. C. Parker and M. G. Dowsett, Plenum, New York, 1982.

33. C. E. C. Wood, III-V Alloy growth by MBE, in *GaInAsP Alloy Semiconductors*, Wiley, New York, 1982.
34. Lester F. Eastman, Use of molecular beam epitaxy in research and development of selected high speed compound semiconductor devices, *J. Vac. Sci. Technol. B* **1**(2), 131 (1983).
35. C. E. C. Wood, MBE doping processes, A review of current understanding, in proceedings of MBE-CST-2 (Second International Symp. on MBE Related Clear Surface Technol.) 27-30 August 1982, Tokyo, Japan.
36. D. L. Smith and V. Y. Pickardt, *J. Appl. Phys.* **46**, 2366 (1975).
37. T. Yao, S. Amano, Y. Makwa, and S. Maekawa, *Jpn. J. Appl. Phys.* **16**, 369 (1977).
38. H. Holloway and J. N. Walpole, MBE techniques for II-VI optoelectronic devices, in *Molecular Beam Epitaxy*, Ed. by B. R. Pamplin, Pergamon, Oxford, England, 1980, pp. 49-94.
39. J. P. Faurie and A. Million, *J. Crystal Growth* **54**, 577 (1981) and **54**, 582 (1981).
40. G. D. Holah, E. L. Meeks, and F. L. Eisele, Molecular beam epitaxial growth of InGaAsP, *J. Vac. Sci. Technol. B.* **1**(2), 182-183 (1983).
41. D. F. Welch, G. W. Wicks, D. W. Woodward, and L. F. Eastman, GaInAs-AlInAs heterostructures for optical devices grown by MBE, *J. Vac. Sci. Technol. B* **1**(2), 202-204 (1983).
42. U. Mishra, E. Kohn, N. J. Kawai, and L. F. Eastman, Permeable base transistor—A new technology, *IEEE Trans Electron Devices.* **ED-29**(10), 1707, Oct. (1982).
43. J. S. Hammis and J. M. Woodcock, *Electron. Lett.* **16**, 319 (1980).
44. A. Y. Cho, H. C. Casey, and P. W. Foy, *Appl. Phys. Lett.* **30**, 397 (1977).
45. P. M. Petroff, Weisbush, *et al.*, Proc. 2nd Int. MBE Workshop, Cornell Univ. 1980.
46. W. T. Tsang and J. A. Ditzenberger, A visible (AlGa)As heterostructure laser grown by molecular beam epitaxy, *Appl. Phys. Lett.* **39**, 193 (1981).
47. W. I. Wang, S. Judaprawira, C. E. C. Wood, and L. F. Eastman, *Appl. Phys. Lett.* **38**, 708 (1981).
48. A. Y. Cho, J. V. DiLorenzo, B. S. Hewitt, W. C. Niehaus, W. Schlosser, and C. Radice, *J. Appl. Phys.* **48**, 336 (1977).
49. S. G. Bandy, D. M. Collins, and C. K. Nishimoto, *Electron. Lett.* **15**, 218 (1979).
50. C. E. C. Wood, D. DeSimone, and S. Judaprawira, *J. Appl. Phys.* **51**, 2074 (1980).
51. R. E. Thorne, S. L. Su, R. J. Fisher, W. F. Kopp, W. G. Lyons, P. A. Miller, and H. Morkoc, Analysis of Camel gate FETs (CAMFETs), *IEEE Trans. Electron Devices.* **ED-30**(3), 212-217 (1983).
52. S. Hiyamizu, T. Minura, and T. Ishikawa, *Jpn. J. Appl. Phys.* **21**, Suppl. 21-1, 161 (1982).
53. H. Morkoc, *Modulation Doped Al_xGa_{1-x}As/GaAs Field Effect Transistors (MODFETs): Analysis, Fabrication, and Performance*, Nijhoff, The Hague, 1983.
54. W. V. McLevice, H. T. Yuan, W. M. Duncan, W. R. Frensley, F. H. Doerbeck, and H. Morkoc, *IEEE Electron Device Lett.* **EDL-3**(2), 43-45 (1982).
55. R. J. Malik, T. R. AuCoin, R. L. Ross, K. Board, C. E. C. Wood, and L. F. Eastman, *Electron. Lett.* **16**, 836 (1980).
56. R. J. Malik, K. Board, L. F. Eastman, T. R. AuCoin, and R. Ross, *Inst. Phys. Conf. Ser.* **56**, 697 (1981).
57. W. Haydl, R. Smith, and R. Bosch, *Appl. Phys. Lett.* **37**, 556 (1980).
58. S. Pham, N. Tung, P. Delecluse, D. Delagebeaudeuf, M. Laviron, J. Chaplart, and N. T. Linh, High speed low power DCFL using planar two-dimensional electron gas FET technology, *Electron. Lett.* **18**(12), 517-518 (1982).
59. D. L. Miller and J. S. Harris, *Appl. Phys. Lett.* **37**, 1104 (1980).
60. Ivor Vrodie and Julius J. Muray, *The Physics of Microfabrication*, Plenum, New York, 1982.
61. A. H. Agajanian, *Semiconducting Devices, A Bibliography of Fabrication Technology, Properties, and Applications*, Plenum, New York, 1976.
62. Method F100-72, Standard Test Method for Thickness of Epitaxial or Diffused Layers in Silicon by the Angle Lapping and Staining Technique, American Society for Testing and Material Standards, 1976.
63. Method F 95-76, Standard Test Method for Thickness of Epitaxial or Diffused Layers in Silicon on Substrates of the Same Type by Infrared Reflection, American Society for Testing and Material Standards, 1976.
64. J. D. Wiley, *C-V* profiling of GaAs FET films, *IEEE Trans. Electron Devices* **ED-25**, 1317-1324 (1978).
65. L. J. van der Pauw, *Philips Res. Rep.* **13**, 1 (1958).
66. L. J. van der Pauw, *Philips Tech. Rep.* **20**, 220 (1958).
67. H. H. Wieder, Electrical and galvanomagnetic measurements on thin films and epilayers, *Thin Solid Films* **31**, 123-138 (1976).
68. S. M. Sze, *Physics of Semiconductor Devices*, Wiley, New York, 1981.

69. D. V. Lang, *J. Appl. Phys.* **45**(7), 3023 (1974).

70. J. S. Blakemore, *Semiconductor Statistics*, Pergamon, New York, 1962, p. 277.

71. E. H. Rhoderick, Transport properties in Schottky diodes, in *Inst. Phys. Conf. Ser., No. 22*, Ed. by K. M. Pepper, Institute of Physics, Manchester, England, 1974, p. 3.

72. E. E. Wagner, D. Hiller, and D. E. Mars, *Rev. Sci. Instrum.* **51**(9), 1205 (1980).

73. G. L. Miller, D. V. Lang, and L. C. Kimerling, *Ann. Rev. Mater. Sci.* **7**, 377–448 (1977).

74. R. S. Muller and T. I. Kamins, *Device Electronics for IC's*, Wiley, New York, 1977, p. 316.

75. K. Yamasaki, M. Yoshida, and T. Sugano, *Jpn. J. Appl. Phys.* **18**(1), 113 (1979).

76. T. R. Ohnstein, Ph.D. thesis., University of Minnesota, 1982.

77. P. D. Kirchner, W. J. Schaff, G. N. Maracas, and L. F. Eastman, The analysis of exponential and non-exponential transients in deep level transient spectroscopy, *J. Appl. Phys.* **52**(11), 6462–6470 (1981).

78. D. V. Lang, A. Y. Cho, A. C. Gossard, M. Ilegems, and W. Wiegman, Study of electron traps in *n*-GaAs grown by molecular beam epitaxy, *J. Appl. Phys.* **47**(6), 2558–2564 (1976).

79. R. E. Honig, Surface and thin film analysis of semiconductor materials, *Thin Solid Films* **31**, 89–122 (1976).

80. D. C. Northrop and E. H. Rhoderick, The physics of Schottky barriers, in *Variable Impedance Devices*, Ed. by M. J. Howes and D. V. Morgan, Wiley, New York, 1978.

81. B. L. Smith and E. H. Rhoderick, *Solid State Electron.* **14**, 71 (1971).

82. G. B. Seiranyan and Y. A. Thorik, *Phys. Status Solidi* **A13**, K115 (1972).

83. C. R. Crowell and S. M. Sze, *Solid State Electron.* **9**, 1035 (1966).

84. A. Padovani and R. Stratton, *Solid State Electron.* **9**, 695 (1966).

85. C. R. Crowell and V. L. Rideout, *Solid State Electron.* **12**, 89 (1969).

86. L. S. Weinman, S. A. Jamison, and M. J. Helix, Sputtered TiW/Au Schottky barriers on GaAs, *J. Vac. Sci. Technol.* **18**(3), 838–840 (1981).

87. E. H. Rhoderick, *Metal–Semiconductor Contacts*, Clarendon Press, Oxford, 1978.

88. B. L. Sharma and S. C. Gupta, Metal-Semiconductor barrier junctions, *Solid State Technol.* **23**, 90–95 (1980).

89. G. Y. Robinson, Schottky diodes and ohmic contacts for the III-V semiconductors, in *Physics and Chemistry of III-V Semiconductor Interfaces*, Ed. by C. W. Wilmsen, Plenum, New York, 1983.

90. N. Braslau, Alloyed Ohmic contacts to GaAs, *J. Vac. Sci. Technol.* **19**(3), 803 (1981).

91. L. L. Chang and G. L. Pearson, The solubilities and distribution coefficients of Zn in GaAs and GaP, *Phys. Chem. Solids* **25**, 23–30 (1964).

92. R. Stall, C. E. C. Wood, K. Board, N. Dandekar, L. F. Eastman, and J. Devlin, A study of Ge/GaAs interface grown by molecular beam epitaxy, *J. Appl. Phys* **52**, 4062–4069 (1981).

93. J. M. Woodall, J. L. Freeouf, G. D. Pettit, T. Jackson, and P. Kirshner, Ohmic contacts to *n*-type GaAs using graded band gap layers of $Ga_x In_{1-x} As$ grown by molecular beam epitaxy, *J. Vac. Sci. Technol.* **19**, 626–627 (1981).

94. N. Braslau, J. B. Gunn, and J. L. Staples, Metal-semiconductor contacts for GaAs bulk effect devices, *Solid-State Electron.* **10**, 381–383 (1967).

95. R. H. Cox and H. Strack, *Solid-State Electron.* **10**, 1213 (1967).

96. D. C. Miller, *J. Electrochem. Soc.* **127**, 467 (1980).

97. A. Christou, *Solid-State Electron.* **22**, 141 (1979).

98. C. P. Lee, B. M. Welch, and W. P. Fleming. Reliability of AuGe/Pt and AuGe/Ni ohmic contacts on GaAs, *Electron. Lett.* **12**, 406–407 (1981).

99. M. Heiblum, M. I. Nathan, and C. A. Chang, *Sol. State Electron.* **25**, 185 (1982).

100. R. P. Gupta and J. Freyer, *Int. J. Electron.* **47**, 459 (1979).

101. K. Heime, U. Konig, E. Kohn, and A. Wortmann, *Sol. State Electron.* **17**, 835 (1974).

102. F. Vidimari, *Electron. Lett.* **15**, 675 (1979).

103. N. Yokoyama, S. Ohkawa, and H. Ishikawa, *Jpn. J. Appl. Phys.* **14**, 1071 (1975).

104. G. Eckhardt, *Laser and Electron Beam Processing of Materials*, Ed. by C. W. White and P. S. Peercy, Academic, New York, 1980, p. 467.

105. G. Badertscher, R. P. Salathe, and W. Luthy, *Electron. Lett.* **16**, 113 (1980).

106. J. G. Werthen and D. R. Scifres, *J. Appl. Phys.* **52**, 1127 (1981).

107. R. L. Mozzi, W. Fabian, and I. J. Piekarski, *Appl. Phys. Lett.* **35**, 337 (1979).

108. Y. I. Nissim, J. F. Gibbons, and R. B. Gold, *IEEE Trans. Electron Devices* **ED-28**, 607 (1981).

109. K. Klohn and Z. Wandinger, *J. Electrochem. Soc.* **116**, 507 (1969).

110. H. Matino and M. Tokunaga, *J. Electrochem. Soc.* **116**, 709 (1969).

111. Y. A. Goldberg and B. V. Tsarenkev, *Soc. Phys. Semicond.* **3**, 551 (1970).

112. W. D. Edwards, W. A. Hartman, and A. B. Torrens, *Solid-State Electron.* **15**, 387 (1972).
113. S. Asai *et al.*, Proc. 5th Conf. on Solid State Devices, Tokyo, p. 442, 1973.
114. G. Y. Robinson, *Solid State Electron* **18**, 331 (1975).
115. H. R. Grinolds and G. Y. Robinson, *Solid-State Electron.* **23**, 973 (1980).
116. A. H. Oraby, K. Murakami, Y. Yuba, K. Gamo, S. Namba, and Y. Masuda, *Appl. Phys. Lett.* **38**, 562 (1981).
117. M. Ogawa, K. Ohata, T. Furutsuka, and N. Kawamura, *IEEE Trans. Micr. Theory Tech*, **MTT-24**, 300 (1976).
118. R. B. Gold, R. A. Powell, and J. F. Gibbons, *Laser–Solid Intersections and Laser Processes*, AIP Conf. Proc. No. 50, New York, 1978, p. 635.
119. S. Margalit, D. Febete, D. M. Pepper, G. P. Leed, and A. Yariv, *Appl. Phys. Lett.* **33**, 346 (1978).
120. T. Inada, S. Kato, T. Hara, and N. Toyada, *J. Appl. Phys.* **50**, 4466 (1979).
121. W. T. Anderson, Jr., A. Christou, and J. F. Giuliani, *IEEE Electron Device Lett.* **EDL-2**, 115 (1981).
122. W. Tseng, A. Christou, H. Day, J. Davey, and B. Wilkins, *J. Vac. Sci. Technol.* **19**, 623 (1981).
123. E. Kuphal, *Solid-State Electron.* **24**, 69 (1981).
124. H. Temkin, R. J. McCoy, V. G. Keramidas, and W. A. Bonner, *Appl. Phys. Lett.* **36**, 444 (1980).
125. S. H. Wemple and W. C. Niehaus, *Inst. Phys. Conf. Ser.* **33b**, 262 (1977).
126. K. Ohata, T. Nozaki, and N. Kawamura, IEEE Trans. Electron Devices, **ED-24**, 1129 (1978).
127. M. Ogawa, *J. Appl. Phys.* **51**, 406 (1980).
128. K. Ohata and M. Ogawa, Proc. 12th Annual Reliability Physics Symposium, IEEE, New York, 1974, p. 278.
129. M. Yoder, *Solid State Electron.* **23**, 117 (1980).
130. M. S. Shur, Resistance of alloyed ohmic contacts to GaAs, unpublished.
131. N. Braslau, presented at GaAs Research Seminar, Department of Electrical Engineering, University of Minnesota, 1981, unpublished.
132. H. J. Gohen and A. Y. C. Yu, Ohmic contacts to epitaxial *p*-GaAs, *Solid-State Electron.* **14**, 515–517 (1971).
133. T. Sanada and O. Wada, Ohmic Contacts to *p*-GaAs with Au/Zn/Au structure, *Jpn. J. Appl. Phys.* **19**, L491–L494 (1980)
134. H. Matino and M. Tokunaga, Contact resistance of several metals and alloys to GaAs, *J. Electrochem. Soc.* **116**, 709–711 (1979).
135. V. L. Rediout, A review of the theory and technology for ohmic contacts to group III–V compound semiconductors, *Solid-State Electron.* **18**, 541–550 (1975).
136. Y. Nakato, S. Takahashi, and Y. Toyoshima, Contact resistance dependence on InGaAsP layers lattice matched to InP, *Jpn. J. Appl. Phys.* **19**, L495–L497 (1980).
137. W. Kellner, Planar ohmic contacts to *n*-type GaAs: Determination of contact parameters using the transmission line model, *Siemens Forsch. Entwickl-Ber* **4**, 137 (1975).
138. I. F. Chang, Contact resistance in diffused resistors, *J. Electrochem. Soc.* **117**, 368 (1970).
139. G. K. Reeves and H. B. Harrison, Obtaining the specific contact resistance from transmission line model measurements, *IEEE Electron Device Lett.* **EDL-3**(5), 111–113 (1982).
140. W. Schockley, Research and investigation of inverse epitaxial UHF power transistors, Report No. Al-TOR-64-207, Air Force Atomic Laboratory, Wright-Patterson Air Force Base, Ohio, September 1964.
141. H. Murrmann and D. Widman, Current crowding on metal contacts to planar devices, *IEEE Trans. Electron Devices* **ED-16**, 1022–1024 (1969).
142. H. H. Berger, Contact resistance on diffused resistors, IEEE ISSCC Digest of Tech. Papers, pp. 160–161, 1969.
143. H. H. Berger, Contact resistance and contact resistivity, *J. Electrochem. Soc.* **119**, 509 (1972).
144. H. H. Berger, Models for contacts to planar devices, *Solid State Electron.* **15**, 145 (1972).
145. R. D. Brooks and H. G. Mathes, Spreading resistance between constant potential surface, *Bell System Tech. J.* **50**, 775–784 (1971).
146. L. E. Terry and R. W. Wilson, Metallization systems for Si integrated circuits, *Proc. IEEE* **57**, 1580–1586 (1969).
147. E. Kuphal, Low resistance ohmic contacts to *n*- and *p*-InP, *Solid-State Electron* **24**, 69–78 (1981).
148. Y. K. Fang, C. Y. Chang, and Y. K. Su, Contact resistance in metal–semiconductor systems, *Solid-State Electron.* **22**, 933–938 (1979).
149. D. V. Morgan, F. H. Eisen, and A. Ezis, Prospects for ion bombardment and ion implantation in GaAs and Inp device fabrication, *IEE Proc.* **128**(4), 109–130 (1981).

150. R. C. Eden and B. M. Welsh, GaAs digital integrated circuits for ultra high speed LSI/VLSI, in *Very Large Scale Integration (VLSI): Fundamentals and applications*, Ed. by D. F. Barbe, Springer-Verlag, Berlin, 1980, pp. 128-177.

151. F. Eisen, C. Kirpartrick, and P. Asbeck, Implantation into GaAs, in *GaAs FET Principles and Technology*, Ed. by J. V. DiLorenzo and D. D. Khandelwal, Artech House, Dedham, Massachusetts, 1982, pp. 117-146.

152. J. F. Gibbons, W. S. Johnson, and S. W. Mylroie, *Projected Range Statistics. Semiconductors and Related Materials*, 2nd Edition, Wiley, New York (1975).

153. J. Lindhard, M. Scharff, and M. E. Schiot, Range concepts and heavy ion ranges, *K. Dan. Vidensk. Seisk. Mat.-Fys. Medd.* **33**(14), 1-42 (1963).

154. J. Lindhard, Influence of crystal lattice on motion of energetic charged particles, *K. Dan. Vidensk. Seisk. Mat.-Fys. Medd.* **34**(14), 64 (1965).

155. J. P. Donnelly, W. T. Lindley, and C. E. Hurwitz, Silicon and selenium ion implanted GaAs reproducibly annealed at temperatures up to 950°C, *Appl. Phys. Lett.* **27**, 41-43 (1975).

156. F. H. Eisen, B. M. Welsh, H. Muller, K. Camo, T. Inado, and J. W. Mayer, Tellurium implantation in GaAs, *Solid-State Electron.* **20**, 219-223 (1977).

157. R. D. Pashify and B. M. Welsh, Tellurium implantation in GaAs, *Solid-State Electron.* **18**, 977-981 (1975).

158. A. A. Immorlica and F. H. Eisen, Capless annealing of ion-implanted GaAs, *Appl. Phys. Lett.* **29**, 94-95 (1976).

159. R. M. Malbon, D. H. Lee, and J. M. Whelan, Annealing of ion-implanted GaAs in a controlled atmosphere, *J. Electrochem. Soc.* **123**, 1413-1415 (1976).

160. J. Kasahara, M. Arai, and N. Watanabe, Capless anneal of ion-implanted GaAs in controlled arsenic vapor, *J. Appl. Phys.* **50**, 541-543 (1979).

161. M. Kuzuhara, H. Kohzu, and Y. Takayama, Infrared rapid thermal annealing of Si-implanted GaAs, *Appl. Phys. Lett,* **41**(8), 755 (1982).

162. B. J. Sealy, S. S. Kular, K. G. Stephens, R. Croft, and A. Palmer, Electrical properties of laser-annealed donor-implanted GaAs, *Electron. Lett.* **14**(22), 720-721 (1978).

163. S. S. Kular, B. J. Sealy, K. G. Stephens, D. R. Chick, Q. V. Davis, and J. Edwards, Pulsed laser-annealing of zinc implanted GaAs, *Electron. Lett.* **14**(4), 85-87 (1978).

164. B. K. Sinh and Y. S. Park, An abrupt dopant-profile in GaAs produced by Te implantation, *J. Electrochem. Soc.* **123**, 1588-1589 (1976).

165. J. P. Donnelly, Ion implantation in GaAs, *Inst. Phys. Conf. Ser.* **33b**, 166-190 (1977).

166. J. M. Woodcock, J. M. Shannon, and D. J. Clark, Electrical and cathodoluminescence measurements on ion implanted donor layers in GaAs, *Solid-State Electron.* **18**, 267-275 (1975).

167. R. G. Hunsprerger and O. J. Marsh, Electrical properties of Cd, Zn, and S ion-implanted layers in GaAs, *Radiat. Eff.* **6**, 263-268 (1970).

168. J. A. Golovchenko and T. N. C. Venkatesan, Annealing of Te-implanted GaAs by ruby laser irradiation, *Appl. Phys. Lett.* **32**, 147-149 (1978).

169. S. U. Campisano, I. Catalano, G. Foti, E. Rimini, F. Eisen, and M. A. Nicolet, Laser reordering of implanted amorphous layers in GaAs, *Solid-State Electron.* **21**, 485-488 (1978).

170. E. Rimini, P. Baeri, and G. Foti, Laser pulse energy dependence of annealing in ion implanted Si and GaAs semiconductors, *Phys. Lett. A* **65**, 153-155 (1978).

171. J. L. Tandon, M. A. Nicolet, W. F. Tseng, F. H. Eisen, S. U. Campisano, G. Foti, and E. Rimini, Pulsed laser annealing of implanted layers in GaAs, *Appl. Phys. Lett.* **34**, 597-599 (1979).

172. S. U. Campisano, G. Foti, E. Rimini, F. Eisen, W. F. Tseng, M. A. Nicolet, and J. L. Tandon, Laser pulse annealing of ion-implanted GaAs, *J. Appl. Phys.* **51**, 295-298 (1980).

173. I. Inada, K. Tokunaga, and S. Taka, Pulsed electron-beam annealing of selenium implanted gallium arsenide, *Appl. Phys. Lett.* **35**, 546-548 (1979).

174. R. L. Mozzi, W. Fabian, and F. J. Piekarski, Non-alloyed ohmic contacts to *n*-GaAs by pulse-electron-beam-annealed selenium implants, *Appl. Phys. Lett.* **35**, 337-339 (1979).

175. R. G. Hunsperger, R. G. Wilson, and D. M. Jamba, Mg and Be ion implanted GaAs, *J. Appl. Phys.* **43**, 1318-1320 (1972).

176. M. A. Littlejohn, J. R. Hauser, and L. K. Monteith, The electrical properties of 60 keV zinc ions implanted into semi-insulating gallium arsenide, *Radiat. Effects* **10**, 185-190 (1971).

177. Y. Yuba, K. Gamo, K. Masuda, and S. Namba, Hall effect measurements of Zn implanted GaAs, *Jpn. J. Appl. Phys.* **13**, 641-644 (1974).

178. R. Heckingbottom and T. Ambridge, Ion implantation in compound semiconductors—An approach based on solid state theory, *Radiat. Effects* **17**, 31-36 (1973).

179. T. Ambridge, R. Heckingbottom, E. C. Bell, B. J. Sealy, K. G. Stephens, and R. K. Surridge, Effect of dual implants into GaAs, *Electron. Lett.* **11**(15), 314–315 (1975).
180. B. J. Sealy, E. C. Bell, R. K. Surridge, K. G. Stephens, T. Ambridge, and R. Heckingbottom, Dual implantation into GaAs, *Inst. Phys. Conf. Ser.* **28**, 75–80 (1976).
181. P. F. Linquist and W. M. Ford, Semi-insulating GaAs Substrates, in *Very Large Scale Integration (VLSI): Fundamentals and Applications*, Ed. by D. F. Barbe, Springer-Verlag, Berlin, 1980, pp. 1–60.
182. J. D. Shpeight, P. Leigh, N. McIntyre, I. G. Grove, S. O'Hara, and P. L. F. Hemment, High efficiency proton-isolated GaAs IMPATT diodes, *Electron. Lett.* **10**(7), 98–99 (1974).
183. J. C. Dyment, L. A. D'Asaro, J. C. North, B. I. Miller, and J. F. Ripper, Proton bombardment formation of stripe geometry heterostructure lasers for 300 K cw operation, *Proc. IEEE* **60**, 726 (1972).
184. H. Stoll, A. Yariv, R. G. Hunsperger, and G. L. Tangonan, Proton implanted optical waveguides detectors in GaAs, *Appl. Phys. Lett.* **23**, 664–665 (1973).
185. D. C. D'Avanzo, Proton isolation for GaAs integrated circuits, *IEEE Trans. Electron Devices* **ED-29**(7), 1051–1059 (1983).
186. P. N. Favennec, Semi-insulating layers of GaAs by oxygen implantation, *J. Appl. Phys.* **47**, 2532–2536 (1976).
187. S. Gecim, B. J. Sealy, and K. G. Stephens, Carrier removal profiles from oxygen implanted GaAs, *Electron. Lett.* **14**(10), 306–308 (1978).
188. K. Gamo, T. Inada, S. Krekfler, J. W. Meyer, F. H. Eisen, and B. M. Welsh, Selenium implantation in GaAs, *Solid-State Electron.* **20**, 213–217 (1977).
189. F. H. Eisen, B. M. Welsh, K. Gamo, T. Inada, H. Mueller, M. A. Nicolet, and J. W. Maeyr, Sulphur, selenium, and tellurium implantation in GaAs, *Inst. Phys. Conf. Ser.* **28**, 64–68 (1976).
190. R. G. Hunsperger and N. Hirsch, *Electron. Lett.* **9**, 1 (1973).
191. K. Ohata, T. Nosaki, and N. Kawamura, *IEEE Trans. Electron Devices* **ED-24**, 1129 (1977).
192. J. A. Higgins, R. A. Kuvas, F. H. Eisen, and D. R. Ch'en, *IEEE Trans. Electron Devices* **ED-25**, 587 (1978).
193. F. H. Dorbeck, H. M. Macksey, G. Brehm, and W. R. Frensley, *Electron. Lett.* **15**, 577 (1979).
194. R. C. Eden and B. M. Welsh, Planar localized fabrication process for fabricating GaAs SDFL, LSI and VLSI circuits, in *GaAs FET Principles and Technology*, Ed. by J. V. DiLorenzo and D. D. Khandelwal, Artech House, Dedham, Massachusetts, 1982, pp. 669–721.
195. K. Lehovec and R. Zuleeg, Direct coupled FET logic (DCFL), in *GaAs FET Principles and Technology*, Ed. by J. V. DiLorenzo and D. D. Khandelwal, Artech House, Dedham, Massachusetts, 1982, pp. 621–668.
196. D. Wilson and D. H. Phillips, Enhancement/depletion GaAs FET logic, in *GaAs FET Principles and Technology*, Ed. by J. V. DiLorenzo and D. D. Khandelwal, Artech House, Dedham, Massachussetts, 1982, pp. 591–597.
197. K. Yamasaki, K. Asai and K. Kurumada, GaAs LSI-directed MESFETs with self-aligned implantation for n^+-layer technology, *IEEE Trans. Electron Devices* **ED-29**, 1772–1777 (1982).
198. K. Yamasaki, Y. Yamane, and K. Kurumada, Below 20 ps/gate operation with GaAs SAINT FETs at room temperature, *Electron. Lett.* **18**(4), 592–593 (1982).
199. R. A. Sadler and L. F. Eastman, High speed logic at 300 K with self-aligned submicrometer-gate GaAs MESFETs, *IEEE Electron Device Lett.* **EDL-4**(7), 215–217 (1983).
200. M. J. Helix, S. A. Jamison, C. Chao, and M. S. Shur, Fan out and speed of GaAs SDFL logic, *IEEE J. Solid State Circuits* **SC-17**(6), 1226–1231 (1982).
201. T. Vu, P. Roberts, R. Nelson, G. Lee, B. Hanzal, K. Lee, D. Lamb, M. Helix, S. Jamison, S. Hanka, J. Brown, and M. S. Shur, A 432-cell GaAs SDFL gate array with on-chip 64-bit RAM, *IEEE Trans. Electron Devices* **ED-31**(2), 144–156 (1984).
202. A. A. Immorlica, Jr. and F. H. Eisen, Planar passivated GaAs hyperabrupt varactor diodes, in Proceedings of Sixth Biennial Cornell IEEE conference, 1977, pp. 151–159.
203. C. O. Bozler, J. P. Donnelly, R. A. Murphy, R. W. Laton, R. W. Sudbury, and W. T. Lindley, High efficiency ion-implanted lo-hi-lo GaAs IMPATT diodes, *Appl. Phys. Lett.* **29**, 123–125 (1976).
204. J. J. Berenz, F. B. Fank, and T. L. Hierl, Ion implanted p-n junction indium phosphide IMPATT diodes, *Electron. Lett.* **14**(21), 683–684 (1978).
205. J. C. C. Fan, R. L. Chapman, J. P. Donnelly, G. W. Turner, and C. O. Bozler, Ion implanted laser annealed GaAs solar cells, *Appl. Phys. Lett.* **34**, 780–782 (1979).
206. H. T. Huan, F. H. Doerbeck, and W. V. Mclevige, Ion implanted GaAs bipolar transistors, *Electron. Lett.* **16**, 637–638 (1980).

4

Ridley–Watkins–Hilsum–Gunn Effect

4-1. INTRODUCTION

In 1963 J. B. Gunn studied the current–voltage characteristics of GaAs and InP devices. He discovered that when the applied electric field

$$F = U/L \qquad (4\text{-}1\text{-}1)$$

(where U is the bias voltage, L is the sample length) was greater than some critical value F_t (~3 kV/cm for GaAs and ~6 kV/cm for InP), spontaneous current oscillations appeared in the circuit [1] (see Fig. 4-1-1). Later Gunn published the results of the detailed experimental study of this effect [2]. Using probe measurements of the potential distribution across the sample he established that a propagating high field domain forms in the sample when $F \geqslant F_t$. It nucleates near the cathode, propagates toward the anode with velocity of the order of 10^5 m/s, and disappears near the anode (see Fig. 4-1-2). Then this process repeats itself. The domain formation leads to a current drop, the domain annihilation results in an increase in the current, and periodic current oscillations exist in the circuit.

In 1963 B. F. Ridley predicted that a domain instability should occur in a semiconductor sample with voltage-controlled negative differential resistance [4]. This could happen if either the carrier velocity or the carrier concentration decreases with the increase of bias. Ridley and Watkins [5] and Hilsum [6] showed that the velocity in n-type GaAs, InP, and some other compound semiconductors should decrease with the electric field if the electric field exceeds a critical value inducing the intervalley electron transitions from a high mobility Γ minimum into higher minima of the conduction band where the effective mass is larger and the scattering rate is higher.

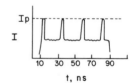

FIGURE 4-1-1. Current waveform for a 2-mm GaAs Gunn diode measured in Ref. 3. The diode is a uniformly n-doped GaAs sample with two ohmic contacts. The dc voltage $U_0 > F_t L$ is applied.

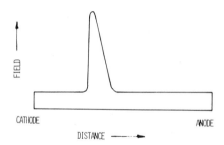

FIGURE 4-1-2. Propagating high field domain.

In 1964 Herbert Kroemer pointed out that all important features of the Gunn effect are consistent with the Ridley–Watkins–Hilsum mechanism [7].

This mechanism can be explained in the frame of a simple two-valley model. When the electric field is low, practically all electrons are in the lowest minimum of the conduction band and the electron drift velocity v is given by

$$v_1 = \mu F \tag{4-1-2}$$

where μ is the low-field mobility and F is the electric field. In the higher electric field electrons are "heated" by the field and some carriets may have enough energy to transfer into upper valleys where the electron velocity

$$v_2 \simeq v_s \tag{4-1-3}$$

Here v_s is the saturation velocity.

The current density is given by

$$j = q n_1(F) v_1 + q n_2(F) v_2 \tag{4-1-4}$$

where n_1 is the electron concentration in the lowest valley and n_2 is the electron concentration in the upper valley:

$$n_1 + n_2 = n_0 \tag{4-1-5}$$

where

$$n_0 = N_D - N_A \tag{4-1-6}$$

N_D is the density of ionized donors, N_A is the density of ionized acceptors. Equation (4-1-4) can be rewritten as

$$j = q n_0 v(F) \tag{4-1-7}$$

where

$$v = \frac{v_1 n_1 + v_2 n_2}{n_0} \tag{4-1-8}$$

or

$$v \simeq \frac{\mu F n_1(F) + v_s n_2(F)}{n_0} \tag{4-1-9}$$

Using the results of a Monte Carlo calculation [see Section (2-5)] one can check that this simplified model is quite adequate for GaAs at 300 K if we assume that the fraction of electrons in the upper valleys $p = n_2/n_0$ is given by

$$p = A\left(\frac{F}{F_s}\right)^t \tag{4-1-10}$$

where $F_{s.} = v_s/\mu$.

Equation (4-1-9) can be then rewritten as

$$v(F) = v_s\left[1 + \frac{F/F_s - 1}{1 + A(F/F_s)^t}\right] \tag{4-1-11}$$

When A, t, and v_s are chosen as the following function of μ [156]:

$$A = 0.6[e^{10(\mu-0.2)} + e^{-35(\mu-0.2)}]^{-1} + 0.01 \tag{4-1-12}$$

$$t = 4[1 + 320/\sin h(40\mu)] \tag{4-1-13}$$

$$v_s = 0.6 + 0.6\mu - 0.2\mu^2 (10^5\,\text{m/s}) \tag{4-1-14}$$

(here μ is in m²/V s), equation (4-1-11) can be used as an interpolation formula for the electron velocity in GaAs (see Fig. 4-1-3). Comparison with the results of the Monte Carlo calculation shows that such an interpolation is accurate within 10%.

From Eq. (4-1-9) we find

$$\frac{dv}{dF} = \mu(1 - p) + (v_s - \mu F)\frac{dp}{dF} \tag{4-1-15}$$

As can be seen from Eq. (4-1-15) when dp/dF exceeds a critical value,

$$\frac{dp}{dF} > \frac{1 - p}{F - F_s} \tag{4-1-16}$$

the differential mobility becomes negative.

As shown below the uniform field distribution becomes unstable when F exceeds the peak field F_p. This instability may lead to the high field domain formation and to the effect observed by J. B. Gunn.

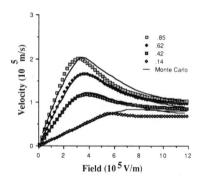

FIGURE 4-1-3. Curves v vs. F in GaAs given by the interpolation formula (4-1-11) for different values of μ (in m²/V s). The curves obtained by Monte Carlo calculations are shown for comparison [156].

4-2. HIGH FIELD DOMAINS

Let us now assume that F is equal to F_p everywhere in the sample except in a small region near the cathode where it is slightly higher owing to a doping nonuniformity. The higher field in this region leads to a smaller electron velocity because of the negative slope of the v vs. F curve (see Fig. 4-2-1). Hence the electrons in front of the high field region and behind the high field region move faster than in the region. It leads to the depletion of electrons in the leading edge and to the accumulation of electrons in the trailing edge. The resulting dipole layer of charge increases the electric field in the region, intensifying the growth of the fluctuation, which propagates toward the anode. If the bias voltage is kept constant the growth of the fluctuation comes at the expense of the electric field F_r outside the domain. The decrease of F_r leads to the current drop during the high field domain formation. The annihilation of the high field domain when it reaches the anode causes the temporary increase in the current. Then the next domain nucleates near the cathode and the process repeats itself.

Most often just one high field domain forms in the sample because the voltage drop across the domain results in a field smaller than F_p everywhere except within the domain.

In most cases (but not always) the domain forms near the cathode and propagates through the sample with a velocity close to the saturation velocity v_s. The current density in the sample with the domain present is then given by

$$j_s = qn_0v_s \qquad (4\text{-}2\text{-}1)$$

During the domain annihilation and the formation of a new domain the current density rises to

$$j_p = qn_0v_p \qquad (4\text{-}2\text{-}2)$$

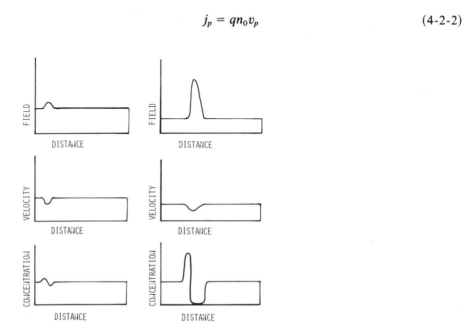

T = 0   T   0   FIGURE 4-2-1. Domain formation.

The domain transit time is given by

$$T \simeq L/v_s \tag{4-2-3}$$

The time of the initial growth of a small fluctuation t_{gr} is determined by the differential dielectric relaxation time τ_{md} :

$$t_{gr} \simeq 3|\tau_{md}| = \frac{3\varepsilon}{q|\mu_d|n_0} \tag{4-2-4}$$

where $\mu_d = dv/dF$.

For the formation of a stable propagating domain t_{gr} should be considerably smaller than the transit time:

$$\frac{3\varepsilon}{q|\mu_d|n_0} < \frac{L}{v_s} \tag{4-2-5}$$

This equation leads to the so-called Kroemer criterion:

$$n_0 L > (n_0 L)_1 = \frac{3\varepsilon v_s}{q|\mu_d|} \tag{4-2-6}$$

For GaAs $\varepsilon = 1.14 \times 10^{-10} \, F/m$, $|\mu_d| \simeq 0.07 \, \text{m}^2/V \, s$, $v_s \simeq 10^5 \, \text{m/s}$, and $(n_0 L)_1 \simeq 3 \times 10^{15} \, \text{m}^{-2}$. When the $n_0 L$ product is close to $10^{16} \, \text{m}^{-2}$ or above, the formation of stable propagating domains takes place when F exceeds F_p. When $(n_0 L)_1 < 10^{16} \, \text{m}^{-2}$ the threshold field of the domain formation F_t exceeds F_p. The value of F_t depends on $n_0 L$ and on the shape of the v vs. F curve.

Once formed the domain does not disappear if the applied voltage is decreased below the threshold voltage during the domain propagation. The reason for that is that the field within the domain remains higher than the peak field F_p even though the average field may be below the peak field unless the voltage drop is too big. For large values of $n_0 L \sim 10^{16} \, \text{m}^{-2}$ the domain sustaining field is close to F_s. The computed curves F_t vs. $n_0 L$ and F_s vs. $n_0 L$ are shown in Fig. 4-2-2 [8].

The difference between F_t and F_s leads to a hysteresis in the domain current–voltage characteristics (see Fig. 4-2-3). It can be utilized for applications of transferred-electron devices in logic circuits (see Fig. 4-2-4).

For $n_0 L \gg (n_0 L)_1$ the electric field outside the stable domain is close to F_s. The domain field depends on $n_0 L$ and the applied voltage and may vary from approximately 30 to 200 kV/cm. It is limited by the critical field of the avalanche breakdown.

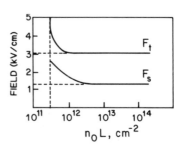

FIGURE 4-2-2. Threshold field of the domain formation F_t and the domain sustaining field F_s vs. $n_0 L$ [8].

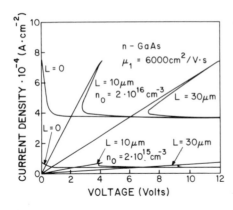

FIGURE 4-2-3. Theoretical I-V curves for Gunn oscillators [9].

The time of the stable domain formation may be considerably larger than the initial growth time $3|\tau_{md}|$ (see Section 4-7). It decreases with the increase in n_0 and may vary from a few picoseconds to a hundred picoseconds.

The domain shape is also dependent on the doping density. When the doping density is large (much larger than $\sim 10^{21}$ m^{-3} for GaAs) a small relative change in the equilibrium electron concentration is sufficient to produce a large space charge supporting a high domain field. In this case the domain shape is nearly symmetrical. In the opposite limiting case ($n_0 \ll 10^{21}$ m^{-3}) the leading edge is totally depleted of carriers creating the positive space charge density $qn_0 = q(N_D - N_A)$. At the same time the density of electrons in the accumulation layer (i.e., in the trailing domain edge) is limited only by the diffusion processes and can exceed n_0 many times. The resulting field distribution is very similar to the field distribution in a p^+-n junction (see Fig. 4-2-5) moving with the domain velocity which is close to the saturation velocity v_s. In the nearly totally depleted leading edge the drift and diffusion currents are small and the current density is determined by the displacement current

$$ j = \varepsilon \frac{\partial F}{\partial t} \simeq \varepsilon v_s \frac{qn_0}{\varepsilon} \tag{4-2-7} $$

so that the total current is

$$ I = qv_s n_0 S \tag{4-2-8} $$

Equations (4-2-7) and (4-2-8) are valid even when n_0 and/or the device cross-section S depend on the space coordinate x, i.e., for a device with a nonuniform

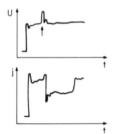

FIGURE 4-2-4. Trigger mode of operation. At the point of time indicated by the arrow the bias voltage exceeds the threshold value. The generated current pulse is caused by a single stable propagating domain [10].

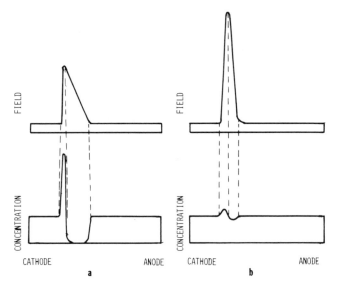

FIGURE 4-2-5. Domain shape. (a) Low doping; (b) high doping.

doping profile and/or with a complex shape. This means that the current waveform should reproduce the $n_0(x)S(x)$ profile:

$$I(t) = qn_0(v_st)S(v_st)v_s \qquad (4\text{-}2\text{-}9)$$

where t is counted from the time of the domain formation (see Fig. 4-2-6). This domain property may be utilized for implementing numerous analog and logic functions [11, 12].

Though we derived Eq. (4-2-9) for a limiting case of relatively low doped samples when the leading edge is totally depleted, it is approximately valid for an arbitrary doping level [8].

4-3. STABLE AMPLIFICATION REGIME

As was shown in Section 4-2, the stable domains develop if $n_0L > (n_0L)_1$. If n_0L is less than $(n_0L)_1$ the transit time is smaller than the characteristic time of the fluctuation growth. In this case the electric field distribution in the sample becomes nonuniform for $F > F_p$ (see Fig. 4-3-1). The field profile is strongly dependent on the boundary condition, especially at the cathode [13-15]. Below we consider a

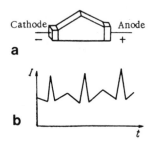

FIGURE 4-2-6. Current waveform in a sample with a nonuniform cross section [8]. (a) Device shape; (b) current waveform. Spikes in the current waveform correspond to the domain annihilation and to the formation of a new domain. The part of the waveform between the spikes reproduces the shape of the device.

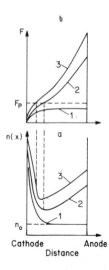

FIGURE 4-3-1. Field and electron concentration distribution in a transferred electron device with $n_0 L < (n_0 L)_1$ [8]. 1, $U/L < F_p$; 2, $U_1/L > F_p$; 3, $U_2 > U_1$.

simple case when the field at the cathode contact is small (an injecting cathode contact). The field distribution in the sample for $U/L < F_p$ is then given by curve 1 of Fig. 4-3-1b. The field is small near the cathode and then increases to

$$F_0 \simeq U/L \tag{4-3-1}$$

Near the cathode the electron concentration is higher so that the current density

$$j = qn(x)\mu E(x) = \text{const} \tag{4-3-2}$$

(we neglect here the diffusion current). Further from the cathode n approaches n_0 (see curve 1 of Fig. 4-3-1a) as F increases and the electron velocity rises.

If $U/L > F_p$ the field and electron concentration distributions near the cathode will be similar to the distributions discussed above until the point X_p is reached where $F = F_p$. Beyond this point the increase in the electric field leads to the decrease of the electron velocity due to the negative differential mobility. As a result the electron concentration has to increase even more to ensure the constant current density along the sample. This increase results in a large field derivative $|dF/dx|$, as can be seen from the Poisson equation

$$\frac{dF}{dx} = \frac{q(n_0 - n)}{\varepsilon} \tag{4-3-3}$$

The increase in the electric field leads to a bigger drop in the electron velocity and the field distribution becomes even steeper closer to the anode.

When the applied voltage increases, the electric field F reaches F_p closer to the cathode, where n is larger. The velocity at this point is still equal to the peak velocity v_p and hence the current density increases with voltage. However, as can be seen from Fig. 4-3-1b, the field distribution is such that a very small increase in the current density requires a large increase in voltage. This means that the current practically saturates at voltages larger than $U_p = F_p L$ (see Fig. 4-3-2). This property makes it possible to use the transferred electron devices as current limiters [16, 17].

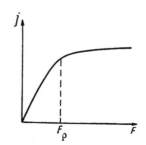

FIGURE 4-3-2. A typical j-F characteristic of a Gunn diode with $n_0L < (n_0L)_1$.

As follows from the above discussion the dc differential resistance is positive at $U > U_p$ owing to the electron injection from the cathode. However, the electron distribution can follow the variation in the external voltage only at frequencies smaller than the transit frequency

$$f_T = v_s/L$$

At frequencies close to f_T and higher the injection of electrons from the cathode will not be able to prevent the growth of a small fluctuation moving from the cathode to the anode and the device may still exhibit the negative differential resistance and can be used as an amplifier.

A crude estimate of the fluctuation growth G can be obtained if we neglect the nonuniformity of the field distribution across the sample:

$$G = \exp\left(\frac{1}{\omega|\tau_{md}|}\right) \tag{4-3-4}$$

where ω is the signal frequency. For $n_0L < (n_0L)_1$ frequency $\omega/2\pi$ is larger than $1/|\tau_{md}|$ and hence

$$G \simeq 1 + \frac{1}{\omega|\tau_{md}|} \tag{4-3-5}$$

Because of the boundary conditions the gain versus frequency curve has maxima at the transit frequency and its harmonics as at these frequencies the sample length is equal to the integer multiple of the space charge wave lengths λ. As shown in Section 4-4

$$\lambda = v_s/f \tag{4-3-6}$$

Thus the wavelength decreases with the increase in frequency, and diffusion effects become more important decreasing the gain.

When the characteristic diffusion time

$$\tau_D = \lambda^2/4\pi^2 D \tag{4-3-7}$$

becomes smaller than $|\tau_{md}|$ the fluctuation decays due to the diffusion. This leads

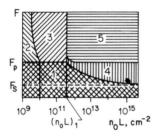

FIGURE 4-3-3. Different regimes of operation of transferred electron
devices (see text) [8].

to the following condition for the stable amplification:

$$n_0\lambda^2 > \frac{4\pi^2\varepsilon D}{q|\mu_d|} \qquad (4\text{-}3\text{-}8)$$

or

$$\frac{n_0}{f^2} > \frac{4\pi^2\varepsilon D}{qv_s^2|\mu_d|} \qquad (4\text{-}3\text{-}9)$$

Because $\lambda \leqslant L$ eq. (4-3-8) also limits the n_0L^2 product

$$n_0L^2 > \frac{4\pi^2\varepsilon D}{q|\mu_d|} \equiv (n_0L^2)_{cr} \qquad (4\text{-}3\text{-}10)$$

is a necessary condition of the stable amplification. For GaAs $\varepsilon = 1.14 \times 10^{-10}$ F/m,
$D = 0.02$ m^2/s, $|\mu_d| \sim 0.07$ m^2/V s and we find from Eq. (4-3-10) $(n_0L^2)_{cr} \simeq 8 \times 10^9$ 1/m.

The different regimes of operation of the transferred electron devices are marked
in Fig. 4-3-3 on the $n_0L - F_0$ plane where $F_0 = U_0/L$. The sample is stable in region
1 and is stabilized by diffusion in region 2. Stable amplification occurs in region 3.
In region 4 (limited by F_t vs. n_0L and F_s vs. n_0L curves) single domains may be
triggered by a short voltage pulse (as shown in Fig. 4-2-4). Region 5[$F_0 > F_t$,
$n_0L > (n_0L)_1$] corresponds to the stable domain propagation.

4-4. SMALL-SIGNAL ANALYSIS

Our one-dimensional analysis of the transferred electron devices is based on
the phenomenological equations described in Section 2-8:

$$\frac{\partial F}{\partial x} = \frac{q(n_0 - n)}{\varepsilon} \qquad (4\text{-}4\text{-}1)$$

$$j = qnv(F) + qD(F)\frac{\partial n}{\partial x} + \varepsilon\frac{\partial F}{\partial t} \qquad (4\text{-}4\text{-}2)$$

Apparently, Eqs. (4-4-1) and (4-4-2) have a solution

$$F = U/L \qquad (4\text{-}4\text{-}3)$$

$$n = n_0 = N_D - N_A \qquad (4\text{-}4\text{-}4)$$

However, this solution may be unstable and/or may not satisfy the boundary conditions. In order to find $F(x, t)$, $n(x, t)$ we should solve Eqs. (4-4-1)–(4-4-2) for the given boundary conditions. Some useful information about the device may, however, be obtained from a small-signal analysis which allows us to investigate the stability of the uniform solution given by Eqs. (4-4-3) and (4-4-4).

We shall seek a solution of Eqs. (4-4-1) and (4-4-2) in the following form:

$$F = F_0 + F_1 e^{i(kx - \omega t)} \tag{4-4-5}$$

$$n = n_0 + n_1 e^{i(kx - \omega t)} \tag{4-4-6}$$

where

$$F_1 \ll F_0 \tag{4-4-7}$$

and

$$n_1 \ll n_0 \tag{4-4-8}$$

Substitution of Eqs. (4-4-5) and (4-4-6) into Eq. (4-4-1) and (4-4-2) yields two uniform algebraic equations for n_1 and F_1 which have a nonzero solution if and only if the system determinant is zero. This leads to the following dispersion equation:

$$\omega = -kv(F_0) - i\left[\frac{1}{\tau_{md}(F_0)} + Dk^2\right] \tag{4-4-9}$$

where

$$\tau_{md} = \frac{\varepsilon}{qn_0 dv/dF} \tag{4-4-10}$$

is the differential dielectric relaxation time.

Equation (4-4-9) may be used to describe the evolution of the sinusoidal fluctuation of the electron concentration which is related to the space charge and electric field fluctuations.

The wavelength (the characteristic size of the fluctuation) λ is related to the wave vector:

$$k = 2\pi/\lambda \tag{4-4-11}$$

The real part of ω

$$\text{Re}(\omega) = -kv(F_0) \tag{4-4-12}$$

determines the frequency and the imaginary part of ω

$$\text{Im}(\omega) = -\frac{1}{\tau_{md}} - Dk^2 \tag{4-4-13}$$

determines the attenuation or growth time constant of the fluctuation.

If $\text{Im}(\omega) < 0$ the fluctuation decays and if $\text{Im}(\omega) > 0$ it grows. This means that if

$$\tau_{md} < 0 \tag{4-4-14}$$

the space charge fluctuations may grow and lead to the instability. The values of k of unstable fluctuations are limited by

$$0 \leqslant k < \frac{1}{(D|\tau_{md}|)^{1/2}} \tag{4-4-15}$$

Equation (4-4-15) is equivalent to Eq. (4-3-8) discussed in Section 4-3.

Equation (4-4-9) can be also used to analyze how the fluctuation induced at a given frequency varies in space. In this case we consider ω real and k complex. Neglecting diffusion we find from Eq. (4-4-9)

$$\mathrm{Re}(k) = -\frac{\omega}{v} \tag{4-4-16}$$

$$\mathrm{Im}(k) = -\frac{1}{\tau_{md} v} \tag{4-4-17}$$

As can be seen from Eqs. (4-4-16) and (4-4-17) the wave grows in the direction of propagation, i.e., from the cathode to anode when $\tau_{md} < 0$. The condition of the stable amplification regime

$$|\mathrm{Im}(k)L| < 1 \tag{4-4-18}$$

is equivalent to the Kroemer criterion.

4-5. SMALL SIGNAL IMPEDANCE

From the Poisson equation (4-4-1) and the equation for the total current density (4-4-2) we can derive an equation for the electric field F:

$$-\varepsilon v \frac{\partial F}{\partial x} + qn_0 v + \varepsilon \frac{\partial F}{\partial t} = j \tag{4-5-1}$$

Here we have neglected diffusion. Assuming

$$j = j_0 + j_1^0 \, e^{-i\omega t} \tag{4-5-2}$$

where $j_1^0 \ll j_0$ we shall seek a solution of Eq. (4-5-1) in the following form:

$$F = F_0 + F_1^0 (1 + a \, e^{ikx}) \, e^{-i\omega t} \tag{4-5-3}$$

where $F_1^0 \ll F_0$ and a is determined by the boundary conditions. Assuming

$$F(0) = F_0 \tag{4-5-4}$$

we find

$$a = -1 \tag{4-5-5}$$

Substitution of Eqs. (4-5-3) and (4-5-5) into (4-5-1) yields

$$qn_0v(F_0) = j_0 \tag{4-5-6}$$

$$-ik = i\frac{\omega}{v(F_0)} - \frac{1}{\tau_{md}v(F_0)} \tag{4-5-7}$$

$$F_1^0 = \frac{j_1^0}{qn_0\mu_d - i\varepsilon\omega} \tag{4-5-8}$$

Equation (4-5-6) is the zero-order solution of Eq. (4-5-1). Equation (4-5-7) coincides with the dispersion equation for $D = 0$. See Eq. (4-4-9). The first-order solution (4-5-8) can be used to evaluate the small-signal impedance:

$$Z(\omega) = \frac{\displaystyle\int_0^L F_1^0(1 - e^{ikx})\,dx}{j_1^0 S} \tag{4-5-9}$$

Substituting Eq. (4-5-8) into Eq. (4-5-9) we find [8]

$$Z(\omega) = -\frac{L^2}{\varepsilon vS}\frac{e^{-\alpha} + \alpha - 1}{\alpha^2} \tag{4-5-10}$$

where

$$\alpha = L\left(-\frac{1}{\tau_{md}v} + i\frac{\omega}{v}\right) \tag{4-5-11}$$

Equation (4-5-11) allows us to investigate the stability of the sample and to establish the boundary between the oscillation and amplification regimes.

The oscillations correspond to the poles of Z in the upper part of the complex plane ω if the sample is connected to a current source ($I_1^0 = 0$). It follows from the equation

$$U_1^0 = Z(\omega)j_1^0 \tag{4-5-12}$$

that if $I_1^0 = 0$ it is necessary to have $Z(\omega) \to \infty$ in order to obtain the finite value of voltage $U_1^0(\omega)$. As can be seen from Eq. (4-5-10) the impedance does not have poles and hence we conclude that the transferred-electron devices connected to a current source ($I_1^0 = 0$) cannot oscillate.

Samples biased by a voltage source [i.e., $U'(\omega) = 0$] will oscillate if the impedance has zeros in the upper half of the ω plane. All zeros of the impedance (with the exception of $\alpha = 0$) coincide with the zeros of the numerator in Eq. (4-5-10)

$$e^{-\alpha} + \alpha - 1 = 0 \tag{4-5-13}$$

Several first zeros α_m of this function are given in Table 4-5-1. For $m \gg 1$

$$\text{Im}(\alpha_m) \simeq \pm 2\pi m \tag{4-5-14}$$

$$\text{Re}(\alpha_m) \simeq \ln(2\pi m) \tag{4-5-15}$$

TABLE 4-5-1. Zeros
of Impedance $Z(\omega)$ [18]

m	α_m
1	$-2.09 \pm i.7.46$
2	$-2.69 \pm i.13.88$
3	$-3.03 \pm i20.22$
4	$-3.22 \pm i26.54$
5	$-3.50 \pm i32.85$

Equation (4-5-11) defining α_m can be written as

$$\alpha_m = \frac{T_0}{\tau_{md}} - i2\pi f T_0 \qquad (4\text{-}5\text{-}16)$$

where

$$T_0 = L/v \qquad (4\text{-}5\text{-}17)$$

is the transit time.

From the comparison of Eq. (4-5-16) with Table 4-5-1 and with the asymptotic solutions given by Eqs. (4-5-14) and (4-5-15) we conclude that the zeros of the impedance (and hence the related instabilities) occur at the frequencies close to the transit frequency $1/T_0$ and its harmonics. For the oscillations to occur the value of τ_{md} should be negative and the absolute value of τ_{md} should be large enough. This condition becomes more and more stringent as the harmonic number m increases. From the condition

$$\mathrm{Re}(\alpha_1) > 2.09$$

(see Table 4-5-1) we find

$$n_0 L < \frac{2.09 \varepsilon v}{q|\mu_d|} \qquad (4\text{-}5\text{-}18)$$

which coincides with the Kroemer criterion discussed in Section 4-3 within a numerical factor.

Equation (4-5-10) implies that the negative differential conductance exists up to very high frequencies. When the nonuniform field distribution and diffusion effects are taken into account a similar calculation shows that the differential conductance at high frequencies may disappear (see Fig. 4-5-1).

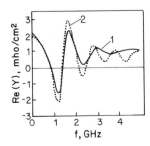

FIGURE 4-5-1. Conductance of a transferred electron device vs. frequency [19]. The nonuniform field distribution is taken into account. Dotted line, $D = 0$; solid line, diffusion is taken into account.

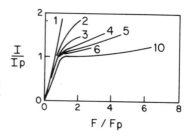

FIGURE 4-5-2. Current–voltage characteristics of a transferred electron device at dc [20]. Numbers indicate the values of $(qn_0L)/(\varepsilon Fp)$. Diffusion has been neglected. Note strong injection at low values of n_0L leading to large ratios I/Ip.

The nonuniform field distribution may also lead to a positive differential conductance at the frequencies below the transit frequency. This effect is related to the electron injection into the sample (see Section 4-3 and Figs. 4-5-1 and 4-5-2).

4-6. ANALYTICAL THEORY OF STABLE HIGH FIELD DOMAINS

The analytical theory of high field domains is based on the set of phenomenological equations discussed in Section 2-7 and used for the small signal analysis in Section 4-4 [Eqs. (4-4-1)–(4-4-2)].

The stable propagating domains can be described by the plane wave solution depending on

$$Z = x - ut \tag{4-6-1}$$

where u is the domain velocity. [For the sign convention used in Eqs. (4-4-1) and (4-4-2) the negative sign of u corresponds to the movement from the cathode to the anode.]

Substitution of Eq. (4-6-1) into Eqs. (4-4-1) and (4-4-2) yields

$$\frac{dF}{dZ} = \frac{q(n_0 - n)}{\varepsilon} \tag{4-6-2}$$

and

$$I = qnv + qD(F)\frac{dn}{dZ} - \varepsilon u \frac{dF}{dZ} \tag{4-6-3}$$

(Here we assume that the diffusion coefficient may be field dependent.) The substitution of n from Eq. (4-6-2) into Eq. (4-6-3) and the introduction of the space charge density

$$\rho = q(n_0 - n) \tag{4-6-4}$$

as a variable results in the following system of equations:

$$\frac{dF}{dZ} = \frac{\rho}{\varepsilon} \tag{4-6-5}$$

$$\frac{d\rho}{dZ} = \frac{qn[v(F) - v(F_r)] - \rho[v(F_r) + u]}{D(F)} \tag{4-6-6}$$

Here F_r is the electric field outside the high field domain where

$$n = n_0 \tag{4-6-7}$$

$$j = qn_0 v(F_r) \tag{4-6-8}$$

Dividing Eq. (4-6-6) by Eq. (4-6-5) we obtain

$$\rho \frac{d\rho}{dF} = \frac{\varepsilon q n_0}{D(F)} \left\{ v(F) - v(F_r) - \frac{\rho}{q n_0} [v(F) + u] \right\} \tag{4-6-9}$$

or

$$\frac{\rho}{1 - \rho/q n_0} \frac{d\rho}{dF} = \frac{\varepsilon q n_0}{D(F)} \left\{ [v(F) - v(F_r)] - \frac{(\rho/q n_0)[v(F_r) + u]}{1 - \rho/q n_0} \right\} \tag{4-6-10}$$

A formal integration of Eq. (4-6-10) yields

$$-\frac{\rho}{q n_0} - \ln\left(1 - \frac{\rho}{q n_0}\right) = \frac{\varepsilon}{q n_0} \left\{ \int_{F_r}^{F} \frac{[v(F') - v(F_r)] \, dF'}{D(F')} \right.$$

$$- [v(F_r) + u]$$

$$\left. \times \int_{F_r}^{F} \frac{\rho/q n_0}{D(F')(1 - \rho/q n_0)} \, dF' \right\}. \tag{4-6-11}$$

For the high field domain

$$\rho(F_m) = 0 \tag{4-6-12}$$

where F_m is the maximum domain field. At this point the right-hand side of Eq. (4-6-11) must be equal to zero. But the sign of the second term in the right-hand side of Eq. (4-6-11) depends on the integration path. For the accumulation layer $\rho_a/q n_0 < 0$ and the integral is negative. For the depletion layer, however,

$$0 < \rho_d/q n_0 < 1 \tag{4-6-13}$$

and the integral is positive. (Here ρ_a and ρ_d are the space charge densities in the accumulation and depletion layer, respectively.) To satisfy Eq. (4-6-12) we have to set

$$u = -v(F_r) \tag{4-6-14}$$

and

$$\int_{F_r}^{F_m} \frac{v(F) - v(F_r)}{D(F)} \, dF = 0 \tag{4-6-15}$$

Equation (4-6-15) determines the maximum domain fields F_m as a function of F_r and hence as a function of the current density [see Eq. (4-6-8)].

If an additional assumption of the field-independent diffusion constant is made Eq. (4-6-15) can be simplified:

$$\int_{F_r}^{F_m} [v(F) - v(F_r)] \, dF = 0 \tag{4-6-16}$$

This equation has a simple geometric interpretation (see Fig. 4-6-1). It is called an "equal area rule" [21].

Substitution of Eq. (4-6-14) into Eq. (4-6-11) leads to

$$-\frac{\rho}{qn_0} - \ln\left(1 - \frac{\rho}{qn_0}\right) = \frac{\varepsilon}{qn_0} \int_{Fr}^{F} \frac{v(F') - v(F_r)}{D(F')} \, dF' \tag{4-6-17}$$

After Eq. (4-6-17) is solved to yield the space charge density in the accumulation and depletion layers ρ_a, ρ_d the domain voltage U_d can be calculated as follows:

$$U_d = \int_{-\infty}^{\infty} (F - F_r) \, dx = \int_{F_r}^{F_m} (F - F_r) \left(\left| \frac{dF}{dx} \right|_a^{-1} + \left| \frac{dF}{dx} \right|_d^{-1} \right) dF$$

$$= \varepsilon \int_{F_r}^{F_m} (F - F_r) \left(\frac{1}{|\rho_a|} + \frac{1}{|\rho_a|} \right) dF \tag{4-6-18}$$

This derivation is based on the theory developed in Ref. 21, 22. However, in a practical case when the fields inside the domain are high and the electron drift velocity and diffusion in high electric fields are nearly saturated, simple expressions determining the domain parameters can be derived [23, 24]. If

$$F_m \gg F_v$$

where F_v is the velocity and diffusion saturation field. The integral in Eq. (4-6-17) can be simplified

$$\int_{F_r}^{F} \frac{[v(F') - v(F_r)]}{D(F')} \, dF' \simeq \int_{F_s}^{F_v} \frac{[v(F) - v_s]}{D(F)} \, dF - \frac{F(F_r - F_s)}{D_s}$$

$$= J - \frac{F(F_r - F_s)}{D_s} \tag{4-6-19}$$

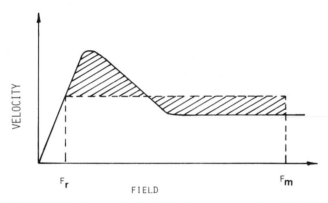

FIGURE 4-6-1. Equal area rule. The shaded areas are equal [see Eq. (4-6-16)].

where D_s is the value of the diffusion coefficient in the high electric field, v_s is the saturation velocity, $F_s = v_s/\mu$, and

$$J = \int_{F_s}^{F_v} \frac{[v(F) - v_s]}{D(F)} \, dF \equiv \frac{\mu F_c^2}{D_s} \qquad (4\text{-}6\text{-}20)$$

where F_c is the characteristic electric field which can be computed once for a given v vs. F curve. The computed dependence of F_c on the low field mobility is shown in Fig. 4-6-2 [24].

When $F = F_m$ and $\rho = 0$ we find from Eqs. (4-6-15) and (4-6-20)

$$D_s J - F_m(F_r - F_s)\mu = 0 \qquad (4\text{-}6\text{-}21)$$

For the field-independent diffusion this equation could be further simplified:

$$F_m(F_r - F_s) = F_c^2$$

where

$$F_c^2 = \frac{\displaystyle\int_{F_s}^{F_v} [v(F) - v_s] \, dF}{\mu} \qquad (4\text{-}6\text{-}22)$$

[see Eq. (4-6-20)]. Using Eq. (4-6-21) we find from Eq. (4-6-17)

$$-\frac{\rho}{qn_0} - \ln\left(1 - \frac{\rho}{qn_0}\right) = \frac{n_{\mathrm{cr}}}{n_0}\left(1 - \frac{F}{F_m}\right) \qquad (4\text{-}6\text{-}23)$$

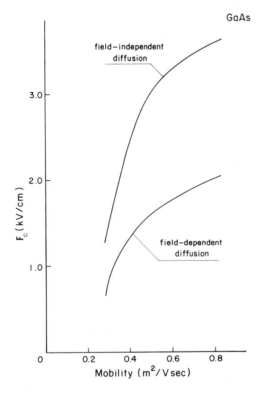

FIGURE 4-6-2. Characteristic field F_c as a function of the low field mobility in GaAs [24].

where

$$n_{cr} = \frac{\varepsilon}{q} J \equiv \frac{\varepsilon \mu F_c^2}{qD_s} \qquad (4\text{-}6\text{-}24)$$

n_{cr} as a function of the low field mobility in GaAs is shown in Fig. 4-6-3.

The domain voltage can then be found from Eq. (4-6-18), which can be rewritten as

$$U_d \simeq \varepsilon \int_{F_s}^{F_m} F\left(\frac{1}{|\rho_d|} + \frac{1}{|\rho_a|}\right) dF \simeq \frac{\varepsilon F_m^2}{qn_o} \int_0^1 W\left(\frac{qn_o}{|\rho_d|} + \frac{qn_o}{|\rho_a|}\right) dW \qquad (4\text{-}6\text{-}25)$$

Here $W = F/F_m$. As can be seen from Eq. (4-6-23) the integral in Eq. (4-6-25) is the universal dimensionless function of n_{cr}/n_o, which does not depend on material parameters nor on the domain field:

$$U_d = \frac{\varepsilon F_m^2}{qn_o} \phi\left(\frac{n_0}{n_{cr}}\right) \qquad (4\text{-}6\text{-}26)$$

We may call ϕ a domain shape function. From Eq. (4-6-26) we find

$$F_m = \left(\frac{qn_0 U_d}{\varepsilon \phi}\right)^{1/2} \qquad (4\text{-}6\text{-}27)$$

In order to determine the domain shape function ϕ let us first consider two limiting

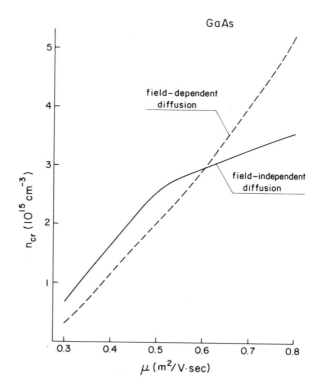

FIGURE 4-6-3. Characteristic electron concentration vs. low field mobility for GaAs [24].

cases $n_o/n_{cr} \ll 1$ and $n_o/n_{cr} \gg 1$ when the integral in Eq. (4-6-25) can be evaluated analytically.

For $n_0/n_{cr} \ll 1$ we find

$$\frac{\rho_d}{qn_0} \simeq 1 \tag{4-6-28}$$

$$\frac{\rho_a}{qn_0} \simeq -\frac{n_{cr}}{n_0}\left(1 - \frac{F}{F_m}\right) \tag{4-6-29}$$

$$\Phi \simeq 1/2 \tag{4-6-30}$$

For $n_0/n_{cr} \gg 1$ we find

$$\frac{\rho_a}{qn_0} = -\left[\frac{2n_{cr}}{n_0}\left(1 - \frac{F}{F_m}\right)\right]^{1/2} \tag{4-6-31}$$

$$\frac{\rho_d}{qn_0} = \left[\frac{2n_{cr}}{n_0}\left(1 - \frac{F}{F_m}\right)\right]^{1/2} \tag{4-6-32}$$

$$\phi \simeq \frac{4\sqrt{2}}{3}\left(\frac{n_0}{n_{cr}}\right)^{1/2} \tag{4-6-33}$$

(see Fig. 4-6-23). Equations (4-6-31)–(4-6-33) are also valid for $n_0/n_{cr} \ll 1$ for the values of F close to F_m, i.e., when $(F_m - F)/F_m \ll n_0/n_{cr}$.

Substitution of Eqs. (4-6-28) and (4-6-33) into Eqs. (4-6-26) and (4-6-27) leads to the formulas identical to those derived in Refs. 21 and 22. It is important, however, that $\phi(n_0/n_{cr})$ can be computed once for all values of n_0/n_{cr} and then can be used to find the explicit relationship between U_d and F_m, n_0 and n_{cr} [23].

The computed dependence ϕ versus n_0/n_{cr} is shown in Fig. 4-6-5. As can be seen from the figure, this dependence can be reasonably well approximated by the following expression:

$$\phi\left(\frac{n_0}{n_{cr}}\right) \simeq 0.363 + 1.67\left(\frac{n_0}{n_{cr}}\right)^{1/2} \tag{4-6-34}$$

if $0.01 < n_0/n_{cr} < 1000$.

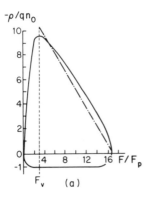

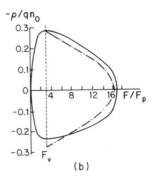

FIGURE 4-6-4. Space charge in the domain walls vs. electric field [8]. (a) $n_0 = 2.25 \times 10^{14}$ cm^{-3}; (b) $n_0 = 4.5 \times 10^{16}$ cm^{-3}. Solid line, numerical calculation; dashed line, analytical calculation using Eqs. (4-6-28) and (4-6-29) for $n_0 = 2.25 \times 10^{14}$ cm^{-3} and Eqs. (4-6-31) and (4-6-32) for $n_0 = 4.5 \times 10^{16}$ cm^{-3}.

For $n_0/n_{cr} \ll 1$ the shape of the high field domain is close to a rectangular triangle. For $n/n_{cr} \gg 1$ the domain practically has a parabolic shape. Using Eqs. (4-6-27) and (4-6-31)–(4-6-33) we find the field profile within the domain in the case when $n_0/n_{cr} \gg 1$.

$$F(x) \simeq F_s + 4(F_m - F_s)\frac{x - x_s}{d_{\text{eff}}}\left(1 - \frac{x - x_s}{d_{\text{eff}}}\right) \qquad (4\text{-}6\text{-}35)$$

Here

$$F_m = F_s + 0.728\left(\frac{qn_0^{1/2}n_{cr}^{1/2}U_d}{\varepsilon}\right)^{1/2} \qquad (4\text{-}6\text{-}36)$$

and

$$d_{\text{eff}} = \frac{3U_d}{2(F_m - F_s)} \simeq 2.06\left(\frac{\varepsilon U_d}{qn_0^{1/2}n_{cr}^{1/2}}\right)^{1/2} \qquad (4\text{-}6\text{-}37)$$

is the effective domain width, x_s is the starting point of the domain, and

$$x_s < x < x_s + d_{\text{eff}}$$

The maximum space charge density in the domain wall is

$$\rho_m = q(2n_0 n_{cr})^{1/2} \qquad (4\text{-}6\text{-}38)$$

and

$$U_d \simeq 1.89\frac{\varepsilon(F_m - F_s)^2}{qn_0^{1/2}n_{cr}^{1/2}} \qquad (4\text{-}6\text{-}39)$$

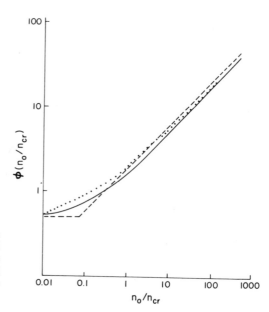

FIGURE 4-6-5. ϕ vs. n_0/n_{cr} [23]. Solid line, computer calculation; dashed line, analytical approximation valid for small and large values of n_0/n_{cr}; dotted line, analytical approximation valid approximation valid for $0.01 < n_0/n_{cr} < 1000$ [see Eq. (4-6-34)].

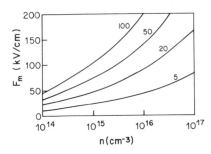

FIGURE 4-6-6. The maximum domain field as a function of the electron concentration for GaAs. Numbers near the curves correspond to the domain voltage U_d in volts. $n_{cr} = 3 \times 10^{15} \text{ cm}^{-3}$.

For arbitrary electron concentration the maximum domain field can be calculated from Eq. (4-6-27) (see Fig. 4-6-6).

The analytical theory of stable high field domains described above is based on highly simplified equations which assume that the electron velocity and diffusion are local instantaneous functions of the electric field. In fact, the width of the accumulation layer in low doped samples and domain size in high doped samples may become quite small and nonlocal transient effects [28, 29] may become important. More accurate simulations of the transferred electron devices which take the transient effects into account have been performed using the Monte Carlo method [25]. A number of alternative models such as the effective electron temperature model [18, 26] and the effective power model [27] have also been used to simulate the high and the effective power model field domains. In many cases, however, the analytical theory based on equations given in this section provides an adequate agreement with the experimental results (see, for example, Figs. 4-6-7 and 4-6-8 [30, 31]).

4-7. DOMAIN DYNAMICS

The simplified treatment of the domain dynamics [32–34] is based on the solution of the equation for the total current density:

$$j = \varepsilon \frac{\partial F}{\partial t} + qnv(F) + qD \frac{\partial n}{\partial x} \tag{4-7-1}$$

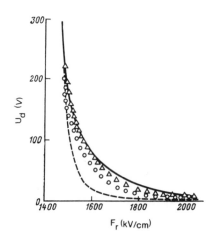

FIGURE 4-6-7. Domain voltage vs. outside domain field [30]. Solid line, experimental curve; dashed line, temperature model; triangles and circles, field-dependent diffusion and velocity with different $D(F)$ curves.

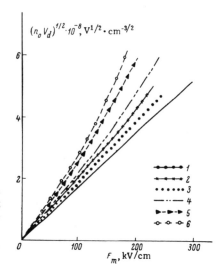

FIGURE 4-6-8. $(n_0 U_d)^{1/2}$ vs. F_m [30]. 1, Analytical calculation assuming the total depletion in the leading domain edge; 2, numerical simulation for the sample with $\rho = 8\,\Omega$ cm; 3, same for $\rho = 4\,\Omega$ cm; 5, 6, 7, calculated curves for $\rho = 2\,\Omega$ cm, $0.5\,\Omega$ cm, and $11\,\Omega$ cm, respectively.

and the Poisson equation

$$\frac{\partial F}{\partial x} = \frac{q}{\varepsilon}(n_0 - n) \tag{4-7-2}$$

which could be combined as

$$j = \varepsilon \frac{\partial F}{\partial t} + q\left(n_0 - \frac{\varepsilon}{q}\frac{\partial F}{\partial x}\right)v(F) - \varepsilon D \frac{\partial^2 F}{\partial x^2} \tag{4-7-3}$$

Integrating Eq. (4-7-3) over x and assuming

$$n(0) = n(L) = n_0$$
$$F(0) = F(L) = F_r \tag{4-7-4}$$

we obtain

$$j(t)L = \varepsilon L \frac{dF_0}{dt} + q\int_0^L\left(n_0 - \frac{\varepsilon}{q}\frac{\partial F}{\partial x}\right)v(F)\,dx \tag{4-7-5}$$

where we defined

$$F_0 = U/L \tag{4-7-6}$$

Here $U(t)$ is the bias voltage. Boundary conditions (4-7-4) correspond to the domain mode of operation (uniform carrier concentration and equal electric field at both ends). The last term in the right-hand part of Eq. (4-7-5) may be rewritten as

$$qn_0\int_0^L v(F)\,dx - \varepsilon \int_{F(0)}^{F(L)} v(F)\,dF \tag{4-7-7}$$

where the second term is equal to zero owing to the boundary conditions assumed [see Eq. (4-7-4)]. Outside the domain the electric field is assumed to be spatially uniform so that

$$j(t) = qn_0 v(F_r) + \varepsilon \frac{dF_r}{dt} \tag{4-7-8}$$

Substituting $j(t)$ from Eq. (4-7-8) into Eq. (4-7-5) and taking into account Eq. (4-7-7) we obtain

$$\frac{dF_r}{dt} = \frac{dF_0}{dt} + \frac{qn_0}{\varepsilon L} \int_0^L [v(F) - v(F_r)] \, dx \tag{4-7-9}$$

or

$$\frac{dF_r}{dt} = \frac{dF_0}{dt} + \frac{qn_0}{\varepsilon L} \int_{F_r}^{F_m} \left\{ [v(F) - v(F_r)] \left[\frac{1}{|(\partial F/\partial x)_a|} + \frac{1}{|(\partial F/\partial x)_d|} \right] \right\} dF \tag{4-7-10}$$

where the subscripts a and d stand for the accumulation and depletion regions, respectively. The knowledge of the charge distribution in the domain walls is required to solve Eq. (4-7-10). This distribution is not known for the transient domain. In this simplified treatment we assume that the shape of the domain during the transient process is similar to the shape of the fully developed domain. The average charge density in the domain walls may then be estimated as

$$\langle \rho \rangle \simeq \frac{\varepsilon F_m}{d} = qn_0/(2\phi) \tag{4-7-11}$$

where d is the domain width. The domain shape function $\phi(n_0/n_{cr})$ is defined by Eq. (4-6-26) and approximated by Eq. (4-6-34). Using Eq. (4-7-11) we can rewrite Eq. (4-7-10) as

$$\frac{dF_r}{dt} = \frac{dF_0}{dt} + \frac{2\phi(n_0/n_{cr})}{L} \int_{F_r}^{F_m} [v(F) - v(F_r)] \, dF \tag{4-7-12}$$

In the limiting cases $n \ll n_{cr}$ and $n \gg n_{cr}$ eq. (4-7-12) may be simplified. When the carrier concentration is small a leading domain edge is totally depleted and the trailing edge (accumulation layer) is narrow. In this case $\phi \simeq 1/2$ [see Eq. (4-6-30)] and Eq. (4-7-12) can be rewritten as

$$\frac{dF_r}{dt} = \frac{dF_0}{dt} + \frac{1}{L} \int_{F_r}^{F_m} [v(F) - v(F_r)] \, dF \tag{4-7-13}$$

In the opposite limiting case of the large electron concentration the domain shape is nearly parabolic [see Eq. (4-6-33)] and Eq. (4-7-12) reduces to

$$\frac{dF_r}{dt} = \frac{dF_0}{dt} + \frac{8\sqrt{2}}{3L} \left(\frac{n_0}{n_{cr}} \right)^{1/2} \int_{F_r}^{F_m} [v(F) - v(F_r)] \, dF \tag{4-7-14}$$

For high field domains $(F_m \gg F_v)$ Eq. (4-7-12) may be further simplified. In this case the integral in the right-hand part of Eq. (4-7-12) may be rewritten as

$$\int_{F_r}^{F_m} [v(F) - v(F_r)] \, dF = \int_{F_s}^{F_v} [v(F) - v_s] \, dF - \int_{F_s}^{F_r} [v(F)$$

$$- v_s] \, dF + v_s F_m - v(F_r) F_m \qquad (4\text{-}7\text{-}15)$$

Using the notation introduced in Section 4-6

$$\int_{F_s}^{F_v} [v(F) - v(F_s)] \, dF = \mu F_c^2 \qquad (4\text{-}7\text{-}16)$$

and assuming $F_r \ll F_v$ we obtain from Eq. (4-7-12)

$$\frac{dF_r}{dt} = \frac{dF_0}{dt} + \frac{2\phi\mu F_c^2}{L} \left[1 - \frac{F_m(F_r - F_s)}{F_c^2} \right] \qquad (4\text{-}7\text{-}17)$$

In the stationary case $dF_r/dt = dF_0/dt = 0$ and Eq. (4-7-17) reduces to the equal area rule

$$F_m(F_r - F_s) = F_c^2 \qquad (4\text{-}7\text{-}18)$$

Substituting F_m from Eq. (4-6-27) into Eq. (4-7-17) we find

$$\frac{dF_r}{dt} = \frac{dF_0}{dt} + \frac{2\phi\mu F_c^2}{L} \left[1 - \left(\frac{qn_0 L}{\varepsilon\phi} \right)^{1/2} \frac{(F_0 - F_r)^{1/2}(F_r - F_s)}{F_c^2} \right] \qquad (4\text{-}7\text{-}19)$$

In the steady state this equation reduces to

$$(F_0^0 - F_r^0)^{1/2}(F_r^0 - F_s) = \left(\frac{\varepsilon\phi}{qn_0 L} \right)^{1/2} F_c^2 \qquad (4\text{-}7\text{-}20)$$

From Eq. (4-7-20) we can calculate the differential resistance of the sample with a propagating domain which is proportional to dF_0^0/dF_r^0:

$$R_d = R_0 + R_- \qquad (4\text{-}7\text{-}21)$$

where $R_0 = L/(q\mu n \cdot S)$ is the low field resistance and

$$R_- = - \frac{2U_{d0}^{3/2}}{(qn_0\varepsilon\phi)^{1/2}\mu F_c^2 S} \qquad (4\text{-}7\text{-}22)$$

or, in a different form

$$R_- = - \frac{2\varepsilon\phi(F_m^0)^3}{q^2\mu n_0^2 F_c^2 S} \qquad (4\text{-}7\text{-}23)$$

Equation (4-7-19) can be rewritten in terms of the domain voltage $U_d(t) = F_0(t)L - F_r(t)L$ as

$$\frac{dU_d}{dt} = \frac{2\mu}{L}\left(\frac{qn_0\phi}{\varepsilon}\right)^{1/2} U_d^{1/2}(U_{d0} - U_d) - 2\mu F_c^2\phi \qquad (4\text{-}7\text{-}24)$$

Here $U_{d0}(t) = F_0(t)L - F_s(t)L$. The second term in the right-hand part of Eq. (4-7-24) is much smaller than the first one if $F_m \gg F_v$ so that

$$\frac{dU_d}{dt} \simeq \frac{2\mu}{L}\left(\frac{qn_0\phi}{\varepsilon}\right)^{1/2} U_d^{1/2}(U_{d0} - U_d) \qquad (4\text{-}7\text{-}25)$$

For the constant bias $[F_0 = \mathrm{const}(t)]$ it is convenient to rewrite this equation in the dimensionless form

$$\frac{du}{dT} = \sqrt{u}(1 - u) \qquad (4\text{-}7\text{-}26)$$

where $u = U_d/U_{d0}$ is the dimensionless domain voltage, $T = t/\tau_f$, and

$$1/\tau_f = \frac{2\mu}{L}\left(\frac{qn_0\phi U_{d0}}{\varepsilon}\right)^{1/2} \equiv 2\phi\frac{\mu}{L}F_m^0 \qquad (4\text{-}7\text{-}27)$$

τ_f is the characteristic time constant of the domain transient. Here F_m^0 is the maximum field of a stable propagating domain. In the limiting case $n_0 \ll n_{cr}$ these expressions become identical to expressions derived in Ref. 33 and 34.

Equation (4-7-27) may also be presented as

$$\tau_f = \frac{1}{2\mu}\left[\frac{\varepsilon L}{qn_0 F_s\phi(F_0/F_s - 1)}\right]^{1/2} \qquad (4\text{-}7\text{-}28)$$

Assuming $\mu = 0.65 \text{ m}^2/\text{V s}$, $\varepsilon = 1.14 \times 10^{-10} \text{ F/m}$, and $F_s \simeq 1.5 \times 10^5 \text{ V/m}$ we find from Eq. (4-7-28)

$$\tau_{f_{(ps)}} \simeq 5.30 \times 10^7\left[\frac{L(\mu m)}{(F_0/F_s - 1)n_0(\text{cm}^{-3})\phi}\right]^{1/2} \qquad (4\text{-}7\text{-}28a)$$

The τ_f vs. n_0 dependence for different values of L for $F_0/F_s = 5$ is shown in Fig. 4-7-1. These results cease be valid when τ_f becomes less than 2–3ps, i.e., comparable to the energy relaxation time.

The solution of Eq. (4-7-26) describing the completion of the domain formation is given by

$$\Delta u \equiv 1 - \mu \simeq 4 \cdot e^{-T} \qquad (4\text{-}7\text{-}29)$$

or

$$\frac{\Delta U_d}{U_{d0}} \simeq 4e^{-t/\tau_f} \qquad (4\text{-}7\text{-}30)$$

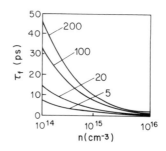

FIGURE 4-7-1. Time constant of the domain transient as a function of the electron concentration n_0. Numbers near the curves correspond to the device length in micrometers. $\mu = 0.65 \text{ m}^2/\text{V s}$, $F_0/F_s = 5$, $F_s = 1.5 \text{ kV/cm}$, $n_{cr} = 3 \times 10^{15} \text{ cm}^{-3}$.

The time constant τ_f may also be expressed as

$$\tau_f = C_D R_0 \tag{4-7-31}$$

where $R_0 = L/(qn_0\mu S)$ is a low field resistance and

$$C_D = \frac{\varepsilon S}{d} \equiv \frac{qn_0 S}{2\phi F_m^0} \equiv \frac{1}{2}\left(\frac{qn_0\varepsilon}{U_{d0}\phi}\right)^{1/2} S \tag{4-7-32}$$

is the domain capacitance [see Eq. (4-7-11)].

Below we shall use Eq. (4-7-19) to evaluate the impedance of a sample with a propagating domain and to describe the equivalent circuit of such a sample. We will then find the solution of Eq. (4-7-26) which describes the process of the domain formation. We will show that the domain transient may be interpreted as charging of the nonlinear capacitance of the domain equivalent circuit via the small signal resistance of the sample.

In order to find the small signal impedance of the sample with a propagating domain we shall assume that

$$F_0 = F_0^0 + F_0^1 e^{i\omega t} \tag{4-7-33}$$

where $F_0^1 \ll F_0^0$ and shall seek a solution of Eq. (4-7-19) as

$$F_r = F_r^0 + F_r^1 e^{i\omega t} \tag{4-7-34}$$

The substitution of Eq. (4-7-34) into Eq. (4-7-19) yields

$$i\omega F_r^1 = i\omega F_0^1 + \frac{1}{R_- C_D}(F_0^1 - F_r^1) - \frac{F_r^1}{R_0 C_D} \tag{4-7-35}$$

We can now find the ratio F_0^1/F_r^1 which can be related to the small-signal impedance Z of a device with unit area cross section using Eq. (4-7-8)

$$Z = \frac{R_0}{1 + i\omega R_0 C_0}\frac{F_0^1}{F_r^1} \tag{4-7-36}$$

where

$$R_0 = \frac{L}{qn_0\mu S} \tag{4-7-37}$$

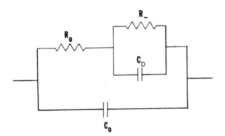

FIGURE 4-7-2. Small signal equivalent circuit of a sample with a stable propagating domain.

is a low field resistance of the sample and

$$C_0 = \varepsilon S / L \qquad (4\text{-}7\text{-}38)$$

is the geometric capacitance. From Eq. (4-7-35) and (4-7-36) we find

$$Z = \frac{R_0 + Z_D}{1 + i\omega R_0 C_0} \qquad (4\text{-}7\text{-}39)$$

where

$$Z_D = \frac{R_-}{1 + i\omega R_- C_D} \qquad (4\text{-}7\text{-}40)$$

is the domain impedance. Equations (4-7-39) and (4-7-40) correspond to the equivalent circuit of the sample with a stable propagating domain shown in Fig. 4-7-2. This equivalent circuit is in good agreement with the numerical simulation and experimental results [35]. Parameters of the domain equivalent circuit R_- and C_D calculated from Eqs. (4-7-22) and (4-7-32) are shown in Figs. 4-7-3 and 4-7-4 as functions of n_0 for different domain voltages.

Let us now consider the process of domain formation using Eq. (4-7-26). At the very beginning of this process the domain voltage is zero, corresponding to the initial condition

$$u\big|_{t=0} = 0 \qquad (4\text{-}7\text{-}41)$$

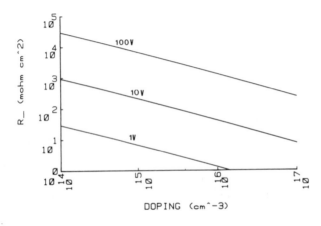

FIGURE 4-7-3. Domain negative differential resistance R_- as a function of the carrier concentration. Numbers near the curves correspond to the domain voltage in volts.

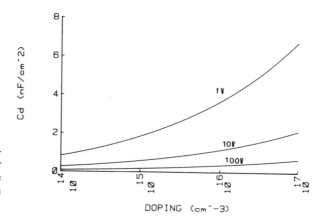

FIGURE 4-7-4. Domain capacitance as a function of the carrier
concentration. Numbers near the
curves correspond to the domain
voltages in volts.

For this initial condition the solution of Eq. (4-7-26) is

$$u = \left(\frac{e^{\tau} - 1}{e^{\tau} + 1}\right)^2 \tag{4-7-42}$$

Computer simulations [36, 37] confirm that Eq. (4-7-42) fairly accurately describes
the domain formation. We should stress, however, that in heavily doped samples
where τ_f becomes comparable to the energy relaxation time the transient effect may
change the domain dynamics quite substantially.

In the case when the high field domain reaches the anode the formation of a
new domain is effected by the process of the previous domain dissappearance in
the anode [38–40]. The domain formation time under such conditions may be
substantially different from $3.5\tau_f$ [40].

Another process of the domain transient to be considered is the dispersal of
the high field domain when the bias drops below the sustaining voltage F_sL. The
computer simulation [41] shows that the domain voltage during the domain annihilation roughly varies as

$$U_d = U_{d0}\, e^{-t/\tau} \tag{4-7-43}$$

τ/τ_f ratio is nearly independent of the sample parameters and is a strong function
of the bias voltage (see Fig. 4-7-5).

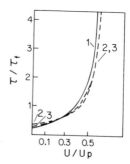

FIGURE 4-7-5. τ/τ_f vs. normalized bias voltage for three samples with
different values of τ_f [41]. 1, $\tau_f = 41$ ps; 2, $\tau_f = 9.5$ ps; 3, $\tau_f = 2.5$ ps.

4-8. ACCUMULATION LAYERS

Stable propagating high field domains typically exist in relatively long samples which are used as the microwave oscillators at frequencies below or about 10 GHz (see Chapter 5). In short samples which operate as the microwave oscillators at frequencies of the order of 50 GHz or above a typical form of instability is a propagation of accumulation layers (see Figure 4-8-1) [18, 42–45]. The propagation velocity of the accumulation layers and their field distribution may be found by solving Eqs. (4-6-2) and (4-6-3), which may be rewritten in the dimensionless form [49] as

$$\frac{d^2y}{d\tau^2} + \frac{1}{\sqrt{\nu}}[w - S(y)]\frac{dy}{d\tau} = S(y) - S(y_1) \qquad (4\text{-}8\text{-}1)$$

Here $y = F/F_p$ is the dimensionless electric field, $\tau = (x - ut)/l$, $l = (\varepsilon D F_p / q n_0 v_t)^{1/2}$, $\omega = |u|/V_p$ is the dimensionless velocity of the accumulation layer, $S(y) = v(F)/v_p$, F_p and v_p are the peak field and peak velocity, respectively, and $\nu = qDn_0/\varepsilon v_p F_p$. Equation (4-8-1) has the same form as Newton's equation for a particle of a unit mass moving under a force $S(y) - S(y_1)$ and experiencing a viscous friction with coordinate-dependent coefficient $[w - S(y)]/\nu$. The general properties of Eq. (4-8-1) and its solutions describing the high field domain and accumulation layers have been studied in Ref. 46–48. The boundary conditions corresponding to the accumulation layer are

$$y(-\infty) = y_1, \qquad y(\infty) = y_2 \qquad (4\text{-}8\text{-}2a)$$

$$\left(\frac{dy}{d\tau}\right) = 0 \qquad (4\text{-}8\text{-}2b)$$

$$\tau = \pm\infty$$

The conditions do not uniquely define the accumulation layer velocity. In particular, solutions corresponding to the electric field $F(Z)$ oscillating in space at the boundary of the accumulation layer are possible [47–49]. Such oscillatory accumulation layers are unstable, however, toward the conversion into the high field domains and have little practical interest [49].

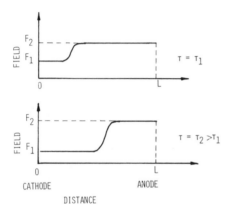

FIGURE 4-8-1. Propagation of an accumulation layer in a GaAs sample.

In order to describe the accumulation layer in the finite sample, Eq. (4-8-1) should be solved together with the equation of the voltage balance

$$F_1 x + F_2(L - x) = U \qquad (4\text{-}8\text{-}3)$$

and the equation demanding the equal electron velocities (equal currents) at both ends of the accumulation layer

$$v(F_1) = v(F_2) \qquad (4\text{-}8\text{-}4)$$

An estimate for the accumulation layer velocity may be obtained from Eq. (4-6-11). Substituting $F = F_2$ and $F_r = F_1$ into Eq. (4-6-11) assuming for simplicity the field-independent diffusion and taking into account that $\rho(F_2) = \rho(F_1) = 0$ we find

$$-[v(F_1) + u]\frac{|\bar{\rho}|(F_2 - F_1)}{|\bar{\rho}| + qn_0} = \int_{F_1}^{F_2} [v(F) - v(F_1)]\, dF \qquad (4\text{-}8\text{-}5)$$

Here $\bar{\rho}$ is the average space charge in the accumulation layer. For large values of $F_2 \gtrsim F_v$ the right-hand side of Eq. (4-8-5) is equal to μF_c^2 and

$$u = -v(F_1) - \frac{\mu F_c^2}{F_2 - F_1}\frac{|\bar{\rho}| + qn_0}{|\bar{\rho}|}. \qquad (4\text{-}8\text{-}6)$$

When $n_0 \ll n_{cr}$ the space charge $|\bar{\rho}|$ is much greater than n_0 and

$$u = -v(F_1) - \frac{\mu F_c^2}{F_2 - F_1} \qquad (4\text{-}8\text{-}7)$$

In the opposite limiting case $n_0 \gg n_{cr}$, $|\bar{\rho}|$ is nearly independent of u and much smaller than n_0:

$$|\bar{\rho}| \simeq (2n_{cr}n_0)^{1/2} \qquad (4\text{-}8\text{-}8)$$

[see Eq. (4-6-32)]. Therefore

$$u = -v(F_1) - \frac{\mu F_c^2}{F_2 - F_1}\left(\frac{n_0}{2n_{cr}}\right)^{1/2} \qquad (4\text{-}8\text{-}9)$$

The results of a more rigorous analysis [49] also lead to Eq. (4-8-7) and (4-8-9) for large values of F_2. These equations are also in agreement with the results of the numerical simulation for a finite sample [41] (see Fig. 4-8-2). As can be seen from

FIGURE 4-8-2. The velocity of the accumulation layer as a function of the field F_2 [49]. Solid lines, results of the analysis based on the solution of Eq. (4-8-1); circles, results of the computer simulation [41]; dashed line, v vs. F curve used in the calculation. 1, $n_0 = 3 \times 10^{15}$ cm^{-3}; 2, $n_0 = 10^{15}$ cm^{-3}; 3, $n_0 = 10^{14}$ cm^{-3}.

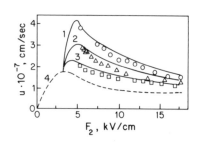

this figure and in agreement with the computer simulation [43], the accumulation layer velocity may be substantially higher than the peak velocity v_p.

The computer simulation [43, 44] also revealed the presence of a dead region near the cathode and the effect of electron thermalization when an accumulation layer disappears and a new accumulation layer nucleates in the sample.

4-9. STATIONARY ANODE DOMAINS

In low doped samples ($n_0 \ll n_{cr}$) the domain propagation is necessary in order to maintain the domain stability. Indeed, in this case the domain leading edge is almost totally depleted of carriers and the displacement current in the depletion edge (induced due to the domain motion) maintains the current continuity. In high doped samples ($n_0 \gg n_{cr}$), however, the space charge in the domain walls is small compared to the electron concentration ($\rho_{max} \lesssim (2n_{cr}n_0)^{1/2}$) and the diplacement current is small compared to the conduction current. Under such conditions a perfectly stable stationary high field domain may appear near the anode [50–57]. The field outside the anode domain is smaller than the domain nucleation field F_p so that the propagating domains do not appear. The computer simulation [56, 58, 59] reveals the following picture of this mode of instability. It exists only when n_0 is close to or larger than n_{cr}. If the bias is just slightly above the Gunn threshold the traveling domains exist in the sample, leading to a conventional Gunn effect. However, when the bias exceeds a critical value (which is typically close to the Gunn threshold) the stable anode domain appears. Close to this new threshold small traveling domains still exist in the sample, but a further increase in bias makes the device perfectly stable. At substantially larger bias voltages the anode domain again disappears and the traveling Gunn domains emerge again leading to the conventional Gunn oscillations. The current–voltage characteristic of the sample with the anode domain is strongly dependent on the anode boundary conditions (see Figure 4-9-11). The possibility of the domain formation near the anode also depends on the cathode boundary conditions (see Section 4-11). The formation of the anode domain is more probable when the electric field at the cathode is low [59].

The characteristic time constant of the anode domain transients is given by

$$\tau_f = C_D R_{od} \qquad (4\text{-}9\text{-}1)$$

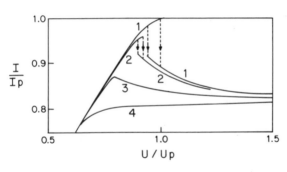

FIGURE 4-9-1. Steady-state I–V characteristics for the sample with the anode domain [59]. $n_0 = 8 \times 10^{15}$ cm^{-3}. The fixed concentration at the anode n_d(cm^{-3}): 1, 8×10^{15}; 2, 1×10^{16}; 3, 1.2×10^{16}; 4, 1.6×10^{16}. The length of sample $L = 4\ \mu$m. Arrows show the switches between conditions without anode domain and "with anode domain" for nucleation and annihilation domain threshold voltages. Currents and voltages are normalized to the peak current and peak voltage.

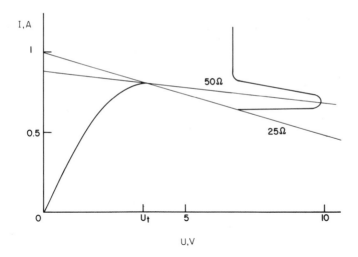

FIGURE 4-9-2. *I-V* characteristic of a sample with a stationary domain [50]. Note *N*-type switching with a 25-Ω load and *S*-type switching with a 50-Ω load.

where C_D is the domain capacitance (see Eq. [4-7-32]) and

$$R_{od} = \frac{L}{qn_0\mu_d(F_r)S} \qquad (4\text{-}9\text{-}2)$$

the differential resistance of the sample for the outside domain field F_r (which is identical to the time constant of a propagating domain). Moreover, in most cases the domain formation process may be fairly well described by Eq. (4-7-42) with the time constant given by Eq. (4-9-1).

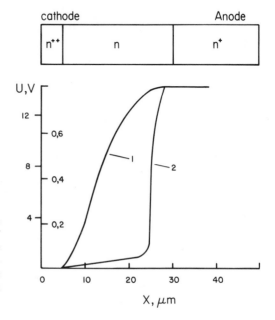

FIGURE 4-9-3. Field distribution along the Gunn diode [60]. 1, Before switching (0.8 V bias); 2, after switching (14 V bias). The state with the anode domain had a high conductivity (*S*-type switching), probably due to impact ionization within the high field domain.

The appearance of the stationary anode domain may lead to switching from a low resistive state without the domain to a high resistive state with the domain (N-type switching) [50, 52, 53]. If the bias voltage is increased even higher the impact ionization in high field domain may lead to the S-type current voltage characteristic [50, 60]; see Figs. 4-9-2 and 4-9-3.

4-10. MULTIDOMAIN REGIME

Under the dc bias the domain nucleation at one of the several possible nucleation sites leads to the decrease of the electric field outside the growing domain to the value smaller than the peak field. Therefore, only one domain nucleates in the sample during each cycle. If, however, the bias increases with time fast enough, the outside domain field may remain constant or even rise. Under such conditions the simultaneous formation of several high field domains is possible [61–63]. The condition of the multidomain regime is

$$\frac{dU}{dt} > \frac{U_d}{\tau_f} \tag{4-10-1}$$

where τ_f is the domain transient time constant. Using Eq. (4-7-27), we can rewrite Eq. (4-10-1) as

$$\frac{dU}{dt} > \frac{2\mu U_d^{3/2}}{L}\left(\frac{qn_0\phi}{\varepsilon}\right)^{1/2} \tag{4-10-12}$$

Using the parameters corresponding to the device studied in [63] $\mu = 0.5 \ \text{m}^2/\text{V s}$, $L \approx 10^{-2} \ \text{m}$, $n_0 = 3 \times 10^{20} \ \text{m}^{-3}$, $\varepsilon = 1.14 \times 10^{-10} \ F/m$, $U_d \approx 300 \ \text{V}$ and assuming $\phi \approx 1/2$ [see Eq. (4-6-30)] we obtain $dU/dt > 2.4 \times 10^{11} \ \text{V/s}$, which is in agreement with the experimental value of $dU/dt \sim 10^{12} \ \text{V/s}$ which was necessary for the nucleation of the multiple domains. The field distributions in the multidomain regime predicted by the computer simulation [64] are shown in Fig. 4-10-1 for the samples with $L = 35 \ \mu\text{m}$ and $L = 70 \ \mu\text{m}$. The samples were divided into regions of $0.5 \ \mu\text{m}$ width with the randomly changing doping with the mean standard deviation of 11%. The simulation included a high-quality parallel resonance LCR circuit leading to the sinusoidal voltage variation across the sample. At a low frequency ($f = 2.7 \ \text{GHz}$) a single traveling domain propagates in the sample during each cycle (see Figure 4-10-1). At $f = 6 \ \text{GHz}$ the multidomain regime with a complex field profile occurs (see Fig. 4-10-1b). At a still higher frequency the electric field in the sample remains larger than the peak field and the current drops due to the negative slope of the v vs. F curve. We should also note that the sample response time in the multidomain mode is faster than for the single domain mode because several domains absorb the bias voltage simultaneouly:

$$\frac{1}{\tau_{\text{eff}}} \simeq \frac{n}{\tau_f} \tag{4-10-3}$$

where n is the number of the domains.

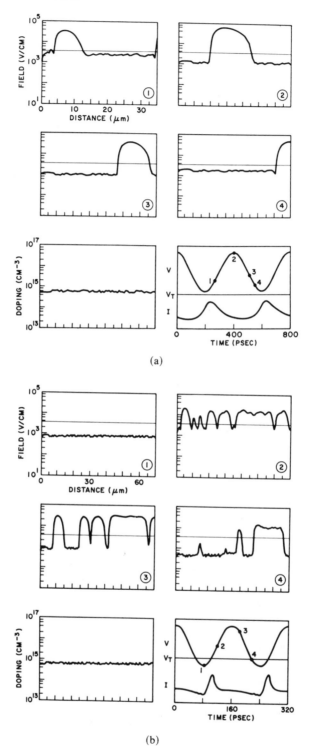

FIGURE 4-10-1. Multidomain regime [64]. (a) $f = 2.7$ GHz, $L = 35$ μm; (b) $f = 6$ GHz, $L = 70$ μm; (c) $f = 14$ GHz, $L = 70$ μm. Field distributions correspond to the moments of time indicated on the voltage waveforms.

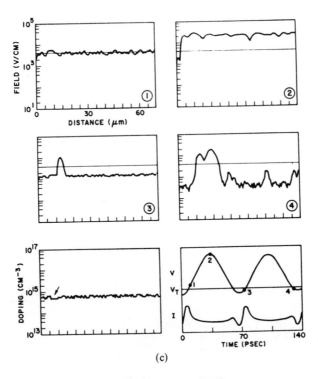

(c)

FIGURE 4-10-1 (*continued*)

4-11. INFLUENCE OF INHOMOGENEITIES AND CONTACTS

Practical Gunn devices may exhibit behavior that is very different from the idealized Gunn effect described in Sections 4-1-4-7. These deviations may be explained by the important role played by the sample inhomogeneities and contacts.

4-11-1. *Influence of Inhomogeneities*

J. B. Gunn was first to point out that a local inhomogeneity near the cathode becomes a domain nucleation site [2]. If there are many doping inhomogeneities the domain may nucleate randomly at different sites leading to the incoherent Gunn oscillations [61].

If the doping concentration somewhere in the sample exceeds the average doping level by a factor of more than F_p/F_s, the electric field within the inhomogeneity may become low enough to suppress the domain propagation. In this case the domain transit length is limited to the distance between the nucleation site and the doping inhomogeneity.

If the variation in the doping level is not too large and the characteristic size of the doping inhomogeneities is bigger than the domain width, the current waveform reproduces the doping profile as has been pointed out in Section 4-2 [see Eq. (4-2-9)]. Assuming that the doping variation is smooth enough so that the diffusion current may be neglected, we find

$$q\mu n_0(x_d)F_r(x_d) = q\mu n_0(x)F_r(x) \qquad (4\text{-}11\text{-}1)$$

where x_d is the domain coordinate (assuming that the domain size is much smaller than the characteristic size of the doping variation). The bias voltage

$$U = U_d + \int_0^L F_r(x)\, dx \qquad (4\text{-}11\text{-}2)$$

The substitution of Eq. (4-11-1) into Eq. (4-11-2) yields

$$U = U_d + n_0(x_d)F_r(x_d)\int_0^L \frac{dx}{n_0(x)} \qquad (4\text{-}11\text{-}3)$$

For the high field domains $(F_m \gg F_v)$ $F_r(x_d) \simeq F_s$, so that

$$U_d = U - n_0(x_d)F_s\beta \qquad (4\text{-}11\text{-}4)$$

where $\beta = \int_0^L dx/n_0(x)$. From this equation we find the domain sustaining voltage

$$U_{ds} = n_0(x_d)F_s\beta \qquad (4\text{-}11\text{-}5)$$

which varies as the domain propagates along the sample because of the variation of x_d and $n_0(x_d)$:

$$x_d = x - ut \qquad (4\text{-}11\text{-}6)$$

where u is the domain velocity. Apparently, for the domain to travel across the whole sample U_{ds} must be smaller than $U_{th} = F_p L$, i.e.,

$$n_{max}F_s\beta < F_p L \qquad (4\text{-}11\text{-}7)$$

If there is just one doping inhomogeneity in the sample which is much shorter than the sample length then $\beta \simeq L/n_0$ and

$$\frac{n_{max}}{n_0} < \frac{F_p}{F_s} \qquad (4\text{-}11\text{-}8)$$

which is in good agreement with the experiment where the carrier concentration has been modulated by a He–Ne laser beam [66]. The computer simulation [67] also confirms the results of the analytical treatment in the case when the domain field is large enough and the domain size is small compared to the size of the, inhomogeneities. When the domain size is much bigger than the inhomogeneities they do not appreciably change the device performance except for the multidomain regime when they could become multiple domain nucleation sites. When the domain size is of the order of the inhomogeneity size the change in doping may change the domain parameters quite substantially [67].

4-11-2. Influence of Contacts

The role of contacts is illustrated by Fig. 4-11-1 [8], where the dc I–V characteristics of the samples with different contacts are presented. As can be seen from the

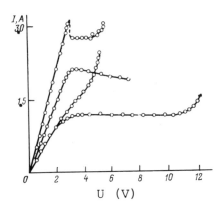

FIGURE 4-11-1. DC I-V characteristics of Gunn diodes with different contacts [8]. All samples are made from the same bulk-grown GaAs with $n_0 = 8 \times 10^{14}$ cm^{-3} and $\mu = 5800$ cm^2/V s. $L = 100\ \mu$m.

figure all kinds of behavior are possible, ranging from the current drop at $F \simeq F_p$ to the superlinear behavior.

Let us first consider the stationary electric field distributions in the samples with different types of contacts. The basic equations are the Poisson equation

$$\frac{dF}{dx} = \frac{q}{\varepsilon}(n_0 - n) \qquad (4\text{-}11\text{-}9)$$

and the equation for the current density

$$j = qnv(F) + qD\frac{dn}{dx} \qquad (4\text{-}11\text{-}10)$$

For simplicity we will limit ourselves to the low carrier concentrations $n_0 \ll n_{cr}$ when the diffusion effects may be neglected everywhere except the narrow boundary region near the anode. We shall choose the fixed value of the electric field at the cathode F_{cb} as a boundary condition [14, 68–71]. When $D \to 0$ Eqs. (4-11-9)–(4-11-10) may be rewritten as

$$\frac{dF}{dx} = -\frac{q}{\varepsilon}\left[\frac{j}{qv(F)} - n_0\right] \qquad (4\text{-}11\text{-}11)$$

The solution of Eq. (4-11-11) is given by

$$X = -\frac{\varepsilon}{q}\int_{F_{cb}}^{F}\frac{dF'}{j/qv(F') - n_0} \qquad (4\text{-}11\text{-}12)$$

General properties of this solution may be illustrated using the $n - F$ plane [69] where

$$n = \frac{j}{qv(F)} \qquad (4\text{-}11\text{-}13)$$

curve is shown together with the $n = n_0$ line (see Fig. 4-11-2). These lines divide the plane into regions with different signs of dF/dx and dn/dx. Indeed when

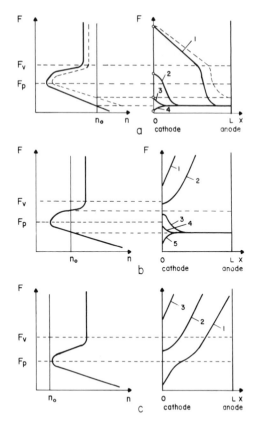

FIGURE 4-11-2. Stationary field distributions in GaAs samples for different current densities [8]. (a) Small current density $j < qn_0v_s$; (b) intermediate current density $qn_0v_s < j < qn_0v_p$; (c) large current density $j > qn_0v_p$.

$n > n_0$ the field derivative $dF/dx < 0$ [see Eq. (4-11-8)]. When $n > j/qv(F)$ the electron concentration derivative $dn/dx < 0$ [see Eq. (4-11-8)]. Therefore, given the cathode field F_{cb} and the current density j we can sketch F vs. x profiles (see Fig. 4-11-2).

The fixed value of the cathode field is not necessarily a realistic boundary condition. Nevertheless, small values of F_{cb} may be related to the ohmic contacts whereas large values of F_{cb} correspond to the Shottky barrier contact. It may be difficult to relate F_{cb} to a given contact in a more accurate way because in a real sample it may not be clear which point should be chosen as the boundary between the sample and the contact. Moreover, even if such a point is chosen in some way, F_{cb} will, generally speaking, be bias dependent.

Stationary field distributions shown in Figure 4-11-2 may be unstable with respect to the domain formation. A detailed computer simulation and experimental study of GaAs samples with different types of contacts [68, 70, 14] relates the sample behavior to the cathode contact field F_{cb}. Samples with $n_0 = 10^{15}$ cm^{-3}, $L = 10^{-2}$ cm, and $\mu = 0.686$ m^2/V s were studied. (In higher doped samples with $n_0 \gtrsim n_{cr}$ the anode boundary condition may be as important as the cathode boundary condition in determining the stationary anode domain formation; see Section 4-8). The study showed that a more or less conventional Gunn effect exists, when the cathode field corresponds to the area hatched in Fig. 4-11-3. The threshold voltage

$$U_{th} \simeq \frac{v(F_{cb})L}{\mu} \qquad (4\text{-}11\text{-}14)$$

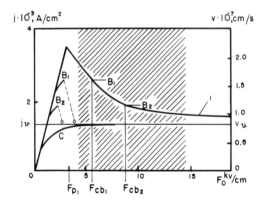

FIGURE 4-11-3. Gunn diode I-V characteristics for different values of the electric field at the cathode [68]. 1, $v(F)$ curve used in the simulation. B_1, B_2, C, I-V characteristics for different values of F_{cb}.

and the threshold current density

$$j_{th} = qn_0 v(F_{cb}) \tag{4-11-15}$$

(see curves B_1 and B_2 of the figure). At bias fields, F_0, smaller than the threshold the current–voltage characteristic tends to saturate and a potential drop develops near the cathode.

At high cathode fields the potential drop near the cathode becomes large and the samples are stable in the resistive circuit (see curve C in Fig. 4-10-1 calculated for $F_{cb} = 24\,\text{kV/cm}$). The sample may, however, become unstable in a resonance circuit.

When the cathode field is small the sample behavior strongly depends on the inhomogeneities and on the external circuit. The domains may be periodically nucleated at a large inhomogeneity with the threshold voltage and current density close to $F_p L$ and $qn_0 v_p$. The current drop is of the order of $qn_0(v_p - v_s)$. This situation is the closest to the "ideal" Gunn effect described in Section 4-1. However, other types of behavior may occur depending on the external circuit, doping profile, and the type of the inhomogeneities. For example, the first high field domain launched may reach the anode and stay there as a stationary domain. In the resistive circuit the current oscillations then cease, however, in the resonance circuit the electric field distribution may still oscillate.

All these types of behavior have been observed experimentally, underscoring the important role which contacts and inhomogeneities play in the Gunn effect [41].

4-12. DOMAINS IN THIN SAMPLES

In a thick sample most of the electric field streamlines created by a space charge within the domain walls or in the space charge wave are in the longitudinal direction. However, if a sample is thin or covered with a dielectric film with a large dielectric constant the transverse flux of the electric field becomes important (see Fig. 4-12-1).

In the one-dimensional case the longitudinal field component created by the electric charge Q is given by

$$F_x = \frac{Q}{2\varepsilon S} \tag{4-12-1}$$

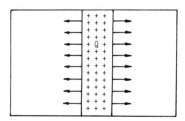

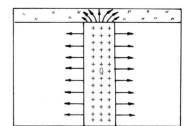

FIGURE 4-12-1. Streamlines of the electric field of a positive
charge in a wide sample, thin sample, and sample covered with
a dielectric film with a high dielectric constant.

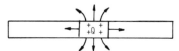

according to Gauss's law. When the transverse component of the electric field F_y
is taken into account Gauss's law states

$$Q = 2\varepsilon S F_x + 2\varepsilon_d F_y \lambda W \qquad (4\text{-}12\text{-}2)$$

where ε_d is the permittivity of the dielectric covering the sample (we assume that
the film is on both surfaces), W is the sample width, and λ is the size of the space
charge region. Assuming for a crude estimate that $F_x \sim F_y$ we find

$$F_x \sim \frac{Q}{2\varepsilon S \left(1 + \dfrac{\varepsilon_d}{\varepsilon} \dfrac{\lambda}{d_0}\right)} \qquad (4\text{-}12\text{-}3)$$

where d_0 is the sample thickness. The growth of the space charge waves in Gunn
devices is due to the decrease of the electron drift velocity with the increase in the
longitudinal electric field. This decrease in the velocity leads to the space charge
build-up, which in turn increases the electric field, etc. The decrease of F_x by a
factor $1 + \varepsilon_d \lambda / \varepsilon d_0$ leads to the corresponding decrease in gain. Therefore the
Kroemer criterion (see Section 4-2) should be modified by this factor

$$n_0 L > (n_0 L)_1 \left(1 + \frac{\varepsilon_d}{\varepsilon} \frac{\lambda}{d_0}\right) \qquad (4\text{-}12\text{-}4)$$

[compare with Eq. (4-2-6)]. For the fundamental component of the space charge
waves $\lambda \sim L$, and when $\varepsilon_d L \gg \varepsilon d_0$ Eq. (4-12-4) reduces to

$$n_0 d_0 > (n_0 L)_1 \frac{\varepsilon_d}{\varepsilon} \qquad (4\text{-}12\text{-}5)$$

When the sample surface is exposed to air we should have

$$n_0 d_0 > \frac{(n_0 L)_1}{12.9} \qquad (4\text{-}12\text{-}6)$$

The domain transient time constant τ_f is affected in a similar way [72]

$$\tau_f = (C_D + C_s) R_0 \simeq \tau_f \left(1 + \frac{\varepsilon_d d}{\varepsilon d_0}\right) \qquad (4\text{-}12\text{-}7)$$

Here

$$C_s \simeq \varepsilon_d W \qquad (4\text{-}12\text{-}8)$$

the fringing capacitance of the domain edge d is the domain width.

More rigorous linear analysis of the space charge waves in thin samples [73–77] leads to the following dispersion relation for the space charge waves [compare with Eq. (4-4-9)]:

$$k = -\frac{\omega}{v} - i\frac{q n_0 \mu_d \omega d_0 (1 - v^2/c_d^2)^{1/2}}{v^2 \varepsilon_d} \qquad (4\text{-}12\text{-}9)$$

Here $c_d = c/(\varepsilon_d \tilde{\mu}_d)^{1/2}$ is the speed of light in the dielectric, the diffusion is neglected, v is the drift velocity of the carriers [$v = v(F)$], and $\tilde{\mu}_d$ is the magnetic permeability of the dielectric. When $c_d \gg v$ Eq. (4-12-9) becomes

$$k = -\frac{\omega}{v} - i\frac{\omega d_0}{\tau_m v^2} \frac{\varepsilon}{\varepsilon_d} \qquad (4\text{-}12\text{-}10)$$

This result confirms that the gain [$\mathrm{Im}(k)$] in a thin sample decreases in $\varepsilon_d \lambda / \varepsilon d_0$ times [compare with Eq. (4-4-9)].

The suppression of the Gunn oscillations in thin samples with and without dielectric cover was reported in Refs. 78–85. The experimental results are in general agreement with Eq. (4-12-4). Figure 4-12-2 illustrates how the device behavior changes depending on the carrier concentration and device thickness [80]. When the $n_0 d_0$ product is large (region I) the Gunn threshold voltage U_{th} is practically the same as in an infinitely thick sample. In region II the threshold voltage depends on the $n_0 d_0$ product, and in region III the Gunn oscillations are absent. The value of $(n_0 d_0)_{\text{cr}}$ could be estimated from the figure as approximately 1.5×10^{11} cm^{-2}.

The disappearance of Gunn oscillations was also caused by covering the samples with different liquids with large dielectric constants [81]. The critical value of $(n_0 d_0)_{\text{cr}}$ is between $(\varepsilon_d / \varepsilon_0) \times (2 \times 10^{10})$ cm^{-2} and $(\varepsilon_d / \varepsilon_0) \times 10^{10}$ cm^{-2} according to that study [81].

If $n_0 d_0$ is larger than $(n_0 d_0)_{\text{cr}}$ the high field domains may form but their formation time may increase substantially [83, 84] [see Eq. (4-12-7)]. In the sample covered with a BaTiO$_3$ film the domain formation time increased from 200 ps to 2 ns [83]. The domain parameters also change (see Fig. 4-12-3). As can be seen from the figure, the dielectric film and the finite sample thickness lead to a smaller domain voltage and larger outside domain field.

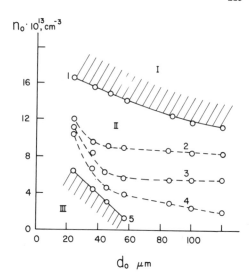

FIGURE 4-12-2. Gunn diode behavior as a function of the $n_0 d_0$ product [80]. Curve 1, threshold voltage U_p; curve 2, threshold voltage $1.1 U_p$; curve 3, threshold voltage $1.3 U_p$; curve 4, threshold voltage $1.5 U_p$; curve 5, no Gunn oscillations.

The quality of the sample surface is also important, even when the surface layer is very small compared to the device dimensions [85].

If the domain nucleates at a small "point" inhomogeneity it has to grow in two dimensions. Computer simulations [86] and simple qualitative arguments [87] show that the domain propagates in the transverse direction with the speed of the order of 10^6 m/s, i.e., about an order of magnitude higher than the speed of the domain propagation in the longitudinal direction.

4-13. BREAKDOWN IN HIGH FIELD DOMAINS

At large domain voltages the domain electric field may become large enough to cause the impact ionization. The light emission from the high field domain is one of the consequences of the impact ionization [88, 89]. The radiation wavelength (0.9 μm) corresponds to the energy gap of GaAs. As a result of the impact ionization within the high field domain the Gunn oscillations may become incoherent [88, 90,

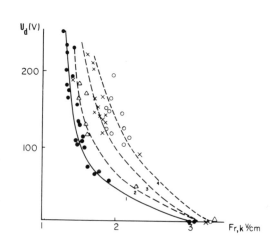

FIGURE 4-12-3. Domain voltage versus outside domain field [83]. 1, Sample surface covered with glass, sample thickness $d_0 = 2000\ \mu$m; 2, 3, 4, sample covered with BaTiO$_3$; 2, $d_0 = 170\ \mu$m; 3, $d_0 = 160\ \mu$m; 4, $d_0 = 130\ \mu$m.

91] and the current–voltage characteristic may become S-type [92–96] leading to the current filamentation [96–98]. When the electron–hole concentration in the current filament is large enough the stimulated light emission is observed [97–99].

The mechanism responsible for these effects was explained in Ref. 100. When the domain electric field is not too large the impact ionization in the high field domain is relatively weak so that the characteristic time of the electron–hole generation is large compared to the domain transit time across its own width. Under such conditions the extra concentration of electrons generated during one cycle is small compared to the total electron concentration. After several cycles, however, the electron concentration may change substantially. After each cycle a new domain appears in the sample with a slightly higher electron concentration, n. If the voltage bias and hence the domain voltage are kept constant the domain field increases with n and the domain width decreases. As a result there may be two states of the device corresponding to the same bias—the state with a low value of n and a wide domain with relatively low domain field and the state with a high value of n and a narrow domain with a higher domain field. This corresponds to the S-type n vs. U curve, i.e., to the S-type I–V characteristic on a time scale large compared to the electron transit time (see Fig. 4-13-1).

The S-type negative resistance may lead to a current filamentation. When the electron–hole density in the filament exceeds the threshold value the stimulated emission from the filament may occur. Such an effect was observed in Gunn diodes doped up to 5×10^{17} cm^{-3} [97]. The sample surfaces were optically polished and served as the mirrors of the resonance cavity. At relatively small currents the Gunn oscillations were accompanied by the spontaneous recombination emission. When the current exceeded a critical value the spectrum of the radiation and the beam narrowed, indicating the stimulated emission. The stimulated radiation was emitted from thin filaments whereas the spontaneous emission was uniform over the cross section. Similar behavior was observed in InP samples doped at 1.4 and 1.8×10^{17} cm^{-3} [98]. If the maximum domain field is small, there is no impact ionization in the domain. With the increase in the domain field, F_m, the impact ionization begins, but the characteristic time of the electron–hole pair generation is large compared with the domain transit time, T, if F_m is still not too high. When the recombination time is also much larger than T, the electron concentration is practically constant along the sample although it can increase substantially during the

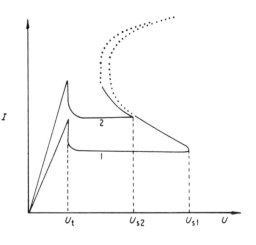

FIGURE 4-13-1. The qualitative current-voltage time-averaged characteristics of two Gunn diodes ($n_{01} > n_{02}$) [102].

many domain transits. In this case we can calculate the domain parameters using the standard theory of stable domains (see Section 4-6). We use the stable domain theory, assuming a field-independent D and a static velocity field characteristic $v(F)$ that saturates beyond the valley field F_v. The validity of this standard approximation will be discussed below. The generation rate G can be determined from Refs. 101–102:

$$G = \frac{1}{qT} \int_0^T \alpha(F) j \, dt \qquad (4\text{-}13\text{-}1)$$

where $\alpha(F)$ is the electron ionization rate and j is the conduction current density (we assume that for the values of the domain electric field the hole ionization rate is smaller and may be neglected). The current density is given by

$$j = qnv(F) + D\frac{\partial n}{\partial x} \qquad (4\text{-}13\text{-}2)$$

and by

$$qn_r v_s = j + \varepsilon\frac{\partial F}{\partial t} \qquad (4\text{-}13\text{-}3)$$

where v_s is the electron velocity outside the domain (assumed to be equal to the saturation velocity) and n_r is the concentration of carriers outside the high field domain. Assuming that the constant domain velocity is close to v_s (so that all variables depend on $x - v_s t$) and using the Poisson equation, we find from Eqs. (4-13-1)–(4-13-3)

$$j = qn(F)v_s \qquad (4\text{-}13\text{-}4)$$

$$G = \frac{\varepsilon F_m \alpha_\infty v_s}{qL} J(a) \qquad (4\text{-}13\text{-}5)$$

where

$$J(a) = \int_0^1 \frac{y(w) \exp(-a^2/w^2) \, dw}{1 - y(w)} \qquad (4\text{-}13\text{-}6)$$

$w = F/F_m$, $y = n/n_{cr}$, $a = F_i/F_m$. In Eq. (4-13-6) we assumed (according to Ref. 103)

$$\alpha(F) = \alpha_\infty \exp(-F_i^2/F^2) \qquad (4\text{-}13\text{-}7)$$

where $\alpha_\infty = 2 \times 10^5 \text{ cm}^{-1}$ and $F_i = 550 \text{ kV/cm}$. The integral in Eq. (4-13-5) can be evaluated in three limiting cases [104]:

$$\frac{n_0}{n_{cr}} \gg 1 \qquad (4\text{-}13\text{-}8)$$

$$\frac{1}{8\pi a^2} \ll \frac{n_0}{n_{cr}} \ll 1 \qquad (4\text{-}13\text{-}9)$$

$$\frac{n_0}{n_{cr}} \ll \frac{1}{8\pi a^2} \qquad (4\text{-}13\text{-}10)$$

where n_{cr} is given by Eq. (4-6-24). Here we limit ourselves to the consideration of the most important case of the intermediate concentration when $1/8\pi a^2 \ll n_0/n_{cr} \ll 1$. As we showed in Section 4-6 the domain field profile is determined by the ratio n/n_{cr}. If $n/n_{cr} \ll 1$, the domain leading edge is fully depleted and the accumulation layer is very narrow so that the field profile is close to the rectangular triangle; but if $n/n_{cr} \gg 1$, the space charge is small compared with qn_0 and the field profile is almost symmetrical. For the typical domain parameters $F_m \ll F_i$. So impact ionization takes place predominantly near the domain top [because of the exponential dependence of $\alpha(F)$ on F]. In GaAs the impact ionization in the domain occurs if $F_m \gg F_v$. In this case we obtain a simple expression for the charge profile near the domain top [see Eq. (4-6-23)]:

$$y - 1 = \left[\frac{2n_{cr}}{n_r}(1 - w)\right]^{1/2} \qquad (4\text{-}13\text{-}11)$$

Substituting (4-13-11) into (4-13-6) we obtain the explicit expression for the integrand of $J(a)$ which is a rapidly varying function of w, so we can estimate $J(a)$ using the standard methods. The resulting formula for G is

$$G = \left(\frac{8\pi n_r^3}{n_{cr}}\right)^{1/2} \frac{\alpha_\infty v_s U_d}{F_0 L} \exp\left(-\frac{\varepsilon F_i^2}{2qn_r U_d}\right) \qquad (4\text{-}13\text{-}12)$$

This can be used for the determination of Gunn diode characteristics in the presence of impact ionization in the domain.

In our calculation we assumed that D is field independent. As can be seen from the above results, the value of $D(F)$ can only alter the value of n_{cr}. Therefore the dependence on $D(F)$ can only change the preexponential coefficient in G by a factor of about unity if $n_{cr} \gg n_r$. In the opposite case ($n_{cr} \ll n$) it can change the value in the exponent by a factor of the same order; but even in this case the dependence on U and n_r remains the same as in the case when $D(F)$ is constant.

The average electron concentration in a sample is determined by the balance between the generation and recombination of electrons and holes. The field outside the domain, F_r, is practically independent of the bias voltage for large enough domain fields ($F_m \gg F_v$). Therefore for $U_0 > U_t = F_p L$ the field dependence of the average external current density $j = qn_r \mu F_r$ is related to the field dependence of the concentration. Thus the dependence $n_r(U)$ reproduces (in the corresponding scale) the current–voltage characteristic.

The recombination of the electron–hole pairs depends on the concentration of recombination acceptor centers n_A and on the level of ionization. The detailed analysis of various recombination mechanisms [104, 105] shows that the most frequent experimental situation corresponds to the case when the recombination via the impurity centers is dominant. Most of the experimental data refer to the critical voltage, U_{0c}, of the beginning of the negative-slope region of the S-type current–voltage characteristic. Here we limit ourselves to the calculation of U_{0c}. The analysis of the form of the current–voltage characteristic for the larger current densities is given in Ref. 104. But the comparison of the corresponding results with experiments is difficult because of the strong influence of contact phenomena, the current filament formation, and the absence of detailed information about the recombination parameters. Also, with very large current densities, the recombination and generation times may become comparable to the transit domain time, with the result that the Gunn oscillations become incoherent and this theory becomes invalid. Also, the experimental data for the high current region are severely limited.

The beginning of the negative-slope region of the current–voltage characteristic occurs at low ionization level, when $(n_r - n_0)/n_0 \ll 1$, where $n_0 = n_D - n_A$ [104].

When $n_D > n_A$ (which is the case in the Gunn samples) all deep centers are occupied by electrons. When the impact ionization occurs the holes are trapped by the centers occupied by the electrons and the electrons are trapped only by the empty centers (by the centers occupied by the holes). Therefore at a low level of the impact ionization the recombination time of the holes is much smaller than that of the electrons. Also, in GaAs the trapping cross section for the holes is much larger than that for the electrons, because the holes are trapped by the negatively charged centers but the electrons are trapped by the neutral centers. So for a low ionization level the number of the empty centers is practically equal to the number of the excess electrons, and the balance between the ionization and recombination rates is given by

$$G = \gamma n_r (n_r - n_0) \tag{4-13-13}$$

where γ is the trapping coefficient of electrons (for GaAs $\gamma \simeq 10^{-8}\,\text{s}^{-1}\,\text{cm}^3$). Equations (4-13-12) and (4-13-13) determine the dependence $n_r(U)$. The critical value U_{0c} can be calculated from the condition $dU/dn_r = 0$. This condition gives two values of U. The larger value corresponds to the threshold voltage. The details of this calculation are given in Ref. 104. The resulting formula for U_{0c} is

$$U_{0c} = \frac{\varepsilon F_i^2}{3qn_0 \ln(n_{ch}/n_0)} \tag{4-13-14}$$

where

$$n_{ch} = \left(\frac{2\pi\alpha_\infty^2 v_s^2 \varepsilon^2 F_i^2}{\gamma^2 q^2 L^2 n_{cr}} \right)^{1/3} \tag{4-13-15}$$

[For GaAs $n_{ch} = 1.2510^{17}/L^{2/3}$ where L is in cm and n_{ch} is in cm^{-3}.] Equation (4-13-14) is valid if $n_0 \ll n_{ch}$, and it determines the dependence of the threshold domain voltage U_{0c} on the sample parameters. U_{0c} increases with the decrease of n_0, and it depends logarithmically on L and γ. The physical meaning of expression (4-13-14) is rather simple: when F_m reaches a relatively small part of the characteristic ionization field, F_i, the ionization becomes so effective that the slope of the characteristic $n_r(U)$ becomes negative.

According to this theory U_{0c} corresponds to the low ionization level when all the trapping centers are occupied by the electrons. The theoretical curves of the critical bias voltage U_{0c}/U_p where $U_p = F_p L$ are compared with experimental data of Kennedy et al. [106] in Fig. 4-13-2. As can be seen, the experimental data agree well with the theory.

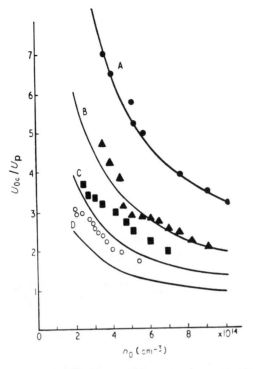

FIGURE 4-13-2. The variation of U_{0c}/U_p with n_0 [105]. The curves are calculated using Eq. (4-13-14) for the following sample thickness: A, 110 μm; B, 200 μm; C, 380 μm; D, 710 μm. The experimental points are taken from Kennedy *et al.* [106].

Owens and Kino [91] studied the impact ionization of Gunn diodes for large values of the bias $U_0 > U_{0c}$. In this case the current density increases with time and after some characteristic time t_c Gunn oscillations become incoherent. When U_0 decreases, t_c increases because the impact ionization becomes weaker. If $U \to U_{oc}$ then $t_c \to \infty$ (if $U < U_{oc}$ Gunn oscillations are coherent). For a sample with $n_0 = 3 \times 10^{14}$ cm^{-3}, $L = 1$ mm, they found that $U_{oc} = 340$ V. The outside domain field was about 2.1 kV cm^{-1} (this estimate being obtained from the time-averaged current–voltage characteristic as well as from the current waveform). The theoretical value of U_{0c} is about 300 V.

Copeland [95] observed S-type switching for a Gunn diode with $n_0 = 3 \times 10^{15}$ cm^{-3}, $L = 20$ μm, and $U_{0c} \sim 60$ V. Our estimate from Eq. (4-13-14) gives $U_{0c} \sim 80$ V, which agrees qualitatively with experiment, although, strictly speaking, Eq. (4-13-14) is only qualitatively valid for concentrations which are close to n_{cr}. The analysis shows that a more accurate theoretical value should be somewhat larger.

The conventional Gunn effect can be realized only if $U_{0c} > U_p$. From this condition and Eq. (4-13-14) we obtain the following criterion:

$$n_0 L < \frac{\varepsilon F_i^2}{3q(F_p - F_s)\ln(n_{ch}/n_0)} \tag{4-13-16}$$

With the parameters listed above and for $L = 1$ mm we find that $n_0 L < 10^{14}$ cm^{-2} (this estimate is valid for other values of length with a logarithmic accuracy because of the dependence of n_{ch} on L). Heeks [88] showed experimentally that the usual Gunn effect could exist only if $n_0 L < 10^{14}$ cm^{-3}, which agrees well with Eq. (4-13-16).

S-type characteristics have also been observed in highly doped Gunn diodes (see, for example, Refs. 90, 108). The same mechanism may be responsible for these characteristics. However, in a highly doped material the impact ionization may become so strong that the quadratic electron–hole recombination dominates and the recombination time $(Rn_0)^{-1}$ becomes comparable to the transit time. This happens when $n_0 L \sim v_s/R \sim 10^{16}$ cm^{-2} for GaAs. Under such conditions the theory described above does not apply.

If $U_0 > U_{0c}$ the stationary state corresponds to the upper branch of the S-type current–voltage characteristics. Thus when the bias voltage $U_0 > U_{0c}$ is applied to the diode, the current increases with time up to the value corresponding to the upper branch or until the sample is destroyed by a very high current. Because of the exponential dependence of G on n_r, the characteristic generation time τ_G can reach a value less than the transit time T. If $\tau_G < T$ the concentration profile becomes inhomogeneous and the Gunn oscillations become incoherent [91, 101] (see Fig. 4-13-3). What happens is this: The first few Gunn pulses (after applying the bias voltage) occur practically at the same level of the electron concentration $n_r \sim n_0$. Then n_r increases abruptly and the coherence is violated during the time interval smaller than the transit time. As shown below, this can be explained by the exponential dependence G on n_r.

When $U_0 \sim U_{0c}$ the generation rate G is of the same order as the recombination rate. As U_0 increases, the value of G increases exponentially. The recombination rate is independent of U_0. Therefore for $U_0 > U_{0c}$ the recombination rate becomes

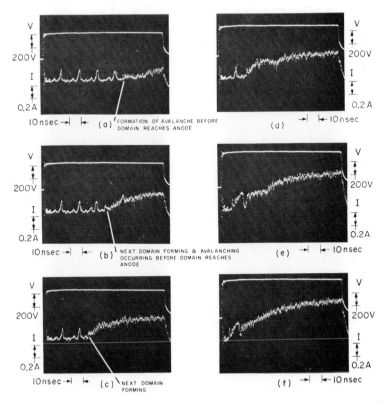

FIGURE 4-13-3. Voltage and current waveforms of Gunn oscillations under the conditions of the impact ionization within the high field domain [91]. Arrows indicate points of time when the Gunn oscillations become incoherent.

negligible compared with the generation rate up to the value of the electron concentration n_c corresponding to the violation of the coherence. Thus we can estimate the time t_c of the coherence violation as follows:

$$t_c = \int_{n_0}^{n_c} \frac{dn}{G(n)} \tag{4-13-17}$$

Using (4-13-12) we can evaluate this integral. The main contribution to this integral is determined by the region of

$$\frac{n - n_0}{n_0} \sim \frac{1}{a^2} < \frac{n_c - n_0}{n_0} \tag{4-13-18}$$

Thus the integral is practically independent of n_c. The calculation of this integral (described in Ref. 101) yields

$$t_c = \frac{n_0}{G(n_0)a_0^2} \tag{4-13-19}$$

where

$$a_0 = a[F_m(n_0, U)] \tag{4-13-20}$$

More detailed calculation of t_c for all values of n_0 is given in Ref. 101. The theoretical dependence t_c on U is compared with the experimental data of Owens and Kino [91] in Fig. 4-13-4. As can be seen from the figure, these calculations agree well with the experimental results.

The current filamentation may be one of the consequences of the S-type current–voltage characteristics. Indeed, if the electron concentration is larger in some small portion of the sample cross section the generation rate is higher there. If the current density corresponds to the region of the S-type current–voltage characteristic with a negative slope, this increase in the generation rate is larger than the concurrent increase in the recombination rate. As a result the carrier concentra-

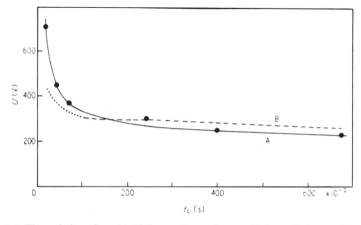

FIGURE 4-13-4. The variation of t_c with U. A, experimental curve; B, theoretical curve. The dotted part of the theoretical curve corresponds to the region $t_c < T$ where the theory is not valid [102].

tion increase will grow even stronger. This increase will stop when the current outside this filament and the applied voltage drop.

In the case when the electron–hole concentration is high and the binary recombination is dominant the filament size is determined by the ambipolar diffusion length $(D\tau_R)^{1/2}$, where τ_R is the recombination time [109]. Therefore the sample width d_0 must be greater than $(D\tau_R)^{1/2}$ for the current filamentation. Here D is an ambipolar diffusion coefficient.

As has been shown in Refs. 110 and 111, the filament should move in the transverse direction in the transverse electric or magnetic field. In a weak transverse electric field F the filament moves due to the ambipolar drift with the velocity

$$U_a = \frac{\mu_n\mu_p(n-p)F}{\mu_n n + \mu_p p} \tag{4-13-21}$$

In the transverse magnetic field the filament moves due to the hole movement in the electron Hall electric field; the electrons follow to maintain the electric neutrality (similar to the motion of electrons and holes due to the Sul effect).

The direction of the movement is determined by the left-hand rule. The velocity of the filament is given by

$$u_B = \mu_n\mu_p F_0 B \tag{4-13-22}$$

The shift of the breakdown channel in the transverse magnetic field in the Gunn diodes was observed in Ref. 112. The direction of the shift corresponded to the predicted direction of the current filament. Thus the breakdown may have been associated with the current filamentation due to the breakdown in the high field domain.

When the current filament moving in the transverse magnetic field reaches the sample boundary it should stop there if the surface recombination velocity S is low

$$S < u_B \tag{4-13-23}$$

or disappear if the surface recombination velocity is large

$$S > u_B \tag{4-13-24}$$

In the latter case a new filament may nucleate in the sample, travel across the sample, etc., leading to periodic current oscillations at the transit frequency

$$f = u_B/d_0 \tag{4-13-25}$$

Low-frequency large-amplitude current oscillations in a transverse magnetic field in Gunn diodes operating in the S-type region of the I–V characteristic were observed in Ref. 113. The oscillation frequency (1.7 MHz for $B = 3.2$ kG and around 10 MHz for $B = 10$ kG) is in agreement with the frequency predicted by Eq. (4-13-25).

A threshold density of electron–hole pairs in the filament which is necessary for the stimulated emission was estimated in Ref. 114. This estimate is in agreement with the experimental data [98, 99].

4-14. INSTABILITIES AND HIGH FIELD DOMAINS IN A SEMICONDUCTOR WITH ELECTRONS AND HOLES

In this section we shall consider a semiconductor with two types of carriers (electrons and holes) and with a current–voltage characteristic of the electrons having a negative slope. Such a situation can be realized in intrinsic semiconductors or under conditions of generation of electron–hole pairs, for example, by illumination, by the impact ionization in transferred electron devices when the transit time of a domain is small compared with the characteristic recombination time of pairs (see Section 4-13), and in GaAs avalanche diodes.

4-14-1. Small Signal Analysis

As was shown in Refs. 115 and 116, when both electrons and holes are present in a sample the ordinary instability, i.e., growing space-charge waves, is accompanied by a new type of instability in the form of quasineutral waves. The quasineutral waves can grow in the region with a positive differential conductance, σ_d, but with a negative electron differential conductance, σ_{nd}, in which the space-charge waves may decay. Thus, the region in which the instability arises is wider than the region with $\sigma_d < 0$ but narrower than the region with $\sigma_{nd} < 0$.

A positive differential conductance of the holes leads to a contraction of the instability region and a sufficiently large increase in the hole density destroys the instability if the electron density remains constant. However, if the electron–hole pairs arise as a result of impact ionization, an increase in the hole density p_0 is accompanied by an increase in the electron density ($n_0 = p_0 + N_D$, where N_D is the donor concentration). Under these conditions, the generation of electron–hole pairs can suppress the instability only if $\mu_p > |v_n/dF|$, where μ_p is the hole mobility, $v_n(F)$ is the electron drift velocity, and F is the electric field strength. If this condition is not satisfied, the instability region simply contracts with increasing pair density to the region with negative differential conductance, which coincides with the region of fields in which $(-\mu_p) > dv_n/dF$ as $p_0 \to \infty$. We shall obtain the condition for the occurrence of the instability in a sample with two types of carriers, assuming that the characteristic excitation frequencies are high compared with the reciprocal times of recombination and generation of holes.

The basic equations are

$$\frac{\partial F}{\partial x} = \frac{q}{\varepsilon}(p - p_0 - n + n_0) \qquad (4\text{-}14\text{-}1)$$

$$q\frac{\partial n}{\partial t} - \frac{\partial j_n}{\partial x} = 0 \qquad (4\text{-}14\text{-}2)$$

$$q\frac{\partial p}{\partial t} + \frac{\partial j_p}{\partial x} = 0 \qquad (4\text{-}14\text{-}3)$$

$$j_n = qnv_n(F) + qD_n\frac{\partial n}{\partial x} \qquad (4\text{-}14\text{-}4)$$

$$j_p = qpv_p - qD_p\frac{\partial p}{\partial x} \qquad (4\text{-}14\text{-}5)$$

Here, n, p, j_n, j_p, $v_n(F)$, $v_p = \mu_p F$, D_n, and D_p are the densities, current densities, drift velocities, and diffusion coefficients of the electrons and holes, respectively, and q is the electronic charge. The linearization of the system ($p = p_0 + p'$, $n = n_0 + n'$, $F = F_0 + F'$) leads to the dispersion equation

$$(\omega + kv_n + iD_n k^2)(\omega - kv_p - iD_p k^2) + i(1/\varepsilon)$$

$$\times [qp_0\mu_p(\omega + kv_n + iD_n k^2) + \sigma_{nd}(\omega - kv_p + iD_p k^2)] = 0 \quad (4\text{-}14\text{-}6)$$

where $\sigma_{nd} = qn_0 dv_n/dF$. The onset of the instability occurs when Im $\omega = 0$. Using this condition, one can obtain an equation for k. For simplicity, we shall write it down for the case $D_p = 0$. In a number of cases, the diffusion of holes can indeed be neglected, for example, under the Gunn-effect conditions, when the electrons are heated more strongly than the holes and have a greater mobility. In addition, as will become clear from the results given below, allowance for the hole diffusion does not alter the qualitative conclusions. When $D_p = 0$, Eq. (4-14-6) yields

$$(\varepsilon k^2 D_n + \sigma_d)^2 + \frac{\varepsilon v^2}{D_n}(\varepsilon k^2 D_n + \sigma_{nd}) = 0 \quad (4\text{-}14\text{-}7)$$

where $\sigma_d = \sigma_{nd} + qp_0\mu_p$, $v = v_n + v_p$.

Equation (4-14-7) is a quadratic equation in k^2. The analysis shows that this equation has at least one positive root when

$$\sigma_d + \frac{\varepsilon v^2}{2D_n} < 0 \quad (4\text{-}14\text{-}8a)$$

or

$$\sigma_d + \frac{\varepsilon v^2}{2D_n} > 0$$

and $\quad (4\text{-}14\text{-}8b)$

$$\sigma_d^2 + \frac{\varepsilon v^2}{D_n}\sigma_{nd} < 0$$

It follows from condition (4-14-8b) that the instability is possible even if $\sigma_d > 0$.

From Eq. (4-14-7) and condition (4-14-8b), we obtain the following result. If the hole density increases, the region of instability in k for $\sigma_d < 0$, which extends from the point $k = 0$ to the value of k determined by Eq. (4-14-7), contracts. For small values of k the solution of Eq. (4-14-6) is given by

$$\omega_1 = -i\frac{\sigma_d}{\varepsilon} - \frac{kq}{\sigma_d}\left(p_0\mu_p^2 F - n_0 v_n \frac{dv_n}{dF}\right) \quad (4\text{-}14\text{-}9)$$

$$\omega_2 = kv_n\frac{\tilde{\sigma}}{\sigma_d} - ik^2\frac{q}{\sigma_d}\left[\mu_p p_0 D_n + n_0\frac{dv_n}{dF}\left(D_p + \frac{\varepsilon q p_0\mu_p v^2}{\sigma_d^2}\right)\right] \quad (4\text{-}14\text{-}10)$$

where

$$\tilde{\sigma} = q\mu_p\left[n_0\frac{d(\ln v_n)}{d(\ln F)} - p_0\right] < 0$$

Equations (4-14-9) and (4-14-10) were derived using the Taylor series expansion in k and their applicability is therefore restricted to the values of k for which the second terms in these equations are small compared with the first. The root ω_1 corresponds to the growth of a fluctuation in time for $\sigma_d < 0$ and to the fluctuation decay for $\sigma_d > 0$. As $\mu_p p_0 \to 0$, the root ω_1 describes the dispersion of the space-charge waves in a semiconductor with one type of carrier. The root ω_2 corresponds to quasineutral oscillations. Their growth is due to a deviation from neutrality. In the case when the electron mobility does not depend on the field the equation for the phase velocity of the quasineutral waves reduces to the well-known expression for ambipolar drift in an electric field [117]. A necessary condition for the root ω_2 to ensure growth when $\sigma_d > 0$ and $dv_n/dF < 0$ is given by

$$\frac{\varepsilon v^2}{D_n} \sigma_{nd} \left(1 + \frac{\sigma_d^2 D_p}{\varepsilon q p_0 \mu_p v^2}\right) + \sigma_d^2 < 0 \tag{4-14-11}$$

Equation (4-14-11) reduces to Eq. (4-14-8b) if $D_p = 0$. If condition (4-14-11) is not satisfied, then both roots correspond to solutions that grow with time for $\sigma_d < 0$.

In the region of large k the solutions of Eq. (4-14-6) are given by

$$\omega_1 = kv_p - i\left(D_p k^2 + \frac{1}{\varepsilon} q p_0 \mu_p\right) \tag{4-14-12}$$

$$\omega_2 = -kv_n - i\left(D_n k^2 + \frac{1}{\varepsilon} \sigma_{nd}\right) \tag{4-14-13}$$

The region of validity of Eqs. (4-14-12) and (4-14-13) depends on the parameters in Eq. (4-14-6). We shall therefore give only the qualitative form of the dependences $\omega(k)$ (see Fig. 4-14-1). Note that for large k the dispersion laws correspond to space-charge waves of holes [Eq. (4-14-12)] and electrons [Eq. (4-14-13)].

The dispersion relations for the quasineutral waves may be interpreted as follows. Let us consider the motion of electrons and holes in the coordinate system which propagates with the velocity u of the drift of the ambipolar fluctuation. The continuity equations are

$$n'(\mu_p F - u) + p\mu_p F' = 0 \tag{4-14-14}$$

$$n'(v_n + u) + nF' \, dv_n/dF = 0 \tag{4-14-15}$$

Here n' is the electron–hole pair concentration in the fluctuation, F' is the fluctuation

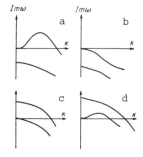

FIGURE 4-14-1. Qualitative behavior of the dispersion curves for different values of the parameters (positive values of Im ω correspond to growing waves). (a) $\sigma_d > 0$, inequality (4-14-11) satisfied; (b) $\sigma_d > 0$, inequality (4-14-11) not satisfied; (c) $\sigma_d < 0$, inequality (4-14-11) satisfied; (d) $\sigma_d < 0$, inequality (4-14-11) not satisfied.

field. We find from Eqs. (4-14-14) and (4-14-15) the velocity u of the drift

$$u = (1/\sigma_d)(\sigma_{nd}v_p - \sigma_p v_n) \tag{4-14-16}$$

where $\sigma_{nd} = qn\,dv_n/dF$, $\sigma_p = qp\mu_p$, $v_p = \mu_p F$, $\sigma_d = \sigma_{nd} + \sigma_p$. (For simplicity we assume that σ_p is not field-dependent.) Equation (4-14-16) reduces to the usual expression for the velocity of the ambipolar drift in the case of the ohmic current-voltage characteristic [117]. In this case the direction of the drift is determined by the sign of $p-n$. In the case when $dv_n/dF < 0$ the ambipolar drift is always directed from cathode to anode. The growth (or damping) of the finite fluctuation of size λ is related to the space charge $q(p' - n') = \rho' \sim \varepsilon F'/\lambda$. When $\rho' \neq 0$ the continuity equations are

$$qp'(\mu_p F - u) + \sigma_p F' = -qJ \tag{4-14-17}$$

$$qn'(v_n + u) + \sigma_{nd}F' = qJ \tag{4-14-18}$$

where J is the electron–hole pair flow which leads to the dissipation (or growth) of the fluctuation. We have neglected diffusion in the simplified equations). If we exclude F' from Eqs. (4-14-17) and (4-14-18) we obtain the relationship between the space charge density ρ' and the flow J:

$$qJ = -\rho' v\sigma_{nd}\sigma_p/\sigma_d^2, \qquad v = v_n + v_p \tag{4-14-19}$$

The characteristic time τ of the fluctuation decay (or growth) is determined from the condition that the decrease of the pairs due to the flow J is equal to the number of pairs in the fluctuation $n'\lambda$:

$$\frac{1}{\tau} \sim \frac{J}{n'\lambda} \sim \frac{\varepsilon\sigma_{nd}\sigma_p v^2}{\lambda^2\sigma_d^3} \tag{4-14-20}$$

Here $v = v_n + v_p$. As can be seen from Eq. (4-14-20) the quasineutral fluctuation grows if $\sigma_{nd} < 0$, $\sigma_d > 0$. The phase velocity and increment of the quasineutral waves [see Eq. (4-14-10)] coincide with u and $1/\tau$ given by Eqs. (4-14-16) and (4-14-17) (when the diffusion is neglected).

An estimate for the case $dv_n/dF = \mu_n = \text{const}$ shows that the field-dependent decay of the ambipolar fluctuation may be much larger than the diffusion decay for typical values of the electron and hole conductivities if the electric field is large enough. Thus for intrinsic semiconductor the field dependent decrement dominates if

$$p < \frac{\varepsilon F^2}{2k_B T} \tag{4-14-21}$$

where T is lattice temperature. (For this estimate we assume $D_n/\mu_n = D_p/\mu_p = k_B T/q$ and $\mu_p \ll \mu_n$.) For F about 1000 V/cm the hole concentration should be less than 10^{14} cm.

It follows from the considerations given above that the measurements of the field-dependent decrement can be used, for example, for the determination of the hole mobility in the intrinsic semiconductor if μ_n and σ are known, for instance, from Hall and conductivity measurements.

4-14-2. High Field Domains

The theory of high field domain in Gunn diodes with electrons and holes was developed in Refs. 118–120.

In GaAs, the hole mobility, $\mu_p \sim 300\text{–}400 \text{ cm}^2/\text{V s}$, is much smaller than the electron mobility in a low electric field, $\mu_n \sim 5000\text{–}8000 \text{ cm}^2/\text{V s}$, and the maximum negative differential mobility of the electrons $(\mu_{nd})_{max} \sim 2000 \text{ cm}^2/\text{V s}$. Nonetheless, at a high electron–hole pair concentration p_0 the influence of the holes on the current–voltage characteristic of the sample is significant in the voltage regions where μ_{nd} is small. This occurs, first, in the region of electric fields close to the threshold, and second, in the region of strong fields, where the drift velocity of the electrons becomes saturated. Accordingly, the holes can influence the threshold field at which the instability occurs as well as the behavior of the already formed high-field domains.

To illustrate the influence of holes at large electron–hole pair concentrations, Fig. 4-14-2 shows the current–voltage characteristic, normalized to the total number of electrons, n_0, in a homogeneous sample of GaAs, for two limiting cases: (1) $p_0 = 0$, $n_0 = N_D$, where N_D is the donor concentration (curve 1), and (2) $p_0 \simeq n_0 \gg N_D$ (curve 2). Curve 1 is taken from Ref. 121, and curve 2 is obtained by adding curve 1 to curve 3 (for $\mu_p \sim 400 \text{ cm}^2/\text{V s}$). The hole mobility in GaAs does not appreciably change with the electric field up to 40–60 kV/cm [122].

As will be shown below, at a low hole concentration the motion of the domain accelerates, and the domain velocity increases with increasing hole concentration. Even at a relatively low hole concentration, there can exist, besides the domain

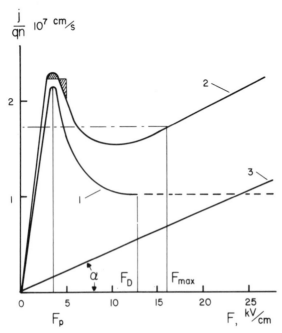

FIGURE 4-14-2. Current-voltage characteristic of a homogeneous GaAs sample, normalized to the total number of electrons. j, Current density; E, bias field; curve 1, $p_0 = 0$, $n_0 = N_D$; curve 2, $p_0 \simeq n_0 \gg N_D$; curve 3, dependence of the drift velocity of the holes in an inhomogeneous sample on the field; $\tan \alpha = \mu_p$. The dashed curve represents an extrapolation of the experimental data obtained in Ref. 3. The shaded areas illustrate the equal area rule for a domain propagating at high velocity u_0 in a sample with holes. The dash-dot line corresponds to the area rule for a trapezoidal domain.

propagating from the cathode to the anode (similar to the domain in a purely electronic semiconductor), a domain of the same shape but propagating in the opposite direction (from the anode to the cathode) with the drift velocity of holes, corresponding to the maximum field in the domain. (At a very low hole concentration, the occurrence of a domain propagating from the anode to the cathode is prevented by hole diffusion.) When the hole concentration is large the domains can propagate from the cathode to the anode and in the opposite direction with equal velocity, determined by the diffusion of the electrons and by the hole dielectric relaxation time ($D_n/u^2 \sim \tau_{mp}$, where D_n is the electron diffusion coefficient, u is the domain velocity, and τ_{mp} is the hole dielectric relaxation time). A positive differential conductivity of holes also leads to the appearance of a rising branch of the current–voltage characteristic in the region of strong fields, thus limiting the domain field.

We shall seek nonlinear solutions of Eqs. (4-14-1)–(4-14-5) that correspond to a stable propagating high field domain when all the quantities depend only on $Z = x - ut$, where u is the domain velocity. In this case the system of Eqs. (4-14-1)–(4-14-5) is transformed into a system of ordinary differential equations, and Eq. (4-14-2) and Eq. (4-14-3) can be integrated:

$$p(\mu_p F - u) - D_p \frac{dp}{dZ} = p_0(\mu_p F_r - u) \tag{4-14-22}$$

$$n[v_n(F) + u] + D_n \frac{dn}{dZ} = n_0[v_n(F_r) + u] \tag{4-14-23}$$

$$\frac{dF}{dZ} = \frac{q}{\varepsilon}(p - p_0 - n + n_0) \tag{4-14-24}$$

Equations (4-14-22)–(4-14-24) can be rewritten as

$$\frac{qD_n}{\varepsilon} \frac{dn}{dF} = \frac{nv_n(F) - n_0 v_n(F_r) + (n - n_0)u}{n - n_0 - p + p_0} \tag{4-14-25}$$

$$\frac{qD_p}{\varepsilon} \frac{dp}{dF} = \frac{\mu_p(pF - p_0 F_r) - u(p - p_0)}{p - p_0 - n + n_0} \tag{4-14-26}$$

Let us consider first the case when it is possible to neglect the hole diffusion. The physical reason why the hole diffusion can be neglected is that the holes have a smaller mobility and a lower temperature than the electrons, since the electrons are heated. It was shown in Ref. 118 that this approximation is valid for a wide range of parameters, including the values corresponding to GaAs.

At $D_p = 0$ we obtain from (4-14-25) and (4-14-26) one first-order equation for the space-charge density $\rho = q(p - p_0 - n + n_0)$:

$$\frac{D_n}{\varepsilon} \rho \frac{d\rho}{dF} = q(N_D + p_0)[v_n(F) - v_n(F_r)]$$

$$- [u + v_n(F_r)]\left[\rho + \frac{qp_0\mu_p(F - F_r)}{\mu_p F - u}\right] + \frac{qD_n p_0\mu_p}{\varepsilon(\mu_p F - u)^2}(u - \mu_p F_r)\rho$$

$$\tag{4-14-27}$$

We confine ourselves first to the case of large stationary electron concentration n_0. As will be shown below, the space-charge density in the domain is small in this case ($\rho/qn_0 \ll 1$). A large stationary electron concentration n_0 can arise in the sample either as a result of a large concentration of electron–hole pairs p_0 produced in the sample by illumination or impact ionization or due to a large doping density $N_D(n_0 = N_D + p_0)$.

We seek the solution of (4-14-27) in the form of a series

$$\rho = \rho^{(0)} + \rho^{(1)} + \cdots$$

$\rho^{(0)}$ and $\rho^{(1)}$ are determined from the equations

$$\frac{D_n}{\varepsilon}\rho^{(0)}\frac{d\rho^{(0)}}{dF} = q(N_D + p_0)[v_n(F) - v_n(F_r)]$$

$$- [v_n(F_r) + u]\frac{qp_0\mu_p(F - F_r)}{\mu_p F - u} \equiv q\phi_0(F, u) \quad (4\text{-}14\text{-}28)$$

$$\frac{D_n}{\varepsilon}\frac{d}{dF}(\rho^{(0)}\rho^{(1)}) = \frac{qD_np_0\mu_p}{\varepsilon(\mu_p F - u)^2}(u - \mu_p F_r)\rho^{(0)} - [u + v_n(F)]\rho^{(0)} \quad (4\text{-}14\text{-}29)$$

Using Eq. (4-14-28), we can rewrite Eq. (4-14-29) in the form

$$\frac{d}{dF}(\rho^{(0)}\rho^{(1)}) = -\frac{2}{3q(N_D + p_0)}\frac{d(p^{(0)})^3}{dF} - \phi_1(F, u)\rho^{(0)} \quad (4\text{-}14\text{-}30)$$

where

$$\phi_1(F, u) = \frac{2\varepsilon}{D_n}\left\{ u + v_n(F_r) - \frac{qD_np_0\mu_p(u - \mu_p F_r)}{\varepsilon(\mu_p E - u)^2} \right.$$

$$\left. + \frac{p_0\mu_p(F - F_r)}{(N_D + p_0)(\mu_p F - u)}[u + v_n(F)] \right\} \quad (4\text{-}14\text{-}30a)$$

From Eqs. (4-14-28)–(4-14-30) we find

$$(\rho^{(0)})^2 = \frac{2\varepsilon q}{D_n}\int_{F_r}^{F} \phi_1(F', u)\, dF' \quad (4\text{-}14\text{-}31)$$

$$\rho^{(1)} = -\frac{2(\rho^{(0)})^2}{3q(N_D + p_0)} - \frac{1}{\rho^{(0)}(F)}\int_{F_r}^{F} dF'\, \rho^{(0)}(F')\phi_1(F', u) \quad (4\text{-}14\text{-}32)$$

In the limiting case $\mu_p \to 0$ or $p_0 \to 0$, the solution (4-14-31) describes a stable strong-field domain in a strongly doped sample (see Section 4-6).

From the Poisson equation it follows that at the maximum value of the electric field in the domain F_m the space-charge density is zero [$\rho^{(0)}(F_m) = 0$]. Hence,

$$\int_{F_r}^{F_m} \phi_1(F, u)\, dF = 0 \quad (4\text{-}14\text{-}33)$$

On the other hand, as is seen from (4-14-32), in order for $\rho^{(1)}$ to remain finite at the point F_m, it is necessary to have

$$\int_{F_r}^{F_m} dF \rho^{(0)}(F)\phi_1(F, u) = 0 \qquad (4\text{-}14\text{-}34)$$

A simultaneous solution of the integral equations (4-14-33) and (4-14-44) determines the maximum value of the domain electric field F_m and the domain velocity u.

In the case of sufficiently low hole concentrations, $\rho^{(0)}$ has the same form as in the absence of holes. It follows from (4-14-33) that F_m is determined in this case (accurate to small corrections) with the aid of the usual "equal area" rule in the theory of high-field domains. Therefore in the zeroth approximation there is likewise no change in the dynamic characteristic of the sample with a high field domain.

It follows from the form of the function $\phi(F, u)$ that Eq. (4-14-34) should have roots

$$u_1 = -v_n(F_r) + \Delta u_1 \qquad (4\text{-}14\text{-}35)$$

$$u_2 = \mu_p F_m + \Delta u_2 \qquad (4\text{-}14\text{-}36)$$

where Δu_1 and Δu_2 are small corrections to the domain velocity which have been evaluated in Ref. 118:

$$-\frac{\Delta u_1}{v_n(F_r)} \simeq \frac{3 D_n p_0}{\varepsilon F_m v_n(F_r)} \qquad (4\text{-}14\text{-}37)$$

$$\frac{\Delta u_2}{\mu_p F_m} = \left\{ \frac{3 \pi D_n q p_0}{4\varepsilon F_m [\mu_p F_m + v_n(F_r)]} \right\}^2 \qquad (4\text{-}14\text{-}38)$$

These equations are valid when the ratios $|\Delta u_1/v_n(F_r)|$ and $\Delta u_2/\mu_p F_m$ are much smaller than unity and the donor concentration N_D is high [118]. The important result is that there may be two kinds of domains in the presence of holes propagating in opposite directions and that the domain velocity for both domain solutions increases with the increase in the hole concentration. One can therefore expect that at sufficiently large hole densities a situation when $u \gg \mu_p F_m + v_n(F_r)$ may be reached. We then find from Eqs. (4-14-28) and (4-14-33)

$$\int_{F_r}^{F_m} \{(N_D + p_0)[v_n(F) - v_n(F_r)] + p_0 \mu_p (F - F_r)\}\, dF = 0 \qquad (4\text{-}14\text{-}33a)$$

Expression (4-14-33a) is a generalization of the equal area rule of the standard theory of strong-field domains. This rule connects the density of the external current $j = q p_0 \mu_p F_r + q n_0 v_n(F_r)$ with the maximum field in the domain F_m (see Fig. 4-14-2). From the equal area rule it follows that the maximum field in the domain F_m cannot be larger than a certain critical value F_{max} that depends on the hole conductivity, even if the characteristic $v_n(F)$ does not impose any limitations on the amplitude of the domain in a purely electronic semiconductor (see Fig. 4-14-2). Thus, in a Gunn diode with two kinds of carriers a "trapezoidal" domain with a constant amplitude F_{max} may exist. The width of such a domain increases with the bias voltage (see Fig. 4-14-3).

FIGURE 4-14-3. Triangular (a) and trapezoidal (b) high field domains.

(a)   (b)

Using the condition $u \gg \mu_p F_m$, we obtain from Eqs. (4-14-30a) and (4-14-34)

$$u = \pm u_0 = \left[\frac{qD_n}{\varepsilon} \mu_p p_0 \right]^{1/2} = \left(\frac{D_n}{\tau_{mp}} \right)^{1/2} \qquad (4\text{-}14\text{-}39)$$

where τ_{mp} is the hole dielectric relaxation time.

Just as in the case of a relatively low hole density, the domain can move not only from the cathode to the anode, but also in the opposite direction. However, when the hole concentration is large the domain velocity is the same in both directions. The concentration of the electron–hole pairs in the domain varies with the direction of the domain motion: in the case of motion from the cathode to the anode, the concentration of the electron–hole pairs in the domain is smaller than outside the domain, while in the case of motion in the opposite direction the electron–hole pair concentration in the domain is higher than outside. At the same time, the distribution profiles of the space charge and of the electric field in the domain do not depend on the direction of the domain motion (see Fig. 4-14-4). The choice of the domain motion direction may be determined by the conditions on the sample contacts.

Let us consider the criteria of the validity of the results obtained for relatively high hole densities. From the conditions $u \gg \mu_p F_m$, $v_n(F_r)$ we obtain lower bounds

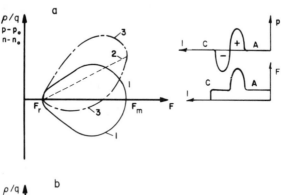

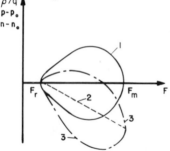

FIGURE 4-14-4. Schematic field dependences of the space charge density ρ/q (curve 1), of the excess hole density $p - p_0$ (curve 2), and excess electron density $n - n_0$ (curve 3) in the domain walls. (a) For a domain propagating from the anode to the cathode; (b) for a domain propagating from the cathode to the anode. In the upper right corner of Fig. 4-14-4a are shown the distribution profiles of the space charge and of the field along the sample, corresponding to the strong-field domain. C, cathode; A, anode. (After Ref. 118.)

for the hole density:

$$p_0 \gg \varepsilon v_n^2(F_r)/qD_n\mu_p \tag{4-14-40a}$$

$$p_0 \gg \varepsilon\mu_p F_m^2/qD_n \tag{4-14-40b}$$

Because in the case of a high hole density the domain field may be limited (see Fig. 4-14-2), we assume for numerical estimates $F_m \sim 30\,\text{kV/cm}$. Then, using the numerical values for the remaining parameters given above, we find from Eq. (4-14-40) $p_0 > 10^{16}\,\text{cm}^{-3}$. Such electron–hole pair concentrations in Gunn diodes can be easily obtained experimentally.

The value of F_m used for the estimates pertains to the case $N_D \ll n_0 \sim p_0$ (see Fig. 4-14-2). If $N_D \gg p_0$, the holes can also strongly alter the domain velocity. The maximum domain field F_m is then approximately the same as in the domain propagating in a purely electronic sample. The hole density, which is necessary to appreciably increase the domain velocity is, in this case, larger and proportional to $\sqrt{N_D}$. The reason for that is a larger value of F_m which increases in proportion to $N_D^{1/4}$ (for $n \gg n_{cr}$, see Section 4-6). The hole charge in the domain

$$q(p - p_0) = \frac{q\mu_p F_m}{u}p_0$$

is, in this case, small compared with the electron space charge $\rho_n \sim q(n_0 n_{cr})^{1/2}$. In strongly doped samples $(n_0 \gg n_{cr})$ the electron space charge $q(n^{(0)} - n_0)$ is small compared to N_D and, in the first approximation, independent of the domain velocity (see Section 4-6). The domain velocity is determined only by the small correction $qn^{(1)}$ to the space charge of the electrons. Therefore in order to change the velocity of the domain, it suffices for the hole charge to be comparable with the small correction $qn^{(1)}$.

From inequalities (4-14-40) we find

$$p_0 \gg \frac{\varepsilon v_n(F_r)F_m}{qD_n} \tag{4-14-40c}$$

[This inequality is obtained by multiplying inequalities (4-14-40a) and (4-14-40b).] It follows from Eq. (4-14-40c) that the domain can propagate with velocity $\pm u_0$ only if $p_0 \gg n_{cr}$, since

$$n_{cr} = \frac{\varepsilon\mu F_c^2}{qD_n} < \frac{\varepsilon v_n(F_r)F_m}{qD_n}$$

[see Eqs. (4-6-22) and (4-6-24)]. The electron and hole density variations in the domain, just as in a highly doped n-type sample $(N_D \gg n_{cr})$, are small in this case in comparison with the stationary densities. When $p_0 \gg N_D$ the charge density ρ in this domain is proportional to $\sqrt{p_0}$, just as $n - n_0 \sim \sqrt{p_0}$ and $p - p_0 \sim \sqrt{p_0}$.

As we show below, in order for Eqs. (4-14-38) and (4-14-39) to be valid it is necessary only to satisfy criteria (4-14-40). The validity of Eqs. (4-14-38) and (4-14-39) is not restricted, for example, by the approximations used in the derivation of Eqs. (4-14-28)–(4-14-30). Indeed, at $u \gg \mu_p F_m$, $v_n(F_r)$ we have from Eq. (4-14-27)

$$\frac{D_n}{2\varepsilon} \frac{d\rho^2}{dF} = q(N_D + p_0)[v_n(F) - v_n(F_r)] + ep_0\mu_p(F - F_r)$$

$$- \rho u \left(1 - \frac{u_0^2}{u^2}\right) \tag{4-14-41}$$

From Eq. (4-14-41) we get

$$\frac{D_n}{2\varepsilon}\rho^2 = q \int_{F_r}^{F} dF'\{(N_D + p_0)[v_n(F') - v_n(F_r)]$$

$$+ \mu_p p_0(F' - F_r)\} - u\left(1 - \frac{u_0^2}{u^2}\right) \int_{F_r}^{F} \rho(F')\, dF' \tag{4-14-42}$$

The high field domain consists of an increasing-field region, in which $\rho > 0$ everywhere, and a decreasing-field region, in which $\rho < 0$. Therefore the sign of the second term in the right-hand side of Eq. (4-14-42) at $u^2 \neq u_0^2$ will depend on the region (of increasing or decreasing field) to which the integration path corresponds. Therefore when $u^2 \neq u_0^2$ it is impossible to satisfy the condition $\rho(F_m) = 0$. Thus we have to put $u = \pm u_0$ [see formula (4-14-39)]. Using the condition $\rho(F_m) = 0$, we then obtain from Eq. (4-14-42) Eq. (4-14-39).

The high-velocity domain motion is easier to describe for the limiting case of a trapezoidal domain, the motion of which constitutes a synchronous motion of two charge layers separated by the width of the flat top of the domain. If the diffusion is neglected, these layers may be considered as discontinuity planes moving through the semiconductor. When the discontinuity plane moves with velocity u, the continuity equations for the electron and hole flows through this plane should be satisfied. Changing over to a coordinate system in which the discontinuity plane is at rest, we obtain, by equating the fluxes of electrons and holes on both sides of the discontinuity surface, the following relations:

$$(u - \mu_p F_m)p_m = (u - \mu_p F_r)p_0 \tag{4-14-43}$$

$$[u + v_n(F_m)]n_m = [u + v_n(F_r)]n_0 \tag{4-14-44}$$

Here n_m and p_m are the electron and hole densities in the domain at a field $F = F_m$. From Eqs. (4-14-43) and (4-14-44) we find that in the case of a discontinuity surface moving with a velocity greatly exceeding the drift velocities of the electrons and the holes, the concentration differences $n_1 = n_m - n_0$ and $p_1 = p_m - p_0$ are equal to

$$p_1 = \frac{\mu_p F_m}{u} p_0$$

$$n_1 = -\frac{v_n(F_m) - v_n(F_r)}{u} n_0$$

with $p_1 = n_1$ since $\rho(F_m) = 0$. Thus,

$$\mu_p F_m = v_n(F_r) - v_n(F_m)$$

The electron diffusion smears the discontinuity surface into a layer, of the width d, which can be estimated by starting from the fact that the diffusion flux in the layer is of the same order as the flux connected with the electron drift:

$$D_n n_1 / d \sim n_1 u$$

Hence $d \sim D_n/u$. As follows from the Poisson equation, the field discontinuity is related to the space charge in the layer $F_m - F_r \sim \rho d/\varepsilon$. Since inside the layer we have $p_1 \sim n_1 \sim (\mu_p F_m/u)p_0$, it follows that $F_m \sim (\mu_p/\varepsilon u)F_m p_0 d$, i.e., $d \sim \varepsilon u/qp_0\mu_p$. Comparing the two relations $d \sim D_n/u$ and $d \sim \varepsilon u/qu_0\mu_p$, we obtain the expression for the velocity with which the discontinuity surface can move through the electron-hole plasma in a semiconductor. The motion of such a discontinuity surface causes an uncompensated charge in the layer into which this surface is smeared out due to the diffusion. In order for the system to remain neutral, it is necessary to introduce another discontinuity surface, with an equal and opposite charge. This is possible only due to the negative differential conductivity of the electrons. The charge ρd on the surface is uniquely connected with the velocity of the surface. Thus, the trapezoidal domain constitutes synchronous motion, with velocity $\pm u_0$, of two discontinuity surfaces that are smeared out by the diffusion.

The picture presented above for the motion of the trapezoidal domain makes it possible to estimate the upper limit of the domain velocity, i.e., the upper value of the hole density, at which this theory is still valid. The smeared-out width of the domain wall cannot be smaller than the electron mean free path l, since the smearing of the discontinuity surface is connected with diffusion of the electrons:

$$d \sim D_n/u > l$$

Substituting the estimate $D_n \sim lv_T$, where v_T is the thermal velocity of the electrons, we obtain $v_T > u$, i.e., the trapezoidal domain cannot move with a velocity exceeding the electron thermal velocity.

An increase of the domain velocity in the presence of holes was reported in Refs. 88, 123, 124. The decrease in the domain field under the impact ionization conditions (which may have been related to the presence of mobile holes) has also been observed [124, 125]. The decrease of the domain field from 120 to 25 kV/cm was reported in Ref. 125.

Trapped holes may also change the domain parameters. In the semi-insulating GaAs the trapping times may be small for both electrons and holes. Under such conditions the domains may form due to both the field dependence of the trapping cross section and negative slope of the $v(F)$ curve [126, 127]. The theory of such domains in the presence of holes was developed in Ref. 120.

A different situation when holes are trapped, but electrons are not, is more typical for the Gunn diodes. This situation was analyzed in Ref. 128. It was shown that two types of high-field domains may exist under such conditions—regular Gunn domains propagating from the cathode to the anode and slow domains moving in the opposite direction with the velocity

$$u \simeq \frac{qD_n\mu_p p_0}{\varepsilon v_n(F_r)}$$

For this slow domain to exist in GaAs the density of the mobile holes should satisfy the following criterion:

$$4 \times 10^{13} L^{1/2} \, (\text{cm}) < p_0 \, (\text{cm}^{-3}) < 2 \times 10^{15} L^{1/2} \, (\text{cm})$$

Which out of two possible domain types will actually occur in a sample is determined by the boundary conditions and by the external circuit.

4-15. INFLUENCE OF MAGNETIC FIELD

The longitudinal magnetic field B (i.e., the magnetic field in the direction of the current) does not have an appreciable influence on the electron intervalley transfer up to $B \simeq 30 \, \text{kG}$ [129, 130].

The influence of the transverse magnetic field is determined by the ratio L/d, where L is the sample length and d is the sample width. When the electric and transverse magnetic fields are turned on, electrons start moving in the direction determined by the ratio of the Lorentz force

$$f_L = qBv \qquad\qquad\qquad (4\text{-}15\text{-}1)$$

and the electrostatic force

$$f_q = qF \qquad\qquad\qquad (4\text{-}15\text{-}2)$$

where v is the electron velocity, F is the electric field ($F = U/L$ in the uniform sample where U is the bias voltage and L is the sample length), q is the elementary charge. In a long sample ($L/d \gg 1$) electrons moving under the Hall angle θ

$$\tan \theta = \frac{qBv}{qF} \qquad\qquad\qquad (4\text{-}15\text{-}3)$$

hit the side surface and lead to the charge accumulation which creates the Hall electric field F_H balancing the Lorentz force

$$F_H = \frac{f_L}{q} \qquad\qquad\qquad (4\text{-}15\text{-}4)$$

As a result the carriers move in the direction parallel to the applied electric field (except a small region near the contacts where the Hall electric field is shorted by the contacts). This situation is depicted in Fig. 4-15-1a.

In the opposite limiting case $L/d \ll 1$ electrons moving under the Hall angle hit the opposite contact (see Fig. 4-15-1b) and the Hall field is almost totally shorted

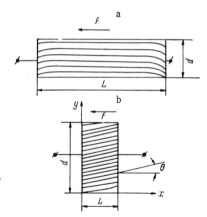

FIGURE 4-15-1. Streamlines of the electric current density in the transverse magnetic field. (a) Long sample ($L/d \gg 1$); (b) short sample ($L/d \ll 1$).

by the contact (see Fig. 4-15-1b). As a result electrons travel a longer distance

$$L_H = L(1 + \tan^2 \theta)^{1/2} \tag{4-15-5}$$

and the component of the electron velocity in the direction of the applied electric field decreases due to the component of the Lorenz force related to the component of the electron velocity parallel to the contacts. Assuming a zero Hall field we find

$$\mathbf{v} = \mu(\mathbf{F} + \mathbf{v} \times \mathbf{B}) \tag{4-15-6}$$

where μ is a low field mobility

$$v_x = \mu(F_x + v_y B) \tag{4-15-7}$$

$$v_y = -\mu v_x B \tag{4-15-8}$$

and, finally,

$$v_x = \frac{\mu F_x}{1 + \mu^2 B^2} \tag{4-15-9}$$

It follows from Eq. (4-15-9) that the effective low field mobility for a sample with $L/d \ll 1$ is given by

$$\mu_B = \frac{\mu}{1 + \mu^2 B^2} \tag{4-15-10}$$

The related increase in the low-field resistivity $\Delta \rho$ in the magnetic field (so-called geometric magnetoresistance) is given by [31]

$$\frac{\Delta \rho}{\rho_0} = (\mu B)^2$$

Here

$$\rho_0 = \frac{1}{q n_0 \mu} \tag{4-15-11}$$

is the low field resistivity and n_0 is the electron concentration.

The magnetic field influence on the Gunn effect in short samples can be interpreted in the frame of the following simple model. We assume that the Gunn threshold is reached when the average electron energy E exceeds the thermal energy by some critical value:

$$E - \frac{3k_BT}{2} = \Delta E_{\text{cr}} \tag{4-15-12}$$

where T is the sample temperature. This increase in the electron energy can be approximately related to the power gained by electrons from the electric field

$$\Delta E_{\text{cr}} \approx q\mu_B F_p^2(B)\tau_T \tag{4-15-13}$$

where τ_T is the effective energy relaxation time, F_p is the peak field of the v vs. F curve. This curve is almost linear up to $F \sim F_p$ and thus

$$v_p(B) \simeq \mu_B F_p(B) \tag{4-15-14}$$

and the peak current density

$$j_p(B) \simeq qn_0\mu_B F_p(B) \tag{4-15-15}$$

From Eqs. (4-15-10) and (4-15-13)–(4-15-15) we find

$$\frac{F_p(B)}{F_p(0)} \simeq (1 + \mu^2 B^2)^{1/2} \tag{4-15-16}$$

and

$$\frac{j_p(B)}{j_p(0)} \simeq (1 + \mu^2 B^2)^{-1/2} \tag{4-15-17}$$

As can be seen from Fig. 4-15-2 this simple model provides reasonable agreement with the experimental data [129, 130] (see also Ref. 132). The same curves computed by the Monte Carlo methods [133] differ by no more than 10% from the curves shown in Fig. 4-15-2. The curves calculated using the Maxellian distribution function [134] are also in good agreement with the simple model described above.

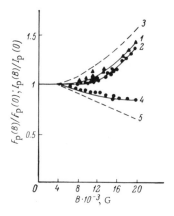

FIGURE 4-15-2. Magnetic field dependences of the relative peak field $F_p(B)/F_p(0)$ and the relative peak current $I_p(B)/I_p(0)$ for short samples $(L/d < 1)$ [129, 130]. 1, $F_p(B)/F_p(0)$ for a sample with $L/d = 0.2$; 2, $F_p(B)/F_p(0)$ for a sample with $L/d = 0.4$; 4, $I_p(B)/I_p(0)$ for a sample with $L/d = 0.4$. Curves 3 and 5 represent calculations in the frame of the simple model discussed in the text.

v-F curves computed by the Monte Carlo method for short $(L/d \ll 1)$ and long $(L/d \gg 1)$ GaAs samples are shown in Fig. 4-15-3 [134]. As can be seen from the figure, the negative differential mobility decreases in the magnetic field. This effect could lead to the prevention of the domain formation as was reported in Ref. 135.

In long samples the influence of the transverse magnetic field is much weaker (see Fig. 4-15-3b) because the Lorentz force is compensated by the Hall electric field. The change in the transport characteristics in the magnetic field is related in this case to the scatter of the electron velocities.

The Lorentz force is larger for faster electrons and therefore the magnetic field tends to "cool" down the electrons, leading to a slight increase electron velocity in large electric field and to a small increase in the peak electric field.

The effect of the magnetic field on the energy distribution for long GaAs samples can be seen from Fig. 4-15-4 [134].

The threshold field of the Gunn effect in a long sample decreases with increasing magnetic field, even though the peak field increases (see Fig. 4-15-36), and the frequency of the Gunn oscillations decreases. Such behavior is explained by the decrease of the Hall electric field near the contacts where the contacts prevent the accumulation of the Hall charge. These narrow regions near the contacts do exhibit the geometric magnetoresistance similar to that in the short sample. As a result the local electric field near the cathode increases in the magnetic field, enhancing the domain formation in this region.

Using the simplified model described above [see Eqs. (4-15-12)–(4-15-13)] we obtain the following expressions for the threshold (not for the peak) electric field and the threshold current in a long sample [8]:

$$F_t(B) = \frac{F_t(0)}{(1 + \mu^2 B^2)^{1/2}} \qquad (4\text{-}15\text{-}18)$$

$$I_t(B) = \frac{I_t(0)}{(1 + \mu^2 B^2)^{1/2}} \qquad (4\text{-}15\text{-}19)$$

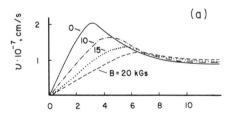

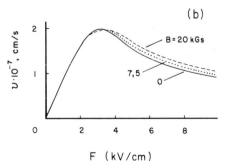

FIGURE 4-15-3. Velocity in the direction of the applied electric field as a function of the applied electric field [133]. (a) $L/d \ll 1$; (b) $L/d \gg 1$.

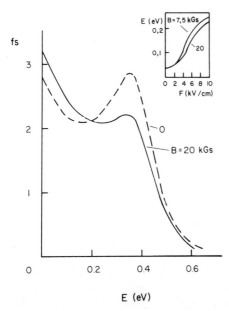

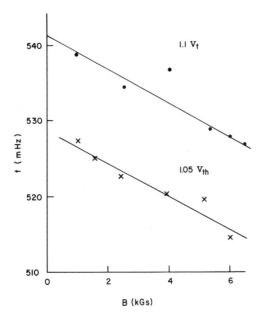

FIGURE 4-15-4. Symmetrical part of the distribution function vs. the electron energy [133]. $F = 18\ \text{kV/cm}$.

Equations (4-15-18) and (4-15-19) are in good agreement with the experimental results [135].

The frequency of the Gunn oscillations vs. magnetic field is shown in Fig. 4-15-5. As can be seen from the figure, a substantial change of frequency can be achieved in the magnetic field of the order of several kilogauss.

The phenomenological theory based on the drift-diffusion model (i.e., v vs. F curve) predicts that in a very high magnetic field the current–voltage characteristic of a short Gunn sample should become S-type due to a very strong geometric magnetoresistance [136, 137]. However, such a behavior has not been observed in

FIGURE 4-15-5. Frequency variation with magnetic field for fixed applied voltages [135].

the magnetic field up to 30,000 G [138]. This may indicate that the drift-diffusion model which seems to be in adequate agreement with many experimental results becomes invalid in a strong magnetic field.

4-16. ULTRASOUND GENERATION

GaAs and many other materials in which the Gunn effect is observed are piezoelectric. This means that an electric field causes a significant deformation of the crystal lattice. Under the Gunn oscillation conditions, when the electric field varies in time and space along the sample, the piezoelectric effect should give rise to a lattice deformation which varies in time and space, i.e., it should give rise to an ultrasonic wave. Grinberg and Kastalskii [139, 140] were the first to draw attention to the possibility of ultrasound generation in Gunn diodes.

A calculation given in Ref. 141 shows that the most effective mechanism for the generation of ultrasound is associated with the formation of high field domains. When a domain is being formed, the field near the cathode (in a region of the order of the domain width d) varies from $F_p \sim 3$ kV/cm to $F_m > 100$ kV/cm at the Gunn oscillation frequency. Thus, a virtual source of a strong alternating field is connected to the cathode region. The piezoelectric effect associated with this alternating field gives rise to forced ultrasonic vibrations at the Gunn oscillation frequency.

The generation of ultrasound in Gunn diodes was observed experimentally and reported in Refs. 142 and 143. Ultrasound was generated at the Gunn oscillation frequency and at its second and third harmonics. In a bias field $F_0 \sim 10$ kV/cm the ultrasonic output power was 3.5 dB mW at 355 MHz, corresponding to a power density of 0.4 W/cm^2.

It is quite likely that Gunn diodes may find practical applications as sources of ultrasound and hypersound. Their principal advantages over other sources would be their small size, wide frequency range extending to the millimeter waves, and small width of the ultrasonic beam.

An interesting modification of a Gunn source of ultrasound was described by Lee and White [144]. A GaAs Gunn diode was bonded to the surface of an LiNbO$_3$ crystal which had very strong piezoelectric properties. The domain field excited in GaAs penetrated, because of the fringe effect, into the adjoining part of the LiNbO$_3$ crystal. Ultrasonic waves were generated in LiNbO$_3$ by the periodic domain transits along the Gunn diode. As was reported in Ref. 144, this source was very promising in several important practical applications.

The maximum density of the ultrasonic power generated by a Gunn diode is estimated at tens of watts per square centimeter. The amplitude of the alternating electric field generated by an ultrasonic wave as a result of the piezoelectric effect may exceed the difference between the peak field F_p and the sustaining field F_s [145]. In this case, the Gunn oscillations may become self-modulated by the ultrasonic wave generated in this way. If the bias field F_0 and the amplitude of the alternating field F_1 generated by ultrasound are such that the field corresponding to the ultrasonic wave peaks exceeds $F_p(F_0 + F_1 > F_p)$ and the field at the troughs of this wave falls to a value below $F_s(F_0 - F_1 < F_s)$, the region in which domains are generated will be determined by the wavelength of ultrasound.

The same effect can be achieved by introducing an ultrasonic wave into a Gunn diode. Estimates show [145] that, depending on the parameters of a Gunn diode, the intensity of the injected ultrasound wave needed for this effect should range from a few tenths to a few hundreds of watts per square centimeter.

4-17. MODULATION OF LIGHT BY A GUNN DIODE

An electric field applied along certain crystallographic directions in GaAs alters the index of refraction because of the linear electro-optical effect. The change in the refractive index Δn is proportional to the first power of the field intensity and is about 10^{-4} for GaAs in the electric field $\sim 40 \, \mathrm{kV/cm}$. When a high field domain crosses an illuminated region the anisotropy of the refractive index resulting from the electro-optical effect can be detected using standard optical instruments. A modulation of the radiation of a He–Ne laser of about $1.5 \, \mu\mathrm{m}$ wavelength by the electro-optical effect induced by high field domains was observed experimentally by Cohen et al. [146].

Light can also be modulated by high field domains using the Franz–Keldysh effect. In this effect the absorption edge of a semiconductor is shifted by a strong electric field. The absorption coefficient of photons whose energy is close to the energy gap depends very strongly on the difference between the photon energy and the energy gap. Illumination of a crystal by photons with energies nearly equal to the energy gap and a simultaneous application of a strong electric field enhances the absorption of light because the Franz–Keldysh effect shifts the absorption edge in the direction of lower energies. Guetin and Boccon-Gibod [147] observed a $\sim 25\%$ increase in the absorption when a high field domain crossed a region illuminated with radiation of wavelength equal to the energy gap of GaAs. This modulation by the Franz–Keldysh effect could be explained by assuming that the average field in a domain was $\sim 38 \, \mathrm{kV/cm}$. The actual domain field should be ~ 1.5–2 times higher, but this discrepancy could be explained by the experimental errors [147].

The Gunn effect in very heavily doped GaAs has a number of special features. In particular, the domain field becomes so high that it causes impact ionization, leading to the generation of electron–hole pairs. The holes in GaAs absorb very strongly an infrared radiation with photon energy of about 0.4 eV. Thus, the excitation of the Gunn oscillations in heavily doped samples enhances strongly the absorption at photon energies close to 0.4 eV. Southgate et al. [148] observed such a modulation of light with photon energy of 0.365 eV which was absorbed in samples with carrier densities $n_0 = (1\text{–}2) \times 10^{17} \, \mathrm{cm}^{-3}$.

A different way of modulating light by a Gunn diode was considered in Ref. 149. Since an electric field can lift a degeneracy of the electron gas, a periodic variation of the field in a Gunn diode can be used to modulate light by the field-induced shift of the fundamental absorption edge related to the lifting of the degeneracy of the electron gas.

In a degenerate semiconductor all the states at the bottom of the conduction band are occupied up to the energies separated from the bottom of the band by the energy of the order of $(E_F - k_B T_e)$, where E_F is the Fermi level. Since the valence-band electrons can be transferred only to the free states in the conduction band, the absorption edge is shifted by $(E_F - k_B T_e)$ (the Moss–Burstein effect).

In the absence of an external electric field the electron temperature is simply equal to the lattice temperature. A strong electric field may "heat" the electron gas so that the condition $k_B T_e > E_F$ may be satisfied for sufficiently high electric fields.

This means that a semiconductor which is degenerate in the absence of the electric field may behave as a nondegenerate material in a strong field. If the carrier density is sufficiently high and the lattice temperature is sufficiently low, one may encounter a situation in GaAs when the semiconductor is degenerate in a field equal to the field outside the domain F_s and nondegenerate when the field reaches the peak value F_p [149]. Under such conditions, a flux of photons whose energy is equal to the energy gap (for example, light emitted by a GaAs laser) is modulated at the frequency of the Gunn oscillations. Such modulation will occur throughout the sample making it possible to modulate fairly wide light beams. As was shown in Ref. 149 this mechanism may also lead to the modulation of light by high field domains.

4-18. INFLUENCE OF ILLUMINATION ON THE RIDLEY–WATKINS–HILSUM– GUNN EFFECT

The influence of illumination [150–155] on the Gunn effect depends on whether the whole sample or just a part is illuminated [8]. If the whole sample is illuminated we can distinguish two cases. If the energy of photons is higher than the energy corresponding to the energy gap, the generated carriers are distributed uniformly across the sample. In this case the relative nonuniformity in the distribution of electrons (associated with a nonuniform doping profile) decreases, and, therefore, the coherence and amplitude of the Gunn oscillations are enhanced. This effect was observed by Haydl and Solomon [150].

The opposite behavior is observed when the photon energy is such that electrons are excited into the conduction band from deep impurity centers. In this case, the nonuniformity in the distribution of the light-generated electrons, associated with a nonuniform distribution of impurity centers, may enhance the nonuniformity of the distribution of the conduction electrons along the sample. This reduces the coherence of the Gunn oscillations [151].

Interesting effects associated with local changes in the conductivity as a result of illumination are observed when only a part of the sample is illuminated. Two different effects may be observed if the region near the anode is illuminated and the light intensity is sufficiently high to increase significantly the conductivity of this region.

Since the field intensity near the anode decreases because of the high illumination-induced conductivity, a high field domain may be extinguished on entering the illuminated region (a virtual anode formation). The length traveled by the domain from the point of the nucleation to the point of quenching decreases and the oscillation frequency increases. Myers et al. [66] used a laser beam (see Fig. 4-18-1) to reduce the effective length of a sample from 600 to 80 μm (the conductivity in the illuminated part of the sample was about eight times higher than the dark conductivity, although the effect could be observed if the conductivity was just doubled or tripled). Figure 4-18-1 shows the dependence of the domain transit time on the effective length of the sample. The shortest effective length at which the oscillations were still observed was 80 μm (see Fig. 4-18-1), which corresponded to the product $n_0 L \sim 1.6 \times 10^{12}$ cm^{-2}. This value was somewhat higher than the critical value $(n_0 L)_1$ given by Kroemer's criterion.

A different effect, reported in Ref. 152, is the switching of the Gunn oscillations by illumination. When a bias voltage is somewhat lower than the threshold value,

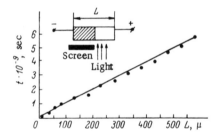

FIGURE 4-18-1. Dependence of the domain transit time on the length of the unilluminated (active) part of a sample [66].

an increase of the conductivity of the anode region by the illumination results in an increase of the electric field near the cathode. When the cathode electric field exceeds the threshold value it leads to domain formation and the resulting Gunn oscillations. If the duration of the light pulse is considerably longer than the domain transit time, oscillations will be observed as long as the sample is illuminated. If the duration of the light pulse is less than the domain transit time, the formed domain reaches the anode and a new domain does not form. Thus, an illumination with a short light pulse generates a single current pulse.

The illumination of the cathode region reduces the field near the cathode and, consequently, suppresses the formation of domains at voltages exceeding the threshold value.

The effects described above may find applications in optoelectronic devices. For example, as reported in Ref. 152, light pulses of 0.3 mW power and 1 ns duration are sufficient to produce (because of the appearance of traveling domains) a 300-mW signal in a 50-Ω load.

We may roughly estimate the power of the light beam required for the switching based on the idealized field and carrier distributions shown in Fig. 4-18-2.

In the dark (see Fig. 4-18-2a) the carrier and field distributions are uniform and the average (bias) field $F_0 < F_p$. When the region near the anode is illuminated (see Fig. 4-18-2b) the carrier concentration there increases and the electric field drops. From the requirement of current continuity we have

$$F_2 n_0 = F_1 n_1 \qquad\qquad (4\text{-}18\text{-}1)$$

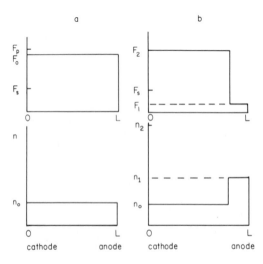

FIGURE 4-18-2. Schematic carrier and field distributions in the device without (a) and with (b) illumination near the anode.

If the bias voltage is kept constant then

$$F_2(L - d_l) + F_1 d_l = F_0 L \qquad (4\text{-}18\text{-}2)$$

Here d is the width of the illuminated region. The electric field F_2 should exceed F_p in order to initiate the domain formation. From this requirement and Eq. (4-18-1)–(4-18-2) we find

$$\Delta n d_l > \frac{(F_p - F_0) n_1 L}{F_p} \qquad (4\text{-}18\text{-}3)$$

where $\Delta n = n_1 - n_0$ is the excess carrier concentration near the anode. The beam power required to create Δn is given by

$$P_L = \frac{\hbar \omega_L \Delta n d_l S}{\tau_n Q_e} \qquad (4\text{-}18\text{-}4)$$

where $\hbar \omega_L$ is the photon energy, S is the cross section, τ_n is the electron lifetime, and Q_e is the quantum efficiency. From Eqs. (4-18-3) and (4-18-4) we find the beam power required for the switching

$$P_L \gtrsim \frac{F_p - F_0}{F_p} \frac{\hbar \omega_L n_1 S L}{\tau_n Q_e} \qquad (4\text{-}18\text{-}5)$$

Assuming for a crude estimate $\hbar \omega_L \sim 1.5$ eV (i.e., 2.4×10^{-19} J), $\tau_n \sim 10^{-8}$ s, $Q_e \sim 0.8$, $n_1 L \sim 10^{16}$ m^{-2}, $S \sim 10^{-8}$ m^2 and $(F_p - F_0)/F_p \sim 0.1$ we find $P \approx 3 \times 10^{-4}$ W. The power of the output pulse is given by

$$P_{\text{out}} = q n_0 v_s F_p L S \qquad (4\text{-}18\text{-}6)$$

Therefore the available gain

$$G = \frac{P_{\text{out}}}{P_L} \sim \frac{q v_s F_p Q_e \tau_n}{\hbar \omega_L} \frac{F_p}{F_p - F_0} \frac{n_0}{n_1} \qquad (4\text{-}18\text{-}7)$$

Assuming $F_p \sim 3.5$ kV/cm, $v_s \sim 10^5$ m/s, $n_0/n_1 \sim 1$ and the values of other parameters to be the same as given above, we find $G \approx 32.7$ dB. Gains up to 30 dB have been observed experimentally.

REFERENCES

1. J. B. Gunn, Microwave oscillations of current in III–V semiconductors, *Solid State Commun.* **1** (4), 88–91 (1963).
2. J. B. Gunn, Instabilities of current in III–V semiconductors, *IBM J. Res. Dev.* **8** (2), 141–159 (1964).
3. P. Guetin, Contribution to the experimental study of the Gunn effect in the long samples, *IEEE Trans. Electron Devices* **ED-14** (9), 552–562 (1967).
4. B. K. Ridley, Specific negative resistance in solids, *Proc. Phys. Soc.* **82** (12), 954–966 (1963).
5. B. K. Ridley and T. B. Watkins, The possibility of negative resistance, *Proc. Phys. Soc.* **78** (8), 293–304 (1961).

6. C. Hilsum, Transferred electron amplifiers and oscillators, *Proc. IRE* **50** (2), 185–189 (1962).
7. H. Kroemer, Theory of the Gunn effect, *Proc. IEEE* **52** (12), 1736 (1964).
8. M. E. Levinstein, Y. K. Pozhela, and M. S. Shur, Gunn Effect, Sov. Radio, Moscow, 1975 (in Russian).
9. H. W. Thim, Linear negative conductance amplification with Gunn oscillators, *Proc. IEEE* **55**, 446–447 (1967).
10. J. S. Heeks, A. D. Wood, and C. P. Sandbank, Coherent high field oscillations in samples of GaAs, *Proc. IEEE* **53** (5), 554–555 (1965).
11. C. P. Sandbank, Synthesis of complex electronic functions by solid state bulk effects, *Solid State Electron.* **10** (5), 369–380 (1967).
12. J. A. Copeland, T. Hayashi and M. Venohara, Logic and memory elements using two valley semiconductors, *Proc. IEEE* **55** (7), 1236–1237 (1967).
13. H. Kroemer, The Gunn effect under imperfect cathode boundary conditions, *IEEE Trans.* **ED-15** (11), 819–837 (1968).
14. M. P. Shaw, H. L. Grubin, and P. R. Solomon, *The Gunn–Hilsum effect*, New York, Academic Press, 1979.
15. K. W. Boer and G. Dohler, Influence of boundary conditions on high field domains in Gunn diodes, *Phys. Rev.* **186** (3), 793–800 (1969).
16. M. E. Levinstein and M. S. Shur, Current stabilizer on Gunn diode, *Sov. Phys. Semicond.* **3** (7), 915 (1970).
17. R. Zuleeg, Possible ballistic effects in GaAs current limiters, *IEEE Electron Device Lett.* **EDL-1** (11), 234–235, (1980).
18. D. E. McCumber and A. G. Chynoweth, Theory of negative conductance amplification and of Gunn instabilities in "two-valley" semiconductors, *IEEE Trans. Electron Devices* **ED-13** (1), 4–21 (1966).
19. H. Kroemer, Detailed theory of the negative conductance of bulk negative mobility amplifiers in the limit of zero ion density, *IEEE Trans. Electron Devices* **ED-14** (9), 476–492 (1967).
20. R. Holmstrom, Small signal behavior of Gunn diodes, *IEEE Trans. Electron Devices* **ED-14** (9), 526–531 (1967).
21. P. N. Butcher, W. Fawcett, and C. Hilsum, A simple analysis of stable domain propagation in the Gunn effect, *Brit. J. Appl. Phys.* **17** (7), 841–850 (1966).
22. B. L. Gelmont and M. S. Shur, Analytical theory of stable domains in high-doped Gunn diodes, *Electron. Lett.* **6** (12), 385–387 (1970).
23. M. S. Shur, Maximum electric field in high field domain, *Electron. Lett.* **14** (16), 521–522 (1978).
24. L. F. Eastman, S. Tiwari, and M. S. Shur, Design criteria for GaAs MESFETs related to stationary high field domains, *Solid State Electron.* **23**, 383–389 (1980).
25. R. A. Warriner, Computer simulation of negative resistance oscillators using a Monte Carlo model of gallium arsenide, *IEE J. Solid State Electron Devices* **1**, 92–96 (1977).
26. K. Blotekjaer, Transport equations for electrons in two-valley semiconductors, *IEEE Trans. Electron Devices* **ED-17** (1), 38–47, 1970.
27. S. I. Anisimov, V. I. Melnikov, and E. I. Rashba, Concerning one model in the theory of the Gunn effect, *Sov. Phys. JETPh Lett.* **7** (7), 196, (1968).
28. A. M. Mazzone and H. D. Rees, Transferred-electron harmonic generators for millimetre band sources, *IEE Proc. I, Solid State Electron Devices* **127** (4), 149–160 (1980).
29. A. M. Mazzone and H. D. Rees, Transferred-electron oscillatiors at very high frequencies, *Electron. Lett.* **17**, 539–540 (1981).
30. J. Owen and G. S. Kino, Theoretical study of Gunn domains and domain avalanching, *J. Appl. Phys.* **42** (12), 5006–5018 (1971).
31. J. Kuru, R. N. Robson, and G. S. Kino, Some measurements of the steady-state and transient characteristics of high-field dipole domains in GaAs, *IEEE Trans. Electron. Devices* **ED-15** (1), 21–29 (1968).
32. K. Kurokawa, Transient behavior of high field domain in bulk semiconductors, *Proc. IEEE* **55** (9), 16–1616 (1967).
33. M. S. Shur, Time of domain parameter changes under small bias variations, *Sov. Phys. Semicond.* **10** (10), 3138–3142 (1968).
34. M. S. Shur, Analytical theory of domain dynamics, *Sov. Phys. Semicond.* **7** (6), 1178–1183 (1973).
35. M. Claasen and E. Reinecker, Equivalent circuit of Gunn devices operating in the monostable switching mode, *IEEE Trans. Electron Devices* **ED-28** (3), 280–284 (1980).
36. M. E. Levinstein and G. S. Simin, Dynamics of changes in a Gunn domain due to changes in the bias voltage, *Sov. Phys. Semicond.* **13** (7), 837–839 (1979).

37. R. B. Robrock II, *IEEE Trans. Electron Devices* **ED-17** (1), 93 (1970).
38. W. Heinle, *Solid State Electron.* **21**, 583 (1968).
39. G. S. Gintsberg, V. I. Efimov, B. A. Kalbekov, and R. I. Kleinerman, *Elektron Tekh. Ser.* 3, *Mikroelektron* **5**, 42 (1973) (in Russian).
40. M. E. Levinshtein and G. S. Simin, Dynamics of disappearance of a Gunn domain in the anode, *Sov. Phys. Semicond.* **13** (7), 529–534 (1979).
41. M. E. Levinshtein and G. S. Simin, Dynamics of dispersal of a Gunn domain at voltages below the disappearance, *Sov. Phys. Semicond.* **13** (12), 1365–1368 (1979).
42. H. Kroemer, *IEEE Trans. Electron Devices* **ED-13**, 27 (1966).
43. D. Jones and H. D. Rees, *Electron. Lett.* **8**, 566 (1972).
44. D. Jones and H. D. Rees, *Electron. Lett.* **8**, 363 (1972).
45. T. G. Ruttan, *Electron. Lett.* **11**, 293 (1975).
46. H. Kroemer, *Solid-State Electron.* **21**, 61 (1978).
47. A. F. Volkov, *Sov. Phys. Solid State* **8**, 2552 (1967).
48. B. W. Knight and G. A. Peterson, *Phys. Rev.* **155**, 393 (1967).
49. M. I. D'yakonov, M. E. Levinshtein, and G. S. Simin, Basic properties of an accumulation layer in a Gunn diode, *Sov. Phys. Semicond.* **15** (11), 1229-1234 (1981).
50. W. Thim, Experimental verification of bistable switching with Gunn diodes, *Electron Lett.* **7** (10), 246–247 (1971).
51. W. Thim, *Proc. IEEE.* **59**, 1285 (1971).
52. S. H. Izadpanah, *Solid State Electron* **19**, 129 (1976).
53. P. Iondrup, P. Jeppesen and B. Jeppsson, *IEEE Trans. Electron Devices* **ED-23**, 1028 (1976).
54. I. Magarshack and A. Mirica, Proceedings of Intern. Conf. Microw. and Opt. Generation and Amplification MOGA-70, Amsterdam, pp. 16.19-16.23, 1970.
55. A. B. Torrens, *Appl. Phys. Lett.* **24**, 432 (1974).
56. A. Aishima, K. Yokoo and S. Ono, *IEEE Trans.* **ED-25**, 640 (1978).
57. K. Yokoo, S. Ono and A. Aishima, *IEEE Trans.* **ED-27**, 208 (1980).
58. S. Hasuo, T. Nakamura, G. Goto, K. Kazetani, H. Ishiwari, H. Suzuki and T. Izobe, *IEEE Trans.* **ED-23**, 1063 (1976).
59. O. A. Kireev, M. E. Levinshtein and S. L. Rumjantsev, Anode domain transient processes in supercritical Gunn diodes, *Solid State Electron.* **27** (3), 233–239 (1984).
60. H. W. Thim and S. Knight, Carrier generation and switching phenomena in *n*-GaAs devices, *Appl. Phys. Lett.* **11** (3), 83 (1967).
61. H. Kroemer, Non-linear space-charge domain dynamics in a semiconductor with negative differential mobility, *IEEE Trans. Electron Devices* **ED-13** (1), 27–40 (1966).
62. H. Thim and M. R. Barber, Observations of multiple high field domains in *n*-GaAs, *Proc. IEEE* **56** (1), 110–111 (1968).
63. M. Ohtomo, Nucleation of high-field domains in *n*-GaAs, *Jpn. J. Appl. Phys.* **7** (11), 1368-1380 (1968).
64. H. W. Thim, Computer study of bulk GaAs devices with random and dimensional doping fluctuations, *J. Appl. Phys.* **39** (8), 3897-3904 (1968).
65. J. B. Gunn, Effect of domain and circuit properties on oscillations in GaAs, *IBM J. Res. Dev.* **10** (4), 310-320 (1966).
66. F. A. Myers, J. McStay, and B. C. Taylor, Variable length Gunn oscillator, *Electron. Lett.* **4** (18), 386–387 (1968).
67. M. Suga and K. Sekido, Effects of doping profile upon electric characteristics of Gunn diodes, *IEEE Trans. Electron Devices* **ED-17** (4), 275–281 (1970).
68. M. P. Shaw, P. R. Solomon, and H. L. Grubin, The influence of boundary conditions on current instabilities in *n*-GaAs, *IBM J. Res. Dev.* **13** (5), 587–590 (1969).
69. E. M. Conwell, Boundary conditions and high field domain in GaAs, *IEEE Trans. Electron Devices* **ED-17** (4), 262–270 (1970).
70. M. P. Shaw, P. R. Solomon, and H. L. Grubin, Circuit-controlled current instabilities in *n*-GaAs, *Appl. Phys. Lett.* **17** (12), 535–537 (1970).
71. P. R. Solomon, M. P. Shaw, H. L. Grubin, and R. Kaul, An experimental study of the influence of boundary conditions on the Gunn effect, *IEEE Trans. Electron Devices* **ED-22** (3), 127–139 (1975).
72. R. Engelmann, Simplified model for the domain dynamics in Gunn effect semiconductors covered with dielectric sheets, *Electron. Lett.* **4** (24), 546–547 (1968).
73. G. S. Kino and P. N. Robson, The effect of small transverse dimensions on the operation of Gunn devices, *Proc. IEEE* **56** (11), 2056-2057 (1968).
74. H. L. Hartnagel, Gunn instabilities with surface loading, *Electron Lett.* **5** (14), 303–304 (1969).

75. P. Gueret, Limits of validity of the 1-dimensional approach in space-charge-wave and Gunn-effect theories, *Electron. Lett.* **6** (7), 197–198 (1970).

76. P. Gueret, Stabilization of Gunn oscillations in layered semiconductor structures, *Electron. Lett.* **6** (20), 637–638 (1970).

77. M. Masuda, N. S. Chang, and Y. Matsuo, Suppression of Gunn-effect domain formation by ferrimagnetic materials, *Electron. Lett.* **6** (19), 605–606 (1970).

78. S. Kataoka, H. Tateno, and M. Kawashima, Improvements in efficiency and tunability of Gunn oscillators by dielectric-surface loading, *Electron. Lett.* **5** (20), 491–492 (1969).

79. P. Kozdon and P. N. Robson, Two-port amplifiers using the transferred electron effect in GaAs, in 8th Int. Conf. on Microwave and Optical Generation and Amplification (MOGA), Amsterdam, paper 16.2, 1970.

80. K. Kumabe, Suppression of Gunn oscillations by a two-dimensional effect, *Proc. IEEE* **56** (12), 2172–2173 (1968).

81. M. E. Levinshtein, Influence of dielectric coatings on oscillation parameters of planar Gunn diodes, *Sov. Phys. Semicond.* **7** (10) 1347–1348 (1974).

82. K. R. Hoffman, Some aspects of Gunn oscillations in thin dielectric-loaded samples, *Electron. Lett.* **5** (11), 227–228 (1969).

83. J. Kuru and Y. Tajima, Domain suppression in Gunn diodes, *Proc. IEEE* **57** (7), 1215–1216 (1969).

84. M. T. Vlaardingerbroek, G. A. Acket, K. Hofmann, and P. M. Boers, Reduced build-up of domains in sheet-type gallium-arsenide Gunn oscillators, *Phys. Lett.* **28A** (2), 97–99 (1968).

85. H. L. Hartnagel, Gunn domains in bulk GaAs affected by surface conditions, *Phys. Stat. Solidi (a)* **7** (2), K99–K103 (1971).

86. K. R. Freeman, C. Sozou, and H. L. Hartnagel, Two-dimensional Gunn-domain growth in bulk GaAs, *Phys. Lett.* **34A** (2), 95–96 (1970).

87. M. Shoji, Theory of transverse extension of Gunn-domains, *J. Appl. Phys.* **11** (2), 774–778 (1970).

88. J. S. Heeks, Some properties of the moving high-field domain in Gunn effect devices, *IEEE Trans.* **ED-13** (1), 68–70 (1966).

89. A. G. Chynoweth, W. L. Feldmann, and D. E. McCumber, Mechanism of the Gunn effect, Proc. Intern. Conf. Phys. Semicond., Kyoto, *J. Phys. Soc. Jpn.* **21** Suppl., 514–521 (1966).

90. P. D. Southgate, Recombination processes following impact ionization by high-field domains in gallium arsenide, *J. Appl. Phys.* **38** (12), 4589–4595 (1967).

91. J. Owens and G. S. Kino, Experimental studies of Gunn domains and avalanching, *J. Appl. Phys.* **42** (12), 5019–5028 (1971).

92. P. D. Southgate, H. J. Pager, and K. K. N. Chang, Modulation of infrared light by holes in pulse-ionized GaAs, *J. Appl. Phys.* **38** (6), 2689–2691 (1967).

93. K. K. Chang, S. G. Liu, and H. J. Prager, Infrared radiation from bulk GaAs, *Appl. Phys. Lett.* **8**, 196–198 (1966).

94. S. G. Liu, Infrared and microwave radiations associated with a current controlled instabilities in GaAs, *Appl. Phys. Lett.* **9** (2), 79–81 (1966).

95. J. A. Copeland, Switching and low-field breakdown in *n*-GaAs bulk diodes, *Appl. Phys. Lett.* **9** (4), 140–142 (1966).

96. M. R. Oliver, A. L. McWhorter, and A. G. Foyt, Current runaway and avalanche effects in CdTe, *Appl. Phys. Lett.* **11** (4), 111–113 (1967).

97. P. D. Southgate, Stimulated emission from bulk field-ionized GaAs, *IEEE J. Quantum Electron.* **QE-4** (4), 179–185 (1968).

98. P. D. Southgate and R. T. Mazzochi, Stimulated emission in field-ionized bulk InP, *Phys. Lett.* **28A** (3), 216–217 (1968).

99. P. D. Southgate, Laser action in field-ionized bulk GaAs, *Appl. Phys. Lett.* **12** (3), 61–63 (1968).

100. B. L. Gelmont and M. S. Shur, Current filamentation in high doped Gunn diodes, *Sov. Phys. JETP Lett.* **11** (7), 350–353 (1970).

101. B. L. Gelmont and M. S. Shur, Characteristic time of the coherency violation of Gunn oscillations, *Sov. Phys. Semicond.* **7** (1), 326–331 (1973).

102. B. L. Gelmont and M. S. Shur, *S*-type current–voltage characteristics in Gunn diodes, *J. Phys. D Appl. Phys.* **6**, 842–850 (1973).

103. R. Hall and J. H. Leck, Avalanche break-down of gallium arsenide *p–n* junctions, *Int. J. Electron.* **25** (6), 529–537 (1968).

104. B. L. Gelmont and M. S. Shur, *S*-type current–voltage characteristics in Gunn diodes with deep levels, Theory, *Sov. Phys. Semicond.* **7** (1), 50 (1973).

105. B. L. Gelmont and M. S. Shur, S-type current-voltage characteristic in Gunn diodes with deep levels. Comparison of theory and experiment, *Sov. Phys. Semicond.* **7** (3), 377-379 (1973).
106. W. K. Kennedy, L. F. Eastman and R. J. Gilbert, LSA operation of large volume bulk GaAs samples, *IEEE Trans.* **ED-14** (9), 500-504 (1967).
107. E. D. Prokhorov and V. A. Shalaev, Influence of a magnetic field on carriers generated within a high field domain under impact ionization in a Gunn diode, *Sov. Phys. Semicon.* **4** (10), 1707 (1970).
108. P. D. Southgate, Role of ionization inhomogeneities in the S-type characteristic of moderately heavily doped n-GaAs, *J. Appl. Phys.* **43** (3), 1038-1041 (1972).
109. B. L. Gelmont and M. S. Shur, Current filamentation in doped Gunn diodes, *Sov. Phys. Semicond.* **4** (9), 1419 (1970).
110. B. L. Gelmont and M. S. Shur, Motion of a current filament movement in a magnetic field under Gunn effect conditions, *Sov. Phys. Semicond.* **4** (9), 1540 (1970).
111. B. L. Gelmont and M. S. Shur, Current filament movement in transverse electric and magnetic fields, *Sov. Phys. Semicond.* **7** (10), 1311-1314 (1971).
112. Z. N. Chigogidze, N. P. Khuchua, L. M. Gutnic, P. G. Kharati, I. V. Varlamov, U. A. Bekirev and A. A. Tyutyun, Mechanism of failure of Gunn diodes *Sov. Phys. Semicond.* **6** (9), 1443 (1973).
113. Sh. G. Suleimanov, Effect of a magnetic field on the current instability in gallium arsenide, *Sov. Phys. Semicond.* **6** (1), 150 (1972).
114. B. L. Gelmont and M. S. Shur, S-type current-voltage characteristic and recombination emission in Gunn diodes, *Electron. Lett.* **6** (16), 531-532 (1970).
115. B. L. Gelmont and M. S. Shur, A quasineutral wave new instability in semiconductors with two kinds of carriers—*Sov. Phys. Semicond.* **5** (6), 955 (1971).
116. B. L. Gelmont and M. S. Shur, The instability of quasineutral waves in a semiconductor with two kinds of carriers, *Phys. Lett.* **35A** (5), 353-354 (1971).
117. R. A. Smith, *Semiconductors*, Cambridge Univ. Press, Cambridge (1959).
118. B. L. Gelmont and M. S. Shur, High field domains in Gunn diodes with two kinds of carriers, *Sov. Phys. JETP* **33** (6), 1234-1239 (1971).
119. B. L. Gelmont and M. S. Shur, High field Gunn domains in the presence of electron-hole pairs, *Phys. Lett.* **36A** (4), 305-306 (1971).
120. B. L. Gelmont and M. S. Shur, High field recombination domains in semiconductors with two types of carriers, *Sov. Phys. JETP* **34** (6), 1295 (1972).
121. B. Fay and G. S. Kino, *Appl. Phys. Lett.* **15**, 337 (1969).
122. V. L. Dalal, Hole velocity in p-GaAs, *Appl. Phys. Lett.* **16** (12), 489-491 (1970).
123. P. A. Borodovskii and I. F. Rosentsvaig, Mobile high field domain in a sample of n-type InSb subjected to an external magnetic field, *Sov. Phys. Semicond.* **6** (12), 1972 (1973).
124. P. M. Boers, Measurements on dipole domains in indium phospide, *Phys. Lett.* **34A** (6), 329-330 (1971).
125. P. Guetin, Contribution to the experimental study of the Gunn effect in long GaAs samples, *IEEE Trans.* **ED-14** (9), 552-562 (1967).
126. B. L. Gelmont and M. S. Shur, Analytical theory of recombination domains, *Sov. Phys. Semicond.* **5** (11), 1841 (1971).
127. B. L. Gelmont and M. S. Shur, An analytical theory of the high field domain in the presence of trapping, *Phys. lett.* **38A** (7), 503-504 (1972).
128. B. L. Gelmont and M. S. Shur, Slow high field domains in Gunn diodes with two kinds of carriers **7** (9), 1279-1286 (1974).
129. M. E. Levinshtein, D. N. Nasledov, and M. S. Shur, Magnetic field influence on the Gunn effect, *Phys. Status Solidi* **33** (2), 897-903 (1969).
130. M. E. Levinshtein, T. V. Lvova, D. N. Nasledov, and M. S. Shur, Magnetic field influence on the Gunn effect II, *Phys. Status Solidi* (a) **1** (1), 177-187 (1970).
131. T. R. Jervis and E. F. Johnson, Geometrical magnetoresistance and Hall mobility in Gunn effect devices, *Solid State Electron.* **13**, 181-189 (1970).
132. H. L. Hartnagel, G. P. Srivastava, P. C. Mathur, K. N. Trepathi, S. K. Lomash, and A. K. Dhall, The influence of transverse magnetic field on Gunn effect threshold, *Phys. Lett.* **49A** (6), 463-465 (1974).
133. A. D. Boardman, W. Fawcett, and J. G. Ruch, Monte Carlo determination of hot electron galvanomagnetic effects in gallium arsenide, *Phys. Status Solidi* (a) **4** (1), 133-141 (1971).
134. W. Heinle, Influence of magnetic field on the Gunn effect characteristic of GaAs, *Phys. Status Solidi* **2** (1), 115-121 (1970).

135. W. K. Kennedy, Variation of the Gunn effect by magnetic field, *Proc. IEEE* 1639–1640, October (1965).
136. M. S. Shur, Turnover of current–voltage characteristic in high magnetic field due to geometric magnetoresistance, Proc. of 13th International Conference on Physics of Semiconductors, pp. 1145–1148, Rome 1977.
137. M. S. Shur, Geometric magnetoresistance and negative differential mobility in semiconductor devices, *Solid State Electron.* **20**, 389–401 (1977).
138. V. B. Gorfinkel, M. E. Levinshtein, and D. V. Mashovets, Influence of a strong transverse magnetic field on the Gunn effect, *Sov. Phys. Semicond.* **13** (3), 331 (1979).
139. A. A. Grinberg and A. A. Kastal'skii, *Phys. Status Solidi*, **26**, 219 (1968).
140. A. A. Kastal'skii, *Sov. Phys. Semicond.* **2**, 995 (1969).
141. N. A. Guschina and M. S. Shur, *Sov. Phys. Semicond.* **4**, 922 (1970).
142. H. Hagakawa, R. Ishiguro, N. Mikoshiba, and M. Kikuchi, *Appl. Phys. Lett.* **14**, 9 (1969).
143. H. Hayakawa, T. Ishiguro, S. Takada, N. Mikoshiba, and M. Kikuchi, *J. Appl. Phys.* **41**, 4755 (1970).
144. R. E. Lee and R. M. White, *Appl. Phys. Lett.* **16**, 343 (1970).
145. A. L. Kazakov and M. S. Shur, *Sov. Phys.—Semicond.* **3**, 1047 (1970).
146. M. G. Cohen, S. Knight, and J. P. Elward, *Appl. Phys. Lett.* **8**, 269 (1966).
147. P. Guetin and D. Boccon-Gibod, *Appl. Phys. Lett.* **13**, 161 (1968).
148. P. D. Southgate, H. J. Prager, and K. K. N. Chang, *J. Appl. Phys.* **38**, 2689 (1967).
149. M. S. Shur, *Phys. Lett.* **29A**, 490 (1969).
150. W. H. Haydl and R. Solomon, The effect of illumination on Gunn oscillations in epitaxial GaAs, *IEEE Trans.* **ED-15** (11), 941–942 (1968).
151. K. G. Sewell and L. A. Boather, Multimode operation in GaAs oscillators induced by cooling and illumination, *Proc. IEEE* **55** (7), 1228–1229 (1967).
152. R. F. Adams and H. J. Schulte, Optically triggered domains in GaAs Gunn diodes, *Appl. Phys. Lett.* **15** (8), 265–267 (1969).
153. R. R. Riesz, Optical interaction with high field domain nucleation in GaAs, *IEEE Trans.* **ED-17** (1), 81–83 (1970).
154. T. Igo *et al.*, Regenerative light pulse detection using Gunn effect, *Jpn J. Appl. Phys.* **9** (10), 1283–1285 (1970).
155. M. E. Levinshtein, Influence of the illumination on the parameters of Gunn diodes, *Sov. Phys. Semicond.* **7** (7), 832 (1973).
156. J. Xu and M. S. Shur, Velocity-Field Dependence in GaAs, unpublished.

5

Transferred Electron Oscillators

5-1. MODES OF OPERATION OF TRANSFERRED ELECTRON OSCILLATORS

Transferred electron devices can be used as oscillators owing to the negative differential resistance related to the intervalley transfer of electrons.

In the stable domain propagation mode the current spikes at the transit frequency (see Fig. 4-1-1) lead to a periodic voltage variation across the resistive load. The efficiency in this regime is, however, very low.

Much higher efficiencies can be obtained when the device is placed into a microwave cavity. In a high Q parallel LRC circuit the voltage across the device is equal to

$$U = U_0 + U_1 \cos \omega t \qquad (5\text{-}1\text{-}1)$$

where U_0 is the dc bias voltage, U_1 is the ac voltage amplitude, and ω is the resonance frequency of the circuit (including the device).

In a series LRC circuit, the current I through the device is controlled by the external circuit

$$I = I_0 + I_1 \cos \omega t \qquad (5\text{-}1\text{-}2)$$

where I_0 is the dc current component and I_1 is the ac current amplitude.

In a practical microwave circuit the voltage or current waveforms may include the second and higher harmonics of the resonance frequency.

For a typical voltage waveform given by Eq. (5-1-1)

$$U_0 - U_1 < F_s L \qquad (5\text{-}1\text{-}3)$$

where F_s is the domain sustaining field so that the device is biased below the peak field F_p for a fraction of each period. During this time the space charge in the device decays. Thus the instabilities are controlled by the applied voltage.

As was first pointed out by Kroemer [1] when the frequency ω is large enough so that

$$\omega \tau_f \gg 1 \qquad (5\text{-}1\text{-}4)$$

where τ_f is the domain formation time, the high-field domains do not form. In this case the current is determined by the electron drift velocity†

$$i = qn_0 v[F(t)] \qquad (5\text{-}1\text{-}5)$$

where

$$F(t) = \frac{U(t)}{L} \qquad (5\text{-}1\text{-}6)$$

Such a regime is called a limited space charge accumulation mode (LSA) [2, 3]. In many cases an accumulation layer (see Section 4-6) propagates in the device during a part of the cycle [4], but the accumulated charge disappears when the voltage drops below the threshold. As can be seen from Eq. (5-1-4) the domain formation should be inhibited for this mode of operation. This may be facilitated by using the inductive external circuit [5–7] or by making thin samples [8].

When the frequency ω is of the order of $1/\tau_f$ the domains start to form but the domain formation cannot be completed during the rf cycle. This mode of operation, which is called the hybrid mode, is probably the most efficient at the frequencies close to 10 GHz [9–12].

If the frequency is low so that

$$\omega \tau_f \ll 1 \qquad (5\text{-}1\text{-}7)$$

the stable domain formation and propagation take place corresponding to the so-called transit regime of operation. A waveform in the transit regime depends on the ratio f/f_T, where $f = \omega/2\pi$ is frequency and

$$f_T = \frac{v_s}{L} \qquad (5\text{-}1\text{-}8)$$

is the transit frequency. If $f > f_T$ the stable domains are quenched before they reach the anode. The current waveform for the *quenched mode* of operation is shown in Fig. 5-1-1.

At the frequencies smaller than the transit frequency the domains travel across the diode and disappear at the anode. If the voltage across the device at the end of the domain transit is smaller than the threshold voltage, a new domain nucleates only after the delay which is necessary to reach the threshold voltage (see Fig. 5-1-2). Such a regime is called the delayed mode of operation. (Both quenched and delayed modes are frequently called transit modes of operation.)

If the doping density is relatively large (of the order of $n_{cr} \geq 2\text{--}3 \times 10^{15}\ cm^{-3}$ and higher) "trapped" domains may form near the anode (see Section 4-8). The diode with a trapped domain exhibits a negative conductance in a microwave cavity. Therefore an oscillation mode with a trapped anode domain may occur [13–15]. In a suitable resonance cavity the diode operates initially in this regime but then the mode of operation changes into the transit mode (with a traveling domain) [16].

At very high frequencies transferred electron devices may operate in a harmonic extraction mode [17–18]. In this regime the field distribution in the device changes

---

† As discussed in Section 2-8, at very high frequencies the electron drift velocity cannot be considered as an instantaneous function of the bias. The retardation effects related to the finite energy relaxation time have to be included for a more realistic analysis of this mode of operation.

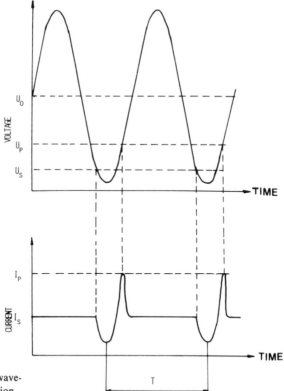

FIGURE 5-1-1. Voltage and current waveforms for the quenched mode of operation.

with the period close to the transit time but the power is extracted at higher (second
or third) harmonics.

At frequencies much smaller than the transit frequency a transferred electron
device may still exhibit a negative differential resistance. Also, the dc I-V characteristic may have a current drop at the threshold voltage due to the domain formation.

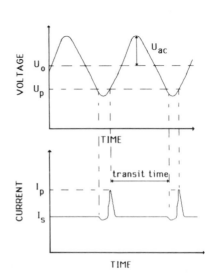

FIGURE 5-1-2. Voltage and current waveforms for the
delayed mode of operation.

As a result low-frequency sinusoidal or relaxation oscillations may occur in resonance circuits connected to the device at frequencies from 1 KHz to the transit frequency [19, 20].

We may also mention the multidomain regime of operation (see Section 4-8), which occurs if the voltage across the diode with doping fluctuations changes very fast (10^{12} V/s or faster) [21, 22].

Thus transferred electron devices operating in different modes (low-frequency oscillations, transit mode, hybrid mode, and LSA mode) may oscillate in a wide frequency range—from several kilohertz to millimeter waves. CW powers from a few milliwatts to a few watts have been generated with efficiencies up to 15%–20%. In the pulsed mode of operation the output powers have reached several kilowatts with efficiencies up to 30%. Low cost, long times before failure (up to several decades), acceptable noise figures, and wide tunability have led to many applications of transferred electron oscillators in microwave systems.

5-2. POWER AND EFFICIENCY OF TRANSFERRED ELECTRON OSCILLATORS

A crude estimate of the device efficiency may be obtained assuming that both voltage and current waveforms are sinusoidal and out of phase:

$$U = U_0 + U_1 \cos \omega t \qquad (5\text{-}2\text{-}1)$$

$$I = I_0 - I_1 \cos \omega t \qquad (5\text{-}2\text{-}2)$$

Then the output rf power

$$P_\sim = \tfrac{1}{2} U_1 I_1 \qquad (5\text{-}2\text{-}3)$$

and the efficiency

$$\eta = \frac{P_\sim}{P_0} = \frac{1}{2} \frac{U_1 I_1}{U_0 I_0} \qquad (5\text{-}2\text{-}4)$$

For large bias voltages $U_0 \simeq U_1 \gg U_t$ and

$$\frac{I_1}{I_0} \leq \frac{I_p - I_s}{I_p + I_s}, \qquad (5\text{-}2\text{-}5)$$

$$I_0 \simeq I_s$$

leading to the following estimate for the efficiency and output power:

$$\eta_{\max} \leq \frac{1}{2} \frac{1 - I_s/I_p}{1 + I_s/I_p} \qquad (5\text{-}2\text{-}6)$$

$$P_\sim = q v_s F_p \eta \frac{U_0}{U_p} n_0 L \qquad (5\text{-}2\text{-}7)$$

Assuming that the peak-to-valley ratio for GaAs $I_p/I_s \sim 2.75$ we obtain $\eta_{\max} \leq$ 23.3%. The efficiency may be even higher if the current and voltage waveforms have a rectangular (not a sinusoidal) shape. In this case we find that the efficiency at the fundamental frequency is given by [23]

$$\eta_{max} \leq \frac{8}{\pi^2} \frac{1 - I_s/I_p}{1 + I_s/I_p} \tag{5-2-8}$$

Thus for GaAs $\eta_{max} \leq 37.7\%$. The actual value of the efficiency depends on the device parameters, bias, and external circuit. A complete solution of the transport equations for the device together with the Kirchoff equations describing the external circuit is required to simulate the device behavior [24, 25].

A simpler approach is based on the assumption that the voltage waveform for the parallel RLC circuit (or the current waveform for the series RLC circuit) is given, which makes it possible to calculate the current waveform (the voltage waveform for the series RLC circuit). After the current waveform is computed the load resistance can be calculated at the fundamental frequency. The impedance at the harmonics can also be calculated if the applied voltage has a complex nonsinusoidal shape.

5-2-1. Transit Mode

Using such an approach the efficiency for the transit modes may be calculated analytically if the domain formation is assumed to be instantaneous [26, 27]. More realistic numerical calculations account for the finite time of the domain transients [9, 10, 28–30].

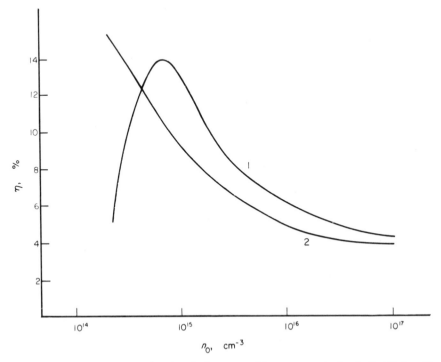

FIGURE 5-2-1. Efficiency of the transferred electron oscillator versus doping [10]. (1) $L = 50\ \mu$m; $f = 2.2$ GHz; Domain formation time, $\tau_f(s) = 10^5/n_0$ (cm^{-3}); (2) $\tau_f = 0$.

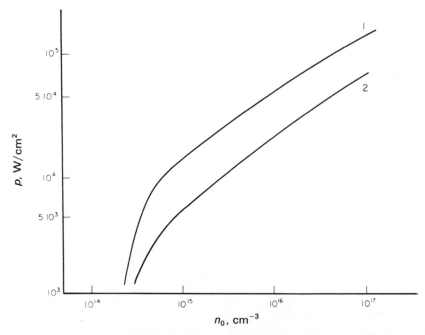

FIGURE 5-2-2. Output power of a transferred electron oscillator versus doping [10]. $f = 2.2\,\text{GHz}$, $L = 50\,\mu\text{m}$, $U_p = F_p L$. The ac field is chosen to give the highest power for each dc bias voltage. (1) $U_0/U_p = 6.82$; (2) $U_0/U_p = 2.73$.

Figure 5-2-1 shows the efficiency (maximized with respect to the bias and load resistance) as a function of doping for a 50-μm-long transferred electron device [10]. As can be seen from the figure the process of the domain formation affects the efficiency quite substantially. The peak efficiency is reached for the $n_0 L$ product $\sim 5 \times 10^{12}\,\text{cm}^{-2}$ (compare with the critical value of $(n_0 L)_1 \sim 10^{12}\,\text{cm}^{-2}$ determining the sample stability).

The output power (maximized with respect to the ac voltage) increases with the doping and bias (see Fig. 5-2-2). The efficiency (for the optimal load) as a function of the bias is shown in Fig. 5-2-3 for the same device ($L = 50\,\mu\text{m}$). The optimal load resistance and optimal ac voltage swing versus bias are shown in Fig. 5-2-4.

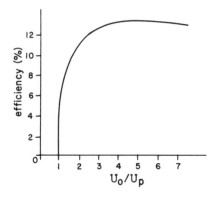

FIGURE 5-2-3. Efficiency of the transferred electron oscillator in a transit mode versus bias [10]. $L = 50\,\mu\text{m}$, $f = 2.2\,\text{GHz}$; $U_p = F_p L$.

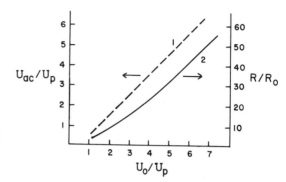

FIGURE 5-2-4. Optimal normalized ac voltage amplitude (1) and load resistance (2) versus bias for the transferred electron oscillator [10]. $L = 50\ \mu m$, $f = 2.2$ GHz.

As can be seen from Fig. 5-2-3 the efficiency peaks out at the bias voltage close to 4–5 threshold voltages and drops with the bias at higher voltages. The difference between the bias and the optimal voltage swing remains approximately constant and equal to the difference between the threshold voltage and domain sustaining voltage. The optimal load resistance increases with the bias. The efficiency close to the maximum value may be achieved in a wide range of load resistances, as can be seen from Fig. 5-2-5. For the higher load resistances the higher bias voltages are necessary to achieve the maximum efficiency. These results are in agreement with the results reported in Ref. 24.

Experimental results confirm that the efficiency first increases with the bias and remains nearly constant at high bias voltages (see Fig. 5-2-6). A sharp increase in the efficiency at the value of the bias close to $5U_p$ shown in the figure is probably due to the change of the mode of operation.

Frequency dependence of the efficiency (see Fig. 5-2-7) reveals that the oscillator may be tuned over a wide frequency range. The efficiency remains high even when the frequency becomes comparable to the inverse domain formation time (which corresponds to the hybrid mode of operation [9-12]). Experimental results also show that the transferred electron oscillators can be tuned over a wide frequency range (see Fig. 5-2-8).

The output power increases with the device length [see Eq. (5-2-7)], i.e., it is inversely proportional to the transit frequency. However, the increase in device length also leads to the increase of the low field resistance. This resistance causes losses due to the domain capacitance discharge. The equivalent circuit of the device in the microwave cavity during the active part of the cycle is shown in Fig. 5-2-9.

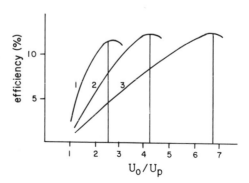

FIGURE 5-2-5. Efficiency versus bias for fixed values of the load resistance [10]. (1) $R/R_0 = 10$; (2) $R/R_0 = 20$; (3) $R/R_0 = 40$.

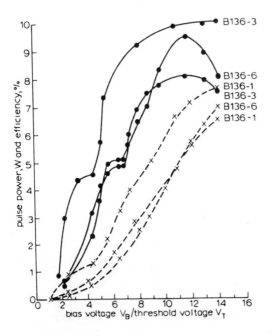

FIGURE 5-2-6. Output power and efficiency versus bias for a GaAs transferred electron oscillator [31]. ———, efficiency; - - -, peak power.

The output power is

$$P_{\sim} = \frac{U_1^2}{2R} \qquad (5\text{-}2\text{-}9)$$

where R is the load resistance. As can be seen from Fig. 5-2-5, fairly large values of R/R_0 are required for efficient operation.

A realistic simulation for the CW mode of operation should include the effects related to a substantial rise in the ambient temperature of the device [28–33], which

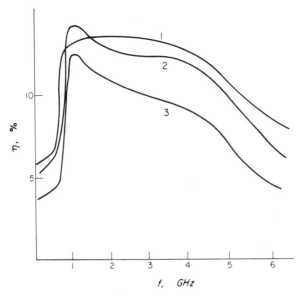

FIGURE 5-2-7. Efficiency of the transferred electron oscillator versus frequency [10]. $n_0 = 10^{15}\,\text{cm}^{-3}$, $\tau_f = 10^{-10}\,\text{s}$, $U_0/U_p = 5.14$. (1) $U_{ac}/U_p = 4.5$; (2) $U_{ac}/U_p = 4.64$; (3) $U_{ac}/U_p = 4.77$.

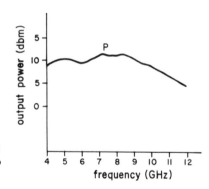

FIGURE 5-2-8. Output power versus frequency for a YIG tuned CW transferred electron oscillator in a microstrip circuit [33].

leads to smaller efficiencies and output powers. As can be seen from Fig. 5-2-10, the efficiency drops almost linearly with the increase in the ambient temperature.

5-2-2. LSA Mode

As was mentioned in Section 5-1, the space charge growth is limited by the applied voltage in the LSA mode. In this mode the ac voltage changes so fast that the high field domains do not form

$$\omega \tau_f > 1 \tag{5-2-10}$$

Using Eq. (4-7-28) for the domain time constant we find from Eq. (5-2-10)

$$\frac{n_0}{f^2} < \frac{\pi^2 \varepsilon L^2}{q \mu^2 \phi (U_0 - U_s)} \tag{5-2-11}$$

where $U_s = F_s L$. Assuming $L \simeq 25\ \mu m$, $\varepsilon \simeq 1.14 \times 10^{-10}\ F/m$, $\mu \simeq 0.6\ m^2/V\,s$, $\phi = 1/2$ (i.e., $n_0 \ll n_{cr}$), $U_0/U_s \sim 6$ we find from (5-2-11)

$$\frac{n_0}{f^2} < 1.6 \times 10^{-5} \left(\frac{s^2}{cm^3} \right)$$

The devices can operate in the LSA mode at smaller frequencies than the frequencies given by Eq. (5-2-10) if the domain formation is inhibited by the external circuit or by making the devices thin (see Section 4-12).

Some accumulation of the space charge part in the LSA mode takes place during the active part of the cycle. The fluctuations of the space charge grow

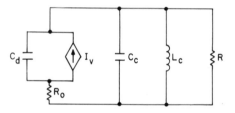

FIGURE 5-2-9. Equivalent circuit of the device in the microwave cavity during the active part of the cycle. C_d, domain capacitance; C_c, circuit capacitance; L_c, circuit inductance; R_0, low field resistance; R, load resistance.

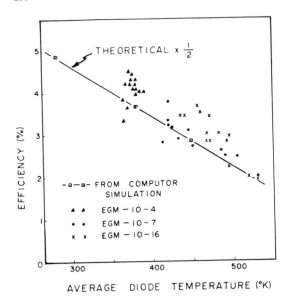

FIGURE 5-2-10. Efficiency versus average diode temperature [33].

proportionally to

$$G_a = \exp\left(\frac{T_a}{|\tau_{md}|}\right) \tag{5-2-12}$$

when the voltage across the sample corresponds to the region with the negative differential mobility, where T_a is the active part of the cycle,

$$|\tau_{md}| = \frac{\varepsilon}{qn_0|\mu_-|} \tag{5-2-13}$$

is the differential dielectric relaxation time and μ_- is the negative differential mobility. This charge should decay during the passive portion of the cycle:

$$G_p = \exp\left(\frac{T_p}{\tau_m}\right) > G_a \tag{5-2-14}$$

where T_p is the passive part of the cycle when the differential mobility is approximately equal to the low field mobility μ and τ_m is the low field dielectric relaxation time:

$$\tau_m = \frac{\varepsilon}{qn_0\mu} \tag{5-2-15}$$

From Eq. (5-2-14) we find

$$\frac{T_p}{T} > \frac{|\mu_-|}{|\mu_-|+\mu} \tag{5-2-16}$$

Criteria (5-2-10) and (5-2-14) are different from criteria

$$n_0/f \leq 2 \times 10^5 \ \text{s/cm}^3 \tag{5-2-17}$$

and

$$n_0/f > 2 \times 10^4 \, \text{s/cm}^3 \tag{5-2-18}$$

of the LSA operation accepted in many papers. Equation (5-2-17) follows from the assumption that the domain formation time may be estimated as

$$\tau_f \sim 2\text{-}3 \times \frac{\varepsilon}{q n_0 |\mu_-|} \tag{5-2-19}$$

whereas the domain formation time constant is given by Eq. (4-7-28). Equation (5-2-18) follows from the requirement that the space charge decay during the passive part of the cycle exceeds the space charge growth during the active part of the cycle by some fixed amount, i.e.,

$$G_p/G_a > e^\delta \tag{5-2-20}$$

where δ is a small positive number [compare with Eq. (5-2-14)], which does not seem to be necessary for the LSA mode.

The low-frequency limit for the operation in the LSA mode is given by Eq. (5-2-10). The upper frequency limit is determined by the energy relaxation time, i.e., by the time scale determining the establishment of the negative differential mobility. The frequency dependence of the efficiency in the LSA mode computed in Ref. 34 shows the efficiency cutoff at the frequencies close to 100 GHz (see Fig. 5-2-11). Higher frequency operation (close to 200 GHz) is possible with the optimized voltage waveforms using higher harmonics.

The simplest theory of the LSA oscillators is based on the assumption that the electric field and electron concentration remain almost uniform during the cycle so that the effects related to the space charge build-up may be neglected. In reality, as was mentioned above and in Section 4-8, the formation and propagation of accumulation layers is quite typical for the LSA mode. Therefore this simple approach may be used only for a crude description of the LSA mode and a more rigorous analysis should include the effects related to the propagation of the accumulation layers [35].

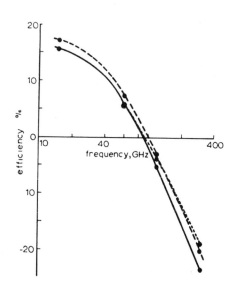

FIGURE 5-2-11. Frequency dependence of efficiency at different operating points [34]. 10 kV/cm dc, 8 kV/cm r.f. (unbroken); 15 kV/cm r.f. (broken); 20 kV/cm d.c., 18.6 V/cm r.f. (dotted).

If the space charge effects are neglected, then the output power per electron and the dc power supplied by the bias are given by

$$\tilde{P}_{\sim} = \frac{qF_{ac}}{T} \int_0^T v \sin \omega t \, dt \tag{5-2-21}$$

and

$$\tilde{P}_0 = \frac{qF_0}{T} \int_0^T v \, dt \tag{5-2-22}$$

respectively. Here $F_0 = U_0/L$ is the bias field, and $F_{ac} = U_{ac}/L$ is the amplitude of the microwave field. (The electric waveform for the LSA mode is shown in Fig. 5-2-12.)

The results of calculations for the LSA mode based on a single carrier approximation [see Eq. (5-21)–(5-22)] are shown in Figs. 5-2-13 and 5-2-14. Figure 5-2-13 shows the efficiency for the optimum load and the optimum load resistance as functions of the bias field [3]. It demonstrates that high efficiency is reached at a fairly large bias, in agreement with the experimental data [39]. Figure 5-2-14 shows the relationship between the ac field F_{ac} and the bias field F_0 in the LSA and transit modes. Large values of F_{ac} correspond to a large passive part of a cycle and lead to the power dissipation. Too small values of F_{ac} correspond to too small a passive part of a cycle so that the space charge accumulated during the active part of the cycle does not dissipate. The space charge build-up leads to hybrid or transit modes of operation [40].

As in the transit modes of operation the oscillation frequency in the LSA mode is smaller than the resonance frequency f_0 of the microwave cavity itself due to the contribution of device capacitance (f is nearly proportional to f_0) [24].

As can be seen from Fig. 5-2-12 a high amplitude of the microwave field F_{ac} is necessary for device operation in the LSA mode, i.e., ac voltage build-up is required in the microwave cavity prior to LSA operation. This may be achieved by utilizing harmonics of a steep bias pulse [41, 42]. Another approach calls for several initial high field domain transits. This sets up oscillations at a relatively low frequency with a higher harmonic build-up in the LSA resonance circuit. Figure 5-2-15 shows a transient voltage waveform for this mode of oscillation. As can be seen from the figure, domain oscillations with a period close to 1.25 ns develop initially. Then the ac voltage build-up with $T \simeq 417$ ps suppresses the domain and the LSA mode takes

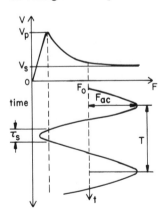

FIGURE 5-2-12. v versus F curve and ac field waveform for the LSA mode.

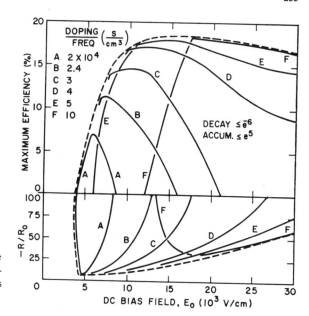

FIGURE 5-2-13. Efficiency for the optimum load and optimum load (relative to the low field resistance) versus bias field [3].

over [43]. This method may lead, however, to a breakdown in the high-field domain and as a result to device failure [39].

Still another possibility is to use a small load resistance for a fast initial ac voltage build-up connecting a half wavelength transmission line between the device and the load [42].

At relatively low frequency ω (comparable to $1/\tau_f$) the device switches from the LSA mode to the hybrid mode of operation. Figure 5-2-16 shows the efficiency in the hybrid and LSA modes as a function of the n_0/f parameter computed in Ref. 11. The experimental dependence of the efficiency on the bias for the hybrid mode of operation is shown in Fig. 5-2-17 [44]. The decrease in efficiency at high bias voltages is due to the impact ionization.

As was pointed out above, the efficiency may be improved substantially if the voltage waveform has a rectangular rather than a sinusoidal form [see Eq. (5-2-8)]. A considerable improvement may be achieved when the applied voltage contains only the fundamental and second harmonic [45–50]:

$$U(t) = U_0 + U_1 \sin \omega t + U_2 \sin(2\omega t + \phi) \qquad (5\text{-}2\text{-}23)$$

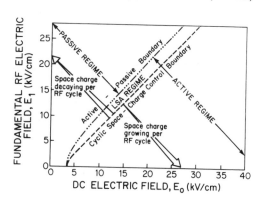

FIGURE 5-2-14. AC field versus bias field for the LSA, hybrid and transit modes [40].

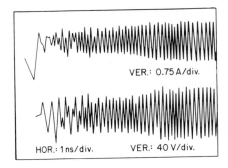

VER.: 0.75 A/div.

HOR.: 1 ns/div.   VER.: 40 V/div.

FIGURE 5-2-15. Transient current (upper trace) and voltage waveforms (lower trace) at the beginning of the LSA oscillations [43].

Microwave cavities which make it possible to establish such a waveform are described, for example, in Ref. 48.

Both simple analytical analysis [45, 46] and computer simulation [47] show that the optimum phase shift ϕ is $3\pi/2$. A $\pm10\%$ deviation from $3\pi/2$ decreases the efficiency to values smaller than in the absence of the second harmonic. The optimum ratio of U_2/U_1 is close to 0.33. The computer simulation predicts that for the optimum waveform the efficiency may increase 1.5 times, in good agreement with the experimental results [48–50].

Let us now consider a transferred electron device operation in a series LRC circuit (see Fig. 5-2-18). Above the threshold voltage the device behaves nearly as a current source shunted by a capacitance. The voltage waveform is then determined by the circuit inductance and effective capacitance (which includes the device capacitance). The duration of the active part of the cycle [8]

$$\tau_A \simeq 2\pi(LC)^{1/2} \tag{5-2-24}$$

Below the threshold the device behaves like a low field resistance R_0. The time

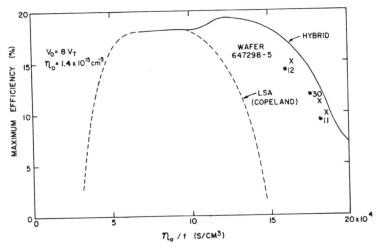

FIGURE 5-2-16. Efficiency versus n_0/f for the hybrid and LSA modes of operation [11]. *, Experimental points.

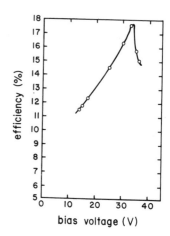

FIGURE 5-2-17. Efficiency versus bias for the hybrid mode of operation [44]. $L = 12\,\mu$m, $f = 8.3$ GHz, $n_0 = 1.3 \times 10^{15}$ cm^{-3}. Maximum output power in the pulsed regime 19 W.

below the threshold is determined by L, R_0, and the bias voltage [8]:

$$\tau_B \simeq \frac{L}{2.5 R(U_0/U_T - 1/2)} \tag{5-2-25}$$

The resulting relaxation voltage waveforms are shown in Fig. 5-2-19. The oscillation frequency

$$f = \frac{1}{\tau_A + \tau_B} \tag{5-2-26}$$

depends on the bias voltage according to Eq. (5-2-25) (see Fig. 5-2-20).

Very high output power (6 kW at 1.75 GHz and 2.1 kW at 5 GHz) from transferred electron devices has been obtained in the LSA relaxation mode [6].

5-2-3. Harmonic Extraction Mode

As was discussed in Chapter 2 (see Fig. 2-7-1), the electron negative differential mobility in GaAs decreases with an increase in the operating frequency. This limits the frequency range of transferred electron devices. However, the frequency range may be extended by extracting power at higher harmonics of the fundamental frequency. This mode of operation (which is sometimes called a harmonic extraction mode) was studied in Refs. 17 and 18. In Ref. 18 it was shown that 1.8- to

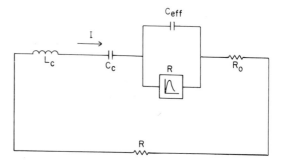

FIGURE 5-2-18. Equivalent circuit of a transferred electron device in a series resonance circuit. C_{eff}, effective device capacitance (C_{eff} is roughly equal to the domain capacitance for the domain modes and is close to the geometric capacitance for the LSA mode).

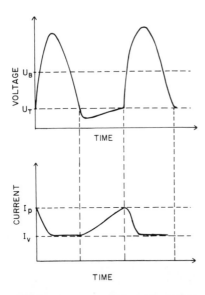

FIGURE 5-2-19. Voltage and current waveforms for the relaxation oscillations in a series LRC circuit.

2-micron-long devices, operating at the second harmonic, and 2.2- to 2.4-micron devices, operating at the third harmonic, can provide microwave power at frequencies close to 100 GHz. A 1.3- to 1.5-micron-long device, operating at the second harmonic, or a 1.7- to 2-micron device, operating at the third harmonic, may provide microwave power at frequencies up to 140 GHz.

5-2-4. Series or Parallel Connection of Transferred Electron Devices

Series or parallel connection of transferred electron devices may be used to increase the output power [50, 53–57]. However, the parallel connection leads to a

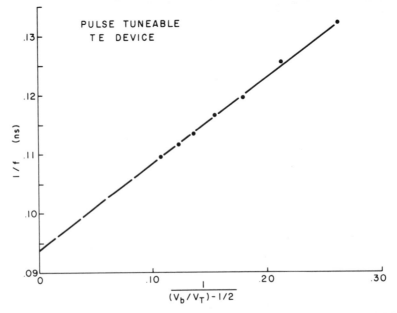

FIGURE 5-2-20. Inverse oscillation frequency as a function of $1/(U_B/U_T - 1/2)$ [4].

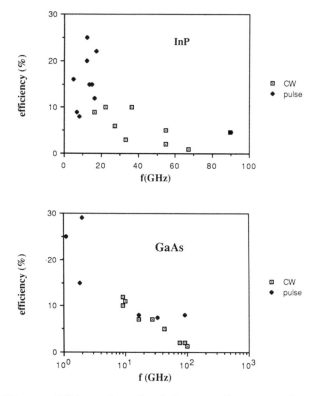

FIGURE 5-2-21. Efficiency of transferred electron oscillators versus frequency.

smaller impedance and does not improve the power–impedance product. For the series connection the devices should be strictly identical (for the transit modes of operation) to prevent an uneven voltage drop leading to the domain formation in the device with the largest resistance. Once a high field domain is formed in such a device it would act as a current limiter and the voltage across the rest of the devices would drop below the threshold.

Enstrom *et al.* [53] grew two nearly identical epitaxial transferred electron devices separated by a highly doped region. The efficiency at 3 GHz was 13.5%.

The computer simulation [54] and experimental studies [55, 56] show that devices with different parameters may still be connected in series if the bias and ac voltage amplitude are large enough and the devices operate in the LSA or hybrid mode.

The output power of 3.4 W at 8.3 GHz was obtained from two devices connected in series for the relative difference in the threshold current of the order of 10% [55]. The efficiency was 28.2%.

Another approach is to separate the devices by a distance which is equal to $\lambda/2$, where λ is the wavelength of the electromagnetic field in the cavity. Four devices connected this way in a coaxial cavity had the output power equal to the sum of the output powers for the individual devices [57].

Efficiency and output microwave power of transferred electron devices versus frequency are shown in Figs. 5-2-21 and 5-2-22 respectively. The data points used

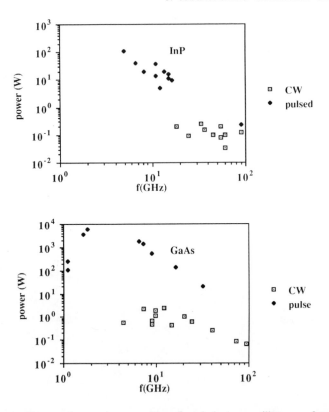

FIGURE 5-2-22. Output microwave power of transferred electron oscillators vs. frequency.

in these figures are taken from Ref. 76 (Chapter 11, Fig. 30) and from Refs. 17, 18, and 77.

5-3. NOISE, FREQUENCY STABILITY, AND TUNING OF TRANSFERRED ELECTRON OSCILLATORS

The noise, frequency stability, tuning, and modulation of any microwave oscillator are interrelated and depend on the loaded quality factor of the resonance circuit, which includes the device susceptance.

In order to reveal this relationship let us consider how the oscillation frequency ω changes when a small ac voltage δV is applied to the oscillator.

For simplicity we assume that the effective negative resistance R_- of the oscillator is frequency-independent and the total circuit reactance X is independent of the current i at the oscillation frequency ω. In this case the active component of the small ac voltage $\delta V \cos \psi$ leads to a change in the current:

$$i \frac{\partial R_-}{\partial i} \delta i = \delta V \cos \psi \qquad (5\text{-}3\text{-}1)$$

and the reactive component $\delta V \sin \psi$ leads to a change in the oscillation frequency

$$i \frac{\partial X}{\partial \omega} \delta \omega = \delta V \sin \psi \tag{5-3-2}$$

Here ψ is the phase angle and the oscillation frequency is determined by

$$X = 0 \tag{5-3-3}$$

From Eq. (5-3-2)

$$\delta \omega = \frac{\delta V \sin \psi}{i \partial X / \partial \omega} \tag{5-3-4}$$

where

$$\frac{\partial X}{\partial \omega} = \frac{\partial}{\partial \omega}\left(\omega L - \frac{1}{\omega C}\right) \tag{5-3-5}$$

Assuming for simplicity that the inductance L and capacitance C are independent of frequency we find

$$\frac{\partial X}{\partial \omega} = L + \frac{1}{\omega^2 C} = 2L = \frac{2QR_L}{\omega} \tag{5-3-6}$$

where we have taken into account that $\omega L = 1/\omega C$ at the oscillation frequency, R_L is the load resistance, and

$$Q = \frac{\omega L}{R_L} \tag{5-3-7}$$

is the loaded quality factor of the circuit. From Eqs. (5-3-4) and (5-3-6) we find

$$\frac{\delta \omega}{\omega} = \frac{1}{2Q} \frac{\delta V \sin \psi}{i_0 R_L} \tag{5-3-8}$$

$\psi = \pi/2$ corresponds to the maximum frequency tuning when

$$\left(\frac{\delta \omega}{\omega}\right)^2 = \frac{1}{4Q^2} \frac{P_{in}}{P_{out}} \tag{5-3-9}$$

where P_{in} is the input power required for the frequency tuning $\delta \omega$ and P_{out} is the oscillator output power. As can be seen from Eq. (5-3-9) a large Q factor leads to a stable oscillation frequency and makes the frequency tuning or injection locking difficult. On the other hand a larger Q factor decreases the frequency variation under the influence of the noise signal, i.e., decreases a frequency modulation noise.

5-3-1. Noise

For a noise signal we may assume that the phase ψ and voltage δV are given by

$$\langle \sin^2 \psi \rangle = 1/2 \tag{5-3-10}$$

$$(\delta V)^2 = 2k_B T_N R_L B \tag{5-3-11}$$

where T_N is the effective noise temperature and B is the bandwidth of one sideband. The noise power is

$$P_N = k_B T_N B \tag{5-3-12}$$

Let us now consider a noise signal at frequency $\omega + \Delta\omega$. We may find the power of the frequency modulated noise $P_{N,FM}$ assuming the modulation of the signal at $\omega + \Delta\omega$ by the noise power P_N:

$$\left(\frac{\Delta\omega}{\omega + \Delta\omega}\right)^2 = \frac{1}{2Q^2}\frac{P_N}{P_{N,FM}} \tag{5-3-13}$$

or, taking into account that $\omega + \Delta\omega \approx \omega$:

$$\frac{P_{N,FM}}{P_N} = \frac{1}{2Q^2}\left(\frac{\omega}{\Delta\omega}\right)^2 \tag{5-3-14}$$

or

$$\frac{P_{N,FM}}{P} = \frac{1}{2Q^2}\left(\frac{\omega}{\Delta\omega}\right)^2 \frac{k_B T_N B}{P} \tag{5-3-15}$$

Equation (5-3-13) is obtained by substituting $P_{N,FM}$ instead of P_{out} and P_N instead of P_{in} into Eq. (5-3-9) and taking into account Eq. (5-3-10). As can be seen from Eq. (5-3-15), the frequency modulated noise is inversely proportional to the squared loaded quality factor and proportional to the effective noise temperature T_N.

The root-mean-square frequency deviation is given by

$$(\delta f)_{rms} = \frac{f}{Q_L}(P_N/P)^{1/2} \tag{5-3-16}$$

We conclude that oscillators with a high value of Q have lower noise but are more difficult to tune [see Eq. (5-3-9)].

The amplitude modulation noise in transferred electron oscillators is smaller because nonlinearities of the device characteristics tend to stabilize the signal amplitude. From Eq. (5-3-1) we get

$$S_L^2(\delta i)^2 R_L^2 = (\delta V)^2/2 \tag{5-3-17}$$

where we have introduced

$$S_L = \frac{i}{R_L}\frac{\delta R_L}{\delta i} \tag{5-3-18}$$

and have taken into account that

$$\langle \cos^2 \psi \rangle = 1/2 \tag{5-3-19}$$

for the noise signal. Using Eq. (5-3-11) we obtain from Eq. (5-3-17)

$$P_{N,AM} \equiv (\delta i)^2 R_L = \frac{k_B TB}{S_L^2} \tag{5-3-20}$$

Thus

$$\frac{P_{N,AM}}{P_{N,FM}} = \frac{2Q^2(\Delta f)^2}{S_L^2 f^2} \tag{5-3-21}$$

For transferred electron oscillators $S_L \sim 3$ or higher [59]. Assuming $Q \approx 100$, $f = 10$ GHz, $\Delta f \approx 100$ kHz, and $S_L = 3$ we find $P_{N,AM}/P_{N,FM} \approx -63.5$ dB, which is in agreement with the experimental results.

The effective noise temperature T_N may be used as a parameter for comparing different devices. For transferred electron devices operating in transit modes one may expect T_N to be fairly high owing to high domain fields. For the LSA mode T_N should be smaller.

The above analysis is based on the simplified assumptions that R_- is independent of frequency and X is independent of current. Generally speaking this is not correct for transferred electron oscillators. Therefore the relationships between the noise,

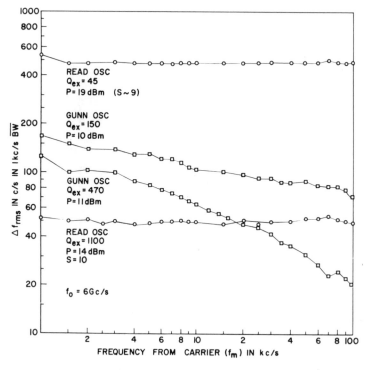

FIGURE 5-3-1. Measured noise spectrum for a transferred electron oscillator and for a Read oscillator [60].

the loaded quality factor, and the effective noise temperature may be more complex for real devices. As an example, Fig. 5-3-1 shows the measured FM noise spectrum for a transferred electron oscillator operating in a transit mode [60]. As can be seen from the figure, $(\delta f)_{\text{rms}}$ is not independent of δf as predicted by Eq. (5-3-16). However, it is a slowly decreasing function of δf and decreases with the increase in Q and P, in agreement with Eq. (5-3-16).

There are several different mechanisms which may contribute to the noise in transferred electron devices such as fluctuations of the domain velocity [61], a jitter in domain nucleation time [62, 63], contact phenomena [64], the device surface [65], and traps [66]. At the bias voltages near the threshold a current noise may be extremely high owing to the plasma instabilities [67, 68].

Injection phase locking may be used to stabilize the oscillation frequency and to decrease the frequency modulated noise of a transferred electron oscillator.

5-3-2. Frequency Tuning

The operating frequency of a transferred electron oscillator is controlled by the microwave circuit. The frequency of the microwave circuit may be changed mechanically, electronically (changing the bias voltage and hence the device reactance or using varactor diodes), or by employing a ferrimagnetic material (YIG spheres) and changing the magnetic field (magnetic tuning).

A description of circuits employed for the mechanical tuning of transferred electron devices may be found in Ref. 29. A wide tuning range is readily achieved. (See, for example, Ref. 69, where tuning from 1.6 to 4.7 GHz was reported.)

The bias dependence of the oscillation frequency is relatively weak for a sinusoidal voltage waveform (2-20 MHz/V depending on the oscillation mode, operating temperature, device parameters, doping profile, etc.). It is considerably more pronounced for the relaxation oscillations and may be used for the frequency tuning (see Fig. 5-2-21).

Figure 5-3-2 illustrates varactor tuning. This type of tuning is relatively fast (sweeping rates in the megahertz range). The tuning range depends on the varactor coupling with the circuit. A strong coupling increases the tuning range, but small changes in the varactor voltage may lead to a substantial variation of the output power. A weak coupling decreases the tuning range. A varactor quality factor which decreases with frequency sets an upper frequency limit for this type of tuning. A tuning range of 1 GHz at 13 GHz was reported [71].

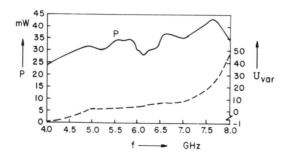

FIGURE 5-3-2. Varactor tuning of a CW transferred electron oscillator [70]. $U_p = 5.77$ V; $U_0 = 11.46$ V; $I_p = 477$ mA; $I_0 = 310$ mA.

p-i-n diodes may be used for the discrete frequency tuning of transferred electron oscillators. The frequency changes when the p-i-n diode switches into a low resistance state. This method is convenient for microstrip circuits.

Magnetic frequency tuning may be achieved if a ferrimagnetic material is inserted into a microwave cavity [72]. The imaginary part of the magnetic permeability decreases at the value of magnetic field close to the values required for a ferrimagnetic resonance. A more suitable approach employs YIG sphere resonators. The resonance frequency

$$\omega = \gamma B \qquad (5\text{-}3\text{-}22)$$

where $\gamma = 2.8\,\text{MHz/G}$ is the gyromagnetic ratio and B is the magnetic field (see Fig. 5-2-8). Both the transferred electron device and the YIG sphere may be a part of a microstrip circuit [74]. Frequency tuning from 5.8 to 13 GHz in a magnetic field changing from 1.8 to 4.5 kG was reported in Ref. 74. The variation of the output power in the tuning range was less than $\pm5\,\text{dB}$. High quality factors of YIG resonators decrease the noise of transferred electron oscillators. Another advantage is a nearly linear dependence between the magnetic field and oscillation frequency.

REFERENCES

1. H. Kroemer, External negative conductance of a semiconductor with negative differential mobility, *Proc. IEEE* **53**(9), 1246 (1965).
2. J. A. Copeland, A new mode of operation for bulk negative resistance oscillators, *Proc. IEEE* **54**(10), 1479–1480 (1966).
3. J. A. Copeland, LSA oscillator diode theory, *J. Appl. Phys.* **38**(8), 3096–3101 (1967).
4. L. F. Eastman, Transferred electron devices, in *Microwave Devices, Device Circuit Interaction*, Ed. by M. J. Howes and D. V. Morgan, Wiley, Chichester, 1978.
5. W. O. Camp, Jr., Computer simulation of multi-frequency LSA oscillations in GaAs, *Proc. IEEE* **57**(6), 220–221 (1969).
6. B. Jeppsson and P. Jeppesen, A high power LSA relaxation oscillator, *Proc. IEEE* **57**(6), 1218–1219 (1969).
7. B. I. Jeppsson and P. Jeppesen, LSA relaxation oscillations in a waveguide iris circuit, *IEEE Trans.* **ED-18**(7), 432–439 (1971).
8. W. O. Camp, Jr., D. W. Woodward, and L. F. Eastman, Bias-tunable c.w. transferred-electron oscillators, Proc. Fourth Cornell Conference, Microwave Semiconductor Devices, pp. 177-183, August 1973.
9. D. P. Dizhur, M. E. Levinstein, and M. S. Shur, Computer calculations of the efficiency of the Gunn generator, *Electron Lett.* **4**(21), 444–446 (1968).
10. K. V. Kuznetsov, M. E. Levinstein, and M. S. Shur, The investigation of the Gunn diode operation in a resonator using the computer model, *Solid State Electron.* **14**(3), 207–220 (1971).
11. H. C. Huang and L. A. MacKenzie, A Gunn diode operated in the hybrid mode, *Proc. IEEE* **56**(7), 1232–1233 (1968).
12. I. W. Monroe and F. M. Briggs, Experimental investigation of hybrid mode operation in oversized Gunn effect devices, *Proc. IEEE* **56**(11), 2066–2068 (1968).
13. H. L. Hartnagel and M. Kawashima, Negative TEO-diode conductance by transient measurement and computer simulation, *IEEE Trans. Microwave Theory Tech.* **MTT-21**, 468–477 (1973).
14. A. Aishima, K. Yokoo, and S. Ono, An analysis of wide band transferred electron devices, *IEEE Trans. Electron Devices* **ED-25**, 640–645 (1978).
15. H. L. Grubin, M. P. Shaw, and P. R. Solomon, On the form and stability of electric-field profiles within a negative differential mobility semiconductor, *IEEE Trans. Electron Devices* **ED-20**, 63-78 (1973).

16. K. Yokoo, S. Ono, and A. Aishima, Experimental observation of large-signal behavior in trapped domain transferred electron devices, *IEEE Trans. Electron Devices* **ED-27**(1), 208–212 (1980).

17. I. G. Eddison and D. M. Brookbanks, Operating modes of millimetre wave transferred electron oscillators, *Electron. Lett.* **17**(3), 112–113 (1981).

18. W. H. Haydl, Harmonic operation of GaAs millimetre wave transferred electron oscillators, *Electron. Lett.* **17**(22), 825–826 (1981).

19. M. E. Levinstein, Dynamics of low frequency oscillations in Gunn effect diodes, *Sov. Phys. JETP Piz.* **7**(9), 248 (1968).

20. W. O. Camp, Jr., Experimental observation of relaxation oscillator waveforms in GaAs from less than transit frequency to several times transit frequency, *Proc. IEEE* **59**(8), 1248–1250 (1971).

21. H. W. Thim, Computer study of bulk GaAs devices with random and dimensional doping fluctuations, *J. Appl. Phys.* **39**(8), 3897–3904 (1968).

22. M. Slater and R. Harrison, An investigation of multiple domain Gunn effect oscillators, *IEEE Trans. Electron Devices* **ED-23**(6), 560–567 (1976).

23. G. S. Kino and I. Kuru, High-efficiency operation of a Gunn oscillator in the domain mode, *IEEE Trans. Electron Devices* **ED-16**(9), 735–748 (1969).

24. J. A. Copeland, Theoretical study of a Gunn diode in a resonant circuit, *IEEE Trans. Electron Devices* **ED-14**(2), 55–58 (1967).

25. M. P. Shaw, H. Grubin, and P. Solomon, *The Gunn–Hilsum Effect*, Academic, New York, 1979.

26. W. Heinle, Determination of current waveform and efficiency of Gunn diodes, *Electron. Lett.* **3**(2), 52–54 (1967).

27. M. E. Levinstein and M. S. Shur, Calculation of the parameters of the microwave Gunn generator, *Electron. Lett.* **4**(11), 233–235 (1968).

28. K. R. Freeman and G. S. Hobson, A survey of CW and pulsed Gunn oscillators by computer simulation, *IEEE Trans. Electron Devices* **ED-20**(10), 891–903 (1973).

29. P. J. Bulman, G. S. Hobson, and B. C. Taylor, *Transferred Electron Devices*, Academic, New York, 1972.

30. M. R. Lakshminarayana and L. D. Partain, Numerical simulation and measurement of Gunn device dynamic microwave characteristics, *IEEE Trans. Electron Devices* **ED-27**, 546–552 (1980).

31. A. L. Edridge, F. A. Myers, B. J. Davidson, and J. C. Bass, Pulsed J band (12.4–18 GHz) Gunn-effect oscillators, *Electron. Lett.* **5**, 103 (1969).

32. D. C. Hanson, YIG-tuned transferred-electron oscillator using thin film microstrip, in Digest of technical papers, IEEE International Solid State Conference, Philadelphia, p. 122, 1969.

33. G. Hasegawa and Y. Aono, Thermal limitation for CW output power of a Gunn diode, *Solid State Electron.* **16**, 337–344 (1973).

34. P. N. Butcher and C. J. Hearn, Theoretical efficiency of the LSA mode for gallium arsenide at frequencies above 10 GHz, *Electron. Lett.* **4**(21), 459–461 (1968).

35. K. T. Ip and L. F. Eastman, Dependence of T.E.O. efficiency on NL product, *Electron. Lett.* **11**(14), 301 (1975).

36. J. A. Copeland, Doping uniformity and geometry of LSA oscillator diodes, *IEEE Trans. Electron Devices* **ED-14**(9), 497–500 (1967).

37. B. G. Bosch and R. W. H. Engelmann, *Gunn-effect Electronics*, Wiley, New York, 1975.

38. J. S. Lamming, Microwave transistor, in *Microwave Devices, Device Circuit Interactions*, Ed. by M. J. Howes and D. V. Morgan, Wiley, New York, 1976.

39. W. K. Kennedy, Jr., L. F. Eastman, and R. J. Gilbert, LSA operation of large volume bulk GaAs samples, *IEEE Trans.* **ED-14**(9), 500–504 (1967).

40. R. J. Harrison, S. P. Denker, and M. L. Hadley, Characteristics ranges for LSA oscillation, *IEEE Trans.* **ED-15**(10), 792–793 (1968).

41. J. F. Dienst and J. J. Thomas, An oversize cavity for exciting the LSA mode in *n*-GaAs, *IEEE Trans.* **ED-15**(8), 615–617 (1968).

42. R. R. Spiwak, Getting LSA started, *Electronics* **42**(4), 89 (1969).

43. T. Ikoma, H. Torizuka, and H. Yanai, Observations of voltage and current waveforms of the transferred-electron oscillators, *Proc. IEEE* **57**, 340 (1969).

44. F. P. Califano, High efficiency X-band Gunn oscillators, *Proc. IEEE* **57**, 251–252 (1969).

45. J. E. Carrol, A Gunn diode self-pumped parametric oscillator, *MOGA-konferenz*, Cambridge, England, IEEE Conf. Publ., No. 27, pp. 309–313, 1966.

46. W. Frey, Influence of a second harmonic voltage component on the operation of Gunn oscillators, In: 8th Int. Conf. on Microwave and Optical Generation and Amplification (MOGA), Amsterdam, paper 2.8, 1970.

47. M. E. Levinstein, L. S. Pushkaroeva, and M. S. Shur, Influence of the second harmonic of a resonator on the parameters of a Gunn generator for transit and hybrid modes, *Electron. Lett.* **8**(2), pp. 31–33 (1972).

48. J. E. Carrol, Hot electron microwave generators, Edwards Arnold Publishers, Ltd., London, England (1970).

49. J. E. Carrol, Resonant-circuit operation of Gunn diodes: A self-pumped parametric oscillator, *Electron. Lett.* **2**(6), 215–216 (1966).

50. J. F. Reynolds, B. E. Berson, and R. E. Enstrom, High-efficiency transferred electron oscillators, *Proc. IEEE* **57**(10), 1692–1693 (1969).

51. W. O. Camp and W. K. Kennedy, Pulse millimeter power using the LSA mode, *Proc. IEEE* **56**, 1105 (1968).

52. J. D. Crowley, J. J. Sowers, B. A. Janis, and F. B. Frank, High efficiency 90 GHz InP Gunn oscillators, *Electron. Lett.* **16**(18), 705–708 (1980).

53. R. E. Enstrom, J. F. Reynolds, and B. E. Berson, Vapour growth of multi-layered GaAs structures for series operation of transferred-electron oscillators, *Electron. Lett.* **5**(26), 714–715 (1969).

54. S. P. Yu, P. J. Shaver, and W. Tantraporn, Direct series operation of Gunn effect diodes with above critical $n_0 L$ products, *Proc. IEEE* **56**(11), 2068–2069 (1968).

55. M. C. Steele, F. P. Califano, and R. D. Larrabee, High efficiency series operation of Gunn devices, *Electron. Lett.* **5**(4), 81–82 (1969).

56. K. M. Baughan and F. A. Myers, Multiple series operation of Gunn-effect oscillators, *Electron. Lett.* **5**(16), 371–372 (1969).

57. J. E. Carrol, Series operation of Gunn diodes for high r.f. power, *Electron. Lett.* **3**(18), 455–456 (1967).

58. S. Boronski, Parallel-fed c.w. Gunn oscillators cascaded in X band waveguide for high microwave power, *Electron. Lett.* **4**(10), 185–186 (1968).

59. M. E. Levinstein, Y. K. Pozhela, and M. S. Shur, Gunn effect, *Soviet Radio*, Moscow, 1975 (in Russian).

60. J. Josenhans, Noise spectra of Read and Gunn oscillators, *Proc. IEEE* **54**(10), 1478–1479 (1966).

61. K. Matsuno, Noise of Gunn effect oscillator, *Proc. IEEE* **56**(1), 108 (1968).

62. G. S. Hobson, Source of f.m. noise in cavity-controlled Gunn effect oscillators, *Electron. Lett.* **3**(2), 63–64 (1967).

63. M. Tanimoto, H. Yanai, and T. Sugeta, Thermally induced FM noise in Gunn oscillators and jitter in Gunn-effect digital devices, *IEEE Trans. Electron Devices* **ED-21**, 258 (1974).

64. A. Ataman, H. Herbst, and W. Harty, The influence of different contact materials on the noise performance of Gunn elements, *Arch. Elektron. Übertragungstechn.* **25**, 396 (1971).

65. P. Kuhn, Noise in Gunn oscillators depending on surface of Gunn diodes, *Electron. Lett.* **6**, 845 (1970).

66. J. A. Copeland, Semiconductor impurity analysis from low frequency noise spectra, *IEEE Trans. Electron Devices* **ED-18**, 50 (1971).

67. B. L. Gelmont and M. S. Shur, Plasma wave instability in the conditions of the intervalley transitions, *Phys. Lett.* **32A**(7), 552–553 (1970).

68. K. Matsumo, Critical fluctuations in GaAs in a d.c. electric field, *Phys. Lett.* **31A**, 335 (1970).

69. Y. S. Narayan and B. E. Berson, High peak power epitaxial GaAs oscillators, *IEEE Trans. Electron Devices* **ED-14**(9), 610–611 (1967).

70. D. Large, Octave band varactor-tuned Gunn diode sources, *Microwave J.* **14**(8), 49 (1970).

71. B. K. Lee and M. S. Hodgart, Microwave Gunn oscillator tuned electronically over 1 GHz, *Electron. Lett.* **4**(11), 233–235 (1968).

72. D. Zeiger, Frequency modulation of a Gunn oscillator by magnetic tuning, *Electron. Lett.* **3**(13), 324–325 (1967).

73. N. S. Chang, T. Hayamizu, and Y. Matsuo, YIG-tuned Gunn effect oscillator, *Proc. IEEE* **55**(9), 1621 (1967).

74. D. A. James, Wide-range electronic tuning of a Gunn diode by a yttrium-iron garnet (y.i.g.) ferrimagnetic resonator, *Electron. Lett.* **4**(21), 451–452 (1968).

75. M. Dydyk, Ferrimagnetically tunable Gunn effect oscillator, *Proc. IEEE* **56**(8), 1363–1364 (1968).

76. S. M. Sze, *Physics of Semiconductor Devices*, John Wiley & Sons, New York, 1981.

77. J. D. Crowley, J. J. Sowers, B. A. Janis, and F. B. Frank, High efficiency 90 GHz InP Gunn oscillators, *Electron. Lett.* **16**(18), 705–706 (1980).

6

Transferred Electron Amplifiers and Logic and Functional Devices

6-1. TRANSFERRED ELECTRON AMPLIFIERS

Transferred electron amplifiers were first developed by Thim *et al.* in 1965 [1], who used subcritically doped devices [$n_0L < (n_0L)_1$; see Section 4-1] to achieve microwave amplification. Since that time other types of transferred electron amplifiers have been developed with the output power up to several tenths of watts. Even higher output powers are possible.

Different types of transferred electron amplifiers include stable amplifiers (where the domain propagation is suppressed), amplifiers utilizing the negative differential resistance of a sample with propagating high field domains, and amplifiers where the field distribution is controlled by the external ac field (similar to the LSA oscillators).

There are several ways to achieve stability of transferred electron devices for their applications in stable amplifiers. Besides using subcritically doped devices with $n_0L < (n_0L)_1$ as originally proposed by Thim *et al.* [1] or with $n_0d < (n_0d)_{cr}$ [2, 3], supercritically doped devices covered by a film with high dielectric constant are used [4–6] (see also Section 4-12). Supercritically doped devices may also be stabilized by a stationary anode domain [7–23] (see Section 4-9), by a special doping profile (a cathode doping "notch" [16, 24–26]), by using a Schottky barrier near the cathode [27–29], or by choosing an appropriate external circuit [30–33].

The dynamic characteristic of a subcritically doped amplifier is shown in Fig. 6-1-1. The noise figure as a function of the gain is shown in Fig. 6-1-2 for the same amplifier. A typical gain versus frequency curve for different bias voltages is presented in Fig. 6-1-3.

A schematic diagram of a transferred electron amplifier stabilized by a dielectric film is shown in Fig. 6-1-4 [6]. The gain of this amplifier as a function of frequency is shown in Fig. 6-1-5. The noise figure was close to 17 dB; the saturation power was around 100 mW at 0.7 GHz and 10 mW at 1.7 GHz.

The maximum output power of a transferred electron amplifier may be estimated as

$$P_\sim = \tfrac{1}{2} V_\sim I = \tfrac{1}{2} q n_0 L \frac{v_\sim}{v_p} \frac{F_\sim}{F_p} v_p F_p S \qquad (6\text{-}1\text{-}1)$$

277

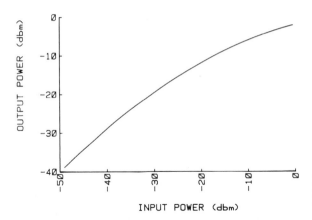

FIGURE 6-1-1. Dynamic characteristic of a subcritically doped reflection type amplifier [34].

where $v_\sim$ is the ac amplitude of the electron velocity. Assuming $v_\sim/v_p \sim \frac{1}{2}$, $F_\sim/F_p \sim \frac{1}{2}$, $F_p = 3.5\,\text{kV/cm}$, and $v_p = 2.2 \times 10^5\,\text{m/s}$, we obtain

$$P \simeq 1.5\,\frac{\text{kW}}{\text{cm}^2} \cdot \left(\frac{n_0 L}{10^{16}\,\text{m}^{-2}}\right) \cdot S(\text{cm}^2) \qquad (6\text{-}1\text{-}2)$$

which is in agreement with the experimental results. This estimate shows that supercritically doped amplifiers with higher values of $n_0 L$ are more practical because of the higher output power. Such amplifiers may be stabilized by a stationary high field domain near the anode (see Section 4-13) which may appear when $n_0 > n_{\text{cr}}$ [7–23]. A field distribution in such a device is shown in Fig. 6-1-6. As can be seen

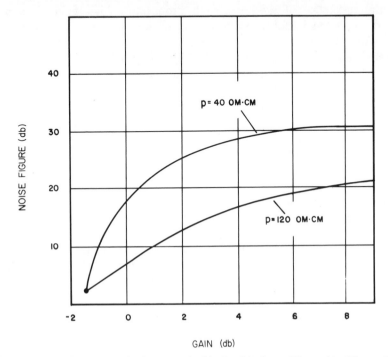

GAIN (db)

FIGURE 6-1-2. Noise figure vs. gain for two subcritically doped amplifiers with different low field resistivities [34].

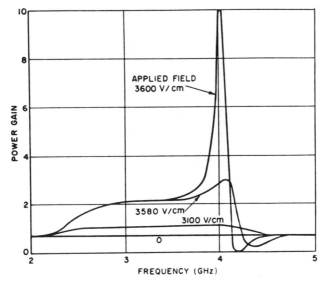

FIGURE 6-1-3. Gain vs. frequency for a subcritically doped transferred electron amplifier [34]. $F_0 = 2F_p$.

from the figure, in a portion of the sample near the cathode the electric field is less than F_p. This part of the sample introduces a parasitic series resistance. Therefore, a wide anode domain is desirable for the efficient operation.

A typical I-V characteristic of a supercritically doped amplifier is shown in Fig. 6-1-7. As can be seen from the figure, the amplification takes place either close to the threshold where the negative differential mobility is not high enough to lead to the domain formation or at large biases.

Higher cathode field enhances the device stability and may be achieved by fabricating a region with a smaller doping (a "doping notch") near the cathode [16, 24–26]. The cathode "notch" causes a larger increase of the electric field near the cathode and a nearly uniform field distribution across the rest of the sample except for a short region near the cathode where the electric field is low. Computed field and electron concentration profiles are shown in Fig. 6-1-8. The doping notch can

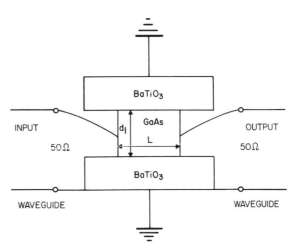

FIGURE 6-1-4. Schematic diagram of a transferred electron amplifier stabilized by a dielectric film [6].

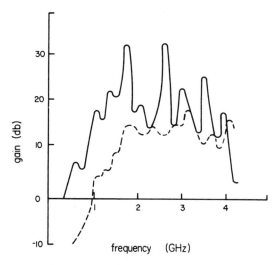

FIGURE 6-1-5. Gain vs. frequency of the transferred electron amplifier stabilized by a dielectric ($BaTiO_3$) film [6]. Solid line, optimum load; dashed line, 50 Ω load; $L = 300\,\mu m$; $d = 200\,\mu m$;   cross section $200\,\mu m \times 600\,\mu m$; $n = 0.6 \times 10^{13}\,cm^{-3}$; $n_0 L = 1.8 \times 10^{11}\,cm^{-2}$; $n_0 d_0 = 1.2 \times 10^{11}$ cm^{-2}; resistivity 170 Ω cm.

be made by an epitaxial growth [16]. An optimum ratio of the doping levels in the notch and in the rest of the sample is approximately $1:3$ [36].

The space charge growth may be prevented if the number of carriers entering the sample is controlled by the cathode (injection limited cathode [37–41]).

A model describing the effect of the injection limited cathode contact on the behavior of a transferred electron amplifier was developed in Ref. 41. A constant electric field distribution was assumed with the cathode controlled current density I_c and electric field at the cathode F_{ct} (limited cathode). The properties of the cathode contact were characterized by a characteristic frequency

$$\omega_{ct} = \frac{\sigma_{cd}}{\varepsilon} \tag{6-1-3}$$

$$\sigma_{cd} = \frac{dI_c}{dF_{ct}}\bigg|_{F_{ct}=F_0} \tag{6-1-4}$$

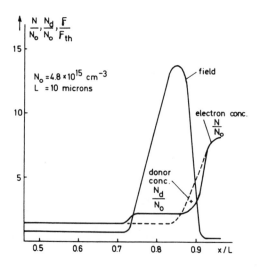

FIGURE 6-1-6. Field, doping, and carrier concentration profiles for supercritically doped transferred electron amplifier with a stable high field domain near the anode [5]. $F_0 = 2F_p$.

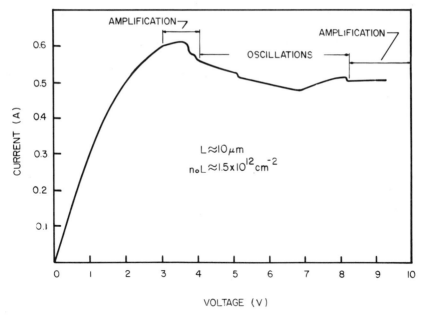

FIGURE 6-1-7. DC current–voltage characteristic of a supercritically doped transferred electron amplifier [3].

When $\omega_{ct} \to \infty$ the expression for the device impedance obtained in the frame of the model reduces to the expression obtained in Section 4-4. However, when ω_{ct} is finite and not too large, a stable amplification may be obtained in a wide frequency range with a relatively low noise figure (of the order of 8 dB for GaAs and 2 dB for InP). A shallow Schottky barrier may be used as an injection limited cathode. The practical difficulties of making such a contact are related to a high density of surface states in GaAs.

A Schottky barrier in the vicinity of the cathode contact may also be used to stabilize a transferred electron amplifier. Such a device may be fabricated as a GaAs field effect transistor with an attached drift region (a "traveling wave transistor"

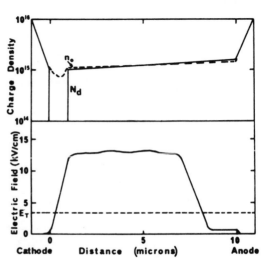

FIGURE 6-1-8. Doping and electric field profiles of a transferred electron amplifier with a cathode doping notch [35].

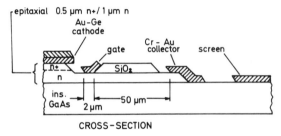

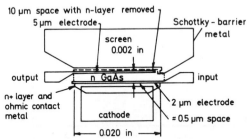

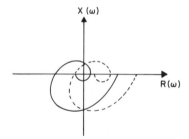

FIGURE 6-1-9. Traveling wave transistor [27–29].

[27–29]; see Fig. 6-1-9). Such an amplifier has a large bandwidth, a gain of 12 dB, and a reverse attenuation about 32 dB [28]. The gain of the device may be varied over approximately 35 db by varying the input gate voltage. The measured noise figure was about 18 dB.

The device stability may be also achieved by choosing an appropriate external circuit. An external resistance has been used [30, 31, 5] to shift the Nyquist diagram toward the right half of the X–R plane (see Fig. 6-1-10) to stabilize a circuit with a supercritically doped diode. Complex loads such as filters may also be used [32, 33].

All devices described above utilize the amplification of the longitudinal space charge waves. So-called active medium propagation amplifiers (AMP amplifiers) employ the amplification of transverse electromagnetic waves in the active GaAs medium with negative differential resistance [42–49]; see Fig. 6-1-11. A practical device configuration features a microstrip line with the gaps narrow enough to prohibit the high field domain propagation in the x–y plane because of the small $n_0 L$ product [46–49]. At the ends of the device the gaps and the line are wider. The tapered lines match 50-Ω coplanar tapers on the alumina ($\varepsilon_r = 9.6$) board which are connected to conventional 50-Ω coaxial launchers. The devices described in Ref. 45 operated at an ambient temperature of $-45°C$ (heat sink simulation) and operating temperature of 110°C. The gain as a function of frequency is shown in Fig. 6-1-12. A typical noise figure was 14–15 dB.

FIGURE 6-1-10. Qualitative Nyquist diagram of a transferred electron amplifier. Solid line: no external load; dashed line: with external resistor [114].

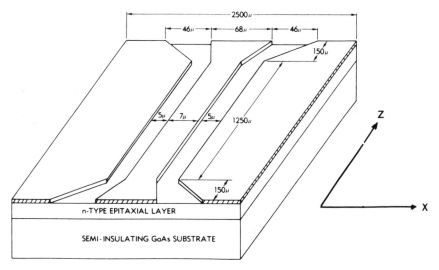

FIGURE 6-1-11. Active medium propagation amplifier [49]. The electromagnetic wave propagates in the Z direction. The electric field of the wave is in the x direction.

The active medium propagation devices can also be used as transmission line switches since they exhibit high loss at zero bias.

Amplifiers with traveling domains (oscillating amplifiers) use the negative differential resistance of a sample with a domain either directly [50-54] or in a parametric scheme [55-59]. The disadvantages of this type of amplifier include a relatively high noise figure (due to high domain fields) and more complex circuitry.

In LSA amplifiers [60-62] the external high-frequency microwave field with frequency

$$\omega \tau_f > 1 \qquad\qquad (6\text{-}1\text{-}5)$$

prevents the domain formation. Such amplifiers may have a somewhat smaller noise figure compared to the oscillating amplifiers, but they have not received much attention.

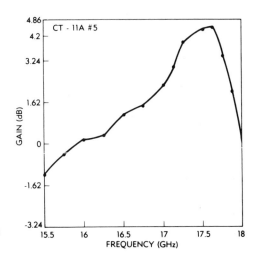

FIGURE 6-1-12. Gain vs. frequency for an active medium propagation amplifier [49].

TABLE 6-1-1. Types of Transferred Electron Amplifiers

Stable	Subcritically doped	$(n_0 L) < (n_0 L)_1$
		$n_0 d < (n_0 d)_{\text{cr}}$
		Active medium propagation
		Covered by a dielectric film
		With a stationary anode domain
	Supercritically doped	Stabilized by external circuit
		Cathode "notch"
		Injection limited cathode
		LSA oscillators
Oscillating	Regular	
	Parametric	

Different types of transferred electron amplifiers are summarized in Table 6-1-1.

All transferred electron amplifiers may be used in a reflection type scheme where a circulator is used to separate the input and output signals (see Fig. 6-1-13). The small-signal (linear) gain of this circuit is given by

$$G = \left| \frac{Z(\omega) - Z_L(\omega)}{Z(\omega) + Z_L(\omega)} \right|^2 \qquad (6\text{-}1\text{-}6)$$

where $Z(\omega)$ and $Z_L(\omega)$ are the device and load impedances, respectively. The gain is bigger than unity if $(Re(Z(\omega)) < 0$ for a resistive load.

That stable transferred electron devices which provide the amplification due to the growth of space charge waves may also be used as traveling wave amplifiers (see Fig. 6-1-14) was first proposed by Robson et al. [63]. Later such amplifiers were fabricated with a Schottky barrier near the cathode contact, to increase the cathode field and to enhance device stability, and in a "traveling wave transistor configuration" [27–29] (see Fig. 6-1-9).

A major drawback of transferred electron amplifiers is a relatively large noise figure. Minimum noise measures of 7 dB for GaAs and 4 dB for InP have been

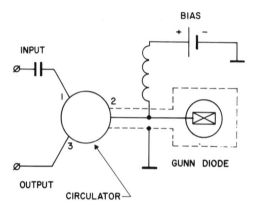

FIGURE 6-1-13. Reflection-type amplifier [114].

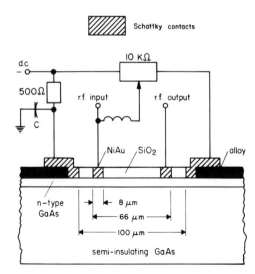

FIGURE 6-1-14. GaAs traveling wave amplifier [64].

predicted [65] (see Fig. 6-1-15 and 6-1-16), but measured noise measures are several decibels higher, with the lowest measured noise figure around 10 dB [24] for GaAs and 7.7 dB for InP [66]. These noise figures favorably compare with IMPATT amplifiers but are considerably larger than what can be achieved with GaAs FETs.

6-2. LOGIC DEVICES

A transferred electron device has two different states—an "ohmic" state and the state with a higher field domain. Thus it may be used as a basic logic element

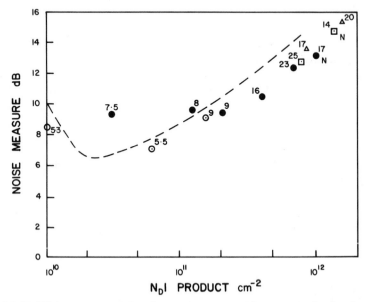

FIGURE 6-1-15. Minimum computed noise measure vs. carrier-concentration-length product for GaAs. The number above each point is the average bias field in kilovolts per centimeter. ● denotes n-n-n^+ at 300 K; ☉, 425 K; △, 500 K; ☐, n^+-p-n-n at 300 K; N is a notch device. Dashed line is the global minimum [65].

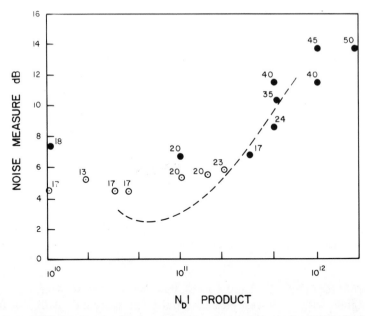

FIGURE 6-1-16. Minimum computed noise measure vs. carrier-concentration-length product for InP. The number above each point is the average bias field in kilovolts per centimeter. ● denotes n^+-n-n^+ at 300 K; ⊙, $n^+-p-n-n^+$ at 300 K. Dashed line is the global minimum [65].

[67–82]. A straightforward way to estimate parameters of a transferred electron device as a logic element is to consider a pulse amplifier operation (see Fig. 6-2-1) [83]. The voltage of the battery U_B is chosen so that the diode voltage $U_0 = U_B - IR_L$ is smaller than U_p but larger than U_s, where U_p and U_s are the domain formation and sustaining voltages, respectively. If an input pulse with the height $U_1 < U_p - U_0$ is applied, the Gunn diode switches to a state with a domain, the current decreases, and the output pulse with the duration T_0 is formed. Such an amplifier is a basic element of Gunn logic circuitry.

The switching time τ_s of the circuit shown in Fig. 6-2-1 is determined by the domain formation time τ_f. For this circuit the voltage across the device

$$U_0(t) = U_B - I(t)R_L \tag{6-2-1}$$

or

$$U_0(t) = U_B - F_r(t)\frac{R_L}{R_0}L \tag{6-2-2}$$

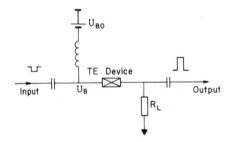

FIGURE 6-2-1. Circuit diagram of a transferred electron pulse amplifier [83].

where $F_r(t)$ is the outside domain field, R_L is the load resistance, $R_0 = L/qn_0\mu S$ is the low field resistance.

Equation (4-7-26) describing the domain dynamics may be now rewritten as

$$\frac{du}{d\tau} = \sqrt{u}\,(1 - u) \tag{6-2-3}$$

where

$$u = \frac{U_d}{U_{d0}} \tag{6-2-4}$$

U_d is the domain voltage,

$$U_{d0} = U_B - F_s L\left(1 + \frac{R_L}{R_0}\right) \tag{6-2-5}$$

and

$$\tau = t/\tau_f \tag{6-2-6}$$

Here

$$\tau_f = \tau_{f0}\left(1 + \frac{R_L}{R_0}\right)^{1/2} \tag{6-2-7}$$

τ_{f0} is the domain formation time constant given by Eq. (4-7-27) for $U_0 = U_p$. Equations (6-2-6) and (6-2-7) are derived assuming

$$U_B \simeq U_p\left(1 + \frac{R_L}{R_0}\right) \tag{6-2-8}$$

i.e., assuming that the voltage across the device in the ohmic state is close to the threshold voltage and that the voltage pulse amplitude is small. These assumptions correspond to typical operation conditions of transferred electron devices in this mode.

Equation (6-2-7) may be easily understood if we relate the process of domain formation to the process of charging the nonlinear domain capacitance C_d through the series load and low-field diode resistances R_L and R_0. The time constant of such a process is $C_d(R_L + R_0)$. C_d is proportional to $\sim 1/(1 + R_L/R_0)^{1/2}$ because C_d is proportional to $1/U_d^{1/2}$ and U_d (in the end of the process) is $(1 + R_L/R_0)$ times larger than the domain voltage corresponding to the constant threshold bias voltage. Indeed, at the end of switching the domain voltage is

$$U_d = U_0 - F_s L = U_B - F_s L\left(1 + \frac{R_L}{R_0}\right) = (F_p - F_s)L\left(1 + \frac{R_L}{R_0}\right) \tag{6-2-9}$$

The height of the output pulse is $\Delta I R_L \simeq I_p R_L/2$. ($\Delta I$ is the current jump corresponding to domain formation.) It increases with an increase of R_L. But as can be seen from Eq. (6-2-9), the voltage applied to the diode after switching also grows with an increase of R_L. Thus starting from the requirements that τ_s should be less than half of the transit time T_0 and that the domain field F_m should not cause impact ionization, we can derive two limitations imposed on the value of R_L. From the condition

$$\tau_s < \frac{T_0}{2} \sim \frac{L}{2v_s} \tag{6-2-10}$$

we obtain

$$1 + \frac{R_L}{R_0} \lesssim \frac{L}{2v_s\tau_{f0}} \tag{6-2-11}$$

When R_L/R_0 tends to zero this expression limits the minimal possible value of the product n_0L (in this case it is analogous to the Kroemer criterion). The allowed value of R_L/R_0 increases with an increase of n_0L. But U_d and the maximal domain field increase at the same time [see Eq. (6-2-9)]. For normal operation in the trigger mode F_m should be smaller than the characteristic threshold field F_i of impact ionization. This results in the following criterion (for $n_0 < n_{cr}$):

$$1 + \frac{R_L}{R_0} \lesssim \frac{\varepsilon F_i^2}{20qn_0L(F_p - F_s)} \tag{6-2-12}$$

Assuming for GaAs $F_i \sim 500 \text{ kV/cm}$, $F_p \approx 3 \text{ kV/cm}$, and $F_s \approx 1.5 \text{ kV/cm}$, we get from Eqs. (6-2-11) and (6-2-12), respectively [83],

$$1 + R_L/R_0 \lesssim n_0L \, (\text{cm}^{-2})/1.25 \times 10^{11}$$

$$1 + R_L/R_0 < 6.8 \times 10^{13}/n_0L \, (\text{cm}^{-2})$$

Here we used Eq. (4-7-28) (in the limit $n_0 \ll n_{cr}$) for τ_{f0} and used the values of the parameters given in Section 4-7. The product of Eq. (6-2-11) and Eq. (6-2-12) gives the simple necessary criterion:

$$1 + \frac{R_L}{R_0} < 24 \tag{6-2-13}$$

Let us now estimate the power required for the switching and the power-delay product. For the applied voltage $U_0 = U_p$ the power dissipated in the device

$$P_1 = qn_0v_pF_pLS \tag{6-2-14}$$

The power dissipated in the load resistance is R_L/R_0 times higher. During the switching, i.e., during the propagation of a stable domain, the dissipated power

$$P_2 \approx qn_0Lv_sF_p\left(1 + \frac{F_p - F_s}{F_p}\frac{R_L}{R_0}\right)S \tag{6-2-15}$$

The power dissipated in the load resistance during the switching is

$$P_{2L} = (qn_0v_sS)^2R_L \tag{6-2-16}$$

Using these expressions we may roughly estimate the minimum switching power.

Taking into account that

$$n_0 L > (n_0 L)_1 \qquad (6\text{-}2\text{-}17)$$

(the Kroemer criterion), we find

$$P_1/S \geqslant q(n_0 L)_1 v_p F_p \qquad (6\text{-}2\text{-}18)$$

The minimum cross section of the device depends on the doping density because at higher doping levels the $n_0 d > (n_0 d)_{cr}$ condition is satisfied at smaller values of the dimension d. Assuming $S_{min} \sim 10^{-10}\,m^2$, $(n_0 L)_1 \simeq 2 \times 10^{11}\,cm^{-2}$, $v_p = 2 \times 10^5\,\mu/s$, $F_p = 3\,kV/cm$, we find

$$P_{1\ min} \simeq 1.9\,mW \qquad (6\text{-}2\text{-}19)$$

with P_2 and P_{2L} in a 10 mW range. The switching speed is determined by the time constant τ_f and may be of the order of several tens of picoseconds, leading to the values of the power-delay product in a picojoule range.

Another important characteristic of the pulse amplifier is the maximum current gain

$$G = \Delta I_p / I_i \qquad (6\text{-}2\text{-}20)$$

where ΔI_p is the output current pulse and I_i is the input current pulse. This gain increases first with the load resistance but then decreases due to the parasitic capacitance from the anode to the ground (see Fig. 6-2-2) [84]. The inherent limitation of the maximum gain is imposed by the thermal noise [78, 79]. Simple estimates show that the maximum gains around 60 may be obtained and the values from 10 to 40 have been measured [78].

The fan-out of the Gunn logic depends on the maximum gain and the load resistance and may be three or higher [85].

The domain nucleation in the trigger mode can be induced by applying a voltage pulse to a Schottky gate placed in the vicinity of the cathode contact [86–95]. A practical design uses planar type devices (see Fig. 6-2-3). The domain formation originates under the Schottky contact depletion layer where the electric field is larger. If the Schottky gate is very short the extra depletion capacitance C_{d_s} under the gate does not appreciably increase the time constant of the circuit (a value of

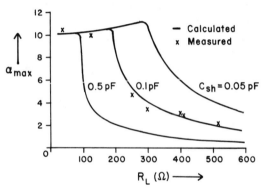

FIGURE 6-2-2. Maximum current gain as a function of load resistance R_L for different values of the parasitic capacitance between the anode and the ground [84].

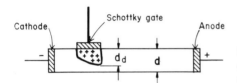

Cathode   Schottky gate
                                        Anode

FIGURE 6-2-3. Schottky gate transferred electron device. Depletion layer under the Schottky gate serves as a domain nucleation site.

C_{d_s} as low as 0.05 pF may be achieved experimentally). The equivalent circuit of the Schottky gate transferred electron device is shown in Fig. 6-2-4. The "gate trigger" capability γ_G is proportional to the device low field conductance σ_o, to the maximum current gain introduced above, and to the gate trigger factor g_t [81–93]

$$\gamma_G = \frac{\Delta I}{(\Delta V_G)_{min}} \simeq G_{max}\sigma_o g_t \qquad (6\text{-}2\text{-}21)$$

The gate trigger factor g_t is a function of the pinch-off voltage

$$V_{po} = \frac{qn_0 d^2}{2\varepsilon} \qquad (6\text{-}2\text{-}22)$$

where d is the device thickness. It also depends on the depletion width under the gate d_d (see Fig. 6-2-5). Experimental values of γ_G reach 40–50 mA/V [89].

A more detailed analysis of the Schottky gate triggered transferred electron element is given in Ref. 94.

The logic functions AND and OR can be performed by a single transferred electron device with two Schottky trigger electrodes perpendicular to the direction of the current flow [95]. A dual Schottky trigger device may also be used as an inhibitor, i.e., as a device where the Gunn domain formation may be inhibited by gate voltages.

A more sophisticated device which includes a Schottky gate-triggered Gunn device integrated with a GaAs MESFET and a resistor was proposed to implement an inhibitor [96–99] and a high-speed carry device operating at the rate of 20 ps/inhibitor [99].

The circuits based on transferred electron devices with Schottky electrodes are capable of processing signals at gigabit rates with delay times below several tens of picoseconds per gate [95–101].

An interesting version of a gate-triggered transferred electron device utilizes the transverse extension of the high field domain in the direction perpendicular to the applied electric field [102–106]. The speed of the transverse domain propagation increases with the bias and may reach 10^7 m/s, i.e., two orders of magnitude faster than the longitudinal domain velocity. This has been demonstrated in Ref. 104, where the velocity of the transverse domain spreading was measured as a function of the bias (see Fig. 6-2-6). The domains were triggered by a pair of capacitive electrodes in the middle of a planar GaAs sample.

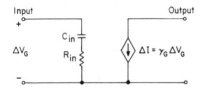

Input   Output

ΔV_G C_{in}
                R_{in} $\Delta I = \gamma_G \Delta V_G$

FIGURE 6-2-4. Equivalent circuit of Schottky gate transferred electron device in the on-state [89].

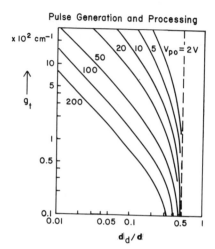

Pulse Generation and Processing

FIGURE 6-2-5. Gate trigger factor g_t as a function of normalized depletion-layer depth for different values of the pinch-off voltage V_{po} [89].

The devices and circuits utilizing transverse spreading of the domain triggered by the Schottky gates were studied in Refs. 105 and 106. Delay times up to 10 ps/stage have been achieved [106].

6-3. FUNCTIONAL DEVICES

As was pointed out in Section 4-2 the current waveform of a sample with a propagating domain reproduces the profile of the doping cross-section product. This makes it possible to generate complex current waveforms (see Fig. 6-3-1).

In the sample shown in Fig. 6-3-2 the electric field decreases towards the anode. Therefore for small bias voltages the domain may propagate only in the portion of the sample adjacent to the cathode where the electric field is larger than the domain

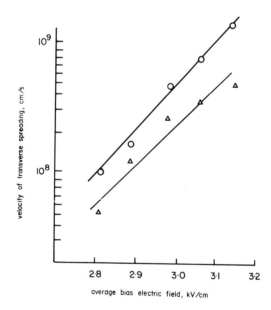

FIGURE 6-2-6. Dependences of the velocity of the transverse spreading of a high-field domain on the average bias electric field [104]. $\bigcirc$, 500-μm-thick sample; $\triangle$, 190-μm-thick sample.

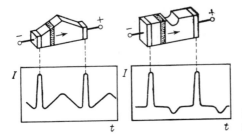

FIGURE 6-3-1. Functional devices [107]. Compare the device shapes and current waveforms.

sustaining field F_s. With the increase in bias the length of this section of the sample increases and the oscillation frequency drops (see Fig. 6-3-2b). At still higher bias voltages the domains reach the anode and the oscillation frequency becomes nearly independent of bias.

A similar device may be realized in a planar geometry using ring contacts (see Fig. 6-3-3). A drop in the oscillation frequency in the resistive circuit from 18 to

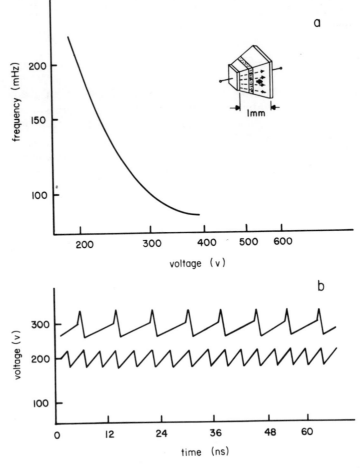

FIGURE 6-3-2. Tapered sawtooth oscillator [108]. (a) Device shape and frequency vs. bias voltage; (b) voltage waveforms.

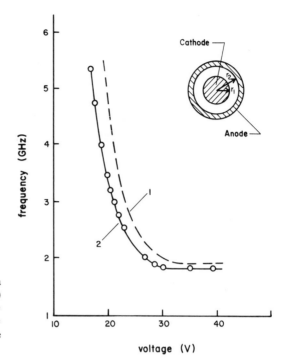

FIGURE 6-3-3. Frequency vs. bias for a planar device with ring contacts [109]. (1) Calculated; (2) measured. $n_0 = 2.3 \times 10^{15}$ cm^{-3}; $\mu = 0.65$ m^2/V s; $r_1 = 48.6\ \mu$m; $r_2 = 55\ \mu$m. Thickness of the active layer 5 μm.

6 GHz was reported for a similar device when the bias voltage changed from U_p to $2.25 \times U_P$ [110]. (The distance between the ring electrodes was 15 μm.)

At small bias voltages the frequency of oscillations of the device shown in Fig. 6-3-1 depends on bias similar to that of the device shown in Fig. 6-3-2. Once, however, the bias exceeds the critical value necessary for the domain to reach the widest section of the sample it can make it across, leading to a drop in frequency (see Fig. 6-3-4).

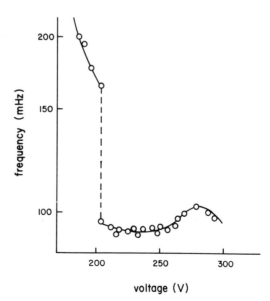

FIGURE 6-3-4. Frequency vs. bias for the sample shown in Fig. 6-3-1 left [107].

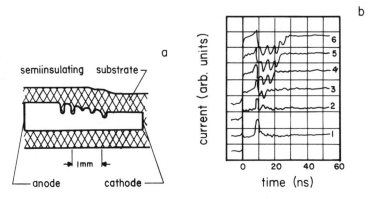

FIGURE 6-3-5. Analog-digital converter [111]. (a) Device shape; (b) current waveforms for different bias voltages.

The analog-digital converter shown in Fig. 6-3-5 is based on a similar idea. Each notch leads to a current pulse, with the number of pulses increasing with bias as the transit domain length increases (see Fig. 6-3-5b).

All these devices in the planar implementation may be used in integrated circuits. Such circuits, called DOFIC (domain oriented functional integrated circuits), are considered in Ref. 111.

The current waveform of a sample with propagating domains may also be changed by covering a part of the device surface with an isolated metal plate [107, 108], by using additional ohmic contacts [107], or by illuminating a part of the device (see Section 4-10).

The changes in the current waveform induced by the extra electrodes are illustrated in Fig. 6-3-6. When the switch K_1 is on, the extra current

$$I_{cm} \simeq C_h \frac{\delta U_d}{\delta t} \qquad (6\text{-}3\text{-}1)$$

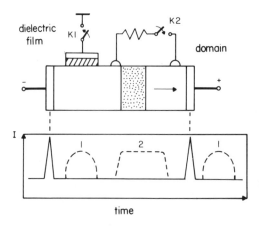

FIGURE 6-3-6. Sample with extra electrodes and current waveforms [111]. Solid line, switches are off; dashed lines, switches are on.

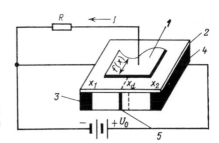

FIGURE 6-3-7. Functional device with a shaped isolated metal electrode [114]. (1) Shaped contact; (2) dielectric film; (3) cathode; (4) anode; (5) high field domain.

is induced. Here U_d is the domain voltage,

$$C_h = \frac{\varepsilon_d S_h}{d_d}$$ (6-3-2)

is the capacitance of the electrode, S_h is the cross section of the electrode, and ε_d and d_d are the dielectric permittivity and thickness, respectively.

An alternative design may include a shaped isolated metal contact and an extra resistor in the charging circuit (see Fig. 6-3-7).

With several additional electrodes rather complex functions may be implemented. As an example a simple multiplexer proposed in Ref. 71 is shown in Fig. 6-3-8 together with the input and output waveforms.

A varying composition in ternary compounds such as $Al_x Ga_{1-x}As$ or $GaAs_x P_{1-x}$ may also be used to change the current waveform [113] and to improve the efficiency of the transferred electron oscillators.

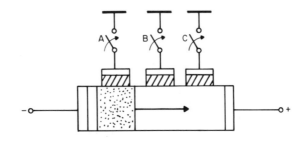

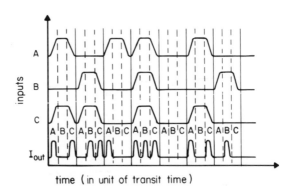

FIGURE 6-3-8. Multiplexer with capacitive electrodes [113].

REFERENCES

1. H. W. Thim, M. R. Barber, B. M. Hakki, S. Knight, and M. Uenohara, Microwave amplification in dc-biased bulk semiconductors, *Appl. Phys. Lett.* **7**(6), 167–168 (1965).
2. P. Kozdon and P. N. Robson, Two-port amplifiers using the transferred electron effect in GaAs, in 8th Int. Conf. on Microwave and Optical Generation and Amplification (MOGA). Amsterdam, 1970, paper 16.2.
3. R. H. Dean, A. B. Dreeben, J. F. Kaminski, and A. Triano, Traveling-wave amplifier using epitaxial GaAs layer, *Electron. Lett.* **6**(24), 775–776 (1970).
4. S. Kataoka, H. Tateno, and M. Kawashima, Improvements in efficiency and tunability of Gunn oscillators by dielectric-surface loading, *Electron. Lett.* **25**(20), 491–492 (1969).
5. H. Pollmann and R. W. H. Engelmann, On supercritical reflection-type amplification and the stability criterion in bulk GaAs devices, MOGA-70, 1970, Summaries, No. 16-5.
6. W. Frey, R. W. H. Engelmann, and B. G. Bosch, Microwave travelling-wave amplification in bulk gallium-arsenide, 8th Int. Conf. on Microwave and Optical Generation and Amplification (MOGA), Amsterdam, 1970, paper 16.3.
7. T. E. Walsch, B. S. Perlman, and R. E. Enstrom, Stabilized supercritical transferred electron amplifiers, *IEEE J. Solid-State Circuits* **SC-4**, 374 (1969).
8. B. S. Perlman, CW microwave amplification from circuit stabilized epitaxial GaAs transferred electron devices, *IEEE J. Solid-State Circuits* **SC-5**, 331 (1970).
9. B. S. Perlman, C. L Upadhyayula, and R. E. Marx, Wide-band reflection-type transferred electron amplifiers, *IEEE Trans. Microwave Theory Techniques* **MTT-18**, 911 (1970).
10. B. S. Perlman, Microwave amplification using transferred-electron devices in prototype filter equalization networks, *RCA Rev.* **32**, 3 (1971).
11. B. S. Perlman and C. L. Upadhyayula, Transferred electron amps challenge the TWT, *Microwaves* **9**(12), 59 (1970).
12. B. S. Perlman, C. L. Upadhyayula, and W. W. Sienkanowicz, Microwave properties and applications of negative conductance TE devices, *Proc. IEEE* **59**, 1229 (1971).
13. C. L. Upadhyayula and B. S. Perlman, Design and performance of transferred electron amplifiers using distributed equalizer networks, IEEE Int. Solid-State Circuits Conference, Philadelphia, PA, 1972; Digest of Technical Papers, p. 40.
14. J. Magarshack and A. Mircea, Wideband CW amplification in X band with Gunn diodes, IEEE Int. Solid-State Circuits Conference, Philadelphia, 1970; Digest of Technical Papers, p. 134.
15. J. Magarshac and A. Mircea, Stabilization and Wideband amplification using overcritically doped transferred-electron diodes, *Proc. 8th Int. Conference MOGA, Amsterdam*, Deventer, Kluwer, 1970, p. 16/19.
16. R. Spitalnik, M. P. Shaw, A. Rabier, and J. Magarshack, On the mechanism for microwave amplification in "supercritically" doped n-GaAs, *Appl. Phys. Lett.* **212**(4), 162–164 (1973).
17. B. I. Jeppsson and P. Jeppesen, On the GaAs supercritical TEA, Paper presented at the 2nd European Solid-State Device Research Conference (ESSDERC), Lancaster, 1970.
18. P. Jeppesen and B. Jeppson, The influence of diffusion on the stability of the supercritical transferred electron amplifier, *Proc. IEEE* **60**(5), 452–454 (1972).
19. H. Pollman and R. W. H. Engelmann, On supercritical reflection-type amplification and the stability criterion in bulk GaAs devices *Proc. 8th Int. Conference MOGA Amsterdam, 1970*, Deventer, Kluwer 1970, p. 16.24.
20. R. W. H. Engelmann, On "supercritical" transferred-electron amplifiers, *Arch. Elecktron. Ubertragungstechn.* **26**, 357 (1972).
21. I. Kuru, performance degradation of Gunn diodes at elevated temperature, *Proc. 2nd Conference on Solid State Devices, Tokyo, Japan, 1970: J. Jpn. Soc. Appl. Phys.* **40** Suppl., 137 (1971).
22. J. W. Monroe and W. K. Kennedy, Amplifiers go solid state at X- and K-band, *Microwave J.* **14**(12), 28 (1971).
23. A. Sene, A wideband CW waveguide Gunn effect amplifier, *IEEE Trans. Microwave Theory Techniques* **MTT-20**, (1972).
24. J. Margashak, A. Rabier, and R. Spitalnik, Optimum design of transferred-electron amplifier devices on GaAs, *IEEE Trans. Electron Devices* **ED-21**, 652–654 (1974).
25. R. M. Raymond, H. Kroemer, and R. E. Hayes, Design of cathode doping notches to achieve uniform fields in transferred electron devices, *IEEE Trans. Electron Devices* **ED-24**(3), 192–195 (1977).

26. C. Berry, G. S. Hobson, M. J. Howard, and P. N. Robson, Design of transferred electron amplifers with good frequency stability, *IEEE Trans. Electron Devices* **ED-24**(3), 270–274 (1977).

27. R. H. Dean and R. J. Matarese, The GaAs travelling wave amplifier as a new kind of microwave transistor, *Proc. IEEE* **60**, 1486–1502 (1972).

28. R. H. Dean, A. B. Dreeben, J. J. Hughes, R. J. Matarese, and L. S. Napoli, Broad-band microwave measurements on GaAs "traveling wave" transistors, *IEEE Trans. Microwave Theory Technique* **MTT-21**, 805–809 (1973).

29. R. H. Dean, R. E. DeBrecht, A. B. Dreeben, J. J. Hughes, R. J. Matarese, and L. S. Napoli, GaAs travelling wave transistor, 1973 IEEE-GMTT International Microwave Symp., IEEE Cat. No. 73 CHO 736-9 MTT, pp. 250–251.

30. D. E. McCumber and A. G. Chynoweth, Theory of negative-conductance amplification and of Gunn instabilities in two-valley semiconductors, *IEEE Trans.* **ED-13**(1), 4–21 (1966).

31. S. Y. Naryaan and F. Sterzer, Stabilization of transferred–electron amplifiers with large $n_0 L$ products, *Electron. Lett.* **5**(2), 30–31 (1969).

32. F. Sterzer, Stabilization of supercritical transferred-electron amplifiers, *Proc. IEEE* **57**(10), 1781–1783 (1969).

33. S. Mahrous and H. L. Hartnagel, Gunn effect domain formation controlled by a complex load, *Brit. J. Appl. Phys. (J. Phys. D)* **2**(1), 1011 (1969).

34. H. W. Thim and M. R. Barber, Microwave amplification in GaAs bulk semiconductor, *IEEE Trans. Electron Devices* **ED-13**(1), 718–719 (1966).

35. R. Charlton and G. S. Hobson, The effect of cathode notch doping profiles on supercritical transferred-electron amplifiers, *IEEE Trans. Electron Devices* **ED-20**, 812–817 (1973).

36. H. W. Thim and W. Hayde, Microwave amplifier circuit considerations, in *Microwave Devices. Device Circuit Interactions*, Ed. by M. J. Howes and D. V. Morgan, Wiley, New York, 1978, pp. 267–313.

37. M. M. Atalla and J. L. Moll, Emitter controlled negative resistance in GaAs, *Solid State Electron.* **12**, 619–129 (1969).

38. H. W. Thim, U.S. Patent 3,537,021, 1970.

39. B. W. Clark, H. G. B. Hicks, and J. S. Heeks, An electronically controlled injection limited cathode for GaAs transferred electron devices, ESSDERC 1973, Nottingham, England, paper A6.2.

40. W. P. Dumke, J. M. Woodall, and V. L. Rideout, GaAs–GaAlAs heterojunction transistor for high frequency operation, *Solid State Electron.* **15**, 1339–1343 (1972).

41. T. Hariu, S. Ono and Y. Shibata, Wide-band performance of the injection limited Gunn diode, *Electron. Lett.* **6**(21), 666–667 (1970).

42. A. C. Baynham, Wave propagation in negative differential conductivity media: *n*-Ge, *IBM J. Res. Dev.* **13**(5), 568–572 (1969).

43. A. C. Baynham, Emission of TEM waves generated within an *n*-type Ge, *Electron. Lett.* **6**(10), 306–307 (1970).

44. A. C. Baynham and D. J. Colliver, New mode of microwave emission from GaAs, *Electron. Lett.* **6**(16), 498–500 (1970).

45. A. C. Baynham, Absolute instability of electromagnetic waves within large subcritically doped gallium arsenide samples, *J. Appl. Phys.* **44**(3), 1247–1250 (1973).

46. P. L. Fleming, The active medium propagation device, *Proc. IEEE (Lett.)* **63**(8), 1253–1254 (1975).

47. P. L. Fleming, U.S. Patent No. 3 975 690, Planar transmission line comprising a material having negative differential conductivity, Aug. 17, 1976.

48. P. L. Fleming and H. E. Carlson, Reflection-mode amplifier utilizing GaAs active-medium-propagation devices *Electron. Lett.* **15**(24), 787–788 (1979).

49. P. L. Fleming, T. Smith, H. Carlson, and W. A. Cox, Continuous wave operation of active medium propagation devices, *IEEE Trans. Electron Devices* **Ed-26**(9), 1267–1272 (1979).

50. H. W. Thim, Linear negative conductance amplification with Gunn oscillators, *Proc. IEEE* **55**(3), 446–447 (1967).

51. H. W. Thim, Linear microwave amplification with Gunn oscillators, *IEEE Trans.* **ED-14**(9), 517–522 (1967).

52. H. W. Thim, Linear negative conductance amplification with Gunn oscillators, *Proc. IEEE* **55**(3), 446–447 (1967).

53. P. Olfs, An oscillating Gunn amplifier with E_{010}-resonator, *Proc. 8th Int. Conference MOGA, Amsterdam, 1970*, Deventer, Kluwer, 1970, p. 2/21.

54. Kh. A. Abdel Fatakh and K. S. Rzhevkin, Amplification of microwave oscillations within GaAs in the presence of domain generation, *Sov. Radio Eng. Electron Phys.* **15**, 1056 (1970).

55. J. E. Carrol, Resonant-circuit operation of Gunn diodes: A Self pumped parametric oscillator, *Electron Lett.* **2**(6), 215–216 (1966).
56. D. J. Vinney, Possible traveling-wave parametric amplifier using Gunn effect, *Electron. Lett.* **2**(10), 357–358 (1966).
57. C. S. Aitchison, Possible Gunn-effect parametric amplifier, *Electron. Lett.* **4**(1), 15–16 (1968).
58. C. S. Aitchison, C. D. Corbey, and B. H. Newton, Self-pumped Gunn-effect parametric amplifier, *Electron. Lett.* **5**(2), 36–37 (1969).
59. H. J. Kuno, Self-pumped parametric amplification with GaAs transferred-electron devices, *Electron. Lett.* **5**(11), 232 (1969).
60. R. R. Spiwak, Frequency conversion and amplification with an LSA diode oscillator, *IEEE Trans. Electron Devices* **ED-15**, 614 (1968).
61. B. Majborn, On the possibility of millimeter wave amplification with an X-band LSA oscillator, *Proc. European Microwave Conference London, 1969*, IEEE Conf. Publication No 587, p. 227.
62. N. Hasizume and S. Katoka, Transferred-electron negative-resistance amplifier, *Electron. Lett.* **6**(2), 34–35 (1970).
63. P. N. Robson, G. S. Kino, and B. Fay, Two-port microwave amplification in long samples of gallium arsenide, *IEEE Trans. Electron Devices* **ED-16**, 612 (1967).
64. R. H. Dean, A. B. Dreeben, J. F. Kaminski, and A. Triano, Traveling wave amplifier using thin epitaxial GaAs layer, *Electron. Lett.* **6**, 775–776 (1970).
65. J. E. Sitch and P. N. Robson, The noise measure of GaAs and InP transferred electron amplifiers, *IEEE Trans. Electron Devices* **23**(9), 1088–1094 (1976).
66. R. M. Corlett, I. Griffith, and J. J. Purcell, A low noise InP reflection amplifier, Proceedings of 5th European Microwave Conference, Hamburg, 1975, paper 10.3.
67. C. P. Sandbank, Synthesis of complex electronic functions by solid state bulk effects, *Solid State Electron* **10**(5), 369–380 (1967).
68. J. A. Copeland, T. Hayashi, and M. Uenohara, Logic and memory elements using two valley semiconductors, *Proc. IEEE* **55**(4), 584–585 (1967).
69. H. L. Hartnagel, Digital logic-circuit applications of Gunn diodes, *Proc. IEEE* **55**(7), 1236–1237 (1967).
70. S. H. Izadpanahj and H. L. Hartnagel, Experimental verification of Gunn-effect comparator, *Proc. IEEE* **55**(10), 1748 (1967).
71. R. S. Engelbrecht, Solid-state bulk phenomena and their application to integrated electronics, *IEEE Trans* **SC-3**(3), 210–212 (1968).
72. H. L. Hartnagel, *Gunn-effect Logic Devices*, American Elsevier, New York, 1971.
73. H. Hartnagel, Some basic logic circuits employing Gunn-effect devices, *Solid State Electron.* **11**(5), 568–572 (1968).
74. T. Sugeta, T. Ikoma, and H. Yanai, Bulk neuristor using the Gunn effect, *Proc. IEEE* **56**(2), 239–240 (1968).
75. R. Engelmann and W. Heinle, Pulse discrimination by Gunn-effect switching, *Solid State Electron.* **14**(1), 1–16.
76. M. Nakamura, H. Kurono, M. Hirao, T. Toyabe, and H. Kodera, High-speed pulse response of planar-type Gunn diodes, *Proc. IEEE* **59**(6), 1039–1040 (1971).
77. H. L. Hartnagel, Theory of Gunn-effect logic, *Sol. St. Electron.* **12**(1), 19–30 (1969).
78. G. White and R. E. Sgams, A 2-GHz multiple Gunn device logic circuit, *Proc. IEEE* **57**(9), 1684–1685 (1969).
79. G. S. Hobson and S. H. Izadpanah, Random domain triggering in Gunn effect pulse regenerators, *Solid State Electron.* **13**(7), 937–942 (1970).
80. R. F. Fisher, Generation of subnanosecond pulse with bulk GaAs, *Proc. IEEE* **55**(12), 2189 (1967).
81. S. H. Izadpanah and H. L. Hartnagel, Pulse gain and analogue-to-pulse conversion by Gunn diodes, *Electron. Lett.* **4**(15), 315–316 (1968).
82. S. H. Izadpanah and H. L. Hartnagel, Memory loop with Gunn-effect pulse diodes, *Electron. Lett.* **5**(3), 53 (1969).
83. M. E. Levinshtein and M. S. Shur, Transient properties of Gunn diodes, *Solid State Electron.* **18**, 983–990 (1975).
84. T. Sugeta, N. Suzuki, M. Tanimoto, and H. Yanai. Gunn-effect functional device, report No. 45-11, Meeting of Japanese IEE Proj. Group on transistors, September 1970.

85. H. Yanai and T. Sujeta, Some features and characteristics of the Gunn effect digital device, Jpn. IECE Nat. Conv. Rec. No. 717, September 1969.
86. T. Sugeta, H. Yanai, and K. Sekido. Schottky-gate Gunn effect digital device, *Proc. IEEE* **59**, 1629 (1971).
87. T. Sugeta, H. Yanai, and T. Ikoma, Switching properties of bulk-effect digital devices, *IEEE Trans. Electron. Devices* **ED-17**, 940 (1970).
88. K. Mause, H. Salow, A. Schlachetzki, K. H. Bachem, and K. Heime, Circuit integration with gate controlled Gunn devices, in Proc. 4th Intern. Symposium on GaAs and related compounds, Boulder, Colorado, 1972, p. 275.
89. H. Yanai, T. Sugeta, and K. Sekido, Schottky-gate Gunn effect digital device, Paper at Int. Electron Devices Meeting, Washington, D.C., Oct. 11-13, 1971.
90. T. Sugeta, H. Yanai, and K. Sekido, Schottky-gate bulk-effect digital device, *Proc. IEEE* **59**, 1629 (1971).
91. T. Sugeta and H. Yanai, Gunn-effect digital functional devices and their performance evaluation, *Trans. IECE (Jpn.)* **55-C**, 437 (1972).
92. T. Sugeta and H. Yanai, Signal-processing of a Gunn-effect digital functional device (in Japanese), Jpn. IECE Techn. Group on Circuits and System Theory, report No. CT-71-71 (1972-02), February 24, 1972, *Trans. IECE (Jpn.)* **55-C**, 445 (1972).
93. T. Sugeta, T. M. Tanimote, and H. Yanai, Gunn effect digital functional device, *J. Fac. Eng. Univ. Tokyo (B)* **31**, 772 (1972).
94. B. G. Bosch and R. W. H. Engelmann, *Gunn-Effect Electronics*, Wiley, New York, 1975.
95. S. Kataoka, N. Hashizume, M. Kawashima, and Y. Komamiya, High field domain functional logic devices with multiple control electrodes, in Proc. 4th Beiennial Cornell Electrical Engineering Conf., pp. 225-234, 1973.
96. N. Hashizume, S. Kataoka, Y. Komamiya, K. Tomizawa, and N. Morisue, GaAs 4 bit gate device of integrated Gunn elements and MESFETs, in Proc. 6th Int. Symp. on Gallium Arsenide and Related Compounds (St. Louis, Mo.), pp. 245-253, 1976.
97. N. Hashizume, M. Kawashima, K. Tomizawa, M. Morisue, and S. Katoka, Gunn effect high-speed carry finding device for 8 bit binary adder, in Tech. Dig. Ind. Electron Device Meet., pp. 209-212, 1977.
98. N. Hasizume, S. Kataoka, and K. Tomizawa, 20 ps/gate Gunn-effect high-speed carry finding device, *Electron. Lett.* **14**, 91-92 (1978).
99. N. Hashizume and S. Kataoka, Gunn-effect inhibitor and its application to high-speed carry finding device, *IEEE Trans. Electron Devices* **ED-26**(3), 183-190 (1979).
100. K. Marese, Multiplexing and demultiplexing techniques with Gunn devices in the gigabit-per-second range, *IEEE Trans. Microwave Theory Technique* **MTT-24**(12), 926-929 (1976).
101. C. L. Upadhyayla, R. E. Smith, J. F. Wilhelm, S. T. Jolly, and J. P. Paczkowski, Transferred electron logic devices for gigabit-rate signal processing, *IEEE Trans. Microwave Theory Techniques* **MTT-24**(12), 920-926 (1976).
102. M. Shoji, Theory of transverse extension of Gunn domains, *J. Appl. Phys.* **41**, 774-0778 (1970).
103. K. Tomizawa, M. Kawashima, and S. Kataoka, New logic functional device using transverse spreading of a high-field domain in *n*-type GaAs, *Electron Lett.* **7**, 239-240 (1971).
104. K. Tomizawa and S. Kataoka, Dependence of transverse spreading velocity of a high-field domain in a GaAs bulk element on the bias electric field, *Electron. Lett.* **8**, 130-131 (1972).
105. S. Hasuo, T. Nakamura, G. Goto, K. Kazetani, H. Ishiwari, H. Suzuki, and T. Isobe, Gunn-effect logic circuits for a high speed computer, in Proc. 5th Biennial Cornell Electrical Engineering Conf., pp. 185-194, 1975.
106. G. Goto, T. Nakamura, S. Hasuo, K. Kazetani, and T. Isobe, Gunn-effect logic device using transverse extension of a high field domain, *IEEE Trans. Electron Devices* **ED-23**, 21-27 (1976).
107. M. Shoji, Functional bulk semiconductor oscillators, *IEEE Trans.* **ED-14**(9), 535-546 (1967).
108. R. S. Engelbrecht, Bulk effect devices for future transmission systems, *Bell. Lab. Rec.* **45**(6), 192-201 (1967).
109. C. O. Newton and G. Bew, Frequency measurements on Gunn effect devices with concentric electrodes, *J. Phys. D: Appl. Phys.* **3**(8), 1189-1198 (1970).
110. G. M. Clark, A. L. Edridge and J. C. Bass, Planar Gunn-effect oscillators with concentric electrodes, *Electron. Lett.* **5**(20), 471-472 (1969).
111. C. P. Sandbank, Synthesis of complex electronic functions by solid state bulk effects, *Solid State Electron.* **10**(5), 369-380 (1967).

112. K. R. Hofmann, Gunn oscillations in thin samples with capacitive surface loading, *Electron. Lett.* **5**, 289 (1969).
113. A. A. Kastalski, E. I. Leonov, and M. S. Shur, Devices with variable energy gap, *Sov. Phys. Semicond.* **4**(8), 1384–1386 (1971).
114. M. E. Levinstein, Y. K. Pozhela, and M. S. Shur, *Gunn Effect*, Soviet Radio, Moscow, 1975 (in Russian).

7

GaAs FETs: Device Physics and Modeling

7-1. INTRODUCTION

Since their introduction in 1970 [1] GaAs field effect transistors have occupied an important niche in the microwave industry. GaAs FET amplifiers, oscillators, mixers, switches, attenuators, modulators, and limiters are widely used and high-speed integrated circuits based on GaAs FETs have been developed. The basic advantages of these devices include a higher electron velocity, leading to smaller transit time and faster response, and semi-insulating GaAs substrates, which allow one to decrease the parasitic capacitances and simplify the fabrication process.

A poor quality of oxide on GaAs and a high density of surface states at the GaAs–insulator interface make it difficult to fabricate GaAs MOSFETs or MISFETs. Schottky barrier metal semiconductor field effect transistors (MESFETs) or junction field effect transistors (JFETs) are examples of practically used GaAs FETs. In many cases these devices are fabricated by direct ion implantation into a GaAs semi-insulating substrate.

In this chapter we consider the basic device physics of GaAs FETs and relate their characteristics to such device parameters as doping density, doping profile, and active layer thickness.

7-2. THE SHOCKLEY MODEL

A field effect transistor was analyzed by W. Shockley in the early 1950s [2, 3]. A schematic diagram showing a field effect transistor is presented in Fig. 7-2-1. The Schottky barrier or p–n junction depletion region under the gate controls the cross section of the conduction channel under the gate, modulating the channel conductivity and hence the drain-to-source current. We consider here an n-type channel.

The depletion region is wider closer to the drain because the reverse bias across the channel-to-gate depletion layer is larger there.

When the gate voltage V_G becomes smaller than

$$V_T = -V_{\text{po}} + V_{\text{Bi}} \qquad (7\text{-}2\text{-}1)$$

301

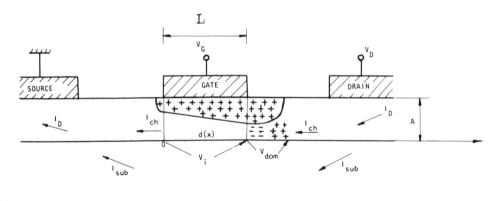

FIGURE 7-2-1. Cross section of a GaAs MESFET. Also shown is a dipole region (a high field domain region with a voltage drop V_{dom}) which may form at the drain side of the gate as discussed further in the text.

where V_T is the threshold voltage, the current practically drops to zero. Here V_{po} is the pinch-off voltage

$$V_{\text{po}} = \frac{qN_D A^2}{2\varepsilon} \qquad (7\text{-}2\text{-}2)$$

V_{Bi} is the built-in voltage, A is the channel thickness width (see Fig. 7-2-1), and N_D is the effective donor density, which we assume to be equal to the electron concentration n_0 in the undepleted portion of the channel. For simplicity we assume a uniform doping profile.

When V_G is greater than V_T the increase in the drain-to-source voltage V_{ds} above the saturation voltage $(V_{\text{ds}})_{\text{sat}}$ leads to current saturation.

This saturation is caused by velocty saturation in the high electric field in the channel. In short-channel GaAs FETs with gate lengths of the order of 0.5–2 μm typical values of the average electric field in the channel are high (5–20 kV/cm) and hot electron effects and the related nonlinearity of the electron drift velocity versus electric field curve are very important.

We will start, however, from the simple model which ignores these effects and simply implies that the drift velocity

$$v = \mu F \qquad (7\text{-}2\text{-}3)$$

is proportional to the longitudinal electric field F up to the point where the channel is pinched off at the drain side of the gate, which happens when

$$V_{\text{gs}} - V_{\text{ds}} \leq V_T \qquad (7\text{-}2\text{-}4)$$

Here μ is the low field mobility. At this point, the longitudinal component of the electric field at the drain side of the gate becomes very large (from the formal point of view it tends to infinity). In reality, velocity saturation occurs at a finite value of the electric field and, hence, prior to the complete pinch-off of the channel. However as we will show in Section 7.3 for devices with large pinch-off voltage, long gates, low field effect mobility, and large electron saturation velocity, we may assume that current saturation occurs exactly at the channel pinch-off at the drain side of the gate and that Eq. (7-2-3) is valid up to the pinch-off.

This model is called the Shockley model after William Shockley, who also proposed the use of the so-called gradual channel approximation [2]. The gradual channel approximation is based on the assumption that the bias of the gate junction is a slowly varying function of position. The gradually varying channel is shown in Fig. 7-2-2. We assume that the conducting channel is neutral, the region under the gate is totally depleted, the electric field F in the channel is in the x direction, the electric field under the gate F_{dep} is in the y direction, the boundary between the neutral channel and the depleted region is sharp, and the potential across the channel varies so slowly that at each point the thickness of the depleted area can be found from the solution of the Poisson equation valid for a one-dimensional junction. Under such assumptions we find the incremental change of the channel potential dV:

$$dV = I_{ch}\, dR = \frac{I_{ch}\, dx}{q\mu N_D W[A - A_d(x)]} \tag{7-2-5}$$

where I_{ch} is the channel current, dR is the incremental channel resistance, x is the coordinate along the channel, A is the thickness of the active layer, $A_d(x)$ is the thickness of the depletion layer (see Fig. 7-2-2), and W is the gate width.

The depletion region width at distance x is given by

$$A_d(x) = \left\{ \frac{2\varepsilon[V(x) + V_{Bi} - V_G]}{qN_D} \right\}^{1/2} \tag{7-2-6}$$

Substituting Eq. (7-2-6) into Eq. (7-2-5) and integrating with respect to x from 0 (the source side of the gate) to L (the drain side of the gate) we derive the fundamental equation of field-effect transistors:

$$I_{ch} = g_0 \left\{ V_i - \frac{2}{3} \frac{[(V_i + V_{Bi} - V_G)^{3/2} - (V_{Bi} - V_G)^{3/2}]}{V_{po}^{1/2}} \right\} \tag{7-2-7}$$

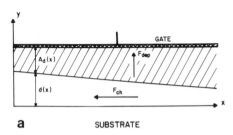

a SUBSTRATE

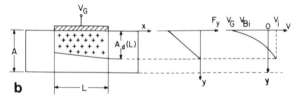

FIGURE 7-2-2. Depletion region and conducting channel under the gate. (a) Section of a gradually varying channel under the gate; (b) charge, electric field, and potential distributions in the depletion region.

where V_i is the voltage drop in the channel across the region under the gate,

$$g_0 = \frac{q\mu N_D WA}{L} \tag{7-2-8}$$

is the conductance of the metallurgical channel, L is the gate length, and V_{po} is the ideal pinch-off voltage given by Eq. (7-2-2). If we neglect the series resistances of the drain-to-gate and gate-to-source regions, including contacts (which is not necessarily a good approximation for the GaAs MESFETs), then $V_i = V_{ds}$, where V_{ds} is the drain-to-source voltage.

Equation (7-2-7) is applicable only up to the point where the neutral channel still exists even in the narrowest spot at the drain side of the channel, i.e.,

$$A_d(L) \equiv A_0 \equiv \left\{ \frac{2\varepsilon [V_i + V_{Bi} - V_G]}{qN_D} \right\}^{1/2} \leqslant A \tag{7-2-9}$$

It is assumed that when $A_d(L) = A$ (the pinch-off condition) the current saturation occurs. That is why the saturation voltage $(V_i)_{\text{sat}}^S$ predicted by the Shockley model is given by

$$(V_i)_{\text{sat}}^S = V_{po} - V_{Bi} + V_G \tag{7-2-10}$$

(the index S stands for Shockley), which is in agreement with Eq. (7-2-4). Substitution of Eq. (7-2-10) into Eq. (7-2-7) leads to the following expression for the saturation current:

$$(I_{ch})_{\text{sat}}^S = g_0 \left[\frac{1}{3} V_{po} + \frac{2}{3} \frac{(V_{Bi} - V_G)^{3/2}}{V_{po}^{1/2}} - V_{Bi} + V_G \right] \tag{7-2-11}$$

A very important characteristic of a field effect transistor is the transconductance

$$g_m = \left. \frac{\partial I}{\partial V_G} \right|_{V_i = \text{const}} \tag{7-2-12}$$

From Eq. (7-2-7) we find that in the linear region

$$g_m = g_0 \frac{(V_i + V_{Bi} - V_G)^{1/2} - (V_{Bi} - V_G)^{1/2}}{V_{po}^{1/2}} \tag{7-2-13}$$

Substituting Eq. (7-2-10) in Eq. (7-2-13) we find the transconductance in the saturation region:

$$(g_m)^S = g_0 \left[1 - \left(\frac{V_{Bi} - V_G}{V_{po}} \right)^{1/2} \right] \tag{7-2-14}$$

For small drain-to-source voltages

$$V_i \ll V_{Bi} - V_G \tag{7-2-15}$$

Eqs. (7-2-7) and (7-2-13) can be simplified

$$I_{ch} \simeq g_0 \left[1 - \left(\frac{V_{Bi} - V_G}{V_{po}} \right)^{1/2} \right] V_i \qquad (7\text{-}2\text{-}16)$$

$$g_m \simeq \frac{g_0}{2} \frac{V_i}{V_{po}^{1/2}(V_{Bi} - V_G)^{1/2}} \qquad (7\text{-}2\text{-}17)$$

The results obtained above can be presented in the universal dimensionless form. Introducing dimensionless variables:

$$i = \frac{I_{ch}}{g_0 V_{po}}$$

$$i_s = \frac{(I_{ch})_{sat}^{S}}{g_0 V_{po}}$$

$$u_i = \frac{V_i}{V_{po}}$$

$$u_G = \frac{V_{Bi} - V_G}{V_{po}} \qquad (7\text{-}2\text{-}18)$$

$$u_S = \frac{(V_i)_{sat}^{S}}{V_{po}}$$

$$G_S = \frac{(g_m)_{sat}^{S}}{g_0}$$

$$u = \frac{V(x)}{V_{po}}$$

we find

$$i = u_i - \tfrac{2}{3}(u_i + u_G)^{3/2} + \tfrac{2}{3}u_G^{3/2} \qquad (7\text{-}2\text{-}19)$$

$$i_s = \tfrac{1}{3} + \tfrac{2}{3}u_G^{3/2} - u_G \qquad (7\text{-}2\text{-}20)$$

$$u_s = 1 - u_G \qquad (7\text{-}2\text{-}21)$$

$$G_s = 1 - \sqrt{u_G} \qquad (7\text{-}2\text{-}22)$$

(see Figs. 7-2-3 and 7-2-4).

We can also find from Eq. (7-2-5) the potential distribution across the channel. Integrating Eq. (7-2-5) with respect to x and taking into account Eq. (7-2-7), we find in terms of the dimensionless variables we introduced

$$u(Z) - \tfrac{2}{3}[u(Z) + u_G]^{3/2} + \tfrac{2}{3}u_G^{3/2} = [u_i - \tfrac{2}{3}(u_i + u_G)^{3/2} + \tfrac{2}{3}u_G^{3/2}]Z \quad (7\text{-}2\text{-}23)$$

where $Z = x/L$. Here we assume

$$u_G < 1 \qquad (7\text{-}2\text{-}24)$$

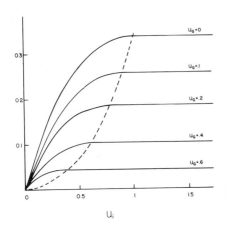

FIGURE 7-2-3. Dimensionless current–voltage characteristics of a field effect transistor predicted by the
Shockley model. Dashed line shows i_s vs. u_s curve.

(if $u_G > 1$, the field effect transistor is off) and

$$u_i \leqslant 1 - u_G \tag{7-2-25}$$

which means that the total voltage drop across the channel is smaller than the
saturation voltage.

The potential distributions across the neutral channel given by Eq. (7-2-23) are
shown in Figs. 7-2-5 and 7-2-6.

The total charge in the depletion layer is given by

$$Q = qN_D W \int_0^L A_d(x)\, dx \tag{7-2-26}$$

Substituting Eq. (7-2-6) into Eq. (7-2-26) we find

$$Q = qN_D WLA \int_0^1 (u + u_G)^{1/2}\, dZ \tag{7-2-27}$$

or

$$Q = qN_D WLA \int_0^{u_i} (u + u_G)^{1/2}\, \frac{dZ}{du}\, du \tag{7-2-28}$$

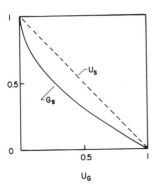

FIGURE 7-2-4. Dimensionless transconductance and saturation voltage u_s vs. dimensionless gate voltage predicted by the Shockley model.

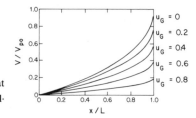

FIGURE 7-2-5. Potential distribution in the neutral channel at the pinch-off predicted by the Shockley model [see Eq. (7-2-23)].

Evaluation of the integral in Eq. (7-2-28) using Eq. (7-2-23) leads to the following equation:

$$Q = Q_0\left(\frac{f_1}{f_2} - 1\right) \tag{7-2-29}$$

where

$$Q_0 = qN_D wLA \tag{7-2-30}$$

$$f_1(u_i, u_G) = u_i - u_i^2/2 - u_G u_i \tag{7-2-31}$$

and

$$f_2 = u_i - \tfrac{2}{3}(u_i + u_G)^{3/2} + \tfrac{2}{3}u_G^{3/2} \tag{7-2-32}$$

The derivatives of Q with respect to voltages V_{gs} and V_{ds} are related to the capacitances C_{gs} and C_{gd} of the small-signal equivalent circuit (see Fig. 7-2-7):

$$dQ = -C_{gs}dV_g + C_{dg}d(V_i - V_g) \tag{7-2-33}$$

and

$$C_{gs} = -\left.\frac{\partial Q}{\partial V_g}\right|_{V_i - V_G = \text{const}} \equiv \frac{1}{V_{po}}\left.\frac{\partial Q}{\partial u_G}\right|_{u_i + u_G = \text{const}} \tag{7-2-34}$$

$$C_{dg} = \left.\frac{\partial Q}{\partial V_i}\right|_{V_G = \text{const}} \equiv \frac{1}{V_{po}}\left.\frac{\partial Q}{\partial u_i}\right|_{u_G = \text{const}} \tag{7-2-35}$$

The substitution of Eq. (7-2-29) into Eqs. (7-2-34) and (7-2-35) yields the following expressions for C_{gs} and C_{dg}:

$$C_{gs} = C_0 \frac{f_{1G}f_2 - f_1 f_{2G}}{f_2^2} \tag{7-2-36}$$

$$C_{dg} = C_0 \frac{f_{1i}f_2 - f_1 f_{2i}}{f_2^2} \tag{7-2-37}$$

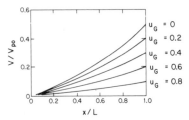

FIGURE 7-2-6. Potential distribution in the neutral channel predicted by the Shockley model. The voltage drop across the channel is one-half of the saturation voltage.

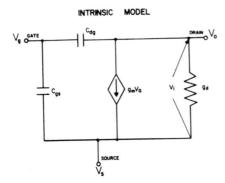

FIGURE 7-2-7. Small signal equivalent circuit of a FET channel.

where

$$f_{1G} = \left.\frac{\partial f_1}{\partial u_G}\right|_{u_i + u_G = \text{const}} = u_G - 1 \tag{7-2-38}$$

$$f_{2G} = \left.\frac{\partial f_2}{\partial u_G}\right|_{u_i + u_G = \text{const}} = u_G^{1/2} - 1 \tag{7-2-39}$$

$$f_{1i} = \left.\frac{\partial f_1}{\partial u_i}\right|_{u_G = \text{const}} = 1 - u_i - u_G \tag{7-2-40}$$

$$f_{2i} = \left.\frac{\partial f_2}{\partial u_i}\right|_{u_G = \text{const}} = 1 - (u_i + u_G)^{1/2} \tag{7-2-41}$$

and

$$C_0 = \frac{2\varepsilon WL}{A} \tag{7-2-42}$$

The computed curves C_{gs}/C_0 vs. u_G and C_{dg}/C_0 vs. $u_G + u_i$ are shown in Figs. 7-2-8 and 7-2-9.

As can be seen from the figures, these curves can be approximated by

$$\frac{C_{gs}}{C_0} = \frac{1}{4u_G^{1/2}} \tag{7-2-43}$$

$$\frac{C_{dg}}{C_0} = \frac{1}{4(u_G + u_i)^{1/2}} \tag{7-2-44}$$

except in the region close to the pinch-off where the Shockley model is not applicable because the extension of the depleted region beyond the gate should be taken into

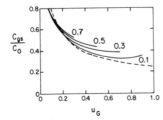

FIGURE 7-2-8. C_{gs}/C_0 vs. u_G. Numbers near the curves correspond to the values of u_i. Dashed line, Eq. (7-2-43).

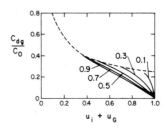

FIGURE 7-2-9. C_{dg} vs. u_G. Numbers near the curves correspond to the values of u_i. Dashed line, Eq. (7-2-44).

account. Equations (7-2-43) and (7-2-44) can be rewritten as

$$C_{gs} = \frac{C_{gs0}}{(1 - V_{gs}/V_{Bi})^{1/2}} \tag{7-2-45}$$

$$C_{gd} = \frac{C_{gs0}}{(1 - V_{dg}/V_{Bi})^{1/2}} \tag{7-2-46}$$

where

$$C_{gs0} = C_{dg0} = \frac{WL}{2}\left(\frac{\varepsilon q N_D}{2 V_{Bi}}\right)^{1/2} \tag{7-2-47}$$

Equations (7-2-45)–(7-2-47) may be explained as follows. At zero drain-to-source voltage and zero gate voltage the total gate capacitance is equal to the capacitance of the space charge region depleted by the built-in voltage:

$$C_{g0} = \frac{\varepsilon WL}{A} = WL\left(\frac{\varepsilon q N_D}{2 V_{Bi}}\right)^{1/2} \tag{7-2-48}$$

This capacitance is equally divided between the source (C_{gs0}) and drain (C_{dg0}) because the space charge distribution is symmetrical. For nonzero V_{ds} and V_{gs} the capacitances C_{gs} and C_{dg} behave almost as capacitances of equivalent Schottky diodes connected between the gate and the source, and the gate and the drain, respectively.

7-3. ANALYTICAL MODELS OF GaAs FETs

7-3-1. Introduction

The rapid development of GaAs technology requires the development of an accurate and simple device model for GaAs field effect transistors. It is desirable to have such a model in a closed analytical form in order to simplify device parameter acquisition and to be able to use it for computer-aided design of GaAs integrated circuits which may include a large number of FETs.

In this section we consider two such models for the calculation of the current–voltage characteristics of GaAs FETs. The first model, which we call a "square law model," provides an accurate description of GaAs FETs with low pinch-off voltages (less than 2 V or so for GaAs devices with a 1-μm gate) and an approximate description of GaAs FETs with higher pinch-off voltages. The second model, which we call a "complete velocity saturation model," accurately describes high pinch-off voltage devices (higher than 3 V or so for GaAs FETs with a 1-μm gate) but considerably overestimates the drain-to-source current in low pinch-off voltage devices. Both models take into account the source and drain series resistance and the output conductance, provide simple analytical expressions for the current voltage characteristics, and are quite suitable for parameter acquisition. The results of the calculation are in good agreement with experimental data.

These models also allow us to gain some insight into the device physics and, in particular, to elucidate the roles played by the electron saturation velocity and low field mobility in determining the comparative performance of GaAs and Si FETs.

7-3-2. Velocity Saturation in the Channel

According to the Shockley model current saturation occurs when a conducting channel is pinched-off at the drain side of the gate. At this point the cross section of the conducting channel predicted by the Shockley model is zero and hence the electron velocity has to be infinitely high in order to maintain the finite drain-to-source current. In reality the electron velocity saturates in a high electric field, and this velocity saturation causes the saturation of the current. The importance of the field dependence of the electron mobility for understanding current saturation in field effect transistors was first mentioned by Dasey and Ross [4]. This concept was later developed in many theoretical models used to describe FET characteristics and interpret experimental results [5–43]. Following Ref. 16 we first consider a very simple approximation for the field dependence of the electron velocity assuming that the velocity is proportional to the electric field until the value of the saturation velocity v_s is reached at $F = F_s$ and then becomes constant:

$$v = \begin{cases} \mu F, & F < F_s \\ v_s, & F \geq F_s \end{cases} \qquad (7\text{-}3\text{-}1)$$

(see Fig. 7-3-1).

The velocity saturation is first reached at the drain side of the gate where the electric field is the highest according to the Shockley model. It occurs when

$$F(L) = F_s \qquad (7\text{-}3\text{-}2)$$

where $F(L)$ is the electric field in the conducting channel at the drain side of the gate. Using the dimensionless variables introduced in Section 7-2, Eq. (7-3-2) may be rewritten as

$$\left. \frac{du}{dZ} \right|_{Z=1} = \alpha \qquad (7\text{-}3\text{-}2a)$$

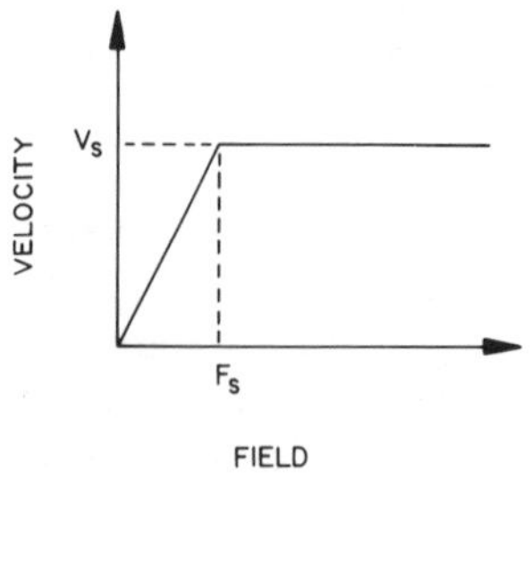

FIGURE 7-3-1. Simplified velocity–electric field dependence.

where

$$\alpha = F_s L / V_{po} \qquad (7\text{-}3\text{-}3)$$

$u = V(x)/V_{po}$ and $z = x/L$. At drain-to-source voltages smaller than the saturation voltage the electric field in the channel may be found from Eq. (7-2-23) obtained in the frame of the Shockley model:

$$\frac{du}{dz} = \frac{u_i - (2/3)(u_i + u_G)^{3/2} + (2/3)u_G^{3/2}}{1 - (u + u_G)^{1/2}} \qquad (7\text{-}3\text{-}4)$$

The dimensionless saturation voltage u_s is then determined from Eq. (7-3-2a):

$$\alpha = \frac{u_s - (2/3)(u_s + u_G)^{3/2} + (2/3)u_G^{3/2}}{1 - (u_s + u_G)^{1/2}} \qquad (7\text{-}3\text{-}5)$$

For large $\alpha \gg 1$ the solution of Eq. (7-3-5) approaches

$$u_s + u_G = 1 \qquad (7\text{-}3\text{-}6)$$

which is identical with the corresponding equation for the Shockley model [see Eq. (7-2-21)]. The opposite limiting case $\alpha \ll 1$ corresponds to the velocity saturation model:

$$u_s = \alpha \qquad (7\text{-}3\text{-}7)$$

For this solution to be valid it is also necessary to have

$$\alpha \ll 2(1 - u_G^{1/2})u_G^{1/2} \qquad (7\text{-}3\text{-}8)$$

The numerical solution of Eq. (7-3-5) is shown in Fig. 7-3-2, where it is compared with a simple interpolation formula suggested in Ref. 42

$$u_s = \frac{\alpha(1 - u_G)}{\alpha + 1 - u_G} \qquad (7\text{-}3\text{-}9)$$

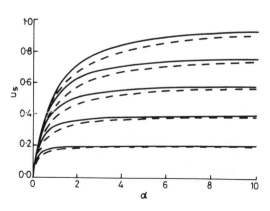

FIGURE 7-3-2. Dimensionless saturation voltage u_s vs. α [42]. Solid line, numerical solution; dashed line, analytical approximation. $u_G = 0$ for the top curve, step 0.2.

The saturation current is given by

$$I_{sat} = qN_d v_s W[A - A_d(L)] \tag{7-3-10}$$

where the depletion width $A_d(L)$ at the drain side of the gate is given by

$$A_d(L) = A(u_G + u_s)^{1/2} \tag{7-3-11}$$

In dimensionless units ($i_s = I_{sat}/g_0 V_{po}$)) Eq. (7-3-10) becomes

$$i_s = \alpha[1 - (u_s + u_G)^{1/2}] \tag{7-3-12}$$

In the limiting case $\alpha \to \infty$ (this corresponds to a long gate device with a small pinch-off voltage) Eq. (7-3-12) reduces to the corresponding equation of the Shockley model [see Eq. (7-2-21)]. In the opposite case $\alpha \ll 1$ (short gate and/or large pinch-off voltage)

$$i_s = \alpha[1 - u_G^{1/2}] \tag{7-3-13}$$

This expression corresponds to a simple analytical model of GaAs MESFETs proposed in Ref. 27.

The computed i_s vs. α curves for different values of u_G are shown in Fig. 7-3-3, where they are compared with a simple interpolation formula:

$$i_s = \frac{\alpha}{1 + 3\alpha}(1 - u_G)^2 \tag{7-3-14}$$

As can be seen from the figure, the agreement is quite good.

Equation (7-3-14) coincides with the equation used for the J-FET saturation current in the SPICE model:

$$I_{sat} = \beta(V_G - V_T)^2 \tag{7-3-15}$$

where

$$V_T = V_{Bi} - V_{po} \tag{7-3-16}$$

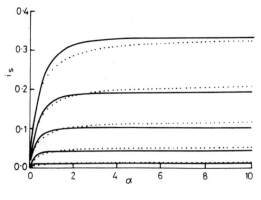

FIGURE 7-3-3. Dimensionless saturation current as a function of α [42]. Solid line, numerical calculation; dashed line, analytical approximation. Top curve for $u_G = 0$, step 0.2.

if we choose the transconductance parameter β as

$$\beta = \frac{2\varepsilon\mu v_s W}{A(\mu V_{po} + 3v_s L)} \tag{7-3-17}$$

This may explain why the SPICE J-FET model may describe fairly well the characteristics of GaAs MESFETs in some cases (see, for example, Ref. 44).
   Equation (7-3-17) can also be rewritten as

$$\beta = \frac{\mu v_s W(2qN_d\varepsilon)^{1/2}}{(\mu V_{po} + 3v_s L) V_{po}^{1/2}} \tag{7-3-18}$$

   The transconductance at the onset of the saturation region is given by

$$g_m = 2\beta(V_G - V_T) \tag{7-3-19}$$

Equations (7-3-17) and (7-3-18) can be used to determine the dependence of β on
the pinch-off voltage. Equation (7-3-17) should be used when A is kept constant
and the pinch-off voltage varies due to the variation of N_d, and Eq. (7-3-18) should
be used when N_d is kept constant and the pinch-off voltage varies due to the
variation of A.
   The variation of β with the pinch-off voltage for these two cases is shown in
Fig. 7-3-4. The most interesting feature of these curves is a dramatic increase in
transconductance at low pinch-off voltages for devices with submicron gates.
   As can be seen from Eqs. (7-3-17) and (7-3-18) the values of β (and, hence,
the values of the transconductance for the same voltage swing) increase with the
decrease of the device thickness and with the increase in doping. This increase of
β is accompanied by a similar increase in the device capacitances C_{gs} and C_{dg} [see
Eqs. (7-2-45)–(7-2-47)). However, the parasitic capacitances do not increase with

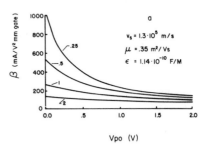

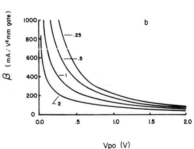

FIGURE 7-3-4. β at the onset of the saturation vs. V_{po}. (a)
V_{po} varies due to the variation in doping (active layer
thickness $A = 0.1 \ \mu$m); (b) V_{po} varies due to the variation
in the thickness of the active layer (doping $N_d = 1.2 \times$
10^{17} cm^{-3}). Numbers near the curves correspond to the gate
lengths in micrometers.

the increase in doping or with the decrease in the device thickness, and hence thin and highly doped layers should lead to a higher speed of operation.

Another important advantage of highly doped devices is a reduction of the active layer thickness for a given value of the pinch-off voltage. This makes it possible to minimize short channel effects which become quite noticeable when $L/A < 3$ [45].

Equation (7-3-18) illustrates the role of the low field mobility, which becomes increasingly important in low pinch-off devices (see Fig. 7-3-5). As can be seen from Fig. 7-3-5, for enhancement mode devices with low pinch-off voltages the high values of low field mobility in GaAs (up to 4500 cm²/V s in highly doped active layers) lead to a substantial improvement in performance even for short channel devices where the velocity saturation effects are very important. The increase in mobility to 8000 cm²/V s which may be achieved in modulation doped devices at room temperature (see Chapter 11) leads to even better performance. However, further increase of the low field mobility, say, to 20,000 cm²/V s (as in modulation doped structures at liquid nitrogen temperatures) does not buy much improvement. The improvement in the MODFET performance at 77 K is probably more due to the higher saturation velocity than due to higher values of the low field mobility.

The higher values of the low field mobility may still help to reduce the series source resistance, especially in devices with conventional (not self-aligned) gates.

The effect of the saturation velocity on the transconductance at the onset of the velocity saturation is shown in Fig. 7-3-6. This figure clearly shows the importance of high values of the electron velocity in short channel devices.

The analysis given above only applies to the values of the drain-to-source current and transconductance at the onset of the current saturation when the electron velocity becomes equal to the electron saturation velocity at the drain side of the

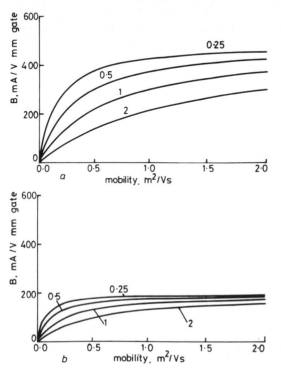

FIGURE 7-3-5. β at the onset of the saturation vs. low field mobility μ [42]. (a) $V_{po} = 0.6$ V; (b) $V_{po} = 1.5$ V. Numbers near the curves correspond to the gate lengths in micrometers.

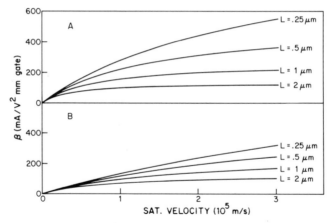

FIGURE 7-3-6. β at the onset of the saturation vs. saturation velocity [42]. (A) $V_{po} = 0.6$ V; (B) $V_{po} = 1.5$ V. Numbers near the curves correspond to the gate lengths in micrometers.

gate. At higher drain-to-source voltages the electric field in the conducting channel increases, leading to velocity saturation in a larger fraction of the channel (the so-called gate length modulation effect). We will consider the implications of gate length modulation based on the ideas developed by Pucel *et al.* [16] and Grebene and Ghandi [11].

We will consider GaAs FETs with self-aligned and non-self-aligned gates shown in Figs. 7-3-7a and 7-3-7b, respectively. In a self-aligned FET high dose n^+ implants for ohmic contacts are made using the gate as a mask. In this case there is no

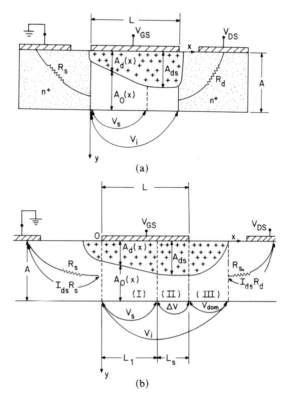

FIGURE 7-3-7. Schematic diagrams of GaAs FETs [43]. (a) Self-aligned FET; (b) non-self-aligned FET.

substantial extension of the depletion region beyond the gate. In non-self-aligned FETs there may be a considerable extension of the depletion region beyond the gate and the formation of the high field region ("high field domain") at the drain side of the channel (see, for example, Refs. 23, 20, 25–27).

In both cases the shape of the depletion region under the gate at low drain-to-source voltages may be described using the Shockley model, which remains valid as long as the longitudinal component of electric field F_x in the conducting channel remains less than the characteristic field F_s of the electron velocity saturation.

When the electric field at the drain side of the gate $F_x(L)$ becomes equal to the saturation field F_s, velocity saturation occurs. At higher drain-to-source voltages the region of velocity saturation extends into the channel (the gate length modulation effect) and, in the case of a non-self-aligned FET, also into the gate-to-drain section (see Fig. 7-3-7b).

Following Pucel et al. [16], we neglect the carrier accumulation and depletion in the velocity saturation region (region II in Fig. 7-3-7). This means that the conducting channel opening $A - A_{ds}$ in this region must be constant to preserve the current continuity:

$$I_{ds} = qn_0(A - A_{ds})v_s W \qquad (7\text{-}3\text{-}20)$$

The potential drop across this region is created by the outside charges, which are charges in the depletion region and in the drain electrode. Therefore the potential distribution in region II may be found from the solution of the Laplace equation. Such a solution, which vanishes at the gate electrode and at the boundary plane $x = L - L_s$, is given by [11]

$$V(x, y) = \sum_{n=0}^{\infty} a_n \cos[(2n + 1)\pi y/2A] \sinh[(2n + 1)\pi(x - L)/2A] \quad (7\text{-}3\text{-}21)$$

Equation (7-3-21) also satisfies the boundary condition

$$F_y(y = A) = 0 \qquad (7\text{-}3\text{-}22)$$

which requires the y component of the electric field to be zero at the active layer–substrate interface. This solution may be approximated by retaining only the first term of the series because of the rapid variation of the exponential terms

$$V(x, y) \approx (2A/\pi)F_s \cos(\pi y/2A) \sinh[\pi(x - L)/2A] \qquad (7\text{-}3\text{-}23)$$

We should add to this potential the potential created by the charges in the depletion region which is found from the solution of the Poisson equation in the depletion region. This additional term is independent of x and y in the conducting channel in region II (where the electron velocity is saturated) and is equal to V_{s1}, where V_{s1} is the potential at the boundary plane with the Shockley region ($x = L_1$). Within the section of the depletion layer adjacent to region II this contribution to the electric potential is independent of x and varies parabolically with y from V_{s1} at $y = A_{ds}$ to $V_G - V_{bi}$ at $y = 0$, where V_{bi} is the built-in voltage of the Schottky barrier and V_G is the potential of the gate metal:

$$\Delta V(L_1 < x < L) = \begin{cases} V_{s1} - (V_{s1} - V_G + V_{bi})(y - A_{ds})^2/A_{ds}^2, & 0 < y \le A_{ds} \\ V_{s1}, & A_{ds} \le y \le A \end{cases}$$

(7-3-24)

Using Eqs. (7-3-23) and (7-3-24) we find the following expression for the voltage drop ΔV_{II} across region II of the conducting channel:

$$\Delta V_{II} \approx (2A/\pi)F_s \sinh(\pi L_s/2A)$$

(7-3-25)

Here L_s is the length of region II. Following Ref. 43 we use an analytical approximation for ΔV which allows us to obtain an approximate analytical solution of the Pucel model:

$$\Delta V_{II} \approx (V_i - V_{s1})K_D$$

(7-3-26)

Here V_i is the voltage drop across the conducting channel and

$$K_D = \Delta V_{II}/(\Delta V_{II} + V_{dom})$$

(7-3-27)

where V_{dom} is the voltage drop across the channel under the extension of the depletion region beyond the gate (see Fig. 7-3-7). In devices with self-aligned gates $K_D \approx 1$. In devices with non-self-aligned gates K_D should depend on the shape of the gate recess and on the length of the drain-to-gate spacing. It may also depend on the gate voltage. In the Pucel model region III is ignored and K_D is always equal to unity. Based on the results of the numerical simulation of GaAs MESFETs (see, for example, Ref. 19) we chose $K_D \approx 0.1$. We should note, however, that the results obtained for practical devices seem to be quite insensitive to the exact value of K_D because of the roughly logarithmic dependence of the length of the velocity saturation region L_s on K_D.

From Eqs. (7-3-25) and (7-3-26) we find

$$L_s = \frac{2A}{\pi} \ln \left\{ \frac{\pi \Delta V}{2F_s A} + \left[\left(\frac{\pi \Delta V}{2F_s A} \right)^2 + 1 \right]^{1/2} \right\}$$

(7-3-28)

As in Ref. 43 we will consider two different analytical models for GaAs FETs. In the first model, which we call a "square law model," we assume that the "square law" dependence of the saturation current on $V_G - V_T$ [see Eq. (7-3-15)] remains approximately valid at voltages above the saturation. As discussed below, this model works quite well for low pinch-off voltage devices (below 2 V or so for 1 μm gate GaAs FETs) and may be adequate (though not as accurate) for higher pinch-off voltage devices as well. In the second model we assume that L_s is close to L at drain-to-source voltages relatively close to the saturation voltage V_s [see Eq. (7-3-9)] so that the electron velocity is saturated nearly everywhere in the channel. We call this model, which was developed in Refs. 27 and 46, a "complete velocity saturation model." This model is quite accurate for high pinch-off voltage devices (larger than 3 V or so for 1 μm GaAs FETs) but may considerably overestimate the drain-to-source current in low pinch-off voltage devices.

7-3-3. "Square Law" Model

If we assume that a "square law" for the drain-to-source saturation current is valid [see Eq. (7-3-15)] we may account for the gate length modulation effect in Eq. (7-3-15) choosing

$$\beta = \frac{2\varepsilon v_s W}{A[V_{po} + 3F_s(L - L_s)]} \tag{7-3-29}$$

as was suggested in Ref. 47. In fact, there is overwhelming experimental evidence that the "square law" is approximately valid for GaAs FETs with low pinch-off voltages (see, for example, Ref. 48). There are also some theoretical considerations leading to the same conclusion [49].

The source and drain series resistances R_s and R_d may be incorporated into the "square law" model as follows. The gate-to-source voltage V_{gs} is given by

$$V_{gs} = V_G + I_{sat}R_s \tag{7-3-30}$$

Substituting Eq. (7-3-30) into Eq. (7-3-15) and solving for I_{sat} we obtain

$$I_{sat} = \frac{1 + 2\beta R_s(V_{gs} - V_T) - [1 + 4\beta R_s(V_{gs} - V_T)]^{1/2}}{2\beta R_s^2} \tag{7-3-31}$$

For the device modeling suitable for a computer aided design one has to model the current–voltage characteristics in the entire range of the drain-to-source voltages, not just above the saturation. We may use a heuristic functional dependence for the drain-to-source current I_{ds} similar to one proposed in Refs. 50, 51:

$$I_{ds} = I_{sat}(1 + \lambda V_i) \tanh(\eta V_i) \tag{7-3-32}$$

where I_{sat} is given by Eq. (7-3-31) and η is chosen in such a way that for $V_{DS} \to 0$ Eq. (7-3-29) converts into a corresponding equation of the Shockley model:

$$\eta = G_{ch}/I_{sat} \tag{7-3-33}$$

where

$$G_{ch} = g_0(1 - u_G^{1/2}) \tag{7-3-34}$$

is the channel conductance at low drain-to-source voltages predicted by the Shockley model,

$$g_0 = qN_d\mu WA/L \tag{7-3-35}$$

is the full channel conductance.

Constant λ in Eq. (7-3-32) is an empirical constant which accounts for the additional output conductance beyond the output conductance related to the gate length modulation and already included in Eq. (7-3-29).

The drain-to-source voltage is given by

$$V_{DS} = V_i + I_{sat}(R_s + R_d) \tag{7-3-36}$$

Eqs. (7-3-29) and (7-3-31)–(7-3-36) form a complete set of equations of this analytical "square law" model. As mentioned above, Eq. (7-3-32) is similar to the empirical equation proposed in Ref. 50 for modeling of GaAs FETs. However, this model includes the effects related to gate length modulation and to the source and drain series resistances. Also, the parameters of the model are related to the device geometry, doping and material parameters such as the saturation velocity, and the low field mobility.

7-3-4. Complete Velocity Saturation Model

As can be seen from Eq. (7-3-29) when

$$V_{po} \gg 3F_s L \tag{7-3-37}$$

the value of β at the onset of velocity saturation is close to the value which corresponds to complete velocity saturation everywhere in the channel. For 1 μm gate GaAs FETs $F_s L$ is close to 0.3–0.4 V and, hence, inequality (7-3-37) is valid for pinch-off voltages higher than approximately 3 V. At such pinch-off voltages the model assuming a complete velocity saturation in the channel [27] becomes valid. This means that at gate voltages not too close to the threshold and at large drain-to-source voltages we can neglect the voltage drop across region 1 (where the electron velocity is not saturated) compared with $V_{bi} - V_G$ (see Fig. 7-3-7):

$$V_{si} \ll V_{bi} - V_G \tag{7-3-38}$$

Hence we may assume that the width of the depletion layer in region 2, A_{ds} (see Fig. 7-3-7), is approximately given by

$$A_{ds} = A[1 - (u_G)^{1/2}] \tag{7-3-39}$$

This leads to the following expression for the drain-to-source saturation current [27]:

$$I_{sat} = qN_d v_s WA[1 - (u_G)^{1/2}] \tag{7-3-40}$$

As can be seen from Eq. (7-3-37) this expression may not be adequate near the threshold when the voltage drop V_s may lead to a substantial relative change in the width of the undepleted region and hence in the drain-to-source saturation current. However, even in this region Eq. (7-3-40) may still be used as an adequate interpolation formula.

For a realistic description of GaAs FETs the effects related to the series source and drain resistances should be incorporated into this model. Substituting

$$V_G = V_{gs} - I_{sat}R_s \tag{7-3-41}$$

[see Eq. (7-3-30)] into Eq. (7-3-40) and solving the resulting equation for I_{sat} we find

$$I_{sat} = I_{fc}[K - (K^2 - 1 + u_{Gs})^{1/2}] \tag{7-3-42}$$

where

$$K = 1 + R_s I_{fc}/(2V_{po}) \tag{7-3-43}$$

and

$$I_{fc} = qN_d v_s WA \tag{7-3-44}$$

is the full channel saturation current, $u_{Gs} = (V_{Bi} - V_{gs})/V_{po}$.

As shown in Section 2, at low drain-to-source voltages the drain-to-source channel current is given by

$$I_{ch} = g_0[1 - (u_G)^{1/2}]V_i \tag{7-3-45}$$

[see Eq. (7-3-33)]. The source series resistance does not change V_{gs} (compared to V_G) at low values of V_i because the values of the channel current and hence of the voltage drop across R_s are small. Therefore we use the same interpolation formula for the drain-to-source current as for the "square law" model [see Eq. (7-3-29)]:

$$I_{ds} = I_{sat}(1 + \lambda V_i)\tanh(\eta V_i) \tag{7-3-46}$$

where I_{sat} is now given by Eq. (7-3-42) and η is chosen in such a way that at $V_{DS} \to 0$ Eq. (7-3-46) converts into a corresponding equation of the Shockley model:

$$\eta = G_{ch}/I_{sat} \tag{7-3-47}$$

where G_{ch} is given by Eq. (7-3-34).

Constant λ in Eq. (7-3-46) is the same empirical constant which is used in the "square law" model to account for the output conductance.

The drain-to-source voltage is given by

$$V_{DS} = V_i + I_{sat}(R_s + R_d) \tag{7-3-48}$$

Equations (7-3-42)–(7-3-48) form a complete set of equations of our analytical "complete velocity saturation" model.

7-3-5. Comparison with Experimental Data

Measured current–voltage characteristics of a GaAs FET with a relatively low pinch-off voltage (close to 1.8 V) and 1.3-μm gate [47] are shown in Fig. 7-3-8a. The current–voltage characteristics calculated using the equations of the "square law" model for the values of the parameters given in Table 7-3-1 are shown in Fig. 7-3-8b. As can be seen from the comparison of Figs. 7-3-8a and 7-3-2b the agreement with the experimental results is quite good. The current–voltage characteristics of the same device calculated using equations of the "complete velocity saturation" model are presented in Fig. 7-3-8c. One can see that the "complete velocity saturation" model considerably overestimates the drain-to-source current. As discussed in Ref. 42 and as can be seen from Eq. (7-3-29), complete velocity saturation in the channel for the devices with low pinch-off voltages may only be achieved if the electron mobility is very high (as in AlGaAs/GaAs modulation doped FETs—see Chapter 10).

In Fig. 7-3-9 the drain-to-source currents calculated as a function of the drain-to-source voltage for a high pinch-off voltage device ($V_{po} = 5.3$ V) in the frame of the "complete velocity saturation" model (Fig. 7-3-9a) and in the frame of the "square

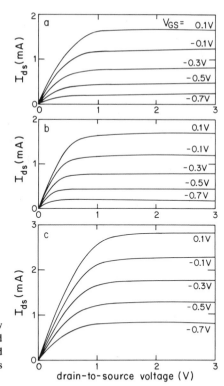

FIGURE 7-3-8. Current–voltage characteristics of a low pinch-off voltage GaAs MESFET [43]. (a) Measured [47]; (b) calculated ("square law" model); (c) calculated ("complete velocity saturation" model). Parameters used in the calculation are given in Table 7-3-1.

law" model (Fig. 7-3-9b) are compared with the experimental data of Pucel *et al.* [16]. The parameters used in the calculation are given in Table 7-3-2. As can be seen from the figure, the agreement with the experimental results is quite good for the "complete velocity saturation" model. For the "square law" model there is still qualitative agreement. However, the deviation of the experimental data from the "square law" model is substantial. This analysis of the current–voltage characteristics of many other devices has shown that such a deviation is typical for high pinch-off voltage FETs. We have also found that using the "square law" model for low pinch-off voltage devices and "complete velocity saturation" model for high pinch-off voltage devices we have been able to simulate current-voltage characteristics of a great variety of GaAs FETs fabricated in different laboratories.

TABLE 7-3-1. Parameters of the Low Pinch-Off Voltage FET[a]

Pinch-off voltage	V_{po}	1.8V
Doping density	N_d	1.81×10^{17} cm^{-3}
Gate length	L	1.3 μm
Gate width	W	20 μm
Built-in voltage	V_{bi}	0.76 V
Saturation velocity	v_s	1.23×10 m/s
Mobility	μ	3060 cm^2/V s
Source series resistance	R_s	68 Ω
Drain series resistance	R_d	68 Ω
Output conductance parameter	λ	0.01
Fraction of voltage drop in region II	K_d	0.1

[a] Reference 47.

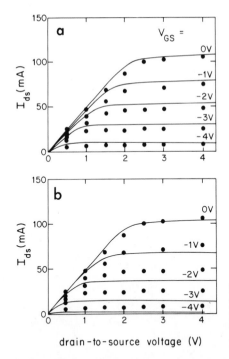

FIGURE 7-3-9. Current–voltage characteristics of a high pinch-off voltage GaAs MESFET [43]. ●, Measured [16]. Solid lines (a) "square law" model; solid lines (b) "complete velocity saturation" model. Parameters used in the calculation are given in Table 7-3-2.

7-3-6. Comparison between GaAs and Si MESFETs

There has been considerable discussion in the literature of the relative potential performances of GaAs and Si devices. The argument has been made that in short gate devices the performance of Si FETs may approach the performance of GaAs FETs if the values of the saturation velocities are comparable. As pointed out in Ref. 42, however, the value of the low field mobility is also very important because it determines whether complete velocity saturation may be achieved. In Figs. 7-3-10 and 7-3-11 we compare current–voltage characteristics of low pinch-off (Fig. 7-3-10) and high pinch-off voltage (Fig. 7-3-11) GaAs and Si FETs. The calculations were done using the "square law" model, which adequately describes the current swing in both low and high pinch-off voltage devices. The parameters used for GaAs FETs are given in Tables 7-3-1 and 7-3-2 for low and high pinch-off voltage devices,

TABLE 7-3-2. Parameters of the High Pinch-Off Voltage FET[a]

Pinch-off voltage	V_{po}	5.3
Doping density	N_d	6.5×10^{16} cm^{-3}
Gate length	L	$1 \, \mu$m
Gate width	W	$500 \, \mu$m
Built-in voltage	V_{bi}	0.75 V
Saturation velocity	V_s	1.2×10 m/s
Mobility	μ	4500 cm^2/V s
Source series resistance	R_s	6.5 Ω
Drain series resistance	R_d	11.3 Ω
Output conductance parameter	λ	0.025
Fraction of voltage drop in region II	K_d	0.1

[a] Reference 16.

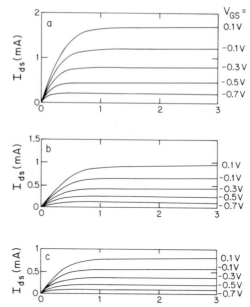

FIGURE 7-3-10. Comparison between low pinch-off voltage GaAs and Si FETs [43]. (a) I-V characteristics of a GaAs FET calculated in the frame of the "square law" model (parameters are given in Table 7-3-1). (b) Same as (a) but $\mu = 1000\ \mathrm{cm^2/V\,s}$. (c) Same as (a) but $\mu = 1000\ \mathrm{cm^2/V\,s}$ and $v_s = 0.7 \times 10^5\ \mathrm{m/s}$.

respectively. The parameters of Si devices were chosen to be the same except for the low field mobility for curves shown in Figs. 7-3-10b and 7-3-11b (for silicon we assumed $\mu = 1000\ \mathrm{cm^2/V\,s}$) and both the low field mobility and saturation velocity for the characteristics shown in Fig. 7-3-10c and 7-3-11c [for these curves we chose $(v_s)_{\mathrm{Si}} = 0.7 \times 10^5\ \mathrm{m/s}$].

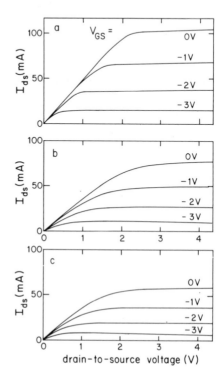

FIGURE 7-3-11. Comparison between high pinch-off voltage GaAs and Si FETs [43]. (a) I-V characteristics of a GaAs FET calculated in the frame of the "square law" model (parameters are given in Table 7-3-2). (b) Same as (a) but $\mu = 1000\ \mathrm{cm^2/V\,s}$. (c) Same as (a) but $\mu = 1000\ \mathrm{cm^2/V\,s}$ and $v_s = 0.7 \times 10^5\ \mathrm{m/s}$.

As can be seen from the figures, a higher mobility in GaAs devices leads to a considerable advantage in low pinch-off voltage devices and to a noticeable advantage in high pinch-off voltage devices. In both cases these advantages are enhanced by a higher effective saturation velocity in GaAs (compare Figs 7-3-10a and 7-3-10c and Figs. 7-3-11a and 7-3-11c). Even greater gains are possible in the short channel GaAs devices where overshoot and ballistic effects may become important (see, for example, Refs. 52 and 53).

Two simple analytical models described in this section allow us to calculate the current–voltage characteristic of GaAs FETs. The "square law model" allows an accurate description of GaAs FETs with low pinch-off voltages (less than 2 V or so for GaAs devices with a 1-μm gate) and an approximate description of GaAs FETs with higher pinch-off voltages. The "complete velocity saturation model" accurately describes high pinch-off voltage devices (higher than 3 V or so for GaAs FETs with a 1-μm gate) but considerably overestimates the drain-to-source current in low pinch voltage devices. These models take into account the source and drain series resistances and the output conductance. The results of the calculations are in good agreement with experimental data.

The relative comparison of GaAs and Si FETs based on the "square law" model shows that a higher electron mobility in GaAs leads to a substantial advantage in device performance, which is further enhanced by a higher effective saturation velocity in GaAs. The proposed models are quite suitable for parameter acquisition and for computer-aided design of GaAs microwave and digital devices and integrated circuits.

7-4. BACKGATING IN GaAs FETs

In our discussion of GaAs FETs we have neglected so far the effects related to the junction between the active channel and the semi-insulating layer beneath the channel (which may be either a semi-insulating substrate in ion implanted FETs or a specially grown buffer layer in epitaxial structures). In fact, a finite depletion region exists at this boundary in the active layer. The width of this layer depends on the density of traps in the substrate and may be found using an "equivalent p-n^+ junction model" illustrated by Fig. 7-4-1 [54].

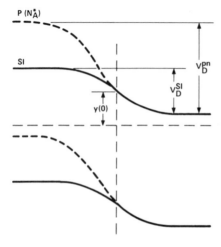

FIGURE 7-4-1. Band diagram for a substrate–active layer junction and for an equivalent p-n^+ junction which leads to the same depletion of the n^+ layer [54]. V_D^{SI} and v_D^{pn} are the diffusion potentials of these junctions [$V_D^{pn} = kT/q \ln(N_d N_a / n_i^2)$].

The energy difference $y(0)$ between the conduction band edge at the interface and the Fermi level may be related to the effective substrate "doping" N_A^* [54]:

$$y(0) + \ln(N_d/N_c) = [\ln(N_d N_A^*/n_i^2) + N_d/N_A^* - 1]/(1 + N_d/N_A^*) \quad (7\text{-}4\text{-}1)$$

The width of the depletion layer in the conducting channel associated with this junction is given by

$$W_n(V) = L_D[\gamma(qV_D^{pn}/kT - 1 + x)$$
$$+ \gamma qV/kT(V_D^{pn}/V_D^{SI})(V_D^{SI} + V)/(V_D^{pn} + V)]^{1/2} \quad (7\text{-}4\text{-}2)$$

Here L_D is the Debye length in the conducting channel, V is the voltage drop across the semi-insulating–conducting channel interface, $x = N_d/N_A^*$, $\gamma = 1/(1 + x)$, and diffusion potentials V_D^{SI} and V_D^{pn} are marked in Fig. 7-4-1.

The variation of W_n with the substrate bias is shown in Fig. 7-4-2. As can be seen from the figure, Eq. (7-4-2) adequately describes the experimental data.

The effects of the backgating on the shape of the depletion region under the gate and on the current–voltage characteristics of a 1-μm gate GaAs FET are shown in Fig. 7-4-3 [55].

In this experiment the gate of a 1×50-μm FET was connected to the source and the substrate was biased via a pad located about 150 μm away from the FET. The drain characteristics under illumination with the substrate voltage changing from 0 to 20 V are shown in Fig. 7-4-3b. The backgating effects were substantially

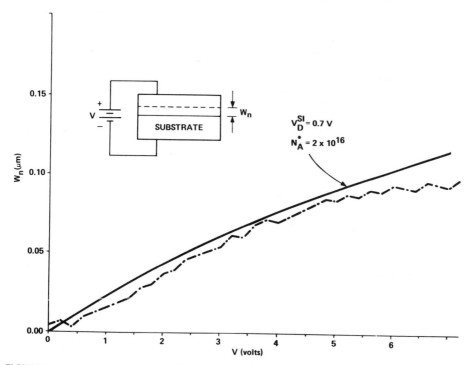

FIGURE 7-4-2. Experimental and calculated (solid line) dependences of the depletion width in the active channel at the substrate boundary on the substrate bias [54].

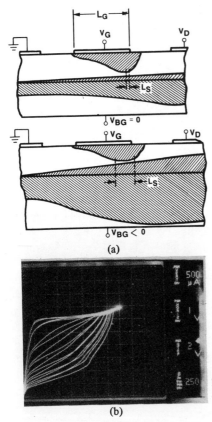

FIGURE 7-4-3. Effect of the backgating on the shape of the depletion region (a) and on the drain characteristics of a 1-μm gate GaAs FET with gate and source common (b) [55].

less pronounced in the dark, which indicates a relationship between backgating effects and light sensitivity related to the traps (see also Refs. 54 and 56). Depending on the substrate properties, illumination may either suppress or enhance backgating.

The gate-to-source and gate-to-drain current–voltage characteristics and series source and drain resistances are also sensitive to the substrate potential (see Fig. 7-4-4 [55]). The observed changes in the series resistances were quite sensitive to processing variations. The devices were fabricated with a plasma deposited silicon nitride layer between the two levels of metallization. The bulk charges in silicon nitride and the charges at the silicon nitride interface may be responsible for the light sensitivity of the series resistances [55].

The variation of the series resistances due to backgating (especially the variation of the source series resistance) has a profound effect on the circuit performance of GaAs FETs [55].

Backgating effects may also be caused by a p-type converted layer beneath the active layer.

An epitaxially grown "buffer layer" between the substrate and the active channel has been shown to decrease the effects of backgating and the light sensitivity of GaAs FETs [54].

The backgating effects have to be taken into account in the design of GaAs ion-implanted FETs and integrated circuits not only because they affect the threshold voltage of ion-implanted MESFETs, but also because they lead to a parasitic

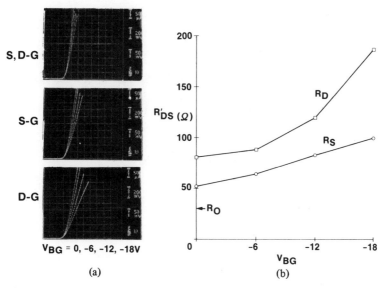

FIGURE 7-4-4. (a) Dependences of the gate-to-source and drain-to-gate characteristics of a GaAs FET on the backgating voltage. (b) Dependence of measured source and drain series resistances on the backgating voltage [55].

interaction in closely located devices [57]. As shown in Ref. 57, such an interaction is smaller for low pinch-off voltage devices.

DLTS studies show that the trapped charge in the substrate, which is responsible for the backgating, is located on the Cr levels and on the so-called EL(2) levels in chromium doped substrates and on the EL(2) levels in high-purity substrates [58]. The backgating effects depend on the degree of compensation in the substrate. Closely compensated substrates should exhibit less backgating [58].

7-5. CURRENT–VOLTAGE CHARACTERISTICS OF GaAs FETs WITH NONUNIFORM DOPING PROFILE

The effects related to the nonuniform doping profile in GaAs FETs, in particular in ion-implanted devices [46, 58–61], may be easily incorporated into a complete velocity saturation model.

For the nonuniform doping profile the channel current may be found from the following equations:

$$\frac{dF_y}{dy} = \frac{qN_d(y)}{\varepsilon} \qquad (7\text{-}5\text{-}1)$$

and

$$I_{\text{ch}} = qv_d W \int_{A_{ds}}^{A} N_d(y)\, dy \qquad (7\text{-}5\text{-}2)$$

Here the electron drift velocity

$$v_d = v_s \qquad (7\text{-}5\text{-}3)$$

in the velocity saturation region and

$$v_d = \mu V_i / L \tag{7-5-4}$$

at low drain-to-source voltages, where V_i is the voltage drop across the conducting channel under the gate (see Fig. 7-3-1).

The depletion width A_{ds} may be related to the gate voltage V_G by integating the Poisson equation (7-5-1) twice:

$$V_{bi} - V_G = (q/\varepsilon) \int_0^{A_{ds}} N_d(y) y \, dy \tag{7-5-5}$$

Differentiating I_{ch} and V_G with respect to A_{ds} we find the device transconductance:

$$g_m = \frac{\partial I_{ch}/\partial A_{ds}}{\partial V_G/\partial A_{ds}} \tag{7-5-6}$$

From Eq. (7-5-6) we obtain

$$g_m = \varepsilon v_d W / A_{ds} \tag{7-5-7}$$

For a variety of doping profiles Eqs. (7-5-2) and (7-5-7) may be solved analaytically [46]. For an arbitrary doping profile it is easy to find a numerical solution [60, 61].

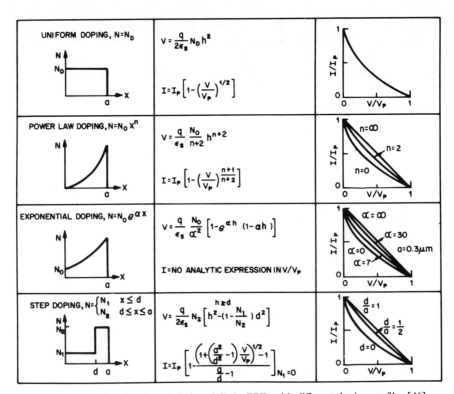

FIGURE 7-5-1. Transfer characteristics of GaAs FETs with different doping profiles [46].

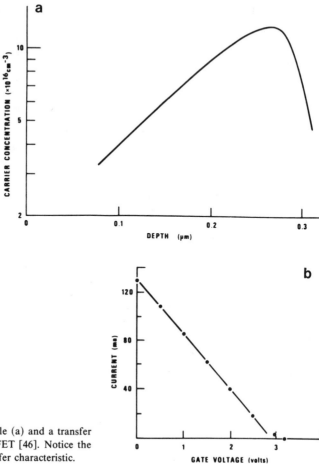

FIGURE 7-5-2. A doping profile (a) and a transfer characteristic (b) of a GaAs FET [46]. Notice the improved linearity of the transfer characteristic.

Transfer characteristics of GaAs MESFETs with different doping profiles are shown in Fig. 7-5-1 [46]. As can be seen from the figure the increase in doping near the substrate–active layer interface allows one to improve the linearity of transfer characteristics. This conclusion is in agreement with experimental results (see Fig. 7-5-2 [46]).

As was shown in Section 7-3, the complete velocity saturation model used here to calculate current–voltage characteristics of GaAs FETs with nonuniform doping profiles is only valid for high pinch-off voltage devices. However, the characteristics of the low pinch-off voltage devices are much less sensitive to variation in the doping profile, as will be further discussed in Section 7-6.

7-6. ANALYTICAL MODELS OF ION-IMPLANTED GaAs FETs

In this section we describe analytical models for the calculation of the current–voltage characteristics of ion-implanted GaAs FETs [59, 63]. The models, which take into account backgating, capping, the source and drain series resistances, and the output conductance, provide simple analytical expressions for the current–voltage characteristics and are quite suitable for parameter acquisition and for computer-aided design of GaAs FETs and ICs. In particular the effective implanted charge and, hence, the activation efficiency may be deduced from the measured pinch-off voltage. The theory may also be used for optimization of doping profiles of GaAs FETs. The results of the calculation are in good agreement with experimental data.

7-6-1. Introduction

Ion implantation has become one of the leading technologies for fabricating GaAs microwave devices and digital ICs. A simple analytical model of ion-implanted GaAs FETs can be quite useful for computer-aided design of GaAs devices and ICs as well as for device parameter acquisition.

The current–voltage characteristics of GaAs FETs with nonuniform doping profiles were analyzed in Section 7-5. However, for the modeling of ion-implanted FETs the important effects of backgating (see Section 7-4) and the change in ion implantation due to capping during the ion-implantation process have to be taken into account. Also, in this section we will obtain analytical expressions for the Gaussian ion-implanted profile.

Our analysis is based on the analytical model of GaAs FETs with the uniform doping profile developed in Section 7-3 (see also Ref. 43), which we modify to include the effects of nonuniform doping, backgating, and capping. As in Section 7-3 we consider two such models for the calculation of the current–voltage characteristics of GaAs FETs. The first one, the square law model, provides an accurate description of GaAs FETs with low pinch-off voltages, using a flat profile with an effective doping concentration and an effective channel thickness. The second one, the complete velocity saturation model, accurately describes high pinch-off voltage devices and takes into account the actual doping profile.

7-6-2. Threshold Voltage, Backgating, and Capping

The doping profile and the energy diagram of an ion-implanted GaAs MESFET with capping depth t_{cap} are shown in Fig. 7-6-1. The doping concentration is described by a Gaussian function:

$$N(x) = N_P \exp\{-[(x - R_P)/(\sigma\sqrt{2})]^2\} \qquad (7\text{-}6\text{-}1)$$

where $N_P = Q/(\sigma\sqrt{2\pi})$, $R_P = R_{po} - t_{cap}$ is the effective projected range of the Gaussian profile and Q, R_{po}, and σ are the dose, the projected range, and the standard deviation of the implant, respectively. Two device doping profiles [47, 63] along with the effective constant doping concentrations which will be discussed later are shown in Fig. 7-6-2.

The decrease in the drain-to-source current when a negative voltage is applied to the substrate is termed backgating (see Section 7-4). The actual pinch-off voltage V_{po} taking into account backgating is defined by

$$V_{po} = \frac{q}{\varepsilon} \int_0^t N(x)x \, dx \qquad (7\text{-}6\text{-}2)$$

where t is the depletion width of the Schottky gate junction when the conducting

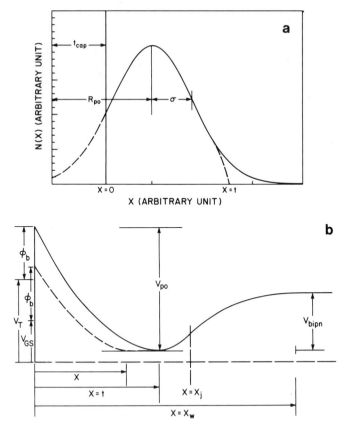

FIGURE 7-6-1. (a) The charge profile in the channel of an ion-implanted MESFET [59]. (b) The potential energy diagram for the MESFET channel: solid and dashed lines represent the potential distributions after and before pinch-off, respectively.

FIGURE 7-6-2. Comparison between actual doping profile and effective flat profile for devices 1 and 2 [63]. Short dashed line, Gaussian profile for device 1; dotted line, Gaussian profile for device 2; solid line, flat profile for device 1; long dashed line, flat profile for device 2.

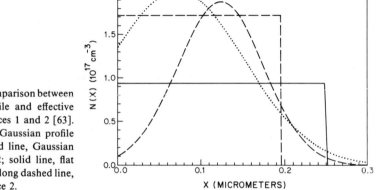

channel is pinched off (see Fig. 7-6-1). Calculating the integral in Eq. (7-6-2) and using Eq. (7-6-1) we find

$$V_{po} = \frac{Qq}{2\varepsilon} \left\{ \frac{\sigma\sqrt{2}}{\sqrt{\pi}} [\exp(-z_p^2) - \exp(-z_t^2)] + R_p[\text{erf}(z_t) + \text{erf}(z_p)] \right\} \quad (7\text{-}6\text{-}3)$$

where

$$z_p = R_p/(\sqrt{2}\sigma)$$

$$z_t = \frac{t - R_p}{\sqrt{2}\sigma}$$

and erf(z) is the error function.

When the conducting channel is pinched off, the total charge in the depletion region of the Schottky gate interface can be obtained by evaluating the integral of the doping concentration $N(x)$ from the metal–semiconductor interface ($x = 0$) to the edge of the depletion region ($x = t$). This gives

$$Q_a = \int_0^t N(x) \, dx \simeq \frac{Q}{2} [\text{erf}(z_t) + \text{erf}(z_p)] \quad (7\text{-}6\text{-}4)$$

Using the approximate relationship between $\exp(-z^2)$ and erf(z) [49]

$$\exp(-z^2) \approx 1 - [\text{erf}(z)]^2 \quad (7\text{-}6\text{-}5)$$

and eliminating erf(z_t) from Eqs. (7-6-3) and (7-6-4), we derive the following expression for the pinch-off voltage:

$$V_{po} = 4V_2 \frac{Q_a}{Q} \left[\frac{Q_a}{Q} + \alpha - \text{erf}(z_p) \right] \quad (7\text{-}6\text{-}6)$$

where

$$V_2 = \frac{qQ\sigma}{\varepsilon\sqrt{2\pi}} \quad \text{and} \quad \alpha = \frac{R_p}{2\sigma} \frac{\sqrt{\pi}}{\sqrt{2}}$$

By inverting Eq. (7-6-6) we can express Q_a in terms of V_{po} as

$$Q_a = \frac{Q}{2} \left(\left\{ [\alpha - \text{erf}(z_p)]^2 + \frac{V_{po}}{V_2} \right\}^{1/2} - [\alpha - \text{erf}(z_p)] \right) \quad (7\text{-}6\text{-}7)$$

Hence, once we know V_{po} from experimental results, Q_a can be calculated from Eq. (7-6-7). Q_a is related to the properties of the substrate through the first term inside the square brackets in the right-hand side of Eq. (7-6-4), erf(z_t). In order to determine erf(z_t), we have to solve the Poisson equation along with the charge neutrality condition for the active channel–substrate junction. The neutrality condition in this case is

$$\int_t^{x_j} N(x) \, dx = N_a(x_w - x_j) \quad (7\text{-}6\text{-}8)$$

where x_j is the channel–substrate junction depth which is the distance from the Schottky gate interface, $x_w - x_j$ is the depletion width on the substrate side, and N_a is the net concentration of negative charge, at equilibrium, on the substrate side (the effective p-doping of a substrate which will cause exactly the same band profile in the active channel as the semi-insulating substrate). The value of N_a depends on the occupancy of the deep traps in the bulk and can be calculated using the results given in Ref. 54 or can be taken as a parameter to fit the measured value of V_{po}. Here we assume that N_a is small when compared to the typical concentration in the channel, i.e.,

$$N_p \gg N_a \qquad (7\text{-}6\text{-}9)$$

and hence

$$\text{erf}(z_j) \approx 1 \quad \text{and} \quad \text{erf}(z_t) \approx 1 \qquad (7\text{-}6\text{-}9a)$$

where $z_j = (x_j - R_p)/(\sigma\sqrt{2}) \approx [\ln(N_p/N_a)]^{1/2}$.

Using the neutrality condition and the boundary condition $F_d(t) = 0$ (where F_d is the electric field in the depletion region), the voltage appearing across the space charge region of the channel–substrate interface can be derived from the Poisson equation as

$$V_{\text{ch-sb}} = \frac{q}{\varepsilon} \int_t^{x_w} N(x)x\, dx \qquad (7\text{-}6\text{-}10)$$

Evaluating the integrals in Eqs. (7-6-8) and (7-6-10) and using Eqs. (7-6-1) and (7-6-5) we obtain

$$\frac{Q}{2}[\text{erf}(z_j) - \text{erf}(z_t)] = N_a(x_w - x_j) \qquad (7\text{-}6\text{-}8a)$$

and

$$V_{\text{ch-sb}} = \frac{q}{\varepsilon} N_a(x_w - x_j)$$

$$\times \left\{ \frac{1}{2}(x_w - x_j) + \frac{\sigma\sqrt{2}}{\sqrt{\pi}}[\text{erf}(z_j) + \text{erf}(z_t)] + R_p \right\} \qquad (7\text{-}6\text{-}10a)$$

Then, substituting Eq. (7-6-8a) into Eq. (7-6-10a) and using Eq. (7-6-9a), we obtain after some algebraic manipulation

$$\text{erf}(z_t) \approx 1 - a_p \qquad (7\text{-}6\text{-}11)$$

where

$$a_p = \frac{V_2}{V_1}(1 + \alpha) \left\{ \left(1 + \frac{V_{\text{bipn}} - \rho V_{\text{bs}}}{V_3} \right)^{1/2} - 1 \right\}$$

$$V_1 = \frac{qQ^2}{8 N_a \varepsilon}$$

$$V_3 = \frac{qN_a[R_p + (\sqrt{2}/\sqrt{\pi})2\sigma]^2}{2\varepsilon}$$

Here

$$\rho = \left(\frac{V_{\text{bip}i}}{V_{\text{bisi}}}\right)\left(\frac{V_{\text{bisi}} - V_{\text{bs}}}{V_{\text{bip}n} - V_{\text{bs}}}\right)$$

is the factor proposed by P. F. Lindquist *et al.* [54] to account for the difference between the semi-insulating substrate and the effective *p*-doping substrate, V_{bs} is the substrate bias voltage, V_{bisi} is the diffusion potential of the channel–semi-insulating junction, and $V_{\text{bip}n}$ is the built-in potential of the effective *p–n* junction at the channel–substrate interface.

Substitution of Eq. (7-6-11) into Eq. (7-6-4) yields

$$Q_a \approx Q_r - \frac{Q}{2} a_p \tag{7-6-12}$$

where

$$Q_r = \frac{Q}{2}[1 + \text{erf}(z_p)] \tag{7-6-12a}$$

is the net charge implanted into substrate. From Eqs. (7-6-6) and (7-6-12) we can see that the pinch-off voltage V_{po} is related to the substrate properties, included in a_p, through Q_a. Therefore, we can determine N_a or $V_{\text{bip}n}$ from the measured value of V_{po} and from the total dose Q. The other constraint which relates N_a, $V_{\text{bip}n}$, and the channel concentration profile can be described approximately by

$$V_{\text{bip}n} \approx V_t \ln\left[\frac{N_a N(t)}{n_i^2}\right] \tag{7-6-13}$$

where n_i is the intrinsic concentration and $V_t = k_B T/q$ is the thermal voltage.

On the other hand, if we know the pinch-off voltage V_{po} from the experimental *I–V* characteristics, then the total dose Q can be calculated. To obtain the expression for the calculation of dose from V_{po}, we eliminate Q_a and Q_r from Eqs. (7-6-7), (7-6-12), and (7-6-12a) and also note that $a_p/(1 + \alpha) \ll 1$ for most *I–V* measurement with floating substrate. Then, we find

$$Q \approx \frac{V_{\text{po}} + 4[q\sigma/(2\varepsilon)]\sqrt{(2/\pi)}b_p(1 + \alpha)}{[q\sigma/(2\varepsilon)]\sqrt{2}\{\exp(-z_p^2)/\sqrt{\pi} + z_p[1 + \text{erf}(z_p)]\}} \tag{7-6-6a}$$

where

$$b_p = \frac{4N_a\sigma}{(2\pi)^{1/2}}(1 + \alpha)\left[\left(1 + \frac{V_{\text{bip}n} - \rho V_{\text{bs}}}{V_3}\right)^{1/2} - 1\right]$$

If we neglect backgating then Eq. (7-6-6a) becomes

$$Q \approx V_{\text{po}}\frac{2\varepsilon}{q\sigma\sqrt{2}}\{\exp(-z_p^2)/\sqrt{\pi} + z_p[1 + \text{erf}(z_p)]\}^{-1} \tag{7-6-6b}$$

The dependences of V_{po} on N_a, V_{bs}, and t_{cap} calculated from Eqs. (7-6-6), (7-6-12), and (7-6-12a) for devices 1 and 2 are shown in Figs. 7-6-3–7-6-5.

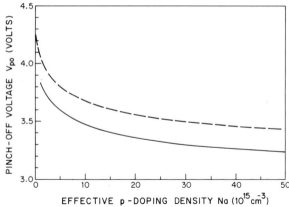

FIGURE 7-6-3. The pinch-off voltage variation with effective p-doping concentration of the semi-insulating substrate [59]. Solid line, device 1; dashed line, device 2 (see Table 7-6-1).

For low pinch-off voltage devices, the current–voltage characteristics can be fairly well described using a uniform profile with an effective doping concentration N_d and an effective channel thickness A[47, 59]. These two parameters can be determined as

$$A = \frac{2\varepsilon V_{po}}{qQ_a} \qquad (7\text{-}6\text{-}14)$$

$$N_d = Q_a/A \qquad (7\text{-}6\text{-}15)$$

Then Eq. (7-6-13) becomes

$$V_{bipn} = V_t \ln\left(\frac{N_a N_d}{n_i^2}\right) \qquad (7\text{-}6\text{-}13a)$$

7-6-3. Current–Voltage Characteristics

To obtain the current–voltage characteristics, we first calculate the drain-to-source saturation current I_{dss}, then we use an interpolation formula, proposed in Ref. 43 and similar to the one proposed in Refs. 50 and 51, for I_{ds}, for the entire

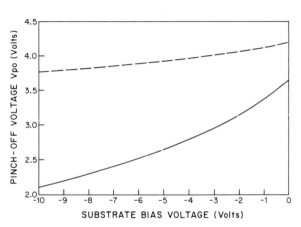

FIGURE 7-6-4. The pinch-off voltage variation with negative substrate bias [59]. Solid line, device 1; dashed line, device 2 (see Table 7-6-1).

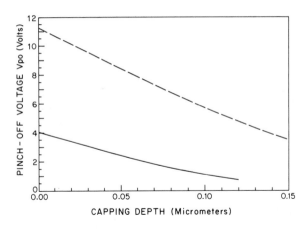

FIGURE 7-6-5. The pinch-off voltage variation with capping depth [59]. Solid line, device 1; dashed line, device 2 (see Table 7-6-1).

range of the drain-to-source voltages V_{ds}:

$$I_{ds} = I_{dss}(1 + \lambda V_{ds}) \tanh(\eta V_{ds}) \qquad (7\text{-}6\text{-}16)$$

where η is chosen in such a way that at $V_{ds} \to 0$ Eq. (7-6-16) converts into the corresponding equation of the Shockley model:

$$\eta = G_{ch}/I_{dss} \qquad (7\text{-}6\text{-}17)$$

Here I_{dss} is the drain-to-source saturation current at the onset of the current saturation and G_{ch} is the channel conductance at $V_{ds} \to 0$. Constant λ in Eq. (7-6-16) is an empirical parameter which accounts for the output conductance in addition to the gate length modulation. (In the square law model the gate length modulation effects will be included in I_{dss}.)

(1) Gaussian Profile. Before calculating the saturation current I_{dss} and the channel conductance G_{ch}, we will first examine the conducting channel charge Q_n, because I_{dss} and G_{ch} are related to Q_n:

$$I_{dss} = qWv_s Q_n(0) \qquad (7\text{-}6\text{-}18)$$

$$G_{ch} = \frac{qW\mu}{L} Q_n(0) \qquad (7\text{-}6\text{-}19)$$

Here $Q_n(0)$ is the conducting channel charge at the source side of the gate, W is the gate width, L is the metallurgical gate length, μ is the low field mobility, and v_s is the electron saturation velocity. Equations (7-6-18) and (7-6-19) are valid for an arbitrary doping profile and are based on the complete velocity saturation model.

The conducting channel charge is equal to the difference between the available charge Q_a taking into account backgating and the charge in the Schottky gate depletion layer Q_d, that is,

$$Q_n = Q_a - Q_d \qquad (7\text{-}6\text{-}20)$$

Since the derivation of Q_d is similar to the determination of Q_a described in Section 7-6-2, we simply replace Q_a, V_{po}, and t in Eq. (7-6-7), by Q_d, $V_{bi} - V_{gs} + V$, and x, respectively, and follow the same procedure. Then we obtain

$$Q_d = \frac{Q}{2}\left\{\left([\alpha - \text{erf}(z_p)]^2 + \frac{V_{bi} - V_{gs} + V}{V_2}\right)^{1/2} - [\alpha - \text{erf}(z_p)]\right\} \quad (7\text{-}6\text{-}21)$$

where V is the channel potential. Substitution of Eq. (7-6-21) into Eq. (7-6-20) yields

$$Q_n(V) = \frac{Q}{2}\left\{\left([\alpha - \text{erf}(z_p)]^2 + \frac{V_{po}}{V_2}\right)^{1/2}\right.$$
$$\left. - \left([\alpha - \text{erf}(z_p)]^2 + \frac{V_{bi} - V_{gs} + V}{V_2}\right)^{1/2}\right\} \quad (7\text{-}6\text{-}22)$$

Then at $V_{ds} \to 0$

$$Q_n(0) = \frac{Q}{2}\left\{\left([\alpha - \text{erf}(z_p)]^2 + \frac{V_{po}}{V_2}\right)^{1/2}\right.$$
$$\left. - \left([\alpha - \text{erf}(z_p)]^2 + \frac{V_{bi} - V_{gs}}{V_2}\right)^{1/2}\right\} \quad (7\text{-}6\text{-}23)$$

(2) *Uniform Profile.* For a uniform profile, the calculation of the conducting channel charge is straightforward. The result is

$$Q_n(V) = Q_a\left[1 - \left(\frac{V_{bi} - V_{gs} + V}{V_{po}}\right)^{1/2}\right] \quad (7\text{-}6\text{-}24)$$

where $Q_a = N_d t$

$$t = \frac{2\varepsilon V_{po}}{qN_d} = d_1\left\{1 - \left[\frac{V_{bipn} - (1-r)V_t - \rho V_{bs}}{(1+r)V_{po1}}\right]^{1/2}\right\} \quad (7\text{-}6\text{-}25)$$

$$d_1/t = (V_{po1}/V_{po})^{1/2} \quad (7\text{-}6\text{-}26)$$

and $r = N_d/N_a$, $V_{po1} = qN_d X_j^2/(2\varepsilon)$.

After determining the conducting channel charge, the drain-to-source saturation current I_{dss} and the channel conductance G_{ch} can be calculated using Eqs. (7-6-18) and (7-6-19). For a low pinch-off voltage device, the channel conductance is still calculated using Eq. (7-6-19); however, the saturation current may be approximated by the square law as proposed in Refs. 42 and 43:

$$I_{dss} = \beta(V_{gs} - V_T)^2 \quad (7\text{-}6\text{-}27)$$

where

$$V_T = V_{bi} - V_{po} \quad (7\text{-}6\text{-}28)$$

$$\beta = \frac{2\varepsilon v_s W}{A(V_{po} + 3F_s L_1)} \quad (7\text{-}6\text{-}29)$$

Here $F_s = v_s/\mu$ is the velocity saturation field and L_1 is the length of the part of the channel where the electrical field is smaller than F_s. L_1 can be determined by

$$L - L_1 = \frac{2A}{\pi} \sinh^{-1}\left[\frac{\pi K_d(V_{ds} - V_{is})}{2AF_s}\right] \qquad (7\text{-}6\text{-}30)$$

where

$$V_{is} = \frac{F_sL(V_T - V_{gs})}{F_sL + V_T - V_{gs}} \qquad (7\text{-}6\text{-}31)$$

is the voltage drop across the channel at the saturation and K_d is defined as

$$K_d = \frac{\text{voltage drop across the high field domain}}{\text{voltage drop across the conducting channel}} \qquad (7\text{-}6\text{-}32)$$

In devices with self-aligned gates, $K_d \approx 1$. In devices with non-self-aligned gates, K_d should depend on the shape of the gate recess and on the length of the drain-to-gate spacing. It may depend on the gate voltage, too. Here we choose $K_d \approx 0.15$ based on the results of the numerical simulation of GaAs MESFETs (see, for examples, Ref. 19). We should note, however, that the results obtained for practical devices seem to be quite insensitive to the exact value of K_d because of a roughly logarithmic dependence of the length of the velocity saturation region on K_d.

Figures 7-6-6 and 7-6-7 show the dependences of the conducting channel charges, calculated from Eqs. (7-6-23) and (7-6-24), on the gate voltages for devices 1 and 2. For device 2, these two curves corresponding to the Gaussian profile and the uniform profile are quite close owing to the deep capping depth. This can be understood by comparing the actual and the effective doping profiles for device 1 and device 2 shown in Fig. 7-6-2.

7-6-4. Role of Source and Drain Series Resistances

For a realistic description of GaAs FETs the effects related to the series source and drain resistances as well as the drain-to-source shunt resistance R_{sh}, which accounts for the additional output conductance beyond the output conductance related to the gate length modulation, should be incorporated into these models.

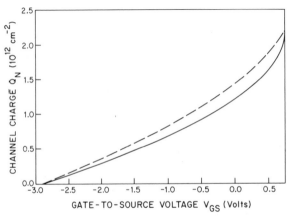

FIGURE 7-6-6. The conducting channel charge as a function of gate bias for device 1 [59]. Dashed line, Gaussian; solid line, flat doping profile. Parameters used in the calculation are given in Table 7-6-1.

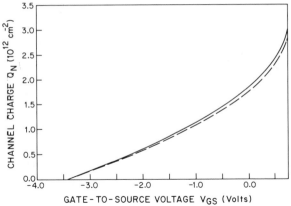

FIGURE 7-6-7. The conducting channel charge as a function of gate bias for device 2 [59]. Dashed line, Gaussian; solid line, flat profile. Parameters used in the calculation are given in Table 7-6-1.

The source and drain series resistances R_s and R_d may be taken into account as follows. The external measurable gate-to-source voltage V_{GS} and drain-to-source voltage V_{DS} are given by

$$V_{GS} = V_{gs} + I_{dss} R_s \tag{7-6-33}$$

and

$$V_{DS} = V_{ds} + I_{dss}(R_s + R_d) \tag{7-6-34}$$

Substituting Eqs. (7-6-33) and (7-6-23) into Eq. (7-6-18) and solving the resulting equation for I_{dss} we find

$$I_{dsso} = \frac{I_o}{2}\left[h_2 - \left(h_2^2 - \frac{4}{V_2}[V_{po} - (V_{bi} - V_{GS})] \right)^{1/2} \right] \tag{7-6-35}$$

where

$$I_o = qWv_s \frac{Q}{2}$$

$$h_2 = 2h_1 + \frac{R_s I_o}{V_2}$$

and

$$h_1 = \left\{ [\alpha - \text{erf}(z_p)]^2 + \frac{V_{po}}{V_2} \right\}^{1/2}$$

Similarly, substituting Eqs. (7-6-33) and (7-6-24) into Eq. (7-6-18) and solving for I_{dss} in the uniform profile case we obtain

$$I_{dsso} = \frac{I_{fc}}{2}\left[h_3 - \left(h_3^2 - \frac{4}{V_{po}}[V_{po} - V_{GS})] \right)^{1/2} \right] \tag{7-6-36}$$

where

$$h_3 = 2 + \frac{R_s I_{fc}}{V_{po}}$$

and

$$I_{fc} = qWv_sQ_a$$

is the full channel saturation current.

Equations (7-6-35) and (7-6-36) are for the high pinch-off voltage devices. For the low pinch-off voltage devices, Eq. (7-6-27) can be applied. The incorporation of R_s and R_d into the "square law" model has already been done in Section 7-3 [43]. The result is

$$I_{dss} = \frac{1 + 2\beta R_s(V_{GS} - V_T) - [1 + 4\beta R_s(V_{GS} - V_T)]^{1/2}}{2\beta R_s^2} \qquad (7\text{-}6\text{-}37)$$

As in the intrinsic device models described in the last section, we use a similar interpolation formula for the drain-to-source current [see Eq. (7-6-16)]:

$$I_{DS} = I_{dss}(1 + \lambda V_{DS})\tanh(\eta V_{DS}) + \frac{V_{DS}}{R_{sh}} \qquad (7\text{-}6\text{-}38)$$

where I_{dss} is given by Eqs. (7-6-35)–(7-6-37), constant λ is the same empirical constant used in Eq. (7-6-16), and R_{sh} is the effective parasitic drain-to-source resistance shunting the intrinsic device.

Parameter η in Eq. (7-6-38) is chosen in the same way as in Eq. (7-6-16), i.e., at $V_{DS} \to 0$ Eq. (7-6-38) converts into the corresponding equation of the Shockley model:

$$\eta = G_{CH}/I_{dsso} \qquad (7\text{-}6\text{-}39)$$

where G_{CH} is the output conductance excluding the contribution due to the shunt resistance $1/R_{sh}$ at $V_{DS} \to 0$:

$$G_{CH} = \lim_{V_{DS} \to 0}\left(\frac{\delta I_{ds}}{\delta V_{DS}}\bigg|_{V_{GS}=\text{const}}\right) \qquad (7\text{-}6\text{-}40)$$

where

$$I_{ds} = \frac{qW\mu}{L}\int_{I_{ds}R_s}^{V_{DS}-I_{ds}R_d} Q_n(V)\,dV \qquad (7\text{-}6\text{-}41)$$

is the channel current in the linear region and $Q_n(V)$ is the conducting channel charge for an arbitrary doping profile. From Eq. (7-6-40) we find

$$G_{CH} = \frac{G_{ch}}{1 + (R_s + R_d)G_{ch}} \qquad (7\text{-}6\text{-}42)$$

where G_{ch} is given by Eq. (7-6-19).

7-6-5. Comparison with Experimental Data

The calculated current–voltage characteristics of the GaAs FET with a high pinch-off voltage are compared with the experimental data [47] in Fig. 7-6-8. The

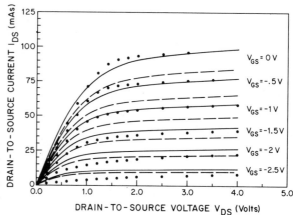

FIGURE 7-6-8. Current–voltage characteristics of device 1 [59]. Solid line, calculated results with a Gaussian profile; dashed line, calculated results with a flat profile; circles, measured results.

parameters used in the calculation are given in Table 7-6-1. As can be seen from this figure, the agreement with experimental results is quite good for the "complete velocity saturation" model with a Gaussian profile (shown in Fig. 7-6-8 by solid lines). The dashed lines in the same figure represent the calculated results for the flat profile model. The agreement with experimental data for this effective uniform model is not as good. However, the agreement can be improved significantly by adjusting the values of F_s and/or v_s. For this particular device, the best fit value of v_s is about 20% greater than that used in the Gaussian profile case ($v_s = 0.88 \times 10^7$ cm/s) (see Fig. 7-6-9).

TABLE 7-6-1. Device Parameter Used in the Calculation[a]

	Symbol	No. 2	No. 1	Unit
Total dose	Q	4×10^{12}	2.4×10^{12}	cm^{-2}
Projected range	R_{po}	0.2	0.123	μm
Standard deviation	σ	0.08	0.051	μm
Capping depth	t_{cap}	0.129	0.0	μm
Gate length	L	1	1.3	μm
Gate width	W	650	490	μm
Built-in voltage	V_{bi}	0.75	0.75	V
Threshold voltage	V_T	−3.45	−2.91	V
Pinch-off voltage	V_{po}	4.2	3.66	V
Saturation velocity	v_s	0.9×10^7	0.88×10^7	cm/s
Saturation velocity field	F_s	3000	4000	V/cm
Source series resistance	R_s	4.5	2.8	Ω
Drain series resistance	R_d	4.5	2.8	Ω
Drain–source shunt resistance	R_{sh}	1.5	5	kΩ
Output conductance parameter	λ	0.0	0.03	
Built-in voltage of channel–semi-insulating substrate interface	V_{bisi}	0.7	0.7	V
Built-in voltage of effective p–n junction at the channel–substrate interface	V_{bipn}	1.14	1.106	V
Effective channel thickness	A	1864.5	2357	Å
Effective uniform channel doping concentration	N_d	1.72	0.938	$\times 10^{17}$ cm^{-3}

[a] Reference [59].

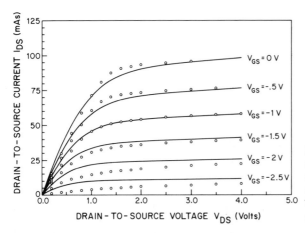

FIGURE 7-6-9. Current-voltage characteristics of device 1 [59]. Solid line, calculated results with a flat profile and parameters given in Table 7-6-1 (except $v_s = 1.05 \times 10^7$ cm/s); circles, measured results.

In Fig. 7-6-10 the drain-to-source current calculated as a function of the drain-to-source voltage for another high pinch-off voltage device in the frame of the "complete velocity saturation" model with a Gaussian profile is compared with the experimental data [59]. The parameters which are chosen to fit the experimental $I-V$ are given in Table 7-6-1. Again, the calculated results for the effective uniform model are shown in the same figure. For this device, these two sets of calculated $I-V$ curves are fairly close because the deep capping makes the resulting profile "flatter."

In both cases the actual implantation dose Q is in good agreement with the dose deduced from Eq. (7-6-6a) using the measured values of V_{po}. (The actual doses and the deduced doses are 2.4×10^{12} cm^{-2} and 2.39×10^{12} cm^{-2} for device 1 and 4.0×10^{12} cm^{-2} and 4.011×10^{12} cm^{-2} for device 2.)

The analytical models for the calculation of the current-voltage characteristics of ion-implanted GaAs FETs discussed in this section take into account backgating, capping, the source and drain series resistances, and the output conductance. They also make it possible to estimate the effective implanted charge and the activation efficiency. The results of the calculation are in good agreement with experimental data. We have also demonstrated that the value of the effective saturation velocity deduced from the GaAs FET characteristics using the effective flat profile may be

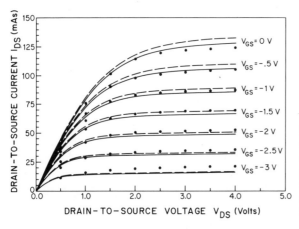

FIGURE 7-6-10. Current-voltage characteristics of device 2 [59]. Solid line, calculated results with a Gaussian profile; dashed line, calculated results with a flat profile; circles, measured results. Parameters used in the calculation are given in Table 7-6-1.

somewhat overestimated because of the error introduced by the "flat" profile approximation.

7-7. CAPACITANCE MODEL FOR GaAs MESFETs

7-7-1. Introduction

Internal device capacitances in GaAs MESFETs play an important role in determining the performance of GaAs devices and integrated circuits. In the JFET model of SPICE, used as an approximate tool for GaAs integrated circuit simulation, the device capacitances are modeled as capacitances of Schottky diodes connected between the gate and the source and the gate and the drain, respectively,

$$C_{gs} = \frac{C_{gs0}}{(1 - V_{gs}/V_{bi})^{1/2}} \tag{7-7-1}$$

$$C_{gd} = \frac{C_{gs0}}{(1 - V_{gd}/V_{bi})^{1/2}} \tag{7-7-2}$$

where V_{bi} is the built-in voltage, V_{gs} is the gate-to-source voltage, V_{gd} is the gate-to-drain voltage, and

$$C_{gs0} = \frac{\varepsilon WL}{2A_0} = \frac{1}{2} WL \left(\frac{\varepsilon q N_d}{2 V_{bi}}\right)^{1/2} \tag{7-7-3}$$

Here A_0 is the depletion width under the gate at zero gate voltage (see Section 7-2). As pointed out in Ref. 47 such a model is completely inadequate for gate voltages smaller than the threshold voltage. Indeed, when the channel under the gate is pinched off, the total charge in the section of the channel under the gate does not change. Hence, the differential gate-to-source and gate-to-drain capacitances are determined by the charge variation in the depletion region extensions beyond the gate (see Fig. 7-7-1). In addition, this oversimplified model fails to account for the nonuniform doping profile in ion-implanted FETs and the possible Gunn domain formation at the drain side of the gate (charges Q_+ and Q_- in Fig. 7-7-1b). The effects related to the high field domains are important for microwave GaAs transistors with high pinch-off voltages [22–27, 63, 64].

In this section we present a new capacitance model [63] which takes into account Gunn domain formation and backgating. The model developed is an analytical model which is suitable for computer-aided design of GaAs devices and integrated circuits. It realistically describes the C-V characteristics of GaAs FETs, in particular the decrease in the gate-to-drain capacitance with the drain-to-source voltage observed in microwave GaAs FETs [24, 63]. This model can be used for ion-implanted GaAs FETs.

7-7-2. Internal Capacitance Model

The calculation of the internal device capacitances is based on the simplified charge distributions shown in Fig. 7-7-1 and the equivalent circuit presented in Fig. 7-7-2.

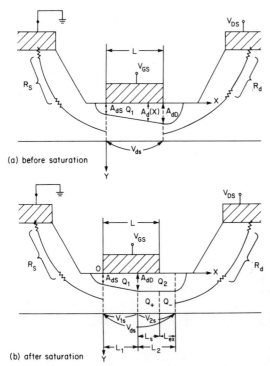

(a) before saturation

(b) after saturation

FIGURE 7-7-1. (a) The charge distribution in a recess gate MESFET in the linear region of the I-V characteristic. (b) The charge distribution in a recess gate planar MESFET in the current saturation region of the I-V curves [63].

Below we derive equations describing C-V characteristics in linear, saturation, and pinch-off regions of the I-V characteristic.

(1) Linear Region. In the linear region, the gradual channel approximation can be applied to the entire channel to determine the position of the depletion layer edge under the metallurgical gate. In order to obtain analytical expressions for the internal gate-to-source capacitance, C_{gs}, and the internal gate-to-drain capacitance, C_{gd}, we assume that the depletion layer edge beneath the gate (Section 1) varies

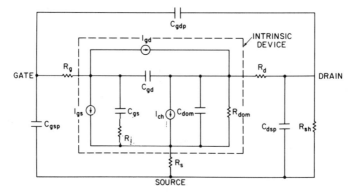

FIGURE 7-7-2. Equivalent circuit of a GaAs MESFET. C_{gs}, internal gate-to-source capacitance; C_{gd}, internal gate-to-drain capacitance; C_{dom}, domain capacitance; C_{gsp}, C_{gdp}, and C_{dsp}, extrinsic parasitic capacitances for the gate-to-source, the gate-to-drain, and the drain-to-source terminals; R_{dom}, domain resistance; R_{ch}, channel resistance; R_{sh}, drain-to-source shunt resistance; R_g, R_s, and R_d, series gate, source, and drain resistance; I_{ch}, drain-to-source channel current; I_{gs}, gate-to-source current; I_{gd}, gate-to-drain current; R_i, input resistance [63].

linearly along the length of the channel as shown in Fig. 7-7-1a. This assumption can be verified by inspecting the depletion layer edge profiles calculated numerically for a device with a uniform doping profile (Fig. 7-7-3a) and for a device with a Gaussian doping profile (Fig. 7-7-3b). In both cases these profiles vary almost linearly with distance.

An assumption of a linear variation of the depletion edge boundary with distance leads to the following expression for the space charge within Section 1, Q_1 (see Fig. 7-7-1a):

$$Q_1 = qWL \int_0^{A_{dS}} N(y)\,dy + \frac{qWL}{A_{dD} - A_{dS}} \int_{A_{dS}}^{A_{dD}} N(y)(A_{dD} - y)\,dy$$

$$= qWL \int_0^{A_{dD}} N(y)\,dy - \frac{qWL}{A_{dD} - A_{dS}} \int_{A_{dS}}^{A_{dD}} N(y)(y - A_{dS})\,dy \quad (7\text{-}7\text{-}4)$$

where $N(y)$ is the doping density and A_{dS} and A_{dD} are the depletion layer widths at the source end and at the drain end. To calculate A_{dS} and A_{dD}, we have to solve Poisson's equation for the gate-channel-substrate (buffer) structure whose potential energy diagram is shown in Fig. 7-7-4. The boundary conditions are

$$V(0) = -(V_{bi} - V_g + V_{ch})$$

$$V(A_d) = -V_m$$

$$F(A_d) = 0 \quad\quad\quad (7\text{-}7\text{-}5)$$

$$V(y \gg A) = -(V_{bisb} + \Delta\phi_p)$$

$$F(y \gg A) = 0$$

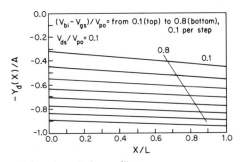

(a) Uniform doping profile

FIGURE 7-7-3. (a) The depletion layer boundary for a MESFET with a uniform doping profile (x, distance measured from the source end of the gate; $y_d(x)$, depletion layer width measured from the Schottky contact; L, gate length; A, active layer thickness). (b) The depletion layer boundary for a MESFET with a Gaussian doping profile (x, distance measured from the source end of the gate; $y_d(x)$, depletion layer width measured from the Schottky contact; L, gate length; σ, standard deviation of the ion implantation) [63].

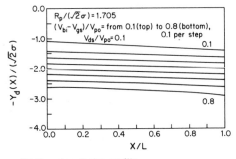

(b) Gaussian doping profile

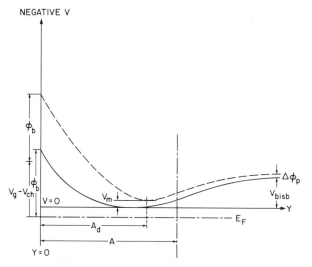

FIGURE 7-7-4. The potential energy diagram for the MESFET channel: solid and dashed lines represent the potential distributions above and at pinch-off, respectively (ϕ_b, potential barrier height of the Schottky contact; A_d, depletion layer width; A, active layer thickness; E_F, Fermi level; V_{bisb}, built-in voltage of the channel–substrate junction) [63].

Here V_g is the gate voltage, V_{ch} is the voltage drop across the channel, V_m is the absolute value of the potential at the depletion layer of the Schottky contact, $\Delta\phi_p$ is the change of the Fermi level in the substrate from its equilibrium position, and V_{bisb} is the built-in voltage of the channel-substrate (buffer layer) junction:

$$V_{\text{bisb}} = \frac{k_B T}{q}\left(\ln\frac{N_{\text{ch}}}{n_i} + \ln\frac{N_{\text{sub}}}{n_i}\right) \tag{7-7-6}$$

Here N_{ch} is the channel doping density ($N_{\text{ch}} = N_d$ for uniform doping profile with a constant doping concentration N_d and $N_{\text{ch}} \approx N_{\text{sub}}$ for nonuniform doping profile), N_{sub} is the effective doping density of the substrate.

The gate-to-source and gate-to-drain voltages, V_{gs} and V_{gd}, are related to the doping profile as follows:

$$V_{\text{bi}} - V_{\text{gs}} - V_{\text{ms}} = \frac{q}{\varepsilon}\int_0^{A_{\text{dS}}} N(y)y\,dy \tag{7-7-7}$$

$$V_{\text{bi}} - V_{\text{gd}} - V_{\text{md}} = \frac{q}{\varepsilon}\int_0^{A_{\text{dD}}} N(y)y\,dy \tag{7-7-8}$$

where $V_{\text{ms}} = V_m[A_{\text{dS}}(V_{\text{gs}})]$ and $V_{\text{md}} = V_m[A_{\text{dD}}(V_{\text{gd}})]$. For a uniform doping profile with a doping density N_d and an active layer thickness A, A_{dS} and A_{dD} are given by

$$A_{\text{dS}} = \left[\frac{2\varepsilon(V_{\text{bi}} - V_{\text{gs}} - V_{\text{ms}})}{qN_d}\right]^{1/2} \tag{7-7-9}$$

$$A_{\text{dD}} = \left[\frac{2\varepsilon(V_{\text{bi}} - V_{\text{gd}} - V_{\text{md}})}{qN_d}\right]^{1/2} \tag{7-7-10}$$

and

$$\frac{\delta V_{\text{ms},d}}{\delta V_{s,d}} = \frac{1}{1 + f(V_g - V_{s,d})} \tag{7-7-11}$$

Here $f(V_g - V_{s,d})$ is a complicated function of $V_g - V_{s,d}$, which is equal to zero when $V_g - V_{s,d} = V_{Tc}$, where

$$V_{Tc} = V_{bi} - V_{bisb} - \frac{qN_dA^2}{2\varepsilon} \tag{7-7-12}$$

Both $V_{ms,d}$ and $\delta V_{ms,d}/\delta V_{s,d}$ are negligibly small when $(A - A_{dS,D})/L_D \gg 1$, where L_D is the Debye length of the active layer, and $V_{ms,d} \approx V_{bisb}$ when $V_g - V_{s,d} = V_{Tc}$, and function $f(V_g - V_{s,d})$ can be approximated by

$$f(V_g - V_{s,d}) \approx \frac{V_{po} - V_{bisb}\sqrt{1 - (A_{dS,D}/A)^4}}{V_{bisb}(A_{dS,D}/A)^2} \tag{7-7-13}$$

Hence

$$V_{ms,d} \approx V_{bisb}[1 - \sqrt{1 - (A_{dS,D}/A)^4}] \tag{7-7-14}$$

Then, from Eqs. (7-7-9), (7-7-10), and (7-7-14) we obtain

$$\frac{V_{ms,d}}{V_{poc}} = \frac{r_1 + (V_{bi} - V_{gs,d})/V_{poc} - [r_1^2 - r_2(r_2 - 2)]^{1/2}}{1 + r_1^2} \tag{7-7-15}$$

where

$$r_1 = V_{poc}/V_{bisb} = (V_{bi} - V_{Tc} - V_{bisb})/V_{bisb} \tag{7-7-16a}$$

and

$$r_2 = (V_{bi} - V_{gs,d})/V_{bisb} \tag{7-7-16b}$$

For a Gaussian profile

$$N(y) = \frac{D_s}{\sigma\sqrt{2\pi}} \exp\left\{ -\left[\frac{y - (R_p - t_{cap})}{\sigma\sqrt{2}} \right]^2 \right\} \tag{7-7-17}$$

with a total implanted dose D_s, a projected range R_p, a standard deviation σ, and a capping thickness t_{cap}. A_{dS} and A_{dD} can be calculated from

$$A_{dS,D} = R_p + \sigma\sqrt{2}\left\{ -\ln\left[\frac{A_1}{A_3}\left(\sqrt{1 + \frac{1 - A_2^2}{A_1^2}} - 1 \right) \right] \right\}^{1/2}$$

$$A_1 = \frac{2A_3A_2 + 1}{2A_3}$$

$$A_2 = \psi_{S,D} - A_3 \exp\left[-\left(\frac{R_p}{\sigma\sqrt{2}} \right)^2 \right] - \mathrm{erf}\left(\frac{R_p}{\sigma\sqrt{2}} \right) \tag{7-7-18}$$

$$A_3 = \sqrt{\frac{2}{\pi}}\frac{\sigma}{R_p}$$

where $\psi_S = 2\varepsilon(V_{bi} - V_{gs} - V_{ms})/(qR_pD_s)$ and $\psi_D = 2\varepsilon(V_{bi} - V_{gd} - V_{md})/(qR_pD_s)$. In deriving Eq. (7-7-18) we have used an approximate expression for the error function, erf (z) [49]

$$\text{erf}^2(z) \approx 1 - \exp(-z^2) \tag{7-7-19}$$

As to the calculations of V_{ms}, V_{md}, $\delta V_{\mathrm{ms}}/\delta V_s$, and $\delta V_{\mathrm{md}}/\delta V_d$, Eqs. (7-7-11)–(7-7-16) may still be used as an approximation if N_d and A in Eq. (7-7-12) are the effective uniform doping density and the effective active layer thickness. N_d and A are defined as [59]

$$\tfrac{1}{2} N_d A^2 = \int_0^{A_j} N(y) y \, dy \tag{7-7-20}$$

$$N_d A = \int_0^{A_j} N(y) \, dy \tag{7-7-21}$$

where A_j is determined from $N(A_j) = N_{\mathrm{sub}}$ for p-type substrate and is assumed to be infinitely large for n-type substrate.

The gate-to-source capacitance C_{gs} is the rate of change of the free charge on the gate electrode, $-Q_g$, with respect to the source bias when the drain and gate potentials are held fixed. Similarly, the gate-to-drain capacitance C_{gd} is the rate of change of the free charge on the gate electrode with respect to the drain bias when the source and gate potential are held fixed. Thus we define [66]

$$C_{\mathrm{gs}} = -\frac{\partial Q_g}{\partial V_s} = \frac{\partial Q_1}{\partial V_s} + \frac{\pi}{2}\varepsilon W \tag{7-7-22}$$

$$C_{\mathrm{gd}} = -\frac{\partial Q_g}{\partial V_d} = \frac{\partial Q_1}{\partial V_d} + \frac{\pi}{2}\varepsilon W \tag{7-7-23}$$

Here the second term, $(\pi/2)\varepsilon W$, in the right-hand side of Eqs. (7-7-22) and (7-7-23) is the side-wall capacitance which is due to the variations of the space charges within Section 2 or Section 3 in Fig. 7-7-1a [67].

Using Eqs. (7-7-4), (7-7-22), and (7-7-23), we obtain

$$C_{\mathrm{gs}} = \frac{\varepsilon WL[1 - (\partial V_{\mathrm{ms}}/\partial V_s)]}{N(A_{\mathrm{dS}})A_{\mathrm{dS}}(A_{\mathrm{dD}} - A_{\mathrm{dS}})^2} \left[A_{\mathrm{dD}} \int_{A_{\mathrm{dS}}}^{A_{\mathrm{dD}}} N(y) \, dy - \frac{\varepsilon}{q}(V_{\mathrm{ds}} + \Delta V_m) \right]$$

$$+ \frac{\pi}{2}\varepsilon W \tag{7-7-24}$$

$$C_{\mathrm{gd}} = \frac{\varepsilon WL[1 - (\partial V_{\mathrm{md}}/\partial V_d)]}{N(A_{\mathrm{dD}})A_{\mathrm{dD}}(A_{\mathrm{dD}} - A_{\mathrm{dS}})^2} \left[\frac{\varepsilon}{q}(V_{\mathrm{ds}} + \Delta V_m) - A_{\mathrm{dS}} \int_{A_{\mathrm{dS}}}^{A_{\mathrm{dD}}} N(y) \, dy \right]$$

$$+ \frac{\pi}{2}\varepsilon W \tag{7-7-25}$$

where $\Delta V_m = V_{\mathrm{ms}} - V_{\mathrm{md}}$. For a uniform profile, Eqs. (7-7-24) and (7-7-25) become

$$C_{\mathrm{gs}} = \frac{\varepsilon WL}{2A_{\mathrm{dS}}}\left(1 - \frac{\partial V_{\mathrm{ms}}}{\partial V_s}\right) + \frac{\pi}{2}\varepsilon W \tag{7-7-26}$$

$$C_{\mathrm{gd}} = \frac{\varepsilon WL}{2A_{\mathrm{dD}}}\left(1 - \frac{\partial V_{\mathrm{md}}}{\partial V_d}\right) + \frac{\pi}{2}\varepsilon W \tag{7-7-27}$$

In the limit of $V_{ds} \to 0$, Eqs. (7-7-24) and (7-7-25) also reduce to Eqs. (7-7-26), (7-7-27). For a Gaussian profile [see Eq. (7-7-17)], by using Eq. (7-7-18), Eqs. (7-7-24) and (7-7-25) may be rewritten as

$$C_{gs} = \frac{\varepsilon WL[1 - (\partial V_{ms}/\partial V_s)]}{N(A_{dS})A_{dS}(A_{dD} - A_{dS})^2}\left\{\frac{A_{dD}D_s}{2}\left[\text{erf}\left(\frac{A_{dD} - R_p}{\sigma\sqrt{2}}\right)\right.\right.$$

$$\left.\left. - \text{erf}\left(\frac{A_{dS} - R_p}{\sigma\sqrt{2}}\right)\right] - \frac{\varepsilon}{q}(V_{ds} + \Delta V_m)\right\} + \frac{\pi}{2}\varepsilon W \qquad (7\text{-}7\text{-}28)$$

$$C_{gd} = \frac{\varepsilon WL[1 - (\partial V_{md}/\partial V_d)]}{N(A_{dD})A_{dD}(A_{dD} - A_{dS})^2}\left\{\frac{\varepsilon}{q}(V_{ds} + \Delta V_m) - \frac{A_{dS}D_s}{2}\right.$$

$$\left. \times \left[\text{erf}\left(\frac{A_{dD} - R_p}{\sigma\sqrt{2}}\right) - \text{erf}\left(\frac{A_{dS} - R_p}{\sigma\sqrt{2}}\right)\right]\right\} + \frac{\pi}{2}\varepsilon W \qquad (7\text{-}7\text{-}29)$$

(2) Saturation Region. In the saturation region the gradual channel approximation can only be applied to part of the channel (Section 1 in Fig. 7-7-1b). The space charge within this section, Q_1, can be calculated from Eq. (7-7-4) with the substitution of L_1 for L, where L_1 is the length of Section 1 in Fig. 7-7-1b:

$$Q_1 = qWL_1 \int_0^{A_{dS}} N(x)\,dx + \frac{qWL_1}{A_{dD} - A_{dS}}\int_{A_{dS}}^{A_{dD}} N(y)(A_{dD} - y)\,dy$$

$$= qWL_1 \int_0^{A_{dD}} N(x)\,dx - \frac{qWL_1}{A_{dD} - A_{dS}}\int_{A_{dS}}^{A_{dD}} N(y)(y - A_{dS})\,dy \qquad (7\text{-}7\text{-}30)$$

Here A_{dS} is the depletion layer width at the source and A_{dD} is the depletion layer width at the boundary of Section 1 and Section 2. A_{dS} is calculated from Eq. (7-7-7), (7-7-9), or (7-7-18). A_{dD} can be determined from Eq. (7-7-8), (7-7-10), or (7-7-18), where V_{gd} should be substituted by $V_{gs} - V_{1s}$ (V_{1s} is the voltage drop across Section 1). For uniform profile, V_{1s} has been found in Refs. 59 and 42 as

$$V_{1s} \approx \frac{(V_{gs} - V_T)F_sL}{F_sL + V_{gs} - V_T} \qquad (7\text{-}7\text{-}31)$$

Equation (7-7-31) will reduce to the saturation drain-to-source voltage predicted by the Shockley model for devices with a very long gate length. Equation (7-7-31) was derived for a uniform doping profile. However, as follows from the results obtained in Ref. 59, it may also be used for devices with ion-implanted (Gaussian) profiles as a reasonable approximation.

If we neglect the small variation of the conducting channel opening in the carrier velocity saturation region (Section 2 in Fig. 7-7-1b), then the space charge within Section 2 can be approximated by

$$Q_2 = qWL_2 \int_0^{A_{dD}} N(y)\,dy \qquad (7\text{-}7\text{-}32)$$

where $L_2 = L_s + L_{ex}$ is the length of Section 2 ($L_s = L - L_1$ is the length of the subsection under the gate and L_{ex} is the length of the extended velocity saturation subsection in the gate-to-drain spacing).

From the definition of the gate-to-source capacitance and gate-to-drain capacitance and referring to Fig. 7-7-1b, we find

$$C_{gs} = \frac{\partial Q_1}{\partial V_s} + \frac{\partial Q_2}{\partial V_s} + \frac{\pi}{2}\,\varepsilon W \qquad (7\text{-}7\text{-}33)$$

$$C_{gd} = \frac{\partial Q_1}{\partial V_d} + \frac{\partial Q_2}{\partial V_d} + \varepsilon W \sin^{-1}\left(\left[\frac{V_{bi} - V_{gs} + V_{1s}}{V_{bi} - V_{gs} + V_{ds}}\right]^{1/2}\right) \qquad (7\text{-}7\text{-}34)$$

where the third term in the right-hand side of Eq. (7-7-34) is an approximate value for the side-wall capacitance [67].

The charge derivatives $\partial Q_{1,2}/\partial V_{d,s}$ depend on the presence or absence of a stationary high field domain which may form at the drain side of the gate. They are evaluated in the Appendix (Section 7-7-A).

(3) Pinch-off Region. If a gate-to-source bias less than the threshold voltage V_T is applied then the conducting channel is totally depleted. Therefore, the contribution to the device capacitances comes only from the variation of the space charges in the side-walls [68]:

$$C_{gs} = \varepsilon W \sin^{-1}\left[\left(\frac{V_{bi} - V_{Tc}}{V_{bi} - V_{gs}}\right)^{1/2}\right] \qquad (7\text{-}7\text{-}35)$$

$$C_{gd} = \varepsilon W \sin^{-1}\left[\left(\frac{V_{bi} - V_{Tc}}{V_{bi} - V_{gd}}\right)^{1/2}\right] \qquad (7\text{-}7\text{-}36)$$

7-7-3. Results of the Calculation

The calculated capacitance–voltage characteristics are shown in Figs. 7-7-5–7-7-8 for a low pinch-off voltage device and in Figs. 7-7-9–7-7-12 for a high pinch-off voltage device. The values of the parameters used in the calculations are given in Table 7-7-1. Both the uniform doping model and the nonuniform doping model were used for the low pinch-off voltage FET. Calculations for the high pinch-off voltage device were done using only the nonuniform doping model because in this

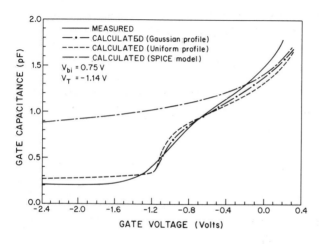

FIGURE 7-7-5. The total gate capacitance (the sum of the gate-to-source capacitance and the gate-to-drain capacitance) of the low pinch-off voltage device at zero drain-to-source voltage as a function of the gate voltage [63].

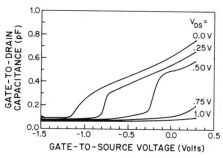

(a) Gaussian doping profile

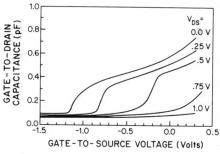

FIGURE 7-7-6. The gate-to-drain capacitance of the low pinch-off voltage device as a function of the gate-to-source voltage. (a) Calculated by using a Gaussian profile; (b) calculated by using an effective uniform doping profile [63].

(b) Uniform doping profile

case the doping nonuniformity strongly affects the device current–voltage characteristics [59]. The measured results [47, 64] are shown in Figs. 7-7-5, 7-7-11, and 7-7-12 for comparison.

As can be seen from Fig. 7-7-5, the calculated gate capacitance is quite close to the experimental data. The error introduced by the uniform profile model is 5%–15% depending on the gate voltage (see Fig. 7-7-7).

The stationary high field domain does not form in the low pinch-off voltage device. As a consequence the C–V characteristics of this device are relatively simple. They may be adequately described by the capacitance model developed in Ref. 16.

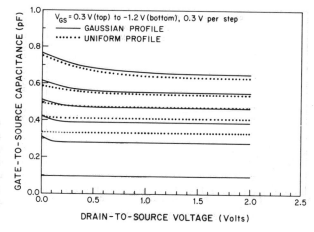

FIGURE 7-7-7. The variation of the gate-to-source capacitance of the low pinch-off voltage device with the drain-to-source voltage. Solid lines, calculated by using a Gaussian profile; dotted lines, calculated by using an effective uniform doping profile [63].

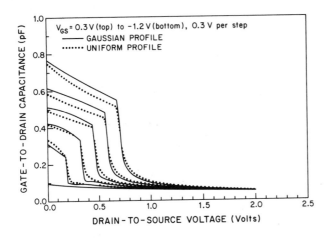

FIGURE 7-7-8. The variation of the gate-to-drain capacitance of the low pinch-off voltage device with the drain-to-source voltage [63].

The formation of the stationary high field domain at the drain side of the channel in the high pinch-off voltage device leads to more complicated dependences which are not described by the model of Ref. 16.

The intervalley scattering from the central valley to the upper valley of the conduction band is affected not only by the electrical field but also by the potential difference. When the voltage across the device is smaller than some critical value, V_{cr}, the electron energy is smaller than the energy gap between the minima of the central valley and the satellite valleys of the conduction band in GaAs. Therefore, the region

$$V_{1s} < V_{ds} < V_{cr} \leqslant 1 \text{ V} \tag{7-7-37}$$

in Figs. 7-7-11 and 7-7-12 cannot be modeled by our equations based on the drift velocity versus electrical field characteristic. In this voltage range the capacitance–voltage characteristics were interpolated by the third degree polynomial. The coefficients of the interpolation polynomial were determined from the values of the capacitance and its derivatives at $V_{ds} = V_{1s}$ and $V_{ds} = V_{cr}$.

The model developed in Ref. 63 and reviewed in this section takes into account the nonuniform doping, backgating, capping, the formation of stationary Gunn domains in high pinch-off voltage devices, and the side wall capacitances. This

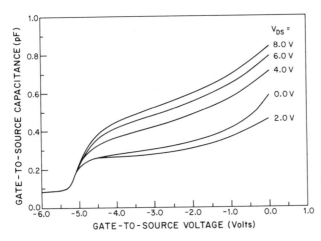

FIGURE 7-7-9. The gate-to-source capacitance of the high pinch-off voltage device as a function of the gate-to-source voltage [63].

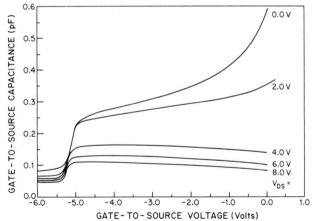

FIGURE 7-7-10. The gate-to-drain capacitance of the high pinch-off voltage device as a function of the gate-to-source voltage [63].

theory explains the complicated voltage dependence of the gate-to-source capacitance and the complicated experimental drain-to-source voltage dependence of the feedback (gate-to-drain) capacitance. The results of the calculation are in good agreement with the measured data. The model presented is quite suitable for computer-aided design of GaAs microwave FETs and integrated circuits.

7-7-A. Appendix

The derivatives of the charges Q_1 and Q_2 may be evaluated using an approximate linear boundary for the depletion edge. For a Gaussian doping profile

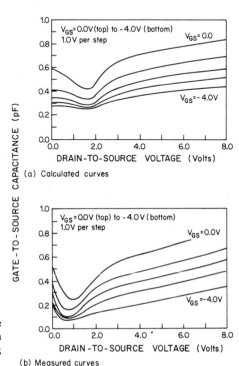

FIGURE 7-7-11. The variation of the gate-to-source capacitance of the high pinch-off voltage device with the drain-to-source voltage. (a) Calculated curves; (b) measured curves [63].

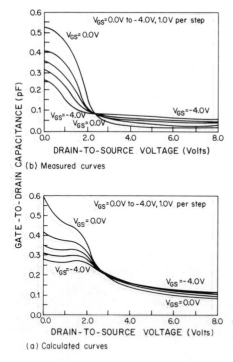

(b) Measured curves

(a) Calculated curves

FIGURE 7-7-12. The variation of the gate-to-drain capacitance of the **high** pinch-off voltage device with the drain-to-source voltage. (a) Calculated curves [63]; (b) measured curves [64].

TABLE 7-7-1. Parameters of Low and High Pinch-Off Voltage Devices

	Symbol	Low	High	Unit
Total dose	D_s	2.2		$\times 10^{12}$ cm^{-2}
Gate width	W	600	600	μm
Gate length	L	1.3	1.7	μm
Projected range	R_p	0.069		μm
Standard deviation	σ	0.0313		μm
Capping thickness	t_{cap}	0.0		μm
Active layer thickness	A		0.32	μm
Active layer doping density	N_d		7.0	$\times 10^{16}$ cm^{-3}
Substrate doping density	N_{sub}	1.0	1.0	$\times 10^{12}$ cm^{-3}
Built-in voltage	V_{bi}	0.75	0.7	V
Threshold voltage	V_T	−1.14	−5.1	V
Pinch-off voltage	V_{po}	1.89	5.8	V
Saturation velocity	v_s	1.0	1.1	$\times 10^7$ cm/s
Saturation velocity field	F_s	4000	5000	V/cm
Domain parameter	K_d		0.4	
Source series resistance	R_s	4	5	Ω
Drain series resistance	R_d	4	5	Ω
Drain-source shunt resistance	R_{sh}	2	3	kΩ
Parasitic capacitance per unit gate width	C_{po}	0.125		fF/μm

we obtain

$$\frac{\partial Q_1}{\partial V_s} = \frac{qW}{A_{\mathrm{dD}} - A_{\mathrm{dS}}} \left(\frac{\partial L_1}{\partial V_s}\right) \left[A_{\mathrm{dD}} Q(A_{\mathrm{dD}}) - A_{\mathrm{dS}} Q(A_{\mathrm{dS}}) - \frac{\varepsilon \Delta V}{q} \right] + \frac{qWL_1}{(A_{\mathrm{dD}} - A_{\mathrm{dS}})^2}$$

$$\times \left\{ \left[A_{\mathrm{dD}} \frac{\partial A_{\mathrm{dS}}}{\partial V_s} - A_{\mathrm{dS}} \frac{\partial A_{\mathrm{dD}}}{\partial V_s} \right] [Q(A_{\mathrm{dD}}) - Q(A_{\mathrm{dS}})] + \left(\frac{\partial A_{\mathrm{dD}}}{\partial V_s} - \frac{\partial A_{\mathrm{dS}}}{\partial V_s} \right) \frac{\varepsilon \Delta V}{q} \right\}$$

$$(7\text{-}7\text{-}A1)$$

$$\frac{\partial Q_2}{\partial V_s} = qW \left[\left(\frac{\partial L_2}{\partial V_s} \right) Q(A_{\mathrm{dD}}) + L_2 \left(\frac{\partial A_{\mathrm{dD}}}{\partial V_s} \right) N(A_{\mathrm{dD}}) \right] \qquad (7\text{-}7\text{-}A2)$$

$$\frac{\partial Q_1}{\partial V_d} = \frac{qW}{A_{\mathrm{dD}} - A_{\mathrm{dS}}} \left(\frac{\partial L_1}{\partial V_d}\right) \left[A_{\mathrm{dD}} Q(A_{\mathrm{dD}}) - A_{\mathrm{dS}} Q(A_{\mathrm{dS}}) - \frac{\varepsilon \Delta V}{q} \right]$$

$$- \frac{\varepsilon W L_1}{(A_{\mathrm{dD}} - A_{\mathrm{dS}})^2 N(A_{\mathrm{dD}}) A_{\mathrm{dD}}} \left\{ \frac{\varepsilon \Delta V}{q} - A_{\mathrm{dS}} [Q(A_{\mathrm{dD}}) - A(A_{\mathrm{dS}})] \right\} \frac{\partial V_{\mathrm{md}}}{\partial V_d}$$

$$(7\text{-}7\text{-}A3)$$

$$\frac{\partial Q_2}{\partial V_d} = qW \left(\frac{\partial L_2}{\partial V_d} \right) Q(A_{\mathrm{dD}}) - \frac{\varepsilon W L_2}{A_{\mathrm{dD}}} \frac{\delta V_{\mathrm{md}}}{\delta V_d} \qquad (7\text{-}7\text{-}A4)$$

$$Q(A_{\mathrm{dS},D}) = \int_0^{A_{\mathrm{dS},D}} N(y)\, dy$$

$$= \frac{D_s}{2} \left[\operatorname{erf}\left(\frac{A_{\mathrm{dS},D} - R_P}{\sigma\sqrt{2}} \right) + \operatorname{erf}\left(\frac{R_P}{\sigma\sqrt{2}} \right) \right] \qquad (7\text{-}7\text{-}A5)$$

$$\frac{\partial A_{\mathrm{dS}}}{\partial V_s} = \frac{\varepsilon}{qN(A_{\mathrm{dS}}) A_{\mathrm{dS}}} \left(1 - \frac{\partial V_{\mathrm{ms}}}{\partial V_s} \right) \qquad (7\text{-}7\text{-}A6)$$

$$\frac{\partial A_{\mathrm{dD}}}{\partial V_s} = \frac{\varepsilon}{qN(A_{\mathrm{dS}}) A_{\mathrm{dS}}} \left(1 + \frac{\partial V_{1s}}{\partial V_s} \right) \qquad (7\text{-}7\text{-}A7)$$

where

$$\Delta V = V_{1s} + V_{\mathrm{ms}} - V_{\mathrm{md}}$$

For a uniform doping profile we find

$$\frac{\partial Q_1}{\partial V_s} = \tfrac{1}{2} qWN_d (A_{\mathrm{dS}} + A_{\mathrm{dD}}) \left(\frac{\partial L_1}{\partial V_s} \right)$$

$$+ C_{g0} \left[\frac{A_0}{A_{\mathrm{dD}}} \left(1 - \frac{\partial V_{\mathrm{ms}}}{\partial V_s} \right) + \left(1 + \frac{\partial V_{1d}}{\partial V_s} \right) \frac{A_0}{A_{\mathrm{dD}}} \right] \qquad (7\text{-}7\text{-}A8)$$

$$\frac{\partial Q_2}{\partial V_s} = qWN_d A_{\mathrm{dD}} \left(\frac{\partial L_2}{\partial V_s} \right) + 2C_{g0} \frac{L_2 A_0}{A_{\mathrm{dD}} L} \left(1 + \frac{\partial V_{1s}}{\partial V_s} \right) \qquad (7\text{-}7\text{-}A9)$$

$$\frac{\partial Q_1}{\partial V_d} = \tfrac{1}{2} qWN_d (A_{\mathrm{dS}} + A_{\mathrm{dD}}) \left(\frac{\partial L_1}{\partial V_d} \right) - \frac{\varepsilon W L_1}{2A_{\mathrm{dD}}} \frac{\partial V_{\mathrm{md}}}{\partial V_d} \qquad (7\text{-}7\text{-}A10)$$

$$\frac{\partial Q_2}{\partial V_d} = qWN_d A_{\mathrm{dD}} \left(\frac{\partial L_2}{\partial V_d} \right) - \frac{\varepsilon W L_2}{A_{\mathrm{dD}}} \frac{\partial V_{\mathrm{md}}}{\partial V_d} \qquad (7\text{-}7\text{-}A11)$$

$$A_{\mathrm{dD}} = \left[\frac{2\varepsilon (V_{\mathrm{bi}} - V_{\mathrm{gs}} + V_{1s} - V_{\mathrm{md}})}{qN_d} \right]^{1/2} \qquad (7\text{-}7\text{-}A12)$$

Here

$$\frac{\partial V_{1s}}{\partial V_s} = -\left(\frac{F_s L}{F_s L + V_{\mathrm{gs}} - V_T} \right)^2 \qquad (7\text{-}7\text{-}A13)$$

and

$$\frac{\partial V_{\mathrm{md}}}{\partial V_d} = \left(\frac{\partial V_{\mathrm{md}}}{\partial V_{1s}} \right)\left(\frac{\partial V_{1s}}{\partial V_d} \right) \qquad (7\text{-}7\text{-}A14)$$

where $\partial V_{\mathrm{md}} / \partial V_{1s}$ is given by Eqs. (7-7-11) and (7-7-13) with the substitution of V_d by $(V_{\mathrm{gs}} - V_{1s})$, and $\delta V_{1s} / \delta V_d$ should decrease from unity at $V_{\mathrm{ds}} = V_{1s}$ to a negligible value when $V_{\mathrm{ds}} > V_{1s}$ [see Eq. (7-7-31)].

The resulting expressions for the capacitances depend on the presence or absence of the Gunn domain in the channel. The criteria of Gunn domain formation in high pinch-off voltage devices are given by [69]

$$A < A_{\mathrm{dm}} \qquad \text{No Gunn domain} \qquad (7\text{-}7\text{-}A15a)$$

$$A_{\mathrm{dm}} < A < 2A_{\mathrm{dm}} \qquad \text{Stationary Gunn domain} \qquad (7\text{-}7\text{-}A15b)$$

$$2A_{\mathrm{dm}} < A \qquad \text{Propagating Gunn domain} \qquad (7\text{-}7\text{-}A15c)$$

where A is the active layer thickness and

$$A_{\mathrm{dm}} = \left[\frac{2\varepsilon (V_{\mathrm{bi}} - V_{\mathrm{gs}} + F_{\mathrm{th}} L_1)}{qN_d} \right]^{1/2} \qquad (7\text{-}7\text{-}A16)$$

Here F_{th} is the threshold field for bulk negative differential mobility. In the nonuniform case, N_d and A in Eqs. (7-7-A15) and (7-7-A16) are the effective uniform doping density and effective active layer thickness.

As can be seen from Eq. (7-7-A15), high field domains do not form in low pinch-off voltage devices ($V_{\mathrm{po}} \le 2V$). Typically pinch-off voltages of digital GaAs FETs are not high enough for the formation of propagating high field domains [see Eq. (7-7-A15c)]. Therefore we do not consider devices with propagating domains. Cases corresponding to Eqs. (7-7-A15a) and (7-7-A15b) are considered below.

(1) No Gunn Domain. In this case, the length of the carrier velocity saturation region (Section 2 in Fig. 7-7-1b), L_2, is only a small fraction of the length of the entire channel. Therefore, the shape of the depletion region edge is insensitive to the source voltage when the drain and gate voltages are held fixed. As a first-order approximation, Eqs. (7-7-24), (7-7-25), and (7-7-28) can still be used to calculate C_{gs} in the saturation region defined in Eq. (7-7-33). As to the calculation of C_{gd}, Eq. (7-7-34) has to be used and L_s is given by [42, 16]

$$L_s = \frac{2A}{\pi} \sinh^{-1}\left[\frac{\pi K_d (V_{\mathrm{ds}} - V_{1s})}{2AF_s} \right] \qquad (7\text{-}7\text{-}A17)$$

and L_{ex} can be approximated by

$$L_{ex} \approx \frac{(1 - K_d) V_{2s}}{F_m} \tag{7-7-A18}$$

where

$$K_d = \frac{V_{2i}}{V_{2s}} \tag{7-7-A19}$$

and

$$F_m = F_s \cosh\left(\frac{\pi L_s}{2A}\right) \tag{7-7-A20}$$

is the maximum field of the velocity saturation region. Here $V_{2s} = V_{ds} - V_{1s}$ and V_{2i} are the voltage drops across Section 2 and across the subsection under the gate, respectively. K_d is determined from the continuity of C_{gd} at $V_{ds} = V_{1s}$ and is nearly independent of the gate bias. then we find

$$\frac{\partial L_1}{\partial V_d} = -\frac{\partial L_s}{\partial V_d} = -\frac{K_d}{F_s \sqrt{1 + b^2}} \tag{7-7-A21}$$

$$\frac{\partial L_2}{\partial V_d} = \left[1 - \frac{1 - K_d}{K_d} \left(\frac{F_s}{F_m} b\right)^2 \right] \left(\frac{\partial L_s}{\partial V_d}\right) + \frac{1 - K_d}{F_m} \tag{7-7-A22}$$

where

$$b = \frac{\pi K_d V_{2s}}{2A F_s} \tag{7-7-A23}$$

(2) Stationary Gunn Domain. If a stationary Gunn domain forms in the channel at the drain side of the gate then Eqs. (7-7-33) and (7-7-34) have to be used to calculate C_{gs} and C_{gd}, because the length of the Gunn domain is equally sensitive to the drain voltage and to the source voltage [64, 69]. In this case L_2 is given by [65]

$$L_2 \simeq 2.06 \left(\frac{\varepsilon V_{2s}}{q \sqrt{n_{cr} N_d}}\right)^{1/2} \tag{7-7-A24}$$

and the charge in the domain walls can be calculated as

$$Q_+ = 0.728 [q\varepsilon \sqrt{n_{cr} N_d} \, V_{2s}]^{1/2} \tag{7-7-A25}$$

Here n_{cr} is the characteristic doping density (a typical value of n_{cr} for GaAs is 3×10^{15} cm^{-3}) [65]. For simplicity we assume

$$L_s \approx K_d L_2 \tag{7-7-A26}$$

where the domain parameter K_d is taken to be a constant.

Differentiating Eqs. (7-7-A24) and (7-7-A25) and using Eq. (7-7-A26), we find

$$\frac{\partial L_2}{\partial V_s} = -\left(1 + \frac{\partial V_{1s}}{\partial V_s}\right)\left(\frac{L_2}{2 V_{2s}}\right) \tag{7-7-A27}$$

$$\frac{\partial L}{\partial V_s} = K_d \left(1 + \frac{\partial V_{1s}}{\partial V_s}\right)\left(\frac{L_2}{2V_{2s}}\right) \qquad (7\text{-}7\text{-}A28)$$

$$\frac{\partial L_2}{\partial V_d} = \frac{L_2}{2V_{2s}} \qquad (7\text{-}7\text{-}A29)$$

$$\frac{\partial L}{\partial V_d} = -K_d \frac{L_2}{2V_{2s}} \qquad (7\text{-}7\text{-}A30)$$

The domain capacitance is given by

$$C_{\text{dom}} = \frac{\delta Q_+}{\delta V_2} = \frac{Q_+}{2V_{2s}} \qquad (7\text{-}7\text{-}A31)$$

7-8. CURRENT–VOLTAGE CHARACTERISTICS OF UNGATED GaAs FETs

Saturated resistors (or ungated FETs) are used as load elements in GaAs logic gates (see, for example, Refs. 32, 71, 72). The current saturation in these structures is a consequence of the velocity saturation of electrons.

The free surface of GaAs is depleted and the surface Fermi level is pinned by a high density of the surface states. The shape and the thickness of this depletion layer determine the low field resistance and saturation current of ungated loads.

In Ref. 73, a new model for GaAs MESFETs was proposed which took into account the change of the surface potential due to the voltages applied to the gate and the drain. The surface potential was assumed to vary linearly along the surface provided that the source-to-gate separation is much larger than the depletion depth, and the electric field is almost perpendicular to the surface. Under these conditions, the resistivity of the channel is modulated not only by the channel potential but also by the surface potential.

We employed this model to characterize ungated GaAs FETs with uniform and nonuniform (ion-implanted) doping profiles [74] and to deduce the electron saturation velocity as a function of device length for 0.5-, 1-, 2-, and 3-μm-long devices.

Figure 7-8-1 shows a schematic cross section of an ungated FET with the applied drain voltage. Using the gradual channel approximation, the depth of the surface depletion layer, $h(x)$, may be found as

$$\frac{qN_d}{2\varepsilon}h^2(x) = V_{\text{Sbi}} + V(x) - \Phi_s(x) \qquad (7\text{-}8\text{-}1)$$

where N_d is the uniform doping density, V_{Sbi} is the surface built-in voltage, $V(x)$ is the channel potential, and $\Phi_s(x)$ is the surface potential. This is equivalent to the structure with a distributed gate whose potential is $\Phi_s(x)$ [51, 75]. The surface potential is assumed to vary linearly along the surface according to

$$\Phi_s(x) = \frac{V_D}{L}x \qquad (7\text{-}8\text{-}2)$$

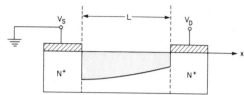

FIGURE 7-8-1. A schematic cross section of an ungated GaAs FET [74].

where V_D is the applied drain voltage and L is the source-to-drain separation. Note that unlike the conventional MESFET model, the conducting channel has the minimum thickness at the source end (Fig. 7-8-1) because of the assumption that the surface potential varies from zero to V_D along the surface.

We use the two-piece linear approximation for the dependence of the electron velocity on the electric field as shown in Fig. 7-3-1. We model the ungated FET first in the linear region of the velocity versus electric field curve assuming constant electron mobility, and then in the saturation region assuming constant saturation velocity.

(1) *Linear Region.* In this region the drain current, I_D, is given by

$$I_D = WqN\mu N_d [A - h(x)] \frac{dV(x)}{dx} \qquad (7\text{-}8\text{-}3)$$

where W is the channel width, μ is the low-field electron mobility, and A is the channel thickness. The boundary conditions for $V(x)$ are

$$V(0) = R_c I_D \qquad (7\text{-}8\text{-}4)$$

$$V(L) = V_D - R_c I_D \qquad (7\text{-}8\text{-}5)$$

where R_c is the ohmic contact resistance under the source and the drain. Integrating Eq. (7-8-3) from $x = 0$ to L leads to the implicit expression for I_D (see Ref. 73)

$$kL = \tfrac{1}{2}[h^2(0) - h^2(L)] - d[h(0) - h(L)]$$
$$+ d(d - A)\ln\frac{h(0) + d - A}{h(L) + d - A} \qquad (7\text{-}8\text{-}6)$$

where

$$h(0) = \left[\frac{2\varepsilon}{qN_d}(V_{Sbi} + R_c I_D)\right]^{1/2} \qquad (7\text{-}8\text{-}7)$$

$$h(L) = \left[\frac{2\varepsilon}{qN_d}(V_{Sbi} - R_c I_D)\right]^{1/2} \qquad (7\text{-}8\text{-}8)$$

$$k = \frac{\varepsilon V_D}{qN_d L} \qquad (7\text{-}8\text{-}9)$$

$$d = \frac{L I_D}{G V_D} \qquad (7\text{-}8\text{-}10)$$

$$G = Wq\mu N_d \qquad (7\text{-}8\text{-}11)$$

Equation (7-8-6) can be solved to find I_D in the linear region for a given V_D. In the limiting case of $R_c = 0$, the thickness of the depletion layer becomes uniform throughout the channel, and under this condition, the drain current is given by

$$I_D = GA_c \frac{V_D}{L} \qquad (7\text{-}8\text{-}12)$$

where A_c is the active channel thickness:

$$A_c = A\left[1 - \left(\frac{V_{Sbi}}{V_{po}}\right)^{1/2}\right] \qquad (7\text{-}8\text{-}13)$$

Note that the current given by Eq. (7-8-12) is identical to the current through the resistor whose resistance, R, is

$$R = R_{ch} = \frac{L}{GA_c} \qquad (7\text{-}8\text{-}14)$$

(2) Velocity Saturation Region. The velocity saturation starts when the electric field reaches the critical value of F_s at the source end of the channel. The corresponding saturation current, I_{Dsat}, is given by

$$I_{Dsat} = WqN_d\left\{A - \left[\frac{2\varepsilon}{qN_d}(V_{Sbi} + R_c I_{Dsat})\right]^{1/2}\right\}v_s \qquad (7\text{-}8\text{-}15)$$

where v_s is the electron saturation velocity. Solving Eq. (7-8-15) for I_{Dsat} we find

$$I_{Dsat} = I_{FC}\left[1 + \frac{\eta}{2} - \left(\frac{\eta^2}{4} + \eta + \zeta\right)^{1/2}\right] \qquad (7\text{-}8\text{-}16)$$

where

$$I_{FC} = WAqN_d v_s \qquad (7\text{-}8\text{-}17)$$

$$\eta = \frac{R_c I_{FC}}{V_{po}}$$

$$\zeta = \frac{V_{Sbi}}{V_{po}}$$

$$V_{po} = \frac{qN_d A^2}{2\varepsilon}$$

The saturation voltage, V_{Dsat}, can be found from Eq. (7-8-6) by substituting I_{Dsat} for I_D and solving for V_D. However, as shown below, an approximate analytical solution for V_{Dsat} may also be found.

In the limiting case of $R_c = 0$, I_{Dsat} is reduced to

$$I_{Dsat} = WqN_d A_c v_s \qquad (7\text{-}8\text{-}18)$$

The results obtained above are valid for the uniform doping profile. For the ion-implanted profile, Eqs. (7-8-1), (7-8-3), and (7-8-15) should be replaced by the following equations:

$$\frac{q}{\varepsilon} \int_0^{h(x)} y N_d(y) \, dy = V_{Sbi} + V(x) - \Phi_s(x) \tag{7-8-19}$$

$$I_D = Wq\mu \frac{dV(x)}{dx} \int_{h(x)}^{y_i} N_d(y) \, dy \tag{7-8-20}$$

$$I_{Dsat} = Wqv_s \int_{h(0)}^{y_i} N_d(y) \, dy \tag{7-8-21}$$

where y_i is the thickness of the depletion layer when the channel is pinched off. The parameter, y_i, is related both to the implantation parameters (such as the implanted dose and the implantation energy), and to the properties of the substrate (such as the deep trap density and the built-in potential across the channel and the substrate) through equations given in Section 7-6 (see also Ref. 59). Equations (7-8-19)–(7-8-21) together with the equations of Section 7-6 should be solved numerically to determine the drain current.

However, we can closely approximate the results replacing the Gaussian profile by an effective uniform profile. The effective uniform doping density, N_d^*, and the effective channel thickness, A^*, are given (see Section 7-6 and Ref. 59) by.

$$A^* = \frac{2\varepsilon V_{po}}{qQ_a} \tag{7-8-22}$$

$$N_d^* = \frac{Q_a}{A^*} \tag{7-8-23}$$

where

$$V_{po} = \frac{V_2}{\alpha_3} \left\{ \frac{\alpha_3 V_{p\infty} + V_{ch-sb}}{V_2} - \frac{2}{\nu} \left[1 + \left(1 + \frac{\nu V_{ch-sb}}{V_1} \right)^{1/2} \right] \right\} \tag{7-8-24}$$

$$Q_a = \frac{Q}{2} \left\{ \left([\alpha - \mathrm{erf}(z_p)]^2 + \frac{V_{po}}{V_1} \right)^{1/2} - [\alpha - \mathrm{erf}(z_p)] \right\} \tag{7-8-25}$$

$$V_1 = \frac{qQ\sigma}{\varepsilon (2\pi)^{1/2}} \tag{7-8-26}$$

$$V_{p\infty} = V_1 \{ \exp(-z_p^2) + 2\alpha [1 + \mathrm{erf}(z_p)] \} \tag{7-8-27}$$

$$V_2 = \frac{qQ^2}{8\varepsilon N_a} \tag{7-8-28}$$

$$\alpha_3 = 1 + \frac{V_2}{V_1} \tag{7-8-29}$$

$$\nu = \frac{\alpha_3}{(1 + \alpha)^2} \tag{7-8-30}$$

$$\alpha = \frac{R_p}{2\sigma} \left(\frac{\pi}{2} \right)^{1/2} \tag{7-8-31}$$

$$V_{ch-sb} = V_{bipn} - \rho V_{bs} \tag{7-8-32}$$

$$V_{\text{bi}pn} = 2V_t \ln\left(\frac{N_a}{n_i}\right) \tag{7-8-33}$$

$$\rho = \frac{V_{\text{bi}pn}(V_{\text{bisi}} - V_{\text{bs}})}{V_{\text{bisi}}(V_{\text{bi}pn} - V_{\text{bs}})} \tag{7-8-34}$$

$$z_p = \frac{R_p}{\sigma\sqrt{2}} \tag{7-8-36}$$

Here, Q is the implanted dose, R_{po} is the projected range, σ is the standard deviation, t_{cap} is the thickness of the capping layer, and N_a is the effective background doping density in the substrate. $V_{\text{ch-sb}}$, $V_{\text{bi}pn}$, V_{bisi}, and V_t are the voltage across the channel–substrate interface, the built-in voltage of the effective p–n junction at the channel–substrate interface, the diffusion potential of the channel–semi-insulating junction, and the thermal voltage ($k_B T/q$ where k_B is the Boltzmann constant and T is the sample temperature), respectively. In Eq. (7-8-25) Q_a represents the implanted dose in GaAs excluding the part in the capping layer. With the measured value of N_a, we may use Eqs. (7-8-22)–(7-8-36) to characterize the ion-implanted device.

Figure 7-8-2 shows I_D–V_D characteristics of the ungated FETs calculated from Eq. (7-8-6). Below the saturation, the ungated FET behaves like a resistor with a resistance

$$R_{\text{off}} = \left(\frac{V_D}{I_D}\right)_{I_D \to 0} = 2R_c + R_{\text{ch}} \tag{7-8-37}$$

where

$$R_{\text{ch}} = \frac{L}{Wq\mu N_D A_c} \tag{7-8-38}$$

Even close to the saturation, the deviation of the slope of the current-voltage characteristics from R_{off} is less than 3%. This may be attributed partly to the assumption that the low field mobility is constant up to the saturation, and partly to the fact that the depletion thickness at the source end increases with the current, while at the drain end it decreases so that the overall change in the shape of the depletion region is not large enough to cause a large variation of the resistivity of the active channel.

As can be seen from Eq. (7-8-16), $I_{D\text{sat}}$ is independent of the source-to-drain separation, provided that the effective electron saturation velocity is independent of the device length.

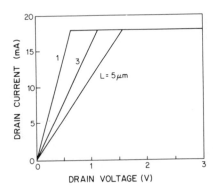

FIGURE 7-8-2. I_D–V_D characteristics of ungated FETs for various source-to-drain separations [74]. $W = 1 \times 10^{-4}$ m, $A = 1.5 \times 10^{-7}$ m, $N_d = 1.5 \times 10^{23}$ m^{-3}, $\mu = 0.4$ m^2/V s, $v_s = 1.2 \times 10^5$ m/s, $R_c = 1\,\Omega$ mm, $V_{\text{Sbi}} = 0.6$ V, $I_{D\text{sat}} = 18.2$ mA.

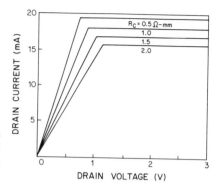

FIGURE 7-8-3. I_D-V_D characteristics of ungated FETs for various contact resistances [74]. The device dimensions and other parameters are the same as in Fig. 7-8-2. $L = 2 \times 10^{-6}$ m.

From the I_D-V_D characteristics, V_{Dsat} can be approximated as

$$V_{Dsat} = R_{off} I_{Dsat} = F_s L + 2 R_c I_{Dsat} \tag{7-8-39}$$

This equation agrees with the equation obtained in Ref. 76.

I_D-V_D characteristics for various contact resistances are shown in Fig. 7-8-3. The corresponding I_{Dsat} and V_{Dsat} as functions of R_c are shown in Figs. 7-8-4 and 7-8-5, respectively. As R_c increases from 0.5 Ω mm to 2.0 Ω mm, I_{Dsat} decreases approximately by 20% while V_{Dsat} increases approximately by 50%. This illustrates the importance of good ohmic contacts for the saturated loads.

Figures 7-8-6 and 7-8-7 show the variation of I_{Dsat} and V_{Dsat} with the doping density assuming that V_{Sbi} is constant due to the high density of the surface states. Both I_{Dsat} and V_{Dsat} increase with N_d.

In Figs. 7-8-8a and 7-8-8b the calculated results are compared with the experimental data for two groups of saturated resistors. Each group consists of four resistors with width of 20 μm and length of 0.5, 1, 2, 3 μm, respectively, on the same chip on a wafer. All the resistors were fabricated using the ion-implantation of Se with energy of 240 keV and dose of 2.4×10^{12} cm^{-2}, and with a Si$_3$N$_4$ capping layer of 200 Å. The effective channel thickness, the effective uniform doping density, and the pinch-off voltage are determined from Eqs. (7-8-22), (7-8-23), and (7-8-24)

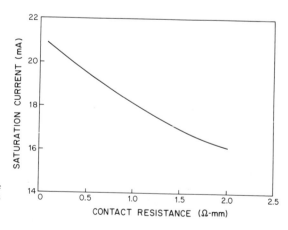

FIGURE 7-8-4. I_{Dsat} as a function of R_c [74]. All the parameters are the same as in Fig. 7-8-2.

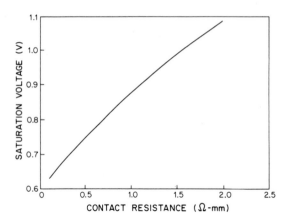

FIGURE 7-8-5. $V_{D\text{sat}}$ as a function of R_c [74]. All the parameters are the same as in Fig. 7-8-2.

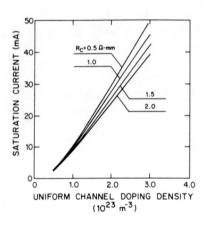

FIGURE 7-8-6. $I_{D\text{sat}}$ as a function of N_d with R_c as a parameter [74]. All other parameters are the same as in Fig. 7-8-2.

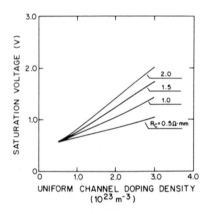

FIGURE 7-8-7. $V_{D\text{sat}}$ as a function of N_d with R_c as a parameter [74]. All other parameters are the same as in Fig. 7-8-2.

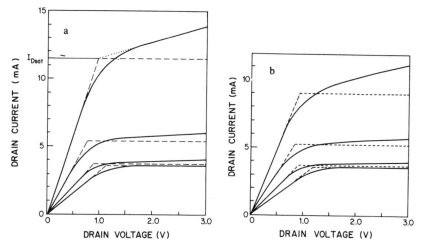

FIGURE 7-8-8. Measured and calculated current-voltage characteristics of ungated loads [74] (solid line, experiment; dashed line, theory). $W = 20 \times 10^{-6}$ m, $A = 1.25 \times 10^{-7}$ m, $N_d = 1.74 \times 10^{23}$ m^{-3}, $\mu_n = 0.35$ m^2/V s.   (a) $V_{\text{Sbi}} = 0.47$ V, $v_s = 1.21 \times 10^5$ m/s, $R_c = 0.74\ \Omega$ mm.   (b) $V_{\text{Sbi}} = 0.46$ V, $v_s = 1.20 \times 10^5$ m/s, $R_c = 0.88\ \Omega$ mm. The top curves are for $L = 0.5\ \mu$m. Other curves (from the top) for $L = 1$, 2, and 3 μm.

(assuming $N_a = 1 \times 10^{15}$ cm^{-3} and $V_{\text{bs}} = 0$):

$$A^* = 0.125\ \mu\text{m}$$

$$N_d^* = 1.74 \times 10^{23}\ \text{m}^{-3}$$

$$V_{\text{po}} = 1.88\ \text{V}$$

The contact resistance of group 1, determined by transmission line model (TLM) measurement, is 37 Ω (or 0.74 Ω mm) and that of group 2 is 44 Ω (or 0.88 Ω mm). We assume $\mu_n = 0.35$ m^2/V s. Using these values, the effective surface built-in voltage can be found fitting R_{off} for a 3-μm device. In doing this, we used the values of N_d^* and A^* given above. Once the effective surface built-in voltage is determined, the electron saturation velocity in a 3-μm device can be deduced fitting the experimental value of the saturation current. (The experimental value of the saturation current is determined at the intersection between the tangent of the measured I_D–V_D characteristics at $V_D = 0$ and the tangent at $V_D = 3$ V; see the dotted line in Fig. 7-8-8a.) The values of V_{Sbi} and v_s deduced in this way from the measured curves for 3-μm saturated resistors are given in Table 7-8-1. As can be seen in Table 7-8-1, these values seem to be quite reasonable.

The saturation currents measured in short devices (see Fig. 7-8-8) are considerably larger than the saturation currents of the 3-μm devices. This may be attributed to the enhancement of the electron saturation velocity due to the ballistic and/or overshoot effects (see Chapter 2 and, for example, Ref. 77) and to the transverse drift of the implanted ions from the n^+ ohmic contact regions to the channel, which results in the increase in the effective doping density for short devices. This is consistent with the smaller values of R_{off} measured for 0.5- and 1-μm devices; see Eqs. (7-8-37) and (7-8-38). Assuming that R_c, V_{Sbi}, and μ_n are the same as for the

TABLE 7-8-1. Parameters of GaAs Saturated Loads[a]

	Group 1				Group 2			
L (μm)	0.5	1	2	3	0.5	1	2	3
R_{off} (Ω)	93.4	145	238	320	105	163	250	331
I_{Dsat} (mA)	11.5	5.43	3.74	3.64	9.0	5.22	3.69	3.56
v_s (10^5 m/s)	2.04	1.64	1.25	1.21	1.35	1.73	1.25	1.20
N_d (10^{23} m^{-3})	2.98	1.92	1.74	1.74	3.29	1.84	1.74	1.74

[a] Reference 74. For the first group, $R_c = 37\ \Omega$ and $V_{Sbi} = 0.47$ V. For the second group, $R_c = 44\ \Omega$ and $V_{Sbi} = 0.46$ V.

long (3-μm) devices and using Eqs. (7-8-37) and (7-8-38), we deduced the effective doping density N_d^* and the electron saturation velocity v_s from the measured characteristics. Figure 7-8-9 shows the increase of the effective doping density for the short devices determined from Eqs. (7-8-37) and (7-8-38) (see also Table 7-8-1). Note that the effective doping density for 0.5 μm is greatly increased, implying that the doping is quite nonuniform along the channel due to the dopant diffusion from the ohmic contact n^+ regions. This is illustrated by a qualitative sketch of the doping profile (see Fig. 7-8-10). In this figure we assumed an exponential decay of the excess dopant concentration with the distance from the n^+ regions and fitted the characteristic decay length to obtain the value of doping density in the center of a 0.5-μm device of group 1 as deduced from our measurements. As can be seen from the figure, the highly nonuniform distribution of dopants in a 0.5-μm device may preclude the application of our model (which is based on the assumption of a uniform doping distribution along the channel) to this device.

An additional increase in the saturation current may come from the enhancement of the effective electron saturation velocity in short devices [77]. As can be seen in Fig. 7-8-11, the effective electron saturation velocity is enhanced from 1.20×10^5 m/s up to 1.73×10^5 m/s as the device length decreases from 3 to 1 μm. The dotted line in the figure is an approximate estimation of the effective electron saturation velocity using the equation given in Ref. 78, which shows the same trend of the enhanced saturation velocity for short devices as this measurement. For 0.5-μm devices, the deduced values have a large variation, indicating that the applicability of this model may be questionable for such a nonuniform doping as mentioned above. We should also notice that the nonuniform doping reduces the effective channel length (see Fig. 7-8-10) by approximately 0.3–0.4 μm.

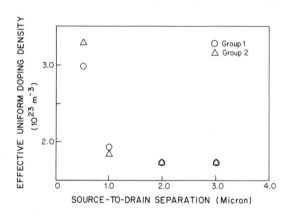

FIGURE 7-8-9. Effective uniform doping density vs. source-to-drain separation [74].

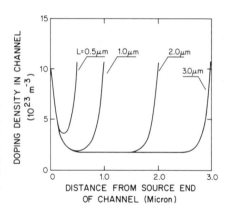

FIGURE 7-8-10. Qualitative channel doping profiles for 3-, 2-, 1-, and 0.5-μm self-aligned ion-implanted loads [74]. We assumed an exponential decay of the doping density with distance from the n^+ contacts. The characteristic decay length was taken to be 0.11 μm. The doping density of the n^+ regions is 1.06×10^{24} m^{-3} and the channel doping density far from the n^+ contacts is 1.74×10^{23} m^{-3}.

This model for the current–voltage characteristics of GaAs saturated resistor loads (or ungated FETs) is in good agreement with our experimental data for 1-, 2-, and 3-μm GaAs ungated FETs. Using this model we determined the value of the electron saturation velocity, v_s, and of the surface built-in voltage from the measured current–voltage characteristics of the long (3-μm) ungated loads. For these devices we obtain $v_s \approx 1.20$–1.21×10^5 m/s and $V_{Sbi} \approx 0.46$–0.47 V, in good agreement with expected values. For 0.5-, 1-, and 2-μm devices we deduced the values of v_s and the effective doping density from the measured current–voltage characteristics. For short (1-μm) devices, the measured values of v_s are considerably higher than for the 3-μm devices (1.64–1.73×10^5 m/s). This may be considered as evidence of velocity enhancement in short structures due to ballistic or overshoot effects.

7-9. DEVICE CHARACTERIZATION

7-9-1. Introduction

Device models considered in previous sections use transistor parameters such as the active layer thickness, the channel doping (or the ion-implantation profile), the built-in voltage, the surface potential, the gate length and width, the source drain and gate series resistance, the electron low field mobility, and the electron saturation velocity. Some of these parameters (such as doping and active layer

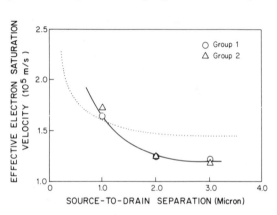

FIGURE 7-8-11. Effective electron saturation velocity vs. source-to-drain separation [74].

thickness, the gate length, and width) may be known from the fabrication procedure; others may be assumed to have some reasonable values (such as the surface potential of 0.5–0.6 V for a free GaAs surface). However, some parameters (such as the source, drain, and gate resistances) may vary from device to device.

One of the most important advantages of simple device models is that they may be used for parameter determination from the experimental data.

In this section we consider parameter acquisition techniques which allow us to determine device parameters from measured device characteristics or from measurements utilizing the special test patterns fabricated on the chip. The device parameters may then be correlated with the fabrication process and/or used for circuit simulation and design. These characterization techniques are based on the device models considered in previous sections of this chapter.

7-9-2. Determination of the Mobility Profile

Different techniques have been proposed to determine the low field mobility profile in active layers of GaAs FETs [79–87]. The simplest technique is based on the Van der Pauw measurements [81] using a Schottky-gate cloverleaf pattern [82]. The low field mobility profile may also be deduced from the C-V and I-V characteristics of fat FETs (i.e., FETs with very long gates) [84] and from magneto-resistance measurements [79]. Here, we consider a technique which is based on measurements of the device transconductance and series resistances at very low drain-to-source voltages when the Shockley model is applicable. It is a direct method which allows one to determine the mobility profile *in situ* for short gate devices [87].

We consider a case when the drain-to-source voltage $V_{ds} \ll V_{bi} - V_g$, where V_{bi} is the built-in voltage and V_g is the voltage drop between the gate and its source side of the channel. Therefore the depletion region boundary is nearly parallel to the boundary between the active channel and the substrate. In effect, under such conditions a short-channel device behaves like a fat FET.

The equivalent circuit of the device between the source and drain may then be presented as a series connection of R_s, R_{ch}, and R_d. Here R_s and R_d are the source and drain series resistances, respectively. The reciprocal of the channel resistance R_{ch} for the channel with nonuniform mobility and doping profiles can be expressed as

$$R_{ch}^{-1} = \frac{qW}{L} \int_X^A \mu(X') N(X') \, dX' \qquad (7\text{-}9\text{-}1)$$

where q is the electronic charge, A is the thickness of the channel, X is the depletion layer thickness, and L is the gate length. Again, (7-9-1) applies only at low V_{ds}.

For a given drain-to-source voltage $V_{ds} = I_d(R_s + R_{ch} + R_d)$ and the device transconductance g_m is given by

$$g_m = \frac{\partial I_d}{\partial V_g} = -\frac{V_{ds}}{R_T^2} \frac{\partial R_{ch}}{\partial V_g} \qquad (7\text{-}9\text{-}2)$$

where

$$R_T = R_s + R_{ch} + R_d$$

and

$$\frac{\partial R_{ch}}{\partial V_g} = R_{ch}^2 \left(\frac{qW\mu N}{L} \right) \frac{\partial X}{\partial V_g} \qquad (7\text{-}9\text{-}3)$$

Here we may also neglect the voltage drop across R_s compared to V_g so that $V_g \simeq V_{gs}$ (the voltage drop $I_d R_s$ may, however, be comparable to V_{ds}).

The relationship between V_g and the depletion layer thickness is given by

$$V_g = V_{bi} - \frac{q}{\varepsilon} \int_0^X X' N(X') \, dX' \qquad (7\text{-}9\text{-}4)$$

where the built-in voltage $V_{bi} \simeq 0.75$ V for GaAs devices. From (7-9-4) we find

$$\frac{\partial X}{\partial V_g} = \left(\frac{\partial V_g}{\partial X}\right)^{-1} = \left(-\frac{q}{\varepsilon} N X\right)^{-1} = -\frac{\varepsilon}{qNX} \qquad (7\text{-}9\text{-}5)$$

Substituting (7-9-5) into (7-9-3), we obtain

$$\frac{\partial R_{ch}}{\partial V_g} = -R_{ch}^2 \cdot \frac{W\mu\varepsilon}{LX} \qquad (7\text{-}9\text{-}6)$$

Combining (7-9-2) and (7-9-6), we get

$$g_m = \frac{\partial I_d}{\partial V_g} = \frac{\varepsilon W V_{ds}}{L} \frac{\mu}{X} \left(\frac{R_{ch}}{R_T}\right)^2$$

$$= \alpha \frac{\mu}{X} \left(\frac{R_{ch}}{R_T}\right)^2 \qquad (7\text{-}9\text{-}7)$$

where

$$\alpha = \frac{\varepsilon W V_{ds}}{L}$$

In the case when $R_s \ll R_{ch}$ and $R_d \ll R_{ch}$ (which is typically not the case for $1 \ \mu m$ GaAs MESFETs with non-self-aligned gates) (7-9-7) coincides with the equation for a fat FET derived by Pucel [85].

The technique described in Ref. 87 involves measuring g_m (V_g) and $R_T(V_g)$ $(R_T = R_s + R_d + R_{ch})$, determining R_s and R_d using the Hower–Bechtel method [90] (which was modified for the nonuniform doping profile), and solving (7-9-7) numerically together with (7-9-4) for the given doping profile. A new technique, of the determination of the series resistances described in Section 7-9-3, may also be used to determine R_s and R_d.

In order to determine the series resistance using the Hower and Bechtel method, we plot the channel resistance R_T versus $(1 - \sqrt{\eta})^{-1}$, where $\eta = (V_{bi} - V_g)/(V_{bi} - V_T)$ and crudely define the sum of the series resistances, from the intercept

$$R_T = R_s^{(0)} + R_d^{(0)} + \frac{1}{G_0(1 - \sqrt{\eta})} \qquad (7\text{-}9\text{-}8)$$

The threshold voltage V_T is adjusted slightly to yield a linear plot.

The next step involves the calculation of the approximate mobility profile as described above. After that, we plot the channel resistance versus $1/\sigma$, where

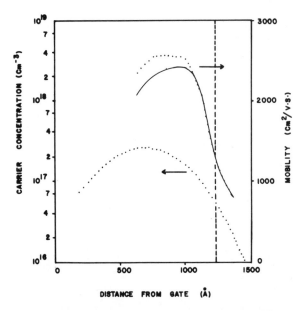

FIGURE 7-9-1. Mobility and channel electron density vs. distance from the gate for an ion-implanted 1-μm gate length MESFET. The dotted line is the zero approximation mobility profile and the solid line is the final mobility profile. In the region to the right of the vertical dashed line our technique does not apply because the Debye length becomes comparable with the characteristic length of the carrier profile variation [87].

$\sigma = L/(WR_{ch})$. The intercept of this dependence yields $R_s + R_d$ and the slope L/W

$$R_T = R_s + R_d + \frac{L}{\sigma W} \qquad (7\text{-}9\text{-}9)$$

Finally, we solve Eqs. (7-9-4) and (7-9-7) again to determine the mobility profile using the new value of $R_s + R_d$ and check for convergence. Typically one iteration is all that is required for the accurate determination of the mobility profile.

The GaAs MESFETs used in Ref. 87 were fabricated using multiple selective implantation into undoped LEC semi-insulating GaAs. Selenium ions were implanted in the active channel and sulfur implantation was used for the n^+ regions under the ohmic contacts. Ohmic contacts were made using AuGe/Ni metals and the Schottky barrier/first-level metal was TiW/Au patterned using a dielectric-assisted liftoff technique. The interlevel dielectric was plasma-enhanced CVD silicon

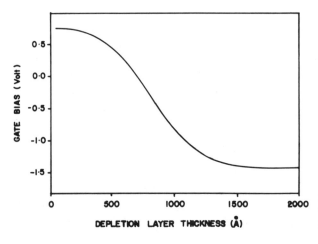

FIGURE 7-9-2. The gate bias vs. the depletion layer thickness [calculated from Eq. (7-9-4)] [87].

oxynitride, and the second-level metal was TiW/Au patterned using ion-beam milling. The threshold voltage was close to -1 V. Doping profiles of the ion-implanted semiconductor can be fairly well approximated by Gaussian distribution [89]. The gate voltage V_g versus X given by Eq. (7-9-4) calculated for the doping profile shown in Fig. 7-9-1 is depicted in Fig. 7-9-2. This curve has been tabulated and used for the numerical solution of (7-9-7). The measured transconductance g_m versus V_g for $V_{ds} = 20$ mV and $V_{ds} = 40$ mV is shown in Fig. 7-9-3. As can be shown from Eq. (7-9-7), the value of $(1/g_m)(\partial g_m/\partial V_g)$ is independent of V_{ds} for low values of V_{ds}. Indeed, the difference in this product measured at 20 and 40 mV was less than 3%.

The channel current I_d measured at low V_{ds} is shown in Fig. 7-9-3b as a function of the gate bias. From this measurement the total resistance was obtained as $R_T = V_{ds}/I_d$. The source and drain resistances R_s and R_d are determined as suggested in Refs. 88 or 15 (see also Ref. 90). Plotting R_T as a function of $(1 - \sqrt{\eta})^{-1}$ [see Eq. (7-9-8)] and slightly adjusting V_T we obtain a linear plot for the data points. This yields both the value of $R_s + R_d$ as the intercept and a more accurate value for V_T than the value obtained from the measurements of the FET characteristics at large V_{ds} ($V_T = -0.995$ V).

Then we solve Eq. (7-9-7) together with (7-9-4). The resulting mobility profile is shown in Fig. 7-9-1 as a dotted line. Plotting R_T versus $1/\sigma$ [see Eq. (7-9-9)] yields a more accurate value of $R_s + R_d = 110 \, \Omega$ (as compared with an approximate value of $120 \, \Omega$) (see Fig. 7-9-4) and the value of $L/W = 0.066$ (the slope), in good agreement with the directly measured value $L/W = 0.065$ (this value was determined optically).

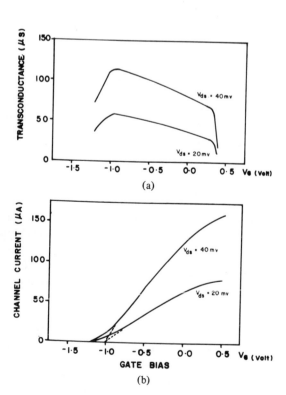

FIGURE 7-9-3. The transconductance (a) and the channel current (b) as functions of gate bias for low V_{ds} for 20-μm-wide devices [87].

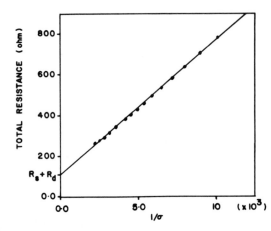

FIGURE 7-9-4. Determination of R_s and R_d. The total resistance is plotted as function of $1/\sigma$, where $\sigma = \int_x^\infty \mu(X')N(X')\,dX'$. This yields $R_s + R_d$ (110 Ω) as the intercept and L/W (0.066) as the slope [87].

Finally, the mobility profile is recalculated using the accurate value of $R_s + R_d$. The final profile is also shown in Fig. 7-9-1 as a solid line. As can be seen from the figure the first iteration is not too far off from the zeroth approximation mobility curve. These results indicate that the mobility may be quite low, especially far from the gate. It is considerably smaller than the measured Hall mobility for the same wafers, which is of the order of 3000–4000 cm^2/V s. A similar apparent decrease of the low field mobility in short devices has been noticed by Fukui [88]. The reasons for this decrease may be related to the device processing. The nonuniform mobility has to be taken into account in modeling and characterization of ion-implanted GaAs devices, especially low pinch-off voltage devices where the low field mobility plays an important role in determining the device transconductance. (See Sections 7-5 and 7-6.)

We should point out that the theory used in this section may become inaccurate when the electron concentration drops below 5×10^{16} cm^{-3} or so (i.e., at a distance farther from the gate than approximately 1200 Å), because in this region the Debye radius (190 Å at 300 K) becomes comparable with the characteristic length of the carrier profile variation. A more elaborate analysis should be used in this region, as suggested in Ref. 91.

7-9-3. Source, Drain, and Gate Series Resistances [92, 93]

The source and drain resistances R_s and R_d are key parameters which determine the performance of field effect transistors. The source series resistance strongly affects the device transconductance and noise figure. Both series resistances increase the power consumption and slow down the device operation. An accurate measurement of the series resistances is crucial for the reliable determination of the effective electron saturation velocity in the channel of a short gate field effect transistor. The values of the series resistances are also required for device and circuit modeling.

Series resistances may vary from device to device because of changes in the contact resistance, which is difficult to calculate theoretically. This makes an experimental determination of series resistances even more important.

Hower and Bechtel [15] obtained the sum of the source (R_s) and drain (R_d) resistances by measuring the small signal drain-to-source resistance r_{ds} at zero drain-to-source voltage as a function of the gate voltage [see Eq. (7-9-8)].

However, this method may not be accurate enough for ion-implanted GaAs FETs because it does not take into account the nonuniform doping and mobility profiles. In Section 7-9-2 this method was extended to include the nonuniform doping and mobility profiles, based on the iterative solution using the depletion approximation [87]. However, this approach only allows us to determine the sum of R_s and R_d, not their values separately.

Fukui [88] suggested determining the difference between R_d and R_s from the measurements of the gate-to-source and gate-to-drain current–voltage characteristics. This approach was combined with the iterative technique to determine R_s and R_d [87]. Recently, it was proposed to use the "end" resistance measurements [94, 95] in order to determine R_s and R_d [93, 96–98] and the electron saturation velocity v_s [98] (see Section 7-9-4).

The basic idea of the "end" resistance measurement technique is illustrated by Fig. 7-9-5. In this scheme the flowing gate current creates a voltage drop across the series resistance R_s and the drain contact is floating so that the drain section of the device acts as a "probe." Hence, the series source resistance was estimated as

$$R_s \approx \frac{V_D}{I_g} \tag{7-9-10}$$

where V_D is the floating drain potential. However, the potential V_D also includes a contribution from the voltage drop across a part of the channel

$$V_D = I_g(R_s + \alpha R_{ch}) \tag{7-9-11}$$

where R_{ch} is the channel resistance and α is some constant. As shown below, Eq. (7-9-10) may lead to a substantial error because the αR_{ch} term may be comparable to or even bigger than R_s.

The distribution of the gate current is described by the following equation:

$$\frac{dI(x)}{dx} = -J_g(x)W \tag{7-9-12}$$

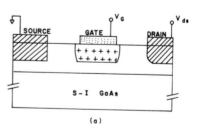

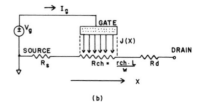

FIGURE 7-9-5. (a) Schematic diagram of a MESFET [93]. (b) "End" resistance measurement technique. R_s, R_{ch}, R_d represent source, channel, and drain series resistances, respectively. $J_g(x)$ is the gate current density. r_{ch} represents a distributed channel resistance.

Here x is the coordinate along the channel, $I(x)$ is the channel current, $J_g(x)$ is the gate current density, and W is the gate width (see Fig. 7-9-5). The potential variation $V(x)$ along the channel is described by the following equation:

$$\frac{dV}{dx} = I(x)\frac{r_{ch}}{W} \tag{7-9-13}$$

where r_{ch} is the channel resistance per square.

Below we assume that r_{ch} is independent of x. For this to be true the depletion region under the gate should have nearly a uniform cross-section, i.e., we should have

$$\frac{\left(\dfrac{V_{bi} - V'_g + V'_s}{V_{po}}\right)^{1/2} - \left(\dfrac{V_{bi} - V'_g + V'_d}{V_{po}}\right)^{1/2}}{1 - \left(\dfrac{V_{bi} - V'_g + V'_s}{V_{po}}\right)^{1/2}} \ll 1 \tag{7-9-14}$$

where V_{bi} is the built-in voltage,

$$V'_g = V_g - I_g R_g \tag{7-9-15}$$

is the gate potential, V_g is the gate voltage, R_g is the gate series resistance, V'_s is the potential at the source side of the channel under the gate, V'_d is the potential of the drain side of the channel under the gate, and V_{po} is the pinch-off voltage. In a typical experimental situation

$$V_d - V_s \ll V_{bi} - V_g \tag{7-9-16}$$

and Eq. (7-9-14) is fulfilled.

The gate current density $J_g(x)$ is given by

$$J_g(x) = J_s \exp\left(\frac{V'_g - V}{nV_T}\right) \tag{7-9-17}$$

where

$$J_s = A^* T^2 \exp\left(-\frac{qV_{bi}}{k_B T}\right) \tag{7-9-18}$$

is the gate saturation current density, A^* is the effective Richardson constant, T is temperature, n is the ideality factor, and $V_T = k_B T/q$ is the thermal voltage. As the gate current density is the exponential function of V whereas r_{ch} is a relatively weak function of V [see Eq. (7-9-14) and the related discussion], we can take into account the nonuniform distribution of the gate current under the gate assuming at the same time that r_{ch} is nearly independent of x. In Eq. (7-9-17) we neglected the reverse gate current, and hence it is valid only for $V'_g - V \gg nV_T$.

The boundary conditions are

$$\frac{dV}{dx} = -I_g\frac{r_{ch}}{W} \qquad \text{at } x = 0 \tag{7-9-19}$$

and

$$\frac{dV}{dX} = 0 \quad \text{at } x = L \tag{7-9-20}$$

(see Fig. 7-9-5).

Differentiating Eq. (7-9-13) with respect to x we find

$$\frac{d^2 V}{dx^2} = \frac{dI}{dx}\frac{r_{ch}}{w} \tag{7-9-21}$$

or using Eq. (7-9-12),

$$\frac{d^2 V}{dx^2} = -J_s r_{ch} \exp\left(\frac{V'_g - V}{n V_T}\right) \tag{7-9-22}$$

The integration of Eq. (7-9-22) with boundary conditions (7-9-19) and (7-9-20) yields

$$I_g^2 = \frac{2 J_s W^2 n V_T}{r_{ch}} \exp\left(\frac{V'_{ds}}{n V_T}\right)\left[1 - \exp\left(-\frac{V}{n V_T}\right)\right] \tag{7-9-23}$$

and

$$L = \left(\frac{2 n V_T}{J_s r_{ch}}\right)^{1/2} \exp\left[\frac{1}{2 n V_T}(V'_{ds} - V'_{gs})\right] \tan^{-1}\left[\exp\left(\frac{V'_{ds}}{n V_T}\right) - 1\right]^{1/2} \tag{7-9-24}$$

Equation (7-9-24), which may be rewritten as

$$l_0 = 2^{1/2} \exp\left(\frac{v_d}{2} - \frac{y}{2}\right) \tan^{-1}[\exp(v_d) - 1]^{1/2} \tag{7-9-25}$$

has to be solved together with Eq. (7-9-23). The resulting equation may be written as

$$i = 2 \exp\left(\frac{v_d}{2}\right)[1 - \exp(-v_d)]^{1/2} \tan^{-1}[\exp(v_d) - 1]^{1/2} \tag{7-9-26}$$

Here

$$l_0 = L\left(\frac{n V_T}{J_s r_{ch}}\right)^{-1/2}, \qquad i = I_g \frac{r_{ch}}{n V_T}, \qquad v_d = \frac{V'_{ds}}{n V_T}, \qquad \text{and} \quad y = V'_{gs}/n V_T$$

At small values of $v_d \ll 1$ Eq. (7-9-26) may be simplified:

$$i \simeq 2 v_d\left(1 + \frac{v_d}{6}\right) \tag{7-9-27}$$

or

$$v_d \simeq \frac{i}{2}\left(1 - \frac{i}{12}\right) \tag{7-9-28}$$

From Eq. (7-8-28) we find that for small values of $V'_{ds} \ll nV_T$

$$R_{end} \equiv \frac{V_{ds}}{I_g} = \frac{R_{ch}}{2} + R_s \qquad (7\text{-}9\text{-}29)$$

Furthermore

$$R'_{end} \equiv \frac{dV_{ds}}{dI_g} = \frac{R_{ch}}{2} + R_s = R_{end} \qquad (7\text{-}9\text{-}30)$$

Equation (7-9-25) for $v_d \to 0$ may be written as

$$y = v_d + \ln\left[v_d\left(1 - \frac{v_d}{6}\right)\right] - 2\ln l_0 + \ln 2 \qquad (7\text{-}9\text{-}31)$$

From Eqs. (7-9-27) and (7-9-31) we find [97, 93]

$$\frac{dV'_{gs}}{dI_g} \simeq \frac{R_{ch}}{3} + \frac{nV_T}{I_g} \qquad (7\text{-}9\text{-}32)$$

Equation (7-9-32) may be rewritten as

$$R^g_{end} \equiv \frac{dV_{gs}}{dI_g} \simeq \frac{R_{ch}}{3} + R_s + R_g + \frac{nV_T}{I_g} \qquad (7\text{-}9\text{-}33)$$

In a general case of an arbitrary gate current, Eqs. (7-9-29), (7-9-30), and (7-9-32) may be rewritten as

$$R_{end} = \alpha(i)R_{ch} + R_s \qquad (7\text{-}9\text{-}34)$$

$$R'_{end} = \alpha'(i)R_{ch} + R_s \qquad (7\text{-}9\text{-}35)$$

$$R^g_{end} = \alpha^g(i)R_{ch} + R_s + R_g + \frac{nV_T}{I_g} \qquad (7\text{-}9\text{-}36)$$

Where $\alpha(i)$, $\alpha'(i)$, and $\alpha^g(i)$ are universal functions of $i = I_gR_{ch}/nV_T$ [97, 93]. For large $v_d \gg 1 (V'_{ds} \gg nV_T)$ we find

$$\alpha(i) \simeq \frac{2}{i}\ln\frac{(i+2)}{\pi} \qquad (7\text{-}9\text{-}37)$$

$$\alpha'(i) \simeq \frac{2}{(i+2)} \qquad (7\text{-}9\text{-}38)$$

and

$$\alpha^g(i) \simeq \frac{1.0}{i} \qquad (7\text{-}9\text{-}39)$$

For small $v_d \ll 1 (V'_{ds} \ll nV_T)$

$$\alpha(i) = \frac{1}{2} - \frac{i}{24} \qquad (7\text{-}9\text{-}40)$$

$$\alpha'(i) = \frac{1}{2} - \frac{i}{12} \qquad (7\text{-}9\text{-}41)$$

$$\alpha^g(i) = \frac{1}{3} - \frac{11}{144}i \qquad (7\text{-}9\text{-}42)$$

Equations (7-9-37)–(7-9-39) apply for $i \geqslant 15$; Eqs. (7-9-40)–(7-9-42) apply for $i \leqslant 0.5$. In the intermediate range of currents $0.5 < i < 15$ the following expressions interpolate α, α', and α^g:

$$\alpha(i) = \frac{1}{2 + 0.166i - 3.12 \times 10^{-4}i^2} \qquad (7\text{-}9\text{-}43)$$

$$\alpha'(i) = \frac{1}{2 + 0.3478i + 7.72 \times 10^{-3}i^2} \qquad (7\text{-}9\text{-}44)$$

$$\alpha^g(i) = \frac{1}{3 + 0.636i + 1.75 \times 10^{-2}i^2} \qquad (7\text{-}9\text{-}45)$$

From Eq. (7-9-25) we find

$$V_{gs} \simeq V_{ds} - \frac{R_{ch}}{6}I_g + nV_T \ln\left(\frac{I_g}{I_s}\right) + R_g I_g \qquad (7\text{-}9\text{-}46)$$

and

$$V_{gs} \simeq V_{ds} + nV_T \ln\left(\frac{I_0 \pi^2}{2I_s}\right) + R_g I_g \qquad (7\text{-}9\text{-}47)$$

for the small and large gate currents, respectively. Here $I_s = J_s WL$ and $I_0 = nV_T/R_{ch}$. If the drain contact is grounded and the source contact floating then

$$R_{end} = \alpha(i)R_{ch} + R_d \qquad (7\text{-}9\text{-}48)$$

$$R'_{end} = \alpha'(i)R_{ch} + R_d \qquad (7\text{-}9\text{-}49)$$

$$R^g_{end} = \alpha^g(i)R_{ch} + R_d + R_g + \frac{nV_T}{I_g} \qquad (7\text{-}9\text{-}50)$$

$$V_{gd} = V_{sd} - \frac{R_{ch}}{6}I_g + nV_T \ln\left(\frac{I_g}{I_s}\right) + R_g I_g \qquad (7\text{-}9\text{-}51)$$

and

$$V_{gd} = V_{sd} + nV_T \ln\left(\frac{I_0 \pi^2}{2I_s}\right) + R_g I_g \qquad (7\text{-}9\text{-}52)$$

The computed and analytical dependences of α, α', and α^g on i agree with better than 1% accuracy (see Fig. 7-9-6) [93]. Equations (7-9-46) and (7-9-47) confirm the validity of the Fukui method because

$$\left(\frac{V_{gs}}{I_g}\right) - \left(\frac{V_{gd}}{I_g}\right) = \frac{V_{ds}}{I_g} - \frac{V_{sd}}{I_g} = R_s - R_d \qquad (7\text{-}9\text{-}53)$$

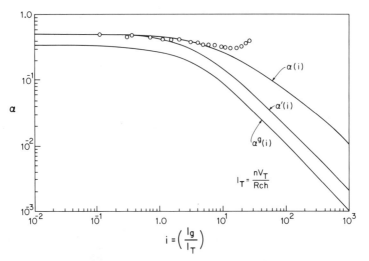

FIGURE 7-9-6. α, α' and α_g as functions of normalized gate current [93]. The dots represent the measured values of $\alpha(i)$ for our $1 \times 20\text{-}\mu\text{m}^2$ transistor.

As can be seen from Eqs. (7-9-34) and (7-9-48) the "end" resistance measurements may be also used to determine $R_s - R_d$:

$$R_{end}^s - R_{end}^d = R_s - R_d$$

Equations (7-9-34) and (7-9-48) offer a new interpretation of the "end" resistance measurements, which is quite different from the conventional approach [see Eq. (7-9-10)]. Below the equations will be used to deduce the series resistance of ion-implanted GaAs FETs and the results will be compared with the modified Fukui method. We should notice that the value of α in Eq. (7-9-11) does not have to be known in order to determine R_s and R_d.

We may also propose a simple measurement to determine the value of $R_s + R_d + R_{ch}$ independently from the "end" resistance measurements [92]. The idea is to bias the drain contact slightly so that the drain current I_d is now flowing, but $I_d \ll I_g$. In this case the drain current does not appreciably change the potential distribution in the channel and the drain-to-source voltage may be found from the superposition principle

$$V_{ds} = (R_s + R_d + R_{ch})I_d + (R_s + R_{ch}/2)I_g \qquad (7\text{-}9\text{-}55)$$

Hence, we may determine $R_s + R_d + R_{ch}$ from the intercept of the V_{ds} vs. I_g curve with the small drain current or, alternatively, as a slope of the V_{ds} vs. I_d curve with the gate current $I_g \gg I_d$. Also we find using Eq. (7-9-46)

$$V_{gs} = nV_t \ln\left(\frac{I_g}{I_s}\right) + \left(R_s + \frac{R_{ch}}{3} + R_g\right)I_g + \left(R_s + \frac{R_{ch}}{3}\right)I_d \qquad (7\text{-}9\text{-}56)$$

A similar approach to the determination of R_s and R_d was discussed in Ref. 96 where Eq. (7-9-55) was derived.

As an example, we consider the results of the measurements reported in Refs. 92, 93.

The fabrication procedure for these devices is described in Section 7-8-2.

The plot of the total resistance as a function of $1/\sigma$ for one of the devices with $W = 20~\mu$m is shown in Fig. 7-9-4. From this measurement we obtained $R_s + R_d = 108~\Omega$, and the slope $L/W = 0.066$, which is in good agreement with the L/W ratio of 0.065 for this device determined by direct optical measurement. The measured end resistances from the V_{ds} vs. I_g curves with the source grounded and the drain floating, and with the drain grounded and the source floating, respectively, are

$$R_s + \frac{R_{ch}}{2} = 102.6~\Omega$$

$$R_d + \frac{R_{ch}}{2} = 126.3~\Omega$$

(7-9-57)

(see Fig. 7-9-7a). Hence, $R_d - R_s = 23.7~\Omega$ and $R_d = 65.8~\Omega$, $R_s = 42.2~\Omega$, $R_{ch} = 121~\Omega$. The sheet resistance for these devices is estimated to be around $1420~\Omega$ per square, leading to a full channel resistance of $92~\Omega$, which is less than the measured resistance. The channel resistance is increased by the partial depletion under the gate.

The plots of dV_{gs}/dI_g vs. $1/I_g$ yield

$$R_s + \frac{R_{ch}}{3} + R_g = 107~\Omega$$

(7-9-58)

$$R_d + \frac{R_{ch}}{3} + R_g = 129~\Omega$$

(7-9-59)

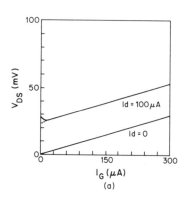

(a)

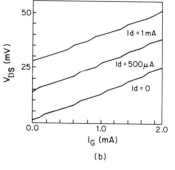

(b)

FIGURE 7-9-7. V_{ds} as a function of the gate current for (a) low pinch-off voltage and (b) high pinch-off voltage [92, 93]. (a) The intercept of curve for 100-μA drain current yield $R_s + R_d + R_{ch} = V_{ds}/I_d = 241~\Omega$ and the slope of curve for zero drain current yield $R_s + R_{ch}/2 = 102.6~\Omega$ [see Eqs. (7-9-29) and (7-9-55)]. (b) The intercept of curve for 500 μA drain current yields $R_s + R_d + R_{ch} = 26~\Omega$. The analysis of the data using the technique described in this section yields $R_s = 10.9~\Omega$, $R_d = 10.1~\Omega$, $R_{ch} = 5.12~\Omega$ and $R_g = 6.8~\Omega$.

TABLE 7-9-1. Values of the Gate Resistance and Device Parameters[a]

Gate current (μA)	$\alpha^g(i)$	$\alpha(i)$	Calculated	R_g (ohm) Eq. (7-9-33)	Eq. (7-9-47)
40	0.3225	0.4939	15.43	22.73	—
50	0.3198	0.4924	15.45	27.50	—
80	0.3117	0.4878	15.50	24.90	—
100	0.3063	0.4847	15.53	24.94	—
—	—	—	—	—	—
1000	—	0.372	17.16	—	16.2
3000	—	0.316	18.40	—	20.4
5000	—	0.372	17.16	—	18.2
6000	—	0.452	15.92	—	16.2

Device parameters
  Gate metal = TiW/Si, thickness = 3000–3500 Å
  Metal sheet resistance = 0.508 ± 0.023 Ω/square
  Gate length = 1.3 μm, gate width = 20 μm
  Parasitic metal length = 80 μm, width = 4 μm
  Diode saturation current (I_s) = 20 pA
  Diode ideality factor (n) = 1.285

[a] Reference 93.

so that $R_d - R_s = 22\,\Omega$, in excellent agreement with the "end" resistance measurement rechnique.

Equations (7-9-33) and (7-9-47) may be used for the determination of the gate series resistance. Equation (7-9-33) is used for small gate currents, and Eq. (7-9-47) for large gate currents. In Table 7-9-1 we compare the values obtained using Eqs. (7-9-33) and (7-9-47) with those calculated from the measured gate metal sheet resistance, the measured value of α, and the measured parasitic metal line resistance between the gate and the bonding pad. (In these devices, the parasitic metal line resistance is much larger than the intrinsic gate metal resistance.)

As can be seen from Table 7-9-1, the gate resistances measured at high gate currents are in good agreement with those calculated using the gate geometry and measured metal sheet resistance. However, at low gate current, the data obtained using Eq. (7-9-33) overestimate the gate resistance. Also from the V_{ds} vs. I_d curve for $I_g = 500\,\mu A \gg I_d$, we find (see Fig. 7-9-8)

$$R_s + R_d + R_{ch} = 224\,\Omega \qquad (7\text{-}9\text{-}60)$$

in good agreement with Eq. (7-9-47) ($R_s + R_d + R_{ch} = 228.9\,\Omega$).

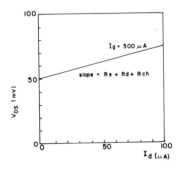

FIGURE 7-9-8. Voltage drop between source and drain, V_{ds}, vs. drain current I_d [92]. The gate current $I_g = 500\,\mu A$. The slope yields $R_s + R_d + R_{ch} = 224\,\Omega$.

Similar studies have been performed for higher pinch-off voltage FETs with wider gates ($W = 198$ μm and $V_{po} \simeq 4$ V) fabricated using a similar ion implantation technique [93]. Reasonable values of the source, drain, and gate resistances were obtained, even though the small values of the resistances made the measurements more difficult (see Fig. 7-9-9 [98]).

7-9-4. Electron Saturation Velocity [93, 98]

The electron saturation velocity can be deduced from end resistance measurement as proposed in Ref. 98. In Fig. 7-9-9 we compare the measured and computed dependences of α on the drain voltage $V'_{ds} = V_{ds} - I_g R_s$ [98].

The values of α first decrease with the increase in V'_{ds} as predicted by the theory (see Fig. 7-9-9). However, with a further increase in V_{ds} the values of α increase sharply. When the voltage drop V'_{ds} becomes comparable with $F_s L_{eff}$, where F_s is the velocity saturation field and $L_{eff} = L$ is the effective length of the section of the channel carrying the gate current, the electron velocity in this section of the channel saturates, leading to the increase of the end resistance R_{end} and, hence, to the increase in the measured value of α. The value of F_s can be estimated from $F_s = (V'_{ds})_{min}/(\alpha_{min} L)$, where α_{min} is the minimum value of α and $(V'_{ds})_{min}$ is the corresponding value of V'_{ds}. The same result is illustrated by Fig. 7-9-10, where the measured value of the electric field

$$F = V'_{ds}/(\alpha L) \tag{7-9-61}$$

is compared with the calculated results obtained using the calculated curve of Fig. 7-9-9. One can clearly see the change in the F vs. V'_{ds} curve when F approaches F_s.

For the samples characterized in Fig. 7-9-9 the low-field mobility μ was determined using the technique described in Section 7-9-2 ($\mu \approx 2500$ cm^2/V s). The saturation velocity was then estimated as $v_s = \mu F_s$. For the devices used in Ref. 98 v_s varied from 1×10^5 to 1.3×10^5 m/s.

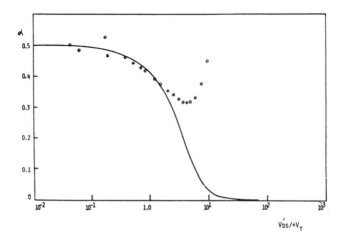

FIGURE 7-9-9. Plot of the calculated α as function of normalized drain-to-source voltage. The dots are the measured values from a 1.3-μm-long gate FET [98].

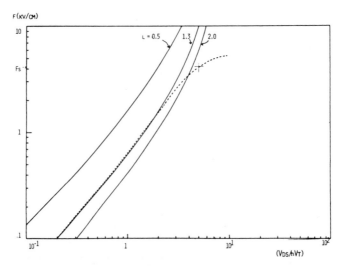

FIGURE 7-9-10. The calculated electric field for 0.5-, 1.3-, and 2.0-μm-gate-length MESFETs versus normalized drain-to-source voltage. The dotted line is the measured one from a 1.3-μm-gate-length MESFET [98].

If the gate current I_g becomes larger than $F_s L_{sg}/R_s$, where L_{sg} is the source-to-gate spacing, the electron velocity may saturate in the source-to-gate spacing, invalidating this technique [93]. This condition may be also rewritten as

$$I_g > I_g^{cr} = \frac{v_s L_{sg}}{\mu R_s} \qquad (7\text{-}9\text{-}62)$$

In the devices discussed above $v_s \approx 1.2 \times 10^5$ m/s, $L_{sg} \approx 1.3\ \mu$m, $\mu = 2500$ cm^2/V s, $R_s \approx 42.2\ \Omega$, and $I_g^{cr} \approx 14.8$ mA, well above the maximum gate current of 5–6 mA used in the measurement (see Fig. 7-9-6).

7-10. SPACE CHARGE LIMITED CURRENT

In this section we will consider a space charge injection into the channel of a field effect transistor which may play an important role in short channel devices.

A schematic diagram of a self-aligned gate GaAs FET [93, 94] is shown in Fig. 7-10-1. This device has a shorter effective gate length and smaller series source and drain resistances than a conventional FET.

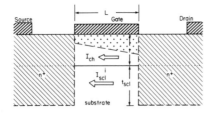

FIGURE 7-10-1. Schematic diagram of a self-aligned GaAs MESFET [113].

As can be seen from Fig. 7-10-1, deep n^+ implants used for the ohmic contacts form a n^+-i-n^+ structure under the active channel in the substrate.

According to our study of the short n^+-n^--n^+ and n^+-p^--n^+ structures [101] (see Chapter 2) the free electron concentration, n_{min}, injected into the center of the structure, due to the electron spill-over at the equilibrium (i.e., at zero drain-to-source bias), may be estimated as

$$n_{min} \sim \frac{2\pi^2 \varepsilon k_B T}{q^2 L^2} \tag{7-10-1}$$

where ε is the dielectric permittivity, k_B is the Boltzmann constant, T is temperature, and q is the electronic charge. For GaAs, $\varepsilon \approx 1.14 \times 10^{-10}$ F/m. Hence

$$n_{min} = \frac{3.6 \times 10^{14}}{L^2(\mu m)} \frac{T}{300 \text{ K}} (\text{cm}^{-3}) \tag{7-10-2}$$

For a self-aligned gate FET with a 0.5-μm gate length, n_{min} is close to 1.4×10^{15} cm^{-3}, which is sufficient to cause a noticeable parallel conduction even at very small drain-to-source voltages. As the device geometry is scaled down, this leakage current increases as $1/L^2$.

When the drain-to-source voltage $V_{DS} > k_B T$ the charge injected into the substrate increases with V_{DS}, leading to the space-charge limited current [102–109]. Below analytical formulas for the space charge limited current in a self-aligned FET are derived [113].

The one-dimensional Poisson equation for the section of the substrate between two parallel implanted n^+ layers (Fig. 7-10-1) is given by

$$dF/dx = -qn(x)/\varepsilon \tag{7-10-3}$$

Here F is the electric field, x is the space coordinate, and n is the free carrier density, injected into the substrate.

As shown in Refs. 102 and 103, the space charge limited current is primarily carried by drift (except in the region near the space charge maximum). Therefore, the space charge limited current density, J, is given by

$$J = qn(x)v(x) \tag{7-10-4}$$

where v is the electron drift velocity. Substituting $n(x)$ from Eq. (7-10-3) into Eq. (7-10-4) we find

$$J = -\varepsilon v(F)(dF/dx) \tag{7-10-5}$$

For the space charge current calculation we approximate the velocity versus electric field dependence as follows:

$$v(F) = v_s \frac{|F|/F_s}{[1 + (|F|/F_s)^2]^{1/2}} \tag{7-10-6}$$

In Fig. 7-10-2 this approximation is compared with the two-piece linear characteristic $v(F)$ and with the approximation for $v(F)$ proposed by Kroemer [110]. As can be seen from the figure, Eq. (7-10-6) provides a reasonable approximation for $v(F)$.

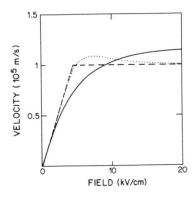

FIGURE 7-10-2. Approximation of $v(F)$ curve given by Eq. (7-10-6) (solid line) [113], $v_s = 1.2 \times 10^5$ m/s, $\mu = 0.4$ m^2/V s. A two-piece linearwise approximation (dashed line) and Kroemer approximation [22]:

$$v(F) = v_s \frac{F/F_s + (F/F_s)^4}{1 + (F/F_s)^4}$$

where $F_s = v_s/\mu$ (dotted line) are shown for comparison. For the dashed and dotted curve $v_s = 1 \times 10^5$ m/s, $\mu = 0.4$ m^2/V s.

At the same time it allows us to derive equations describing the space charge limited current in the entire range of drain-to-source voltages. It should also be noticed that in short devices the velocity of the electrons injected into the substrate is saturated for most relevant drain-to-source voltages. In the saturation region any approximation of the velocity curve which approaches the value of the saturation velocity with the increase in the electric field is adequate.

Substituting Eq. (7-10-6) into Eq. (7-10-5) we find

$$J dx = -\varepsilon v_s \frac{|F|/F_s}{[1 + (|F|/F_s)^2]^{1/2}} dF \tag{7-10-7}$$

Integrating Eq. (7-10-7) with respect to x and solving the resulting equation for the electric field strength $F(x)$, we obtain

$$F(x) = -F_s \left[\left(1 + \frac{Jx}{\varepsilon v_s F_s} \right)^2 - 1 \right]^{1/2} \tag{7-10-8}$$

The boundary condition used above is

$$F(0) = 0 \tag{7-10-9}$$

This boundary condition may be justified by a very large electron concentration in the n^+ implanted region. The drain-to-source voltage can be obtained by integrating $F(x)$ from $x = 0$ to $x = L$:

$$u = (2j)^{-1}((1+j)[(1+j)^2 - 1]^{1/2} - \ln\{(1+j) + [(1+j)^2 - 1]^{1/2}\}) \tag{7-10-10}$$

Here $u = V_{ds}/F_s L$ is the normalized drain-to-source voltage and $j = J/(\varepsilon v_s F_s/L)$ is the normalized current density. For large j ($j \gg 1$), the solution of this equation is given by

$$j \approx 2(u + 1) \tag{7-10-11}$$

which reduces to the equation for the constant velocity case when $u \gg 1$:

$$j \approx 2u \tag{7-10-12}$$

In the opposite limiting case $j \ll 1$, the solution of Eq. (7-10-10) reduces to the well-known Mott–Gurney law [111]:

$$j \approx (9/8)u^2 \qquad (7\text{-}10\text{-}13)$$

In the intermediate current range, Eq. (7-10-10) has to be solved numerically. However, we have found the following interpolation formula, which provides a very good approximation to the exact solution of Eq. (7-10-10):

$$j = \begin{cases} (2/3)u^2 - (2/27)u^3 & \text{for } u \leqslant 3 \\ 2(u-1) & \text{for } u > 3 \end{cases} \qquad (7\text{-}10\text{-}14)$$

(see Fig. 7-10-3).

In some cases the substrate may have a residual p-type (p^-) doping. This is, for example, the case when devices are fabricated by ion implantation into an MBE grown unintentionally doped buffer layer. (The saturated GaAs loads in modulation doped circuits described in [72] were fabricated that way.) We will examine the influence of the substrate impurity concentration N_a on the space-charge limited leakage current: in the most interesting limiting case of relatively high drain-to-source voltages the electrons in the substrate move with the saturation velocity, v_s. Replacing $n(x)$ in Eq. (7-10-3) by $n(x) + N_a(x)$ and $v(x)$ by v_s in Eq. (7-10-4) and repeating the derivation given above, we obtain

$$j = 2(u - u_{\text{ofs}}) \qquad (7\text{-}10\text{-}15)$$

where

$$u_{\text{ofs}} = V_{\text{ofs}}/F_s L \qquad (7\text{-}10\text{-}16)$$

is the dimensionless punch-through voltage. Comparing Eq. (7-10-12) with Eq. (7-10-15), we conclude that for a doped substrate, the J–V curve is simply displaced by voltage V_{ofs}, where

$$V_{\text{ofs}} = \frac{qN_aL^2}{2\varepsilon} \qquad (7\text{-}10\text{-}17)$$

is the punch-through voltage.

The space-charge limited leakage current, I_{scl}, may be found by multiplying j by the unit of the current, I_s, where

$$I_s = \frac{\varepsilon v_s F_s}{L}(Wt_{\text{scl}}) \qquad (7\text{-}10\text{-}18)$$

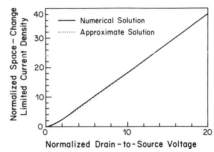

FIGURE 7-10-3. Numerical solution and approximate analytical interpolation for normalized space charge current density [113].

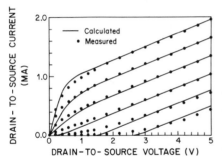

FIGURE 7-10-4. Current–voltage characteristics of the self-aligned 0.55-μm gate MESFET [113]. Experimental data from Refs. 99, 100. Top curve, $V_{gs} = 0.8$ V; step, -0.2 V; bottom curve, $V_{gs} = -0.2$ V. Parameters used in the calculation are given in Table 7-10-1.

Here t_{scl} is the depth of the implanted n^+ region beyond the channel–substrate junction (see Fig. 7-10-1).

When the gate voltage is smaller than the threshold voltage, the potential profile in the substrate is affected by the gate voltage, especially within the depletion region of the channel–substrate junction. In this case the two-dimensional solution of the Poisson equation is required. However, this situation is similar to a vacuum tube triode with the effective gate voltage $V_{gs} - V_T$, and the current–voltage characteristics may be found from Eq. (7-10-14), where u should be redefined as

$$u = [G_v(V_{gs} - V_T) + V_{ds} - V_{ofs}]/(F_s L) \qquad (7\text{-}10\text{-}19)$$

Here G_v is the amplification factor.

Equations (7-10-14)–(7-10-19) are used to describe the space charge limited current in the updated version of the DOMES program [47].

The I–V characteristics for a self-aligned gate MESFET, calculated taking into account the space-charge limited leakage currents, are compared with the experimental data [99, 100] in Fig. 7-10-4.

The current–voltage characteristics were calculated using an expression for the drain-to-source saturation current in submicron MESFETs proposed in Ref. 112:

$$I_{dsat} = \frac{I_{fc}(V_{gs} - V_T)}{V_{po} + R_s I_{fc}} \qquad (7\text{-}10\text{-}20)$$

TABLE 7-10-1. Parameters Used for the Calculated I–V Curves Shown in Fig. 7-10-4

Threshold voltage $V_T = 0.2$ V
Built-in voltage $V_{bi} = 0.74$ V
Saturation field $F_s = 4000$ kV/cm
Saturation velocity $v_s = 1 \times 10^5$ m/s
Source series resistance $R_s = 65\ \Omega$
Drain series resistance $R_d = 65\ \Omega$
Gate series resistance $R_g = 120\ \Omega$
Richardson constant $A^{**} = 120$ A/cm^2 K
Diode ideality factor $n = 1.18$
n^+ implant depth $t_{scl} = 0.24\ \mu$m
Amplification factor $G_v = 6$
Channel implantation does $Q = 1 \times 10^{12}$ cm^{-2}
Channel implantation range $R_p = 850$ Å
Standard deviation for the channel implant $\sigma = 440$ Å
n^+ implantation dose $Q = 2 \times 10^{13}$ cm^{-2}
n^+ implantation range $R_p = 1070$ Å
Standard deviation for the n^+ implant $\sigma = 525$ Å

Here V_{po} is the pinch-off voltage ($V_{po} = -V_T + V_{bi}$) and $I_{fc} = qWv_sQ/2$, where Q is the implantation dose. As can be seen from the figure, this model provides a very good fit of the measured current–voltage characteristics.

REFERENCES

1. K. Drangeid, R. Sommerhalder, and W. Walter, *Electron. Lett.* **6**, 228 (1970).
2. W. Shockley, *Bell Syst. Tech. J.* **30**, 990–1034 (1951).
3. W. Shockley, A unipolar field-effect transistor, *Proc. IRE* **40**, 1365 (1952).
4. G. C. Dacey and I. M. Ross, *Bell Syst. Tech. J.* **34**, 1149–1189 (1955).
5. F. N. Trofimenkoff, *Proc. IEEE* **53**, 1765–1766 (1965).
6. K. Tarney, *Proc. IEEE* **54**, 1077–1078 (1966).
7. R. Zuleeg, *Solid-State Electron.* **10**, 559–576 (1967).
8. H. R. Winteler and A. Steinemann, Proc. Int. Symp. Gallium Arsenide, 1st, 1966 Paper 30A, pp. 228–232, 1967.
9. J. R. Hauser, *Solid-State Electron.* **10**, 577–587 (1967).
10. T. Turner and B. Wilson, Implications of carrier velocity saturation in a gallium arsenide field-effect transistor, in Proc. 1968 Symp. GaAs (Inst. Phys., Conf. Series No. 7, London, 1969), pp. 195–204.
11. A. B. Grebene and S. K. Ghandhi, *Solid-State Electron.* **12**, 573–589 (1969).
12. K. Lehovec and R. Zuleeg, Voltage–current characteristics of GaAs J-FET's in the hot electron range, *Solid State Electron.* **13**, 1415–1426 (1970).
13. H. Himsworth, A two-dimensional analysis of gallium arsenide function field effect transistors with long and short channels, *Solid State Electron.* **15**, 1353–1361 (1972).
14. M. Reiser, A two-dimensional FET model for dc, ac, and large-signal analysis, *IEEE Trans. Electron Devices* **ED-20**, 35–45 (1973).
15. P. Hower and G. Bechtel, Current saturation and small-signal characteristics of GaAs field-effect transistors, *IEEE Trans. Electron Devices* **ED-20**, 213–220 (1973).
16. R. Pucel, H. Haus, and H. Statz, Signal and noise properties of gallium arsenide microwave field-effect transistors, in *Advances in Electronics and Electron Physics*, Vol. 38, Academic, New York, 1975, pp. 195–205.
17. J. J. Barnes, R. J. Lomax, and G. I. Haddad, Finite-element simulation of GaAs MESFET's with lateral doping profiles and submicron gates, *IEEE Trans. Electron Devices* **ED-23**, 1042–1048 (1976).
18. K. Yamaguchi, S. Asar, and H. Kodera, Two-dimensional analysis of stability criteria of GaAs FET's, *IEEE Trans. Electron. Devices* **ED-23**, 1283–1290 (1976).
19. K. Yamaguchi and H. Kodera, Drain conductance of junction gate FETs in the hot electron range, *IEEE Trans. Electron Devices* **ED-23**, 545–553 (1976).
20. H. L. Grubin, Large signal numerical simulation of field effect transistors, Abstracts of Technical Papers of Sixth Biennial Conf. on Active Microwave Semiconductor Devices and Circuits, Cornell University, Ithaca, New York, p. 45, 1977.
21. S. P. Yu and W. Tantraporn, Time domain two-dimensional computer simulation of three-terminal semiconductor devices, Abstract of Technical Papers of Sixth Biennial Conference on Active Microwave Semiconductor Devices and Circuits, Cornell University, Ithaca, New York, p. 44, 1977.
22. R. W. H. Engelmann and C. A. Liechti, Gunn domain formation in the saturated current region of GaAs MESFETs, *IEDM Tech. Dig.* **Dec**, 351–354 (1976).
23. R. A. Warriner, Computer simulation of gallium arsenide field-effect transistors using Monte Carlo methods, *Solid-State Electron Devices* **1**, 105 (1977).
24. H. A. Willing and P. de Santis, Modeling of Gunn domain effects on GaAs MESFET's, *Electron. Lett.* **13**(18), 537–539 (1977).
25. R. W. H. Engelmann and C. A. Liechti, Bias dependence of GaAs and InP MESFET parameters, *IEEE Trans. Electron Devices*, **ED-24**, 1288–1296 (1977).
26. M. S. Shur and L. F. Eastman, Current-voltage characteristics, small-signal parameters and switching times of GaAs FET's, *IEEE Trans. Electron Devices* **ED-25**, 606–617 (1978).
27. M. S. Shur, Analytical model of GaAs MESFET's, *IEEE Trans. Electron Devices* **ED-25**, 612–618 (1978).
28. T. Wada and S. Frey, Physical basis of short-channel MESFET operator, *IEEE Trans. Electron Devices* **ED-26**, 476–490 (1979).
29. A. Madjar and F. J. Rosenbaum, A practical ac large-signal model for GaAs microwave MESFETs, in Proc. IEEE MTT-S 1979 Int. Microwave Symp. (Orlando, FL, IEEE Cat. No. 79CH1439-9 MTT-S), pp. 399–401, 1979.

30. C. Rauscher and H. A. Willing, Quasi-static approach to simulating nonlinear GaAs FET behavior, in Proc. IEEE MTT-S 1979 Int. Microwave Symp. (Orlando, FL, IEEE Cat. No. 79CH1439-9 MTT-S), pp. 402-404, 1979.

31. H. Fukui, Determination of the basic device parameters of a GaAs MESFET, *Bell Syst. Tech. J.* **58**(3), 771-797 (1979).

32. K. Lehovec and R. Zuleeg, Analysis of GaAs FET's for integrated logic, *IEEE Trans. Electron Devices* **ED-27**, 1074-1091 (1980).

33. H. L. Grubin, D. K. Ferry, and K. R. Gleason, Spontaneous oscillations in gallium arsenide field effect transistors, *Solid-State Electron.* **23**, 157 (1980).

34. H. L. Grubin, Switching characteristics of nonlinear field effect transistors: Gallium arsenide versus silicon, *IEEE Trans. Microwave Theory Tech.* **MTT-28**, 442 (1980).

35. M. Ino and M. Ohmori, Intrinsic response of normally off MESFET's of GaAs, Si and InP, *IEEE Trans. Microwave Theory Tech.* **MTT-28**, 456 (1980).

36. D. E. Norton and R. E. Hayes, Static negative resistance in calculated MESFET drain characteristics, *IEEE Trans. Electron Devices* **ED-27**, 570 (1980).

37. W. R. Curtice and Y. H. Yun, A temperature model for the GaAs MESFET, *IEEE Trans. Electron Devices* **ED-28**, 954 (1981).

38. P. Bonjour, R. Castagne, J.-F. Pne, J-P. Courat, G. Bert, G. Nuzillat, and M. Peltier, Saturation mechanism in 1 μm gate GaAs FET with channel–substrate interfacial barrier, *IEEE Trans. Electron Devices* **ED-27**, 1019 (1980).

39. S. E. Laux and R. J. Lomax, Effect or mesh spacing on static negative resistance in GaAs MESFET simulation, *IEEE Trans. Electron Devices* **ED-28**, 120 (1981).

40. V. Faricelli, J. Frey, and J. P. Krusius, Physical basis of short-channel MESFET operation II: Transient behavior, *IEEE Trans. Electron Devices* **ED-29**(3), 377 (1982).

41. R. K. Cook and J. Frey, Two-dimensional numerial simulation of energy transport effects in Si and GaAs MESFET's, *IEEE Trans. Electron Devices* **ED-29**(6), 970-976 (1982).

42. M. S. Shur, Low field mobility, effective saturation velocity and performance of submicron GaAs MESFETs, *Electron. Lett.* **18**(21), 909-911 (1982).

43. M. S. Shur, Analytical models of GaAs FETs, *IEEE Trans. Electron Devices* **32**(1), 18-28 (1985).

44. M. J. Helix, S. A. Jamison, C. Chao, and M. Shur, Fanout and speed of GaAs SDFL logic, *Solid-State Circuits* **SC-17**(6), 1126-1232 (1982).

45. H. Dambkes, W. Brokerhoff, and K. Heime, GaAs MESFETs with highly doped (10^{18} cm^{-3}) channels—An experimental and numerical investigation, *IEDM Tech. Digest* 621-624 (1983).

46. R. E. Williams and D. W. Shaw, Guided channel FETs: Improved linearity and noise figure, *IEEE Trans. Electron Devices* **ED-25**(6), 600-605 (1978).

47. T. H. Chen, M. S. Shur, B. Hoefflinger, K. W. Lee, T. T. Vu, P. C. T. Roberts, and M. J. Helix, DOMES—GaAs IC simulator, in Proceedings of IEEE Biennial Conference on High Speed Devices, Cornell University, Ithaca, New York, p. 327-337, 1983.

48. N. Yokoyama, H. Onodera, T. Ohnishi, and A. Shibatomi, Orientation effect of self-aligned source/drain planar GaAs Schottky barrier field-effect transistors, *Appl. Phys. Lett.* **42**(3), 270-271, (1983).

49. G. W. Taylor, H. M. Darley, R. C. Frye, and P. K. Chatterjee, A device model for an ion-implanted MESFET, *IEEE Trans. Electron Devices* **ED-26**(3), 172-182 (1979).

50. W. R. Curtice, A MESFET model for use in the design of GaAs integrated circuits, *IEEE Trans. Microwave Theory Tech.* **MTT-28**(5), 448-456 (1980).

51. C. L. Chen and K. D. Wise, Transconductance compression in submicrometer GaAs MESFETs, *IEEE Electron Device Lett.* **EDL-4**(10), 341-343 (1983).

52. M. S. Shur and L. F. Eastman, Ballistic transport in semiconductors at low temperatures for low power high speed logic, *IEEE Trans. Electron Devices*, **ED-26**(11), 1677-1683 (1979).

53. A. Cappy, B. Carnes, R. Fauquembergues, G. Salmer, and E. Constant, Comparative potential performance of Si, GaAs, GaInAs, InAs submicrometer-gate FETs, *IEEE Trans. Electron Devices* **ED-27**(11), 2158-2168 (1980).

54. P. F. Lindquist and W. M. Ford, Semi-insulating GaAs substrates, in *GaAs FET Principles and Technology*, ed. by J. V. DiLorenzo and D. D. Khandewal, Artech House, Dedham, Massachusetts, 1982.

55. H. Goronkin, M. S. Birrittella, W. C. Seelbach, and Vaitkus, Backgating and light sensitivity in ion-implanted GaAs integrated circuits, *IEEE Trans. Electron Devices* **ED-29**(5), 845-850 (1982).

56. J. Graffeuil *et al.*, Light induced effects in GaAs FETs, *Electron Lett.* **15**(14), 439-440 (1979).

57. M. S. Biritella, W. C. Seelbach, and H. Goronkin, The effect of backgrating on design and performance on GaAs digital integrated circuits, *IEEE Trans. Electron Devices* **ED-29**(7), 1135–1142 (1982).

58. C. Kocot and C. A. Stolte, Backgating in GaAs MESFETs, *IEEE Trans. Electron Devices* **ED-29**(7), 1059–1064 (1982).

59. T. H. Chen and M. S. Shur, Current–voltage characteristics of ion-implanted GaAs MESFET's, *IEEE Trans. Electron Devices* **ED-32**(1), 18–27 (1985).

60. M. S. Shur and L. F. Eastman, I–V characteristics of GaAs MESFETs with non-uniform doping profile, *IEEE Trans. Electron Devices* **ED-27**(2), 455–461 (1980).

61. N. McIntyre, Calculation of I/V characteistics for ion-implanted GaAs MESFETs, *Electron. Lett.* **18**(5), 208–210 (1982).

62. C. H. Hyun, M. S. Shur, and A. Peczalski, Analysis of noise margin and speed of GaAs MESFET DCFL using UM-SPICE, *IEEE Trans. Electron Devices* **ED-33**(10), Oct. (1986).

63. T. H. Chen and M. S. Shur, Capacitance model of GaAs MESFETs, *IEEE Trans. Electron Devices* **ED-32**(5), 883–891 (1985).

64. Harry A. Willing, Christen Rauscher, and Pietro de Santis, A technique for predicting large-signal performance of a GaAs MESFET, *IEEE Trans. MTT* **MTT-26**(12), 1017–1023 (1978).

65. M. S. Shur, Small-signal nonlinear circuit model of GaAs MESFET, *Solid-State Electron.* **22**, 723–728 (1979).

66. C. R. Viswanathan and Miguel E. Levy, Modelling interelectrode capacitances in a MOS transistor, *IEDM Tech. Digest* 38–41 (1979).

67. T. Takada, K. Yokoyama, M. Ida, and T. Sudo, A MESFET variable-capacitance model for GaAs integrated circuit simulation, *IEEE Trans. MTT* **MTT-30**(5), 719–723 (1982).

68. H. F. Cooke, Small-signal GaAs FET problems, Report Number ECOM-76-C-1340-I, Appendix I, March 1977, Document request to Commander, U.S. Army Electronics Command, ATTN: DRSEL-TL-IC, Fort Monmouth, New Jersey, 07703.

69. R. E. Neidert and C. J. Scott, Computer program for microwave GaAs MESFET modelling, Report Number WRL 8561, February 1982, Naval Research Lab., Washington, DC 20375.

70. L. F. Eastman, S. Tiwari, and M. S. Shur, Design criteria for GaAs MESFETs, *Solid-State Electron.* **23**, 383–389 (1980).

71. C. P. Lee, B. M. Welch, and R. Zucca, Saturated resistor loads for GaAs integrated circuits, *IEEE Trans. Electron Devices* **ED-29**(7), 1103–1109 (1982).

72. N. C. Cirillo, Jr., J. K. Abrokwah, and M. S. Shur, A self-aligned gate process for ICs based on modulation doped (Al, Ga)As/GaAs FETs, Presented at the Device Research Conference, June 1984.

73. Takashi Hariu, Kazuhito Takahashi, and Yukio Shibata, New modeling of GaAs MESFETs, *IEEE Trans. Electron Devices* **ED-30**(12), 1743–1749 (1983).

74. J. Baek, M. S. Shur, K. Lee, and T. Vu, Current–voltage characteristics of ungated GaAs FETs, *IEEE Trans. Electron Devices* **ED-32**(11), 2426–2430 (1985).

75. C. L. Chen, Characterization of GaAs MESFETS fabricated using ion-beam etching technology, Ph.D. dissertation, University of Michigan, Ann Arbor, 1982.

76. H. J. Boll, J. E. Iwersen, and E. W. Perry, High-speed current limiter, *IEEE Trans. Electron Devices* **ED-13**(12), 904–907 (1966).

77. M. S. Shur and L. F. Eastman, Near ballistic electron transport in GaAs devices at 77°K, *Solid-State Electron.* **24**, 11–18 (1981).

78. M. S. Shur and D. Long, Performance prediction for submicron GaAs SDFL logic, *IEEE Electron Devices Lett.* **EDL-3**(5), 124–127 (1982).

79. J. S. Sites and H. H. Wider, Magnetoresistance mobility profiling of MESFET channels, *IEEE Trans. Electron Devices* **ED-27**(12), 2277–2281 (1980).

80. A. A. Immorlica, Jr., D. R. Decker, and W. A. Hill, A diagnostic pattern for GaAs FET material development and process monitoring, *IEEE Trans. Electron Devices* **ED-27**(12), 2285–2291 (1980).

81. L. J. Van der Pauw, A method of measuring specific resistivity and Hall effect of disc of arbitrary shape, *Phillips Res. Rep.* **13**, 1–9 (1958).

82. T. L. Tansley, AC profiling by Schottky-gate cloverleaf, *J. Phys. E: Sci. Instr.* **8**, 52–54 (1975).

83. L. J. Van der Pauw, A method of measuring the resistivity and Hall coefficient on lamellae of arbitrary shape, *Phillips Res. Rep.* **20**, 220–224 (1958/59).

84. K. Lehovec, Determination of impurity and mobility distribution in epitaxial semiconducting films on insulating substrate by C–V and Q–V analysis, *Appl. Phys. Lett.* **25**, 29–281 (1974).

85. R. A. Pucel and C. F. Krumm, Simple method of measuring drift mobility profiles in thin semiconductor films, *Electron. Lett.* **12**, 240–244 (1976).

86. J. D. Wiley and G. L. Miller, Series resistance effects in semiconductor *C-V* profiling, *IEEE Trans. Electron Devices* **ED-22**, 265–273 (1975).

87. Kang Lee, M. S. Shur, Kwyro Lee, Tho T. Vu, P. C. T. Roberts, and M. J. Helix, Low field mobility profile in GaAs ion-implanted FETs, *IEEE Trans. Electron Devices* **ED-31**(3), 390–393 (1984).

88. H. Fukui, Determination of the basic device parameters of a GaAs MESFET, *Bell Syst. Tech. J.* **58**(3), 771–797 (1979).

89. J. F. Gibbons, W. S. Johnson, and S. W. Myrlie, Projected range statistics, in *Semiconductors and Related Material*, 2nd ed., Dowen, Hutchinson, and Ross, Stroudsburg, Pennsylvania, 1975.

90. K. W. Lee, Characterization, modeling, and circuit design of GaAs MESFETs, Ph.D. thesis, University of Minnesota, 1984.

91. H. Kroemer and W. Chien, On the theory of Debye averaging in the *C-V* profiling of semiconductors, *Solid-State Electron.* **24**, 655–660 (1981).

92. K. Lee, M. Shur, K. W. Lee, T. Vu, P. Roberts, and M. Helix, New interpretation of "End" Resistance Measurements, *IEEE Electron Device Lett.* **EDL-5**(5), 5–7 (1984).

93. K. Lee, M. Shur, T. Vu, P. Roberts, and M. Helix, Source, drain and gate series resistances and electron saturation velocity in ion implanted GaAs FETs, *IEEE Trans. Electron Devices*, **ED-32**(5), 987–992 (1985).

94. W. Shockley, Research and investigation of inverse epitaxial UHF power transistors, Report No. A1-TOR-64-207, Air Force Atomic Laboratory, Wright-Patterson Air Force Base, Ohio, September 1964.

95. G. K. Reeves and H. B. Harrison, Obtaining the specific contact resistance from transmission line model measurements, *IEEE Electron Device Lett.* **EDL-3**(5), 111–113 (1982).

96. P. Urien and D. Delagebeaudeuf, New method for determining the series resistances in a MESFET or TEGFET, *Elec. Lett.* **19**, 702–703 (1983).

97. S. Chaudhuri and M. Das, On the determination of source and drain series resistances of MESFET's, *IEEE Electron Device Lett.* 244–246 (1984).

98. Kang Lee, Michael S. Shur, and Tho T. Vu, New technique for measurement of electron saturation velocity in GaAs MESFETs, *IEEE Electron Device Lett.* **EDL-5**(10), 426–427 (1984).

99. R. A. Sadler and L. F. Eastman, High-speed logic at 300 K with self-aligned submicron-gate GaAs MESFET's, *IEEE Electron Devices Lett.* **EDL-4**, 215–217 (1983).

100. R. A. Sadler and L. F. Eastman, Self-aligned submicron GaAs MESFET's for high speed logic, Device Research Conference, June 1983.

101. A. Van Der Ziel, M. S. Shur, K. Lee, T. H. Chen, and K. Amberiadis, Carrier distribution and low-field resistance in short n^+-n^--n^+ and n^+-p^--n^+ structure, *IEEE Trans. Electron Devices*, **ED-30**(2), (1983).

102. S. M. Sze, *Physics of Semiconductor Devices*, 2nd edition, Wiley and Sons, New York, 1981.

103. W. Shockley and R. C. Prim, Space-charge limited emission in semiconductors, *Phys. Rev.* **90**, 753 (1953).

104. R. Zuleeg, A silicon space-charge-limited triode and analog transistor, *Solid-State Electron.* **10**, 449–460 (1967).

105. C. De Blasi, G. Micocci, A. Rizzo, and A. Tepore, One-carrier space-charge-limited-currents in non-homogeneous solids under cylindrical flow, *Solid-State Electron.* **36**(11), 1095–1099 (1983).

106. Z. Van der Ziel, Space-charge-limited solid-state diodes, *Semiconductors and Semimetals*, Vol. 14, Academic, New York, 1979.

107. G. C. Dacey, Space-charge limited hole current in Germanium, *Phys. Rev.* **90**(5), 759 (1953).

108. E. De Chambost, Estimate of substrate influence on space-charge-limited current, *Electron. Lett.* **9**, 351 (1973).

109. L. F. Eastman and M. S. Shur, Substrate current in GaAs MESFETs, *IEEE Trans. Electron Devices* **ED-26**(9), 1359–1361, 1979.

110. H. Kroemer, The Gunn effect under imperfect boundary conditions, *IEEE Trans. Electron Devices* **ED-15**(11), 819–837 (1968).

111. Mott and Gurney, *Electronic Processes in Ionic Crystals*, Clarendon Press, Oxford, 1940.

112. J. Grafteuil and P. Rossel, Semiempirical expression for direct transconductance and equivalent saturated velocity in short-gate-length MESFETs, *IEE Proc.* **129**, Pt. 1(5), pp. 185–188 (1982).

113. T. H. Chen, High speed GaAs device and integrated circuit modeling and simulation, Ph.D. thesis, University of Minnesota, 1984.

8

GaAs FET Amplifiers
and Microwave Monolithic
Integrated Circuits

8-1. INTRODUCTION

One of the most important applications of GaAs FETs is in small signal amplifier components. High-frequency low-noise GaAs FETs are used in phase-array radars, signal processors, space based electronic detection systems, tracking devices, and digital transmitter–receivers. In particular, GaAs low-noise amplifiers are used in communication equipment for the 3.7–4.2-GHz television receive-only (TVRO) band and for the 12-GHz direct broadcast satellite (DBS) band. The DBS receivers represent a large potential market for GaAs components.

Different GaAs FETs operate in the frequency range from dc to 40 GHz and above, with higher frequency and lower noise figure generally achieved in shorter gate structures, with the best results achieved for submicron gate devices.

GaAs power amplifiers are very convenient for airborne and space applications because of their small size, weight, radiation hardness, and reliability. Broadband GaAs FETs are used in instrumentation and in electronic countermeasure systems. GaAs microwave monolithic circuits are especially useful in applications where a potentially low cost is an advantage or where many identical devices are utilized (as in phased array radars).

In this chapter we will consider low-noise GaAs FETs, power GaAs amplifiers, and microwave monolithic integrated circuits (MMIC).

8-2. NOISE FIGURE OF MICROWAVE GaAs FETs

An important figure of merit characterizing the noise performance of GaAs FETs is the noise figure. The noise figure NF is defined as the ratio of the noise output power in a given bandwidth to the noise power output that would be obtained in the same bandwidth if the only noise came from the internal resistance R_{source} of the signal source. Typically the noise figure is measured in decibels:

$$NF = 10 \log_{10} \frac{\text{total noise power output}}{\text{noise power output due to } R_{source}} \qquad (8\text{-}2\text{-}1)$$

It may be considered as a parameter comparing a real device to an idealized noiseless amplifier.

The noise figure measurements are based on the linear relationship between the output noise power and the equivalent noise temperature of the noise source. For a resistor R_{source} the equivalent noise temperature T_n is equal to the ambient temperature and is related to the thermal noise voltage V_n as follows:

$$T_n = V_n^2 / 4k_B R_{source} B \qquad (8\text{-}2\text{-}2)$$

Here k_B is the Boltzmann constant and B is the bandwidth. As can be seen from this expression, the thermal (or Johnson) noise has the same power in any given bandwidth independently of the central frequency (it is also called white noise).

The linear dependence of the output noise power on the source temperature is shown in Fig. 8-2-1. If the noise temperature had been zero all the output noise power would come from the noise generated by the device (P_0 in Fig. 8-2-1). At any finite temperature T_s the source noise has a power density of $k_B T_s$. It is common to measure the noise figure at 290 K where the input noise power density is -174 dB m/Hz and the output noise power is equal to P_{290} (see Fig. 8-2-1). The slope of the characteristic shown in Fig. 8-2-1 is proportional to the device gain, G, and hence the output noise power is given by

$$P = k_B T_s GB + P_0 \qquad (8\text{-}2\text{-}2)$$

In this equation GB is the gain–bandwidth product of the device. According to the definition given above the noise figure at 250 K may be found as

$$\mathrm{NF} = \frac{P_{290}}{P_{290} - P_0} \qquad (8\text{-}2\text{-}3)$$

Consequently, the noise figure can be measured by finding the ratio Y of the noise power output of the device under test when the noise source impedance is hot ($T = T_h$) and cold ($T = T_c$). These effective noise temperatures are created by a special solid state noise source or by a tube or (at low frequencies) by a hot and cold resistor. The noise temperature of the noise source in the hot state is defined by the so-called excessive noise ratio (ENR) of the noise source [1]:

$$\mathrm{ENR} = (T_h - 290 \text{ K})/290 \text{ K} \qquad (8\text{-}2\text{-}4)$$

ENR is frequently measured in decibels:

$$\mathrm{ENR_{dB}} = 10 \log_{10} \frac{T_h - 290 \text{ K}}{290 \text{ K}} \qquad (8\text{-}2\text{-}5)$$

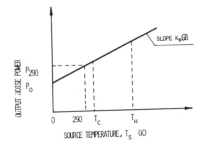

FIGURE 8-2-1. Output noise power vs. source resistance temperature (see text).

TABLE 8-2-1. Typical Noise-Measurement Uncertainties
of the HP8970a/HP346 Measuring Set at the X Band[a]

Instrumentation uncertainty	± 0.1 dB
ENR uncertainty	± 0.15 dB
Uncertainty due to second stage contribution	± 0.1 dB
Mismatch uncertainty	± 0.15 dB
Total measurement uncertainty	± 0.25 dB

[a] Reference 1.

From Eqs. (8-2-3) and (8-2-4) we find[1]

$$NF = \frac{ENR - Y(T_c/290 - 1)}{Y - 1} \qquad (8\text{-}2\text{-}6)$$

which can be simplified if the temperature T_c of the cold noise source is equal to 290 K (not a typical case):

$$NF = ENR/(Y - 1)$$

More details of the noise figure measurements may be found in Ref. 1, which describes a microprocessor controlled noise figure and gain measurement using the HP8970A noise figure meter and the HP346B noise source. Typical noise-measurement uncertainties of the HP8970a/HP346 measuring set at the X band are given in Table 8-2-1 [1].

The theory of noise in GaAs field effect transistors was first developed by van der Ziel and was based on the gradual channel approximation [2, 3]. A good review of earlier work on noise properties of GaAs FETs was published by Liechti [4].

Pucel et al. [5] derived an expression for the minimum noise figure in terms of parameters of the noise equivalent circuit (see Fig. 8-2-2):

$$F_{\min} = 1 + 2(2\pi f C_{gs}/g_m)\{K_g[K_r + g_m(R_g + R_s)]\}^{1/2}$$
$$+ 2(2\pi f C_{gs}/g_m)^2[K_g g_m[R_g + R_s + K_c R_i)] + \cdots \qquad (8\text{-}2\text{-}7)$$

where f is the operating frequency, C_{gs} is the gate-to-source capacitance, g_m is the

FIGURE 8-2-2. FET Equivalent circuit used in the analysis of noise [5, 6]. i_{ng}, the induced gate noise current; i_{nd}, the drain circuit noise current; e_{ng}, the thermal noise of the gate metallization resistance R_g; e_{ns}, the thermal noise of the source series resistance R_s; e_s, the source signal voltage; Z_s, the source impedance. Parasitic resistances R_s and R_g are relatively independent of the bias voltages. Other parameters of the equivalent circuit are complicated functions of the gate and drain bias voltages.

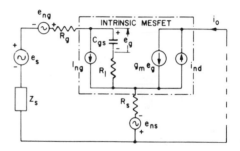

device transconductance,

$$K_g = P[(1 - C\sqrt{R/P})^2 + (1 - C^2)R/P] \qquad (8\text{-}2\text{-}8)$$

$$K_r = \frac{R(1 - C^2)}{(1 - C\sqrt{R/P})^2 + (1 - C^2)R/P} \qquad (8\text{-}2\text{-}9)$$

$$K_c = \frac{1 - C\sqrt{R/P}}{(1 - C\sqrt{R/P})^2 + (1 - C^2)R/P} \qquad (8\text{-}2\text{-}10)$$

Here R_i is the gate-to-source internal channel resistance, P is the drain noise coefficient, R is the gate noise coefficient, C is the noise correlation coefficient, R_g is the ac gate series resistance, and R_s is the source series resistance. Parameters R_s and R_g may be considered fairly bias independent. However, other parameters vary with the gate and drain voltages in a complicated fashion. As pointed out by Fukui [6], the minimum value of the noise figure is usually obtained at relatively large reverse gate bias and at a relatively low drain bias in the saturation region of the current voltage characteristic.

Fukui [5] proposed several empirical relationships which allow a simple evaluation of the minimum noise figure in terms of the equivalent circuit elements evaluated at zero gate bias in the saturation region (typically at drain-to-source voltages close to 5 V).

According to Fukui [6], the optimum value of the minimum noise figure F_0 at zero gate bias is given by

$$F_0 = 1 + 2\pi K_f f C_{gs}[(R_g + R_s)/g_m]^{1/2} \times 10^{-3} \qquad (8\text{-}2\text{-}11)$$

where K_f is the fitting parameter, which is approximately equal to 2.5 and is dependent on the material quality (in this equation and in equations below f is in GHz and capacitances are in pF). Equation (8-2-11) may be recovered from Eq. (8-2-7) in the limiting case of $C = 1$ and $R = 0$ when all higher-order terms are neglected.

Equation (8-2-11) may be rewritten in terms of the cutoff frequency f_T:

$$f_T = g_m/(2\pi C_{gs}) \times 10^3 \qquad (8\text{-}2\text{-}12)$$

where C_{gs} is pF and f_T is in GHz. Using Eq. (8-2-12) we find

$$F_0 = 1 + K_f(f/f_T)[g_m(R_g + R_s)]^{1/2} \qquad (8\text{-}2\text{-}13)$$

Taking into account that f_T is inversely proportional to the gate length, this equation can be further rewritten as

$$F_0 = 1 + K_1 L f[g_m(R_g + R_s)]^{1/2} \qquad (8\text{-}2\text{-}14)$$

where K_1 is a fitting factor (close to 0.27) and L is in microns.

Fukui proposed several empirical relationships which link the device transconductance, the gate metallization resistance, and the source series resistance in terms of geometrical and material parameters of a device [7]:

$$g_m \approx K_m W(N/AL)^{1/3} \ (\Omega^{-1}) \qquad (8\text{-}2\text{-}15)$$

where W is the gate width in mm, A is the effective channel thickness in μm, N is the effective channel doping density in 10^{16} cm^{-3}, and $K_m \approx 0.023$ is a fitting constant.

$$R_g \approx 17z^2/(WhL_g) \; (\Omega) \tag{8-2-16}$$

where h is the gate metallization height in μm, z is the unit gate width in mm, and L_g is the average gate metallization length in μm.

For a recessed gate Fukui proposed the following empirical formula for the series resistance R_s:

$$R_s = R_1 + R_2 + R_3 \tag{8-2-17}$$

where R_1 is the source contact resistance, and R_2 and R_3 are partial resistances of the channel between the source and gate electrodes [7]:

$$R_1 \approx 2.1/(Wa_1^{1/2}N_1^{0.66}) \; (\Omega) \tag{8-2-18}$$

$$R_2 \approx 1.1L_2/(Wa_2N_2^{0.82}) \; (\Omega) \tag{8-2-19}$$

$$R_3 \approx 1.1L_3/(Wa_3N_3^{0.82}) \; (\Omega) \tag{8-2-20}$$

Here a_1 is the effective channel thickness under the source contact, in μm, N_1 is the effective carrier concentration in the channel under the source contact, in 10^{16} cm^{-3}; L_2, L_3 are effective lengths of each sectional channel between the source and the gate contacts, in μm; a_2, a_3 are effective thicknesses of the sectional channel, in μm; N_2, N_3 are effective carrier concentrations of the sectional channel, in 10^{16} cm^{-3}. The effective channel thicknesses should take into account the surface depletion layer thickness A_s in GaAs, which is approximately given by

$$A_s \approx (2\varepsilon V_s/qN)^{1/2} \tag{8-2-21}$$

In the case of a device with an n^+ layer between the ohmic metal and n-GaAs, L_2 is approximately equal to the distance between the source electrode and the n^+ layer, and L_3 is approximately equal to the distance between the edge of the n^+ region and the gate electrode. As an example, the values of the parameters for five different GaAs FETs are given in Table 8-2-2 [6]. Devices A and B did not have n^+ layers under the ohmic contacts; devices marked C, D, and E had such layers. Parameters L_g and h were determined from the SEM measurements. The calculated values of g_m, R_g, R_1, R_2, R_3, and R_s are given in Table 8-2-3 [6], where the experimental results are also given. As can be seen from the table the agreement with the experimental results is very good. It can also be seen that the n^+ layer decreases the source series resistance by about 40%. It is also clear that the gate metallization in device C is too thin, resulting in a very large gate resistance.

Substituting the expressions for the resistances into Eq. (8-2-13) for the minimum noise figure we obtain

$$F_0 = 1 + kf(NL^5/A)^{1/6}[17z^2/(hL_g) + 2.1/(a_1^{1/2}N_1^{0.66})$$
$$+ 1.1L_2/(a_2N_2^{0.82}) + 1.1L_3/(a_3N_3^{0.82})]^{1/2} \tag{8-2-22}$$

Here

$$k = K_1\sqrt{K_m} \approx 0.040 \tag{8-2-23}$$

TABLE 8-2-2. Predetermined Values of Geometrical and Material Parameters
for Sample GaAs MESFETs[a]

Parameter		Device				
Symbol	Units	A	B	C	D	E
N	10^{16} cm^{-3}	7.4	8.1	3.5	6.6	9.8
N_1	10^{16} cm^{-3}	7.4	8.1	100	100	100
N_2	10^{16} cm^{-3}	7.4	8.1	100	100	100
N_3	10^{16} cm^{-3}	7.4	8.1	3.5	6.6	9.8
a	μm	0.155	0.138	0.273	0.165	0.140
a_1	μm	0.35	0.45	0.15	0.15	0.15
a_2	μm	0.25	0.35	0.10	0.10	0.10
a_3	μm	0.155	0.138	0.273	0.165	0.140
L	μm	0.46	0.51	0.57	0.51	0.50
L_g	μm	0.8	0.8	0.8	0.8	0.8
L_2	μm	0.75	0.75	0.85	0.75	0.85
L_3	μm	0.4	0.3	0.4	0.3	0.4
h	μm	0.65	0.65	0.2	0.65	0.65
z	mm	0.25	0.25	0.25	0.25	0.25
k		0.04	0.04	0.04	0.04	0.04

[a] Reference 6.

As can be seen from Eq. (8-2-22) F_0 is independent of the device width but varies
with the unit gate width z.

At higher operating frequencies the skin effect increases the gate metallization
resistance and the expression for the minimum noise figure becomes

$$F_0 = 1 + kf(NL^5/A)^{1/6}[17z^2/(hL_g) + 1.3z^2(f/hL_g)^{1/2}$$
$$+ 2.1/(a_1^{1/2}N_1^{0.66}) + 1.1L_2/(a_2N_2^{0.82})$$
$$+ 1.1L_3/(a_3N_3^{0.82})]^{1/2} \tag{8-2-24}$$

TABLE 8-2-3. Comparison between the Calculated and Measured Values of
Transconductance and Parasitic Resistances for Sample GaAs MESFETs[a]

	Parameter		Device				
	Symbol	Units	A	B	C	D	E
Calculated	g_m	m℧	54	56	33	49	61
Measured	g_m	m℧	54	56	33	48	59
Calculated	R_g	Ω	4.1	4.1	13.3	4.1	4.1
Measured	R_g	Ω	4.5	3.8	13.7	3.8	4.0
Calculated	R_1	Ω	1.89	1.57	0.52	0.52	0.52
	R_2	Ω	1.28	0.85	0.68	0.59	0.68
	R_3	Ω	1.13	0.86	1.15	0.86	1.00
	R_s	Ω	4.3	3.3	2.4	2.0	2.2
Measured	$\dfrac{R_s + R_d}{2}$	Ω	4.3	3.2	2.7	1.7	2.3

[a] Reference 6.

TABLE 8-2-4. Comparison between the Predicted and Measured Values of Optimal Noise
Figure for Sample GaAs MESFETs[a]

	Parameter		Device				
	Symbol	Units	A	B	C	D	E
Measured	F_0	dB	1,75	1.76	2.22	1.51	1.74
	L	μm	0.46	0.51	0.57	0.51	0.50
	g_m	m℧	54	56	33	48	59
	R_g	Ω	4.5	3.8	13.7	3.8	4.0
	R_s	Ω	3.8	2.9	2.7	1.5	2.3
Predicted	F_0	dB	1.75	1.76	2.21	1.50	1.73

[a] Reference 6.

The calculated values of the noise figure are compared with the experimental results
in Table 8-2-4 [6]. As can be seen from the table, the agreement with experimental
results is very good.

The frequency dependence of an optimum noise figure predicted by Eq. (8-2-22)
is shown in Fig. 8-2-3 [6]. As can be seen from the figure, a submicron gate GaAs
FET with an appropriately scaled gate unit width may have a superior noise
performance at the microwave frequencies through the K band. As was pointed
out by Fukui [6], the noise performance of GaAs FETs should be much better than
that of bipolar junction transistors.

The frequency dependences of the noise figure calculated by Fukui [6] for
GaAs with parameters given in Table 8-2-5 are shown in Fig. 8-2-3. As can be seen
from the figure, the improvement in the noise figure results from the decrease in

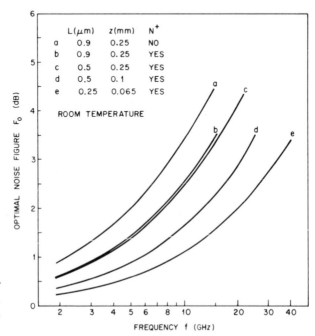

FIGURE 8-2-3. Calculated optimum noise figure as a function of
frequency for GaAs MESFETs [6].
Device parameters are given in
Table 8-2-5.

TABLE 8-2-5. Design Parameters of Five Representative GaAs MESFETs Used for Calculation of the Optimal Noise Figure as a Function of Frequency, as Shown in Fig. 8-2-3[a]

Parameter		Device				
Symbol	Units	a	b	c	d	e
L	μm	0.9	0.9	0.5	0.5	0.25
L_g	μm	0.9	1.2	0.8	0.8	0.4
L_2	μm	1.0	0.75	0.75	0.75	0.4
L_3	μm	0	0.4	0.3	0.3	0.2
h	μm	0.5	1.0	0.65	0.65	0.4
N	10^{16} cm^{-3}	7	4	8	8	18
N_1	10^{16} cm^{-3}	7	200	200	200	200
N_2	10^{16} cm^{-3}	7	200	200	200	200
N_3	10^{16} cm^{-3}	—	4	8	8	18
a	μm	0.3	0.27	0.15	0.15	0.1
a_1	μm	0.3	0.15	0.15	0.15	0.15
a_2	μm	0.17	0.12	0.12	0.12	0.12
a_3	μm	—	0.27	0.15	0.15	0.1
z	mm	0.25	0.25	0.25	0.1	0.065
z_m	mm	0.24	0.23	0.14	0.14	0.065

[a] Reference 6.

the gate length and from choosing the unit gate width z sufficiently short to minimize the gate resistance.

Fukui estimated the noise figure at the zero gate bias. In fact, a noise figure is very much dependent on the drain current and is typically measured at the noise bias (15% of I_{dss}). Podell [8] demonstrated that the noise performance of a GaAs FET can be adequately described by two uncorreleted noise sources. One source, at the input of the FET, is the thermal noise generated in the various resistances in the gate–source loop. This frequency-dependent noise may be related to the parameters of the FET equivalent circuit. The second noise source, at the output of the FET, is frequency-independent and is a function of the drain current and voltage. This output noise may be related to the stationary high field domain at the drain side of the gate [9–12] as well as to the substrate current [13–15].

Podell's calculation is based on the simple noise equivalent circuit proposed by Bruncke and Van der Ziel [3, 16] for a silicon FET (see Fig. 8-2-4). However, the equivalent noise resistance given by

$$R_n = Q/g_m \qquad (8\text{-}2\text{-}25)$$

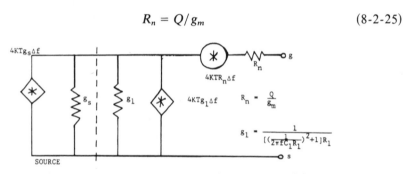

FIGURE 8-2-4. Noise equivalent circuit for a GaAs FET [8].

where

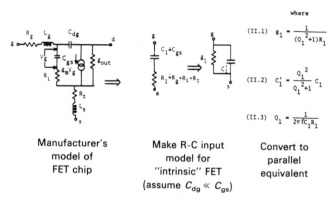

(II.1) $\quad g_1 = \dfrac{1}{(Q_1^{\,2}+1)R_1}$

(II.2) $\quad C_1' = \dfrac{Q_1^{\,2}}{Q_1^{\,2}+1}C_1$

(II.3) $\quad Q_1 = \dfrac{1}{2\pi f C_1 R_1}$

Manufacturer's model of FET chip	Make R-C input model for "intrinsic" FET (assume $C_{dg} \ll C_{gs}$)	Convert to parallel equivalent

FIGURE 8-2-5. Simplified FET input model [8].

in the Bruncke–Van der Ziel theory, where $Q \approx 0.7$, is determined from the empirical relationship

$$R_n = K_0 \exp(K_2 I)/g_m \qquad (8\text{-}2\text{-}26)$$

where g_m is the device transconductance, $I = I_{DS}/I_{DSS}$, I_{DS} is the drain-to-source current, I_{DSS} is the value of I_{DS} at zero gate-to-source voltage, and K_0 and K_2 are empirical constants. The minimum noise figure is then found from the expression predicted by the Bruncke–Van der Ziel theory:

$$F_{min} = 1 + 2A + 2(A + A^2)^{1/2} \qquad (8\text{-}2\text{-}27)$$

where $A = g_1 R_n$ and the value of g_1 is found as shown in Fig. 8-2-5 [8].

The comparison between the measured and calculated noise figures is shown in Fig. 8-2-6 [8]. Figure 8-2-7 illustrates the dependence of the noise figure on the drain-to-source current for five different devices [8].

Once the minimum noise figure for the optimum source admittance (with respect to the noise performance) is known, the device noise figure can be found as

$$F = F_{min} + (R_n/g_s)[(g_s - g_{s\,opt})^2 + (b_s - b_{s\,opt})^2] \qquad (8\text{-}2\text{-}28)$$

Here the optimum source admittance

$$Y_{sopt} = g_{sopt} - jb_{sopt}$$

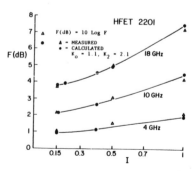

FIGURE 8-2-6. Comparison of measured and calculated noise figures [8].

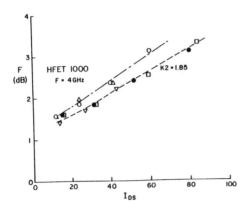

FIGURE 8-2-7. Dependence of noise figure on drain-to-source current for five different FETs [8].

is given by

$$g_{sopt} = g_1(F_{min} + 1)/(F_{min} - 1) \qquad (8\text{-}2\text{-}29)$$

$$b_{sopt} = -\omega C_1' = -Q_1^2 \omega C_1/(Q_1^2 + 1) \qquad (8\text{-}2\text{-}30)$$

(see Fig. 8-2-5). C_1 is approximately equal to the FET input capacitance measured at a relatively low frequency (say 2 GHz), where inductances L_s and L_g (see Fig. 8-2-5) may be neglected.

The optimum source impedance for noise performance is different from the optimum source impedance for power matching. As a consequence, trade-offs must be made in practical FET amplifier design.

8-3. DESIGN AND FABRICATION OF LOW-NOISE FETs

As can be seen from the results discussed in Section 8-2, short gate lengths, low series source and gate resistances, and optimized unit gate widths are necessary prerequisites for low-noise device performance. More detailed calculations which link the noise figure to such device parameters as the gate metallization thickness, active layer thickness, and source-to-drain separation based on the Fukui model are given in Ref. 17, where device design and fabrication are also considered.

A chemical vapor deposition process was used in Ref. 17. The reactor was equipped with two doping lines to allow n^+/n structures to be grown without a pause in the growth sequence. A mercury Schottky barrier probe was used to determine the carrier profile. Mobility profiles were determined from the Hall measurements utilizing a standard Van der Pauw cloverleaf pattern with a Schottky diode over the central active area. The Schottky diode was reverse biased and the Hall data were measured as a function of the reversed voltage, making it possible to determine the Hall mobility as a function of the distance from the surface. Also, the buffer resistivity was measured in the dark and under illumination.

The back face of the epitaxial slice was metallized by a Au/Ge alloy to allow the soldering of the final transistor chip into the circuit. Mesa regions were defined by conventional photolithography and adjacent device areas were isolated using a chemical etch. Source and drain contacts were defined by an evaporation process with an In/Au/Ge alloy used for low-resistance contacts. The gate area was defined

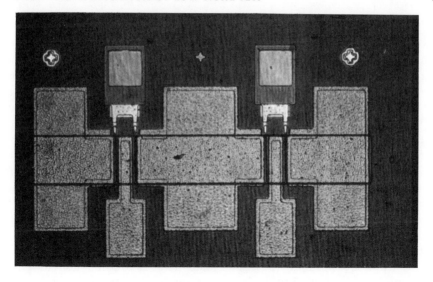

FIGURE 8-3-1. Low-noise GaAs FET [17].

by an electron beam lithography process. The channel region was chemically etched until the monitored source–drain saturation current reached a predetermined value related to final drain-to-source device saturation current. After this the aluminum gate was evaporated. The fabricated 0.3-μm gate GaAs FET is shown in Fig. 8-3-1 [17].

The predicted noise figure and associated gain (based on the Fukui model) are compared with the experimental results in Fig. 8-3-2 [17]. As can be seen from the figure, the agreement is close.

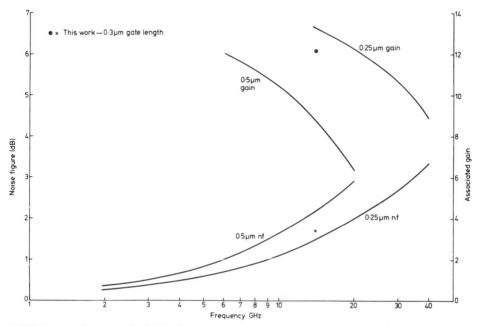

FIGURE 8-3-2. Calculated (solid lines) and measured noise figure and associated gain of the low-noise GaAs amplifier [17].

Different layout patterns for low-noise GaAs FETs are shown in Fig. 8-3-3 [18]. In Ref. 18 the dependences of the noise figure on the carrier concentration in the channel, on the drain-to-source current, on doping profiles in the channel, and on the recessed depth and frequency were studied experimentally for 0.5- and 1-μm gate FETs.

Half-micrometer and 0.3-μm gate GaAs devices are now produced commercially. Performance specifications of NEC microwave GaAs FETs are given in Table 8-3-1 as an example [19]. Typical noise figure and associated gain for NEC67863 transistors are shown in Fig. 8-3-4 [19].

The development effort is concentrated on even shorter gate devices, which require, in most cases, electron beam lithography [17, 20, 21]. However, optically defined quarter-micron gate FETs with superior noise performance have also been reported [22]. The devices described in Ref. 22 have a T gate geometry with multiple gate feeds (see Fig. 8-3-5). The small gate pads on either side are connected to a central bonding pad by a plated airbridge which acts as a low loss bus line. The gatewidth was chosen to provide a good match to 50 Ω over the desired frequency band. Gate widths of 250 and 75 μm were chosen for 4-18-GHz and 18-40-GHz bands, respectively.

Mesas were isolated by chemical etching. Alloyed AuGe/Ni/Au drain and source ohmic contacts were fabricated. A 0.25-μm gate was created using an electrom beam generated mask with contact printing lithography and angle evaporated metal. The gate metal was Ti/Pt/Au. The SEM microphotograph of the gate cross section is shown in Fig. 8-3-6 [22]. As can be seen from the figure the gate tapers toward the bottom. This shape reduces the gate series resistance while maintaining a small gate length.

The rf results achieved in Ref. 22 are given in Table 8-3-2. The device equivalent circuit is shown in Fig. 8-3-7 [22]. The parameters of this equivalent circuit determined from the S-parameter measurements are given in Table 8-3-3. The calculations

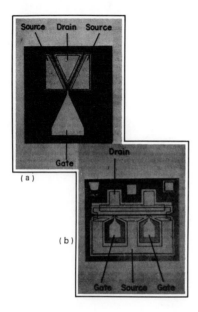

FIGURE 8-3-3. Examples of low-noise GaAs FET patterns [18]. (a) 1-μm gate FET with 150-μm unit gate width (NE244); (b) 0.5-μm gate FET with two gate pads (NE388).

TABLE 8-3-1. Performance Characteristics of NEC Low-Noise GaAs FETs[a]

	NE part number: EIAJ registered number: Package code:		NE67300 Chip			NE67383/NE67383-4 2SK407 83		
Symbols	Parameters and conditions	Units	Min	Typ	Max	Min	Typ	Max
f_{max}	Maximum frequency of oscillation at $V_{DS} = 3$ V, $I_{DS} = 30$ mA	GHz		100			100	
MAG	Maximum available gain at $V_{DS} = 3$ V, $I_{DS} = 30$ mA							
	$f = 8$ GHz	dB		15			15	
	$f = 12$ GHz	dB		12			12	
	$f = 18$ GHz	dB		8.5			8.5	
NF_{opt}	Optimum noise figure at $V_{DS} = 3$ V, $I_{DS} = 10$ mA							
	$f = 4$ GHz, $\Gamma_{opt} = 0.64\angle 69°$, $R_n = 0.38$	dB		0.4			0.4	0.6
	$f = 8$ GHz, $\Gamma_{opt} = 0.55\angle 115°$, $R_n = 0.20$	dB		0.8			0.8	
	$f = 12$ GHz, $\Gamma_{opt} = 0.48\angle 155°$ $R_n = 0.20$	dB		1.4	1.6		1.4	1.6
	$f = 18$ GHz, $\Gamma_{opt} = 0.46\angle -33°$, $R_n = 0.40$	dB		1.9				
	$f = 26$ GHz	dB		3.3				
G_a	Associated gain at optimum noise figure at $V_{DS} = 3$ V, $I_{DS} = 10$ mA							
	$f = 4$ GHz	dB		14.5		12	14.5	
	$f = 8$ GHz	dB		11.5			11.5	
	$f = 12$ GHz	dB	8.5	10.0		8.5	10.0	
	$f = 18$ GHz	dB		8.0				
	$f = 26$ GHz	dB		6.0				
P_{1dB}	Output power at 1 dB compression point at $V_{DS} = 3$ V, $I_{DS} = 30$ mA							
	$f = 12$ GHz	dB m		14.5			14.5	

[a] Reference 19.

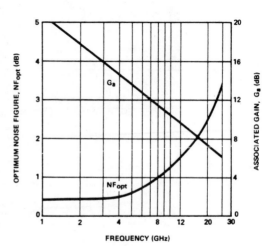

FIGURE 8-3-4. Performance characteristics of NEC microwave GaAs FETs [19].

FIGURE 8-3-5. GaAs FET with airbridge gate feeds [22]. Bonding pads are labeled. Gate width is 75 μm.

based on this equivalent circuit project that these devices should yield a reasonable performance at frequencies up to 60 GHz [22].

8-4. GaAs POWER AMPLIFIERS

8-4-1. Classes of Amplifiers

GaAs low-noise amplifiers considered in Sections 8-2 and 8-3 are small-signal amplifiers. They typically operate under the conditions when the output current and voltage swings correspond to relatively small deviations from the operating point. Power amplifiers may operate in this regime (class A amplifiers) but they may also have output waveforms such that the output current flows only during a part of the cycle (class AB, class B, and class C amplifiers; see Fig. 8-4-1).

The amplifier efficiency η is defined as

$$\eta = \frac{\text{ac output power}}{\text{dc input power}} \times 100\% = \frac{P_{\text{ac}}}{P_{\text{dc}}} \times 100\% \qquad (8\text{-}4\text{-}1)$$

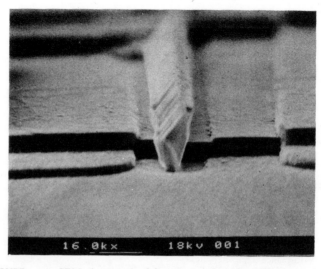

FIGURE 8-3-6. SEM photograph of the cross section of a 0.25 μm gate [22].

TABLE 8-3-2. RF Performance of 0.25-μm FETs[a]

Device No.	Gate width (μm)	Frequency (GHz)	NF (dB)	Associated gain (dB)
A	250	12	0.95	11.5
B	250	18	1.72	8.5
C	75	18	1.55	12.3
D	75	21.7	1.98	10.55
E	75	32	2.60	7.2

[a] Reference 22.

For a class A amplifier the maximum current swing is roughly equal to $I_Q = I_{fc}/2$ and the maximum voltage swing is approximately equal to $V_{dd}/2$, where V_{dd} is the voltage of the dc power supply, I_{fc} is the full channel current.

$$P_{ac} = (I_{fc}/2)*(V_{dd}/2)/2 \tag{8-4-2}$$

The dc power supplied by the voltage source V_{dd} is given by

$$P_{dc} = I_Q V_{dd} = (I_{fc}/2) V_{dd} \tag{8-4-3}$$

and hence the maximum efficiency of the class A amplifier is around 25%. The power dissipated in the transistor is

$$P_D = P_{dc} - I_Q^2 R_L = I_Q V_{dd}/2 = 2P_{ac} \tag{8-4-4}$$

For comparison in a class B amplifier the current flows only during half of the period (see Fig. 8-4-1) and the maximum efficiency for a push–pull amplifier is $(\pi/4) \times 100\%$ or 78.5%. The power dissipation P_D in each transistor of the push-pull class B amplifier is $P_{ac}/5$. Below, however, we will consider class A amplifiers.

8-4-2. Estimates of Output Power per Millimeter of Gate Periphery

In order to obtain the maximum power from a class A amplifier the load line should be chosen as shown in Fig. 8-4-2 where I_{fc} is the full channel current, V_A is

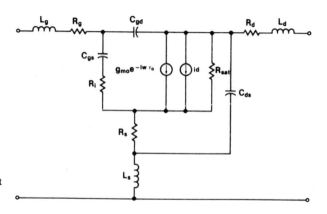

FIGURE 8-3-7. Equivalent circuit of the 0.25-μm gate FET [22].

TABLE 8-3-3. Circuit Element Values
for 0.25-μm FET[a]

g_m	19 mS	C_{gs}	0.07 pf
τ_0	2.5 ps	C_{gd}	0.003 pf
R_s	4Ω	C_{ds}	0.02 pf
R_D	10 Ω	L_s	0.15 nH
R_i	2 Ω	L_g	0.5 nH
R_{sat}	700 Ω	L_d	0.5 nH

[a] Reference 22.

the avalanche breakdown voltage, V_s is the saturation voltage, and $I_Q = I_{fc}/2$. The maximum ac power which may be obtained is then given by

$$P_{ac} = I_{fc}(V_A - V_s)/8 \qquad (8\text{-}4\text{-}5)$$

In a typical power FET amplifier the highest voltage drop is reached between the gate and the drain when the channel is pinched off (i.e., the gate voltage is below threshold; see Fig. 8-4-3). At this point no drain-to-source current flows unless the avalanche breakdown occurs in the channel. This breakdown limits the maximum voltage swing and hence the maximum output power P_{max}.

A crude estimate of the breakdown voltage can be obtained using an analytical model describing the electric field distribution in GaAs FETs near the pinch-off [23]. This model is based on the device physics revealed by the numerical simulation of the recessed gate FETs (see Fig. 8-4-3) [23]. As can be seen from this figure, the electric field at the gate edge at the drain side of the gate diverges. Using a simplified model which was solved using the conformal mapping technique, W. Frensley showed that the electric field at this edge behaves as

$$F(x) = V_c f(B/A)/(Ax)^{1/2} \qquad (8\text{-}4\text{-}6)$$

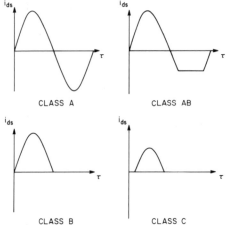

FIGURE 8-4-1. Qualitative waveforms for class A, class AB, class B, and class C amplifiers.

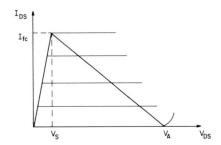

FIGURE 8-4-2. Schematic current–voltage characteristics and the load line for class A GaAs power FET
amplifier.

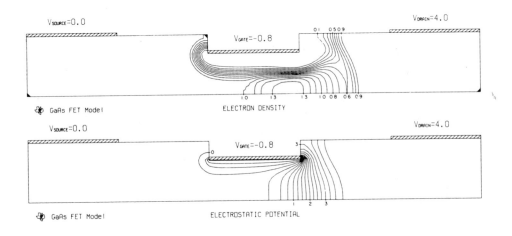

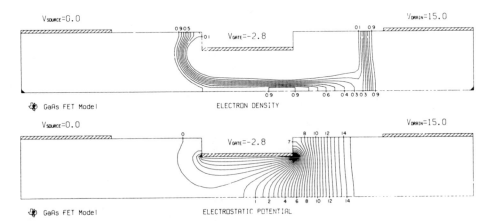

FIGURE 8-4-3. Results of the numerical simulation of a recessed gate GaAs FET [23]. The gate potential
indicated in the figures includes −0.8 built-in voltage. $V_{\text{gate}} = -2.8$ V and $V_{\text{drain}} = 15$ V correspond to
the breakdown condition. Notice electron depletion and accumulation at the drain side of the gate for
$V_{\text{gate}} = -0.8$ V corresponding to the high field domain and very high values of the electric field at the
drain side of the gate contact.

where

$$V_c = (qA/2\varepsilon) \int_0^A N(y)\, dy \qquad (8\text{-}4\text{-}7)$$

($V_c = V_{po}$ for the uniform doping profile), B is the length of the depletion layer extension beyond the gate, A is the thickness of the active channel, and $f(B/A)$ is a function shown in Fig. 8-4-4. The extension of the depletion region beyond the gate is found from

$$V_{dg} = V_c g(B/A) \qquad (8\text{-}4\text{-}8)$$

where function $g(B/A)$ is shown in Fig. 8-4-4. One may crudely assume $g \approx f^2$ [23]. The breakdown voltage is found from the approximate avalanche condition

$$\int_0^\infty \alpha\, dx = 1 \qquad (8\text{-}4\text{-}9)$$

where

$$\alpha = A_i \exp[-(F_0/F)^2] \qquad (8\text{-}4\text{-}10)$$

is the ionization rate, F is the electric field, and F_0 is the characteristic electric field of the impact ionization. The evaluation of this integral taking into account Eq. (8-4-6) yields the following expression for the breakdown voltage:

$$V_A = 2\varepsilon F_0^2/(qA_i Q_d) \qquad (8\text{-}4\text{-}11)$$

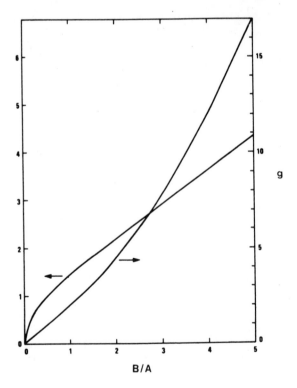

FIGURE 8-4-4. f and g vs. the depletion layer aspect ratio B/A [23]. f describes the magnitude of the singularity of the electric field at the gate edge; g is related to the potential at the edge of the depletion region.

where

$$Q_d = \int_0^A N_d(y)\, dy$$

is the total donor charge in the channel per unit of gate area. Equation (8-4-11) may be rewritten as

$$V_A = K/Q_d \tag{8-4-11a}$$

where the numerical constant K may be determined from the experimental data [24, 25]; see Fig. 8-4-5 ($K = 5.3 \times 10^{13}$ V/cm^2).

Substituting Eq. (8-4-11a) into Eq. (8-4-5) and taking into account that the knee voltage, V_s, is typically much smaller than the breakdown voltage we find

$$P_{ac} \approx qv_s KW/8 \tag{8-4-12}$$

which for $v_s = 1.5 \times 10^5$ m/s leads to $P_{ac} \approx 1.6$ W/mm gate periphery. The output power obtained in the best devices is generally close to $1 - 1.4$ W/mm gate periphery. We should also mention that mechanisms other than the avalanche breakdown at the drain may limit the breakdown voltage and the output power (see Section 8-4-3).

8-4-3. Design and Performance of GaAs Power FET Amplifiers

Basic GaAs FET structures utilized in power amplifiers are shown in Fig. 8-4-6 [26]. They include devices with and without the gate recess and with and without ion-implanted n^+ region under the ohmic contacts, which reduces the series source resistance. As can be seen from Fig. 8-4-7 [26] the presence and shape of the gate recess strongly influence the drain breakdown voltage. The drain-to-source breakdown voltage is also dependent on the drain-to-gate spacing. It may substantially increase if the drain-to-gate spacing is large enough to house a stationary high field domain forming at the drain side of the gate [9–12, 27–28]. For this reason there

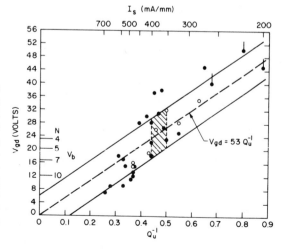

FIGURE 8-4-5. Experimental values of dc gate avalanche voltages for a fixed value of the current (1 mA/mm) vs. reciprocal donor charge in the channel [24]. ($Q_u = N_d A$, where the donor concentration N_d is in the range of 10^{16} cm^{-3} and the channel thickness A is in the range of micrometers.) The top scale is obtained by using an approximate relation for the channel current: I_s (mA/mm) $= 176 Q_u$. Open circles present data for a nonuniform wafer with $N_d = 10^7$ cm^{-3}.

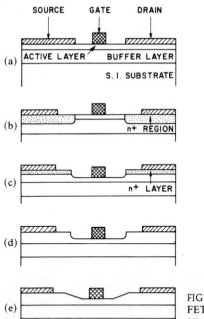

FIGURE 8-4-6. GaAs power FET structures [26]. (a) Flat type FET, (b) FET with n^+ implants for ohmic contacts, and (c), (d), (e) FETs with different gate recess.

may be some advantage in increasing the gate-to-drain spacing. As shown in Ref. 27 the high-field domain size may be roughly estimated as

$$d_{\text{dom}} \approx 2.06[\varepsilon V_{\text{dbr}}/(qN_d^{1/2}n_{\text{cr}}^{1/2})]^{1/2}$$

where V_{dbr} is the drain-to-source breakdown voltage, N_d is the doping density in the channel, and n_{cr} is the characteristic doping density ($\approx 3 \times 10^{15}$ cm^{-3} for GaAs).

The device structure shown in Fig. 8-4-6b may have the smallest source series resistance. The device structure of Fig. 8-4-6e has the highest breakdown voltage (see Fig. 8-4-7 [26]).

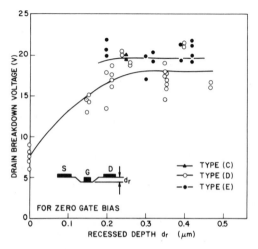

FIGURE 8-4-7. Dependence of the drain-to-source breakdown voltage on the recessed depth of the gate region at zero gate bias for different structures shown in Fig. 8-4-6 [26].

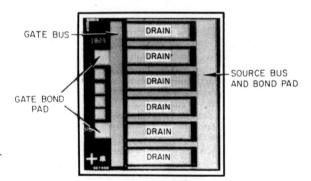

FIGURE 8-4-8. A 6-mm GaAs FET
[26].

As has been discussed in Section 8-3 with regard to low-noise GaAs amplifiers, the total gate periphery should be divided into several "unit gates" or gate fingers in order to minimize the series gate resistance and related gain degradation (see for example Fig. 8-4-8 [25]). As shown in Ref. 26 a transmission line model may be used in order to estimate the gain degradation:

$$G_d = \int_0^{z_u} \exp(-2\alpha z) \cos^2 \beta z \, dz \qquad (8\text{-}4\text{-}13)$$

where z_u is the unit gate width and α and β are propagation constants of a parallel transmission line:

$$
\begin{aligned}
\alpha &= \omega C_{gs}\{-L_g + [(L_g/C_{gs})^2 + (\omega L_g/g_s + r_g/\omega C_{gs})^2]^2\}^{1/2} \\
\beta &= \omega C_{gs}\{L_g + [(L_g/C_{gs})^2 + (\omega L_g/g_s + r_g/\omega C_{gs})^2]^2\}^{1/2}
\end{aligned} \qquad (8\text{-}4\text{-}14)
$$

Here $\omega = 2\pi f$ is the angular frequency, f is the operating frequency, C_{gs} is the gate-to-source capacitance, L_g is the gate inductance, g_s is the gate-to-source conductance, r_g is the gate resistance. Equation (8-4-14) was derived assuming $g_s \gg \omega C_{gs}$ and $L_g g_s \gg r_g C_{gs}$. The calculated dependences of the gain degradation on the unit gate width z_u are shown in Fig. 8-4-9 [26].

Depending on how the unit gates are connected, the power FET designs may be divided into two groups—with the cross-over of the connecting metal over the contacts and without the cross-over. The isolation between the connecting metal

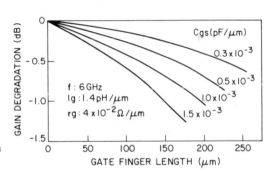

FIGURE 8-4-9. Calculated gain degradation
vs. gate finger width [26].

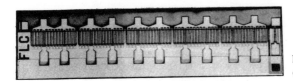

FIGURE 8-4-10. Dielectric isolation cross-over FET structure (NEC FET) [26].

and the device contacts which are crossed in devices with a cross-over design may be achieved by using either a suitable dielectric (typically CVD deposited SiO_2) or air bridge technology (see Figs. 8-4-10 and 8-4-11 [26]). Air bridges may be made by using a thick gold plating on top of a thick photoresist. The photoresist is then removed to make an air bridge. A slight advantage of this structure over dielectric isolation is a smaller parasitic capacitance. However, the parasitic capacitance of the interconnects is typically small compared to the parasitic capacitance of the gate, drain, and source pads.

Power FET structures without cross-overs include a wire bonding construction, the flip-chip configuration, and the via hole connection [26]. The wire bonding construction is shown in Fig. 8-4-12 [29]. This is perhaps the simplest design; however, it may lead to increased parasitic components.

The fabrication process of a "flip-chip" power FET is shown in Fig. 8-4-13 [30]. The posts are used in order to mount the chip on a heat sink. The technique results in a low thermal resistance and in a low common source inductance of the transistor. A chip carrier with a 1-W X-band device is shown in Fig. 8-4-14 [30]. The thermal resistance of the structure was 17°C/W. An infrared scan confirmed the temperature uniformity of the device, achieved because of a good thermal mount. The flip-chip design was first introduced by RCA [31] and later used by Microwave Semiconductor Corporation [30] and Mitsubishi [32].

The via hole source connect structure is shown in Fig. 8-4-15 [26, 33, 34]. This design results in a low parasitic source inductance, in a low parasitic capacitance, and in good heat-sinking because of the electroplated heat sink.

The thermal design of GaAs power FETs was reviewed in Ref. 35. In the analytical treatment the heat sources introduced by FETs were approximated by planar line heat sources (or by equivalent half cylinder line heat sources) on an infinite dielectric slab (see Fig. 8-4-16 [35]). This approach led to the following simple expression for the thermal impedance θ:

$$\theta = (\pi k)^{-1} \ln [8C/\pi d)]$$

(8-4-15)

where θ is in °C mm/W, k is the substrate thermal conductivity in W/°C mm, C is the substrate thickness, and d is the width of the thermal source (see Fig. 8-4-16). This expression is valid when $C \gg d$ (which is a practically important case) and in agreement with a computer solution when parameter d is adjusted to provide a good fit. The dependence of θ on d for various values of C is shown in Fig. 8-4-17

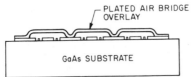

AIR BRIDGE OVERLAY SOURCE CONNECTIONS   FIGURE 8-4-11. Air bridge crossover FET structure [26].

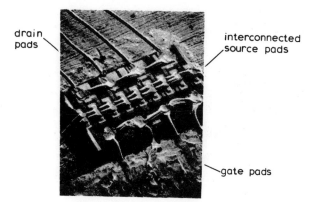

FIGURE 8-4-12. Power GaAs FET with wire bonding connections (Texas Instruments FET) [29].

[35] for $k = 0.038$ W/°C mm, which corresponds to the thermal conductivity of GaAs at 60°C [35].

For a practical design the results of the numerical solution which takes into account the finite width of the gate fingers, etc. may be used. Such results, reported in Ref. 35, are presented in Figs. 8-4-18 through 8-4-20. Different techniques for the measurement of the temperature distribution in GaAs FETs are also reviewed in Ref. 35. The measured temperature profile of a five-cell four-gate GaAs FET is shown in Fig. 8-4-21 as an example. The total gate periphery of this device was

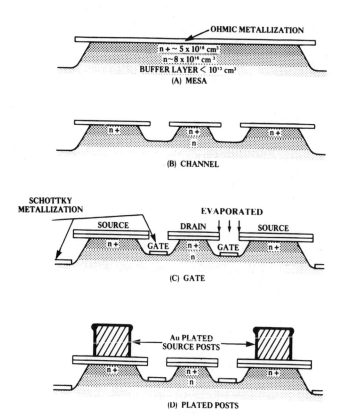

FIGURE 8-4-13. Fabrication of GaAs flip-chip FET [30].

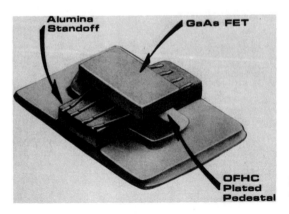

FIGURE 8-4-14. Chip carrier mounted GaAs FET [30].

3 mm and the peak temperature of 85°C corresponded to the thermal resistance of 10°C/W ($\theta = 57$°C mm/W) [35]. Experimental data agree within 10%–20% with the results of the numerical simulation. Both theory and experiment predict the value of $\theta \approx 70$–80°C mm/W for the thick backside mounted design, $\theta \approx$ 50–60°C mm/W for flip-chip configurations, and $\theta \approx 20$–40°C mm/W for the thin chip backside mounted design. As pointed out in Ref. 35, the thermal impedance shown in Figs. 8-4-18 through 8-4-20 is numerically close to the actual maximum temperature in the channel. Hence, it is possible to have a power GaAs FET with the maximum temperature only 20–60°C above ambient temperature.

As was discussed in Section 8-4-2 the output power is limited by the breakdown voltage. One possible mechanism which seems to be consistent with the experimental results is the avalanche breakdown at the edge of the gate closer to the drain which was considered in Section 8-4-2. However, alternative mechanisms such as buffer layer conduction [36], high field domains [9–11, 37], traps in the interface region [38], thermal burnout [39], and the properties of the channel-substrate interface [40] have also been considered.

An interesting new design of a GaAs power FET is based on the idea that the depletion layer at the substrate–active channel interface limits the output power [40]. As a consequence a castellated gate GaAs FET has been designed in order to ensure that the depletion of the active channel is controlled by the Schottky barrier at the gate–channel interface (see Figs. 8-4-22 and 8-4-23) [40]. As shown in the figures, the gate periodically touches the substrate, eliminating the drain-to-source leakage current and enhancing the breakdown voltage as a consequence. The performance of a conventional GaAs MESFET and of a GaAs MESFET with a castellated gate are compared in Fig. 8-4-24 [40]. As can be seen from the figure the castellated gate FET has a constant value of the power-added efficiency whereas the power-added efficiency of the conventional FET degrades with the drain bias.

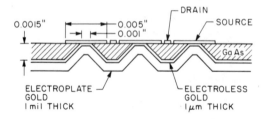

FIGURE 8-4-15. GaAs FET structure with via hole source connections [26].

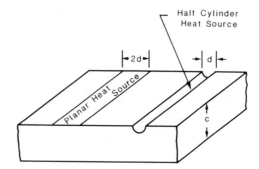

FIGURE 8-4-16. Equivalent planar and half-cylinder sources on an infinite dielectric slab of thickness C [35].

Typical values of the maximum output power in the best GaAs power FETs are in excess of 1 W/mm gate periphery. Output power up to 1.4 W/mm gate has been reported by Macksey and Derbeck [41] (see Fig. 8-4-25). The output power drops with frequency (see, for example, Fig. 8-4-26 [42] and Table 8-4-1 [43]). An output power of 710 mW with 4.5 db gain and 17.7% power-added efficiency has been achieved at 21 GHz for GaAs FETs fabricated using molecular beam epitaxy [44]. The output power per millimeter of gate periphery was 563 mW/mm gate periphery. The largest power has been achieved for InP MISFETs (4.2 W/mm gate at 9 GHz with 40% power added efficiency and 4 dB gain [45]). These devices, however, suffer from the carrier injection into the insulator of the metal–insulator–semiconductor gate [45].

8-5. MONOLITHIC MICROWAVE INTEGRATED CIRCUITS (MMICs)

The concept of monolithic microwave integrated circuits (MMICs) was first explored in 1965 [46]. However, the poor quality of semi-insulating silicon substrates did not allow fabrication of reliable circuits. In 1968 Mehal and Wacker [47] reported an integrated 94-GHz receiver front end implemented on a semi-insulating GaAs substrate. The circuit utilized GaAs Gunn diodes and Schottky diodes. In 1976 monolithic broadband GaAs MESFET amplifiers were reported [48], and since that time GaAs FETs have emerged as prime contenders for the monolithic microwave integrated circuit technology [49–88].

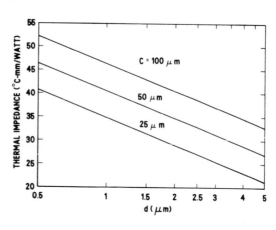

FIGURE 8-4-17. Calculated thermal impedance [see Eq. (8-4-15)] as a function of the equivalent heat source dimension (see Fig. 8-4-16) for different substrate thicknesses [35].

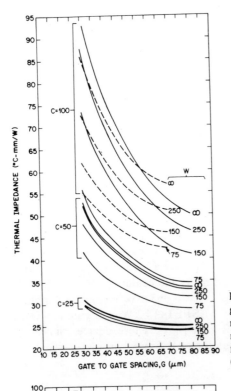

FIGURE 8-4-18. Thermal impedance as a function of gate-to-gate spacing (G) for various substrate thicknesses (C) and gate finger widths (W). Backside mounted (solid lines) and flip-chip (dashed lines) configurations are included [35]. The value of $k = 0.038$ W/°C mm was used in this calculation.

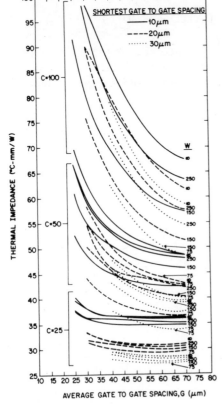

FIGURE 8-4-19. Thermal impedance as a function of average gate-to-gate spacing (G) for various substrate thicknesses (C), gate finger widths (W), and shortest gate-to-gate spacing (gate spacings need not be equal). The results are for backside mounted chips [35]. The value of $k = 0.038$ W/°C mm was used in this calculation.

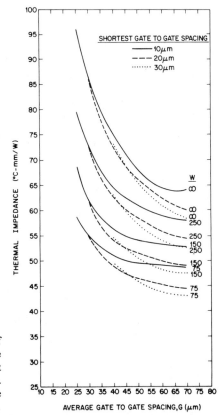

FIGURE 8-4-20. Thermal impedance as a function of average gate-to-gate spacing (G) for various substrate thicknesses (C), gate finger widths (W), and shortest gate-to-gate spacing (gate spacings need not be equal). The results are for flip-chip configuration [35]. The value of $k = 0.038$ W/°C mm was used in this calculation.

The key advantages of this technology include broad band performance, improved reliability and reproducibility, circuit design flexibility, multifunction performance on a chip, small size and weight, and potentially low cost [49].

There are also disadvantages, such as small device-to-chip area ratio, difficulties of circuit trouble-shooting and tweaking, a possibility of crosstalk between different elements of the circuit, and difficulty of integrating high power sources such as IMPATTs [49]. The performance of MMICs may also suffer compared to the performance of discrete FETs because of additional processing steps involved in MMIC fabrication. In particular, thermal cycles associated with the dielectric

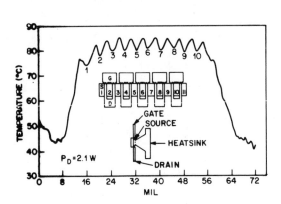

FIGURE 8-4-21. Temperature profile of a five-cell four-gate FET [35].

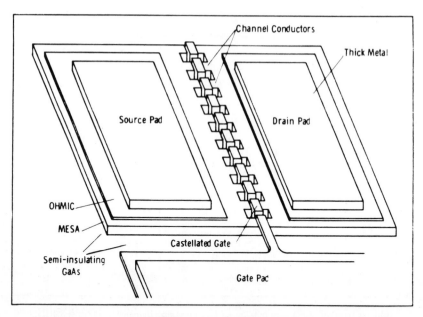

FIGURE 8-4-22. A schematic diagram of a castellated gate GaAs power FET [40].

deposition may degrade ohmic contacts. Exposed active layers (especially between gate and source) may be damaged or etched during dielectric plasma deposition and reactive ion etching. Also, the high dielectric constant material (such as silicon nitride) deposited over the FETs increases the gate-to-source and gate-to-drain capacitances [55].

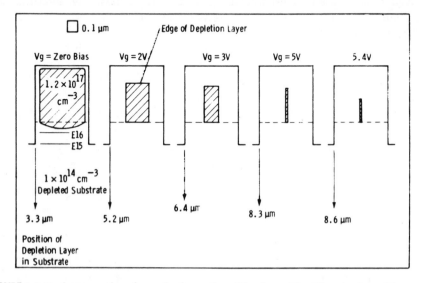

FIGURE 8-4-23. A cross section of a conducting section of the channel for different values of the negative gate voltages [40]. Numbers in micrometers represent the depth of the depletion layer in the substrate.

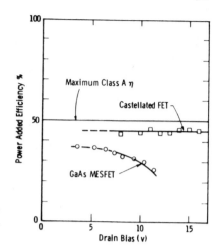

FIGURE 8-4-24. The power added efficiency vs. drain-to-source voltage for a conventional GaAs FET and for a castellated FET [40].

Difficulties in tweaking and trouble-shooting may be overcome by using computer aided design (CAD) techniques in designing and trouble shooting MMICs [88]. Also, a technological process may be chosen to minimize the effects of additional processing steps on the characteristics of discrete devices.

An example of the MMIC fabrication process described in Refs. 54, 55, and 87 is illustrated by Fig. 8-5-1. As can be seen from the figure, FET fabrication is complete after step 3. The remaining steps are required for the fabrication of metal–insulator–metal (MIM) capacitances, rf and dc circuitry, via holes, etc. in order to complete a MMIC. In addition to a reactively sputtered silicon nitride cap

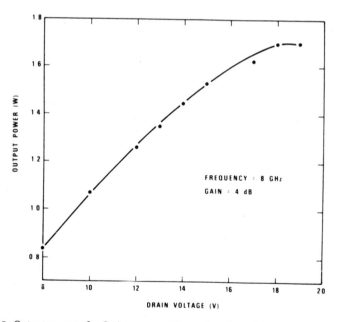

FIGURE 8-4-25. Output power of a GaAs power FET as a function of the drain voltage for the device with a gate periphery of 1.2 mm [41].

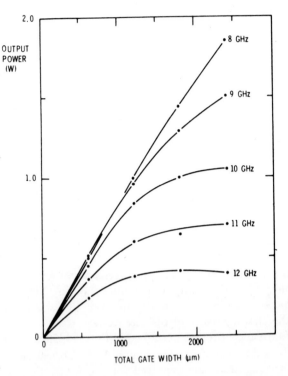

FIGURE 8-4-26. Maximum output power (with 4 dB gain) of a 2-μm gate length GaAs FET vs. the total gate width at several frequencies [42]. The drain bias is 8 V in all cases.

used in the ion implantation process and not shown in Fig. 8-5-1, two silicon nitride layers are used in this process. The first layer is deposited after the ohmic contact alloying and is used for dielectric-aided lift-off of FET gates of the first metallization layer which formed ohmic contact overlays and the bottom contact of MIM capacitors. The second dielectric layer is used as an insulator in MIM capacitors [55].

TABLE 8-4-1. GaAs Power FET Performance[a]

Frequency (GHz)	Output power (W)	Gain (dB)	Power-added efficiency (%)	Gate length (μm)	Total gate width (mm)	Drain voltage (V)	Number of cells
8	5.1	5.0	35	1.0	6.4	13	4
8	3.9	7.0	39	0.8	4.8[b]	10	4
8	1.7	4.0	37	2.5	1.2	18	2
9	5.3	4.2	19	1.0	9.6	13	8
10	4.0	3.6	21	1.0	4.8	11	4
10	3.9	6.0	39	0.8	4.8[b]	10	4
10	0.87	6.0	45	0.7	1.2	8	1
12	3.2	4.0	25	0.7	4.8	9	4
15	1.8	4.0	21	0.7	2.4[b]	8	4
16	1.6	4.0	20	0.7	2.4[b]	8	4
18	0.85	4.0	24	0.5	1.2[b]	10	4
20	1.0	4.0	17	0.5	2.7	8	2

[a] Reference 43.
[b] Low source-lead inductance configurations.

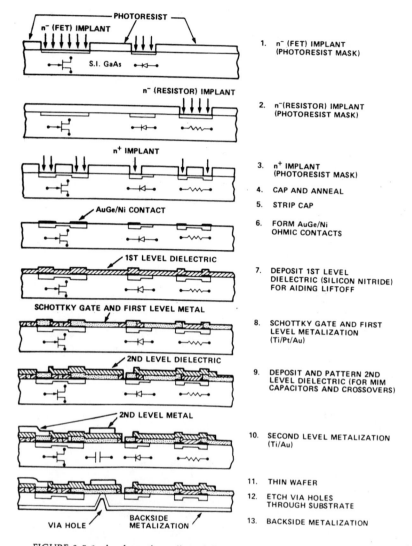

1. n⁻ (FET) IMPLANT
   (PHOTORESIST MASK)

2. n⁻(RESISTOR) IMPLANT
   (PHOTORESIST MASK)

3. n⁺ IMPLANT
   (PHOTORESIST MASK)

4. CAP AND ANNEAL

5. STRIP CAP

6. FORM AuGe/Ni
   OHMIC CONTACTS

7. DEPOSIT 1ST LEVEL
   DIELECTRIC (SILICON NITRIDE)
   FOR AIDING LIFTOFF

8. SCHOTTKY GATE AND FIRST
   LEVEL METALIZATION
   (Ti/Pt/Au)

9. DEPOSIT AND PATTERN 2ND
   LEVEL DIELECTRIC (FOR MIM
   CAPACITORS AND CROSSOVERS)

10. SECOND LEVEL METALIZATION
    (Ti/Au)

11. THIN WAFER

12. ETCH VIA HOLES
    THROUGH SUBSTRATE

13. BACKSIDE METALIZATION

FIGURE 8-5-1. A schematic outline of the MMIC fabrication process [87].

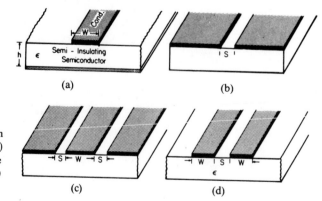

FIGURE 8-5-2. Four propagation modes for monolithic circuits. (a) Microstrip (MS); (b) slot line (SL); (c) coplanar waveguide; (d) coplanar strips (CS) [49].

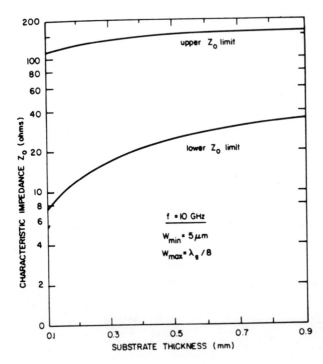

FIGURE 8-5-3. Range of characteristic impedance of microstrip on GaAs substrate as a function of substrate thickness [49]. W is the microstrip width; λ is the wavelength.

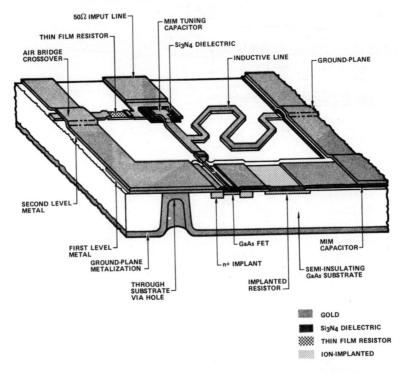

FIGURE 8-5-4. Different elements used in GaAs MMICs [87].

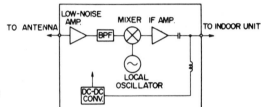

FIGURE 8-5-5. Typical block diagram of an outdoor unit of a DBS receiver [58].

The back-side metallization is required in order to provide microstrip connections between the elements of the circuit. Several alternative propagation modes for monolithic circuits are compared in Fig. 8-5-2 [49]. The characteristic impedance of a typical connection—microstrip—depends on the substrate thickness and varies approximately between 10 and 100 Ω (see Fig. 8-5-3 [49]).

When the ground connection is required on the top plane of the chip it may be provided using via holes etched in the GaAs wafer (see Fig. 8-5-1). The estimated inductance of a via hole is approximately 40–60 pH/mm of substrate thickness [49].

In addition to MIM capacitors (see Fig. 8-5-4), planar inductors and planar resistors may be fabricated. A schematic illustration of different elements used in MMICs is shown in Fig. 8-5-4 [87]. More detailed discussion of the relevant properties of thin insulating and resistive films used in MMICs may be found in Ref. 49.

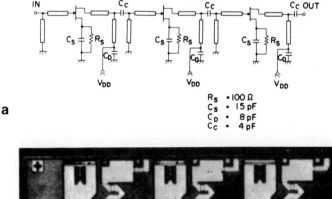

a

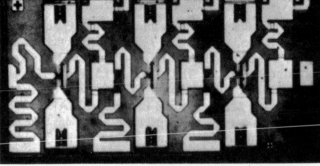

b

FIGURE 8-5-6. (a) Circuit diagram of a three-stage low-noise amplifier [58]. (b) Three-stage low-noise amplifier chip (1.5 × 3.0 mm) [58].

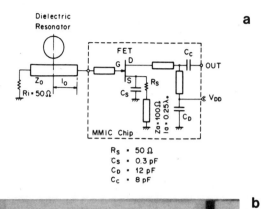

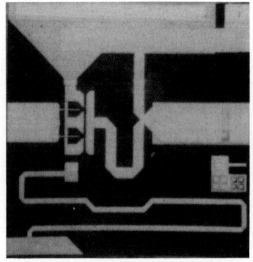

R_S = 50 Ω
C_S = 0.3 pF
C_D = 12 pF
C_C = 8 pF

FIGURE 8-5-7. (a) Circuit diagram of a dielectric resonator oscillator [58]. (b) Dielectric resonator oscillator chip (1.5 × 1.5 mm) [58].

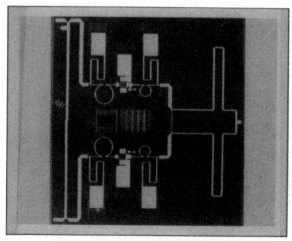

FIGURE 8-5-8. GaAs monolithic 90-degree phase shifter. Sized 4.5 × 4.5 × 0.1 mm, the chip uses air bridges for connecting the interdigitated conductors of the coupler as well as the two second gate-pads of the dual-gate FET. Si_3N_4 MIM rf bypass capacitors are fabricated on the same chip [72].

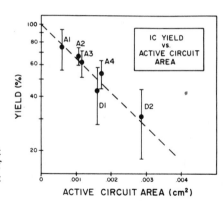

FIGURE 8-5-9. GaAs IC yield for six different circuit designs (*A*, amplifier type; *D*, digital type). Average of individual wafer yields is plotted against active circuit area. Bars indicate one standard deviation [52].

Some recent examples of MMICs include GaAs low-noise amplifiers, monolithic ICs for direct broadcast satellite (DBS) receivers, and monolithic phase shifters.

MMIC chips described in Ref. 58 included a 12-GHz low-noise three-stage amplifier, a 1-GHz IF amplifier, and an 11-GHz dielectric resonator oscillator. These MMICs were incorporated into a DBS receiver (see Fig. 8-5-5), which exhibited an overall noise figure less than 4 dB for frequencies from 11.7 to 12.2 GHz.

The three-stage low-noise MMIC amplifier reported in Ref. 58 exhibited a 3.4-dB noise figure and a 19.5-dB gain over 11.7–12.2 GHz. The negative-feedback-type three-stage IF amplfier showed a 3.9-dB noise figure and a 23-dB gain over 0.5–1.5 GHz. The dielectric resonator oscillator provided 10 mW output power at 10.67 GHz with a frequency stability of 1.5 mHz over a temperature range from −40 to 80°C. The circuit diagrams and photographs of the three-stage low-noise MMIC amplifier and of the MMIC dielectric resonator are shown in Figs. 8-5-6 and 8-5-7.

A GaAs monolithic 90-degree phase shifter which includes a 90-degree interdigitated coupler, two dual gate FETs, and in-phase coupler is shown in Fig. 8-5-8 [72].

One of the most important problems facing GaAs MMICs is the problem of manufacturing yield. In Ref. 52 an automated rf test system was used to collect functional yield data for a variety of linear and digital GaAs ICs. The data collected from 23 tested wafers are presented in Fig. 8-5-9 [52]. The functional yield percentages (full-wafer, including edges) are plotted against active circuit area (for devices and wiring) for four different amplifier circuits (A1–A4) and two digital circuits (D1–D2). Functional yields ranging from 13% to 97% are included. This study showed that the manufacturing yield may be substantially increased by using larger wafers and better lithographic techniques [52]. However, even present yields are adequate for economic manufacturing of small-to-medium scale GaAs ICs [52].

The examples of microwave monolithic integrated circuits given above clearly demonstrate that GaAs MMIC technology has matured and MMIC circuits have been used for a variety of microwave applications.

REFERENCES

1. M. Cuevas, A microprocessor mediates the noise figure debate, *Microwaves* **December**, 51 (1981).
2. A. van der Ziel, Thermal noise in field effect transistor, *Proc. IRE* **50**, 1808–1812 (1962).
3. A. van der Ziel, Gate noise in field effect transistors at moderately high frequencies, *Proc. IRE* **51**, 462–467 (1963).

4. C. A. Liehti, Microwave field-effect transistors, *IEEE Trans. Microwave Theory Technique* **MTT-24**, 279-300 (1976).

5. R. A. Pucel, H. A. Haus, and H. Statz, Signal and noise properties of gallium arsenide microwave field-effect transistors, in *Advances in Electronics and Electron Physics*, Vol. 38, Academic, New York, 1975, pp. 195-265.

6. H. Fukui, Optimum noise figure of microwave GaAs MESFETs, *IEEE Trans. Electron Devices* **ED-26**, 1032-1037 (1979).

7. H. Fukui, Determination of the basic device parameters of a GaAs MESFET, *Bell Syst. Tech. J.* **58**, 771-797 (1979).

8. A. Podell, A Functional GaAs FET noise model, *IEEE Trans. Electron Devices* **ED-28**(5), 511-517 (1981).

9. R. W. H. Engelmann and C. A. Liechti, Gunn domain formaton in the saturated region of GaAs MESFETs, *IEDM Tech.* Digest **Dec**. 351-354 (1976).

10. R. W. H. Engelmann and C. A. Liechti, Bias dependence of GaAs and InP MESFET parameters, *IEEE Trans. Electron Devices* **ED-24**(11), 1288-1296 (1977).

11. M. S. Shur and L. F. Eastman, Current–voltage characteristics, small-signal parameters and switching times of GaAs FETs, *IEEE Trans. Electron Devices* **ED-25**(6), 606-611 (1978).

12. R. E. Neidert and C. J. Scott, Computer program for microwave GaAs MESFET modeling, NRL Report 8561, Naval Research Laboratory, February 12, 1982.

13. M. Reiser, Two-dimensional analysis of substrate effects in junction FETs, *Electron. Lett.* **6**, 493-494 (1970).

14. L. F. Eastman and M. S. Shur, Substrate current in GaAs MESFETs, *IEEE Trans. Electron Devices* **ED-26**, 1359-1361 (1979).

15. R. A. Kiehl and G. C. Osborn, Physics of short gate GaAs MESFETs from hydrostatic pressure studies, *IEEE Trans. Electron Devices* **ED-28**(8), (1981).

16. W. C. Bruncke and A. Van. der Ziel, Thermal noise in junction gate field effect transistor, *IEEE Trans. Electron Devices* **ED-13**, 323-329 (1966).

17. J. A. Turner, R. S. Butlin, D. Parker, R. Bennet, A. Peake, and A. Hughes, The noise and gain performance of submicron gate length GaAs FETs, in *GaAs FET Principles and Technology*, Ed. by J. V. DiLorenzo and D. D. Khandelwal, Arctech House, Dedham, Massachusetts, 1982, pp. 151-174.

18. F. Hasegawa, Low Noise GaAs FETs, in *GaAs FET Principles and Technology*, Ed. by J. V. DiLorenzo and D. D. Khandelwal, Arctech House, Dedham, Massachusetts, 1982, pp. 177-193.

19. NEC application notes, published by California Eastern Labs., Inc., exclusive sales agent for NEC Corporation, Santa Clara, California (1983).

20. K. Kamei, H. Kawasaki, T. Chigua, T. Nakanier, T. Kawabuchi, and M. Yoshimi, Extremely low-noise MESFET's fabricated by metelorganic chemical vapor deposition, *Electron. Lett.* **17**, 450-451 (1981).

21. C. H. Oxley, A. H. Peake, R. H. Bennet, J. Arnold, and R. S. Butlin, Q-band (26-40 GHz) GaAs FETs, *IEDM Tech.* Digest **Dec**, 680-683 (1981).

22. P. W. Chye and C. Huang, Quarter micron low noise GaAs FET's, *IEEE Electron Device Lett.* **EDL-3**(12), 401-403 (1982).

23. W. R. Frensley, Power limiting breakdown effects in GaAs MESFETs, *IEEE Trans. Electron Devices* **ED-28**(8), 962-970 (1981).

24. S. H. Wemple, W. C. Niehaus, H. M. Cox, J. V. DiLorenzo, and W. O. Schlosser, Control of gate-drain avalanche in GaAs MESFETs, *IEEE Trans. Electron Devices* **ED-27**, 1013-1018 (1980).

25. W. C. Niehaus, S. H. Wemple, L. A. D'Asaro, H. Fukui, J. C. Irvin, H. M. Cox, J. V. DiLorenzo, J. C. M. Hwang, and W. O. Schlosser, GaAs power FET design, in *GaAs FET Principles and Technology*, Ed. by J. V. DiLorenzo and D. D. Khandeiwal, Arctech House, Dedham, Massachusetts, 1982, pp. 279-306.

26. F. Hasegawa, Power GaAs FETs, in *GaAs FET Principles and Technology*, Ed. by J. V. DiLorenzo and D. D. Khandelwal, Arctech House, Dedham, Massachusetts, 1982, pp. 219-255.

27. L. F. Eastman, S. Tiwari, and M. S. Shur, Design criteria for GaAs MESFETs related to stationary high field domains, *Solid State Electron.* **23**, 383-389 (1980).

28. M. S. Shur, L. F. Eastman, S. Judraprawira, J. Gammel, and S. Tiwari, Design Criteria for GaAs MESFETs related to stationary high field domains, *IEDM Tech.* Digest **Dec**, 381-383 (1978).

29. H. Macksey, R. L. Adams, D. N. McQuiddy, and W. R. Wisseman, X-band performance of GaAs power FETs, *Electron. Lett.* **12**(2), (1976).

30. I. Drukier, Power GaAs FETs, in *GaAs FET Principles and Technology*, Ed. by J. V. DiLorenzo and D. D. Khandelwal, Arctech House, Dedham, Massachusetts, 1982, pp. 202-217.

31. I. Drukier *et al.*, *Electron. Lett.* **11**, 104 (1975).

32. Y. Mitsui *et al.*, Europian Microwave Conference Technical Digest, p. 272, 1979.

33. L. A. D'Asaro, J. V. DiLorenzo, and H. Fukui, *IEDM Tech. Digest*, 370 (1977).

34. B. S. Hewitt *et al.*, European Microwave Conference Technical Digest, p. 265, 1979.

35. S. H. Wemple and H. C. Huang, Thermal Design of Power GaAs FETs, in *GaAs FET Principles and Technology*, Ed. by J. V. DiLorenzo and D. D. Khandelwal, Arctech House, Dedham, Masachusetts 1982, pp. 309–347.

36. S. Tiwari, L. F. Eastman, and L. Rathburn, Physical and material limitations on burnout voltage of GaAs power MESFETs, *IEEE Trans. Electron Devices* **ED-27**, 1045 (1980).

37. T. Furutsuka, T. Tsuji, and F. Hasegawa, *IEEE Trans. Microwave Theory Technique* **MTT-24**, 512 (1978).

38. P. Ladbrooke and A. L. Martin, Material and structure factors affecting the large signal operation of GaAs MESFETs, International Conference on Semi-insulating GaAs, p. 313, France.

39. K. Morizane, M. Dosen, and Y. Mori, A mechanism of source–drain burnout in GaAs MESFETs, *Inst. Phys. Conf. Ser.* **45**, 287 (1979).

40. R. C. Clarke, A high-efficiency castellated gate power FET, in Proceedings of IEEE/Cornell Conference on High-Speed Semiconductor Devices and Circuits, IEEE Cat. No. 83ch1959-6, pp. 93–111, Ithaca, New York, 1983.

41. H. M. Macksey and F. H. Doerbeck, GaAs FETs having high output power per unit gate width, *IEEE Electron Device Lett.* **EDL-2**(6), 147–148 (1981).

42. H. M. Macksey, R. L. Adams, D. N. McQuiddy, D. W. Shaw, and W. R. Wisseman, Dependence of GaAs power MESFET microwave performance on device and material parameters, *IEEE Trans. Electron Devices* **ED-24**(2), 113–122 (1977).

43. H. M. Macksey, GaAs power FET design, in *GaAs FET Principles and Technology*, Ed. by J. V. DiLorenzo and D. D. Khandelwal, Arctech House, Dedham, Massachusetts, 1982, pp. 257–275.

44. P. Saunier and H. D. Shih, State-of-the-art K-band GaAs power field effect transistors prepared by molecular beam epitaxy, *IEEE Trans. Elecron Devices* **ED-30**(11), 1599 (1983).

45. M. Armand, D. V. Bui, J. Chevrier, and N. T. Linh, High power microwave amplification with InP MISFETs, in Proceedings of IEEE/Cornell Conference on High-Speed Semiconductor Devices and Circuits, IEEE Cat. No. 83ch1959-6, pp. 218–225, Ithaca, New York, 1983.

46. T. M. Hyltin, Microstrip transmission on semiconductor substrates, *IEEE Trans. Microwave Theory Tech.* **MTT-13**, 777–781 (1965).

47. E. Mehal and R. W. Wacker, GaAs integrated microwave circuits, *IEEE Trans. Microwave Theory Tech.* **MTT-16**, 451–454 (1968).

48. R. S. Pengelly and J. A. Turner, Monolithic broadband GaAs FET amplifiers, *Electron. Lett.* **12**, 251–252 (1976).

49. R. A. Pucel, Design considerations for monolithic microwave circuits, *IEEE Trans. Microwave Theory Techniques* **MTT-29**(6), 513–534 (1981).

50. C. Kermarrec, J. Gaguet, P. Harrop, and C. Tsironis, Monolithic circuits for 12 GHz direct broadcasting satellite reception, in Proc. 1982 Microwave and Millimeter-Wave Monolithic Circuits Symp., June 1982, Dallas, p. 5.

51. W. C. Petersen, A. K. Gupta, and D. R. Decker, A monolithic GaAs dc to 2 GHz feedback amplifier, *IEEE Trans. Electron Devices* **ED-30**, 27-29 (1983).

52. R. L. Van Tuyl, V. Kumar, D. C. D'Avanzo, T. W. Taylor, V. E. Peterson, D. P. Hornbuckle, R. A. Fisher, and D. B. Estreich, A manufacturing process for analog and digital gallium arsenide integrated circuits, *IEEE Trans. Electron Devices* **ED-29**(7), 1032–1037 (1982).

53. A. K. Gupta, W. C. Petersen, and D. R. Decker, Yield considerations for ion implanted GaAs MMICs, *IEEE Trans. Electron Devices* **ED-30**, 16–20 (1983).

54. L. R. Decker, W. C. Petersen, and A. K. Gupta, Monolithic GaAs microwave analog integrated circuits, Electronics Technology and Devices Lab., Technical report No. DELET-TR-78-2999-F Final Report, Aug. 1982.

55. A. K. Gupta, D. P. Siu, and K. T. Ip, Low-noise MESFETs for ion-implanted GaAs MMICs, *IEEE Trans. Electron Devices* **ED-30**(12), 1850–1854 (1983).

56. V. Sokolov, J. J. Geddes, A. Contolatis, P. E. Bauhahn, and C. Chao, A Ku-band GaAs monolithic phase shifter, *IEEE Trans. Electron Devices* **ED-30**(12), 1855–1861 (1983).

57. T. Sugiura, H. Itoh, T. Tsuji, and K. Honjo, 12 GHz-band low-noise GaAs monolithic amplifiers, *IEEE Trans. Electron Devices* **ED-30**(12), 1861–1866 (1983).

58. S. Hori, K. Kaprei, K. Shibata, M. Tatematsu, K. Mishima, and S. Okana, GaAs monolithic MIC's for direct broadcast satellite receivers, *IEEE Trans. Electron Devices* **ED-30**(12), 1867–1874 (1983).

59. G. Avery, The GaAs IC industry structure—Present and future, GaAs IC Symposium Technical Digest, p. 3, Phoenix, Arizona, October 1983.

60. T. Sugiura, K. Honjo, and T. Tsuji, 12 GHz-band GaAs dual-gate MESFET monolithic mixers, GaAs IC Symposium Technical Digest, pp. 3-6, Phoenix, Arizona, October 1983.

61. S. Moghe, T. Andrade, G. Policky, and C. Huang, A wideband two stage miniature amplifier, GaAs IC Symposium Technical Digest, pp. 7-10, Phoenix, Arizona, October 1983.

62. G. Kaelin, J. Seligman, and A. Gupta, 20 GHz two stage low noise monolithic amplifier, GaAs IC Symposium Technical Digest, pp. 11-12, Phoenix, Arizona, October 1983.

63. B. Considine and D. Wandrei, X-band receive module using monolithic GaAs MMIC's, GaAs IC Symposium Technical Digest, pp. 13-15, Phoenix, Arizona, October 1983.

64. H. Finlay, J. Jenkins, R. Pengelly, and J. Cockrill, Accurate coupling predictions and assessments in MMIC networks, GaAs IC Symposium Technical Digest, pp. 16-19, Phoenix, Arizona, October 1983.

65. M. Le Brun, P. Jay, C. Rumelhard, G. Rey, and P. Delescluse, Monolithic microwave ampifier using a two-dimensional electron GAS FET—A comparison with GaAs, GaAs IC Symposium Technical Digest, pp. 20-24, Phoenix, Arizona, October 1983.

66. C. Suckling, M. Williams, T. Banbridge, R. Pengelly, K. Vanner, and R. S. Butlin, An S-band phase shifter using monolithic GaAs circuits, GaAs IC Symposium Technical Digest, pp. 102-105, Phoenix, Arizona, October 1983.

67. Y. Ayasli, R. Mozzi, T. Tsukii, and L. Reynolds, 6-19 GHz GaAs FET transmit-receive switch, GaAs IC Symposium Technical Digest, pp. 106-108, Phoenix, Arizona, October 1983.

68. E. Strid, A monolithic 10 GHz vector modulator, GaAs IC Symposium Technical Digest, pp. 109-112, Phoenix, Arizona, October 1983.

69. G. Kaelin, J. Seligman, and A. Gupta, A wide band medium-power monolithic microwave amplifier, GaAs Symposium Technical Digest, pp. 113-114, Phoenix, Arizona, October 1983.

70. J. Dormail, Y. Tajima, R. Mozzi, M. Durschlag, S. McOwen, and A. Morris, A 2-8 GHz 2-watt monolithic amplifier, GaAs IC Symposium Technical Digest, pp. 115-118, Phoenix, Arizona, October 1983.

71. W. Petersen and A. Gupta, A three-stage power amplifier for a 20 GHz monolithic transmit module, GaAs IC Symposium Technical Digest, pp. 119-122, Phoenix, Arizona, October 1983.

72. L. C. Upadhyayula, M. Kumar, and H. C. Huang, GaAs MMICs could carry the waves of the high-volume future, *Microwave Syst. News* **July**, 58 (1983).

73. H. Q. Tserng and H. M. Macksey, Performance of monolithic GaAs FET oscillators at J-band, *IEEE Trans. Electron Devices* **ED-28**, 163-165 (1981).

74. M. Kumar, G. C. Taylor, and H. C. Huang, Monolithic dual-gate FET amplifier, *IEEE Trans. Electron. Devices* **ED-28**, 197-198 (1981).

75. M. Kumar, S. N. Subbarao, R. J. Menna, and H. C. Huang, Monolithic GaAs interdigitated couplers, *IEEE Trans. Microwave Theory Techniques* **MMT-31**(1), 29-32 (1983).

76. M. Kumar, S. N. Subbarao, R. J. Menna, and H. C. Huang, Broadband active phase shifter using dual-gate MESFET, *IEEE Trans. Microwave Theory Techniques* **MTT-29**, 1098-1102 (1981).

77. R. L. Vantuyl and C. Liechti, High-speed GaAs MSI, ISSCC Digest of Technical Papers, pp. 20-21, February 1976.

78. L. C. Upadhyayula, GaAs FET comparators for high-speed analog-to-digital conversion, in Digest of Technical Papers, GaAs IC Symp., Lake Tahoe, NE, Sept. 1979.

79. L. C. Upadhyayula, W. R. Curtice, and R. Smith, Design, fabrication and evaluation of 2- and 3-bit GaAs MESFET analog-to-digital converter ICs, *IEEE Trans. MTT* **MTT-31**(1), 2 (1983).

80. W. C. Petersen, D. R. Decker, A. K. Gupta, J. Dully, and D. R. Ch'en, A monolithic GaAs 0.1 to 10 GHz amplifier, IEEE MTT-S Symposium Digest, No. 81CH1592-5, pp. 354-355, June 1981.

81. Y. Ayasli, R. Mozzi, L. Hanes, and L. D Reynolds, An X-band 10 W monolithic transmit-receive GaAs FET switch, 1982 Monolithic Circuits Symposium Digest, IEEE catalog No. 82CH1784-8, pp. 42-46.

82. M. Kumar, S. N. Subbarao, R. J. Menna, and H. C. Huang, Monolithic GaAs interdigitated 90° hybrids with 50- and 25-Ohm impedance, 1982 Monolithic Circuits Symposium Digest, IEEE catalog No. 82Ch1794-9, pp. 50-53.

83. G. E. Brehm and R. E. Lehmann, Monolithic GaAs FET low-noise amplifiers for X-band applications, *Microwave J.* **25**(11), 103-107 (1982).

84. M. C. Driver, G. W. Eldridge, and J. E. Degenford, Broadband monolithic integrated power amplifiers in GaAs, *Microwave J.* **25**(11), 87-94 (1982).

85. A. Contolatis, C. Chao, S. Jamison, and C. Butter, *Ku*-band monolithic GaAs balanced mixers, 1982 Monolithic Circuits Symposium Digest, IEEE catalog No. 82CH1784-8, pp. 28–30.
86. B. N. Scott and G. E. Brehm, Monolithic voltage controlled oscillator for *X* and *Ku*-bands, IEEE MTT-S Symposium Digest, No. 82CH1705-3, pp. 482–485, June 1982.
87. D. R. Decker, Are MMICs a fad or fact?, *Microwave Syst. News* **13**(7) (1983).
88. L. Besser, Synthesize amplifiers exactly, *Microwave Syst. News* **Oct.**, 28–40 (1979).

9

GaAs Digital Integrated Circuits

9-1. INTRODUCTION

One of the most promising applications of GaAs technology is in ultrafast digital integrated circuits [1–206]. Gate delays as short as 15 ps for logic based on self-aligned GaAs MESFETs [124] at 300 K and of 11.6 ps at 300 K [200, 201] and 5.8 ps at 77 K [73] for logic based on modulation doped AlGaAs–GaAs transistors (also called HEMTs) have been achieved, making GaAs circuits the fastest solid state circuits. Circuits as complicated as 16×16 mutipliers [116, 117], 1-kb RAMs [122, 123, 143, 172], 4-kb static RAMs [172, 173], 16-kb static RAMs, and gate arrays [126, 127] have been built, and GaAs medium-scale integration (MSI) ICs have been fabricated with reasonable yields.

An approximate sketch of the propagation delay per gate for different solid state technologies is shown in Fig. 9-1-1.

Figure 9-1-2 shows a projected system gate delay for GaAs MESFET logic and (Al, Ga)As modulation doped FET logic [111] [also called high electron mobility transistor (HEMT) logic and selectively doped heterostructure transistor logic (SDHL)]. For comparison, the best expected system delay per chip for Si MOS transistors would be close to 500 ps.

Most GaAs digital ICs are fabricated by direct ion implantation into a semi-insulating GaAs substrate. Steady progress in substrate quality, fabrication process control, and GaAs circuit design since the first introduction of GaAs ICs

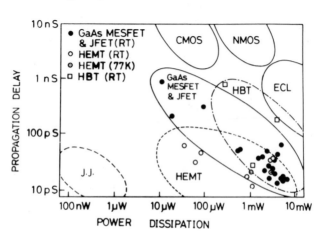

FIGURE 9-1-1. Propagation delay per gate versus power dissipation per gate for different solid state logic families [205].

431

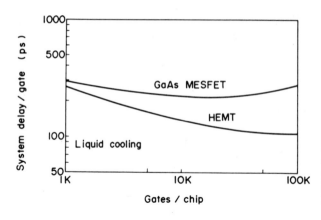

FIGURE 9-1-2. Predicted system delay per gate versus number of gates per chip for GaAs MESFET logic and HEMT (or MODFET) logic [111].

[1] makes it virtually certain that they will find many applications in gigabit communication systems and superfast computers.

In Section 9-2 we describe different GaAs logic families. In Sections 9-3 and 9-4 we consider two widely used GaAs logic families—direct coupled field effect transistor logic (DCFL) and Schottky diode field effect transistor logic (SDFL). In Section 9-5 we discuss modulation doped field effect transistor DCFL circuit design and simulation. In Section 9-6 we briefly review GaAs IC fabrication. Some examples of GaAs digital ICs are given in Section 9-7.

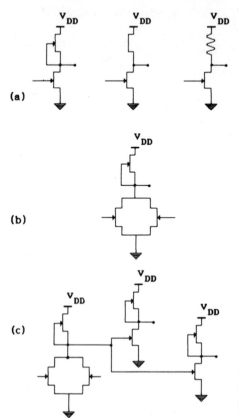

FIGURE 9-2-1. Direct coupled field effect transistor logic (DCFL). (a) Basic inverter with a FET load, an ungated FET load, and a resistive load. (b) NOR gate with fan-in 2. (c) NOR gate with fan-in 2 and fan-out 2.

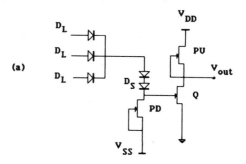

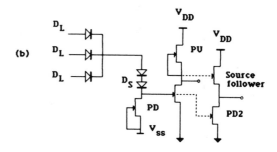

FIGURE 9-2-2. Schottky diode field effect transistor logic (SDFL). (a) Standard SDFL gate. Q, switching transistor; PU, pull-up transistor; PD, pull-down transistor; D_L, logic diode; D_S, level-shifting diode; OR function is implemented using logic Schottky diodes. (b) SDFL gate with an optional push-pull output buffer [129]. (c) SDFL OR/NAND gate [129].

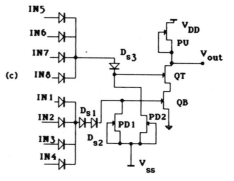

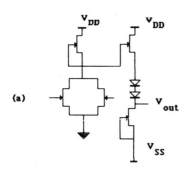

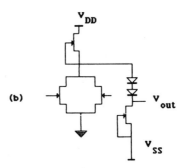

FIGURE 9-2-3. Buffered FET logic (BFL). (a) Basic inverter circuit with a source follower. (b) Basic inverter circuit without a source follower.

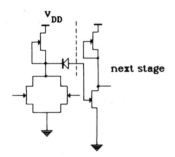

next stage

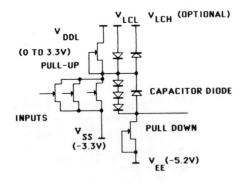

V_{LCL} V_{LCH} (OPTIONAL)

V_{DDL}

(0 TO 3.3V)

PULL-UP

CAPACITOR DIODE

INPUTS

V_{SS}
(-3.3V)

PULL DOWN

V_{EE} (-5.2V)

SWITCHING FETs 20 MICRON WIDE

PULL-UP 16.5 MICRON WIDE

PULL-DOWN 1.5 MICRON WIDE

FIGURE 9-2-4. (a) Capacitive coupled logic (CCL). (b) Capacitive diode-coupled FET logic (CDFL) three-input NOR logic gate with output level shifted [163].

9-2. GaAs LOGIC FAMILIES

Circuit diagams of different basic inverter logic gates implemented using GaAs FETs are shown in Figs. 9-2-1–9-2-7.

9-2-1. Direct Coupled Field Effect Transistor Logic (DCFL)

DCFL gates with a FET load, an ungated FET load, and a resistive load (see Fig. 9-2-1) have an advantage of circuit simplicity and very few circuit elements per gate. This results in lower interconnect parasitics, higher density, lower power consumption, and higher speed than for other GaAs logic families.

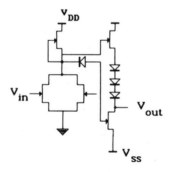

V_{DD}

V_{in}

V_{out}

V_{SS}

FIGURE 9-2-5. Composite logic [105].

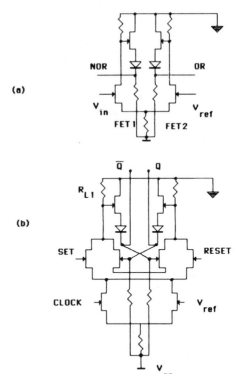

FIGURE 9-2-6. Source coupled FET logic (SCFL)
[87, 152]. (a) Inverter circuit. (b) Clocked SCFL R-S
flip-flop.

The operation of a DCFL gate may be described as follows: when the voltage V_{in} applied to the gate of the switching transistor Q1 is low (smaller than the threshold voltage V_T of this transistor), the transistor is off. For the zero fan-out case, i.e., no logic gates are connected to the output, the voltage V_{out} is nearly equal to V_{DD}. In a more realistic situation when other gates are connected to the output, V_{out} is determined by the input characteristics of the next stage. If GaAs MESFETs or JFETs are used, V_{out} is limited by the turn-on voltage of the gate junction (see Fig. 9-2-8).

When the input voltage V_{in} is high the switching transistor is on and the output voltage V_{out} is low (see Fig. 9-2-8). When the input voltage is low (less than V_T) the output voltage is high. Since the lowest output voltage in the circuit is zero, enhancement mode drivers ($V_T > 0$) should be used in DCFL inverters. The high value of the output voltage is limited by the barrier height of the gate diode. As a consequence DCFL circuits have a relatively small voltage swing and small noise margins.

The processing technology for DCFL circuits is difficult because of the stringent requirements on the active-layer doping and thickness which are necessary in order to maintain the threshold voltage uniformity. Also, multiple ion implants are required for enhancement mode drivers and depletion mode loads.

9-2-2. Schottky Diode Field Effect Transistor Logic (SDFL)

Two power supplies are required for SDFL circuits: a negative power supply voltage V_{ss} which is smaller than the threshold voltage of the normally-on switching transistor and the regular power supply voltage V_{DD} (see Fig. 9-2-2). A pull-up

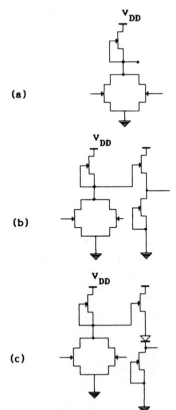

(a)

(b)

(c)

FIGURE 9-2-7. Comparison between DCFL logic and quasi-normally-off logic. (a) DCFL gate (fan-in 2); (b) buffered DCFL gate (fan-in 2); (c) quasi-normally-off logic gate (fan-in 2).

transistor (PU) serves as a load; a pull-down transistor (PD) connects the gate of the switching transistor to the negative power supply. A level shifting diode D_s (or several level shifting diodes) decreases the voltage on the gate of the switching transistor Q1 so that when the input voltage V_{in} is low the switching transistor Q1 is turned off. A logic "OR" function is implemented using Schottky logic diodes (diodes D_L in Fig. 9-2-2).

The source follower with or without the connection between the pull-down transistors of the switching stage and the source follower stage (see Fig. 9-2-2) may be used in order to increase the driving capability of an SDFL gate and the maximum fan-out it can handle.

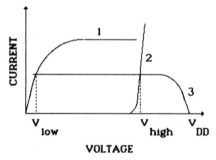

FIGURE 9-2-8. Operating points for a DCFL gate. 1, Current–voltage characteristic of the driver FET in parallel with the Schottky diode of the next stage for the high value of the input voltage. 2, Current–voltage characteristic of the driver FET in parallel with the Schottky diode of the next stage for the low value of the input voltage. 3, Current–voltage characteristic of the load transistor. V_{low}, low output voltage; V_{high}, high output voltage.

The normally-on switching transistors are used in the SDFL circuits. This leads to a larger voltage swing and noise margins and, as a consequence, leads to a higher yield and better reliability at the penalty of higher power consumption. Also, a larger number of transistors and diodes per gate may lead to a somewhat smaller speed compared to DCFL circuits.

SDFL processing technology is less demanding than for DCFL circuits because larger variations of the threshold voltage may be tolerated and because only normally-on FETs are fabricated.

9-2-3. Buffered FET Logic (BFL)

A BFL gate introduced by Van Tuyl et al. is shown in Fig. 9-2-3. Just like DCFL and SDFL, a BFL inverter is static (i.e., it has a low pass filter response). BFL circuits require two bias voltages. A basic inverter includes a logic branch and a driver/voltage shifter branch. Depletion mode MESFETs are used in BFL circuits and a level shift is required to make the input and output voltage levels of the logic gate compatible. This level shift is provided by Schottky diodes incorporated into the output buffer (similar to what is done in SDFL circuits). The required number of diodes is determined by the pinch-off voltage of the switching transistor and in turn determines the magnitude of the logic swing. Circuits utilizing MESFETs with smaller pinch-off voltages have fewer level-shift diodes and exhibit smaller power consumption, smaller logic swing, and smaller noise margins.

Power dissipation of a BFL gate may be reduced by removing the source follower as shown in Fig. 9-2-3b.

This circuit configuration has a smaller speed–power product and about 40% less power dissipation compared with the BFL gate with a source follower. However, BFL gates with source followers have a higher speed, especially for large fan-outs.

Using electron beam lithography, Greiling et al. [23] fabricated submicrometer gate BFL circuits and reported a delay time of 34 ps and 30–40 mW power dissipation per gate. NAND/NOR logic gates with one micron FETs exhibit 100 ps propagation delay with 40 mW power consumption yielding 4 pJ speed–power product.

Frequency dividers using master-slave flip-flops operating at counting rates ranging from dc to 4 GHz were reported [12].

High power dissipation (40 to 50 mW per gate) makes it difficult to fabricate large-scale integrated circuits using buffered depletion mode FET logic.

9-2-4. Capacitive Coupled Logic (CCL)

In CCL circuits proposed by Livingstone and Mellor [58] a diode is used as a capacitor which provides a dc isolation between the states. The power dissipation is considerably less than for the BFL logic. However, this inverter must be initialized for proper operation and it does not work at dc. This may be inconvenient for many applications.

Recently a better version of this approach was used by Gigabit Logic [163]. In these circuits the capacitance of the reverse-biased diode is charged by a very small current flowing through the chain of several small Schottky diodes. As an example a three-input NOR gate with an output level shifter is shown in Fig. 9-2-4-b. Using this approach a frequency divider, providing divide by 2, by 4, by 8, by 16, by 32, by 64, and by 128, was implemented [163]. This circuit has approximately 100 gates, operates up to frequencies close to 3 GHz, and dissipates about 600 mW.

9-2-5. Composite Logic

A composite circuit which operates as a static gate but still has a smaller power consumption than the BFL logic is shown in Fig. 9-2-5 [105]. It is called forward-feed static (FFS) logic. Experimental results presented in Ref. 105 show that this circuit requires about 30% of the power of a comparable BFL circuit and is somewhat faster (propagation delays of 76.7 ps were obtained compared to propagation delays of 104 ps for the BFL ring oscillators). Still, the power consumption was quite high (18.8 mW per stage).

9-2-6. Source Coupled FET Logic (SCFL)

The spread in the threshold voltages of GaAs FETs in different logic gates of the same circuit is an important factor limiting the integration scale and device yield. This factor is important even in SDFL circuits where the FET pinch-off voltages are larger than in DCFL circuits and the relative variation of pinch-off voltages is smaller. In the source coupled field effect transistor logic (SCFL) [87, 152, 154, 156] the circuitry of the logic gate is such that only the relative variation of the threshold voltages is important (the absolute values of the threshold voltages still determine the switching speed).

The SCFL circuit includes a FET differential amplifier and a pair of buffer stages (see Fig. 9-2-6). The operation of the SCFL gate may be explained as follows. Let us first consider two cases [157]:

(a) The transistor FET1 is on and the transistor FET2 is off. Then the following conditions should be fulfilled:

$$V_{ina} - V_s \geq V_{t1} + V_{on1}(0) \tag{9-2-1}$$

$$V_{ref} - V_s < V_{t2} \tag{9-2-2}$$

(b) The transistor FET1 is off and transistor FET2 is on. Then we have

$$V_{inb} - V_s < V_{t1} \tag{9-2-3}$$

$$V_{ref} - V_s \geq V_{t2} + V_{on2}(0) \tag{9-2-4}$$

Here V_{ina} and V_{inb} are the input voltages required for switching in cases (a) and (b), respectively, V_{t1} and V_{t2} are the threshold voltages of FET1 and FET2, V_s is the source voltage (common to FET1 and FET2), voltages $V_{oni}(0)$ $(i = 1, 2)$ are defined as

$$V_{oni}(0) = V_{gsi} - V_{ti} \qquad (i = 1, 2) \tag{9-2-5}$$

where V_{gsi} $(i = 1, 2)$ is the gate-to-source voltage, required in order to turn the transistor on.

Substituting Eq. (9-2-2) into Eq. (9-2-1) and Eq. (9-2-4) into Eq. (9-2-3) we obtain conditions required for a proper operation of an SCFL gate:

$$V_{ina} > V_{ref} + V_{on1}(0) + V_{t1} - V_{t2} \tag{9-2-6}$$

$$V_{inb} < V_{ref} + V_{on2}(0) + V_{t1} - V_{t2} \tag{9-2-7}$$

Hence, the input level required for switching is only dependent on the difference between the threshold voltages of the FETs. In most cases the threshold voltage difference, $V_{t1} - V_{t2}$, between neighboring FETs on a chip is small.

Output voltages of SCFL gates calculated as functions of the FET threshold voltage V_T are shown in Fig. 9-2-9 [154]. These results were obtained using SPICE circuit simulation. The input parameters of FETs with different threshold voltages were scaled assuming, for simplicity, a constant active channel thickness. Delay times, power dissipation, and power–delay product versus FET threshold voltage are shown in Fig. 9-2-10. The simulation results show that the optimum threshold voltage range is from -0.6 V to 0.0 V. This demonstrates that SCFL circuits operate over a wide range of FET threshold voltages.

The power consumption of an SCFL inverter is considerably larger than the power consumption of inverters implemented using other logic families because of a more complicated circuit configuration and large values of V_{ss} required for an optimum output voltage swing. However, total power consumption for an SCFL circuit may compare favorably with those for other logic families because a flip-flop, for example, may be implemented using a single SCFL inverter with two complementary outputs (see Fig. 9-2-6). Still, SCFL circuits may not be a suitable choice for very low power VLSI circuits. They seem to be more appropriate for very high speed and moderate power SSI and MSI circuits [154].

Another advantage of SCFL circuits is that they are compatible with bipolar ECL logic. FET differential amplifiers usually exhibit smaller voltage gain than bipolar amplifiers. Nevertheless, SCFL circuits may be quite fast because the FETs operate in the saturation region of the current–voltage characteristic where the feedback drain-to-gate capacitance is smaller.

An SCFL frequency divider operating at 4 GHz with 25 mW power dissipation is described in Ref. 156.

9-2-7. Low Pinch-off Voltage FET Logic (LPFL)

LPFL (also called quasi-normally-off logic) [61, 153] (see Fig. 9-2-7c) is a logic family which occupies an intermediate position between the SDFL and DCFL circuits. The switching transistor has a threshold voltage close to zero (either slightly positive or slightly negative). Because of larger pinch-off voltages (compared to DCFL) LPFL circuits are more tolerant to the variations in the threshold voltages than DCFL circuits. In practice, an acceptable variation of the threshold voltages for LPFL circuits may be two times larger than for DCFL circuits.

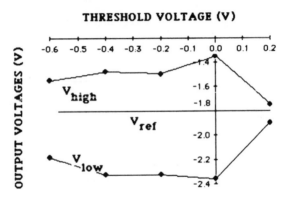

FIGURE 9-2-9. Output voltages of SCFL gates versus FET threshold voltages [154].

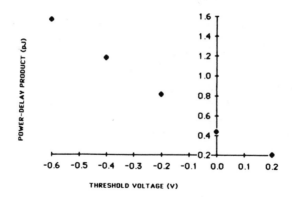

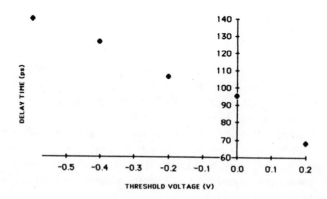

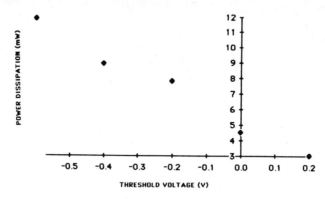

FIGURE 9-2-10. (a) Power-delay product, (b) delay time, and (c) power dissipation of SCFL gates versus FET threshold voltages [154].

An LPFL gate consists of two stages and is similar to a BFL gate though it requires only one power supply. LPFL circuits may be expected to have larger power dissipation, larger delay time, and higher fan-out capabilities than DCFL circuits.

9-2-8. Tunnel Diode FET Logic (TDFL) [177–179]

A tunnel diode FET logic cell is shown in Fig. 9-2-11, where the computed characteristics of the enhancement mode driver FET and of the tunnel diode are

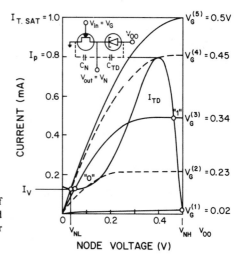

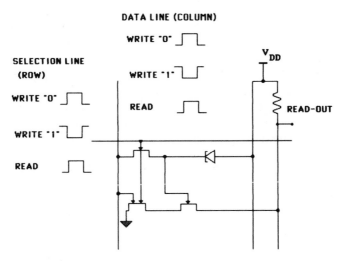

FIGURE 9-2-11. Current–voltage characteristics of an enhancement mode FET with tunnel diode load line [178]. The insert shows the basic TDFL inverter circuit.

also shown [177]. As can be seen from the figure, the negative resistance of the tunnel diode allows us to achieve a smaller power dissipation in the stable states, providing at the same time a large switching current. The predicted delay time is larger than for a comparable DCFL inverter. A delay time of 280 ps for fan-out of 3 and a dc power dissipation of 25 μW/gate were projected, resulting in a power-delay product of 1 fJ/gate [178]. The advantages of TDFL include a larger voltage swing and an ability to handle larger fan-outs.

A proposed TDFL static random access memory cell is shown in Fig. 9-2-12 [178]. A depletion mode switching transistor, an enhancement mode dual gate FET, and an enhancement mode buffer FET are used in this particular design. Various combinations of the voltages applied to the selection line and data line provide the write and erase voltages for the read-out as shown in the figure. However, the required voltage levels are quite small, which imposes stringent requirements for parameters of tunnel diodes which may be beyond the scope of the state-of-the-art technology [178].

FIGURE 9-2-12. Tunnel diode field effect transistor memory cell [178].

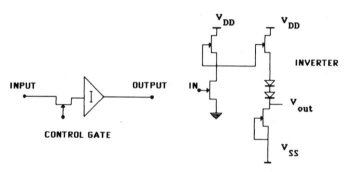

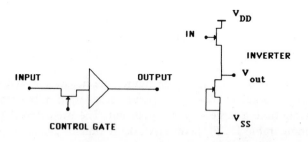

TRANSMISSION GATE INVERTING CONFIGURATION

TRANSMISSION GATE NON-INVERTING CONFIGURATION

FIGURE 9-2-13. Transmission gate configurations [114].

9-2-9. Transmission Gates

Transmission gates are an equivalent of a bidirectional switch. They can be used along with inverters in both inverting and noninverting configurations (see Fig. 9-2-13) [114].

The control signals should satisfy the following inequalities:

$$V_{GL} \leq V_L + V_T \tag{9-2-8}$$

$$V_H + V_T \leq V_{GH} \leq V_L + 0.7 \text{ V} \tag{9-2-9}$$

where V_{GL} and V_{GH} are the low and high control voltages, V_L and V_H are the low and high logic levels, and V_T is the MESFET threshold voltage. In Fig. 9-2-13 the transmission gates are used with the BFL inverters but similar circuits may be designed with inverters implemented using other logic families.

The transmission gates may be used quite effectively in dynamic circuits. As an example, a normally-on dynamic frequency divider circuit [114] is shown in Fig. 9-2-14. The timing diagram for this circuit is shown in Fig. 9-2-15. This circuit will

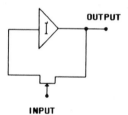

FIGURE 9-2-14. Dynamic frequency divider by two [114].

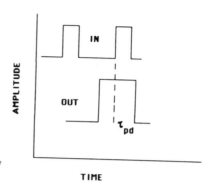

FIGURE 9-2-15. Timing diagram for dynamic frequency divider [114].

operate as a frequency divider by two up to frequencies of the order of $1/\tau_{pd}$ where τ_{pd} is the gate propagation delay. The implementation of this divider using BFL inverters is shown in Fig. 9-2-16. Simulation results for 40-μm-wide inverter transistors and 8-μm-wide transmission gate transistors showed that this circuit should operate in a frequency range between approximately 500 MHz and 10 GHz [114].

Circuits were fabricated using direct selenium ion implantation into a semi-insulating substrate. A boron implant was used for the isolation. The fabrication process was planar. The gates were self-aligned with the drain and source using an underetching technique. The gate length was from 0.7 to 1 μm. Five masks were required for the fabrication process. The yield was quite high (80%).

Measured values were in excellent agreement with the simulation results [114].

A normally-off dynamic frequency divider [114] is shown in Fig. 9-2-17. Two inverters in series are required to restore the logic levels because the level degradation is large compared to the voltage swing and the inverter gain is relatively small. The maximum frequency for this circuit is roughly $1/(3\tau_{pd})$ because the loop delay is $3\tau_{pd}$. The complementary clock generator required for the operation of the circuit is shown in Fig. 9-2-18.

A normally-off dynamic flip-flop shown in Fig. 9-2-19 was simulated in Ref. 114. The results of the simulation show that the circuit should operate up to approximately 2.5 GHz (see Fig. 9-2-20). The minimum operating frequency is about 100 MHz. The circuit has a substantially smaller power dissipation than a comparable static frequency divided by two circuit (see Fig. 9-2-20).

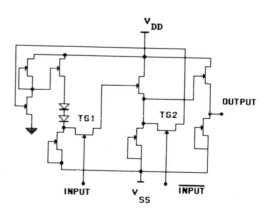

FIGURE 9-2-16. BFL implementation of two-phase dynamic frequency divider by two [114]. TG1, TG2, transmission gates.

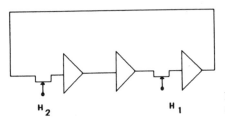

FIGURE 9-2-17. Normally-off dynamic frequency divider by two [114].

Transmission gates may be also used in MSI or even LSI GaAs circuits. As an example of possible implementations, a semidynamic shift register and a four-bit dynamic shift register are shown in Figs. 9-2-21 and 9-2-22 [114]. The results reported in Ref. 114 show that transmission gates can be used to reduce the gate count and power consumption in MSI and LSI GaAs circuits.

9-2-10. GaAs JFET Logic

Most GaAs digital circuits mentioned above have been implemented using GaAs MESFETs. GaAs junction field effect transistor logic has been developed in parallel with GaAs MESFET technology [19, 41, 43, 81, 133, 180–185].

Enhancement mode GaAs JFETs are fabricated using an ion-implantation of the active n-type channel followed by a p^+ implant under the gate to form a p^+-n junction (see Fig. 9-2-23). Alternatively, the p^+-n junction may be formed using a diffusion process [183, 184]. The normally-off JFETs are used as drivers. The depletion mode ion-implanted two-terminal devices (gateless FETs) may be used as loads in DCFL circuits which are similar to GaAs MESFET DCFL circuits. Power dissipation as low as 50 μW/gate and propagation delays of 45 ps have been observed [184]. This offers a potential for developing VLSI circuits with 10,000 gates and more. Further reduction in power (accompanied, however, by reduction in speed) may be achieved using double-implanted GaAs complementary JFETs. Using this approach one may fabricate static GaAs RAMs dissipating 50–200 nW per cell [133].

GaAs JFET logic exhibits a good radiation hardness with the ring oscillator operation not affected by doses as high as 10^7 rad [180].

One of the advantages of GaAs JFET technology is a larger built-in voltage (which may be close to the energy gap of GaAs, i.e., close to 1.4 V). This makes it possible to obtain a larger voltage swing and better noise margins than for GaAs MESFETs.

There are also some disavantages of GaAs JFET logic. First of all, the GaAs JFET logic has a somewhat lower speed than GaAs MESFET logic. This is because

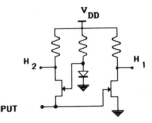

FIGURE 9-2-18. Complementary clock-pulse generator [114].

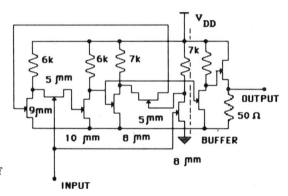

FIGURE 9-2-19. Schematic diagram of normally-off dynamic T flip-flop [114].

the rise time of the enhancement mode MESFET inverter is reduced by the gate current flowing when the gate of the driver FET is forward biased. Also, the formation of p^+ regions is a difficult technological step. As pointed out in Ref. 184, however, the junction depth or the pinch-off voltage may be monitored on chip and adjusted by an additional drive-in diffusion.

9-2-11. Modulation Doped Field Effect Transistor Logic

Modulation doped (Al, Ga)As field effect transistors (MODFETs), also called high electron mobility transistors (HEMTs) (see Chapter 10) have attracted considerable interest for ultra-high-speed applications. DCFL ring oscillators, frequency dividers, multiplexers, RAMs, and other circuits have been implemented using the MODFET technology. Gate propagation delays of 8.5 ps at 77 K and 11.6 ps at 300 K [200–201] have been obtained for MODFET ring oscillators. 4-kb RAMs have recently been described [173]. This RAM has a minimum address time of 2.0 ns at 77 K. A 1-kb MODFET RAM exhibited a subnanosecond performance [146]. MODFET frequency dividers operated at up to 10.1 GHz dividing frequency [137].

MODFET logic circuits have been fabricated using a self-aligned ion-implantation process [135, 150, 176, 186–188].

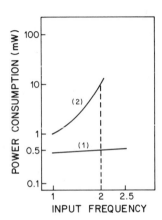

FIGURE 9-2-20. Power consumption versus input frequency for dynamic (1) and static (2) frequency dividers by two [114].

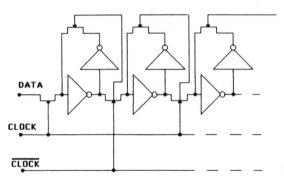

DATA

CLOCK

$\overline{CLOCK}$

FIGURE 9-2-21. Semi-dynamic implementation of GaAs shift register [114].

Recently, complementary MODFETs and complementary MODFET ICs (similar to CMOS) have been fabricated, opening new possibilities for ultra-high-speed low-power logic [142, 143].

Different MODFET devices (such as regular MODFET, inverted MODFET, MODFET with an n^+ GaAs gate, MODFETs with undoped AlGaAs layer, MODFETs with graded heterointerface, etc.; see Chapter 10) have been studied and may find applications in digital circuits.

Basic advantages of MODFETs include a higher electron mobility and velocity (leading to a higher speed and device transconductance), closer distance from the conducting channel to the gate (which also enhances the device transconductance and speed), high current swing, and a larger voltage swing (compared to GaAs MESFETs).

The turn-on voltage of the Schottky gates used in MODFET circuits may be in excess of 1 V (compared to 0.7 V or so for GaAs MESFETs). It leads to better noise margins and also makes it possible to make NAND gates using two MODFETs in series or a dual gate MODFET [137] (see Fig. 9-2-24). A frequency divider designed with AND-NOR gates showed about 50% improvement in noise margins compared to a conventional DCFL design with NOR gates.

The comparison of speed–power performance of MODFET ring oscillators and frequency dividers with other GaAs technologies is shown in Figs. 9-2-25 and 9-2-26 [137].

A disadvantage of MODFET logic is that it requires an MBE technology for growing modulation doped structures. Whereas MBE technology is an excellent laboratory tool, its application to production is still unproven. Recently, however, MOCVD epitaxy has been used for (Al,Ga)As–GaAs integrated circuit fabrication [147]. Eventually MOCVD may develop into a production tool for MODFET integrated circuits.

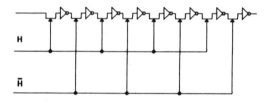

H

$\bar{H}$

FIGURE 9-2-22. A four-bit dynamic shift register [114].

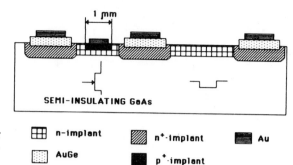

FIGURE 9-2-23. GaAs JFET inverter fabricated by an ion implantation process [41].

⊞ n–implant	▨ n⁺-implant	▦ Au	
▒ AuGe	■ p⁺-implant		

SEMI-INSULATING GaAs

9-2-12. AlGaAs–GaAs Heterojunction Bipolar Transistor Logic

Heterojunction bipolar transistors (HBTs) [95] have a number of advantages over conventional bipolar transistors (see Chapter 11). Because of a larger energy gap in the emitter region the charge injection from the base into the emitter region is suppressed, resulting in nearly unity injection efficiency and high current gain. Current gains in excess of 3000 have been achieved for AlGaAs/GaAs HBTs. In addition, there is the freedom to choose to decrease the emitter doping and to increase the base doping, leading to smaller base spreading resistance and to lower emitter–base capacitance. All these factors contribute to a higher speed of operation.

The HBT digital integrated circuits may have several advantages such as high threshold voltage uniformity and high current driving capabilities [95, 108].

Small-scale integrated I^2L and ECL circuits have been fabricated [189, 190, 122, 148, 138].

Logic circuits reported in Ref. 138 were based on current mode logic (CML). In this logic the current from an appropriate source is steered between alternate paths depending on the input voltages (see Fig. 9-2-27). CML logic provides complementary outputs. It also maintains constant output current minimizing inductive transients on power supply lines [138].

Using this approach, ring oscillator propagation delays close to 40 ps were obtained and frequency dividers operating at 8.5 GHz input frequency were fabricated with excellent transistor uniformity and good circuit yield (as high as 90%) [138]. A four-bit pattern generator operating with a clock frequency up to 4 Gbit/s

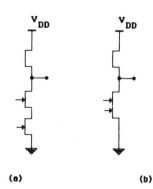

FIGURE 9-2-24. DCFL NAND gates. (a) Two driver transistors connected in series. (b) Dual gate driver.

(a)   (b)

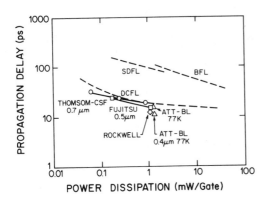

FIGURE 9-2-25. Comparison of speed–power performance of MODFET ring oscillators with other GaAs logic families (dashed lines) [137].

was also demonstrated. Better performance with higher speed and smaller power dissipation has been projected (see Fig. 9-2-28) [138].

HBTs used in Refs. 122, 138, and 148 were fabricated using MBE grown layers and Be implantation to form p^+ base regions for the contacts with the base. Oxygen implants were used to decrease the extrinsic base–collector capacitance. Isolation was achieved using the boron bombardment of the surface outside the active transistor areas (see Fig. 9-2-29).

This technology is more difficult than GaAs MESFET or AlGaAs/GaAs MODFET technologies. That probably means that the integration scale will remain relatively small, at least in the near future.

9-2-13. GaAs MOSFET Logic

MOSFET circuits have the advantage of a large voltage swing. In spite of difficulties related to the formation of a stable native oxide on GaAs, logic circuits with enhancement mode drivers and enhancement or depletion mode loads ((E/E) or (E/D)) have been successfully fabricated using GaAs MOSFETs [38, 191].

Thirteen-gate ring oscillators were fabricated using a low-temperature oxidation technique for gate insulation. Devices with a gate length of 1.5 μm were employed in E/E circuits with the gate driver width of 100 μm and load width of 10 μm. Transistors with 2 μm gate lengths were used in E/D ring oscillators. In the latter case, the driver width was 200 μm and the load width was 80 μm.

For the E/D ring oscillators, a minimum propagation delay of 110 ps was obtained with power-delay product of 2 pJ. For the E/E ring oscillators the gate propagation delay was 385 ps with a minimum power–delay product of 26 fJ.

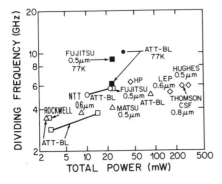

FIGURE 9-2-26. Comparison of the frequency dividers implemented using different GaAs logic families. Open and solid circles, MODFET DCFL AND-NOR gate logic. Open squares and solid squares, MODFET DCFL NOR logic. Open diamonds, GaAs BFL AND-NOR gate logic. Open triangles, GaAs DCFL NOR gate logic.

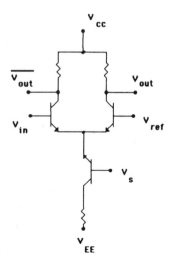

FIGURE 9-2-27. Current mode logic (CML) gate [138].

9-3. DESIGN AND SIMULATION OF GaAs MESFET DCFL CIRCUITS [192]

9-3-1. Introduction

GaAs DCFL technology has become a leading contender for ultrafast digital integrated circuits. Gate delays as short as 15 ps for logic based on self-aligned GaAs MESFETs [128] and 8.5 ps at 77 K for logic based on modulation doped AlGaAs–GaAs transistors [200, 201] have been achieved. Circuits as complicated as a 16×16 multiplier [120] have been fabricated and reasonable yields have been demonstrated for small-scale GaAs integrated circuits [151].

Direct coupled field effect transistor logic (DCFL) has a higher speed and a smaller power consumption than another popular GaAs logic family—Schottky diode field effect transistor logic (SDFL) (see Sections 9-2 and 9-4). The disadvantages of DCFL circuits include a smaller logic swing, smaller noise margins, and more stringent requirements for threshold voltage uniformity.

In this section we consider the design of a DCFL logic gate. We derive a set of analytical equations which relate the inverter parameters, such as propagation delay, inverter gain, switching voltage, output voltage levels, and noise margins, to the parameters of the switching transistor and saturated transistor load, such as the load saturation current, the switching transistor threshold voltage, and the load and switching transistor output conductances. These results allow us to establish

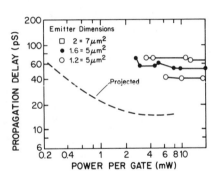

FIGURE 9-2-28. Propagation delay time versus power obtained and projected for CML ring oscillators [138].

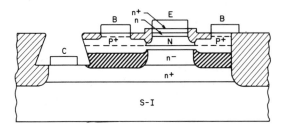

Be DOPED P REGIONS
OXYGEN-IMPLANTED REGIONS
IMPLANT DAMAGE ISOLATION REGIONS   FIGURE 9-2-29. HBT cross section [138].

trade-offs between the noise margins, propagation delay, and power consumption for different fan-ins and fan-outs.

We then compare our experimental data for GaAs DCFL inverters and ring oscillators and demonstrate that they are in good agreement with the results of our analytical model.

9-3-2. GaAs MESFET Model

The saturation current–voltage characteristics of GaAs MESFETs with low pinch-off voltages (below 2 V or so) can be accurately described by the "square law" model (see Chapter 7):

$$I_{d\text{sat}} = \beta_0(V_{\text{GS}} - V_T)^2 \tag{9-3-1}$$

Here $I_{d\text{sat}}$ is the drain-to-source saturation current, V_{GS} is the gate-to-source voltage, V_T is the threshold voltage, and

$$\beta_0 = \frac{2\varepsilon\mu v_s W}{A(\mu V_{\text{po}} + 3v_s L)} \tag{9-3-2}$$

Here ε is the dielectric permittivity, v_s is the electron saturation velocity, W is the gate width, μ is the low field mobility, A is the effective active channel depth, L is the gate length,

$$V_{\text{po}} = qN_d A^2/2\varepsilon \tag{9-3-3}$$

is the pinch-off voltage, and N_d is the active channel doping.

Effects related to a source series resistance R_s are neglected in Eq. (9-3-1). In order to include these effects the gate-to-source voltage V_{GS} in Eq. (9-3-1) should be replaced by $V_{\text{GS}} - I_{\text{DS}}R_s$. The resulting expression for the saturation drain-to-source current is given by

$$I_{d\text{sat}} = \frac{1 + 2\beta_0 R_s V_{\text{GT}} - (1 + 4\beta_0 R_s V_{\text{GT}})^{1/2}}{2\beta_0 R_s^2} \tag{9-3-4}$$

where $V_{\text{GT}} = V_{\text{GS}} - V_T$. This expression is rather complicated for an analytical modeling of a DCFL inverter. Therefore we account for the source series resistance in an empirical fashion substituting β_0 in Eq. (9-3-1) by β, where

$$\beta = I_{d\text{sat1}}/(V_{\text{bi}} - 0.1 - V_T)^2 \tag{9-3-5}$$

Here $I_{d\,\text{sat}1}$ is the value of the drain-to-source saturation calculated using Eq. (9-3-4) for $V_{\text{GS}} = V_{\text{bi}} - 0.1$ V and V_{bi} is the built-in potential.

In Fig. 9-3-1 we compare an exact solution (solid curve) with our approximation given by Eq. (9-3-5) (dashed curve) for a typical GaAs MESFET (see Table 9-3-1). As can be seen from the figure the agreement is quite good, corroborating the validity of the approximate expression. The results of the calculation also agree well with the experimental data.

The drain-to-source current–voltage characteristics in both the linear and the saturation regimes are described by the following heuristic equation (see Chapter 7):

$$I_{\text{DS}} = I_{d\,\text{sat}}(1 + \lambda V_{\text{DS}}) \tanh(V_{\text{DS}}/V_{\text{ss}}) \tag{9-3-6}$$

where V_{DS} is the drain-to-source voltage, λ is an empirical parameter describing the output conductance in the saturation region, and $V_{\text{ss}} = I_{d\,\text{sat}}/g_{\text{ch}}$. Here g_{ch} is the effective channel conductance at $V_{\text{DS}} \to 0$ defined as

$$g_{\text{ch}} = \frac{g_{\text{ch}i}}{1 + g_{\text{ch}i}(R_s + R_d)} \tag{9-3-7}$$

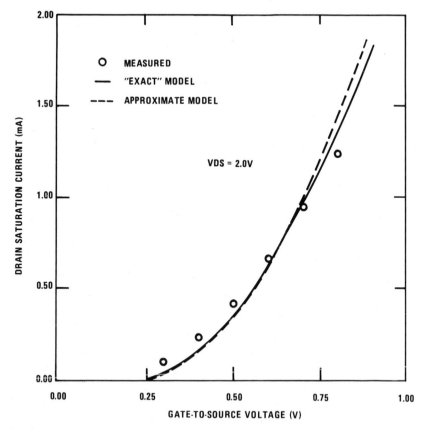

FIGURE 9-3-1. Transfer characteristics of a GaAs MESFET at $V_{\text{DD}} = 2.0$ V [192]. Parameters used in the calculation are given in Table 9-3-1.

TABLE 9-3-1. Parameters Used in the Calculation of Figure 9-3-1

	Symbol	Unit	Value
A. Switching transistor			
Schottky diode ideality factor	n	—	1.44
Schottky diode saturation current density	J_s	A/m^2	0.255
Thermal voltage	V_{th}	eV	0.02584
Electron saturation velocity	v_s	m/s	1.3×10^5
Low field mobility	μ	m^2/V s	0.25
Source series resistance	R_s	Ω	50
Drain series resistance	R_d	Ω	50
Channel doping density	N_d	cm^{-3}	7.24×10^{16}
Channel thickness	A	m	1.0×10^{-7}
Built-in voltage	V_{bi}	V	0.72
Dielectric permittivity	ε	F/m	1.14×10^{-10}
Gate width	W	m	2×10^{-5}
Gate length	L	m	0.7×10^{-6}
Output conductance parameter	λ_1	—	0.15
Intrinsic transconductance parameter	β_0	mA/V^2	3.7
Effective transconductance parameter	β_1	mA/V^2	3.23
Threshold voltage	V_{T1}	V	0.21
Effective node capacitance per unit area per fan-out	C_1/LW	F/m^2	10
B. Load			
Output conductance parameter	λ_2	—	0.027
Saturation current per unit width	I_{Lsw}	A/m	250
Low field resistance per millimeter	R_{offw}	Ω mm	2.24
C. Interconnects			
Effective node interconnect per unit area per fan-out	C_{IN}	fF	10

The intrinsic channel conductance g_{chi} is given by

$$g_{chi} = \frac{q\mu N_d A W}{L}\left[1 - \left(\frac{V_{bi} - V_{GT}}{V_{po}}\right)^{1/2}\right] \tag{9-3-8}$$

for the case when $V_{bi} > V_{GT}$, otherwise $g_{chi} = q\mu N_d A W/L$.

Figure 9-3-2 compares the calculated I_{DS} vs. V_{DS} characteristics with the experimental data. The MESFET device parameters used in all the model calculations were determined from the measured I-V characteristics using the following procedure. First, a suitable λ is chosen from the I_{DS} vs. V_{DS} curve for one value of V_{GS}. Then the saturation currents are extrapolated to $V_{DS} = 0$ for all values of V_{GS}, keeping λ constant. To obtain the source resistance R_s we plot $\sqrt{I_{DS}}$ vs. $V_{GS} - I_{DS}R_s$ for different values of R_s until a best least-squares fit is obtained. The slope and intercept of this line give us β_0 and V_T, respectively. From the gate I-V characteristic we can derive V_{bi}. The channel thickness and doping are determined using Eq. (9-3-3) and the implant dose data. Assuming $R_s = R_d$ we can get the mobility from the slope of the I_{DS} vs. V_{DS} characteristic in the linear region at large V_{GS}. Once μ is known, the saturation velocity can be calculated using Eq. (9-3-2).

The ungated saturated load device characteristics are described using a velocity saturation model (see Chapter 7). The load I-V characteristics may be approxi-

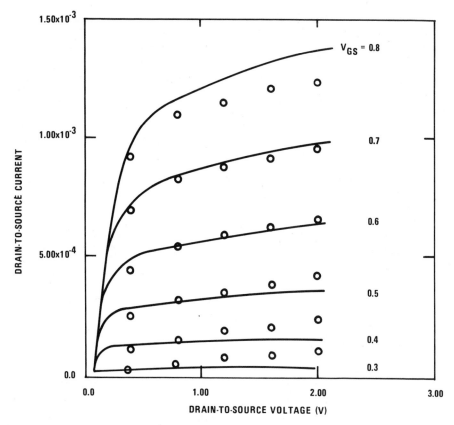

FIGURE 9-3-2. Drain-to-source current versus voltage characteristics of the GaAs MESFET with parameters given in Table 9-3-1. Solid lines, analytical model; dots, measured data.

mated by an expression of the form of Eq. (9-3-6) as follows:

$$I_L = I_{Ls} \tanh(V_L/V_{Lss})(1 + \lambda_L V_L) \qquad (9\text{-}3\text{-}9)$$

where λ_L is the load device output conductance parameter, V_L is the drain-to-source voltage across the load, and V_{Lss} is an empirical parameter given by

$$V_{Lss} = I_{Ls} R_{off} \qquad (9\text{-}3\text{-}10)$$

The parameters I_{Ls} and R_{off} may be either derived from the I–V characteristics or calculated from physical and process parameters as discussed in Chapter 7. The experimental and calculated I–V characteristics of a saturated load device are compared in Fig. 9-3-3. For a gated load, the model identical to the switching FET model can be used.

9-3-3. DCFL Inverter Transfer Curves

The circuit diagrams of DCFL inverters are shown in Fig. 9-2-1. A measured dc inverter transfer curve for an inverter using an ungated FET load device is shown in Fig. 9-3-4. As can be seen, the transfer curve can be approximated well by a

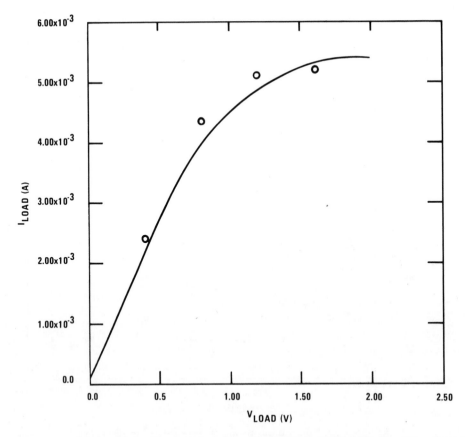

FIGURE 9-3-3. Current–voltage characteristics of the ungated GaAs FET load device with a contact separation of 2 μm and width of 20 μm. FET parameters are given in Table 9-3-1.

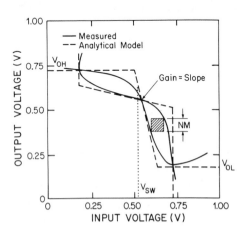

FIGURE 9-3-4. Measured dc transfer curve of a DCFL inverter (solid line) and the piecewise linear approximation of the transfer curve (dotted line) [206].

piecewise linear model. Using this approximation we define the inverter noise margin as shown in Fig. 9-3-5. In reality noise margins are somewhat smaller because the piecewise linear approximation of the transfer curve exaggerates the gain near the boundaries of the transition region of the inverter transfer curve (see Fig. 9-3-4). However, the SPICE computer simulations show that noise margins defined based

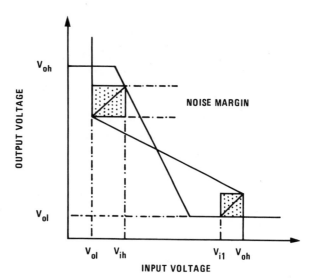

FIGURE 9-3-5. Determination of noise margins based on the "largest square" concept using the linear piecewise transfer characteristic [192].

on the piecewise linear transfer curve scale in the same way with device parameters and bias voltage as the computed values of the noise margins.

Figure 9-3-6 illustrates the operation of a DCFL gate. In the case of a zero fan-out (unloaded gate) the output high voltage V_{oh} depends on the power supply voltage V_{DD}. However, if the fan-out is greater than zero the output high voltage V_{oh} is clamped by the Schottky gate diode of the next stage. We will only consider the latter case when V_{oh} is given by

$$V_{oh} = nV_{th} \ln\left(\frac{I_{Lh}}{J_s WLF_O}\right) + \left(\frac{1}{F_O} + \frac{1}{F_I}\right) I_{Lh}R_s \qquad (9\text{-}3\text{-}11)$$

Here n is the ideality factor of the Schottky gate junction, $V_{th} = k_B T$ is the thermal voltage, J_s is the gate saturation current density, F_O is the fan-out, F_I is the fan-in, and I_{Lh} is the current through the load device when the output voltage is

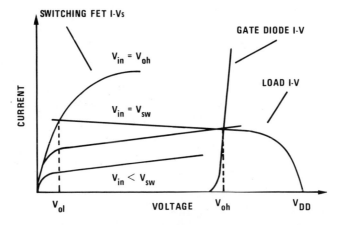

FIGURE 9-3-6. Current–voltage characteristics illustrating switching of a DCFL inverter [192]. Dotted curves show the variation of the current–voltage characteristics of the switching FET with the input voltage V_{in}. Notice that a drastic change in the output voltage occurs when $V_{in} = V_{sw}$.

high. The first term in Eq. (9-3-11) is the diode voltage drop. The second term accounts for the voltage drop across R_s. Normally the drop across the diode is much larger, except at very high load currents; thus the accuracy of V_{oh} is not significantly affected by assuming the driver current flowing through R_s to be I_{Lh}. Then the load device saturation current I_{Ls} can be related to the parameters of the load transistor.

Assuming that in the transition region the gate current can be neglected, we simply equate the driver and load currents to derive the switching voltage, output low voltage, and gain. Hence we find

$$I_{Ls}[1 + \lambda_L(V_{DD} - V_{out})] \tanh\left(\frac{V_{DD} - V_{out}}{V_{Lss}}\right)$$

$$= \beta(V_{in} - V_T)^2(1 + \lambda V_{out}) \tanh\left(\frac{V_{out}}{V_{ss}}\right) \qquad (9\text{-}3\text{-}12)$$

where V_{DD} is the power supply voltage, and V_{in} and V_{out} are the input and output voltages.

The switching voltage V_{sw} is the input voltage at which the onset of the transition from the output high to low voltage occurs. By evaluating Eq. (9-3-12) at $V_{out} = V_{oh}$ and solving for $V_{in} = V_{sw}$, we get

$$V_{sw} = V_T + \left[\frac{I_{Lh}}{\beta F_I(1 + \lambda V_{oh}) \tanh(V_{oh}/V_{ss})}\right]^{1/2} \qquad (9\text{-}3\text{-}13a)$$

In cases when both the driver and the load transistors are in saturation and λ is small, Eq. (9-3-13a) simplifies to

$$V_{sw} = V_T + \left(\frac{I_{Lh}}{\beta F_I}\right)^{1/2} \qquad (9\text{-}3\text{-}13b)$$

The output low voltage V_{ol} is determined in a similar manner by solving Eq. (9-3-12) at $V_{in} = V_{oh}$ for $V_{out} = V_{ol}$:

$$V_{ol} = \frac{V_{ss}}{2} \ln\left(\frac{1 + \alpha}{1 - \alpha}\right) + \frac{I_{Lh}}{F_O} R_s \qquad (9\text{-}3\text{-}14)$$

where $\alpha = I_L(V_{ol})/[I_{ds}(V_{ol})F_I]$. Because α depends on V_{ol}, an iterative process is required to accurately determine V_{ol}. However, only a few iterations are required to get convergence in most cases. An initial guess for V_{ol} can be obtained using the expression

$$V_{ol} \approx V_{ss}I_{Ls}/I_{ds1} + I_{Lh}R_s/F_O \qquad (9\text{-}3\text{-}15)$$

where $I_{ds1} = \beta(V_{oh} - V_T - I_{Lh}R_s/F_O)^2$.

In order to obtain the inverter gain $G = dV_{out}/dV_{in}$ we differentiate both sides of Eq. (9-3-12) with respect to V_{out}. The resulting expression for the gain, G, can be expressed as

$$G = -\frac{\partial I_D / \partial V_{in}}{\partial I_L / \partial V_{out} - \partial I_D / \partial V_{out}} \qquad (9\text{-}3\text{-}16)$$

The maximum gain is obtained for large values of V_{DD} and small load saturation currents when both the driver and load devices are in full saturation. In this case Eq. (9-3-16) reduces to

$$G_{max} = -\frac{2(\beta F_I / I_{Ls})^{1/2}}{(\lambda + \lambda_L)} \qquad (9\text{-}3\text{-}17)$$

The inverter gain corresponding to the slope of our piecewise linear model of Fig. 9-3-4 is calculated using Eq. (9-3-16) with $V_{out} = (V_{oh} + V_{ol})/2$, which roughly represents the point of highest gain on the real transfer curve.

The preceding analysis (see Eqs. (9-3-12)–(9-3-16)) is valid only in the operating regime where the condition

$$I_D(V_{gs} = V_{ol}, V_{ds} = V_{oh}) < I_L(V_L = V_{DD} - V_{oh}) \qquad (9\text{-}3\text{-}18)$$

is satisfied. However, this is not a severe limitation in practice since acceptable noise margins needed to operate an inverter in circuits other than ring oscillators can only be obtained in this regime.

The noise margin derivation is based on the widely used "largest square" definition (see Fig. 9-3-5). The diagonal corresponding to the largest square is defined by the two points on the normal and mirror transfer curves where the slopes are equal. Hence, from simple geometrical considerations, the noise margin is given by

$$NM = V_{ih} - V_{ol} \qquad \text{for } V_{sw} \leqslant V_{ih} \qquad (9\text{-}3\text{-}19a)$$

or

$$NM = V_{oh} - V_{il} \qquad \text{for } V_{il} \leqslant V_{isw} \qquad (9\text{-}3\text{-}19b)$$

The intercept voltages V_{ih} and V_{il} (see Fig. 9-3-5) are defined as

$$V_{ih} = \frac{V_{oh} + GV_{sw} + V_{ol} - V_{isw}}{1 + G} \qquad (9\text{-}3\text{-}20)$$

$$V_{il} = \frac{2V_{oh} + (G - 1)V_{sw}}{1 + G} \qquad (9\text{-}3\text{-}21)$$

where $V_{isw} = V_{sw} + (V_{oh} - V_{ol})/G$.

The transient switching behavior of the inverter may be approximated by considering the charging and discharging time of an effective input capacitance C_N. This capacitance is related to the voltage dependent MESFET gate capacitance and to interconnect capacitances. The total average capacitance over the range of the voltage swing, $V_{oh} - V_{ol}$, can be roughly estimated as

$$C_N = C_o F_o \frac{(V_{oh} - V_T)}{(V_{oh} - V_{ol})} + C_{IN} \tag{9-3-22}$$

where C_{IN} is the effective interconnect capacitance and

$$C_o = \frac{\varepsilon L W}{A} \tag{9-3-23}$$

is the effective MESFET capacitance.

The inverter turn-off time τ_{OFF} is determined by the average current I_{OFF} available to discharge C_N given by

$$I_{OFF} = \frac{1}{2}\left[\beta \left(V_{oh} - V_T - \frac{I_{Lh}R_s}{F_O} \right)^2 (1 + \lambda V_{oh}) - I_{Lh} \right] \tag{9-3-24}$$

Then τ_{OFF} can be estimated as

$$\tau_{OFF} = \frac{(V_{oh} - V_{ol})C_N}{I_{OFF}} \tag{9-3-25}$$

The turn-on current to charge C_N is supplied by the load device. In inverters with low threshold drivers ($V_T < V_{ol}$) this current can be substantially reduced, limiting the speed attainable. The turn-on time based on an average turn-on current is given by

$$\tau_{ON} = \frac{2(V_{oh} - V_{ol})C_N}{I_{Ll} - \beta(V_{ol} - V_T)^2(1 + \lambda V_{ol}) \tanh(V_{ol}/V_{ss})} \tag{9-3-26}$$

where I_{Ll} is calculated for the voltage $V_{Lds} = V_{DD} - V_{ol}$ across the load device. The average propagation delay is given by

$$\tau_D = (\tau_{ON} + \tau_{OFF})/2 \tag{9-3-27}$$

The inverter power consumption may be estimated as follows:

$$P = I_2 V_{DD} + C_N(V_{oh} - V_{ol})^2 f \tag{9-3-28}$$

where f is the frequency at which the inverter stage is operating and $I_2 = (I_{Lh} + I_{Ll})/2$. For the case of a N-stage ring oscillator circuit $f = 1/(2\tau_D N)$. The first term in the right-hand side of Eq. (9-3-28) represents the static component and the second term represents the dynamic component.

9-3-4. Calculation Results

The calculated dependences of the output high voltage, switching voltage, and output low voltage on the load saturation current are shown in Fig. 9-3-7, where they are compared with the experimental data ($F_I = F_O = 1$) measured on devices

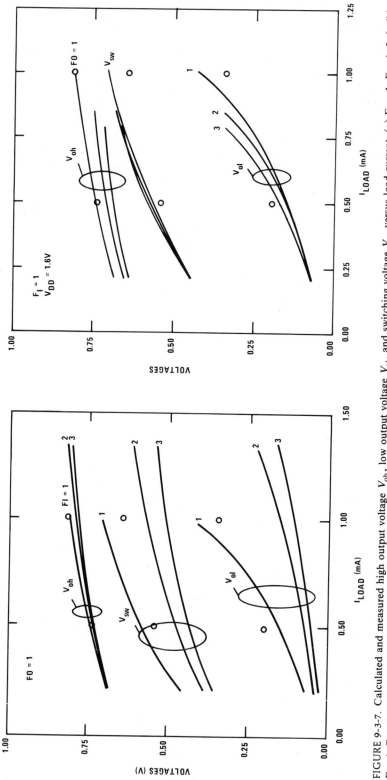

FIGURE 9-3-7. Calculated and measured high output voltage V_{oh}, low output voltage V_{ol}, and switching voltage V_{sw} versus load current. (a) $F_O = 1$, $F_I = 1, 2, 3$; (b) $F_I = 1$, $F_O = 1, 2, 3$. Parameters used in the calculation are given in Table 9-3-1.

fabricated using the self-aligned gate process. As can be seen from the figure, there is good agreement except at high load saturation currents where our analytical model is no longer valid. Figure 9-3-7 also illustrates the effects of different fan-ins (Fig. 9-3-7a) and fan-outs (Fig. 9-3-7b). Figure 9-3-7a shows that for circuits with inverter stages of different fan-ins the worst case noise margin at large load currents is determined by the difference between the switching voltage of the inverter with the largest F_I and the output low voltage of the inverter with $F_I = 1$ and not by Eq. (9-3-19).

The calculated gain is shown in Fig. 9-3-8 and the noise margin in Fig. 9-3-9 as functions of the load saturation current. As mentioned in Section 9-3-3, the piecewise linear model of the inverter transfer curve tends to exaggerate the noise margin, but as can be seen from Fig. 9-3-9 the analytical model correctly predicts the overall trend.

The results of the analytical calculation and of the computer simulation (using SPICE) are compared in Fig. 9-3-10. (The GaAs MESFET model used in UM-SPICE is described in Ref. 74.) The results of the analytical calculation are close to the results of the computer simulation. Hence we conclude that the derived analytical expressions describe the operation of a DCFL gate quite accurately.

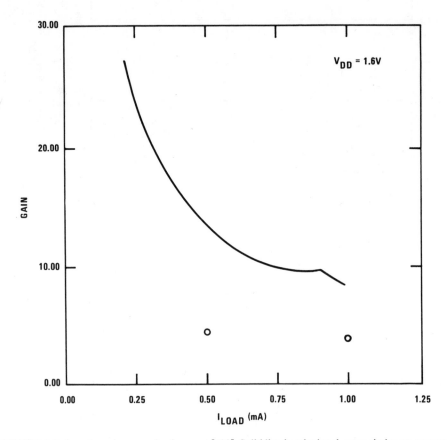

FIGURE 9-3-8. Inverter gain versus load current [192]. Solid line is calculated, open circles are measured data.

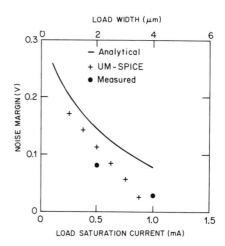

FIGURE 9-3-9- Noise margins versus the load saturation current [74].

The experimental and analytically calculated propagation delay times are compared in Fig. 9-3-11. In order to study the effects of fan-in on the gate propagation delay times we will consider the worst case situation when the voltage swing is the largest. In this case the low output voltage is the smallest (this corresponds to the case when all parallel transistors are on) and the high output voltage is the largest [i.e., when $F_O = 1$ in Eq. (9-3-11)]. As can be seen from Fig. 9-3-12, for $F_I > 1$ the minimum propagation delay times are obtained when the ratio $I_{d\,sat}/I_{LS} \approx 2$.

The effects of the driver threshold voltage on noise margin and delay times are illustrated in Fig. 9-3-13. In these calculations the threshold voltage was varied by changing the doping density N_d. The change in the transconductance parameter β with N_d was also taken into account [see Eqs. (9-3-2) and (9-3-3)]. The minimum delay time decreased with decreasing threshold voltages and overall noise margin increased. However, at small load currents (i.e., in low power inverters), the delay time is almost independent of the threshold voltage. In this case the propagation delay is dominated by the turn-on time τ_{on}, which is nearly independent of the threshold voltage.

The effects of different power supply voltages V_{DD} on the propagation delay are shown in Fig. 9-3-14. Because of the clamping of the output high voltage there is only a small variation of the delay time with respect to V_{DD}.

The results of the analytical model of the GaAs MESFET DCFL inverter are in reasonable agreement with the experimental data, and most importantly the

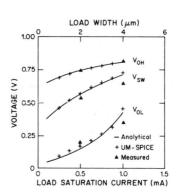

FIGURE 9-3-10. Comparison of the analytical model results with UM-SPICE simulations and experimental data [74].

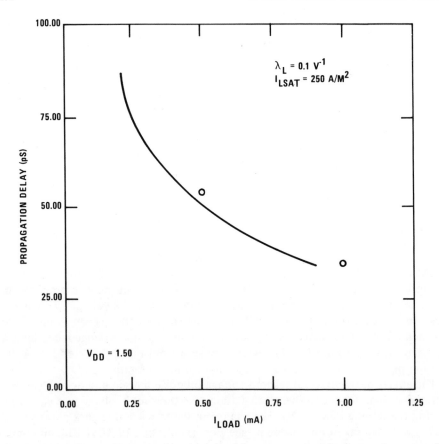

FIGURE 9-3-11. Ring oscillator circuit propagation delay per gate versus load saturation current [192].

model correctly predicts the dependence of the inverter performance characteristics on the device parameters of the driver and load transistors. This allows one to quickly establish design trade-offs and approximate process guidelines to achieve the desired circuit performance.

9-3-5. Circuit Simulation of Ultra-High-Speed DCFL Inverters [199]

Device models described in Chapter 7 and in Section 9-3-2 were implemented in the GaAs circuit simulator (DOMES) (see a description of an older DOMES version in Ref. 121) and, more recently, in a customized version of SPICE called University of Minnesota SPICE (UM-SPICE). Below we describe the DOMES circuit simulation results for self-aligned DCFL [199].

The simulated inverters consisted of 11.5-μm-wide enhancement mode MESFETs with self-aligned T-gates in series with linear resistors [127, 128]. The gate length was 0.55 μm. The dc characteristics of the enhancement mode MESFET driver are shown in Fig. 7-10-4. They were simulated taking into account the space charge current. Using the DOMES program, a five-stage ring oscillator was simulated for different values of load resistances, different values of power supply voltage, and different fan-outs. The results of the simulation are compared with measured

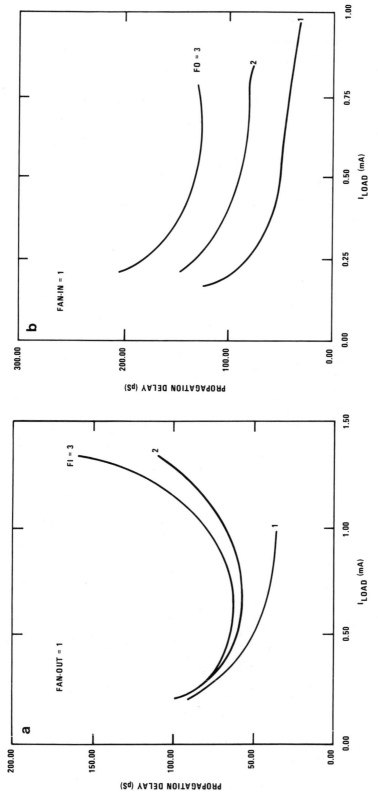

FIGURE 9-3-12. Worst case propagation delays for (a) different fan-ins with $F_O = 1$, and (b) different fan-outs with $F_I = 1$ [192].

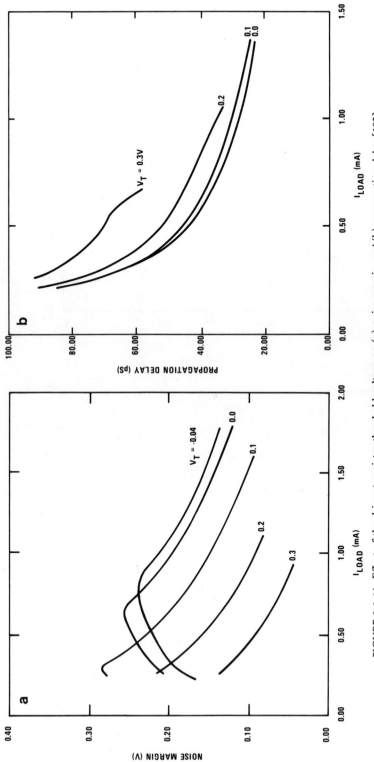

FIGURE 9-3-13. Effect of the driver transistor threshold voltage on (a) noise margin, and (b) propagation delay [192].

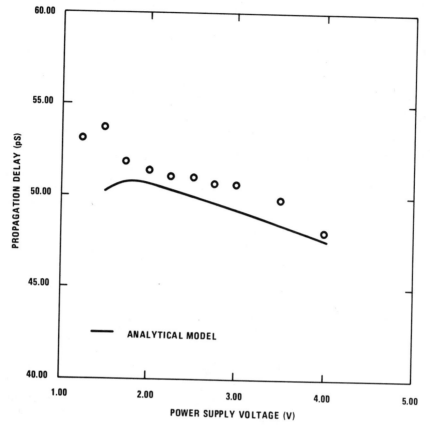

FIGURE 9-3-14. Calculated and measured dependence of the propagation delay as a function of the power supply voltage [192].

data [127, 128] in Figs. 9-3-15 amd 9-5-16. The agreement is very good, which shows that the upgraded version of DOMES (or an equivalent but more efficient UM-SPICE simulator) may be used as an accurate simulation and computer design tool for GaAs integrated circuits.

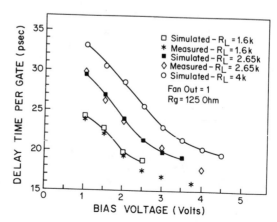

FIGURE 9-3-15. Propagation delay per gate versus supply voltage for a five-stage ring oscillator [199]. Experimental points from Refs. 127 and 128.

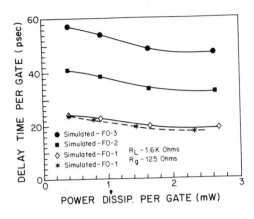

FIGURE 9-3-16. Propagation delay per gate versus power dissipation for a five-stage ring oscillator [199]. Experimental points from Refs. 127 and 128.

9-4. FAN-OUT, FAN-IN, POWER DISSIPATION, AND SPEED OF GaAs SDFL LOGIC [103]

9-4-1. Fan-out Limitations

In this section we will discuss the fan-out and speed of SDFL inverters relating them to the parameters of the switching transistor and of the pull-up and pull-down FETs (see Fig. 9-2-2).

We will first consider the output voltages in the off and on states of the SDFL inverter.

The output voltage of the SDFL gate depends on the fan-out. If the fan-out is larger than some critical value, the output, with the switching transistor in the off state, drops to a value which is too low to turn on the switching transistors in the next stages, and the circuit malfunctions.

The dc equivalent circuit of the output loop of the gate in the off state loaded by the pull-down transistors of the next stages can be represented by one of the equivalent circuits shown in Fig. 9-4-1.

Three V_d corresponds to the voltage drop across the switching diode D_L and the level shifting diode (or diodes) D_S. V_{dd} is the voltage of the positive power supply; V_s is the voltage of the negative power supply ($V_s > 0$ with this sign convention). Other parameters of the equivalent circuits are explained in Fig. 9-4-2, where the simplified $I-V$ characteristics of the FETs are presented. When the fan-out

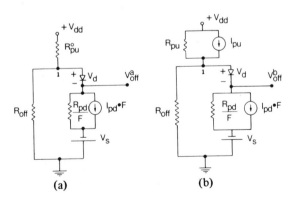

FIGURE 9-4-1. Equivalent circuits of the output loop of the SDFL gate loaded by F gates in the off state [103]. (a) With the pull-up transistor in the linear regime. (b) With the pull-up transistors in the saturation mode.

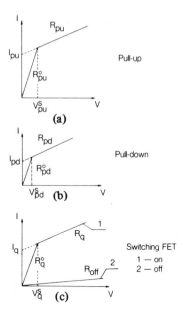

FIGURE 9-4-2. Linear-piecewise approximations of the I-V characteristics of the SDFL FETs [103].

is zero (unloaded gate) or low, almost all the voltage (V_{dd}) is across the switching transistor in the off state and the equivalent circuit of Fig. 9-4-1a is applied. At higher fan-outs, however, the voltage drop across the switching transistor is smaller, the pull-up transistor is saturated, and the equivalent circuit of Fig. 9-4-1b is relevant. The latter case limits the fan-out of the SDFL gate at dc.

Voltage V_{off} at the gates of the switching transistors in the next stages can be found from the solution of the nodal equation at node 1 of the circuit shown in Fig. 9-4-2b:

$$\frac{V_{off}^b + V_d}{R_{off}} + \frac{(V_{off}^b + V_{ss})F}{R_{pd}} + \frac{V_{off}^b - V_{dd} + V_d}{R_{pu}} + I_{pd}F - I_{pu} = 0 \qquad (9\text{-}4\text{-}1)$$

Hence, we find

$$V_{off}^b = \left(I_{pu} - I_{pd}F + \frac{V_{dd} - V_d}{R_{pu}} - \frac{V_d}{R_{off}} - \frac{V_s F}{R_{pd}} \right) R_1^b \qquad (9\text{-}4\text{-}2)$$

where

$$R_1^b = \left(\frac{1}{R_{pu}} + \frac{F}{R_{pd}} + \frac{1}{R_{off}} \right)^{-1}$$

This expression is valid when

$$V_{dd} - V_{off}^b \geq V_{pu}^s \qquad (9\text{-}4\text{-}3)$$

where V_{pu}^s is the saturation voltage for the pull-up transistor. Otherwise, the equivalent circuit of Fig. 9-4-1a should be used, leading to the following expression:

$$V_{off}^a = \left(-I_{pd}F + \frac{V_{dd} - V_d}{R_{pu}^0} - \frac{V_d}{R_{off}} - \frac{V_s F}{R_{pd}} \right) R_1^a \qquad (9\text{-}4\text{-}4)$$

where

$$R_1^a = \left(\frac{1}{R_{\text{pu}}} + \frac{F}{R_{\text{pd}}} + \frac{1}{R_{\text{off}}}\right)^{-1} \tag{9-4-5}$$

(All notations are explained in Fig. 9-4-2.)

The critical fan-out can be found from (9-4-2) by demanding

$$V_{\text{off}}^b \approx 0 \tag{9-4-6}$$

so that the next stages may be turned on:

$$F_{cr} \approx \frac{I_{\text{pu}} + \dfrac{V_{\text{dd}} - V_d}{R_{\text{pu}}} - \dfrac{V_d}{R_{\text{off}}}}{I_{\text{pd}} + \dfrac{V_s}{R_{\text{pd}}}} \tag{9-4-7}$$

When the next stages are turned on (i.e., $V_{\text{off}}^b > 0$), the gate current of the switching transistors in the next stages should be included in (9-4-1) and (9-4-4). The major effect of the gate current is to limit V_{off}^b at

$$V_{\text{off}}^b \approx V_{\text{bi}} + R_{\text{gs}}(I_{\text{ds}} + I_{\text{gs}})$$

where V_{bi} is the built-in voltage, I_{ds} is the drain-to-source current, I_{gs} is the gate-to-source current, and R_{gs} is the source series resistance. Assuming that $V_d/R_{\text{off}} \ll I_{\text{pu}}$ and that the FET current and resistance scale with device width, we find

$$F_{cr} \approx \frac{W_{\text{pu}}}{W_{\text{pd}}} \frac{1 + \dfrac{V_{\text{dd}} - V_d}{I_{\text{pd}} R_{\text{pd}}}}{1 + \dfrac{V_s}{I_{\text{pd}} R_{\text{pd}}}} \tag{9-4-8}$$

where W_{pu} and W_{pd} are widths of the pull-up and pull-down transistors, respectively.

For typical values ($I_{\text{pd}} \approx 0.2$ mA, $R_{\text{pd}} \approx 25$ KΩ, $I_{\text{pd}}R_{\text{pd}} \approx 5$ V), the critical value of the fan-out seems to be of the order of the ratio $W_{\text{pu}}/W_{\text{pd}}$. However, as can be seen from (9-4-8), the fan-out capabilities may be somewhat improved by increasing V_{dd} and decreasing V_s (the absolute value of the negative power supply). However, too large a decrease in V_s may result in an unacceptable value of the output voltage when the switching transistor is on (see below).

An increase in the fan-out may be achieved by increasing the ratio $I_{\text{pu}}/I_{\text{pd}}$. This approach, however, may lead to an increase in the time delay because of the smaller current of the pull-down transistor available to discharge the gate-to-source capacitance of the switching transistor (see Section 9-4-4).

In the analysis of the fan-out given above, we assumed that the gate is loaded by identical gates. However, the fan-out is determined by the total widths of all switching transistors of the loading gates. This means that each gate with a 20-μm-wide switching transistor can be replaced by two gates with 10-μm-wide switching transistors, etc., allowing the additional measure of flexibility in the SDFL design.

The fan-out capabilities can be greatly enhanced using source followers. This approach is analyzed in Ref. 196.

When the switching transistor is in the on state, the output loop of the SDFL gate can be represented by one of the equivalent circuits shown in Fig. 9-4-3. Voltage V_{on} at the gate of the switching transistors in the next stages can be found from the nodal equation at node 1 of these circuits. For the circuit shown in Fig. 9-3-3a, we find

$$\frac{V_{on} + V_d}{R_q} + I_q - \frac{V_{dd} - V_{on} - V_d}{R_{pu}} + \frac{(V_{on} + V_s)F}{R_{pd}^o} - I_{pu} = 0 \tag{9-4-9}$$

$$V_{on} = \left[I_{pu} - I_q + \frac{(V_d - V_s)F}{R_{pd}^o} + \frac{V_{dd}}{R_{pu}} \right] R_2 - V_d \tag{9-4-10}$$

where

$$R_2 = \left(\frac{1}{R_q} + \frac{1}{R_{pu}} + \frac{F}{R_{pd}^o} \right)^{-1} \tag{9-4-11}$$

(All notations are explained in Fig. 9-4-2.)

The solution is valid if $V_{on} + V_d$ is greater than V_q^s. In the opposite case, the equivalent circuit of Fig. 9-4-3b is valid, and V_{on} can be found from

$$V_{on} = \left[I_{pu} - \frac{(V_d - V_s)F}{R_{pd}^o} + \frac{V_{dd}}{R_{pu}} \right] R_2^o - V_d \tag{9-4-12}$$

where

$$R_2^o = \left(\frac{1}{R_q^o} + \frac{1}{R_{pu}} + \frac{F}{R_{pd}^o} \right)^{-1} \tag{9-4-13}$$

In order to switch off the switching transistors of the next gates, V_{on} should be negative and of the order of V_t, where V_t is the threshold voltage:

$$V_{on} \leq V_t \tag{9-4-14}$$

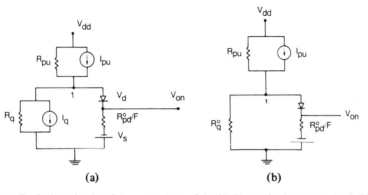

(a) (b)

FIGURE 9-4-3. Equivalent circuits of the output loop of the SDFL gate in the on state loaded by F gates [103]. (a) Switching transistors in saturation region. (b) Switching transistors in the linear region.

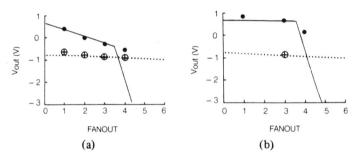

FIGURE 9-4-4. Voltages V_{off} and V_{on} at the gates of the switching transistors in the next stages versus fan-out predicted by the analytical model [103]. —, switching transistor off. · · ·, switching transistor on. Also shown are the results of the SPICE simulation ($\oplus$ off state, $\bullet$ on state). (a) $V_{dd} = 2$ V, $V_s = -1.2$ V. (b) $V_{dd} = 5$ V, $V_s = -1.2$ V.

As can be seen from Eqs. (9-4-10) and (9-4-12), this condition is easier to fulfill when the fan-out is larger. One can also see that the condition can be met either by increasing V_s or by increasing the ratio I_q/I_{pu}. Typical dependences of V_{on} and V_{off} on the fan-out predicted by the model are shown in Fig. 9-4-4a and 9-4-4b for different values of V_{dd}. The results of the approximate SPICE calculations described below are also shown for comparison. The agreement seems to be quite good, indicating that the analytical model gives an accurate estimate of the fan-out as a function of the circuit parameters and supply voltages.

9-4-2. Fan-in Limitations

The fan-in capabilities of the SDFL logic are limited by the gate-to-source series resistance R_{gs} of the switching FET. The maximum input gate current in the on state is

$$I_{\text{in}} = F_{\text{in}} I_{\text{pu}} - I_{\text{pd}} \tag{9-4-15}$$

where F_{in} is the fan in, I_{pu} is the pull-up current, and I_{pd} is the pull-down current. This current leads to the voltage drop

$$\Delta V_{\text{out}} \approx I_{\text{in}} R_{\text{gs}} \tag{9-4-16}$$

and hence raises the output voltage in the on state, decreasing the noise margin. For typical values of $R_{\text{gs}} \approx 100\,\Omega$, $I_{\text{pu}} \approx 1$ mA, and $I_{\text{pd}} \approx 0.3$ mA, a fan-in of 5 would reduce the noise margin by 0.5 V.

This analysis assumes that the pull-up transistor is in the saturation regime. When the voltage drop across the logic diode, the level-shifting diodes, and the input gate become large, the pull-up transistor operates in a linear regime. Under such conditions, the input current drops and, hence, the SDFL gate may have a better fan-in than indicated by (9-4-15) and (9-4-16), but somewhat reduced speed.

9-4-3. Transfer Characteristics

As an example, we consider SDFL circuits fabricated using multiple selective ion implantation directly into qualified Cr-doped GaAs substrates [103]. Selenium

TABLE 9-4-1. Parameters of the Pull-Up FET Used in the SPICE
Simulation[a,b]

Threshold voltage	-1.4 V
β	0.473×10^{-3} A/V^2
Built-in voltage V_{bi}	0.7 V
Gate-to-source capacitance C_{gs0}	8 fF
Gate-to-drain capacitance C_{gd0}	8 fF

[a] Reference 103.
[b] The parameters of the pull-down and switching FETs were scaled proportionally to the gate width (14, 20, and 4 μm for the pull-up, switching, and pull-down FETs, respectively).

ions were implanted in the active channel and sulfur ions under ohmic contacts and Schottky diodes. The wafers were annealed at 850°C for 30 min. Au/Ge/Ni metals were used to form ohmic contacts with a resulting ohmic line resistance $R_c \approx 0.3 \, \Omega$ mm. TiW/Au Schottky gates with $\Phi_b = 0.76$ eV were sputter-deposited to form the 1-μm gates and first-level metallization. The interlevel dielectric was reactively sputtered Si_3N_4 with via defined by plasma etching. Sputtered Ti/Au patterned with ion etching formed the second level metallization. The fabricated MESFETs had a threshold voltage ≈ -1.4 V. The measured current–voltage characteristics were well approximated by the SPICE JFET model with the parameters given in Tables 9-4-1 and 9-4-2. The measured and calculated transfer curves of a single gate are compared in Fig. 9-4-5. A relatively good agreement indicates that the SPICE simulation gives quite reasonable results for the GaAs logic at dc. The difference in the transfer curves at large values of V_{dd} is due to the inaccuracy in describing the transfer characteristic of the switching FET.

A graphical analysis of the measured transfer curve in Fig. 9-4-5a leads to the worst case noise margin of 0.5 V.

In order to investigate how the transfer characteristics vary with the fan-out, the chains of gates shown in Fig. 9-4-6 were simulated. As can be seen from the transfer characteristics shown in Figs. 9-4-7 and 9-4-8, the critical fan-out value is smaller than 3, in good agreement with the predictions of the analytical model.

Measured and simulated response of the chains of ten inverters with a fan-out of 1 and fan-out of 2 are shown in Fig. 9-4-9. The good agreement indicates again that the simulation using SPICE gives reasonable results for GaAs digital circuits. As can be seen from Fig. 9-4-9, for a fan-out of 2, there is a considerable change in response compared to a fan out of 1, which agrees with the results of the simulation described above (see Figs. 9-4-6 and 9-4-7).

TABLE 9-4-2. Parameters of the Diodes Used in the SPICE Simulation[a]

	Switching diode	Level-shifting diode
Series resistance (Ω)	170.00	170.00
Built-in voltage (V)	0.65	0.65
Zero-bias capacitances (fF)	3.00	18.00

[a] Reference 103.

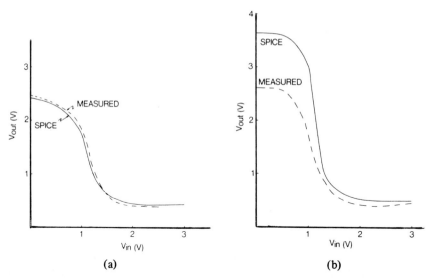

FIGURE 9-4-5. Transfer characteristics of the SDFL gate for different supply voltages [103]. (a) $V_{dd} =$ 2.5 V, $V_s = -1.2$ V. (b) $V_{dd} = 4$ V, $V_s = -0.7$ V.

9-4-4. Fan-out and Speed of SDFL Ring Oscillators

The computed and experimental values of the time delay for seven-stage ring oscillators with fan-outs of 1 and 2 are given in Table 9-4-3. When the fringing capacitances are not taken into account, the computed period of oscillations is about half of the experimentally observed value.

The exact values of the fringing capacitances depend on the layout because the mutual fringing capacitance between two metal lines decreases very slowly with the separation. According to Van Tuyl [197], this capacitance can be estimated as

$$C_c = \frac{1.39 \times 10^{-2}(\varepsilon_r + 1) W}{\ln[4(1 + d/l)]}\,(\text{fF}) \qquad (9\text{-}4\text{-}17)$$

for $l/d \le 0.75$ and

$$C_c = 2.82 \times 10^{-3}(\varepsilon_r + 1) \ln[4(1 + 2l/d)] W\,(\text{fF}) \qquad (9\text{-}4\text{-}18)$$

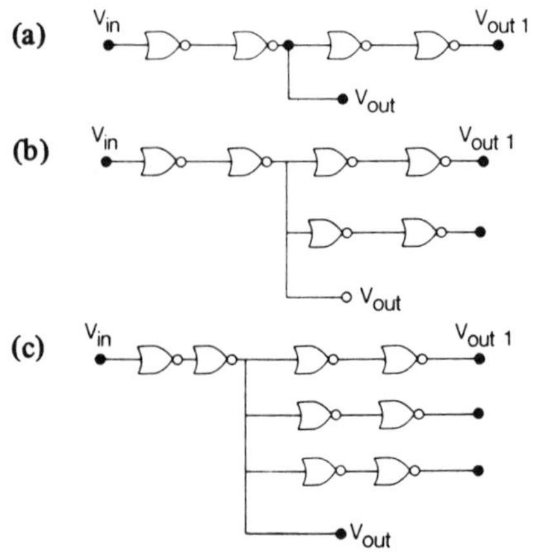

(a) V_{in} ▷○—▷○—▷○—▷○— $V_{out\,1}$
      • V_{out}

(b) V_{in} ▷○—▷○—▷○—▷○— $V_{out\,1}$
      ▷○—▷○—
      ○ V_{out}

(c) V_{in} ▷○▷○—▷○—▷○— $V_{out\,1}$
      ▷○—▷○—
      ▷○—▷○—
      • V_{out}

FIGURE 9-4-6. Chains of gates used in the SPICE simulation [103]. (a) Fan-out of 1. (b) Fan-out of 2. (c) Fan-out of 3.

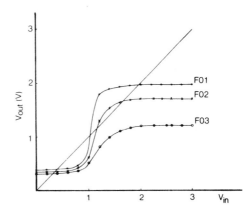

FIGURE 9-4-7. Transfer characteristics of chains of gates shown in Fig. 9-4-6 [103]. $V_{dd} = 2.5$ V, $V_s = -1.2$ V.

for $l/d \geqslant 0.75$. Here dimensions are in microns, and symbols are explained in Fig. 9-4-10. The calculated curve C_c versus d/l is shown in Fig. 9-4-11. As can be seen from the figure, for the exact calculation of the fringing capacitances, the mutual capacitances between practically all lines in the circuit have to be taken into account. A substantial rise of the mutual fringing capacitance when $d/l \leqslant 3$ leads to a design rule which requires setting parallel lines apart by approximately $2d$-$4d$.

In Ref. 103 two different models were used in order to take the fringing capacitance into account. In the first model, the line capacitances to the ground were calculated using the Van Tuyl formulas [197]:

$$C_{GR} = 5.56 \times 10^{-2}\left[\frac{W\varepsilon^*(w)}{\ln(8h/l)} + \frac{l\varepsilon^*(l)}{\ln(8h/W)}\right] - 8.85 \times 10^{-3}\frac{\varepsilon_r l W}{h} \text{ (fF)} \quad (9\text{-}4\text{-}10)$$

where

$$\varepsilon^*(l) = \frac{\varepsilon_r + 1}{2} + \frac{\varepsilon_r - 1}{2}\left(1 + 12\frac{h}{l}\right)^{-1/2} \quad (9\text{-}4\text{-}20)$$

$$\varepsilon^*(W) = \frac{\varepsilon_r + 1}{2} + \frac{\varepsilon_r - 1}{2}\left(1 + 12\frac{h}{W}\right)^{-1/2} \quad (9\text{-}4\text{-}21)$$

Here h, W, and L are in microns.

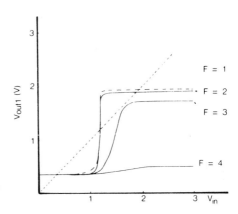

FIGURE 9-4-8. Transfer characteristics of the chain of gates shown in Fig. 9-4-6 (the output voltage V_{out1} is amplified by two gates with fan-outs 1 and 0, respectively; see Fig. 9-4-7) [103]. $V_{dd} = 2.5$; V, $V_s = -1.2$ V.

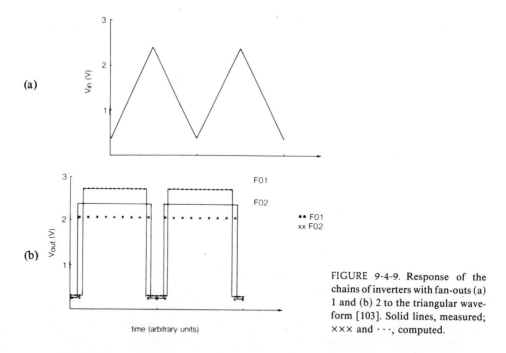

FIGURE 9-4-9. Response of the chains of inverters with fan-outs (a) 1 and (b) 2 to the triangular waveform [103]. Solid lines, measured; ××× and · · ·, computed.

As shown in Ref. 103 this formula agrees within 10%–15% of the empirical formula suggested in Ref. 198. The limitations of this approach are apparent from Fig. 9-4-11, which shows that the contacts located farther from the gate may still give considerable contribution to the effective fringing capacitance.

Another approach used in Ref. 103 was simply to assign the fringing capacitance to the ground, which is proportional to the periphery of the contact. The fringing capacitance per unit length 0.07 fF/μm periphery (0.14 fF/μm length) provided a good agreement with the calculations using the mutual fringing capacitances and with the experimental data.

As can be seen from Table 9-4-3, the speed decreases with an increase of the fan-out. This result can be interpreted using the following simplified mode.

The total time delay per stage is determined by

$$\tau_D \approx (\tau_{on} + \tau_{off})/2 \tag{9-4-22}$$

where τ_{on} is the time required to switch the switching transistor on and τ_{off} is the time required to switch the switching transistor off. These times can be roughly

TABLE 9-4-3. Delay per Gate for SDFL Ring Oscillators

Fan-out	τ_D (ps) Computed with no fringing capacitances	τ_D (ps) Computed with fringing capacitances	τ_D (ps) Measured
1	59.3	105	107
2	—	157	178

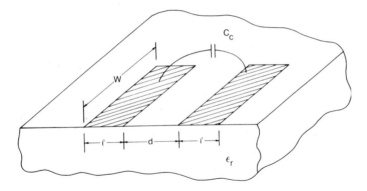

FIGURE 9-4-10. Planar metal contacts [103].

estimated as follows:

$$\tau_{\rm on} \approx \frac{QF}{K_{\rm on}I_{\rm pu}} \tag{9-4-23}$$

where Q is the effective logic gate charge (including fringing) and $K_{\rm on}$ is the fraction of the pull-up current available to charge C_g (part of the current goes through the pull-down):

$$\tau_{\rm off} \approx \frac{C_g \Delta V}{I_{\rm pd}} \tag{9-4-24}$$

where ΔV is the logic swing (assuming that the transistor discharges primarily through the pull-down transistor). Hence

$$\tau_D = \frac{C_g \Delta V}{2 I_{\rm pu}}\left(\frac{F}{K_{\rm on}} + \frac{I_{\rm pu}}{I_{\rm pd}}\right) \tag{9-4-25}$$

For $I_{\rm pu} \approx 1$ mA, $I_{\rm pu}/I_{\rm pd} = 3.75$, $\Delta V \approx 1.5$ V, and from the comparison of (9-4-25) to the results in Table 9-4-3, we find $K_{\rm on} = 0.27$ and $C_g = 19$ fF.

This analytical and computer simulation shows that in the first order, the fan-out of SDFL logic is limited by the ratio of the pull-up and pull-down transistor current. More detailed analysis shows that the fan-out is also dependent on the supply voltages and drain-to-source conductances of the transistors. The time delay of the SDFL logic increases with the fan-out. For simulated ring oscillators, the critical

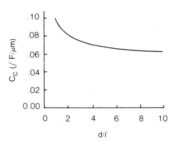

FIGURE 9-4-11. Mutual fringing capacitances of two planar metal contacts as a function of distance between the contacts [103]. $W = 1\ \mu$m.

fan-out is close to two. However, larger fan-outs may be tolerable for individual gates. Because the fan-out is primarily limited by the ratio of the currents of the pull-up and pull-down transistors, using different implants for pull-ups and pull-downs may allow for an extra measure of flexibility in the SDFL design.

9-5. DESIGN AND SIMULATION OF MODFET DCFL CIRCUITS [187, 188]

9-5-1. Introduction

Modulation-doped (AlGa)As/GaAs field effect transistors (MODFETs), also called HEMTs, have been fabricated offering excellent ultra-high-speed performance. Propagation delays as low as 8.5 ps at 77 K and 11.6 ps at 300 K [200, 201] have been obtained in MODFET DCFL ring oscillator circuits. A 4 kb MODFET RAM has been demonstrated [173].

In this section we describe a MODFET circuit simulator based on a customized version of SPICE called UM-SPICE [187, 188]. This circuit simulator implements an extended charge control model, formulated in Chapter 10, to describe the current–voltage characteristics of MODFETs. This model includes the effects of the finite interface carrier density of the two-dimensional electron gas (2D gas) at high gate biases as well as the parallel conduction through the parasitic MESFET on the transfer characteristics. The capacitance–voltage characteristics are calculated using an empirical equation derived by fitting the numerically calculated gate capacitance curve. UM-SPICE also implements the velocity saturation model for the ungated GaAs FET [193] used as the load devices in MODFET inverter stages.

Based on the results of the MODFET inverter circuit simulations and experimental data, a set of analytical equations which allows us to obtain design guidelines for MODFET inverters is derived. These guidelines relate the inverter parameters such as output voltage levels, switching voltages, inverter gain, noise margins, and switching times to the parameters of the driver and load devices.

Finally we present experimental results obtained for self-aligned MODFET ring oscillators with gate propagation delays of 11.6 ps at 300 K with a power dissipation of 2.5 mW and demonstrate that these results are in good agreement with the UM-SPICE circuit simulation.

9-5-2. MODFET Current–Voltage Characteristics

A schematic diagram of a MODFET is shown in Fig. 9-5-1. As shown in chapter 10, a two-dimensional electron gas is formed in the unintentionally doped GaAs buffer layer at the heterointerface. The charge density of the two-dimensional electron gas versus the gate voltage is given by

$$n_s = \frac{C_0}{q}[V_{gs} - V_{TO} - V(X)] \tag{9-5-1}$$

where V_{gs} is the intrinsic gate to source voltage, $V(X)$ is the voltage at X in the channel, and q is the electronic charge. The threshold voltage is defined as

$$V_{TO} = \Phi_b - \Delta E_c - V_{P2} \tag{9-5-2}$$

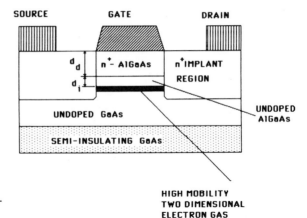

FIGURE 9-5-1. Self-aligned modulation doped field effect transistor.

Here, Φ_b is the built-in surface barrier, ΔE_c is the energy band discontinuity at the (AlGa)As/GaAs interface, and

$$V_{P2} = \frac{qN_d d_a^2}{2\varepsilon} \qquad (9\text{-}5\text{-}3)$$

is the pinch-off voltage, ε is the dielectric constant of (AlGa)As, N_d is the doping concentration in the (AlGa)As layer, and d_a is its thickness. The geometric gate to channel capacitance per area C_0 is defined as

$$C_0 = \varepsilon/(d + \Delta d) \qquad (9\text{-}5\text{-}4)$$

where $d = d_a + d_i$, d_i is the thickness of the undoped AlGaAs buffer layer, and $\Delta d \approx 80\ \text{Å}$ is the correction term to account for the quantization of the 2D gas in the potential well (see Chapter 10) in the direction normal to the heterointerface plane.

Using the gradual channel approximation (see Chapter 7) and integrating over the channel length, we find for the current–voltage characteristic below the saturation voltage

$$I_{ds} = \beta\left(V_{gst}V_{ds} - \frac{V_{ds}^2}{2}\right) \qquad (9\text{-}5\text{-}5)$$

where V_{ds} is the drain-to-source voltage, $V_{gst} = V_{gs} - V_{TO}$ and

$$\beta = \mu C_0(W/L) \qquad (9\text{-}5\text{-}6)$$

Here μ is the low field electron mobility in the channel, and W and L are the width and length of the gate.

Assuming that the drain current saturates when the electric field at the drain side of the channel exceeds the critical field $F_s = v_s/\mu$, where v_s is the electron saturation velocity, the saturation current is given by

$$I_{dss} = \beta V_{sl}^2\{[1 + (V_{gst}/V_0)^2]^{1/2} - 1\} \qquad (9\text{-}5\text{-}7)$$

when V_{ds} is equal to the saturation voltage

$$V_{dss} = V_{gst} + V_O - (V_{gst}^2 + V_O^2)^{1/2} \tag{9-5-8}$$

where $V_O = F_s L$.

For devices with $n_{so}/N_d > d_d$, where n_{so} is the equilibrium 2D gas interface carrier density, three different operating regimes can be observed with increasing gate bias (devices with $n_{so}/N_d < d_d$ behave as MOSFETs when $V_{gst} > (V_{PO})_{2D}$). The simple charge control model described above is valid for gate voltages $V_{TO} < V_{gs} < V_{TO} + (V_{PO})_{2D}$; here the 2D gas pinch-off voltage is given by

$$(V_{PO})_{2D} = qn_{so}(d + \Delta d)/\varepsilon \tag{9-5-9}$$

For the range of gate voltages $V_{TO} + (V_{PO})_{2D} \leq V_{gs} < V_{T3}$, where

$$V_{T3} = V_{TO} + (V_{PO})_{2D} + V_O \tag{9-5-10}$$

is the effective threshold voltage of the parasitic MESFET, the drain saturation current is reduced from the value given by Eq. (9-5-7) owing to the finite value of n_{so} (see Chapter 10). In this regime the saturation current is given by

$$I_{dss} = I_2 \frac{(V_{PO})_{2D} + 2V_{g2}}{V_O + [V_O^2 + 2(V_{PO})_{2D} V_{g2} + (V_{PO})_{2D}^2]^{1/2}} \tag{9-5-11}$$

where $V_{g2} = V_{gs} - V_{TO} - (V_{PO})_{2D}$, and

$$I_2 = qn_{so}v_s W \tag{9-5-12}$$

is the maximum current through the 2D gas channel.

In the gate bias range $V_{gs} \geq V_{T3}$, we have the onset of the parasitic MESFET which provides a second parallel conduction path from drain to source in the (AlGa)As layer. Thus, the total saturation current is given by

$$I_{dss} = I_2 + \beta_0(V_{gs} - V_{T3})^2 \tag{9-5-13}$$

where

$$\beta_0 = \frac{2\varepsilon v_{s3} W}{t_d[(V_{PO})_{3D} + 3v_{s3}L/\mu_3]} \tag{9-5-14}$$

$$t_d = d_d - \frac{n_{so}}{N_d} \tag{9-5-15}$$

is the MESFET channel thickness, and v_{s3} and μ_3 are the carrier saturation velocity and mobility in AlGaAs. The MESFET pinch-off voltage is given by

$$(V_{PO})_{3D} = \frac{qN_d t_d^2}{2\varepsilon} \tag{9-5-16}$$

So far we have treated the intrinsic device equations. When the source and drain series resistance R_s and R_d are included, Eq. (9-5-7) and (9-5-8) should be modified to account for the additional voltage drops. The extrinsic 2D gas pinch-off voltage can be expressed as

$$V_{P1} = (V_{PO})_{2D} + I_1 R_s \qquad (9\text{-}5\text{-}17)$$

where

$$I_1 = I_2 \frac{(V_{PO})_{2D}}{V_O + [V_O^2 + (V_{PO})_{2D}^2]^{1/2}} \qquad (9\text{-}5\text{-}18)$$

Then for the range $0 < V_{gst} < V_{P1}$ we have

$$I_{dss} = \beta V_{sl}^2 \frac{[1 + 2\beta R_s V_{gst} + (V_{gst}/V_{sl})^2]^{1/2} - 1 - \beta R_s V_{gst}}{1 - (\beta R_s V_{sl})^2} \qquad (9\text{-}5\text{-}19)$$

$$V_{dss} = V_{gst} + V_O - (V_{gst}^2 + V_O^2)^{1/2} + I_{dss}(R_s + R_d) \qquad (9\text{-}5\text{-}20)$$

The extrinsic MESFET threshold voltage is given by

$$V_{T3'} = V_{TO} + V_{P1} + V_{OR} \qquad (9\text{-}5\text{-}21)$$

where $V_{OR} = V_O + (I_2 - I_1)R_s$. Then for gate voltages $V_{P1} < V_{gst} < V_{T3}$, the drain saturation current can be expressed as

$$I_{dss} = I_2 \frac{(V_{PO})_{2D} + 2V_{g2'}}{V_O + [V_O^2 + 2(V_{PO})_{2D}V_{g2'} + (V_{PO})_{2D}^2]^{1/2}} \qquad (9\text{-}5\text{-}22)$$

here $V_{g2'} = V_{g2}V_O/V_{OR}$. And for the range $V_{gst} > V_{T3'}$, Eq. (9-5-13) should be replaced by

$$I_{dss} = I_2 + \beta_{MES}(V_{gs} - V_{T3})^2 \qquad (9\text{-}5\text{-}23)$$

where the extrinsic transconductance parameter β_{MES} may be approximated as follows:

$$\beta_{MES} = I_{d\,sat1}/(V_{bi} - 0.1 - V_T)^2 \qquad (9\text{-}5\text{-}24)$$

where $I_{d\,sat1}$ is given by

$$I_{d\,sat1} = \frac{1 + 2\beta_0 R_s V_{GT3} - (1 + 4\beta_0 R_s V_{GT3})^{1/2}}{2\beta_0 R_s^2} \qquad (9\text{-}5\text{-}25)$$

here $V_{GT3} = V_{bi} - 0.1\,V - V_{T3}$.

Equations (9-5-17)–(9-5-23) are approximate equations but the numerical calculations show that they agree with the exact solutions over a wide range of series resistances R_s and R_d.

The drain-to-source current I_{ds} as a function of V_{ds} is described using an empirical equation (see Chapter 7) of the form

$$I_{ds} = I_{dss}(1 + \lambda V_{ds}) \tanh\left(\frac{g_{ch}V_{ds}}{I_{dss}}\right) + \frac{V_{ds}}{R_{sh}} \qquad (9\text{-}5\text{-}26)$$

in both the linear and saturation regimes. Here R_{sh} is the gate-voltage dependent shunt resistance [188], λ is the output conductance parameter to account for channel length modulation in the saturation region and g_{ch} is the channel conductance at low V_{ds} given by

$$g_{ch} = \frac{g_{chi}}{1 + g_{chi}(R_s + R_d)} \qquad (9\text{-}5\text{-}27)$$

where $g_{chi} = \beta V_{gst}$. Equations (9-5-26) reduces to the Shockley model as $V_{ds} \to 0$.

9-5-3. UM-SPICE MODFET and Ungated Load Models

The large signal equivalent circuit of the MODFET implemented in UM-SPICE is illustrated in Fig. 9-5-2. The intrinsic MODFET model, shown enclosed in dashed lines, is represented by a nonlinear current source in parallel with a drain-to-source conductance g_{ds}. The conductance g_{ds} is included in the model through the empirical parameter λ. Then from Eq. (9-5-26), the current source is described by

$$I_{ds'} = I_{dss} \tanh(V_{ds'}/V_{ss'}) \qquad (9\text{-}5\text{-}28)$$

where I_{dss} is given by Eqs. (9-5-8), (9-5-11), and (9-5-13) and $V_{ss'} = I_{dss}/g_{chi}$.

The gate current is modeled by two Schottky diodes connected from the gate to the source and to the drain, respectively. Using the well-known diode equation we find for the total gate current

$$I_g = J_s W_L \left[\exp\left(\frac{V_{gs}}{n V_{th}}\right) + \exp\left(\frac{V_{gd}}{n V_{th}}\right) - 2 \right] \qquad (9\text{-}5\text{-}29)$$

where V_{th} is the thermal voltage, n is the diode ideality factor, and J_s is the reverse saturation current density.

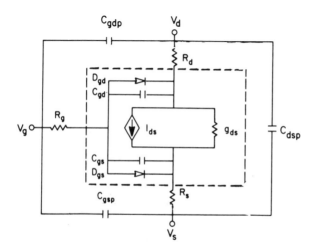

FIGURE 9-5-2. Nonlinear equivalent circuit of the MODFET implemented in UM-SPICE [188].

The nonlinear capacitances $C_{gs}(V_{gs}, V_{ds})$ and $C_{gd}(V_{gd}, V_{ds})$ represent channel charge storage effects. These capacitances are calculated using a modified Meyer's model [195] as implemented in the SPICE MOSFET model and are given by

$$C_{gs} = \frac{2}{3} C_g \left[1 - \frac{(V_{dss} - V_{ds})^2}{(2V_{dss} - V_{ds})^2} \right] \qquad (9\text{-}5\text{-}30)$$

and

$$C_{gd} = \frac{2}{3} C_g \left[1 - \frac{V_{dss}^{\ 2}}{(2V_{dss} - V_{ds})^2} \right] \qquad (9\text{-}5\text{-}31)$$

in the linear region whereas in the saturation region we have $C_{gs} = 2C_g/3$ and $C_{gd} = 0$. The capacitance C_g is the total gate to channel capacitance at $V_{ds} = 0$. An approximate analytical expression for C_g of the form

$$C_g = C_0 \left[1 - \frac{(\alpha - 1)^2 V_{CO}^{\ 2}}{(-\alpha V_{CO} + V_{gst} + \Delta V_T)^2} \right] \qquad (9\text{-}5\text{-}32)$$

is used to fit the numerically calculated C_g vs. V_{gs} curve (see Chapter 10), which includes the Fermi level dependence on n_s. Here, $\alpha = 2 + \sqrt{2}$, $\Delta V_T = 0.03$ V, and $V_{CO} = -3V_{th}$. As can be seen from Fig. 9-5-3, there is good agreement over the range of V_{gs} where C_g is significant. The constant capacitances C_{gsp}, C_{gdp}, and C_{dsp} are the interelectrode capacitances.

We also implemented the saturation velocity model for ungated FETs [193] in UM-SPICE. The current–voltage characteristics are calculated using an expression of the form of Eqn. (9-5-26). Substituting the corresponding parameters for the ungated FET, we find

$$I_L = I_{Ls}(1 + \lambda_L V_{Lds}) \tanh(V_{Lds}/V_{Lss}) \qquad (9\text{-}5\text{-}33)$$

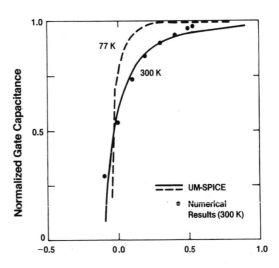

FIGURE 9-5-3. Comparison of the normalized gate capacitance versus gate voltage calculated numerically and using UM-SPICE [188].

where λ_L is the output conductance parameter, V_{Lds} is the drain-to-source voltage, and V_{Lss} is an empirical parameter given by

$$V_{Lss} = I_{Ls}R_{OFF} \tag{9-5-34}$$

The parameters I_{Ls} and R_{OFF} are derived as shown in Ref. 193 (see Section 7-8). Since both I_{Ls} and R_{OFF}^{-1} are linearly proportional to the gate width W_L, the value of V_{Lss} is independent of W_L.

9-5-4. Analytical Inverter Model [188]

Below we formulate a set of analytical equations describing the performance of MODFET inverters and NOR gates using ungated FET loads. Our approach here is similar to our analysis for GaAs DCFL MESFET inverters [192] (see Section 9-3).

We propose the following approximate expression to describe the drain-to-source saturation current versus the gate voltage

$$I_{d\,sat} = g_m(V_{gs} - V_{Teff}) \tag{9-5-35}$$

in the range of gate bias $V_{Teff} < V_{gs} < V_{T3}$, where V_{Teff} is the intercept at $I_{d\,sat} = 0$, and the transconductance g_m is defined as

$$g_m = \frac{I_{dss}(V_{T3}) - I_{dss}(V_{P1}/2 + V_{TO})}{V_{T3} - (V_{P1}/2 + V_{TO})} \tag{9-5-36}$$

where I_{dss} is given by Eq. (9-5-7). Figure 9-5-4 compares our approximate equation versus the I-V curve computed using Eqs. (9-5-7), (9-5-11), and (9-5-13). Equation (9-5-35) gives an adequate fit over the range of gate voltages where it is most likely to operate in an inverter circuit.

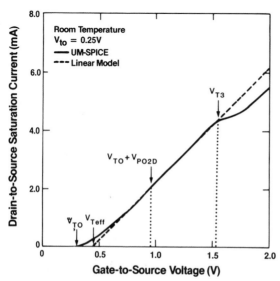

FIGURE 9-5-4. Linear model of the MODFET drain saturation current used in the analytical inverter model compared with the UM-SPICE model [188].

The MODFET inverter dc transfer curve can be adequately approximated by a piecewise linear model similar to one used for GaAs MESFET (see Fig. 9-5-4). In order to describe the transfer curve we need to determine equations for the output high voltage, the switching voltage, the output low voltage, and the gain. The detailed derivations of these equations for the MESFET DCFL inverter are given in Section 9-3 [192]. The corresponding equations for the MODFET inverter are obtained by replacing the MESFET I–V equation by Eq. (9-5-35).

For fan-out greater than zero, the output high voltage V_{oh} is mostly determined by the clamping action of the gate-to-source diode of the next stage. Hence, if the load current at $V_{out} = V_{oh}$ is specified, the output high voltage can be expressed as

$$V_{oh} = nV_{th} \ln\left(\frac{I_{Lh}}{J_s WLF_O}\right) + \left(\frac{1}{F_O} + \frac{1}{F_I}\right)I_{Lh}R_s \qquad (9\text{-}5\text{-}37)$$

where F_O is the fan-out, F_I is the fan-in, and I_{Lh} is the current through the load device when the output voltage is high. Then the load device saturation current I_{Ls} can be easily determined from Eq. (9-5-33).

Assuming that in the transition region, $V_{SW} < V_{in} < V_{ISW}$, the gate current can be neglected, we simply equate the driver and load currents to derive the switching voltage, output low voltage, and gain as illustrated in Fig. 9-3-6. Hence, using Eqs. (9-5-33) and (9-5-35) we find

$$I_{Ls}[1 + \lambda_L(V_{DD} - V_{out})]\tanh\left(\frac{V_{DD} - V_{out}}{V_{Lss}}\right)$$

$$= g_m F_i(V_{in} - V_{Teff})(1 + \lambda V_{out})\tanh\left(\frac{V_{out}}{V_{ss}}\right) \qquad (9\text{-}5\text{-}38)$$

where V_{DD} is the power supply voltage.

Solving Eq. (9-5-38) for the switching voltage V_{SW}, defined as the input voltage at which the onset of the transition from the output high to low voltage occurs, we get

$$V_{SW} = V_{Teff} + \frac{I_{Lh}}{g_m F_I(1 + \lambda V_{oh})\tanh(V_{oh}/V_{ss})} \qquad (9\text{-}5\text{-}39)$$

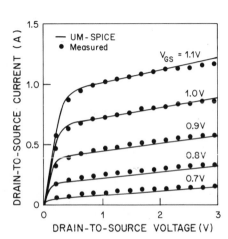

FIGURE 9-5-5. Current–voltage characteristics of MODFET with $V_{TO} = 0.62$ V. Solid lines, simulated; ●, measured data. The gate length of the MODFET is 1 μm; the gate width is 20 μm [188].

The output low voltage V_{ol} is given by

$$V_{\text{ol}} = \frac{V_{ss}}{2} \ln\left(\frac{1+\alpha}{1-\alpha}\right) + \frac{I_{Lh}}{F_O} R_s \qquad (9\text{-}5\text{-}40)$$

where $\alpha = I_L(V_{\text{ol}})/[I_{ds}(V_{\text{ol}})F_I]$. To accurately determine V_{ol}, an iterative solution is required (see Section 9-3). However, as in the case of MESFETs, only a few iterations are required to get convergence in most cases. An initial guess for V_{ol} can be obtained using the expression

$$V_{\text{ol}} \approx V_{ss} I_{Ls}/I_{ds1} + I_{Lh} R_s/F_O \qquad (9\text{-}5\text{-}41)$$

where $I_{ds1} = g_m(V_{\text{oh}} - V_{T\text{eff}} - I_{Lh} R_s/F_O)$.

The inverter gain $G = dV_{\text{out}}/dV_{\text{in}}$ can be expressed as

$$A = -\frac{\partial I_D/\partial V_{\text{in}}}{\partial I_L/\partial V_{\text{out}} - \partial I_D/\partial V_{\text{out}}} \qquad (9\text{-}5\text{-}42)$$

The maximum gain is obtained for large values of V_{DD} when both the driver and load devices are in full saturation. In this case Eq. (9-5-42) reduces to

$$G_{\text{max}} = \frac{g_m}{I_{Ls}(\lambda + \lambda_L)} \qquad (9\text{-}5\text{-}43)$$

The inverter gain corresponding to the slope of our piecewise linear model is obtained by evaluating Eq. (9-5-42) with $V_{\text{out}} = (V_{\text{oh}} + V_{\text{ol}})/2$, which roughly represents the point of highest gain on the real transfer curve.

Throughout this section the noise margin is defined as the "largest square" enclosed by the normal and mirror transfer curves (see Fig. 9-3-5). For the piecewise linear model, the noise margin is given from simple geometrical considerations by

$$\text{NM} = V_{ih} - V_{\text{ol}} \qquad \text{for } V_{sw} \leq V_{ih} \qquad (9\text{-}5\text{-}44)$$

or

$$\text{NM} = V_{\text{oh}} - V_{il} \qquad \text{for } V_{il} \leq V_{sw} \qquad (9\text{-}5\text{-}45)$$

The intercept voltages V_{ih} and V_{il} (see Fig. 9-3-5) are defined as

$$V_{ih} = \frac{V_{\text{oh}} + GV_{sw} + V_{\text{ol}} - V_{Isw}}{1 + G} \qquad (9\text{-}5\text{-}46)$$

$$V_{il} = \frac{2V_{\text{oh}} + (G-1)V_{sw}}{1 + G} \qquad (9\text{-}5\text{-}47)$$

where $V_{Isw} = V_{sw} + (V_{\text{oh}} - V_{\text{ol}})/G$ (see Eqs. (9-3-20)–(9-3-21)). In reality the noise margins are somewhat smaller than those given by Eq. (9-5-44) or (9-5-45) because the piecewise linear model of the transfer curve exaggerates the gain near the boundaries of the switching region.

The transient switching behavior of the inverter may be approximated by considering the charging and discharging time of an effective input capacitance C_N.

This capacitance must include the voltage-dependent MODFET gate capacitance plus any interconnect capacitances. The total average capacitance over the range of the voltage swing, $V_{oh} - V_{ol}$, can be roughly estimated as

$$C_N = \frac{C_0 F_O}{1.5} \frac{(V_{oh} - V_{TO})}{(V_{oh} - V_{ol})} + C_{IN} \tag{9-5-48}$$

where C_{IN} is the effective interconnect capacitance.

The inverter turn-off time τ_{OFF} is determined by the average current I_{OFF} available to discharge C_N given by

$$I_{OFF} = \frac{1}{2} g_m \left(V_{oh} - V_{Teff} - I_{Lh} \frac{R_s}{F_O} \right)(1 + \lambda V_{oh}) - I_{Lh} \tag{9-5-49}$$

Then τ_{OFF} can be estimated as

$$\tau_{OFF} = \frac{(V_{oh} - V_{ol})C_N}{I_{OFF}} \tag{9-5-50}$$

The turn-on current to charge C_N is supplied by the load device. In inverters with low threshold drivers ($V_{TO} < V_{ol}$), this current can be substantially reduced, limiting the attainable speed. The turn-on time is given by

$$\tau_{ON} = \frac{2(V_{oh} - V_{ol})C_N}{I_{L1} - g_m(V_{ol} - V_{Teff})(1 + \lambda V_{ol}) \tanh(V_{ol}/V_{ss})} \tag{9-5-51}$$

where I_{L1} is given by Eq. (9-5-33) with $V_{Lds} = V_{DD} - V_{ol}$. Equations (9-5-50) and (9-5-51) apply to the case of $F_I = 1$. When $F_I > 1$ we only consider the worst case analysis which corresponds to the largest voltage swing. This happens when all the drivers in a gate switch simultaneously and are driven by separate outputs. Thus, the worst case voltage swing is $V_{oh}(F_O = 1) - V_{ol}(F_I > 1)$.

The average switching time, defined as

$$\tau_D = (\tau_{ON} + \tau_{OFF})/2 \tag{9-5-52}$$

roughly determines the maximum frequency at which the inverter stage can be operated while maintaining the maximum noise margin. The highest operating speed is determined by the average propagation delay time which is obtained from the measured or simulated ring oscillator waveforms.

The inverter power dissipation consists of two components: the static dc term plus the dynamic component. Hence, the average power consumption may be estimated as follows:

$$P = I_{Ls}V_{DD} + C_N(V_{oh} - V_{ol})^2 f \tag{9-5-53}$$

For an N-stage ring oscillator $f = (2\tau_D N)^{-1}$.

9-5-5. Simulation Results

The measured and simulated current–voltage characteristics of a high threshold ($V_{TO} = 0.62$ V) MODFET are shown in Fig. 9-5-5. The measured and derived

TABLE 9-5-1. MODFET and Ungated FET Device Parameters Used in the Simulated Results Shown Here[a]

Description	Symbol	Value	Units
Gate length	L	1	μm
Gate width	W	20	μm
Doped layer	d_d	284	Å
Buffer layer	d_i	60	Å
Low field mobility	μ	2900	cm^2/V s
Saturation velocity	v_s	1.2×10^5	m/s
Conduction band discontinuity	ΔE_c	0.318	eV
Built-in potential	ϕ_b	1.0	eV
Electron mobility in (AlGa)As	μ_3	1000	cm^2/V s
Saturation velocity in (AlGa)As	v_{s3}	1×10^5	m/s
Doping concentration in (AlGa)As	N_d		
$V_{\text{TO}} = 0.25$ V		7×10^{17}	cm^{-3}
$V_{\text{TO}} = 0.62$ V		1×10^{17}	cm^{-3}
Output conductance parameter	λ	0.07	V^{-1}
Source series resistance	R_s	100	Ω
Drain series resistance	R_d	100	Ω
Gate series resistance	R_g	60	Ω
Ungated FET contact separation	L_L	2	μm
Ungated FET saturation current	I_{LS}	173	A/m
Ungated FET linear region resistance	R_{OFF}	5	Ω mm
Ungated FET output conductance parameter	L	0.012	V^{-1}

[a] Ref. 188.

device parameters used in our simulations are listed in Table 9-5-1. The measurements were performed on devices fabricated using a self-aligned, completely planar technology [176]. The results of the inverter simulations are illustrated in Figs. 9-5-6–9-5-9. As can be seen in Fig. 9-5-6, the simulated transfer curves exhibit good agreement with the measured curves over a wide range of power supply voltages. Small differences between them may be attributed to some variation of the device parameters across the wafer. The noise margins and ring oscillator propagation delays as functions of the load device gate width for two different MODFET drivers with threshold voltages of 0.25 and 0.62 V are compared in Fig. 9-5-7. The inverter with the higher threshold driver offers higher speeds with increasing load currents

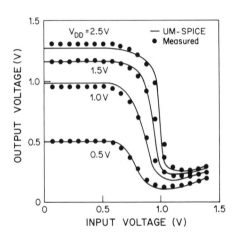

FIGURE 9-5-6. Inverter transfer characteristics for different power supply voltages. Solid lines, simulated; +, measured. The gate length and width of the MODFET drivers are 1 μm and 20 μm, respectively. The width of the ungated FET load is 4 μm and the separation between the contacts is 2 μm.

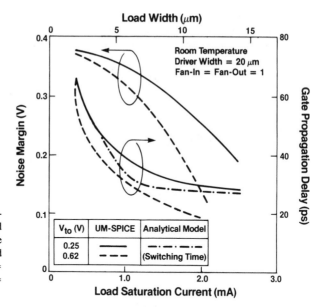

V_{to} (V)	UM-SPICE	Analytical Model
0.25	——	— · — · — · —
0.62	- - - -	(Switching Time)

FIGURE 9-5-7. UM-SPICE simulation of MODFET inverters and ring oscillators. These results are for fan-in and fan-out of one. Solid lines are for drivers with $V_{TO} = 0.25$ V and dashed lines for $V_{TO} = 0.62$ V.

but is accompanied by a faster decrease in the noise margin compared to the low threshold driver.

The simulated output waveform of an 11-stage ring oscillator circuit is shown in Fig. 9-5-8. The measured and simulated gate delay versus power dissipation (power supply voltage) curves are presented in Fig. 9-5-9. The small change in the propagation delay with power dissipation is due to the clamping of the output high voltage by the gate diode and the saturation of the load device.

Figures 9-5-10–9-5-13 compare the results of our analytical model with the output of UM-SPICE. As can be seen, the analytical model offers a good agreement over a wide range of load currents. The worst case analysis of the different fan-out and fan-in loading effects on the switching times are shown in Figs 9-5-13a and

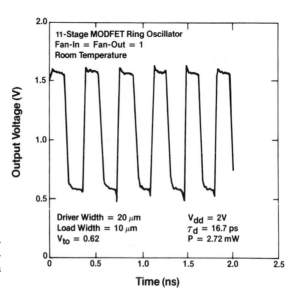

FIGURE 9-5-8. Simulated output waveform of an 11-stage ring oscillator circuit. The delay time is 16.7 ps with a power dissipation of 2.7 mW.

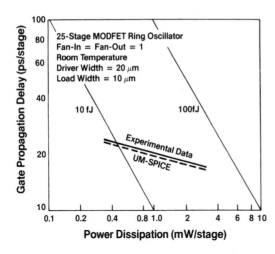

FIGURE 9-5-9. Comparison of experimental and simulated propagation delay versus power dissipation per stage for 25-stage ring oscillators. The minimum experimental delay time is 17.6 ps.

9-5-13b. The noise margin versus load current curves for different driver threshold voltages (i.e., for different doping concentrations in the doped AlGaAs layer) are illustrated in Fig. 9-5-11. The highest noise margin over the widest range of load currents is obtained with drivers with a threshold voltage of ≈ 0.4 V.

The simulation results demonstrate that UM-SPICE may be used as a reliable tool for the simulation of MODFET integrated circuits.

9-6. FABRICATION OF GaAs INTEGRATED CIRCUITS

Most GaAs ICs are fabricated by direct ion implantation into a semi-insulating GaAs substrate grown by the Czochralski method. Material growth, characterization, ion implantation, and ohmic and Schottky contact fabrication are discussed in Chapter 3.

Below we briefly describe several different fabrication processes.

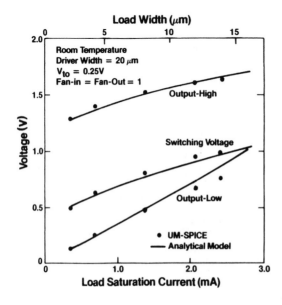

FIGURE 9-5-10. Output and switching voltages versus load currents calculated using the analytical inverter model.

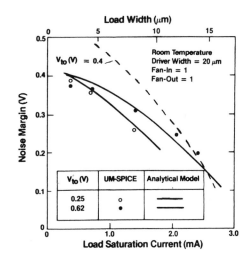

FIGURE 9-5-11. Noise margin versus load current derived from analytical model for different driver threshold voltages.

The first GaAs ICs [11] were fabricated using mesa-implanted depletion mode MESFETs (see Fig. 9-6-1). In this approach the active devices are isolated by etching through an epitaxial or ion-implanted layer. The process is similar to the discrete MESFET fabrication with a second layer of metal and dielectric layer added in order to connect the circuit elements. Alternatively, plated air bridges are used to isolate the first and second level metal at the crossings. Initially epitaxial active n-type GaAs layers were used. Now an ion-implantation into a semi-insulating substrate is used to produce active channels. The process is appropriate for small and medium scale BFL integrated circuits.

A similar process was applied to fabricate epitaxial enhancement mode MESFETs for DCFL circuits. Epitaxial layers were grown using vapor epitaxy, and the enhancement mode devices were fabricated using a recessed gate structure (see Fig. 9-6-2). As a result the source and drain series resistances were reduced, but reproducibity and threshold voltage uniformity were difficult to control. Both contact photolithography and electron beam lithography were used to fabricate circuits of this type [112].

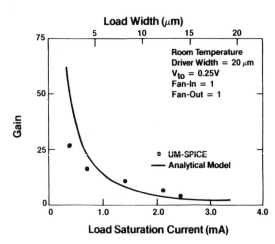

FIGURE 9-5-12. Inverter dc gain versus load width.

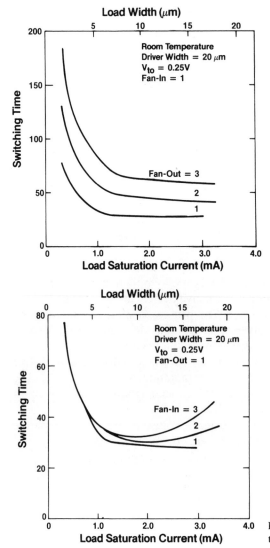

FIGURE 9-5-13. Worst case inverter switching times for different (a) fan-outs and (b) fan-ins.

Better threshold voltage uniformity and reproducibility are achieved using a planar fabrication process. An example of a planar GaAs SDFL IC is shown in Fig. 9-6-3 [16]. In this fabrication process, a GaAs substrate is coated with a thin Si_3N_4 film, which remains throughout all subsequent processing steps. Two localized implantation steps are carried out by implanting through this layer. The first implant is optimized for the FET channels; the second deeper implant is designed for Schottky barrier switching diodes. Both implants are used for ohmic contacts and for the level shifting diodes. Additional implantations may be incorporated into the process if needed. After the implantation steps, an additional insulator is deposited and the implants are annealed at 850°C. This approach provided a good threshold voltage uniformity with the threshold voltage standard deviation between 20 and 40 mV. MSI/LSI circuits (of 1000 gates complexity) have been demonstrated using this technology [112].

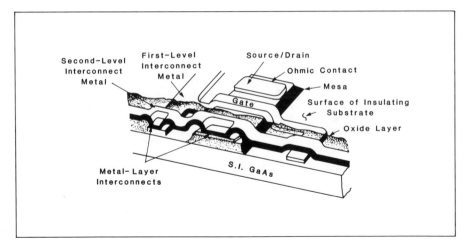

FIGURE 9-6-1. Mesa-implanted depletion mode MESFET [11].

As discussed in Chapters 7 and 8, surface depletion of the active channel may substantially increase the parasitic series resistances (see Fig. 9-6-2). A self-aligned gate fabrication process (see Fig. 9-6-4 [111]) solves this problem, reducing the series resistances and making the fabrication process planar.

A self-aligned gate MESFET IC process using refractory metal–silicon gates was developed for LSI circuits at gigabit speeds [175]. A cross section of a self-aligned gate IC at different stages of this fabrication process is shown in Fig. 9-6-5. The circuits were fabricated by selective ion-implantation into 3-in. LEC substrates. The FET channel and load implants were annealed using a Si_3N_4 cap. Following the anneal, the cap is stripped and the refractory metal–silicon gate is sputter-deposited and patterned using reactive ion etching. This metal serves as the implant mask for the source and drain implants. The n^+ implant is also annealed using a Si_3N_4 cap. The devices are completed using a AuGe ohmic contact formed by liftoff. Interconnect metallization begins with the deposition of a Si_3N_4 layer which assists the liftoff. Au-based (approximately $6000\,Å$ thick) metal is deposited and lifted off using the plasma-etched dielectric. The SiO_2 interlevel dielectric is then deposited. The formation of the second interconnect level is quite similar. Both interconnect levels have sheet resistances less than $0.07\,\Omega/\text{square}$. Excellent threshold voltage uniformity obtained with this process is illustrated by Fig. 9-6-6 [175].

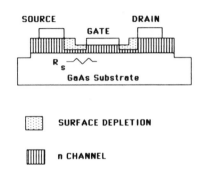

FIGURE 9-6-2. Recessed gate MESFET.

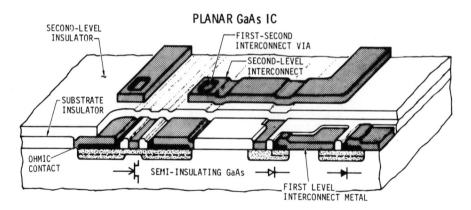

FIGURE 9-6-3. GaAs planar ion-implanted ICs (SDFL) [16]. Two implants, insulators, metallizations, and second level interconnect lines are shown.

Transconductances as high as 216 mS/mm were obtained for 0.7-μm gate length devices. Fabricated ring oscillators achieved speeds as fast as 30 ps and speed-power products as low as 1.1 fJ. Divide by four circuits operated up to 2.6 GHz with a power dissipation per gate of 1.18 mW [175].

A similar process was used to fabricate MODFET self-aligned gate ICs [135, 150, 176, 187, 188]. The modulation doped heterostructures were grown on semi-insulating GaAs substrates using MBE. The cross-sectional view of a self-aligned gate MODFET IC is shown in Fig. 9-6-7, where fabrication steps are explained. An extrinsic transconductance as high as 249 mS/mm at room temperature was observed with an output conductance less than 10 mS/mm. For the average threshold voltage of 0.44 V the standard deviation of 11.9 mV was obtained [176]. Saturated loads were fabricated by direct implantation of GaAs ungated FETs.

High speed for GaAs MESFET circuits (15 ps ring oscillator propagation delay [127,128]) was achieved with the so-called T-gate process [116, 117, 127, 128]. The basic fabrication steps are shown in Fig. 9-6-8 [127]. Active layers were formed by direct Si$^+$ implantation into undoped LEC substrates and annealed at 830°C with a SiO$_2$ cap. After removing the SiO$_2$, TiW was deposited over the entire wafer. The gate-level electron-beam lithography was used to define Al or Ni mask in order to form the T-gates by CF$_4$ reactive ion etching. After a self-aligned n^+ implant, the

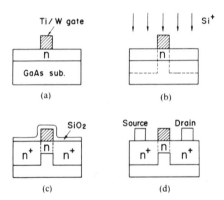

FIGURE 9-6-4. Fabrication process for the self-aligned GaAs MESFETs [111]. (a) Gate metallization (Schottky TiW gate metal, stable at high annealing tempeatures). (b) n^+ implant. (c) SiO$_2$ deposition and annealing. (d) Ohmic metallization (AuGe).

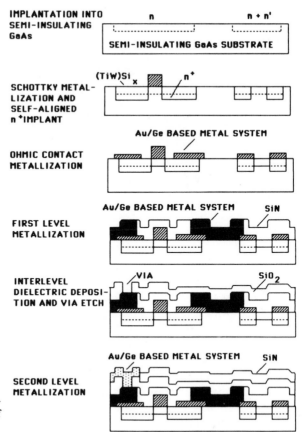

FIGURE 9-6-5. Cross section of self-aligned gate IC at various stages of fabrication process [175].

etch mask was stripped and the wafer was caplessly annealed at 800°C in an arsenic vapor. The device isolation was achieved by a 70-keV boron implantation. The third photolithography level provided patterns for ohmic contacts and interconnect metallization for the circuits. The resulting device structure is shown in Fig. 9-6-9. It has a very short electrical gate length with separations that ensure lower interelectrode capacitances and higher breakdown voltages.

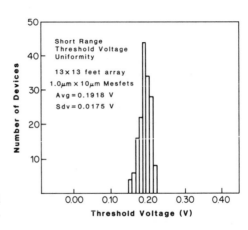

FIGURE 9-6-6. Histogram of short-range threshold voltage uniformity over a 1.0×0.6 mm^2 area [175].

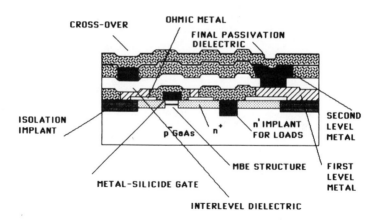

FIGURE 9-6-7. Cross-sectional view of self-aligned gate MODFET IC [176].

A self-aligned gate technology with implanted n^+ layers (SAINT) was described in Refs. 88, 92–94, and 136. The diagram of the process and a SAINT MESFET cross section are shown in Fig. 9-6-10. This process has been used to fabricate GaAs MESFETs with transductances higher than 300 mS/mm. Using SAINT technology the largest GaAs IC—a 16 kb RAM—was implemented [136].

Electron beam photolithography was used to fabricate the fastest solid state circuits (5.8 ps MODFET ring oscillators) [207]. The modulation doped layers were grown by MBE. All features were patterned using conventional optical photolithography except the ring oscillator driver gates, which were written by e-beam. The devices had a recessed gate structure with the threshold voltage adjusted by a wet chemical etch.

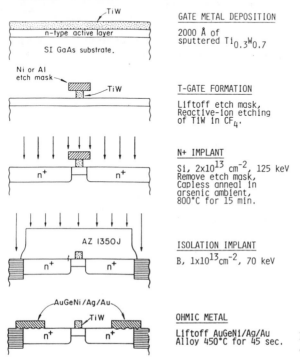

GATE METAL DEPOSITION

2000 Å of
sputtered $Ti_{0.3}W_{0.7}$

T-GATE FORMATION

Liftoff etch mask,
Reactive-ion etching
of TiW in CF_4.

N+ IMPLANT

Si, 2×10^{13} cm^{-2}, 125 keV
Remove etch mask,
Capless anneal in
arsenic ambient,
800°C for 15 min.

ISOLATION IMPLANT

B, 1×10^{13} cm^{-2}, 70 keV

OHMIC METAL

Liftoff AuGeNi/Ag/Au
Alloy 450°C for 45 sec.

FIGURE 9-6-8. Self-aligned T-gate process steps [127].

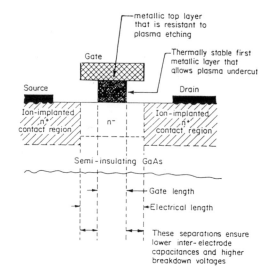

FIGURE 9-6-9. T-gate structure [204].

A self-aligned MODFET fabrication process which produces both enhancement mode and depletion mode MODFETs was used to fabricate 4-kb MODFET RAMs [173]. A selective dry etching technology was applied. The process sequence is explained in Fig. 9-6-11. First, the active region is isolated by a shallow mesa step (180 nm). The source and drain contacts for enhancement and depletion mode MODFETs are metallized with AuGe eutectic alloy and Au overlay. Then the ohmic

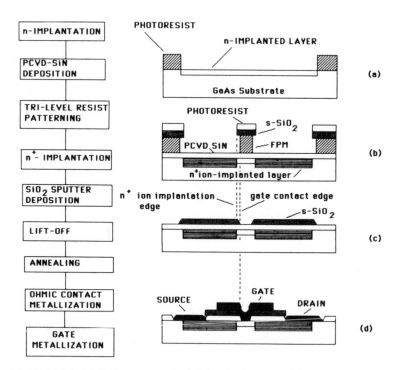

FIGURE 9-6-10. SAINT fabrication process [88]. (a) n-Implantation. (b) Formation of tri-level resist and n^+ implantation. (c) Liftoff of sputtered SiO_2. (d) Metallization.

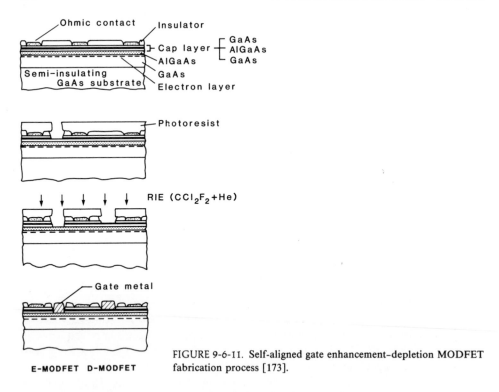

FIGURE 9-6-11. Self-aligned gate enhancement–depletion MODFET fabrication process [173].

contacts are alloyed. Fine gate patterns are formed for enhancement mode MODFETs, and the top GaAs layer and thin $Al_{0.3}Ga_{0.7}As$ stopper are etched off by non-selective chemical etching. Using the same resist, after the formation of gate patterns for depletion mode MODFETs, the top GaAs layer is removed for depletion mode devices and the GaAs layer under the stopper layer is removed for the enhancement mode FETs. This is followed by the deposition of Al Schottky contacts, which are self-aligned with the GaAs cap layer used for the ohmic contacts. Ti/Pt/Au interconnecting metal is attached to the device terminals through the contact holes etched to the device terminals through the contact holes etched in a crossover insulator film. A low-temperature annealing process is used to remove damage due to selective dry recess etching.

This fabrication process has provided an excellent threshold voltage uniformity (20 mV and 12 mV standard deviation of the threshold voltages for the normally-on and normally-off FETs, respectively)[173].

9-7. GaAs DIGITAL ICs—EXAMPLES

9-7-1. GaAs LSI Circuits

As has been mentioned above, a 16-kbit RAM [136] is the largest GaAs IC. A schematic diagram of the 16-kbit static RAM circuit is shown in Fig. 9-7-1. The memory has a 409-word × 4-bit organization. In order to shorten the word-line, the memory cells were divided into two 8-kb memory cell areas with an X-address decoder and word-line drivers located in between. Each memory cell area had one

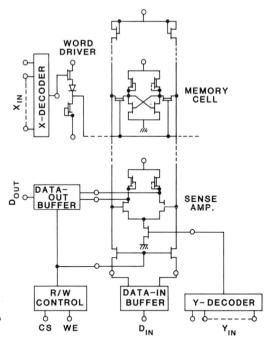

FIGURE 9-7-1. Schematic diagram of 16-kb
GaAs static RAM [136].

word-line driver. The memory cell was designed with six transistors and included
one enhancement–depletion-type cross-coupled flip-flop and two transfer gates. The
gate length was 1 μm, and the gate width was 12 μm for the enhancement mode
devices and 2 μm for the depletion mode load FETs and the transfer gate FETs.
The threshold voltage of enehancement FETs was 85 mV and the threshold voltage
of the depletion mode FETs was −409 mV. The average FET transconductance was
143 mS/mm. Measurements on 23 stage ring oscillators showed a delay time per
stage of 115 ps with power dissipation of 0.11 mW for a bias voltage of 1 V.

The minimum address access time measured was 4.1 ns. The total power
dissipation was 2.52 W, of which memory cell circuits dissipated 0.77 W. The 4-kb
RAM circuits fabricated by the same group [136] had a minimum address time of
2 ns with 0.9 W power dissipation.

The simulations predicted 2.06 ns access time for the 16-kb RAM [136]. A
larger measured access time (4.1 ns) was explained by the scatter in device threshold
voltages which was related to the dislocations in LEC-grown material [136]. Using
a material with a drastically reduced dislocation density, a fully functional 4 kb
RAM was demonstrated [136].

The performance of GaAs memories is summarized in Table 9-7-1 [136]. A
4-kb GaAs RAM (with address access time of 2.6 ns at power dissipation of 1.8 W)
was also described in Ref. 172.

A 4-kb MODFET RAM was designed and fabricated using DCFL logic [173].
A novel fabrication process used for this RAM is described in Section 9-6. The
memory cell was a six-transistor cross-coupled flip-flop circuit. Switching FETs had
a gate length of 2 μm. Switching FETs used in peripheral devices had a gate length
of 1.5 μm. The memory cell was 55 × 39 μm. The chip contained 26864 MODFETs
and had a size of 4.76 × 4.35 mm.

TABLE 9-7-1. Performance of GaAs RAMs[a]

	16 kb	4 kb
Organization	4096×4	1024×4
Chip size (mm × mm)	7.18×6.24	4.13×3.56
Access time (ns)	4.1	2.0
Power dissipation (W)	2.52	0.89
Circuit	DCFL	
Device	SAINT MESFET	
Gate length	$1\,\mu$m	
Line/space	$1.5\,\mu$m/$1.5\,\mu$m	
Via hole	$1.5\,\mu$m × $1.5\,\mu$m	
Cell size	$41\,\mu$m × $32.5\,\mu$m	

[a] Reference 136.

A minimum address access time was estimated at 2 ns at 77 K. At 300 K the access time was 4.4 ns with power dissipation of 0.86 mW.

Another example of a GaAs LSI circuit is a 16 × 16 bit multiplier [119, 120], which included 3168 DCFL gates implemented using self-aligned gate technology. Full adders (FA) and half adders (HA) are based on NOR gates (see Fig. 9-7-2). The block diagram of the multiplier is shown in Fig. 9-7-3.

9-7-2. GaAs MSI and SSI Circuits

One of the fastest frequency dividers was implemented using (Al, Ga)As/GaAs MODFET DCFL circuits [137]. The circuit included both NOR and NAND gates which decreased the gate count and increased the frequency of operation (see Fig. 9-7-4). NAND gates were implemented using a dual gate MODFETs and two MODFETs connected in series. Such a design is possible because of a larger voltage swing of MODFETs compared to GaAs MESFETs. A maximum dividing frequency of 10.1 GHz was achieved at 77 K. The maximum dividing frequency at 300 K was 5.5 GHz. The total power dissipation at 300 K was 34.8 mW at 1.97 V bias. The power dissipation at 77 K was 49.9 mW at 1.67 V Bias. This is a considerably smaller power dissipation than for GaAs BFL circuits which achieve similar dividing frequencies (see Section 9-2-3).

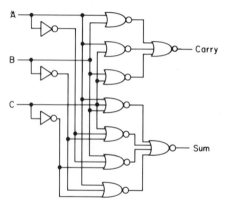

FIGURE 9-7-2. Logic circuit of a full adder with NOR gates [119, 120].

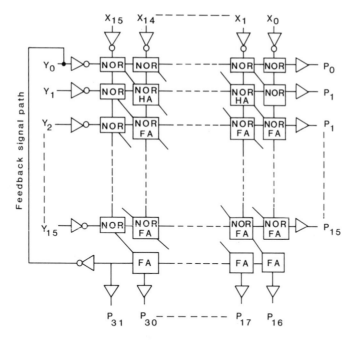

FIGURE 9-7-3. Circuit block diagram of 16×16 GaAs multiplier [119, 120].

Another example of a GaAs SSI circuit is frequency dividers implemented using heterojunction bipolar junction transistors (HBTs) [138]. These dividers are based on current mode logic (CML) discussed in Section 9-2-12. A circuit diagram of a CML bi-level latch used to implement frequency dividers is shown in Fig. 9-7-5 [138]. The maximum input frequency of operation for divide by four circuits was 8.6 GHz, with a total power consumption of 210 mW (exclusive of output drivers).

9-7-3. GaAs Gate Arrays [129, 130]

One of the examples of an SDFL circuit is a GaAs gate array such as described in Refs. 129 and 130. This array featured 432 programmable Schottky diode field-

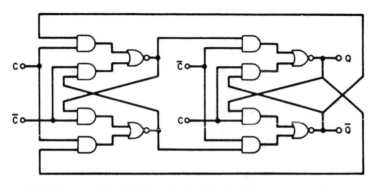

FIGURE 9-7-4. Circuit design of a dual-clocked frequency divider [137].

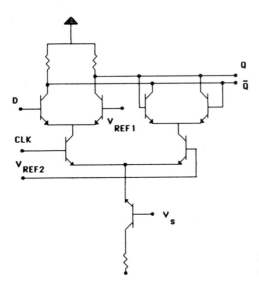

FIGURE 9-7-5. CML bi-level (series gated) latch used to implement frequency dividers [138].

TABLE 9-7-2. Features of GaAs SDFL Cell Array[a]

Uncommitted cell array	
Cells/gate, max. F1/F0	432 cells ≡ 1296 GEC, 8/6
I/O, Max FO of input buffer	32 I/O, 10
Basic gate logic functions	NOR, OR/NAND
Design rule	3 μm
Cell area	$120 \times 112 \ \mu m^2$
Tracks/cell	8 horizontal × 6 vertical
Mask layers	9
Programmable layers	3
Chip area	147 × 185 mils including 64-bit RAM
Package	144 pins maximum
Expected % of cell utilization	70–80%
Performance	
Signal voltage swing (nominal)	1.5 V
Gate delay (minimum)	150 ps
Gate power, speed × power	1.5 mW, 0.2 PJ
I/O Interface options	ECL, T^2L, CMOS SDFL
I/O Power	75 mW (ECL), 130 mW (T^2L)
I/O Clock rate	500 MHz (ECL), 250 MHz (T^2L)
Power supplies	2.5 V, −1.5 V, 5 V (T^2L), −5.2 V (ECL)
Chip power	3 W including 64-bit RAM
Figure of merit	7.4×10^{11} gates cm^{-2} Hz
Expected radiation hardness	
Neutron fluence (n/cm^2)	1×10^{15}
Total dose (rads)	5×10^7
Dose rate (rad/s)	1×10^{11}
Operating temperature	−55 to 125°C

[a] 432 SDFL + RAM, non-self-aligned, depletion mode.

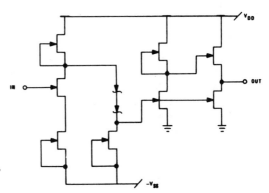

FIGURE 9-7-6. Input buffer of ECL to SDFL interface [130].

effect-transistor logic (SDFL) cells, 32 programmable interface input/output (I/O) buffers, and four on-chip 4 × 4 bit static random access memories (RAM) (see Table 9-7-2). Each SDFL cell can be programmed as a NOR gate with as many as eight inputs with any unbuffered or buffered output. The cell can be configured also as a dual OR-NAND gate with four inputs per side. The interface I/O buffer can be programmed for ECL, TTL, CMOS, and SDFL logic families. Each 4 × 4 bit RAM was fully decoded using depletion mode MESFETs and SDFL circuit approach. The chip size was 147 × 185 mils, and the total power dissipation of the whole chip was less than 3 W. Testing an associated test chip gave yield figures of 90%, 85%, and 50%, respectively, for FETs and diodes, logic gates, and a 4:1 multiplexer circuit. The best speed performance of the unbuffered and buffered SDFL NOR gates as measured with a 15-stage ring oscillator was 90 ps and 110 ps with a fan-out of 1 and 3, respectively. The 4:1 multiplexer circuit was operated up to 1.9 GHz in a microstrip test fixture.

Each cell could be programmed as an eight-input unbuffered or buffered NOR gate or as a dual OR-NAND gate with four inputs per side as shown in Fig. 9-2-2. The basic unbuffered SDFL NOR gate was used for a fan-out of 1 to 3. The buffered gate with a push-pull driver stage featured good tolerance of fan-out and capacitive load. Cells were programmed using the first metal layer and vias. All gate inputs and outputs lay on a 7 × 7 μm uniform grid that allowed adjacent vias in both directions. First and second metal interconnect lines were 4 μm wide with 3 μm spacing. Vias were 2 × 2 μm with 1-μm overlap of first and second metals. There were eight horizontal second-metal and six vertical first-metal routing tracks for

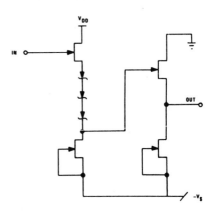

FIGURE 9-7-7. Output buffer of SDFL to ECL interface [130].

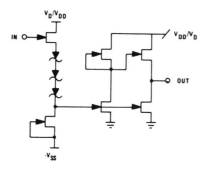

FIGURE 9-7-8. TTL/CMOS-SDFL input and output buffers [130].

each cell. The vertical to horizontal ratio of each gate was 8:6, identical to the cell array ratio of 24 columns to 18 rows. This allowed for excellent routeability. The cell array had two halves, each containing 216 cells in a 12×18 array. Average dissipation was 1.5 mW per cell using $V_{DD} = 2.5$ V and $V_{SS} = -1.5$ V with $V_T = -1.0$ V. The chip had 32 programmable interface I/O buffers. Each could be programmed for ECL, TTL, CMOS, and SDFL logic families. Figures 9-7-6 and 9-7-7 show the ECL-SDFL interface input and output circuit configurations. Figure 9-7-8 presents TTL/CMOS-SDFL input and output buffers. Each input buffer has a maximum fan-out of 10, and each output buffer operates up to 500 MHz using a standard load for each circuit family. Average power dissipations were 75 mW for ECL buffers using $V_s = -5.2$ V and 130 mW for TTL and CMOS buffers using $V_D = 5$ V.

Static RAMs on chip were organized in a 4×4 bit configuration and were fully decoded using depletion mode MESFETs with SDFL circuit approach. The RAM had two decoding addresses, four direct data inputs and ouputs, and a write-enable line. Figure 9-7-9 shows the circuit schematic of the SDFL static RAM cell. Average

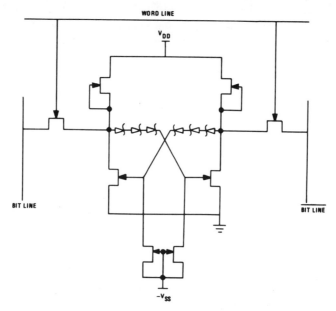

FIGURE 9-7-9. SDFL static RAM cell [130].

power dissipation is 0.5 mW per memory cell adding to a total of 38 mW per 16-bit RAM.

In addition to the electrical testing, some radiation testing was done with this circuit. Recovery times from pulsed x-ray tests at 6×10^{10} rad/s with a 20-ns pulse were of the order of 1.3 μs.

Further work on development of low power 2K GaAs gate arrays using a self-aligned process is described in Ref. 206.

REFERENCES

1. R. L. Van Tuyl and C. A. Liechti, High-speed integrated logic with GaAs MESFETs, in ISSCC Digest of Technical Papers, p. 114, 1973.
2. R. L. Van Tuyl and C. Liechti, High-speed integrated logic with GaAs MESFETs, *IEEE J. Solid-State Circuits* **SC-9**, 269–276 (1974).
3. R. L. Van Tuyl and C. A. Liechti, Gallium arsenide digital integrated circuits, Air Force Avionics Laboratory, Contract F33615-73-C-1242, Technical Report AFAL-TR-74-40, March 1974.
4. S. Yanagisawa, O. Wada, and H. Takanashi, Gigabit rate Gunn-effect shift register, 1975 Int. Electron Meeting, Technical Digest, December 1975, pp. 317–319.
5. G. Nuzillat, C. Arnado, and J. P. Puron, A subnanosecond integrated switching circuit with MESFETs for LSI, *IEEE J. Solid-State Circuits* **SC-11**, 385–394 (1976).
6. K. Mause, E. Hesse, and A. Schlachetzki, Shift register with Gunn devices for multiplexing techniques in the gigabit-per-second range, *Solid-State Electron* **1**, 17–23 (1976).
7. H. Muta *et al.*, Femto Joule logic circuit with enhancement-type Schottky barrier gate FET, *IEEE Trans. Electron Devices* **ED-23**, 1023–1027 (1976).
8. D. Boccon-Gibod and G. Durand, GaAs integrated circuits for high-speed logic, in 1976 European Solid-State Circuits Conf., Digest of Technical Papers, pp. 88–89.
9. H. Beneking and W. Filensky, "The GaAs MESFET as a pulse regenerator in the gigabit per second range, *IEEE Trans. Microwave Theory Tech.* **MTT-24**, 385–386 (1976).
10. R. L. Van Tuyl, C. A. Liechti, and C. A. Stolte, Gallium arsenide digital integrated circuits, Air Force Avionics Laboratory, Contract F33165-73-C-1242, Technical Report AFAL-TR-76-264, April 1977.
11. C. A. Liechti, GaAs FET logic, 1976 Intern. GaAs Symp., *Inst. Phys. Conf. Ser.* **33a**(5), 227–236 (1977).
12. R. L. Van Tuyl, C. A. Liechti, R. E. Lee, and E. Gowen, GaAs MESFET Logic with 4-GHz clock rate, *IEEE J. Solid-State Circuits* **SC-12**, 485–496 (1977).
13. T. G. Mills, D. H. Claxton, J. M. Crishal, and L. M. Tichauer, GaAs monolithic integrated circuit and device development, Proceedings of 6th Biennial Cornell Electric. Eng. Conf., 1977, pp. 347–357.
14. H. Ishikawa, H. Kusakawa, K. Suyama, and M. Fukuta, Normally-off type GaAs MESFET for low-power high-speed logic circuits, 1977 Int. Solid State Circuits Conf., Digest of Technical Papers, February 1977, pp. 200–201.
15. G. Bert, G. Nuzillat, and C. Arnodo, Femtojoule logic circuit using normally-off GaAs MESFETs, *Electron. Lett.* **13**, 644–645 (1977).
16. B. M. Welch and R. C. Eden, Planar GaAs integrated circuits fabricated by ion implantation, 1977 Int. Electron Device Meeting, Technical Digest, December 1977, pp. 205–208.
17. W. Filensky, H. J. Klein, and H. Beneking, The GaAs MESFET as a pulse regenerator, amplifier and laser modulator in Gbit/S Range, *IEEE J. Solid-State Circuits* **SC-12** (1977).
18. R. C. Eden, B. M. Welch, and R. Zucca, Low power GaAs digital IC's using Schottky diode-FET logic, in Digest of Technical Papers, 1978 Int. Solid-State Circuits Conf., February 1978, pp. 68–69.
19. R. Zuleeg, J. K. Notthoff, and K. Lehovec, Femtojoule high speed planar GaAs E-JFET logic, *IEEE Trans. Electron Devices* **ED-25**, 628–639 (1978).
20. R. C. Eden, B. M. Welch, and R. Zucca, Planar GaAs IC technology: Applications for digital LSI, *IEEE J. Solid-State Circuits* **SC-13**, 419–426 (1978).
21. P. M. Solomon, The performance of GaAs logic Gates in LSI, in IEDM Technical Digest, Washington, DC, December 1978, pp. 201–204.

22. M. Fukuta, K. Suyama, and H. Kusakawa, Low power GaAs digital integrated circuits with normally off MESFETs, *IEEE Trans. Electron Devices* **ED-25**, 1340 (1978).

23. P. T. Greiling, C. F. Krumm, F. S. Ozdemir, L. H. Hackett, and R. F. Lohr, Jr., Electron-beam fabricated GaAs inverter, *IEEE Trans. Electron Devices* **ED-25**, 1340 (1978).

24. D. O. Wilson, W. R. Scoble, R. P. Mandal, and H. L. Petersen, A comparison of high-speed enhancement and depletion mode GaAs MESFETs, International Electron Devices Meeting, Washington, DC, December 5, 1978, pp. 601–603.

25. M. Cathelin and G. Durand, Logic IC's using GaAs FET's in a planar technology, *L'onde Electrique* **58**, 218–221 (1978).

26. R. C. Eden, GaAs integrated circuits, MSI status and VLSI prospects, 1978 Int. Electron Devices Meeting, Technical Digest, December 1978, pp. 6–11.

27. T. Mizutani, M. Ida, and M. Ohmori, An 80 ps normally-off GaAs MESFET logic, presented at the 1st Specialty Conf. Gigabit Logic for Microwave Syst., Orlando, Florida, May 3–4, 1979.

28. J. A. Maupin, P. T. Greiling, and N. G. Alexopoulos, Speed–power tradeoff in GaAs FET integrated circuits, in Proc. 1st Specialty Conf. Gigabit Logic for Microwave Syst., Orlando, Florida, May 3–4, 1979, pp. 82–89.

29. R. E. Lundgren, C. F. Krumm, and R. L. Pierson, Fast enhancement-mode GaAs MESFET logic, presented at 37th Ann. Dev. Research Conf., Boulder, Colorado, June 25–27, 1979.

30. R. C. Eden and I. Deyhimy, Application of GaAs integrated circuits and c.c.d.s. for high speed signal processing, Paper at IEEE acousto-optic bulk wave devices meeting (SPIE), November 1979.

31. K. Suyama, H. Kusakouva, and M. Fukuta, GaAs integrated logic with normally-off MESFET, *Jpn J. Appl. Phys.* **18**, pp. 145–152, Suppl. 18-1 (1979).

32. M. Gloanec, G. Nuzillat, C. Arnodo, and M. Peltier, A GaAs integrated edge-triggered *D*-type flip-flop, Digest of the First Specialty Conference on Gigabit Logic for Microwave Systems, Orlando, Florida, pp. 114–119, May 1979.

33. P. T. Greiling, F. S. Ozdemir, C. F. Krumm, and B. F. Lohr, Jr., Electron beam fabricated GaAs integrated circuits, 1979 IEDM Technical Digest.

34. B. M. Welch, Y. D. Shen, R. Zucca, and R. C. Eden, Planar high yield GaAs IC processing techniques, 1979 IEDM Technical Digest.

35. B. M. Welch, Y. D. Shen, R. Zucca, R. C. Eden, and S. I. Long, LSI processing technology for planar GaAs integrated circuits, in *IEEE Trans. Electron Devices*, pp. 1116–1124, Special Issue, June, 1980.

36. R. E. Lundgren, C. F. Krumm, and R. L. Pierson, Fast enhancement-mode GaAs MESFET logic, 37th Ann. Dev. Research Conf., Boulder, Colorado, 25–27, June 1979.

37. R. E. Lundgren, C. F. Krumm, and R. F. John, Jr., Enhancement-mode GaAs MESFET logic, Presented at the IEEE GaAs IC Symposium, Lake Tahoe, Nevada, September 1979.

38. N. Yokoyama, T. Mimura, H. Kusakawa, K. Suyama, and M. Fukuta, Low power, high-speed integrated logic with GaAs MOSFETs, presented at Int'l Conf. on Solid State Devices, Tokyo, Japan, 1979.

39. D. H. Phillips, Outlook for the development of gallium arsenide circuits, Paper No. 1, IEEE Gallium Arsenide Integrated Circuit Symposium, Lake Tahoe, Nevada, September 27, 1979.

40. D. H. Phillips and H. L. Petersen, High-speed gallium arsenide integrated circuit development for satellite systems, published by New York University and AIAA, *Progr. Astronautics Aeronautics* **67**, 391 (1979).

41. G. Troeger, A. Behle, P. Friebertshauser, K. Hu, and S. Watanabe, Fully ion implanted planar GaAs E-JFET process, IEDM Digest of Technical Papers, 1979, 21.7, pp. 497–500.

42. G. Bert, T. Pham Ngu, G. Nuzillat, and M. Gloanec, Quasi-normally-off MESFET logic for high performance GaAs ICs, Paper 7 in Research Abstracts of the First Annual Gallium Arsenide Integrated Circuit Symposium, Lake Tahoe, Nevada, September 17, 1979.

43. J. K. Nottoff and C. H. Vogelsang, Gate Design for DCFL with GaAs E-JFETs, Paper 10 in Research Abstracts of First Annual Gallium Arsenide Integrated Circuit Symposium, Lake Tahoe, Nevada, September 27, 1979.

44. R. C. Eden, F. S. Lee, S. I. Long, B. M. Welch, and R. Zucca, Multi-level logic gate implementation in GaAs ICs using Schottky diode-FET logic, 1980 Int. Solid State Circuits Conf., Digest of Technical Papers, February 1980.

45. S. I. Long, B. M. Welch, R. C. Eden, F. S. Lee, and R. Zucca, MSI high speed low power GaAs integrated circuits, 1979 Int. Conf. on Solid State Devices, Digest of Technical Papers, Tokyo, August 1979, pp. 29–30.

46. R. C. Eden, B. M. Welch, R. Zucca, and S. I. Long, The prospects for ultrahigh-speed VLSI GaAs digital logic, *IEEE J. Solid-State Circuits* **ED-26**, 299–317 (1979).

47. R. Zucca, B. M. Welch, C. P. Lee, R. C. Eden, and S. I. Long, Process evaluation test structures and measurement techniques for a planar GaAs digital IC technology, *IEEE Trans. Electron Devices* **ED-27**, 2292–2298 (1980).

48. M. Peltier, G. Nuzillat, and M. Gloanec, A monolithic GaAs decision circuit for Gbit/s PCM transmission systems, 1980 International microwave symposium, Washington, DC, May 28–30, 1980, pp. 107–110.

49. B. G. Bosch, Device and circuit trends in gigabit logic, *Proc. Inst. Elec. Eng.* **127**, 254–265 (1980).

50. R. C. Eden and B. M. Welch, GaAs digital integrated circuits for ultra high speed LSI/VLSI, in *Springer Series in Electrophysics*, Vol. 5: *Very Large Scale Integration (VLSI) Fundamentals and Applications*, Ed. by D. F. Barbe, Springer-Verlag, Berlin, 1980, Chap. 5.

51. F. S. Lee, E. Shen, G. R. Kaelen, B. M. Welch, R. C. Eden, and S. I. Long, High speed LSI GaAs digital integrated circuits, presented at the 2nd GaAs Symp., Las Vegas, Nevada, November 1980, paper No. 3.

52. B. Welch, Y. Shen, R. Zucca, R. Eden, and S. Long, LSI processing technology for planar GaAs integrated circuits, *IEEE Trans. Electron Devices* **ED-27**, 1116–1123 (1980).

53. R. V. Tuyl, Monolithic GaAs IC's in instruments, presented at 1980 MIT short course on Monolithic GaAs Integrated Circuits for Microwave Systems, March 1980, unpublished.

54. N. Yokoyama, T. Mimura, and M. Fukuta, Planar GaAs MOSFET integrated logic, *IEEE Trans. Electron Devices* **ED-27**, 1124–1127 (1980).

55. M. Cathelin, G. Durand, M. Gavant, and M. Rocci, 5 GHz binary frequency division of GaAs, *Electron. Lett.* **16**, 535–536 (1980).

56. N. Yokoyama, GaAs MOSFET high-speed logic, *IEEE Trans.* **MTT-28**, 483–486 (1980).

57. K. Lehovec and R. Zuleeg, Analysis of GaAs FET for integrated logic, *IEEE Trans.* **ED-27**, 1074–1091 (1980).

58. A. W. Livingstone and P. J. T. Mellor, Capacitor coupling of GaAs depletion mode FET's, 1980 GaAs IC Symposium Abstracts, paper No. 10.

59. T. Mizutani, Gigabit logic operation with enhancement-mode GaAs MESFET IC's, presented at IEEE MTT-S Workshop Gigabit Logic for Microwave Systems, May 1980.

60. P. T. Greiling, R. E. Lundgren, C. F. Krumm, and R. F. Lohr, Jr., Why design logic with GaAs and How?, *Microwave Systems News*, 48–60, Jan. (1980).

61. G. Nuzillat, G. Bert, T. P. Ngu, and M. Gloanec, Quasi-normally-off MESFET logic for performance GaAs IC's, *IEEE Trans.* **ED-27**, 1102–1109 (1980).

62. T. Mizutani, N. Kato, M. Ida, and M. Ohmori, High-speed enhancement-mode GaAs MESFET logic, *IEEE Trans. Microwave Theory Tech.* **MTT-28**, 479–483 (1980).

63. M. Gloance, G. Nuzillat, C. Arnodo, and M. Peltier, An e-beam fabricated GaAs *D*-type flip-flop i.c., *IEEE Trans.* **MTT-28**, 472–478 (1980).

64. M. Ida, T. Mizutani, K. Asai, M. Uchida, K. Shimada, and S. Ishida, Fabrication technology for an 80 ps normally-off GaAs MESFET logic, *IEEE Trans. Electron Devices* **ED-28**, 489–493 (1981).

65. N. Hashizume, H. Yamada, and K. Tomizawa, Schottky-barrier coupled Schottky-barrier gate GaAs FET logic, *Electron. Lett.* **17**(1), 51–52 (1981).

66. N. Yokoyama, T. Mimura, M. Fukuta, and H. Ishikawa, A self-aligned source/drain planar device for ultrahigh-speed GaAs MESFET VLSI's, IEEE International Solid-State Circuits Conference, Digest of technical papers, 1981, pp. 218–219.

67. G. M. Metze, H. M. Levy, D. W. Woodard, C. E. C. Wood, and L. F. Eastman, GaAs integrated circuits by selective molecular beam epitaxy, *Solid State Technol.* **Aug**, 127–130 (1981).

68. R. Yamamoto and A. Higashisaka, High speed GaAs digital integrated circuit with clock frequency of 4.1 GHz, *Electron. Lett.* **17**(8), 291–292 (1981).

69. N. Hashizume, H. Yamada, T. Kojima, K. Matsumoto, and K. Tomizawa, Low-power gigabit logic by GaAs CSFL, *Electron. Lett.* **17**(16), 553–554 (1981).

70. F. Damay-Kavala, M. Gloanec, M. Peltier, G. Nuzillat, and C. Arnodo, Speed power performances in sequential GaAs logic circuits, GaAs and Related Compounds Symp., Osio, Japan, September 1981.

71. F. Katano, T. Furutsuka, and A. Higashisaka, High speed normally-off GaAs MESFET integrated circuits, *Electron. Lett.* **17**(6), 236–239 (1981).

72. F. Damay-Kavala, G. Nuzillat, and C. Arnodo, High-speed frequency dividers with quasi-normally-off GaAs MESFETs, *Electron. Lett.* **17**(25), 968–970 (1981).

73. N. J. Shah, S. S. Pei, and C. W. Tu, Gate-length dependence of the speed of SSI circuits using submicrometer selectively doped heterostructure transistor technology, *IEEE Trans. Electron Devices* **ED-33**, 543 (1986).

74. C. H. Hyun, M. Shur, A. Peczalskí, Analysis of noise margin and speed of GaAs MESFET DCFL using UM-SPICE, *IEEE Trans. Electron Devices*, **ED-33**, 1421–1426, Oct. (1986).

75. T. Mimura, S. Hiyamizu, H. Ishikawa, and T. Misugi, An enhancement-mode high electron mobility transistor for VLSI, in Proc. High speed Digital Technologies Conf., San Diego, paper III-5, January 14, 1981.

76. H. T. Yuan, GaAs bipolar gate array technology, in Technical Digest, 1982 GaAs IC Symp., p. 100.

77. R. C. Eden and B. M. Welch, Ultra high speed GaAs VLSI: Approaches, potential and progress, in *VLSI Electronics: Microstructue Science*, Vol. 3, Ed. by N. Einspruch, Academic, New York, 1981.

78. T. Mimura, K. Joshin, S. Hiyamizu, K. Hikosaka, and M. Abe, High electron mobility transistor logic, *Jpn. J. Appl. Phys.* **20**, L598–L600 (1981).

79. G. Nuzillat, GaAs IC's for gigabit logic applications, in Proc. 7th European Solid-State Circuits Conf., Frieburg, Germany, September 1981, pp. 65–74e.

80. P. M. Asbeck, D. L. Miller, R. A. Milano, J. S. Harris, Jr., G. R. Kaelin, and R. Zucca, (GaAl)As/GaAs bipolar transistors for digital integrated circuits, in IEDM Technical Digest, Washington, DC, December 1981, pp. 629–632.

81. R. Zuleeg, J. K. Notthof, and K. Lehovec, Effects of total dose ionizing radiation on GaAs JFETs and ICs, presented at the GaAs IC Symp., San Diego, October 1981, paper 14.

82. D. Ankri, A. Scavennec, and C. Vivier, GaAlAs–GaAs bipolar transistors for high-speed digital circuits, in Research Abstracts, 1981 GaAs IC Symp., p. 18.

83. M. Ohmori, T. Mizutami, and N. Kato, Very low power gigabit logic circuits with enhancement-mode GaAs MESFET's, 1981 IEEE MTT-S International Microwave Symposium Digest, pp. 188–190, Los Angeles, June 1981.

84. S. I. Long *et al.*, GaAs large scale integrated circuits, *J. Vac. Sci. Technol.* **19**, 531–536 (1981).

85. W. Filensky, F. Ponse, and H. Beneking, Applications of single and dual gate GaAs MESFETs for Gbit/s optimal data transfer systems, *IEEE J. Solid State Circuits* **SC-16**(2), 93–99 (1981).

86. Pham N. Tung, D. Delagebeaudeuf, M. Laviron, P. Delescluse, J. Chaplart, and N. T. Linh, High-speed two-dimensional electron–Gas FET Logic, *Electron. Lett.* **18**(3), Feb. (1982).

87. S. Katsu, S. Nambu, A. Shimano, and G. Kano, A GaAs monolithic frequency divider using source coupled FET logic, *IEEE Electron Device Lett.* **EDL-3**(8), 197–199 (1982).

88. K. Yamasaki, K. Asai, T. Mizutani, and K. Kurumada, Self-aligned implantation for n^+-layer technology (Saint) for high-speed GaAs ICs, *Electron. Lett.* **18**(3), 119–121 (1982).

89. N. Toyoda *et al.*, An application of Pt–GaAs solid phase reaction to GaAs IC, Proc. Int. Conf. on GaAs and Related Comp., Inst. Phys. Conf. Ser. No. 63, p. 521, 1982.

90. R. Kolbas, J. Carney, J. Abrokwah, E. Kalweit, and D. Hitchell, Planar optical sources and detectors for monolithic integration with GaAs MESFET electronics, in *SPIE Integrated Optics II* **321**, 94–102 (1982).

91. J. K. Carney, M. J. Helix, R. M. Kolbas, S. A. Jamison, and S. Ray, Monolithic optoelectronic/electronic circuits, in Technical Digest GaAs IC Symp., New Orleans, 1982, pp. 38–41.

92. K. Yamasaki, K. Asai, and K. Kurumada, GaAs LSI-directed MESFETs with self-aligned implantation for n^+-layer technology, *IEEE Trans. Electron Devices* **ED-29**, 1772–1777 (1982).

93. K. Yamasaki, N. Kato, Y. Matsuoka, and K. Ohwada, EB-writing n^+ self-aligned GaAs MESFETs for high-speed LSIs, in IEDM Technical Digest, 1982, pp. 166–169.

94. K. Yamasaki, Y. Yamane, and K. Kurumada, Below 20 ps/gate operation with GaAs SANIT FETs at room temperature, *Electron. Lett.* **18**(4), 592–593 (1982).

95. H. Kroemer, Heterostructure bipolar transistors and integrated circuits, *Proc. IEEE* **70**, 13–25 (1982).

96. G. Bert, J. Morin, G. Nuzillat, and C. Arnodo, High-speed GaAs static random access memory, *IEEE Trans. Electron Devices* **ED-29**(7), 1110–1115 (1982).

97. M. S. Shur and D. Long, Performance prediction for submicron GaAs SDFL Logic, *IEEE Electron Device Lett.* **EDL-3**(5), 124–127 (1982).

98. D. Kinnell, A 320 gate GaAs logic gate array, in GaAs IC Symp. Technical Digest, pp. 17–20, 1982.

99. N. Toyoda, T. Terada, M. Mochizuki, K. Kamazawa, T. Mizoguchi, and A. Hojo, Capability of GaAs DCFL for high-speed gate array, in IEDM Technical Digest, pp. 602–637, 1982.

100. C. P. Lee, B. M. Welch, and R. Zucca, Saturated resistor loads for GaAs integrated circuits, *IEEE Trans. Electron Devices* **ED-29**(7), 1103–1109 (1982).

101. M. Ino, M. Hirayama, K. Ohwada, and K. Kurumada, GaAs 1KB static RAM with E/D MESFET DCFL, in GaAs IC Symp. Technical Digest pp. 2-5, 1982.

102. F. S. Lee *et al.*, A high-speed LSI GaAs 8 × 8 bit parallel multiplier, *IEEE J. Solid State Circuits* **SC-17**, 638-647 (1982).

103. N. Helix, S. Jameson, C. Chao, and M. Shur, Fan-out and speed of GaAs SDFL logic, *IEEE J. Solid-State Circuits* **SC-17**(6), 1226-1232 (1982).

104. N. Yokoyama *et al.*, TiW silicide gate self-alignment technology for ultra-high-speed GaAs MESFET LSI/VLSI's, *IEEE Trans. Electron Devices* **ED-29**, 1541-1547 (1982).

105. M. R. Namordi and W. A. White, A novel low-power static GaAs MESFET logic gate, *IEEE Electron Devices Lett.* **EDL-3**(9), 264-267 (1982).

106. C. Liechti *et al.*, A GaAs MSI word generator operating at 5 Gbits/s data rate, *IEEE Trans. Electron Devices* **ED-29**(7), 1094-1102 (1982).

107. M. R. Namordi and W. M. Duncan, The effect of logic cell configuration, gatelength, and fan-out on the propagation delays of GaAs MESFET logic gates, *IEEE Trans. Electron Devices* **ED-29**, 402 (1982).

108. P. Solomon, Semiconductor devices for high-speed logic, *Proc. IEEE* **70**(5), 489-509 (1982).

109. T. Mizutani, N. Kato, K. Osafune, and M. Ohmori, Gigabit logic operation with enhancement-mode MESFET ICs, *IEEE Trans. Electron Devices* **ED-29**, 199-204 (1982).

110. R. C. Eden, Comparison of GaAs device approaches for ultra-high-speed and expected circuit performance, *Proc. IEEE* **70**, 5-12 (1982).

111. M. Abe, T. Mimura, N. Yokoyama, and H. Ishikawa, New technology toward GaAs LSI/VLSI for computer applications, in a joint special issue of *IEEE Trans. Electron Devices* and *IEEE Trans. Microwave Theory Tech.*, July 1982.

112. S. I. Long, B. M. Welch, R. Zucca, P. M. Asbeck, C.-P. Lee, C. G. Kirkpatrick, F. S. Lee, G. R. Kaelin, and R. C. Eden, High speed GaAs integrated circuits, *Proc. IEEE* **70**, 35-45 (1982).

113. D. Wilson, N. Frick, J. Kwiat, S. Lo, J. Churchill, and J. Barrera, Package study for high speed (GHz) commerical GaAs products, 1982 GaAs IC Symposium Digest, pp. 13-16.

114. M. Rocchi and B. Gabillard, GaAs digital dynamic ICs for applications up to 10 GHz, *J. Solid State Circuits* **SC-18**(3), 369-376 (1983).

115. T. Hsu, O. Wilson, and J. Barrera, An overview of GaAs MESFET digital IC technology, presented at the 1983 VLSI Symposium, Taiwan.

116. H. M. Levy, R. E. Lee, and R. A. Sadler, A submicron self-aligned GaAs MESFET technology for digital integrated circuits, Device Research Conf., paper IVB-3, June 1982; also abstract published in *IEEE Trans. Electron Devices* **ED-29**, 1687 (1982).

117. H. M. Levy and R. E. Lee, Self-aligned submicron gate digital GaAs integrated circuits, *IEEE Electron Device Lett.* **EDL-4**, 102-104 (1983).

118. C. P. Lee and W. I. Wang, High-performance modulation-doped GaAs integrated circuits with planar structures, *Electron. Lett.* **19**, 155-157 (1983).

119. H. Shimizu, K. Suyama, Y. Nakayama, N. Yokoyma, A. Shibatomi, and H. Ishikawa, GaAs 16 × 16 bit parallel multiplier, presented at the National Convention of the Institute of Electronics and Communication Engineers of Japan, 1983.

120. Y. Nakayama *et al.*, A GaAs 16 × 16 bit parallel multiplier, *IEEE J. Solid State Circuits* **SC-18**, 599-603 (1983).

121. T. Chen, M. Shur, B. Hoefflinger, K. Lee, T. Vu, P. Roberts, and M. Helix, DOMES-GaAs IC simulator, in Proc. IEEE Biennial Conf. on High Speed Devices and Circuits, Ithaca, 1983.

122. P. M. Asbeck, D. L. Miller, R. J. Anderson, and F. H. Eisen, Emitter-coupled logic circuits implemented with heterojunction bipolar transistors, in Technical Digest, 1983 GaAs IC Symp., p. 170.

123. S. Yamakoshi, T. Sanada, O. Wada, T. Fujii, and T. Sakurai, Low threshold AlGaAs/GaAs MQW laser integrated with FETs, in Technical Digest, Fourth Int. Conf. Integrated Opt., Opt. Fiber Commun., Tokyo, Japan, 1983, pp. 140-141.

124. C. H. Vogelsang, J. L. Hogin, and J. K. Notthoff, Yield analysis methods for GaAs ICs, 1983 IEEE GaAs IC Symp. Technical Digest, pp. 149-152, October 1983.

125. K. Asai, K. Kurumada, M. Hiroyama, and M. Ohmori, 1kb static RAM using self-aligned FET technology, in ISSCC Technical Digest, pp. 46-47, 1983.

126. N. Yokoyama, T. Ohnishi, Ho. Onodera, T. Shinoki, A. Shibatomi, and H. Ishikawa, A GaAs 1K static RAM using tungsten-silicide gate self-alignment technology, in ISSCC Technical Digest, pp. 44-45, 1983.

127. R. A. Sadler and L. F. Eastman, Self-aligned submicron GaAs MESFET's for high speed logic, Device Research Conference, June 1983.

128. R. A. Salder and L. F. Eastman, High-speed logic at 300K with self-aligned submicron-gate GaAs MESFET's, *IEEE Electron Devices Lett.* **EDL-4**, 215–217 (1983).

129. T. Vu, P. Roberts, R. Nelson, G. Lee, B. Hanzal, K. Lee, D. Lamb, M. Helix, S. Jamison, S. Hanka, J. Brown, and M. S. Shur, A 432-Cell GaAs SDFL gate array with on-chip 64-bit RAM, Proc. 1983 Custom Integrated Circuits Conf., pp. 32–36, Rochester, New York, May 1983.

130. T. Vu, P. Roberts, R. Nelson, G. Lee, B. Hanzal, K. Lee, N. Zafar, D. Lamb, M. Helix, S. Jamison, S. Hanka, J. Brown, Jr., and M. S. Shur, A gallium arsenide SDFL gate array with on-chip RAM, *IEEE Trans. Electron Devices* **ED-31**(2), 144–156 (1984).

131. R. A. Kiehl, M. D. Feuer, R. H. Hendel, J. C. M. Hwang, V. G. Keramidas, C. L. Alyn, and R. Diryle, Selectively doped heterostructure frequency dividers, *IEEE Electron Device Lett.* **EDL-4**(10), 377 (1983).

132. C. P. Lee, D. L. Miller, D. Hou, and R. J. Anderson, Ultra-high speed integrated circuits using GaAs/GaAlAs high electron mobility transistors, presented at Device Research Conf., June 1983, Burlington, Vermont, see also *IEEE Trans. Electron Devices* **ED-30**(11), 1569 (1983).

133. R. Zuleg, J. K. Notthoff, and G. L. Troeger, Double-implanted GaAs complementary JFET, *IEEE Electron Devices Lett.* **EDL-5**, 21, 23 (1984).

134. P. G. Flahive, W. J. Clementson, P. O'Connor, A. Dori, and S. C. Shunk, A GaAs DCFL chip set for multiplex and de-multiplex applications at gigabit/sec data rates, presented at the 1984 GaAs IC Symp. Boston, October 23–25, 1984.

135. N. C. Cirillo, Jr., J. K. Abrokwah, and M. S. Shur, A self-aligned gate process for ICs based on modulation doped (Al, Ga)As/GaAs FETs, presented at the Device Research Conference, June 1984.

136. Y. Ishii, M. Ino, M. Idda, M. Hirayama, and M. Ohmori, Processing technologies for GaAs memory LSIs, IEEE GaAs IC Symposium Technical Digest, pp. 121–123, October 1984.

137. S. Pei, N. Shah, R. Hendel, C. Tu, and R. Dingle, Ultra high speed integrated circuits with selectively doped heterostructure transistors, 1984 IEEE GaAs IC Symposium Technical Digest, pp. 129–132, October 1984.

138. P. Asbeck, D. Miller, R. Anderson, R. Deming, R. Chen, C. Liehti, and F. Eisen, Application of heterojunction bipolar transistors to high-speed, small-scale digital ICs, 1984 IEEE GaAs IC Symp. Technical Digest, pp. 133–136, October 1984.

139. S. I. Long, VLSI gallium arsenide—What are the limitations, how will they be overcome?, Proc. Int. Conf. on Computer Design, p. 267, October 1984.

140. S. A. Rooslid, DARPA plans and pilot production line project, Proc. 1984 Int. Conf. on Computer Design, Rye, New York, pp. 251–257, October 1984.

141. R. Kiehl, H. L. Strormer, K. Baldwin, A. C. Gossard, and W. Wiegman, Modulation doped field effect transistors and logic gates based on two-dimensional hole gas, presented at 42nd Annual Device Res. Conf., Santa Barbara, California, June 18–20, 1984.

142. R. A. Kiehl and A. C. Gossard, *p*-Channel modulation doped logic gates, *IEEE Electron Device Lett.* **EDL-5**, Oct. (1984).

143. R. A. Kiehl and A. C. Gossard, Complementary *p*-MODFET and N-HB MESFET (Al, Ga)As transistors, *IEEE Electron Device Lett.* **EDL-5**(12), 521–523 (1984).

144. R. H. Hendel, S. S. Pei, R. A. Kiehl, C. W. Tu, M. D. Feuer, and R. Dingle, A 10-GHz frequency divider using selectively doped heterostructure transistors, *IEEE Electron Device Lett.* **EDL-5**(10), 406–408 (1984).

145. S.S. Pei, R. H. Henel, R. A. Kiehl, C. W. Tu, M. d. Feuer, and R. Dingle, Selectively doped heterostructure transistors for ultrahigh speed integrated circuits, presented at 1984 Device Research Conf., Santa Barbara, California, June 18–20, 1984.

146. K. Nishiuchi, N. Kobayashi, S. Kuroda, S. Notomi, T. Nimura, M. Abe, and M. Kobayashi, A subnanosecond HEMT 1 Kb SRAM, in IEEE Intl. Sol. State Circ. Conf., Technical Digest, 1984, pp. 48–49.

147. M. E. Kim, C. S. Hong, D. Kasemset, and R. A. Milano, GaAs/GaAlAs selective MOCVD epitaxy and planar ion-implantation technique for complex integrated optoelectronic circuit applications, *IEEE Electron Devices Lett.* **EDL-5**(8), 306–309 (1984).

148. P. M. Asbeck, D. L. Miller, R. J. Anderson, L. D. Hou, R. Deming, and F. Eisen, Nonthreshold logic ring oscillators implemented with GaAs/(GaAl)As heterojunction bipolar transistors, *IEEE Electron Devices Lett.* **EDL-5**(5), 181–183 (1984).

149. P. M. Solomon and H. Morkoc, Modulation-doped GaAs/AlGaAs heterojunction field-effect transistors (MODFET's), ultrahigh-speed devices for supercomputers, *IEEE Trans. Electron Devices* **ED-31**(8), 1015–1027 (1984).

150. N. C. Cirillo, Jr., J. K. Abrokwah, and M. S. Shur, Self-aligned modulation doped (Al, Ga)/GaAs field-effect transistors, *IEEE Electron Device Lett.* **EDL-5**(4), 129-131 (1984).

151. R. L. Tyul, V. Kumar, D. C. D'Avanzo, T. M. W. Taylor, V. E. Peterson, D. P. Hornbuckle, R. A. Fisher, and D. B. Estreisch, A manufacturing process for analog and digital gallium arsenide integrated circuits, *IEEE Trans. Electron Devices* **ED-29**(7), 1032-1037 (1982).

152. A. Shimano, S. Katsu, S. Nambu, G. Kano, A 4 GHz 25 mW GaAs IC using source coupled FET logic, in Proceedings of ISSCC 83, p. 42, February 1983.

153. G. Nuzillat, G. Bert, F. Damay-Kavala, and C. Arnodo, High speed low power logic ICs using quasi-normally-off MESFETs, *IEEE J. Solid State Circuits*, No. 3, pp. 226-232, June 1981.

154. Kang Lee, Characterization, modeling, and circuit design of GaAs MESFETs, Ph.D. thesis, University of Minnesota, October 1984.

155. V. I. Staroselskii and A. N. Sapelnikov, Limiting properties of gallium arsenide integrated circuits using field effect transistors, *Mikroelectronika* **11**(2), March-April (1982).

156. A. Shimano, S. Katsu, S. Nambu, and G. Kano, A 4 GHz 25 mW GaAs IC using SCFL, ISSCC 1983, February 1984.

157. M. Idda, T. Takada, and T. Sudo, Analysis of high speed GaAs FET logic circuits, *IEEE Trans. MTT.* **MTT-32**(1), 5-10 (1984).

158. M. Cathelin, M. Gavant, and M. Rocchi, A 3-5 GHz self-aligned single-clocked binary frequency divider on GaAs, *Proc. Inst. Elec. Eng.* **127**, 270-277 (1980).

159. M. Ohmori, T. Mizutani, and N. Kato, Very low power gigabit logic circuits with E-mode GaAs MESFET's, in Proc. 1981 IEEE MTT-S Int. Microwave Symp., Los Angeles, June 15-19, 1981, pp. 188-190.

160. Y. Ikawa *et al.*, A 1K-gate GaAs gate array, IEEE ISSCC, pp. 40-41, 1984.

161. R. Kiehl, P. Flahive, S. Wemple, and H. Cox, DCFL GaAs ring oscillators with self-aligned gates, *IEEE Electron Device Lett.* **EDL-3**(11), 325-326 (1982).

162. A. Mitonneau, M. Rocchi, I. Talmud, and J.-C. Maudmit, Direct experimental comparison of sub-micron GaAs and Si MOS MSI digital ICs, 1984 IEEE GaAs IC Symp. Technical Digest, pp. 3-6, October 1984.

163. R. Eden, Capacitor diode FET logic (CDFL) circuit approach for GaAs D-MESFET ICs, 1984 IEEE GaAs IC Symp. Technical Digest, pp. 11-14, October 1984.

164. R. Deming, P. Griffith, R. Zucca, and R. Vahrenkamp, A gallium arsenide configuration cell array using buffered FET logic, 1984 IEEE GaAs IC Symp. Technical Digest, pp. 15-18, October 1984.

165. B. Gilbert, D. Schwab, G. Heimbigner, E. Walton, and S. Roosild, Investigation of optimum interchip signal transmission protocols using GaAs BFL MSI components with various I/O peripheries, 1984 IEEE GaAs IC Symp. Technical Digest, pp. 19-22, October 1984.

166. K. Yoshihara, T. Sudo, A. Aida, T. Miyagi, S. Ooe, and T. Saito, Cross-talk predictions and reducing techniques for high speed GaAs digital ICs, 1984 IEEE GaAs IC Symp. Technical Digest, pp. 24-26, October 1984.

167. L. Pengue, G. McCormack, E. Strid, D. Smith, and T. Bowman, The "quick chip": A depletion mode digital/analog array, 1984 IEEE GaAs IC Symp. Technical Digest, pp. 27-30, October 1984.

168. P. Greiling, R. Lee, H. Winston, A. Hunter, J. Jensen, R. Beaubien, and R. Bryan, 1984 IEEE GaAs IC Symp. Technical Digest, pp. 31-34, October 1984.

169. T. Gheewala, Packages for ultra-high speed GaAs ICs, 1984 IEEE GaAs IC Symp. Technical Digest, pp. 67-70, Oct. 1984.

170. Y. Kawakami, K. Tanaka, M. Tsunotani, H. Nakamura, and Y. Sano, Advanced GaAs DCFL constructions with 8 × 8 bit parallel multipliers, 1984 IEEE GaAs IC Symp. Technical Digest, pp. 107-110, October 1984.

171. T. Hayashi, A. Masaki, H. Tanaka, H. Yamashita, N. Masuda, T. Doi, N. Hashimoto, N. Kotera, J. Shigeta, T. Lohashi, and S. Takahashi, ECL-compatible GaAs SRAM circuit technology for high performance computer application, 1984 IEEE GaAs IC Symp. Technical Digest, pp. 111-114, October 1984.

172. S. Lee, R. Vahrenkamp, R. Deming, F. Chang, R. Zucca, and C. Kirpartrick, 1K GaAs low-power, high speed static RAM with depletion mode MESFETs, 1984 IEEE GaAs IC Symp. Technical Digest, pp. 115-116, October 1984.

172. T. Mizoguchi, N. Toyoda, K. Kanazava, Y. Ikawa, T. Terada, M. Mochizuki, and A. Hojo, A GaAs 4K bit static RAM with normally-on and normally-off combination circuit, 1984 IEEE GaAs IC Symp. Technical Digest, pp. 117-120, October 1984.

173. S. Kuroda, T. Mimura, M. Suzuki, N. Kobayshi, K. Nishiuchi, A. Shibatomi, and M. Abe, New device structure for 4kb RAM, 1984 IEEE GaAs IC Symp. Technical Digest, pp. 125–128, October 1984.

174. S. Roosild, Survey of manufacturing capabilities for GaAs digital integrated circuits, 1984 IEEE GaAs IC Symp. Technical Digest, pp. 155–158, October 1984.

175. M. Helix, S. Hanka, P. Vold, and S. Jamison, A. low power gigabit IC fabrication technology, 1984 IEEE GaAs IC Symp. Technical Digest, pp. 163–166, October 1984.

176. N. Cirillo, Jr., J. Abrokwah, and S. A. Jamison, A self-aligned gate modulation doped (Al, Ga)As/GaAs FET IC process, 1984 IEEE GaAs IC Symp. Technical Digest, pp. 167–170, October 1984.

177. K. Lehovec, GaAs enhancement mode FET-tunnel diode ultra-fast low power inverter and memory cell, *IEEE J. Solid State Circuits* **SC-14**, 797 (1979).

178. K. Lehovec, VLSI GaAs tunnel diode-FET logic and memory cell, *Jpn. J. Appl. Phys.* **19**, Suppl. 19-1, 335–338 (1979).

179. K. Lehovec, presented at the IEEE/MTT-S First Specialty Conference on Gigabit Logic for Microwave Systems, Orlando, Florida, 3–4 May 1979.

180. R. Zuleeg and K. Lehovec, Radiation effects in GaAs JFETs, *IEEE Trans. Nucl. Sci.* **NS-27**, 1343–1354 (1980).

181. Y. Kato, M. Dohsen, J. Kasahara, and N. Wanatabe, Planar GaAs normally-off JFET for high speed logic circuits, *Electron. Lett.* **16**, 821 (1980).

182. J. Kasahara, K. Tayra, Y. Kato, M. Dohsen, and N. Wanatabe, Fully implanted GaAs ICs using normally-off JFETs, *Electron. Lett.* **17**, 621 (1980).

183. M. Dohsen, J. Kasahara, Y. Karo, and N. Wanatabe, GaAs JFETs formed by localized Zn diffusion, *IEEE Electron Device Lett.* **2**, 157 (1981).

184. Y. Kato, M. Dohsen, J. Kasahara, K. Taira, and N. Wanatabe, Planar GaAs normally-off JFET for high speed logic circuits, *Electron. Lett.* **17**, 951 (1981).

185. J. K. Notthoff and G. L. Troeger, All-ion-implantation JFET MSI performance, 2-d International GaAs IC symposium, Las Vegas, 1980.

186. N. C. Cirillo, A. Fraasch, H. Lee, L. F. Eastman, M. S. Shur, and S. Baier, Novel multilayer modulation doped (Al, Ga)As/GaAs structures for self-aligned gate FETs, *Electron. Lett.* **20**(21), 854–855 (1984).

187. M. S. Shur, T. H. Chen, C. H. Hyun, P. N. Jenkins, and N. C. Cirillo, Jr., Simulation and design of self-aligned modulation doped AlGaAs/GaAs integrated circuits, International Solid State Circuit Conference Technical Digest, paper FAM 18.3, pp. 264–265, IEEE, 1985.

188. C. H. Hyun, M. S. Shur, and N. C. Cirillo, Jr., Design and simulation and fabrication of modulation doped DCFL circuits, *IEEE Trans. on CAD*, **CAD-5**(2), 284–292 (1986).

189. H. T. Yuan, GaAs bipolar gate array technology, 1982 GaAs Symposium Technical Digest, p. 100, New Orleans, 1982.

190. W. V. McLevige, H. T. Yuan, W. M. Duncan, W. R. Frensley, F. H. Doerbeck, H. Morkoc, and T. J. Drummond, GaAs/AlGaAs heterojunction bipolar transistors for integrated circuit applications, *Electron Device Lett.* **EDL-3**(43), (1982).

191. N. Yokoyama, T. Mimura, H. Kusakawa, K. Suyama, and M. Fukuta, GaAs MOSFET high-speed logic, *IEEE Trans. Microwave Theory Tech.* **MTT-28**(5), 483–486, 43–45 (1980).

192. A. Peczalski, M. S. Shur, C. H. Hyun, K. Lee, and T. Vu, Design and simulation of GaAs direct coupled field effect transistor logic, *IEEE Trans. on CAD*, **CAD-5**(2), 266–273 (1986).

193. J. Baek, M. S. Shur, K. Lee, and T. Vu, Current–voltage characteristics of ungated GaAs FETs, *IEEE Trans. Electron Dev.*, **ED-32**(11), 2426–2440 (1985).

194. J. E. Meyer, MOS models and circuit simulation, *RCA Rev.* **32**, 42–63 (1971).

195. C. F. Hill, Noise margin and noise immunity in logic circuits, *Microelectronics* **1**, 15–21 (1968).

196. M. J. Helix, S. A. Jamison, S. A. Hanka, R. P. Vidano, P. Ng, and C. Chao, Improved logic gate with a push-pull output for GaAs digital ICs, in 1982 GaAs IC Symposium Technical Digest, 1982.

197. R. V. Van Tuyl, Monolithic GaAs IC's in instruments, presented at 1980 M.I.T. Short Course on Monolithic GaAs Integrated Circuits for Microwave Systems, March 1980.

198. A. Higashika and F. Hasegawa, Estimation of fringing capacitance of electrodes on S.I. GaAs substrate, *Electron. Lett.* **16**, 411–412 (1980).

199. T. H. Chen and M. S. Shur, Simulation of GaAs self-aligned enhancement mode logic, unpublished.

200. N. C. Cirillo, Jr. and J. K. Abrokwah, 8.5-picosecond ring oscillator gate delay with self-aligned gate modulation doped n^+-(Al, Ga)As FETs, presented at 43rd Device Research Conference, June 1983.

201. C. H. Hyun, M. S. Shur, J. H. Baek, and N. C. Cirillo, Comparative study of MODFET integrated circuits operating at 77 and 300 K, in Proceeding of Cornell IEEE Conference on Advanced Concepts in High Speed Devices, pp. 220–229, Ithaca, New York, July 1985.

202. N. C. Cirillo, Jr., M. S. Shur, P. Vold, J. K. Abrokwah, R. R. Daniels, and O. N. Tufte, Complementary heterojunction insulated gate field effect transistors, IEDM 1985 Technical Digest, IEEE, pp. 317–320, 1985.

203. N. C. Cirillo, Jr., M. S. Shur, P. Vold, J. K. Abrokwah, and O. N. Tufte, Realization of n-channel and p-channel high-mobility (Al, Ga)As/GaAs heterostructure insulating gate FETs on a planar water surface, *IEEE Electron Dev. Lett.* **EDL-6**(12), 645–647 (1985).

204. R. A. Sadler and L. F. Eastman, A performance study of GaAs direct-coupled FET logic by self-aligned ion implantation, in Proceedings of IEEE/Cornell conference on high-speed semiconductors and circuits, Ithaca, New York, August 1983, pp. 267–276.

205. H. Hasegawa, M. Abe, P. M. Asbeck, A. Higashizaka, Y. Kato, and M. Ohmori, GaAs LSI/VLSIs: Advantages and applications, in Extended Abstracts of the 16th International Conference on Solid State Devices and Materials, KOBE, pp. 413–414, 1984.

206. T. Vu *et al.*, Low power logic circuits and 2K gate array using GaAs self-aligned MESFETs, *IEEE J. Sol. State Circuits*, to be published.

207. N. J. Shah, S. S. Pei, and C. W. Tu, Gate-length dependence of the speed of SSI circuits using submicrometer selectively doped heterostructure transistor technology, *IEEE Trans. Electron Devices*, **ED-33**, 543 (1986).

10

Modulation Doped Field Effect Transistors

10-1. INTRODUCTION

Modulation doped field effect transistors (MODFETs), also called high electron mobility transistors (HEMTs) and selectively doped heterojunction transistors (SDHTs), have recently emerged as the fastest solid state devices. Ring oscillator propagation delay times as low as 10.2 ps at 300 K and 5.8 ps at 77 K have been demonstrated. MODFET frequency dividers have operated at up to 10.1 GHz input frequency. A 4-kb MODFET RAM has also been fabricated. (A more detailed discussion of the MODFET IC performance may be found in Chapter 9.)

A modulation doped transistor is a heterojunction device which utilizes the high mobility and high velocity of the two-dimensional electron gas formed at the heterointerface, typically at the heterointerface between (Al, Ga)As and undoped GaAs in the GaAs region.

The first heterojunction device was proposed by William Shockley in 1951 [1]. Also, in 1951 A. I. Gubanov developed a theory describing heterojunctions [2]. In 1957 H. Kroemer published a pioneering paper on bipolar junction transistors with a wide gap emitter [3]. In 1960 R. A. Anderson predicted that an accumulation layer may appear at the interface of a heterojunction [4]. In 1960 he reported the results of an experimental study of Ge–GaAs heterojunctions and proposed a simple model which has become a starting point for most discussions of heterojunction behavior [5]. Alternative models have been proposed by Adams and Nussbaum [6] and by von Roos [7] (see also the discussion by Kroemer [8]), and by Heime and Nussbaum [9]. Books and review articles on heterojunctions have been published by Milnes and Feucht [10], Sharma and Purihit [11], Casey and Panich [12], Taunsley [13], Kressel and Butler [14, 15], and others.

In 1969 L. Esaki and R. Tsu pointed out that the mobility of the two-dimensional electron gas at the heterointerface should be enhanced [16]. In 1978 R. Dingle, H. L. Stormer, A. C. Gossard, and W. Wiegman were first to observe the enhanced electron mobility of the two-dimensional electron gas [17]. In 1980 T. Mimura, S. Hiyamizu, T. Fujii, and K. Nanbu fabricated the first heterojunction field effect transistor which utilized the mobility enhancement of the two-dimensional electron gas [18]. Since that time considerable progress has been achieved in improving the performance and increasing the integration scale for these devices.

In Section 10-2 we will consider the original Anderson model [4, 5]. In Section 10-3 we will discuss the two-dimensional electron gas which forms at the heterointerface. The density of the two-dimensional gas will be calculated in Sections 10-4 and 10-5. In Sections 10-6 and 10-7 we will consider the low field mobility and saturation velocity of electrons in the two-dimensional gas. This will be followed by the charge control model (Section 10-8). This model allows us to describe current–voltage characteristics and capacitance–voltage characteristics of MODFETs (Section 10-9). In Section 10-10 we will discuss the so-called inverted GaAs–AlGaAs modulation doped structure. In Section 10-11 we will consider the maximum device transconductance and the maximum current swing. The improved extended charge control model which allows us to consider MODFETs at large gate biases will be discussed in Section 10-12. Device characterization techniques will be briefly discussed in Section 10-13. Nonideal phenomena related to traps, such as the shift of the threshold voltage with temperature, transient capacitance response, and persistent photoconductivity, will be considered in Sections 10-14 and 10-15. Finally, in Sections 10-16, 10-17, and 10-18 we will discuss new MODFET devices—a self-aligned MODFET, a GaAs gate MODFET, a p-channel MODFET, and complementary heterostructure insulated gate FETs.

10-2. ANDERSON MODEL

Band diagrams of two separated semiconductors of different composition with the vacuum level chosen as a point of reference are shown in Fig. 10-2-1a. The electron affinities X_1 and X_2 are defined as energies required to promote an electron from the bottom of the conduction band into vacuum. Work functions ϕ_1 and ϕ_2 are equal to the energies separating the Fermi level and the vacuum level.

When the semiconductor materials are joined to form a heterojunction the discontinuities in the conduction and valence band edges appear at the heterojunction interface (see Fig. 10-2-1b). According to the Anderson model the conduction band discontinuity is

$$\Delta E_c = X_1 - X_2 \tag{10-2-1}$$

i.e., is equal to the difference between the electron affinities. The valence band discontinuity is given by

$$\Delta E_v = \Delta E_g - \Delta E_c \tag{10-2-2}$$

where ΔE_g is the difference in the energy gaps (the values of the energy gaps, electron affinities, and lattice constants for different semiconductors are given in Table 10-2-1). Using this assumption we may find the built-in voltage from Fig. 10-2-1b:

$$V_{Bi} = E_{g1} - \Delta E_n - \Delta E_p + \Delta E_c \tag{10-2-3}$$

Here we reserve the subscript 1 for a narrow gap material which we assume to be doped p-type (notation is explained in Fig. 10-2-1).

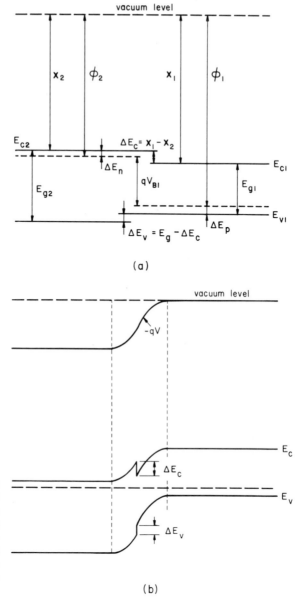

FIGURE 10-2-1. (a) Band diagrams for two different semiconductor materials (*n*-type Al$_{0.35}$Ga$_{0.65}$As and *p*-type GaAs). Vacuum level is chosen as the reference energy. V_{bi} is the built-in voltage. (b) Band diagram of a *p–n* heterojunction under zero bias according to the Anderson model [4, 5]. V is the electric potential.

Using the depletion approximation we find how V_{Bi} is divided between the *p*- and *n*-type regions

$$V_{Bi1} = \frac{\varepsilon_2 N_D}{\varepsilon_2 N_D + \varepsilon_1 N_A} V_{Bi} \qquad (10\text{-}2\text{-}4)$$

$$V_{Bi2} = \frac{\varepsilon_1 N_A}{\varepsilon_2 N_D + \varepsilon_1 N_A} V_{Bi} \qquad (10\text{-}2\text{-}5)$$

TABLE 10-2-1. Energy Gaps, Electron Affinities, and Lattice Constants of Different Semiconductor Compounds[a]

	E_g (eV)	X (eV)	a_0 (Å)
GaAs	1.424	4.07	5.654
AlAs	2.16	2.62	5.661
GaP	2.2	4.3	5.451
AlSb	1.65	3.65	6.135
GaSb	0.73	4.06	6.095
InAs	0.36	4.9	6.057
InSb	0.17	4.59	6.479
Ge	0.66	4.13	5.658
Si	1.11	4.01	5.431
ZnTe	2.26	3.5	6.103
CdTe	1.44	4.28	6.477
ZnSe	2.67	3.9	5.667
InP	4.38	5.34	5.869
CdS	2.42	4.87	4.137

[a] From Kressel and Butler [14].

The depletion widths x_1 and x_2 are given by

$$x_1 = \left[\frac{2 N_A \varepsilon_1 \varepsilon_2 V_{Bi}}{q N_D (\varepsilon_2 N_D + \varepsilon_1 N_A)} \right]^{1/2} \tag{10-2-6}$$

$$x_2 = \left[\frac{2 N_D \varepsilon_1 \varepsilon_2 V_{Bi}}{q N_A (\varepsilon_2 N_D + \varepsilon_1 N_A)} \right]^{1/2} \tag{10-2-7}$$

and the depletion capacitance is given by

$$C = \left[\frac{N_D N_A \varepsilon_1 \varepsilon_2}{2 (\varepsilon_2 N_D + \varepsilon_1 N_A) V_{Bi}} \right]^{1/2} \tag{10-2-8}$$

When a reverse or small forward ($V < V_{Bi}$) bias is applied, the voltage drops across the depletion region will be distributed between the semiconductor regions as follows:

$$V_{Bi1} - V_1 = \frac{\varepsilon_2 N_D}{\varepsilon_2 N_D + \varepsilon_1 N_A} (V_{Bi} - V) \tag{10-2-9}$$

$$V_{Bi2} - V_2 = \frac{\varepsilon_1 N_A}{\varepsilon_2 N_D + \varepsilon_1 N_A} (V_{Bi} - V) \tag{10-2-10}$$

Equations (10-2-6)–(10-2-8) should be then changed to

$$x_1 = \left[\frac{2 N_A \varepsilon_1 \varepsilon_2 (V_{Bi} - V)}{q N_D (\varepsilon_2 N_D + \varepsilon_1 N_A)} \right]^{1/2} \tag{10-2-11}$$

$$x_2 = \left[\frac{2 N_D \varepsilon_1 \varepsilon_2 (V_{Bi} - V)}{q N_A (\varepsilon_2 N_D + \varepsilon_1 N_A)} \right]^{1/2} \tag{10-2-12}$$

$$C = \left[\frac{N_D N_A \varepsilon_1 \varepsilon_2}{2(\varepsilon_2 N_D + \varepsilon_1 N_A)(V_{Bi} - V)}\right]^{1/2} \qquad (10\text{-}2\text{-}13)$$

Anderson calculated the current–voltage characteristics of a heterojunction by combining the proposed band diagram with a diode emission theory (see Ref. 12). He obtained the saturation currents by using the diffusion theory as for a homojunction. This led, however, to large discrepancies between the theory and the experiment. A more realistic theory of the I–V characteristics of p–n heterojunctions based on a classical emission model and on the assumption of the quasi-equilibrium (i.e., on the assumption of constancy of electron and hole quasi-Fermi levels) was developed by S. S. Perlman and D. F. Feucht [19].

10-3. TWO-DIMENSIONAL ELECTRON GAS AT (Al, Ga)As–GaAs INTERFACE

When a doped (Al, Ga)As layer is grown on top of an undoped GaAs layer a two-dimensional electron gas is formed at the interface owing to the difference in the electron affinity of the two layers (see Fig. 10-3-1). The electron motion in the two-dimensional gas is quantized in the direction perpendicular to the heterointerface, because the de Broglie wavelength is larger than the width of the potential well (the de Broglie wavelength of a thermal electron is approximately 260 Å at room temperature and longer at lower temperatures). This quantization effect is well known for Si inversion layers [21], but it is more pronounced in the (Al, Ga)As–GaAs structure because of a smaller electron effective mass in GaAs.

Based on the effective mass approximation, the motion of the electrons in the two-dimensional (2D) gas may be characterized by the envelope function [21]:

$$F(x, y, z) = \phi(z) \exp(i q_e \cdot r) \qquad (10\text{-}3\text{-}1)$$

where r is a 2D vector in the interface plane, q_e is a 2D wave vector for the motion parallel to the interface, and z is the distance from the interface into the GaAs layer. The wave function $\phi_i(z)$ satisfies the Schrödinger equation

$$\frac{\hbar^2}{2m} \frac{d^2 \phi_i}{dz^2} + [E_i - V(z)]\phi_i = 0 \qquad (10\text{-}3\text{-}2)$$

where m is the effective mass of an electron in the conduction band of the bulk GaAs, E_i is the quantized energy for the ith subband, and $V(z)$ is the potential energy (corresponding to the band bending near the interface). The boundary conditions are $\phi(\infty) = 0$ and $\phi(-\infty) = 0$. The potential energy $V(z)$ is found from the solution of Poisson's equation:

$$\frac{d^2 V}{dz^2} = q\rho(z)/\varepsilon \qquad (10\text{-}3\text{-}3)$$

The space charge density is determined by the electron concentration at the interface

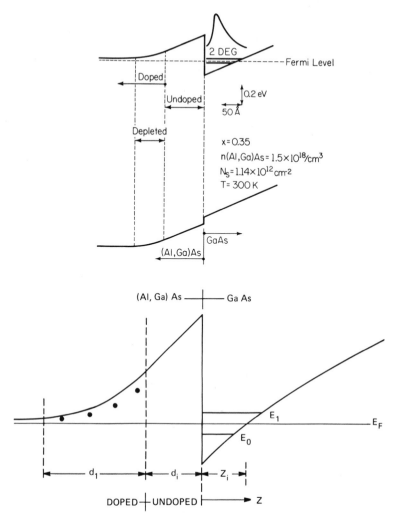

FIGURE 10-3-1. (a) Band diagram of an $Al_xGa_{1-x}As$-GaAs n^+-p^- heterojunction with the electron inversion layer (two-dimensional electron gas) at the heterointerface in GaAs [50]. Also shown, quantum energy levels of electrons in the two-dimensional gas (2DEG) and the qualitative distribution of electrons in the potential well. (b) Conduction band edge for the same structure [29, 30]. Dots represent ionized donors. Z_i is the effective width of the two-dimensional electron gas.

and by the doping of GaAs:

$$\rho(z) = q(N_{D1} - N_{A1}) - q \sum_{i=0}^{\infty} n_i |\phi_i|^2(z) \tag{10-3-4}$$

$$n_i = (mk_BT/\hbar^2) \ln\{1 + \exp[q(E_F - E_i)/k_BT]\} \tag{10-3-5}$$

Here ε is the permittivity of the GaAs layer, N_{D1} and N_{A1} are the concentrations of ionized donors and acceptors in the GaAs layer, E_F is the Fermi level, q is the electronic charge, k_B is the Boltzmann constant, and $\hbar$ is the reduced Planck constant.

For the inversion layer in silicon structures Eqs. (10-3-1)–(10-3-5) were solved both numerically and analytically [21–23]. The analytical solution is based on the assumption of the infinite barrier height for $z < 0$ (see Fig. 10-3-1) and on the linear approximation for the potential energy in the vicinity of the heterointerface

$$V = qF_s z$$

where F_s is the surface electric field (the triangular potential well approximation). This approach yields a quite satisfactory solution for the surface carrier density as a function of the Fermi potential. It leads to the well-known Airy equation and to the subband energies [21]

$$E_i \approx (\hbar^2/2m)^{1/3}[3qF_s\pi(i + \tfrac{3}{4})/2]^{2/3} \qquad (10\text{-}3\text{-}6)$$

Equation (10-3-6) approximates the exact solution for the triangular potential well with a 2% error.

The surface field F_s is related to the surface carrier density n_s by Gauss's law

$$\varepsilon F_s = qn_s + Q_{B1} \qquad (10\text{-}3\text{-}7)$$

where

$$n_s = \sum_{i=0}^{\infty} n_i \qquad (10\text{-}3\text{-}8)$$

n_i is given by Eq. (10-3-5) and

$$Q_{B1} = q \int_0^{w_{\text{dep}}} (N_{D1} - N_{A1})\, dz \qquad (10\text{-}3\text{-}9)$$

is the total surface charge in GaAs related to the ionized impurities. The GaAs layer in modulation doped structures is typically undoped in order to have a high electron mobility. Typically the GaAs layer is p-type with the concentration of the ionized acceptors of the order of 10^{14} cm^{-3} leading to values of Q_{B1} of the order of 4×10^{10} cm^{-2}, which is negligible compared to the typical values of n_s of the order of 10^{11}–10^{12} cm^{-2}. Hence we find from Eq. (10-3-7)

$$\varepsilon F_s = qn_s \qquad (10\text{-}3\text{-}10)$$

Substitution of Eq. (10-3-10) into Eq. (10-3-6) yields

$$E_0 = \gamma_0(n_s)^{2/3} \qquad (10\text{-}3\text{-}11)$$

$$E_1 = \gamma_1(n_s)^{2/3} \qquad (10\text{-}3\text{-}12)$$

Parameters γ_0 and γ_1 may be calculated using the value of the effective mass for GaAs in Eq. (10-3-6). However, some adjustment of these parameters is necessary to account for the bulk charge and for the deviation of the potential distribution from a triangular shape. The values of γ_0 and γ_1 were estimated in Ref. 24 from Shubnikov–de Haas and cyclotron resonance data. For the undoped GaAs

$$\gamma_0 = 2.5 \times 10^{-12}\text{V m}^{4/3} \qquad (10\text{-}3\text{-}13)$$

$$\gamma_1 = 3.2 \times 10^{-12}\text{V m}^{4/3} \qquad (10\text{-}3\text{-}14)$$

We do not consider higher energy levels, which seems to be a satisfactory approximation in most practical cases [24]. In this case

$$n_s = Dk_BT/q \sum_{i=0}^{1} \ln\{1 + \exp[q(E_F - E_i)/k_BT]\} \qquad (10\text{-}3\text{-}15)$$

Here $D = qm/\pi\hbar^2$ is the density of states of the two-dimensional electron gas. It is determined from the measured cyclotron effective mass [24]

$$D = 3.24 \times 10^{17}\,\mathrm{m^{-2}\,V^{-1}} \qquad (10\text{-}3\text{-}16)$$

In silicon MOSFETs, the surface electron concentration may be evaluated using a simpler approach based on the following assumptions:

1. The Maxwell–Boltzmann (rather than the Fermi–Dirac) distribution function may be used.

2. The density of states in the potential well near the interface may be assumed to be continuous.

Our analysis shows that both assumptions are typically justified for Si MOSFETs but *not* for GaAs modulation doped structures. This is illustrated by Fig. 10-3-2 [25], where the interface carrier densities are plotted as functions of the Fermi level positions using the "three-dimensional" Joyce–Dixon approximation [26] (dashed lines) and a more accurate "two-dimensional gas" analysis based on solving Eqs. (10-3-5), (10-3-6), and (10-3-9). We also show a simple linear approximation for the n_s vs. E_F curves, which we use in our analytical analysis of modulation doped stuctures (see Sections 10-6 and 10-7).

In silicon there are six equivalent minima of the conduction band contributing to the density of states effective mass. The effective mass in each minimum is also high ($m/m_e = 0.3$). Hence, there are many levels in the potential well near the interface and the conventional "three-dimensional" approach is justified. Moreover, the Fermi level at the silicon–oxide interface is well below the bottom of the conduction band (about 0.1 eV or so). As a consequence the Maxwell–Boltzmann distribution may be used.

In AlGaAs–GaAs modulation doped structures there are only two relevant energy levels in the potential well near the interface, and the discrepancy between the three-dimensional (continuous) and the two-dimensional (discrete) models is

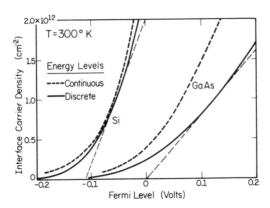

FIGURE 10-3-2. Interface carrier density versus Fermi potential for Si and GaAs at 300 K. Solid lines are from assuming two lowest discrete energy levels and the heavy dashed lines are from assuming continuous energy levels using the Joyce–Dixon approximation. The light dashed lines are linear approximations. The energy reference is the bottom of the conduction band at the interface [25].

quite large. Also, the Fermi level is in the potential well and, hence, the Fermi–Dirac distribution function should be used.

An important consequence is a much stronger dependence of the Fermi level position on the surface carrier concentration (compared to the inversion layers in Si MOSFETs). This should be taken into account in modeling modulation doped structures and devices.

10-4. DENSITY OF THE TWO-DIMENSIONAL ELECTRON GAS AS A FUNCTION OF THE FERMI LEVEL POSITION

When a doped (Al, Ga)As layer is grown on top of an undoped GaAs layer, a two-dimensional electron gas is formed at the interface owing to the difference in the electron affinities of these layers (see Fig. 10-3-1). The amount of charge transfer across the interface is found by equating the charge depleted from the (Al, Ga)As to the charge accumulated in the potential well. A solution is then found such that the Fermi level is constant across the heterointerface. The derivation below follows Ref. 27.

We start from the one-dimensional Poisson equation,

$$\frac{d^2 V}{dx^2} = -\frac{\rho(x)}{\varepsilon} \tag{10-4-1}$$

describing the potential distribution in (Al, Ga)As in the direction perpendicular to the heterointerface. Here V is the electrostatic potential. The space charge density $\rho(x)$ is given by

$$\rho(x) = q[N_d^+(x) - n(x)] \tag{10-4-2}$$

where $n(x)$ is the free electron concentration and

$$N_d^+(x) = \frac{N_d}{1 + g \exp[(E_d + qV)/k_B T]} \tag{10-4-3}$$

is the concentration of the ionized donors. Here N_d is the total donor density, g is the degeneracy factor of the donor level, E_d is the donor activation energy, k_B is the Boltzmann constant, and T is the lattice temperature. We will use a simple analytical approximation for $n(V)$ which describes quite accurately both the nondegenerate and degenerate cases [28] and allows one to find the solution of Eq. (10-4-1) in the analyical form,

$$n(V) = N_c \frac{\exp(qV/k_B T)}{1 + \exp(qV/k_B T)/4} \tag{10-4-4}$$

where N_c is the equivalent density of states of the conduction bands and the Fermi level, E_F, is chosen as the origin of the energy scale ($E_F = 0$).

Combining equations (10-4-1)–(10-4-4), we obtain

$$\frac{d^2 V}{dx^2} = -\frac{qN_c}{\varepsilon} \left[\frac{N_d'}{1 + g' \exp(qV/k_B T)} - \frac{\exp(qV/k_B T)}{1 + \exp(qV/k_B T)/4} \right] \tag{10-4-5}$$

where $N'_d = N_d/N_c$ and $g' = g \exp(E_d/k_B T)$. The integration of Eq. (10-4-5) from $V(-\infty)$ to $V(0)$ with respect to V using the boundary condition $F(-\infty) = 0$, where F is the electric field, yields

$$F^2(0) = \frac{2k_B T N_c}{\varepsilon}\left\{N'_d \ln\frac{g' + \exp[-qV(0)/k_B T]}{g' + \exp[-qV(-\infty)/k_B T]} + 4\ln\frac{4 + \exp\{qV(0)/k_B T\}}{4 + \exp\{qV(-\infty)/k_B T\}}\right\}$$

$$(10\text{-}4\text{-}6)$$

The constant $V(-\infty)$ can be found from the requirement of space charge neutrality at $x = -\infty$, i.e., from

$$\rho(-\infty) = -\frac{qN_c}{\varepsilon}\left\{\frac{N'_d}{1 + g'\exp[qV(-\infty)/k_B T]} - \frac{\exp[qV(-\infty)/k_B T]}{1 + \exp[qV(-\infty)/k_B T]/4}\right\}$$

$$= 0 \qquad\qquad (10\text{-}4\text{-}7)$$

The solution of Eq. (10-4-7) is given by

$$y = \exp[qV(-\infty)/k_B T] = \frac{-(1 - N_d/4) + [(1 - N_d/4)^2 + 4g'N_d]^{1/2}}{2g'} \quad (10\text{-}4\text{-}8)$$

As can be seen from Fig. 10-4-1, $V(-\infty)$ is simply equal to the difference between the Fermi level and the bottom of the conduction band away from the heterojunction. This value is shown in Fig. 10-4-1 as a function of doping for the classical Boltzmann statistics (solid line) and for the approximate Fermi–Dirac statistics (dotted line). This comparison demonstrates that the deviation from Boltzmann statistics is quite noticeable even at a relatively low doping level ($N_d \approx 10^{17}$ cm^{-3}) and becomes quite large at doping densities commonly used ($1 - 2 \times 10^{18}$ cm^{-3}).

The equilibrium interface carrier concentration n_{s0} is determined by the surface field:

$$n_{s0} = \frac{\varepsilon}{q}F(0) = \frac{\varepsilon}{q}F(d_i^-) \qquad\qquad (10\text{-}4\text{-}9)$$

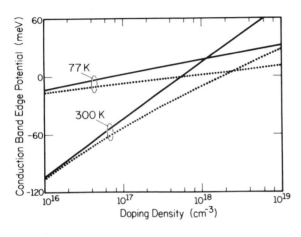

FIGURE 10-4-1. Potential of the conduction band edge $V(-\infty)$ versus doping density for the bulk Al$_{0.3}$Ga$_{0.7}$As at 300 and 77 K. Solid lines, Boltzmann statistics; dotted lines, Fermi Dirac statistics [27].

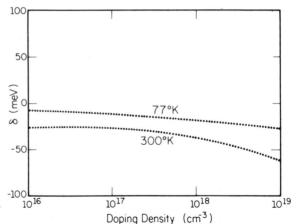

FIGURE 10-4-2. δ versus doping density at 300 and 77 K [27].

Equation (10-4-9) follows from Gauss's law if the doping density in the GaAs is small enough so that the total bulk charge in the depletion layer of GaAs is much smaller than qn_{s0}. The expression for $F(0)$ is given by Eq. (10-4-6), which may be simplified when the inequality

$$\exp\left[-\frac{qV(0)}{k_BT}\right] \gg 1 \tag{10-4-10}$$

is taken into account as follows:

$$F^2(0) = \frac{2qN_d}{\varepsilon}\left\{-V(0) + V(-\infty) - \frac{k_BT}{q}\left[\ln(1 + g'y) + \frac{4}{N'_d}\ln\left(1 + \frac{y}{4}\right)\right]\right\} \tag{10-4-11}$$

Substitution of Eq. (10-4-11) into Eq. (10-4-9) yields

$$n_{s0} = \left\{\frac{2\varepsilon N_d}{q}[-V(d_i^-) + V(-\infty) + \delta] + N_d^2 d_i^2\right\}^{1/2} - N_d d_i \tag{10-4-12}$$

where

$$\delta = -\frac{k_BT}{q}\left[\ln(1 + g'y) + \frac{4}{N'_d}\ln(1 + y/4)\right] \tag{10-4-13}$$

The relation

$$V(d_i^-) = V(0) - F(0)d_i \tag{10-4-14}$$

has been used to derive Eq. (10-4-12). The coordinate $x = d_i^-$ corresponds to the (Al, Ga)As side of the heterointerface.

Equation (10-4-12) differs from a similar equation in Ref. 24 by the factor δ in the right-hand part, which is subtracted from total band bending $-V(d_i^-) + V(-\infty)$.

This contribution to the band-bending is shown in Fig. 10-4-2 as a function of the doping density N_d in (Al, Ga)As for 77 and 300 K. As can be seen from the figure, this correction is quite important because it represents a significant fraction of ΔE_c.

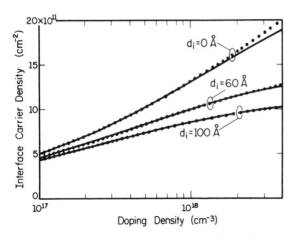

FIGURE 10-4-3. Interface carrier density n_{s0} versus doping density with the undoped $Al_{0.3}Ga_{0.7}As$ layer thickness d_i as a parameter. Dotted line, numerical solution of Eqs. (10-4-12)–(10-4-18); solid line, Eq. (10-4-20) [27].

In order to determine n_{s0}, we should solve Eq. (10-4-12) together with the equation

$$n_{s0} = D\frac{k_B T}{q} \ln\left[\left(1 + \exp\left\{\frac{q}{k_B T}[V(d_i^+) - E_0]\right\}\right)\left(1 + \exp\left\{\frac{q}{k_B T}[V(d_i^+) - E_1]\right\}\right)\right]$$

$$(10\text{-}4\text{-}15)$$

In Eq. (10-4-15) n_{s0} is expressed in terms of the density of states D and the two lowest energy levels in the potential well in the GaAs at the heterointerface [24], given by

$$E_0 = \gamma_0 n_{s0}^{2/3} \qquad (10\text{-}4\text{-}16)$$

$$E_1 = \gamma_1 n_{s0}^{2/3} \qquad (10\text{-}4\text{-}17)$$

Here $\gamma_0 \approx 2.5 \times 10^{-12}\ \text{eV cm}^{4/3}$ and $\gamma_1 \approx 3.2 \times 10^{-12}\ \text{eV cm}^{4/3}$. E_0 and E_1 in Eqs. (10-4-15)–(10-4-17) are measured from the bottom of the conduction band in the GaAs at the heterointerface (see Fig. 10-3-1). Here

$$V(d_i^+) = \Delta E_c + V(d_i^-) \qquad (10\text{-}4\text{-}18)$$

(see Fig. 10-4-1). Equations (10-4-13)–(10-4-18) are solved numercially to find n_{s0} (dotted lines in Fig. 10-4-3). However, a fairly accurate analytical approximation to the exact computer solution may be obtained if equations (10-4-15)–(10-4-17) are linearized with respect to $V(d_i^+)$ (as was done in Ref. 29). According to Ref. 29

$$V(d_i^+) = \Delta E_{F0}(T) + an_{s0} \qquad (10\text{-}4\text{-}19)$$

where $a = 0.125 \times 10^{-16}\ \text{V m}^{-2}$ and $\Delta E_{F0} = 0$ at 300 K and $\Delta E_{F0} = 0.025$ eV at 77 K and below. Using Eq. (10-4-19) instead of Eqs. (10-4-15)–(10-4-17) leads to the following formula for n_{s0}:

$$n_{s0} = \left(\frac{2N_d\varepsilon}{q}[\Delta E_c - E_{F0}(T) + \delta + V(-\infty)] + N_d^2(d_i + \Delta d)^2\right)^{1/2}$$

$$- N_d(d_i + \Delta d) \qquad (10\text{-}4\text{-}20)$$

where $\Delta d = (\varepsilon a)/q \approx 80$ Å [29].

In Figure 10-4-3 the analytical solution of Eq. (10-4-20) is compared with the exact numerical calculation n_{s0} based on the solution of Eqs. (10-4-13)–(10-4-19). As can be seen from the figure, Eq. (10-4-20) is indeed a very good approximation.

The expression for n_{s0} given in Ref. 24 is recovered if one assumes $\delta \to 0$ and $\Delta E_{F0}(T) \to 0$. The actual values of δ and ΔE_{F0} at 77 K are close to -0.025 V and 0.025 eV, respectively, resulting in a total correction of 0.05 eV. At 300 K $\delta \approx 0.05$ eV and $\Delta E_{F0} \approx 0$ leading again to a correction about 0.05 eV.

A rather large difference between the solution based on the depletion approximation and the Boltzmann statistics for the electrons in (Al, Ga)As and the results given above may be explained as follows. The depletion approximation is definitely applicable in (Al, Ga)As near the heterointerface where the electrostatic potential is much larger than the thermal voltage [see Eq. (10-4-10)]. However, at the edge of the depletion region where the electrostatic potential is less than the thermal voltage, the depletion approximation does not apply. The space charge density there is smaller than qN_d because of the finite electron concentration. If the Boltzmann statistic applies, the length of this region is of the order of the Debye radius

$$L_D = \left(\frac{\varepsilon k_B T}{q^2 N_d}\right)^{1/2} \tag{10-4-21}$$

In a degenerate (or nearly degenerate) case this distance, while still proportional to L_D, is substantially larger and depends on the position of the Fermi level. This is reflected in the expression for the electric field at the interface [see Eq. (10-4-11)], which can be rewritten as

$$F(0) = \frac{qN_D}{\varepsilon} (W_0^2 - 2L_D^2 f)^{1/2} \tag{10-4-22}$$

where

$$W_0 = \left\{\frac{2\varepsilon[-V(0) + V(-\infty)]}{qN_d}\right\}^{1/2} \tag{10-4-23}$$

is the width of the space charge region given by the depletion approximation and

$$f = \ln(1 + g'y) + \frac{4}{N_d'} \ln\left(1 + \frac{y'}{4}\right) \tag{10-4-24}$$

may be appreciably larger than unity, leading to smaller values of $F(0)$ and hence n_{s0}. Alternatively, we may say that whereas the correction to the voltage drop across the space charge region is equal to the thermal voltage $k_B T/q$, when Boltzmann statistics is applicable, it may be substantially larger in the degenerate or nearly degenerate case.

In Fig. 10-4-4 we compare the results of our calculation with experimental data given in Ref. 30. As can be seen from the data, our theory explains very well the general trend in the n_{s0} vs. d_i curve. The large change in the measured n_{s0} with respect to temperature not predicted by the theory is a result of a relatively thick ($0.15\text{-}\mu$m) layer of doped (Al, Ga)As conducting parallel to the two-dimensional electron gas. At low temperatures the electron gas mobility typically increases by a factor of 10. Consequently, electrons remaining in the (Al, Ga)As make a much smaller contribution to the measured sheet carrier concentration [31].

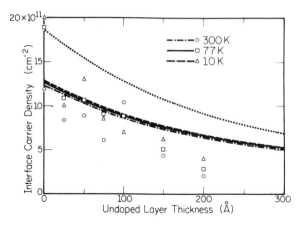

FIGURE 10-4-4. Interface carrier concentration n_{so} versus undoped $Al_{0.33}Ga_{0.67}As$ layer thickness for different temperatures. Experimental points are for heterostructures with 0.15 μm of $Al_{0.33}Ga_{0.67}As$ doped to $N_d = 7 \times 10^{17}$ cm^{-3}. Solid lines, Eq. (10-4-20); dotted line, using depletion approximation with $\Delta d = 0$ at 300 K [25].

10-5. DENSITY OF TWO-DIMENSIONAL ELECTRON GAS IN MODULATION DOPED STRUCTURE WITH GRADED INTERFACE [32, 33]

In the previous section we have considered modulation doped layers with abrupt heterointerfaces. As was shown in Refs. 32 and 33, grading of the heterointerface may lead to an improvement in device characteristics at small grading lengths. The interface grading leads to a wider potential well for the 2D electron gas. This, in turn, decreases the electron subband energies and, hence, increases the concentration of the 2D electron gas for a fixed Fermi level position. On the other hand, grading decreases the magnitude of the interface barrier (see Fig. 10-5-1), decreasing the concentration of the 2D electron gas at large grading lengths. As a result an optimum grading length corresponding to the maximum concentration of the 2D gas should exist, as discussed in Ref. 25 (and in Section 10-11), and hence an increase in the concentration n_{so} of the 2D electron gas leads to larger device transconductance and current swing.

The determination of the energy levels in the interface potential well requires a self-consistent solution of the Schrödinger and Poisson equations (see Section 10-3 and Refs. 22 and 34). However, a constant field (constant slope of the potential well) approximation and a related variational approach with one parameter have been found to be quite accurate, especially at large values of n_{so} [22].

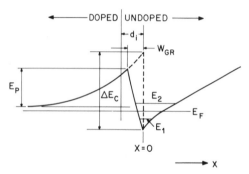

FIGURE 10-5-1. Energy band diagram of a graded AlGaAs/GaAs heterostructure at a thermal equilibrium [32].

In case of a graded heterointerface the shape of the triangular well is determined by two effective electric fields, both dependent on n_{s0}. The slope of the potential well in GaAs, F^+, is equal to the electric field, $F(0)$, in GaAs at the heterointerface. The slope of the potential well in AlGaAs, F^-, is given by

$$F^- = -F(0) - F_{GR} = -F^+ + \Delta E_c/(qW_{GR}) \qquad (10\text{-}5\text{-}1)$$

where we assumed that the conduction band variation in the graded region in AlGaAs is a linear function of distance. In Eq. (10-5-1) ΔE_c is the total conduction band discontinuity and W_{GR} is the grading length. The electron wave function Ψ_j and energy E_j for the jth subband in the graded potential well are found from the solution of the Schrödinger equation:

$$\Psi(x) = \begin{cases} D \, \text{Ai}(-\gamma_+) \, \text{Ai}(-\zeta_-)/\text{Ai}(-\beta^2\gamma_+) & \text{for } x \leqslant 0 \\ D \, \text{Ai}(\zeta_+) & \text{for } x \geqslant 0 \end{cases} \qquad (10\text{-}5\text{-}2)$$

$$E_j = (\hbar^2/2m)^{1/3}(qF^+)^{2/3}\gamma_+^{\,j} \qquad (10\text{-}5\text{-}3)$$

where D is a normalization constant, m is the electron effective mass, q is the electronic charge, and

$$\zeta_\pm = \alpha_\pm(x - E/qF^\pm), \qquad \alpha_\pm = (2mqF^\pm/\hbar^2)^{1/3}$$
$$\gamma_\pm = \alpha_\pm E/qF^\pm, \qquad \beta = (F^+/F^-)^{1/3} \qquad (10\text{-}5\text{-}4)$$

Ai (x) is an Airy function and $\gamma_+^{\,j}$ are roots of the following equation:

$$\beta \, \text{Ai}'(-\gamma_+) \, \text{Ai}(-\gamma_+\beta^2) + \text{Ai}(-\gamma_+) \, \text{Ai}'(-\gamma_+\beta^2) = 0 \qquad (10\text{-}5\text{-}5)$$

The lowest roots for $\beta < 1$ may be approximated as

$$\gamma_+^1(\beta) = 2.338 - 1.506\beta + 0.188\beta^2$$
$$\gamma_+^2(\beta) = 4.088 - 1.163\beta - 0.585\beta^2 \qquad (10\text{-}5\text{-}6)$$

with 1% accuracy.

For $\beta > 1$ the roots of Eq. (10-5-5) $\gamma_+^{1,2}(\beta)$ are given by

$$\gamma_+^{1,2}(\beta) = \gamma_+^{1,2}(1/\beta)/\beta^2 \qquad (10\text{-}5\text{-}7)$$

Integration of the Poisson equation (using the depletion approximation for the space charge region in the AlGaAs layer) yields

$$F(0) = F^+ = (q/\varepsilon)\{N_d L - [\beta^3/(1+\beta^3)]\sum N_j\} \qquad (10\text{-}5\text{-}8)$$

where L is the thickness of the space charge layer (excluding the undoped spacer layer with thickness d_i), and ε is the dielectric permittivity, which we assume to be constant. It can be seen from Eq. (10-5-8) and Gauss's law that the ratio of the surface concentration of electrons, N^-, located in the region $x < 0$, to the surface concentration of electrons, N^+, located in the region $x > 0$ (see Fig. 10-5-1), is equal to β^3, i.e., $N^-/N^+ = F^+/F^-$.

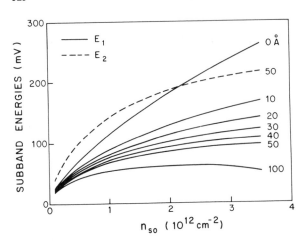

FIGURE 10-5-2. The lowest energy subband in the interface potential well versus the surface concentration of the 2D electron gas for different grading lengths [32]. (a) $x = 0.3$; (b) $x = 0.37$. Dotted lines show the second lowest subband for $W_{GR} = 50$ Å.

Assuming that the total donor charge $N_d L$ in the depletion layer in AlGaAs is equal to the electron charge in the 2D gas ($n_{so} = \sum N_j$), we find from Eqs. (10-5-4) and (10-5-8)

$$\beta = (q^2 W_{GR} n_{so} / \varepsilon \Delta E_c)^{1/3} \qquad (10\text{-}5\text{-}9)$$

$$F^+ = q n_{so} / [\varepsilon (1 + \beta^3)] \qquad (10\text{-}5\text{-}10)$$

Substituting Eq. (10-5-9) into Eq. (10-5-6) and using Eq. (10-5-3) we can calculate the subband levels as functions of n_{so} for different grading lengths (see Fig. 10-5-2). With the increase in grading length the subband levels approach the bottom of the potential well. A noticeable change takes place even at 10 Å grading. This is because of large changes in the effective electric field with grading at large values of n_{so}. For small n_{so}, $F_{GR} \gg F^+$ and the energy levels are nearly the same as for the abrupt heterointerface (see Fig. 10-5-2). The decrease of the energy levels becomes substantial when F^- becomes comparable to or smaller than F^+.

Expressing the position of the Fermi level as a function of the doping level, N_d, in AlGaAs as was done in Section 10-4 we can find n_{so} as a function of N_d. (We have assumed a donor binding energy of 11 meV.) The n_{so} vs. N_d curves for

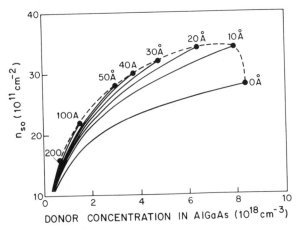

FIGURE 10-5-3. Surface concentration of the 2D gas versus doping in $Al_{0.37}Ga_{0.63}As$ layer for different grading lengths [32]. $\Delta E_c = 0.39$ eV; $d_i = 0$. The curves are terminated when the maximum of interface potential to the Fermi level is equal to $2kT$. The envelope of the termination points yields dependence of the maximum achievable 2D concentration on doping of the AlGaAs layer.

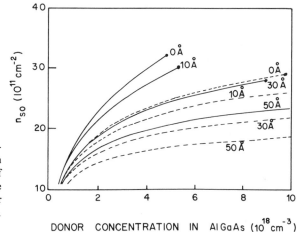

FIGURE 10-5-4. Surface concentration of the 2D gas versus doping in $Al_{0.37}Ga_{0.63}As$ for grading length of 30 Å for different thicknesses of the undoped spacer layer. Dashed lines for $W_{GR} = 0$; solid lines for $W_{GR} = 30$ Å [32].

$d_i = 0$ are shown in Fig. 10-5-3. All the curves are terminated when the Fermi level approaches the interface peak E_p in the conduction band so that $E_p - E_F = 50$ mV ($\approx 2 k_B T$). E_p decreases with grading and, hence, at large grading lengths the maximum value of n_{s0} starts to decrease with grading. For small values of W_{GR} the decrease in the subband energy is larger than the decrease in the conduction band peak and n_{s0} increases with grading. As can be seen from Fig. 10-5-3, the value of W_{GR} corresponding to the maximum value of n_{s0} is approximately 20 Å. It is worth mentioning that the maximum value of n_{s0} occurs at smaller doping densities in AlGaAs as the grading length increases. In practical devices it may be difficult to dope AlGaAs more than 2×10^{18} cm^{-3}. With this constraint, a larger grading length (of the order of 70 Å) corresponds to the maximum value of n_{s0}.

The dependences of n_{s0} on the doping density in AlGaAs for $W_{GR} = 0$ and $W_{GR} = 30$ Å and different thicknesses of the spacer layer d_i are shown in Fig. 10-5-4. As can be seen from the figure, there is a considerable increase in n_{s0} even when $d_i > W_{GR}$. In this case no ionized donors are present in the 2D potential well and, hence, no substantial mobility degradation should be expected compared with the abrupt modulation doped layers.

10-6. LOW FIELD MOBILITY OF A TWO-DIMENSIONAL ELECTRON GAS

10-6-1. Impurity Scattering [35]

Impurity scattering by remote donors plays a dominant role in determining the mobility of the two-dimensional electron gas (2DEG) in modulation doped layers at low temperatures [35–50].

Impurity potential is screened by electrons in the two-dimensional gas and, as shown in Ref. 35, is also affected by the electrons in the neutral section of the doped AlGaAs layer (see Fig. 10-6-1). The latter effect may be interpreted as the effect related to the image charge of the ionized donor induced in the neutral AlGaAs layer. In an idealized limit of extremely large carrier concentration in the neutral doped AlGaAs layer this image charge is equal to $-qZ$, where qZ is the effective charge of the ionized donor and its location mirrors the location of the ionized donor with respect to the neutral region boundary. In this limiting case impurity

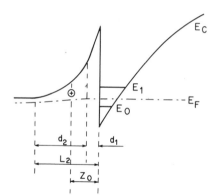

FIGURE 10-6-1. Band diagram of (Al, Ga)As–GaAs
modulation doped layer [35].

scattering is caused by the resulting dipole potential and, hence, is diminished. For
practical values of the carrier concentration in the doped AlGaAs layer the image
charges are further removed from the depletion region boundary by an additional
distance of the order of the screening length of carriers in AlGaAs. Hence, the
effects of the image charge are smaller but they are still quite important.

The theory developed in Ref. 35 and considered below is a generalization of
the theory of impurity scattering of electrons in the two-dimensional gas as developed
by Stern and Howard [34]. It takes into account the effect of the finite width of the
depletion layer in AlGaAs on the impurity scattering potential. This is equivalent
to taking into account the image charge of the scattering center reflected with respect
to the boundary of the neutral layer.

The ionized impurity potential $\phi(r, z)$ is defined by Poisson's equation:

$$\nabla^2\phi(r, z) - 2S|\Psi|^2\hat{\phi}(r) = -(q/\varepsilon_1)\delta(z - z_0)\delta(r), \qquad z > 0 \quad (10\text{-}6\text{-}1a)$$

$$\nabla^2\phi(r, z) = -(q/\varepsilon_2)\delta(z - z_0)\delta(r), \qquad -L_2 \leqslant z \leqslant 0 \quad (10\text{-}6\text{-}1b)$$

$$\nabla^2\phi(r, z) - \phi(r, z)/L_{sc}^2 = 0, \qquad -\infty < z \leqslant -L_2 \quad (10\text{-}6\text{-}1c)$$

where

$$\hat{\phi}(r) = \int_0^\infty \phi(r, z)|\Psi|^2 \, dz \qquad (10\text{-}6\text{-}2)$$

is the electron potential, averaged over coordinate z, Ψ is the 2DEG wave function,
r is a two-dimensional vector in the plane of the two-dimensional gas, and z is the
coordinate perpendicular to the plane ($z = 0$ corresponds to the heterointerface;
see Fig. 10-6-1), z_0 is the z-coordinate of the impurity center,

$$S = (2/R_B)/[1 + \exp(-E_F/k_BT)] \qquad (10\text{-}6\text{-}3)$$

is the screening factor of the 2DEG [34, 43], E_F is Fermi level of the 2DEG relative
to the bottom of the first subband, T is temperature, R_B is the effective Bohr radius
in GaAs ($R_u = 4\pi\varepsilon_1\hbar^2/m_1q^2$), m_1 and ε_1 are the electron effective mass and the
dielectric constant in GaAs, ε_2 is the dielectric constant in AlGaAs, and

$$L_{sc} = \left[\frac{1.5q^2n_0}{\varepsilon_2}\frac{\int E^{1/2}(\partial f/\partial E)\, dE}{\int E^{3/2}(\partial f/\partial E)\, dE}\right]^{-1/2} \qquad (10\text{-}6\text{-}4)$$

L_{sc} is the screening length here, f is the electron distribution function, n_0 is the electron concentration in the neutral region of the doped AlGaAs, $L_2 = d_1 + d_2$ is the distance between the heterointerface and the boundary of the neutral region in AlGaAs, d_1 and d_2 are undoped spacer layer and depletion layer thicknesses, respectively (see Fig. 10-6-1).

Here we assume that electrons occupy only the first subband. This limits the validity of our theory to relatively small values of the 2DEG concentration $n_{s0} <$ 10^{12} cm^{-2} (see, for example, Ref. 36).

Equation (10-6-1c) for the impurity potential in the region $z \leqslant -L_2$ is valid only if the absolute value of the potential energy $q\phi$ is smaller than $k_B T$. Otherwise, the image effect is underestimated.

We will define the Fourier component of the potential of the ionized impurity in the form

$$\phi(k, z) = \int \phi(r, z) \exp(-ik \cdot r) \, d^2 r \qquad (10\text{-}6\text{-}5)$$

Performing the Fourier transform of Eqs. (10-6-1a)–(10-6-1c) taking into account the boundary conditions

$$\phi(k, \pm\infty) = 0 \qquad (10\text{-}6\text{-}6)$$

and the requirements of the continuity of the electric flux and potential at $z = 0$ and $z = -L_2$ one may obtain the solution of Eqs. (10-6-1a)–(10-6-1c).

The impurity scattering is determined by $\hat{\phi}(k)$. The equations for $\hat{\phi}(k)$ are obtained by multiplying $\phi(k, z)$ by $|\Psi|^2$ and integrating with respect to z:

$$\hat{\phi}(k) = \frac{q}{2\varepsilon_1 k} \frac{[\chi(k, 0) R_1(-\lambda) \exp(-kz_0) + \chi(k, z_0) R_1(\lambda)]}{P(k)} \qquad (10\text{-}6\text{-}7)$$

for $z_0 \geqslant 0$,

$$\hat{\phi}(k) = \frac{q}{\varepsilon_1 k} \frac{\chi(k, 0)\{\exp(kz_0) - \gamma \exp[-k(2L_2 + z_0)]\}}{P(k)} \qquad (10\text{-}6\text{-}8)$$

for $-L_2 \leqslant z_0 \leqslant 0$. Here

$$\chi(k, z_0) = \int_0^\infty |\Psi|^2 \exp(-k|z - z_0|) \, dz \quad , \quad \hat{F}(k) = \int_0^\infty |\Psi|^2 F(z) \, dz$$

$$P(k) = [1 - \hat{F}(k)] R_1(\lambda) - \chi(k, 0)\{F'(0)[1 - \gamma \exp(-2kL_2)]/k$$
$$- \lambda F(0)[1 + \gamma \exp(-2kL_2)]\}$$

$F'(z)$ is the derivative of $F(z)$ with respect to z, $R_1(\lambda) = \lambda + 1 + \gamma(\lambda - 1)$ $\exp(-2kL_2)$, $\lambda = \varepsilon_2/\varepsilon_1$, and

$$\gamma = (1 - k/k^*)/(1 + k/k^*) \qquad (10\text{-}6\text{-}9)$$

The limiting case $\gamma \to 0$ corresponds to the negligible effect of the image charges.

Assuming that the wave function Ψ is given by [22, 23]

$$\Psi(z) = \frac{\alpha^{3/2}}{2} \exp(\alpha z/2) \tag{10-6-10}$$

where α is related to the sheet electron concentration n_{so}:

$$\alpha = (33\pi n_{so}/2R_B)^{1/3}$$

we find

$$F(z) = \frac{S\alpha^3}{(\alpha^2 - k^2)^3} [2(3\alpha^2 + k^2) + 4\alpha(\alpha^2 - k^2)z + (\alpha^2 - k^2)^2 z^2] \exp(-\alpha z) \tag{10-6-11}$$

$$\chi(k, z_0) = \frac{\alpha^3 \exp(-\alpha z_0)}{(k+\alpha)^3} \left(\frac{(k+\alpha)^2 z_0^2}{2} + (k+\alpha)z_0 + 1 \right.$$

$$\left. - \frac{(k+\alpha)^3}{(k-\alpha)^3} \{\exp[-(k-\alpha)z_0] - 1 + (k-\alpha)z_0 - (k-\alpha)^2 z_0^2/2\} \right) \tag{10-6-12}$$

$$\chi(k, 0) = [\alpha/(k+\alpha)]^3 \tag{10-6-13}$$

$$F(0) = \frac{2S\alpha^3(3\alpha^2 + k^2)}{(\alpha^2 - k^2)^3}, \qquad F'(0) = -\frac{2S\alpha^4(\alpha^2 + 3k^2)}{(\alpha^2 - k^2)^3} \tag{10-6-14}$$

$$\hat{F}(k) = \frac{S\alpha}{8(\alpha^2 - k^2)^3} (15\alpha^4 - 10\alpha^2 k^2 + 3k^4) \tag{10-6-15}$$

$$P(k) = 1 + \lambda - R_2(-\lambda) - \gamma[1 - \lambda - R_2(\lambda)] \exp(-2kL_2) \tag{10-6-16}$$

$$R_2(\lambda) = \frac{S\alpha}{8k(\alpha+k)^6} [(\lambda - 1)k(3k^4 + 18\alpha k^3 + 44k^2\alpha^2 + 54k\alpha^3 + 33\alpha^4) - 16\alpha^5] \tag{10-6-17}$$

For scattering centers located at the heterointerface we obtain from Eq. (10-6-7)

$$\hat{\phi}(k) = \frac{q\alpha^3[1 - \gamma \exp(-2kL_2)]}{\varepsilon_1 k(k+\alpha)^3 P(k)} \tag{10-6-18}$$

In the above equations the screening factor S was assumed to be given by Eq. (10-6-3). This equation for the inverse screening length is only valid for a strong degenerate case and when $k \leq 2k_F$. In a general case the dielectric formalism should be used to account for the screening effects [43, 51] and in the above equations S should be replaced by

$$S \rightarrow \left[\frac{\varepsilon(k)}{\varepsilon_1} - 1 \right] \frac{k^2}{2} \tag{10-6-19}$$

where $\varepsilon(k)$ is the wave-vector-dependent longitudinal static dielectric constant of a 2DEG:

$$\frac{\varepsilon(k)}{\varepsilon_1} = 1 + \frac{4mq^2}{\hbar^2 k \varepsilon_1}\left(1 - \exp(E_{F0}/k_B T)\right.$$

$$\left. - \frac{1}{2k_B T E_k^{1/2}}\int_0^{E_k/4}\frac{\mu^{1/2}\,d\mu}{\cosh^2\{(E_k/4 - \mu)/k_B T - (1/2)\ln[\exp(E_{F0}/k_B T) - 1]\}}\right)$$

Here $E_k = \hbar^2 k^2/2m_1$.

The surface current density is given by

$$J = \frac{q\hbar}{2\pi^2 m_1}\int (\partial f/\partial E_{kp})g(E_{kp})(k_p \cdot n)^2 d^2 k_p \qquad (10\text{-}6\text{-}20)$$

where

$$g(E_{kp}) = 8\pi\hbar^4 k_p^3 F_0/[qm_1^2 G_I(k_p)] \qquad (10\text{-}6\text{-}21)$$

$$G_I(k_p) = 4\int\frac{k^2\,dk}{(1 - k^2/4k_p^2)^{1/2}}\int |\hat{\phi}(k)|^2 N(z_0)\,dz_0 \qquad (10\text{-}6\text{-}22)$$

F_0 is the electric field and n is the unit vector in the direction of the electric field.

For a strongly degenerate 2DEG we obtain from Eqs. (10-6-20)–(10-6-22) the following expression for the mobility:

$$\mu_I = \frac{8\varepsilon_1^2 E_F^2}{q^3 \hbar n_{s0}\tilde{G}_I(k_F)} \qquad (10\text{-}6\text{-}23)$$

where

$$\tilde{G}_I(k_F) = \left(\frac{4\varepsilon_1 k_F}{q}\right)^2\int_0^1\frac{\xi^2\,d\xi}{(1 - \xi^2)^{1/2}}\int |\hat{\phi}(\xi)|^2 N(z_0)\,dz_0 \qquad (10\text{-}6\text{-}24)$$

$N(z_0)$ is the volume concentration of the impurity charges, function $\hat{\phi}(\xi)$ is obtained from function $\hat{\phi}(q)$ by substituting q by $2k_F\xi$, and letting

$$\xi = q/2k_F, \qquad \tilde{\alpha} = \alpha/2k_F, \qquad \tilde{S} = S/2k_F, \qquad \tilde{L}_2 = 2k_F L_2 \quad (10\text{-}6\text{-}25)$$

Using Eqs. (10-6-7), (10-6-8), (10-6-23), and (10-6-24) one can calculate the mobility for different locations of the scattering centers relative to the 2DEG.

A. Scattering by Interface Charge Centers. If N_I is the surface concentration of the charged scattering centers located at the heterointerface then $N(z_0) = N_I\delta(z_0)$. If we assume that the 2DEG wavefunction is localized at the heterointerface ($\alpha \to \infty$) (as was done in Ref. 39), we find

$$\tilde{G}_I(k_F) = N_I\int_0^1\left[\frac{\xi[1 - \gamma\exp(-2\xi\tilde{L}_2)]^2}{\xi + \tilde{S} - \tilde{S}\gamma\exp(-2\xi\tilde{L}_2)}\right]^2 d\xi \qquad (10\text{-}6\text{-}26)$$

If $L_2 \gg 1$ (this condition is required in order to neglect the effects related to image charges) then we obtain from Eq. (10-6-26)

$$\tilde{G}_I(k_F) = N_I\{\pi/2 - \tilde{S}[1 + (\tilde{S}^2 - 2)Q(\tilde{S})]/(\tilde{S}^2 - 1)\} \qquad (10\text{-}6\text{-}27)$$

where

$$Q(\tilde{S}) = \ln\{[1 + (1 - \tilde{S}^2)^{1/2}]/\tilde{S}\}/(1 - \tilde{S}^2)^{1/2} \qquad \text{for } S \leqslant 1$$

$$Q(\tilde{S}) = \tan^{-1}(\tilde{S}^2 - 1)^{1/2}/(\tilde{S}^2 - 1)^{1/2} \qquad \text{for } S \geqslant 1$$

The dependences of the mobility on the spacer layer thickness calculated using Eqs. (10-6-26)–(10-6-27) for a finite localization of the wave function and for $\alpha \to \infty$ are shown in Fig. 10-6-2.

As can be seen from the figure the approximation $\alpha \to \infty$ substantially overestimates scattering by the interface states. This result has an important implication. As the thickness of the spacer layer, d_1, increases, the measured values of the mobility tend to a constant value at large values of d_1. This result was explained in Ref. 36 by taking into account scattering by the interface states. However, as can be seen from Fig. 10-6-2, this scattering is considerably smaller than predicted by the model used in Ref. 36 (which was based on the assumption $\alpha \to \infty$).

In order to explain the constant value of the mobility at large values of d_1 we take into account scattering by compensated impurities in the spacer layer in addition to the scattering by the remote donors.

B. Scattering by the Remote Donors and Compensated Impurities in the Spacer Layer. Scattering by the remote donors is also overestimated by all calculations where the effects related to the image charges are neglected.

For this type of scattering we obtain the following expression for $G_I(k_F)$ from Eq. (10-6-26):

$$\tilde{G}_I(k_F) = \frac{(2\tilde{a})^6}{4k_F} \int \frac{\xi W_2(\xi)}{(1 - \xi^2)^{1/2} W_1(\xi)} d\xi \qquad (10\text{-}6\text{-}28)$$

where

$$W_2(\xi) = N_d[y_1 y_2 (1 - y_2)(\gamma^2 + 1/y_2^2) - 2\gamma\lambda_2]$$
$$+ n_{sl}[(1 - y_1)(1 + y_1 y_2^2 \gamma^2) - 2y_1 y_2 \gamma\lambda_1]$$

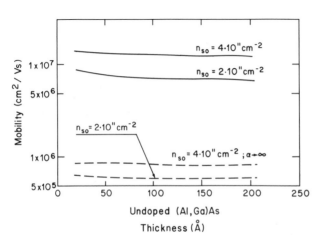

FIGURE 10-6-2. Mobility limited by interface state scattering versus spacer layer thickness [35]. Solid curves have been calculated taking into account the finite width of the wave function of the two-dimensional gas; dashed curves have been computed using the wave function localized at the heterointerface. Doping density in AlGaAs $N_d = 10^{18}$ cm^{-3}. Surface concentration of the charged interface states $N_I = 10^9$ cm^{-2}. Other parameters used in the calculation are given in Table 10-6-1.

and $\lambda_1 = 4k_F d_1 \xi$, $\lambda_2 = 4k_F d_2 \xi$, $y_1 = \exp(-\lambda_1)$, $y_2 = \exp(-\lambda_2)$. Here N_d is donor concentration in the depletion layer of the AlGaAs and n_{sl} is total (positive and negative) concentration of the compensated impurity centers in the spacer layer.

The computed dependences of the mobility limited by remote donor scattering and scattering by compensated impurities in the spacer layer are shown in Fig. 10-6-3a. These curves are calculated for a strongly degenerate case ($T = 10$ K) when

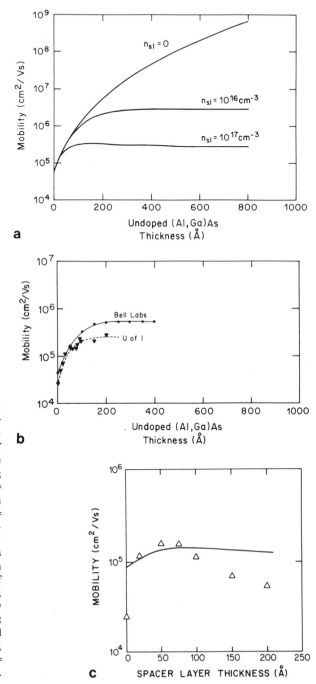

FIGURE 10-6-3. Impurity scattering mobility versus spacer layer thickness [35]. (a) Calculated curves for different concentrations of compensated impurities in the undoped spacer layer. Doping density in AlGaAs $N_d = 10^{18}$ cm^{-3}. Surface concentration of the charged interface states $N_I = 0$. Surface density of the two-dimensional gas $n_{s0} = 4 \times 10^{11}$ cm^{-2}. Other parameters used in the calculation are given in Table 10-6-1. (b) Examples of measured curves: Bell Labs [52]; U of I [36]. (c) Example of the low field mobility decreasing with d_1 at large values of d_1. Experimental points [30], $N_d = 7 \times 10^{17}$ cm^{-3}; solid curve, calculated for $N_d = 7 \times 10^{17}$ cm^{-3}, $n_{s1} = 3 \times 10^{17}$ cm^{-3}.

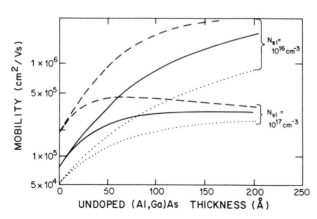

FIGURE 10-6-4. Impurity scattering mobility versus spacer layer thickness [35]. Solid curves have been calculated taking the image charges into account; dashed curves have been calculated assuming image charges as in metal; dotted curves have been computed neglecting the effects related to the image charges. Doping density in AlGaAs $N_d = 10^{18}$ cm^{-3}. Surface concentration of the charged interface states $N_I = 0$. Surface density of the two-dimensional gas $n_{s0} = 4 \times 10^{11}$ cm^{-2}. Other parameters used in the calculation are given in Table 10-6-1.

the effect of the phonon scattering is quite small [38]. For $n_{sl} = 0$ the mobility rises with increasing d_1 as expected. In this case the predicted values of μ_I are too large. The values of n_{sl} between 10^{16} and 10^{17} cm^{-3} seem to correspond to typical experimental data (see Fig. 10-6-3b [36, 52]). As can be seen from Fig. 10-6-3a for these values of n_{sl} the mobility tends to a constant value at large values of d_1. Moreover, there is a small decrease of the mobility at large d_1 for $n_{sl} = 10^{17}$ cm^{-3}. It should be noted that some experimental results seem to indicate that the low-temperature mobility first rises with increasing d_i and then drops [30]. As can be seen from Fig. 10-6-3c, the observed drop is even sharper than expected from our theory.

The effect of the image charges in the doped AlGaAs layer is illustrated by Fig. 10-6-4, where mobility versus d_1 curves are compared for three different cases: no image charges, image charges as in metal, and for the semiconductor case. As can be seen from the figure, the effect of the image charges increases the mobility several times. Hence, this effect has to be taken into account in all calculations of impurity scattering in the two-dimensional gas.

The dependences of the mobility on the donor concentration in AlGaAs (for $d_1 = 0$) are shown in Fig. 10-6-5. As expected the effects of the image charges are more important at large doping densities. Another important result is that at large doping densities the mobility is quite insensitive to n_{s0}, contary to what may be expected from the calculation neglecting the image charges.

10-6-2. Phonon Scattering

At temperatures above 100 K or so polar optical scattering is a dominant photon scattering mechanism for the electrons in the two-dimensional gas. Strictly speaking, a momentum relaxation time cannot be defined for the polar optical scattering. A

TABLE 10-6-1. Parameters Used in Calculation [35]

	Symbol	Value
Dielectric constants GaAs and AlGaAs	$\varepsilon_1, \varepsilon_2$	1.14×10^{-10} F/m
Effective mass of the electrons in GaAs	m_1	$0.067 \times 9.1 \times 10^{-31}$ kg
Electron concentration in AlGaAs	n_0	10^{18} cm^{-3}

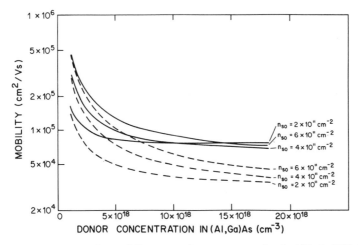

FIGURE 10-6-5. Impurity scattering mobility versus donor concentration in AlGaAs [35]. Solid curves have been calculated taking the image charges into account; dotted curves have been computed neglecting the effects related to the image charges. Surface concentration of the charged interface states $N_I = 0$. Spacer layer's thickness $d_1 = 0$. Other parameters are given in Table 10-6-1.

simple approximate expression for the effective relaxation time for this scattering mechanism was introduced in Chapter 2. However, screening and localization of electrons in the two-dimensional gas may play an important role in determining the polar optical phonon scattering of electrons in the two-dimensional gas so that the effective momentum relaxation time may be quite different.

The results of the calculation of the phonon limited mobility of electrons in the two-dimensional gas at 77 and 300 K were reported in Ref. 53. The Boltzmann equation was solved using an iterative technique similar to one used by Rode [54] (see Chapter 2). The scattering rate calculated in Ref. 53 is shown in Fig. 10-6-6. The results of the calculation are shown in Fig. 10-6-7. They indicate that the mobility is nearly independent of n_{s0} at room temperature. However, at 77 K it decreases with increasing n_{s0} from 210000 cm^2/V s at $n_{s0} \approx 0$ to approximately 150000 cm^2/V s at $n_{s0} = 6 \times 10^{11}$ cm^{-2}. The highest observed values of mobility in the two-dimensional electron gas at 77 K (see Refs. 55–58 and Fig. 10-6-8 [59]) are in agreement with the results of this calculation.

A simple empirical expression for the mobility limited by the polar optical scattering was proposed in Ref. 36:

$$\mu_{po} = 0.883(300/T)^2 + 57(77/T)^6 \ (m^2/V\,s) \qquad (10\text{-}6\text{-}29)$$

As can be seen from Fig. 10-6-7, combined acoustic (deformation potential) and piezoelectric scattering is stronger than the polar optical scattering at 77 K. In this temperature range the mobility μ_{pa} limited by these scattering mechanisms is approximately proportional to $1/T$. At low temperatures ($T < 10$–20 K) the mobility is limited by the impurity scattering, which is only weakly dependent on temperature in this temperature range. As a result we propose the following approximate expression for the low field mobility:

$$1/\mu = 1/\mu_I + 1/\mu_{po} + 1/\mu_{pa} \qquad (10\text{-}6\text{-}30)$$

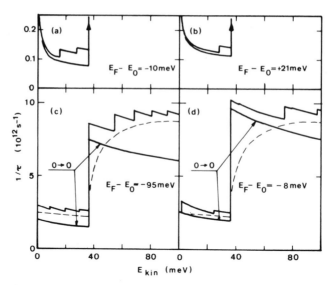

FIGURE 10-6-6. Phonon scattering rates for electrons in the lowest subband in the two-dimensional gas versus kinetic energy [53]. (a) $T = 77$ K, $n_{s0} = 4 \times 10^{10}$ cm^{-2}; (b) $T = 77$ K, $n_{s0} = 6.5 \times 10^{11}$ cm^{-2}; (c) $T = 300$ K, $n_{s0} = 4 \times 10^{10}$ cm^{-2}; (d) $= T = 300$ K, $n_{s0} = 6.2 \times 10^{11}$ cm^{-2}. Also shown, the three-dimensional polar optical scattering rate (dashed line) and the scattering rate for the intraband scattering alone (marked $0 \rightarrow 0$).

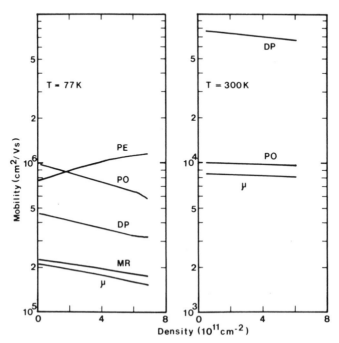

FIGURE 10-6-7. Phonon limited mobility of electrons in the two-dimensional gas at 77 and 300 K [53]. PO, polar optical scattering; DP, deformation potential scattering; PE, piezoelectric scattering; MR, mobility calculated using Matthiessen's rule; μ, total phonon limited mobility.

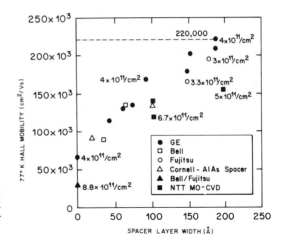

FIGURE 10-6-8. Low field monility of electrons in the two-dimensional gas at 77 K versus the thickness of the undoped spacer layer [59].

where μ_I is a temperature independent parameter for the impurity limited mobility, μ_{po} is given by Eq. (10-6-29), and

$$\mu_{pa} = 50(T/77) \; (m^2/V \, s)$$

in crude agreement with the calculation [53].

At low temperature ($T \leqslant 20$ K or so) μ_{pa} may increase very rapidly with decreasing temperature. However, in this temperature range the contribution from the impurity scattering is dominant.

The temperature dependence of the mobility, calculated using Eq. (10-6-30), is shown in Fig. 10-6-9. The experimental points marked by dark dots are from Ref. 52. As can be seen from the figure, the value of $\mu_I = 250 \; m^2/V \, s$ results in a good fit to the experimental data.

10-6-3. Contribution of the Parallel Conduction in AlGaAs

In many modulation doped samples the thickness of the doped AlGaAs layer is so large that the electrons in this layer may substantially contribute to the total

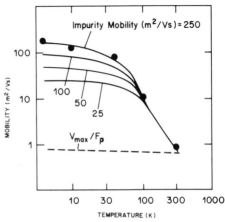

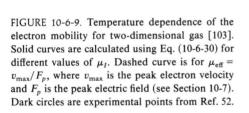

FIGURE 10-6-9. Temperature dependence of the electron mobility for two-dimensional gas [103]. Solid curves are calculated using Eq. (10-6-30) for different values of μ_I. Dashed curve is for $\mu_{eff} = v_{max}/F_p$, where v_{max} is the peak electron velocity and F_p is the peak electric field (see Section 10-7). Dark circles are experimental points from Ref. 52.

conductance at room temperature, and, hence, to the measured value of the low field mobility. This is because at room temperature the mobility of electrons in the two-dimensional gas is relatively small. At very low temperatures the electron mobility of the two-dimensional gas is so large that the AlGaAs layer does not contribute substantially to the overall conductance. The Hall effect for the modulation doped structure may be estimated using the two carrier model as follows:

$$n_H \mu_H = n_{s0} \mu + n_3 \mu_3 \tag{10-6-31}$$

$$n_H \mu_H^2 = n_{s0} \mu^2 + n_3 \mu_3^2 \tag{10-6-32}$$

Here μ_3 is the electron mobility in AlGaAs, n_3 is the surface concentration of electrons in AlGaAs, n_H and μ_H are the Hall concentration (per unit area) and the Hall mobility, n_{s0} is the surface concentration of the two-dimensional gas and μ is the mobility of the two-dimensional gas. In most samples

$$n_{s0} \mu^2 \gg n_3 \mu_3^2 \tag{10-6-33}$$

even at room temperature. In addition, at low temperatures when μ is very large

$$n_{s0} \mu \gg n_3 \mu_3 \tag{10-6-34}$$

Therefore at low temperature (let us say $T = 10$ K) $n_{s0} \approx n_H$ and $\mu \approx \mu_H$. At room temperature we have [see Eq. (10-6-32)]

$$\mu(300\ \text{K}) = \mu_H(300\ \text{K})[n_H(300\ \text{K})/n_H(10\ \text{K})]^{1/2} \tag{10-6-35}$$

The values of $\mu(300\ \text{K})$ calculated from Eq. (10-6-35) are shown in Fig. 10-6-10 together with measured points [31]. As can be seen from the figure, there is good agreement between the measured data and Eq. (10-6-35). The measured values of the electron mobility for the two-dimensional gas at room temperature and the values of $\mu(300\ \text{K})$ deduced from Eq. (10-6-35) (i.e., corrected for the contribution

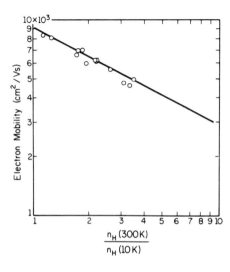

FIGURE 10-6-10. Measured Hall mobility μ_H (300 K) versus n_H (300 K)/n_H (10 K). Solid line represents Eq. (10-6-35) with μ (300 K) = 9200 cm^2/V s [31].

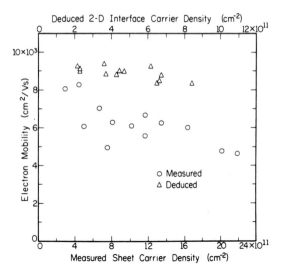

FIGURE 10-6-11. Measured Hall Mobility μ_H versus measured sheet carrier density (open circles) and corresponding deduced values of mobility μ (300 K) versus interface density of the two-dimensional gas (open triangles) [31].

from the parallel conduction in AlGaAs) are shown in Fig. 10-6-11. As can be seen from the figure, μ(300 K) is practically independent of n_{s0}, in agreement with the theoretical predictions [36, 53].

10-7. ELECTRON VELOCITY IN THE TWO-DIMENSIONAL ELECTRON GAS

With voltage swings of the order of a volt and a gate length of the order of a micrometer, the average electric field in the channel of a modulation doped FET may be in excess of 10 kV/cm. Therefore the MODFET characteristics are strongly dependent on the electron velocity in large electric fields.

The field dependence of the electron mobility for the two dimensional gas and the velocity versus electric field dependence were studied both experimentally and theoretically (see, for example, Refs. 60–66).

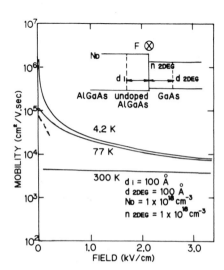

FIGURE 10-7-1. Electron mobility versus electric field for the two-dimensional electron gas. Solid lines, Monte Carlo simulation [66]; dashed line, measured data [63]. A qualitative sketch of the device structure is also shown.

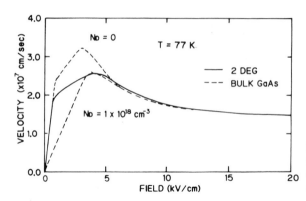

FIGURE 10-7-2. Velocity versus electric field curves for electrons in the two-dimensional gas and for the highly doped and undoped GaAs [66].

A crude estimate of the mobility dependence on the electric field may be obtained by studying nonlinear current–voltage characteristics of modulation doped devices at low temperatures [63]. The experimental data [63] and the Monte Carlo simulations of the two-dimensional electron gas [66] show a rapid decrease in the electron mobility with the electric field (see Fig. 10-7-1). This decrease is related to the polar optical phonon scattering of the hot electrons with energies high enough for the emission of the polar optical phonons.

The velocity versus electric field curve for the electrons in the two-dimensional electron gas, computed by the Monte Carlo technique, is compared with similar curves for undoped and highly doped GaAs in Fig. 10-7-2 [66]. Compared to the highly doped GaAs sample it offers the advantage of a faster velocity rise with the electric field, which is quite important for the device performance for two reasons: First of all it makes it possible to reach high velocities even in relatively long devices (such as 2-μm gate FETs) at relatively small voltage swings (of less than a volt) which may be typical of the MODFET logic. Secondly, it may result in a higher velocity enhancement in short gate devices due to the ballistic or overshoot effects (see Chapter 2). The latter effect is clearly illustrated by Fig. 10-7-3, where the velocity transients for the two-dimensional electron gas and highly doped GaAs are compared.

Hall mobility measurements and Shubnikov–de-Haas effect measurements were used to determine the velocity versus electric field curve [65]. Quantitative results

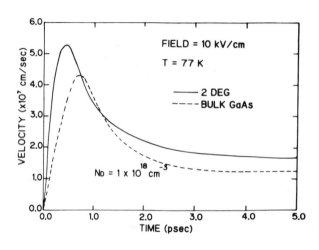

FIGURE 10-7-3. Velocity transients for a uniform electric field at 77 K for the two-dimensional electron gas and for the highly doped GaAs sample [66].

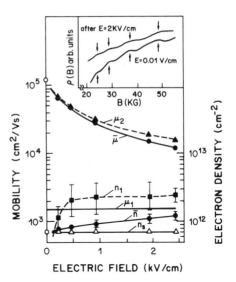

FIGURE 10-7-4. Electric field dependences of transport parameters in GaAs/n-Al$_{0.2}$Ga$_{0.8}$As heterostructure [65]. μ_2, the electron mobility in the two-dimensional electron gas; $\bar{\mu}$, average electron mobility; μ_1, electron mobility in AlGaAs; n_s, concentration of electrons in the two-dimensional gas; $\bar{n}$, average surface electron concentration; n_1, surface concentration of electrons in AlGaAs. The insert shows two traces of SdH oscillations observed at the initial condition of the sample and just after the pulsed Hall measurement at 2 kV/cm.

were obtained by analyzing the contribution of the electrons in the parallel AlGaAs layer to the total conductivity of the structure [31, 65].

The results of this study are presented in Figs. 10-7-4 and 10-7-5. Figure 10-7-4 shows an average Hall mobility and the mobility for the two-dimensional gas as well as the average electron concentration and the concentration of electrons in the two-dimensional gas and in the AlGaAs layer. The decrease in the mobility with the electric field is more gradual than predicted by the Monte Carlo simulation (see Fig. 10-7-1). However, this may be related to the fact that the experimental value of the mobility in the zero electric field is much less and, hence, and additional polar optical phonon scattering in high electric field leads to a relatively smaller change of the overall scattering rate.

The measured electron velocity in the two-dimensional gas is compared with the velocity versus electric field curve for an n-type GaAs sample in Fig. 10-7-5 [65]. The maximum velocity in the two-dimensional gas ($\approx 3.6 \times 10^5$ m/s) is slightly higher than what is expected for a GaAs sample at low temperatures [65]. Nevertheless, the velocty versus electric field curves for undoped GaAs (see Chapter 2) seem to be a useful guide for the velocity of electrons in the two-dimensional gas.

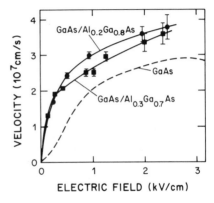

FIGURE 10-7-5. Velocity versus electric field dependences for electrons in the two-dimensional gas at 4.2 K [65]. Velocity versus electric field curve for n-type GaAs at 4.2 K determined by the microwave heating technique is shown for comparison.

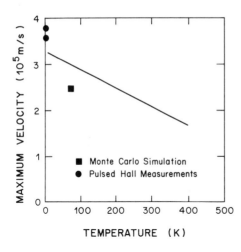

FIGURE 10-7-6. Temperature dependence of the maximum electron velocity in GaAs. Monte Carlo results from [66], Hall measurement data from [65].

The peak electron velocity for undoped GaAs may be estimated using an empirical equation by Blakemore [67]:

$$v_p = (3.3 - 0.04 \times T) \times 10^5 \ (\text{m/s}) \qquad (10\text{-}7\text{-}1)$$

with the peak electric field corresponding to the peak velocity given by

$$F_p = 4.7 - T/215 \ (\text{kV/cm}) \qquad (10\text{-}7\text{-}2)$$

v_{max} vs. T is shown in Fig. 10-7-6, where the maximum values of the electron velocities determined from Figs. 10-7-2 and 10-7-5 are shown for comparison. The ratio $\mu_{\text{eff}} = v_{\text{max}}/F_{\text{max}}$ is compared with the low field electron mobility in the two-dimensional gas in Fig. 10-6-9.

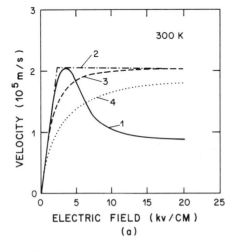
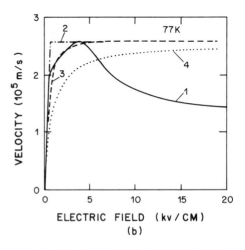

FIGURE 10-7-7. Different approximations of velocity versus electric field curve [103]. a, 300 K; b, −77 K; 1, −$v(F)$ curve for undoped GaAs at 300 K and $v(F)$ curve for the two-dimensional electron gas at 77 K [66]; 2, piecewise linear approximation; 3, $v(F) = \mu F/[1 + (\mu F/v_s)^2]^{1/2}$ [68]; 4, $v(F) = \mu F/(1 + \mu F/v_s)$ [69].

Very little is known at the present time about the magnitude of the negative differential mobility in the two-dimensional gas (see Fig. 10-7-2) and about the effects it may have on device performance. One may only speculate about the possibility of related instabilities in the two-dimensional gas.

A simple piecewise linear model with the saturation velocity region is frequently used in device modeling to account for velocity saturation in high electric fields. This approximation as well as two other approximations used in different MODFET models,

$$v(F) = \mu F / [1 + (\mu F / v_s)^2]^{1/2} \qquad (10\text{-}7\text{-}3)$$

and

$$v(F) = \mu F / (1 + \mu F / v_s) \qquad (10\text{-}7\text{-}4)$$

are compared with the velocity versus electric field curve for undoped GaAs at 300 K and with the velocity versus electric field curve for the two-dimensional electron gas in Fig. 10-7-7.

10-8. CHARGE CONTROL MODEL

In this section we present a simple charge control model which describes the change in the concentration of the 2D gas with the gate voltage [29].

Two possible device structures are shown in Fig. 10-8-1. The structure shown in Fig. 10-8-1a has yielded high values of the transconductance when used in MODFETs and high values of the 2D electron gas mobility. We refer to this device as a "normal" structure, as opposed to the inverted structure shown in Fig. 10-8-1b. Here we consider a charge control model for a normal structure. An inverted

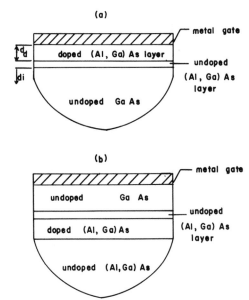

FIGURE 10-8-1. Normal (a) and inverted (b) (Al, Ga)As–GaAs structures with a gate contact (not to scale).

structure which may also be quite promising for the device applications is dealt
with in Section 10-10.

Placing a Schottky gate on the (Al, Ga)As layer results in a certain amount of
depletion beneath the gate. If the (Al, Ga)As layer is thin enough, or a sufficiently
large negative gate voltage is applied, the gate depletion and junction depletion
regions will overlap (see Fig. 10-8-2).

In this case the surface carrier concentration n_s of the two-dimensional gas is
given by [24]

$$n_s = \frac{\varepsilon}{qd} [V_g - (\phi_b - V_{P2} + V(d_i^+) - \Delta E_c)] \qquad (10\text{-}8\text{-}1)$$

where ϕ_b is the Schottky barrier height, V_g is the gate voltage, and

$$V_{P2} = q N_d d_d^2 / 2\varepsilon$$

is the pinch-off voltage of the doped AlGaAs layer. Here d_d is the thickness of the
doped (Al, Ga)As beneath the gate and $d = d_d + d_i$.

Equation (10-8-1) simply gives the charge induced in a parallel plate capacitor
formed by the gate and by the two-dimensional electron gas separated by the
(Al, Ga)As layer.

Equation (10-8-1) should be solved together with Eq. (10-4-15) (where n_{s0}
should now be replaced by n_s) in order to yield the electron concentration n_s in
the potential well. This solution (valid for n_s greater than zero and less than the
equilibrium n_{s0}) is given in Section 10-4.

If the doped (Al, Ga)As layer is too thick, or a sufficiently large positive gate
voltage is applied, a parallel conduction path in the (Al, Ga)As is created. For a
given d_d the maximum gate threshold voltage which affects the two-dimensional
electron gas can be obtained by equating (10-4-12) (which is valid in the equilibrium)

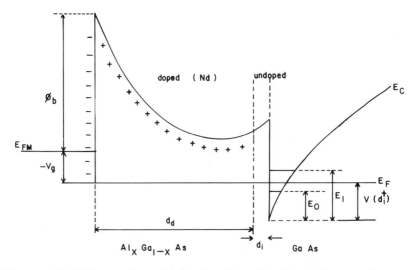

FIGURE 10-8-2. Band diagram of a modulation doped layer. A sufficiently large negative gate voltage
is applied so that the doped (Al, Ga)As layer is totally depleted and the two-dimensional electron gas
is partially depleted. E_{FM} is the Fermi level in the gate metal.

and Eq. (10-8-1) and by using the equilibrium value of the Fermi level at the interface. This condition implies that the (Al, Ga)As underneath the gate is fully depleted but the gate and junction depletion depths do not interpenetrate.

The relationship between the Fermi potential $V(d_i)$ and n_s may be found from Eq. (10-4-15), which is a quadratic equation with respect to $\exp[qV(d_i)/k_BT]$.

The solution of Eq. (10-4-15) is given by

$$\exp\left[\frac{V(d_i^+)}{V_T}\right] = -\frac{a_0 + a_1}{2} + \left[\frac{(a_0 + a_1)^2}{2} - a_0 a_1[1 - \exp(n_s/DV_T)]\right]^{1/2} \quad (10\text{-}8\text{-}2)$$

where $a_0 = \exp(E_0/V_T)$, $a_1 = \exp(E_1/V_T)$, and $V_T = k_BT/q$ is the thermal voltage.

The Fermi potential $V(d_i)$ versus temperature given by Eq. (10-8-2) is shown in Fig. 10-8-3 for 300, 77, and 4 K. The calculated density n_s for 300 K is about half of the density predicted by a less accurate three-dimensional gas model which neglects the quantization in the potential well and uses the Joyce–Dixon approximation. For the values of n_s between 5×10^{11} cm^{-2} and 1.5×10^{12} cm^{-2} the dependencies shown in Fig. 10-8-3 can be approximated as (shown in dashed lines)

$$V(d_i^+) = \Delta E_{F0}(T) + an_s \quad (10\text{-}8\text{-}3)$$

where $a \approx 0.125 \times 10^{-16}$ V m^2, and $\Delta E_{F0} \approx 0$ at 300 K and 0.025 V at 77 K and below. Substituting Eq. (10-8-3) into Eq. (10-8-1), we find the equation of the charge control model

$$n_s = \frac{\varepsilon}{q(d + \Delta d)}(V_g - V_{off}) \quad (10\text{-}8\text{-}4)$$

where

$$V_{off} = V_{off'} + \Delta E_{F0} \quad (10\text{-}8\text{-}5)$$

$$V_{off'} = \phi_b - \Delta E_c - V_{P2} \quad (10\text{-}8\text{-}6)$$

and

$$\Delta d = \varepsilon a/q \approx 80 \text{ Å} \quad (10\text{-}8\text{-}7)$$

Thus Δd gives an important correction especially for a normally-off device where d may be of the order of 300 Å or less.

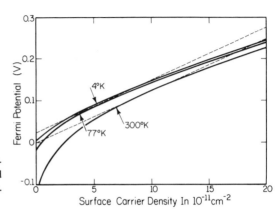

FIGURE 10-8-3. Fermi potential versus surface carrier density at 300, 77, and 4 K (solid lines). Linear approximations [see Eq. (10-8-5)] are shown as dashed lines [29].

In Fig. 10-8-4 the charge control model given by Eq. (10-8-4) is compared with the exact solution of Eqs. (10-4-15) and (10-8-1). A simplified solution

$$n_s = \frac{\varepsilon}{qd}(V_g - V_{\text{off}'}) \tag{10-8-8}$$

is also shown in Fig. 10-8-4 for comparison [24]. We can see from this figure that if we introduce a threshold voltage V_{off} [see Eq. (10-8-4)] similar to the threshold voltage of a MOSFET, a substantial "subthreshold" charge exists leading to the subthreshold current. However, our charge control model given by Eq. (10-8-4) is quite adequate for the analytical device modeling. The estimate for Δd given by Eq. (10-8-7) is in good agreement with the experimental data reported in Ref. 70. The "apparent mobility" for the two-dimensional gas calculated in Ref. 61 from the drain conductance versus gate voltage curve is $55,000 \text{ cm}^2/\text{V s}$ at 77 K. The measured value of the Hall mobility was $61,000 \text{ cm}^2/\text{V s}$. We estimate that the value of Δd required to explain this difference is about 100 Å, in good agreement with Eq. (10-8-6). This value also has the same order as the thickness of the two-dimensional gas (see, for example, Ref. 50) which can be expected on the basis of the device physics.

In some modulation doped structures an optional thin n^+-GaAs layer is grown between the gate and the doped AlGaAs layer. In this case Eq. (10-8-6) should be replaced by [68]

$$V_{\text{off}'} = \phi_b - \Delta E_x - V_{P20} \tag{10-8-6a}$$

where

$$V_{P20} = qN_t d_t^2/2\varepsilon_t + qN_d d_d^2/2\varepsilon + qN_d d_t d_d/\varepsilon_t$$

$$\Delta E_x = \begin{cases} \Delta E_c & \text{if } d_t = 0 \\ 0 & \text{if } d_t > 0 \end{cases}$$

and N_t, ε_t, and d_t are the doping concentration, dielectric permittivity, and thickness of the optional top n^+-GaAs layer.

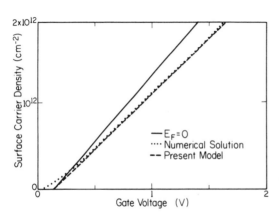

FIGURE 10-8-4. Surface carrier density versus voltage difference between gate and channel ($V_{\text{off}}' = 0.16$ V and $d = 400$ Å) [29]. Dotted line, numerical solution; dashed line, charge control model; solid line, simplified charge control model ($\Delta d = 0$).

The capacitance C of the modulation doped structure shown in Fig. 10-8-1a may be found as [91]

$$C = \frac{\partial(qn_s)}{\partial V_g} = 1 \bigg/ \left(\frac{\partial V_g}{\partial(qn_s)}\right) = 1 \bigg/ \left(\frac{d}{\varepsilon} + \frac{1}{q}\frac{\partial V(d_i^+)}{\partial n_s}\right) \qquad (10\text{-}8\text{-}9)$$

where $\partial V(d_i^+)/\partial n_s$ can be determined from Eq. (10-8-2):

$$\frac{\partial V(d_i)}{\partial n_s} = \frac{2}{3n_s}\left\{\frac{3n_s/D + 2(E_0 + E_1)}{4C/(C+4)}[1 + (1 + C/A^2)^{-1/2}] - B\right\} \qquad (10\text{-}8\text{-}10)$$

where

$$A = \exp(E_0/V_T) + \exp(E_1/V_T)$$

$$B = E_0 \exp(E_0/V_T) + E_1 \exp(E_1/V_T)$$

$$C = 4\{\exp[(n_s/D + E_0 + E_1)/V_T] - 1\}$$

The C vs. $V_g - V_{off}$ curve may now be found from the solution of Eqs. (10-8-9), (10-8-10), and (10-8-2). The resulting C-V_g plots are shown in Fig. 10-8-5. As can be seen from the figure, at gate voltages larger than the threshold voltage given by the charge control model [see Eq. (10-8-5)] the capacitance per unit area approaches the value of

$$C_0 = \frac{\varepsilon}{d + \Delta d} \qquad (10\text{-}8\text{-}11)$$

However, near the threshold C-V_g characteristics are temperature dependent. At

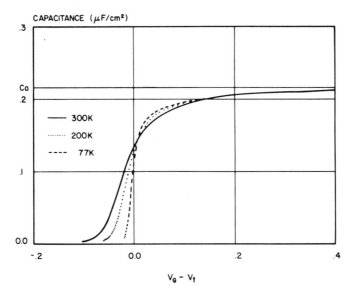

FIGURE 10-8-5. Capacitance of the modulation doped structure as a function of $V_g - V_t$ [71]. $C_0 = \varepsilon/(d + \Delta d)$.

low temperature these characteristics are closer to the step function predicted by the charge control model.

10-9. CURRENT–VOLTAGE AND CAPACITANCE–VOLTAGE CHARACTERISTICS OF MODULATION DOPED FIELD EFFECT TRANSISTORS

Current–voltage characteristics of a modulation doped field effect transistor may be found based on the charge control model, using the gradual channel approximation [72]. This implies that the surface carrier concentration in the channel is given by

$$n_s(x) = \frac{\varepsilon}{d + \Delta d} [V_g - V_{\text{off}} - V(x)] \qquad (10\text{-}9\text{-}1)$$

where x is the space coordinate along the channel and $V(x)$ is the channel potential. Equation (10-9-1) should be solved together with the equation

$$I_{ds} = q n_s v(F) W \qquad (10\text{-}9\text{-}2)$$

Equation (10-9-2) relates the drain-to-source current I_{ds} to the electron velocity $v(F)$ in the channel, which is assumed to be a function of the electric field F in the channel. Here W is the gate width. Using the gradual channel approximation we assume that the electric field in the channel is parallel to the heterointerface (it is directed from the drain to the source) and neglect the diffusion current.

We consider two models—a very simple two-piece linear approximation for the electron velocity as a function of the electric field and a more realistic three-piece linear approximation (Fig. 10-9-1); see Section 10-7 for a more detailed discussion of $v(F)$.

When the drain-to-source voltage V_{ds} is small ($V_{ds}/L < F_s$ for the two-piece linear approximation and $V_{ds}/L < F_1$ for the three-piece linear approximation [see Fig. 10-9-1]), the electron velocity is simply proportional to the electric field:

$$v = \mu F \qquad (10\text{-}9\text{-}3)$$

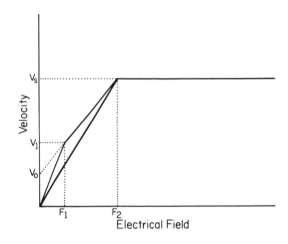

FIGURE 10-9-1. Two- and three-piece linear approximations for the velocity-field characteristic [72].

where μ is the effective low field mobility. Substituting Eq. (10-9-3) into Eq. (10-9-2), replacing F in Eq. (10-9-2) by $-dV/dx$, and integrating with respect to x from 0 to L, where L is the gate length, we obtain a conventional Shockley model equation describing the current–voltage characteristics of the MODFETs at low drain-to-source voltages V_{ds}:

$$I_{ds} = \beta(V'_g V_{ds} - V_{ds}^2/2) \tag{10-9-4}$$

where

$$V'_g = V_g - V_{off} \tag{10-9-5}$$

and

$$\beta = \frac{\varepsilon \mu W}{(d + \Delta d)L} \tag{10-9-6}$$

The drain-to-source saturation current I_{ds}^s is found assuming, as suggested in Refs. 75, 29, and 72, that the current saturation occurs when the electric field at the drain side of the gate exceeds the velocity saturation field $F_s = v_s/\mu$ and using the Shockley model in order to describe the longitudinal field distribution in the channel where the potential is below the saturation voltage.

For the two-piece linear approximation of $v(F)$ such an approach leads to the following expressions [72]:

$$I_{ds} = \beta V_0^2 \frac{(1 + 2\beta R_s V'_g + V_g'^2/V_0^2)^{1/2} - 1 - \beta R_s V'_g}{1 - \beta^2 R_s^2 V_0^2} \tag{10-9-7}$$

and

$$V_{ds}^s = V'_g + V_0 - (V_g'^2 + V_0^2)^{1/2} + I_{ds}^s(R_s + R_d) \tag{10-9-8}$$

Here V'_g is given by Eq. (10-9-5), β is given by Eq. (10-9-6),

$$V_0 = F_s L \tag{10-9-9}$$

R_s is the source series resistance, R_d is the drain series resistance, and V_{ds}^s is the drain-to-source saturation voltage.

The normalized drain saturation current for $R_s = 0$ ($i_{ds} = I_{ds}/\beta V_0^2$) versus normalized gate voltage ($v_g = V_g/V_0$) is depicted in Fig. 10-9-2 (dotted line). Also shown is the drain saturation current found from the exact solution of Eqs. (10-4-15) and (10-8-1) using the two-piece model (dashed line). The solid line in Fig. 10-9-2 shows the asymptote of Eq. (10-9-7) for large gate voltages and $R_s = 0$:

$$I_{ds}^s = \frac{\varepsilon W}{d + \Delta d}(V'_g - V_0)v_s \tag{10-9-10}$$

As can be seen in Fig. 10-9-2, the agreement is quite good at high gate voltages (above threshold), but there is a considerable subthreshold current not predicted by the charge control model. For V_g greater than V_0 and d of around 300 Å, the simplified charge control model seems to be quite adequate.

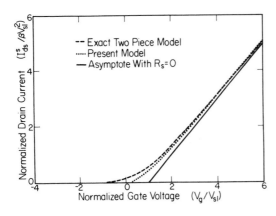

FIGURE 10-9-2. Normalized drain saturation current versus normalized gate-to-source voltage [72]. $R_s = 0$. Dotted line, Eq. (10-9-7); dashed line, exact solution using two-piece model; solid line, asymptote of Eq. (10-9-7), i.e., Eq. (10-9-10).

A more realistic (but also more complicated) three-piece linear model for the velocity versus electric field (see Fig. 10-9-1) was considered in Ref. 72. The results of the calculation using both models are compared below.

The calculated and experimental drain saturation currents at room temperature as a function of the gate voltage are shown in Figs. 10-9-3 and 10-9-4 [72]. The following values of parameters were used in the calculation: $N_d = 1 \times 10^{18}$ cm^{-3}, $\mu = 0.68$ m^2/V s, $v_s = 2 \times 10^5$ m/s, $F_1 = 1$ kV/cm, $F_s = 3.5$ kV/cm, $R_s = 7\,\Omega$ for the normally-on and $10\,\Omega$ for the normally-off FET, respectively. The low field mobility was obtained from the Van der Pauw–Hall measurements of the particular wafer on which the FET was fabricated. The gate width was $145\,\mu$m and the gate length was $1\,\mu$m. The heterostructures used to fabricate these FETs were grown by molecular beam epitaxy on Cr-doped semi-insulating substrates. The structure consisted of a 1-μm undoped GaAs buffer layer, a 60-Å undoped Al$_x$Ga$_{1-x}$As layer, a 600-Å n-type (Al, Ga)As layer doped with Si to a level of 1×10^{18} cm^{-3}, and the undoped Al$_x$Ga$_{1-x}$As layer of thickness $d_i = 60$ Å. The gate was recessed with the thickness of the doped (Al, Ga)As remaining beneath the gate about 250 and 350 Å to obtain normally-off and normally-on devices, respectively.

The solid lines in Figs. 10-9-3 and 10-9-4 correspond to the three-piece velocity model, and dotted lines to the two-piece velocity model. The experimental values are marked by the dots. The measured transconductance was 225 mS/mm for the normally-on device at 300 K.

Even though $10\,\Omega$ was measured as R_s for the normally-off FET, $12\,\Omega$ was used in the calculation to have the best fit to the experimental data. A possible

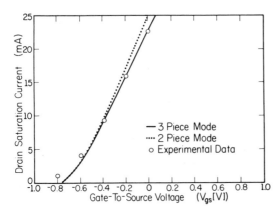

FIGURE 10-9-3. Drain saturation current versus gate-to-source voltage for normally-on FET [72].

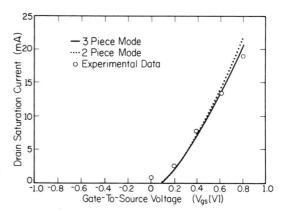

FIGURE 10-9-4. Drain saturation current versus gate-to-source voltage for normally-off FET [72].

explanation is the increase in R_s at large values of I_{ds} due to the velocity saturation in the source-to-gate region. Indeed, this 1-μm-long region may be thought of as an ungated FET. This means that R_s should increase because in the normally-off devices the voltage drop across R_s should be larger owing to the relatively larger resistance of the gate-to-source region.

As can be seen from Figs. 10-9-3 and 10-9-4 the agreement between the model and the experimental data is quite good except near the threshold. (This region is not described by the model.) The two-piece model overestimates the current predicted by the three-piece model by approximately 10%–20%.

The I_{ds}–V_{ds} characteristics for the normally-on FET are shown in Fig. 10-9-5. As can be seen from the figure, the agreement between the measured and calculated characteristics is quite adequate.

For the normally-off FET, the agreement is also good but, as mentioned above, the value of R_s had to be adjusted to give a best fit. The slightly smaller current measured at $V_{gs} = 0.8$ V may be due to the fact that the model becomes invalid at gate voltages higher than $V_{off} + (V_{po})_{2D}$, where

$$(V_{po})_{2D} = \frac{qn_{s0}(d + \Delta d)}{\varepsilon}$$

(10-9-11)

(see Section 10-12).

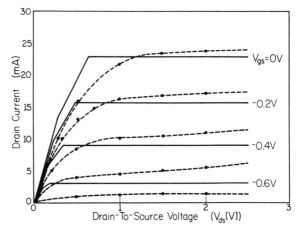

FIGURE 10-9-5. I_{ds}–V_{ds} characteristics of a normally-on FET [72]. Solid line, three-piece model; dots, measured points.

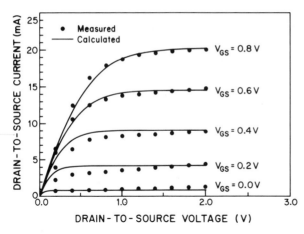

FIGURE 10-9-6. $I_{ds}-V_{ds}$ characteristics of a normally-on FET[68]. Smooth approximation for $v(F)$ curve. Dots, measured points.

As can be seen from Fig. 10-9-5, the major drawback of a piecewise linear model is that the calculated current–voltage characteristics change abruptly at the onset of the velocity saturation in the channel. This can be corrected by using a smooth approximation for the velocity versus electric field curve (as shown in Fig. 10-7-7 (see also Refs. 68, 69)). The current–voltage characteristics calculated using curve 3 of Fig. 10-7-7 are shown in Fig. 10-9-6. As can be seen from the figure, the agreement with the experimental results is much better.

Another way to achieve a smooth transistion between the linear and saturation region of the current–voltage characteristic is to use the results of the two-piece linear model for the linear region and for the saturation region and use an interpolation function, for example, the hyperbolic function as described in Section 9-5; see Eq. (9-5-27) [74]. This approach is very convenient for the applications in integrated

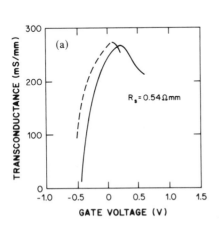

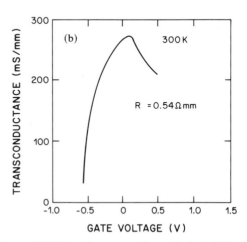

FIGURE 10-9-7. Transconductance versus gate voltage. Solid line, charge control model; dashed line, two-dimensional simulation [75]. $V_{ds} = 1.7$ V. Other parameters used in the two-dimensional simulation are given in Table 10-9-1. The value of $R_s = 0.54\ \Omega$ mm was used in the charge control model calculation. (a) The threshold voltage for the charge control model calculation is given by Eq. (10-8-5). (b) The threshold voltage for the charge control model calculation is shifted [see Eq. (10-9-12)]. In this case the results of the two-dimensional simulation and of the charge control model practically coincide. Therefore only the solid line is shown. The decrease in the transconductance at large gate voltages is due to the parallel conduction through the doped AlGaAs layer (see Section 10-13, where this decrease is discussed).

TABLE 10-9-1. Parameters Used in the Two-Dimensional
Simulation[a]

Al fraction	0.3
AlGaAs doped layer	5×10^{17} cm^{-3}, 500 Å
Undoped AlGaAs spacer layer	1×10^{13} cm^{-3}, 60 Å
Gate length	$1\,\mu$m
Source–gate length	$1\,\mu$m
Drain–gate length	$1\,\mu$m
Schottky barrier height	0.8 V
p-GaAs	1×10^{14} cm^{-3}

[a] Reference 75.

circuit simulators and also provides very good agreement with the experimental results (see Fig. 9-5-5).

It is interesting to compare the results predicted by the charge control model with the results of the two-dimensional modeling of modulation doped FETs [75] (see Figs. 10-9-7a and 10-9-7b). As can be seen from the figures, the agreement is excellent when a small shift of the threshold voltage is introduced into the charge control model equation (10-9-24) changing V_{off} to V_{off1}, where

$$V_{off1} = V_{off} - \Delta V_{off} \tag{10-9-12}$$

This shift ΔV_{off} is related to the subthreshold effects and may be crudely estimated as $\Delta V_{off} \approx 3k_B T$.

Let us now briefly discuss small signal device capacitances.

For simplicity, we calculate the small signal gate capacitance using the two-piece model at small drain-to-source voltages $V_{ds} < V_{ds}^s$. The total charge Q_T in the Shockley regime is given by

$$Q_T = W \int_0^L q n_s \, dx$$

$$= W \int_{V_s}^{V_d} q n_s (dV/dx)^{-1} dV$$

$$= (2/3) C_0^1 \frac{V_{gs}^3 - V_{gd}^3}{V_{gs}^2 - V_{gd}^2} \tag{10-9-13}$$

where

$$C_0^1 = \varepsilon WL/(d + \Delta d)$$

Then

$$C_{gs} = \frac{\partial Q_T}{\partial V_{gs}}$$

$$= \frac{2}{3} C_0^1 \frac{V_{gs}(V_{gs} + 2V_{gd})}{(V_{gs} + V_{gs})^2}$$

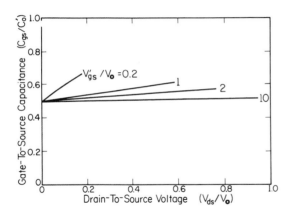

FIGURE 10-9-8. Normalized gate-to-source capacitance versus normalized drain-to-source voltage. Numbers near the curves correspond to the different values of the normalized gate-to-source voltage [72]. (All voltages are normalized with respect to $V^1_{gs} = V_{gs} - V_{off}$.)

and

$$C_{gd} = \frac{\partial Q_T}{\partial V_{gd}}$$

$$= \frac{2}{3} C^1_0 \frac{V_{gd}(V_{gd} + 2V_{gd})}{(V_{gs} + V_{gd})^2}$$

In Figs. 10-9-8 and 10-9-9 normalized capacitances C_{gs}/C_0 and C_{gd}/C_0 are plotted against normalized drain-to-source voltage V_{ds}/V_0 using normalized gate voltage V^1_{gs}/V_0 as parameter. The calculation is done up to the saturation point. As can be seen from Figs. 10-9-8 and 10-9-9, in the linear region each capacitance value is nearly one half of the total gate capacitance. We expect the capacitances to stay nearly constant in the saturation regime with $C_{gs} \approx C_0$ and $C_{gd} \ll C_{gs}$.

A more accurate capacitance description may be obtained using an empirical approach similar to that used in MOSFET modeling; see Eq. (9-5-32) and Fig. 9-5-3 [74]. This approach based on the gate capacitance calculations of Section 10-8 (see Fig. 10-8-5) takes into account an important subthreshold region and allows one to follow the transition between the linear and saturation regimes.

In Fig. 10-9-10 the capacitance–voltage characteristic calculated using Eq. (9-5-32) is compared with the results of the two-dimensional simulation [75]. As can be seen from the figure, the agreement is good, indicating that our approach adequately describes the device physics. An additional rise in the device capacitance

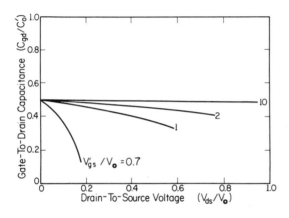

FIGURE 10-9-9. Normalized gate-to-drain capacitance versus normalized drain-to-source voltage [72]. Numbers near the curves correspond to the different values of the normalized gate-to-source voltage. (All voltages are normalized with respect to V_0.)

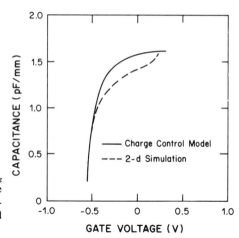

FIGURE 10-9-10. Small-signal gate capacitance C_g versus gate voltage. Solid line, analytical model [see Eq. (9-5-32)]; dashed line, two-dimensional simulation [75]. Parameters used in the simulation and analytical calculation are given in Table 10-9-1.

at high gate voltages predicted by the two-dimensional simulation is due to the contribution of electrons induced into the doped AlGaAs layer, which is neglected in our simplified model (see Section 10-13).

10-10. CHARGE CONTROL MODEL OF AN "INVERTED" GaAs–AlGaAs MODULATION DOPED STRUCTURE

As we mentioned in Section 10-8, most of the recent work on modulation doped field effect transistors (MODFETs) has concentrated on metal–AlGaAs–GaAs structures. However, the mobility enhancement in the 2D electron gas was also found in metal–GaAs–AlGaAs layers ("inverted" structures) [76]. The "inverted" MODFET was fabricated [77] and the first-order theory for such a device was also reported [78].

There are several potential advantages of the inverted structures. Among them is a stable surface of GaAs compared to the AlGaAs surface. This may lead to higher yield and reproducibility in device fabrication, which is especially important for MODFET IC development. Another possible advantage is related to the fact that the transconductance of the inverted MODFET is nearly independent of the AlGaAs doping, for reasons explained in this section. As has been pointed out in Ref. 25, for a given threshold voltage the transconductance in "normal" MODFETs decreases with reduced doping. This is because the thickness of the doped AlGaAs layer has to be increased to maintain the same threshold voltage. As a consequence the channel is farther away from the gate, which results in a drop of the transconductance. For relatively low levels of doping in AlGaAs (below approximately 5×10^{17} cm^{-3}), the inverted device may have better performance.

The inverted layers must also be a part of any multilayered structures where several 2D electron gas layers are involved. Hence, understanding the inverted structure is the first necessary step toward the modeling of multilayered modulation doped devices. Finally it is easier to fabricate Schottky barriers and ohmic contacts on GaAs than on AlGaAs.

In this section we present a charge control model for the inverted modulation doped structure [79] similar to the charge control model of a "normal" modulation doped structure given in Sections 10-8 and 10-9. In this case the formalism describing the 2D gas developed in the quantum well cannot be used, because the potential in the GaAs layer should be known. Therefore we use the three-dimensional degenerate statistics. As can be shown based on the charge control model for the "normal" structure, this approximation may lead to some error in the calculated value for the equilibrium surface concentration of the 2D gas n_{s0}, but should give a very close estimate for the surface concentration n_s controlled by the gate voltage.

In order to solve the equations of the charge control model we use the first-order Joyce–Dixon approximation for the Fermi integral

$$q\phi/k_BT = \ln(n/N_c) + n/(N_c\sqrt{8}) \qquad (10\text{-}10\text{-}1)$$

where ϕ is the potential with respect to the Fermi potential, q is the electronic charge, k_B is the Boltzmann constant, T is the lattice temperature in K, n is the volume electron density, and N_c is the effective density of states in the conduction band. Equation (10-10-1) is accurate within 2% error for $q\phi/k_BT < 7$ [26].

Our further discussion depends on whether there is a conduction path in the AlGaAs layer shunting the 2D gas layer. Two limiting cases are illustrated in Fig. 10-10-1. The structure shown in Fig. 10-10-1a has a doped AlGaAs layer thick enough so that the bottom part of this layer is undepleted even at large gate voltages. The structure shown in Fig. 10-10-1b has a relatively thin doped AlGaAs layer which is always depleted. In both cases an undoped AlGaAs layer of thicknerss d_i may be included to enhance the low field mobility.

The optimum structure for the device applications should correspond to the "borderline" second case, the doped AlGaAs layer thick enough to provide enough

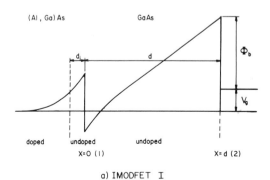

a) IMODFET I

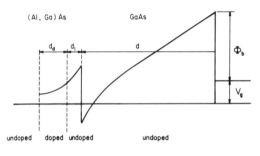

b) IMODFET II

FIGURE 10-10-1. Energy band diagrams for two kinds of "inverted" MODFETs (IMOD-FETs). IMODFET-I (a) has a thick doped AlGaAs layer and IMODFET-II (b) has a relatively thin doped AlGaAs layer [79].

carriers to the 2D electron gas but thin enough to eliminate the parallel conduction path with a low electron mobility of the doped AlGaAs layer.

The GaAs layer should be undoped or low doped in order to take advantage of the high electron mobility and velocity in the 2D electron gas.

The Poisson equation for the undoped GaAs layer is

$$\frac{d^2\phi}{dx^2} = \frac{qn}{\varepsilon} \tag{10-10-2}$$

where x is the coordinate and ε is the dielectric permittivity of the GaAs layer. Integrating Eq. (10-10-2) with respect to x and using Eq. (10-10-1), we can find the electric field F as a function of n:

$$F = \{F_0^2[(n/N_c) + (n/N_c)^2/(2\sqrt{8})] + C^2\}^{1/2} \tag{10-10-3}$$

where $F_0 = (2N_c k_B T/\varepsilon)^{1/2}$ and C is an integration constant. At $x = d$ (at the metal–semiconductor interface):

$$F = F_2 \tag{10-10-4}$$

$$n = n_2 \approx 0 \tag{10-10-5}$$

$$\phi = \phi_2 \tag{10-10-6}$$

At $x = 0$ (at the heterointerface)

$$F = F_1 \tag{10-10-7}$$

$$n = n_1 \tag{10-10-8}$$

$$\phi = \phi_1 \tag{10-10-9}$$

From Eqs. (10-10-3), (10-10-4), and (10-10-5)

$$F = \{F_0^2[(n/N_c) + (n/N_c)^2/(2\sqrt{8})] + F_2^2\}^{1/2} \tag{10-10-3}$$

Then

$$d = -\int_{\phi_1}^{\phi_2} \frac{d\phi}{F}$$

$$= L_c \int_{n_2/N_c}^{n_1/N_c} \frac{(1/z + 1/\sqrt{8})\,dz}{(z + z^2/2\sqrt{8} + F_2^2/F_0^2)^{1/2}} \tag{10-10-11}$$

where $L_c = (\varepsilon k_B T/2q^2 N_c)^{1/2}$ is the effective Debye length. In order to evaluate the integral in Eq. (10-10-11) we divide the integration region into two parts: small $z < z_0 \ll 1$ (in this region the integral is evaluated analytically) and $z > z_0$, where we calculate the integral numerically:

$$d = L_c \left\{ \int_{n_2/N_c}^{n_3/N_3} \frac{(1/z + 1/\sqrt{8})\,dz}{[z + (1/2\sqrt{8})z^2 + F_2^2/F_0^2]^{1/2}} \right.$$

$$\left. + \int_{n_3/N_c}^{n_1/N_c} \frac{(1/z + 1/\sqrt{8})\,dz}{[z + (1/2\sqrt{8})z^2 + F_2^2/F_0^2]^{1/2}} \right\} \tag{10-10-12}$$

We choose $z_0 = n_3/N_c = 0.0182$, which corresponds to $q\phi_3/k_BT = -4$, so that $z + z^2/2\sqrt{8}$ can be neglected with respect to F_2^2/F_0^2 for the first integral in Eq. (10-10-12). Here ϕ_3 is the potential at the point where $n = n_3 = 0.0182N_c$.

This approximation corresponds to a constant electric field in the region where $n < n_3$, i.e., in the region where the potential (counted from the Fermi level) is less than $-4k_BT$. Of course this approximation may be used only for gate voltages $V_g < \phi_b - 4k_BT$, where ϕ_b is the Schottky barrier height.

Then we have

$$\phi_2 = -F_2d + \phi_3 + F_2L_c \int_{0.0182}^{n_1/N_c} \frac{(1/z + 1/2\sqrt{8})\,dz}{[z + (1/2\sqrt{8})z^2 + F_2^2/F_0^2]^{1/2}} \quad (10\text{-}10\text{-}13)$$

Furthermore,

$$\phi_2 = V_g - \phi_b \quad (10\text{-}10\text{-}14)$$

(see Fig. 10-10-1). Now the total sheet carrier density n_s in GaAs layer can be found from Gauss's Law:

$$n_s = \varepsilon(F_1 - F_2)/q \quad (10\text{-}10\text{-}15)$$

Combining Eqs. (10-10-13), (10-10-14), and (10-10-15), we find the gate voltage dependence of n_s if n_1/N_c and F_1 are known.

The relationship between F_1 and n_1 is different for the structures shown in Figs. 10-10-1a and 10-10-1b. For the first structure (IMODFET-I; see Fig. 10-10-1a) this relationship is the same as for the normal structure where AlGaAs is grown on top of the GaAs layer (see Section 11-4):

$$F_1 = \frac{q}{\varepsilon}\left\{\left[N_d^2d_i^2 + (\Delta E_{c,\text{eff}} - \phi_1)\frac{2\varepsilon N_d}{q}\right]^{1/2} - N_dd_i\right\} \quad (10\text{-}10\text{-}16)$$

Here d_i is the thickness of the undoped AlGaAs layer, $\Delta E_{c,\text{eff}}$ is the effective conduction band discontinuity, and

$$\phi_1 = \frac{k_BT}{q}\left[\ln\left(\frac{n_1}{N_c}\right) + \frac{1}{\sqrt{8}}\frac{n_1}{N_c}\right] \quad (10\text{-}10\text{-}17)$$

as defined by Eq. (10-10-1).

For the second structure (IMODFET-II; see Fig. 10-10-1b) the electric field F_1 is constant, because the potential of the bulk AlGaAs is floating and all donors in the doped AlGaAs layer are ionized.

Figure 10-10-2 shows n_s vs. V_g dependences for MODFET-I with $d_i = 20$ Å and with the GaAs thickness d as a parameter. Figure 10-10-3 shows the same dependence as Fig. 10-10-2 for $d_i = 60$ Å. The doping density of the AlGaAs layer is 10^{18} cm^{-3}. Similar dependences for IMODFET-II are depicted in Figs. 10-10-4 and 10-10-5. The total ionized sheet doping density $N_s = N_dd_d$ is assumed to be $N_s = 10^{12}$ cm^{-2} for Fig. 10-10-4 and $N_s = 1.5 \times 10^{12}$ cm^{-2} for Fig. 10-10-5, respectively. In the same figures, a simple capacitor-type approximation

$$n_s = N_s + \frac{\varepsilon(V_g - \phi_b)}{qd} \quad (10\text{-}10\text{-}18)$$

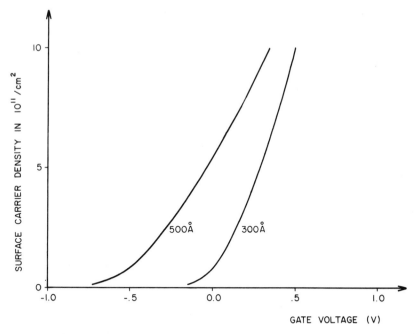

FIGURE 10-10-2. Interface carrier density n_s versus gate voltage for IMODFET-I with $d_i = 20$ Å for different thicknesses of the undoped GaAs layer [79]. Composition $x = 0.3$ and $N_d = 10^{18}$ cm^{-3} are used in the calculation.

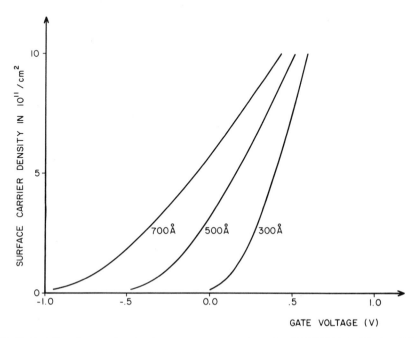

FIGURE 10-10-3. Interface carrier density n_s versus gate voltage for IMODFET-I with $d_i = 60$ Å for different thicknesses of the undoped GaAs layer [79]. Composition $x = 0.3$ and $N_d = 10^{18}$ cm^{-3} are used in the calculation.

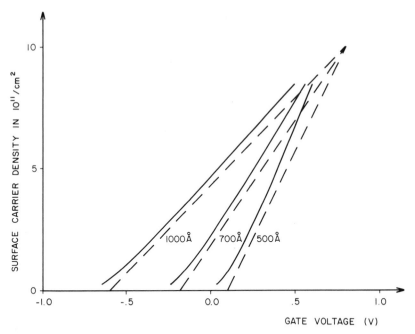

FIGURE 10-10-4. Interface carrier density n_s versus gate voltage for IMODFET-II with the thickness of the undoped layer as a parameter [79]. $N_s = 10^{12}\,\mathrm{cm}^{-2}$.

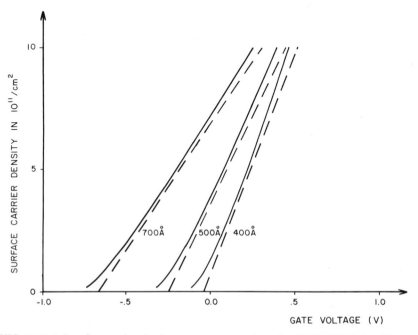

FIGURE 10-10-5. Interface carrier density n_s versus gate voltage for IMODFET-II for different thicknesses of the undoped GaAs layer [79]. $N_s = 1.5 \times 10^{12}\,\mathrm{cm}^{-2}$.

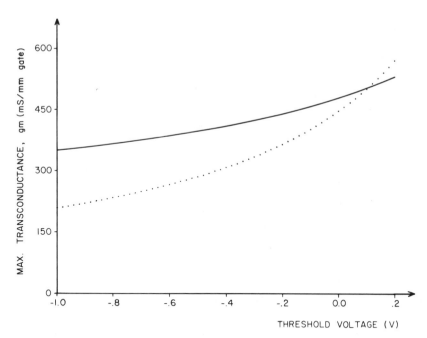

FIGURE 10-10-6. Maximum transconductances versus threshold voltage for "normal" structure (solid line) and for IMODFET-II (dotted line) [79].

is also shown. As can be seen from these figures, Eq. (10-10-18) is quite a good approximation for IMODFET-II. But for IMODFET-I, the effective distance between the channel and the gate is longer than d owing to the dependence of the Fermi level on the sheet electron concentration in the 2D gas related to the band-bending in the doped AlGaAs layer.

The maximum transconductances for the "normal structure" and IMODFET-II are compared in Fig. 10-10-6 using the $I-V$ model developed in Section 10-9. They are comparable for normally-off devices. The "normal structure" has somewhat higher transconductance for normally-on devices for the parameters used in the calculation, $N_d = 10^{18} \, \text{cm}^{-3}$, $x = 0.3$, $d_i = 20 \, \text{Å}$, $\mu = 0.7 \, \text{m}^2/\text{V s}$, $v_s = 2 \times 10^5 \, \text{m/s}$, and the gate length $L = 1 \, \mu\text{m}$.

As can be seen from these calculations the electron concentration in the 2D gas for the inverted structure is comparable to the values of n_s in the normal structure for comparable distances from the 2D gas to the gate and comparable threshold voltages. Thus the "inverted" modulation doped devices may exhibit as good a performance as the "normal" modulation doped devices. They may have, however, a number of technological advantages, discussed in the beginning of this section.

As shown in this section, the optimum device design of the inverted structure calls for a sheet concentration of the doped AlGaAs layer close to the maximum electron concentration in the 2D gas.

As can be seen from Fig. 10-10-6, for enhancement mode transistors the transconductance is actually larger for inverted FETs. This is because the 2D gas is actually closer to the gate in inverted structures than in normal structures (for equal distances between the gate and the heterointerface). This advantage becomes especially important in devices with very small distances between the gate and the heterointerface. Inverted MODFETs with a separation of only 100 Å between

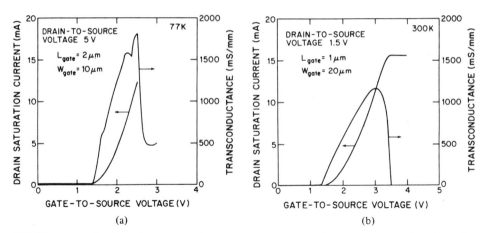

FIGURE 10-10-7. Drain-to-source saturation current and transconductance versus gate voltage for enhancement mode inverted MODFETs [71].

the gate and heterointerface were described in Ref. 71. Transconductances as high as 1810 mS/mm at 77 K and 1180 mS/mm at 300 K were observed (see Fig. 10-10-7). These values of the extrinsic (measured) transconductance were primarily limited by source series resistances, with intrinsic transconductance being even higher.

10-11. MAXIMUM TRANSCONDUCTANCE AND CURRENT SWING

As has been shown in Chapter 7, the device transconductance in FETs with high values of the low field mobility and low pinch-off voltages is primarily determined by the electron saturation velocity. High values of the effective electron saturation velocity in modulation doped devices (up to 2×10^5 m/s at 300 K and up to 3×10^5 m/s at 77 K) lead to a very high device transconductance and current swing. The extremely high values of the low field mobility calculated and measured for modulation doped structures can be obtained only at very low electric fields (see Sections 10-6 and 10-7). As shown in Section 10-7, the differential mobility at low temperatures decreases very fast with the increase in the electric field (see also Fig. 10-11-1 [80]). This reduction starts at fields as low as 10 V/cm. The values of the electric field in the conducting channel of a modulation doped transistor are of the order of several kV/cm. Hence, the saturation velocity is more important in determining the device characteristics. Higher values of the low field mobility are

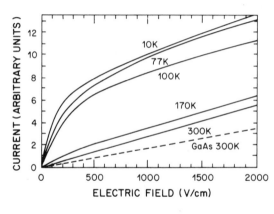

FIGURE 10-11-1. Current versus electric field characteristics up to a field of 2 V/cm in a lattice temperature range of 10–300 K associated with MODFET structure. A bulk GaAs layer with equivalent real electron concentration is also shown for comparison [80].

important because they ensure velocity saturation in most of the channel even at low drain-to-source voltage. However, as will be shown below, the increase of the low field mobility beyond 30,000 cm^2/V s or so does not dramatically increase the device transconductance of 1 μm or submicron gate devices. It follows from this discussion that a mobility value at low temperatures is not an appropriate "figure of merit" for modulation doped devices even though it may be a good indication of the quality of the modulation doped layer. As shown in this section the decrease in the undoped "spacer" layer thickness d_i, for example, may improve the device performance because of a larger charge transfer across the heterointerface even though the maximum value of the low field mobility decreases with decreasing d_i.

The charge control model, based on a two-piece linear approximation for the electron velocity versus electric field curve, leads to the following expression for the maximum intrinsic (i.e., $R_s = 0$) transconductance of a modulation doped transistor [25]:

$$(g_m)_{\max} = \frac{q\mu n_{s0}}{L\{1 + [q\mu n_{s0}(d + \Delta d)/\varepsilon v_s L]^2\}^{1/2}} \tag{10-11-1}$$

Here $(g_m)_{\max}$ is the intrinsic transconductance per unit gate length, and

$$d = d_d + d_i \tag{10-11-2}$$

is the thickness of the AlGaAs layer.

The gate voltage $(V'_g)_{\max}$, where highest transconductance can be obtained, is the pinch off voltage of the 2D gas given by

$$(V'_g)_{\max} = (V_g - V_{off})_{\max} = (V_{po})_{2D} = q(d + \Delta d)n_{s0}/\varepsilon \tag{10-11-3}$$

Equations (10-11-1) and (10-11-2) are obtained using the charge control model described in Sections 10-8 and 10-9 [25].

For a given threshold voltage and doping of the doped AlGaAs layer the thickness d_d of the doped AlGaAs layer is given by

$$d_d = [2(V_{bi} - V_{off})\varepsilon/qN_d]^{1/2} \tag{10-11-2a}$$

Here $V_{bi} = \phi_b - \Delta E_c$.

As can be seen from Eq. (10-11-2) a higher doping of the (Al, Ga)As reduces the thickness of the doped (Al, Ga)As beneath the gate leading to a higher transconductance.

We can rewrite Eq. (10-11-1) in a slightly different form:

$$(g_m)_{\max} = \frac{\varepsilon\mu(V_{po})_{2D}}{(d + \Delta d)L\{1 + [\mu(V_{po})_{2D}/v_s L]^2\}^{1/2}} \tag{10-11-1a}$$

Even though $(V_{po})_{2D}$ depends on n_{s0} and $d + \Delta d$ for a crude comparison between different devices it may be taken as a constant. Indeed, n_{s0} increases with doping but for a given threshold voltage d decreases with doping [see Eq. (10-11-2a)] so that the product is less dependent on doping. We choose $(V_{po})_{2D} = 0.8$ V as a representative value for the device comparison.

The dependences of g_{max} on doping in the AlGaAs layer for different values of d_i and for two threshold voltages corresponding to normally-on and normally-off devices are shown in Fig. 10-11-2. As can be seen from the figure, the maximum transconductance increases with increasing threshold voltage, with increasing doping of AlGaAs layer, and with decreasing space layer thickness. The reason for this is that all these factors bring the channel closer to the gate. This means that the gate capacitance, C_g, increases accordingly and the g_{max}/C_g ratio remains constant. However, in practical circuits the total capacitance is determined by the device and the interconnect and fringing capacitance, C_i, so that the figure of merit is

$$f = g_{max}/(C_g + C_i)$$
$$= (g_{max}/(C_g))/(1 + C_i/C_g)$$

This figure of merit increases with the increase in $(g_m)_{max}$ and corresponding increase in C_g.

The dependences of $(g_m)_{max}$ on temperature for 0.5-, 1-, and 2-μm devices are shown in Fig. 10-11-3. In this calculation we assumed that the temperature depen-

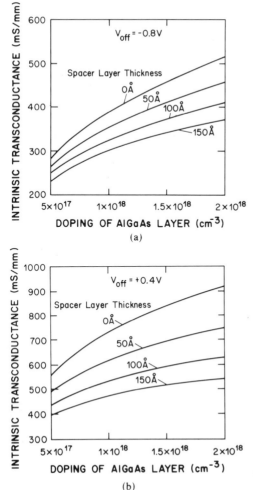

FIGURE 10-11-2. Maximum intrinsic transconductance versus doping level in doped AlGaAs layer for different thicknesses of the spacer layer [103]. (a) $V_{off} = -0.8$ V; (b) $V_{off} = 0.4$ V. $\mu = 0.5$ m^2/V s, $V_{bi} = \phi_b - \Delta E_c = 0.7$ V, $v_s = 2.2 \times 10^5$ m/s, $R_s = 0$.

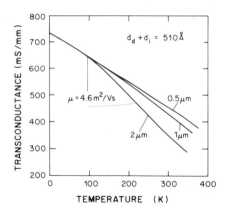

FIGURE 10-11-3. Maximum intrinsic transconductance versus temperature for different device lengths [103].

dence of the saturation velocity is given by Eq. (10-7-1) (see Fig. 10-7-6) and the temperature dependence of the low field mobility is given by Eq. (10-6-30) with $\mu_I = 25 \text{ m}^2/\text{V s}$. As can be seen from the figure, there is a substantial increase in transconductance at low temperatures because of the velocity saturation at lower voltages for higher values of the low field mobility and because of the velocity increase with decreasing temperatures. The latter factor is most important at low temperatures. It may also be seen that the transconductance at low temperatures is quite insensitive to the gate length because the mobility is high enough to saturate the electron velocity at very low drain-to-source voltages (see also Fig. 10-11-4). This means that the MODFET circuits operating at low temperatures may have longer gates with relatively little effect on the transconductance and with the benefit of smaller short-channel effects (smaller output conductance in the saturation regime). Even at room temperature transconductance becomes nearly independent of the gate length at very small gate lengths owing to the velocity saturation. In reality, however, an additional enhancement of the transconductance in short gate structures may still be possible owing to ballistic and overshoot effects (see Section 2-9).

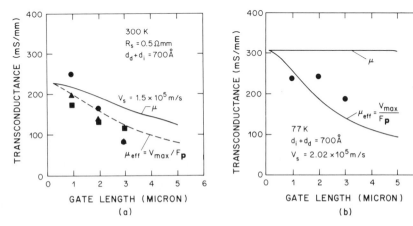

FIGURE 10-11-4. Maximum intrinsic transconductance versus gate length. (a) 300 K; (b) 77 K. Curves marked μ_{eff} were calculated assuming $\mu_{\text{eff}} = v_{\text{max}}/F_p$ where v_{max} is the peak electron velocity in GaAs and F_p is the corresponding electric field [103].

For very short gates when

$$\mu (V_{po})_{2D}/(v_s L) \gg 1 \qquad\qquad (10\text{-}11\text{-}4)$$

we find

$$(g_m^{\text{short}})_{\text{max}} = \varepsilon v_s/(d + \Delta d) \qquad\qquad (10\text{-}11\text{-}5)$$

This expression sets an upper limit for the transconductance of short gate MODFETs.

Because of the higher velocity at low temperature, very high transconductance can be obtained at 77 K. The transfer characteristics at 77 K for one of the normally-off MODFETs are shown in Figure 10-11-5. The calculated curve is shown as a solid line and the measured data are indicated by circles (device width $W = 290\ \mu\text{m}$) [81]. For the theoretical calculation, a relationship between the saturation current and the gate voltage was derived assuming a two-piece linear approximation for the electron velocity–electric field characteristic. The exact dependence of the Fermi level in the 2D gas on gate voltage was also taken into account.

For a gate voltage of +0.6 V the transconductance was 225 mS/mm at 300 K and 400 mS/mm at 77 K. At a gate voltage of +0.8 V for the device at 300 K, the 2D gas is entirely undepleted and an additional parallel conduction path through the $Al_{0.33}Ga_{0.67}As$ is formed, resulting in the obseved decrease in transconductance (see Section 10-12).

At 300 K a source resistance of $R_s = 4\ \Omega$ and an electron saturation velocity of $v_s = 2 \times 10^5$ m/s were used. At 77 K the value of R_s decreased to 2.5 Ω and v_s increased to 3×10^5 m/s. Also presented in Figure 10-11-5 are the results for another

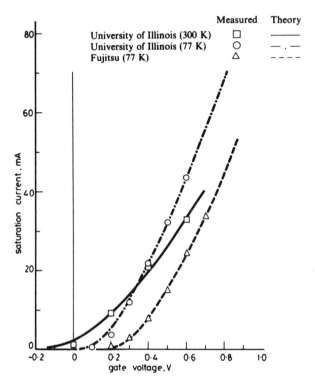

FIGURE 10-11-5. I_{ds}^s vs. V_g calculated for the University of Illinois devices at 300 and 77 K and the device fabricated by Fujitsu Ltd. at 77 K [81].

MODFET operating at 77 K [82]. The data, represented as triangles, were taken from the published drain I–V characteristic. The dotted line was then calculated to fit to the data. This device [82] had a gate length of 2 μm, a doping level of $N_d = 2 \times 10^{18}$ cm^{-3} in the $Al_{0.3}Ga_{0.7}As$, and no undoped spacer layer. To model the device at 77 K the reported mobility of 20,000 cm^2/V s was used with $R_s = 3.8\ \Omega$ and $v_s = 3 \times 10^5$ m/s.

The transconductances of each device, calculated numerically from the curves in Figure 10-11-5 ($g_m = \partial I_{dss}/\partial V_{gs}$), are plotted in Fig. 10-11-6. As can be seen, the maximum transconductance of the University of Illinois device at 77 K for a gate voltage of +0.8 V is 450 mS/mm at 77 K, as compared to a value of 400 mS/mm obtained for the Fujitsu device at 77 K [82]. Calculating the maximum intrinsic transconductance (zero source resistance) for the U of I device, we find $g_{m0} = 352$ mS/mm at 300 K and 668 mS/mm at 77 K.

Both devices show nearly the same value of threshold voltage shift ($\approx$0.2 V) when they cool down to 77 K. This will be discussed in Section 10-13.

At room temperature when the mobility is only a weak function of d_i, the transconductance should increase both with a decrease in d_i (in agreement with experimental results) and with a decrease in gate length. This reduction in d_i leads to several consequences. First, it increases both the capacitance and transconductance. Second, it increases n_{s0} (see Section 10-4), the maximum voltage swing [Eq. (10-11-3)], and the maximum drain saturation current at $(g_m)_{max}$ [see Eq. (10-11-1)]. Finally, it reduces the value of source series resistance due to the increase in n_{s0}. (The source resistance includes the contribution from the resistance of the source-to-gate region which decreases with the increase in n_{s0}.)

To investigate the role of the $Al_xGa_{1-x}As$ spacer layer, a series of structures were grown and devices were fabricated [83]. The spacer layer thickness was decreased from 100 to 20 Å. The decrease in undoped $Al_xGa_{1-x}As$ thickness was to increase the charge transfer. As a result, the gate capacitance remained nearly constant while the maximum saturation current almost doubled. The maximum transconductance increased with the increase in the current.

The transistor characteristics were modeled using the equations derived in Section 10-9 except that the exact dependence of the interface carrier density on the gate voltage was accounted for. The saturation velocity was chosen to be 1.7×10^5 m/s.

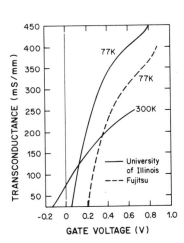

FIGURE 10-11-6. Transconductance against V_g for the University of Illinois device at 300 and 77 K and for the Fujitsu device at 77 K [81].

TABLE 10-11-1. Measured and Theoretical Maximum Values of Transconductances and Saturation Current[a]

	g_m (mS/mm)		$(I^s_{ds})_{max}$ (mA)		R_s (Ω)	
d_i (Å)	N-on	N-off	N-on	N-off	N-on	N-off
20	210 (290)	250 (275)	7.5 (27)	19 (19)	5	5.5
40	230 (245)	155 (175)	24 (27.5)	13 (14.5)	5	16.5
60	235 (240)	205 (210)	23 (26.5)	18.8 (16)	5.5	12.5
80	210 (225)	160 (170)	14 (19)	8.4 (11)	4	13
100	145 (170)	125 (135)	7.5 (11)	8.5 (8.5)	13	22

[a] Theoretical values are in parentheses. The values of source resistances are those employed in fitting the theoretical curve to the data [83]. $L = 1$ μm, $W = 145$ μm.

The two parameters adjusted to fit the data were the doped $Al_xGa_{1-x}As$ layer thickness beneath the gate, d_d, and the source resistance R_s. Once the I^s_{ds} vs. V_g characteristic had been determined, the transconductance and gate capacitance were calculated numerically as $g_m = \partial I^s_{ds}/\partial V_g$ and $C_g = \partial Q/\partial V_g$. While the values obtained for the transconductance are an accurate reflection of the experimental data, the capacitance values are good only as estimates.

Table 10-11-1 lists the source resistances used to model each device as well as the maximum saturation currents and transconductances. The maximum theoretical values of the sauration current and transconductances are given in parentheses. The thickness of the undoped $Al_xGa_{1-x}As$ spacer, d_i, the doped $Al_xGa_{1-x}As$ beneath the gate, d_d, the equilibrium junction depletion depth, d_1, and the open channel gate capacitance are given in Table 10-11-2. The large discrepancy in the parameters of the 20-Å normally-on FET relative to the other normally-on devices results from it being only quasi-normally-on. The gate voltage swing ran from -0.4 V to $+0.6$ V. The current and transconductance values for this device are listed in Table 10-11-2 for $V_g = 0$ V. The normally-off device with $d_i = 20$ Å is anomalous in that the doped layer is too thin and the source resistance too small to be consistent with the other normally-off devices.

In Table 10-11-2 the gate capacitance is nearly independent of spacer layer thickness. This is a result of the particular parameters used in the calculation rather than being characteristic of MODFETs in general. The capacitance values could be reduced further by increasing the Al mole fraction and by decreasing the doping

TABLE 10-11-2. Calculated Parameters of Modulation Doped Transistors with Different "Spacer" Layer Thicknesses[a]

		d_d (Å)		$d_i + d_d$ (Å)		C (pF)	
d_i (Å)	d_1 (Å)	N-on	'N-off	N-on	N-off	N-on	N-off
20	138	358	283	378	293	0.73	0.88
40	127	460	307	500	347	0.55	0.78
60	117	441	281	501	341	0.57	0.80
80	109	420	268	500	348	0.56	0.78
100	101	382	245	482	345	0.58	0.78

[a] Reference 83.

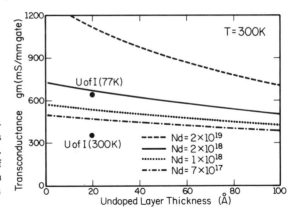

FIGURE 10-11-7. Intrinsic transconductance per millimeter of gate width as a function of undoped layer thickness. Theoretical curves are shown for N-off devices and experimental points for both normally-on and normally-off devices [83].

level in the $Al_xGa_{1-x}As$. The open channel capacitances given in Table 10-11-2 represent an upper limit for devices operating in a current saturation regime. The data imply that the maximum operating frequency of these devices should easily exceed 10 GHz.

Reducing the spacer thickness from 100 to 20 Å resulted in doubling both the maximum saturation current and transconductance, which should lead to faster operating speeds. The increased current is a result of the larger charge transfer possible with the thinner spacer layer. The maximum transconductances obtained (for a 145-μm gate width) were 250 mS/mm for a normally-off device with a 20-Å spacer layer and 230 mS/mm for a normally-on device with a 40-Å spacer layer. In general, for a given spacer thickness, larger transconductances were obtained for normally-on devices. This is a direct result of the lower source resistances, typically one-third to one-half of the value obtained for the normally-off devices.

In Fig. 10-11-7 the maximum intrinsic transconductances are compared with theory. Although the functional dependence of the experimental g_m on d_i is in good agreement with the theory, the theoretical values are slightly larger. This may be attributed to a small uncertainty in the electron velocity and doping level in the $Al_xGa_{1-x}As$.

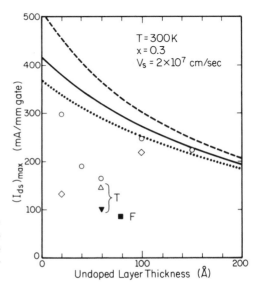

FIGURE 10-11-8. Maximum drain-to-source saturation current in modulation doped structures versus undoped AlGaAs spacer layer thickness [25]. T, Tompson CSF device; ○ and ◇, University of Illinois devices.

The high current swing is even more important than the high transconductance in logic devices geared for maximum speed because the current determines the time necessary to charge the effective gate capacitance. The maximum current carried by the two-dimensional electron gas is given by

$$(I_{ds}^s)_{max} = qn_{s0}v_s W \tag{10-11-6}$$

Using the results given in Section 10-4, we calculated $(I_{ds}^s)_{max}$ as a function of d_i for $N_d = 7 \times 10^{17}$ cm^{-3}, 10^{18} cm^{-3}, and 2×10^{18} cm^{-3} [25]. The results of this calculation are shown in Figure 10-11-8. As can be seen from the figure, the $(I_{ds}^s)_{max}$ variation with d_i agrees well with the experimental data.

Because $(I_{ds}^s)_{max}$ is only a function of n_{s0} and v_s, and independent of R_s, more detailed studies of the maximum current swing may yield important information about these crucial parameters.

10-12. EXTENDED CHARGE CONTROL MODEL AND MODFET $I-V$ CHARACTERISTICS AT LARGE GATE BIAS [86, 87]

The charge control model developed in Sections 10-8 and 10-9 implies that the conduction between the source and drain occurs only via the two-dimensional electron gas. This is correct when the gate voltage swing is less than the pinch-off voltage of the two-dimensional electron gas

$$(V_{po})_{2D} = \frac{qn_{s0}(d + \Delta d)}{\varepsilon} \tag{10-12-1}$$

At more positive gate voltages the doped (Al, Ga)As layer becomes partially undepleted and contributes to the conduction. A qualitative sketch of the interface carrier density n_s as a function of the gate voltage is shown in Fig. 10-12-1. According to the charge control model presented in Section 10-6

$$qn_{s2} = C_{2D}(V_g - V_{t2}) \tag{10-12-2}$$

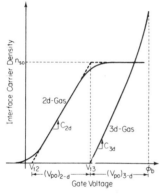

FIGURE 10-12-1. Qualitative sketch showing carrier dependence on gate bias [86]. The interface carrier density of the two-dimensional electron gas (2D-gas) and the sheet carrier density in the (Al, Ga)As MESFET region (3D-gas) are plotted versus gate voltage. Zero drain-to-source bias is assumed.

when V_{t2} is the threshold voltage, V_g is the gate voltage,

$$C_{2D} = \frac{\varepsilon}{d + \Delta d} \tag{10-12-3}$$

is the capacitance of the two-dimensional electron gas per unit area. Here ε is the dielectric permittivity of (Al, Ga)As, $d = d_a + d_i$, d_a is the thickness of the doped (Al, Ga)As layer, and d_i is the thickness of the undoped "spacer" layer,

$$d_a = [2\varepsilon(\phi_b - \Delta E_c/q - V_{t2})/(qN_a)]^{1/2} \tag{10-12-4}$$

where ϕ_b is the Schottky barrier height, ΔE_c is the conduction band discontinuity at the heterointerface, and N_d is the doping density in (Al, Ga)As. The dependence of n_s on V_g predicted by Eq. (10-12-2) is shown in Fig. 10-12-1 by a dashed line.

At gate voltages higher than

$$V_{t3} = V_{t2} + (V_{po})_{2D} \tag{10-12-5}$$

$n_{s2} \approx n_{s0}$ and the parallel conduction from the (Al, Ga)As MESFET has to be included. The charge qn_{s3} of the electrons in the MESFET channel is given by

$$qn_{s3} = C_{3D}(V_g - V_{t3}) \tag{10-12-6}$$

where

$$C_{3D} = \varepsilon/t_d \tag{10-12-7}$$

and

$$t_d = d_a - \frac{n_{s0}}{N_d} \tag{10-12-8}$$

Equation (10-12-6) is applicable when (see Fig. 10-12-1)

$$V_g - V_{t3} \ll (V_{po})_{3D} \tag{10-12-9}$$

where

$$(V_{po})_{3D} = \frac{qN_dt_d^2}{2\varepsilon} \tag{10-12-10}$$

is the pinch-off voltage of the doped (Al, Ga)As layer.

The qualitative distribution of electrons in the two-dimensional electron gas and in the doped (Al, Ga)As layer is shown in Figure 10-12-2. In the gate voltage range

$$V_{t2} < V_g < V_{t3} \tag{10-12-11}$$

the two-dimensional gas is partially depleted everywhere in the channel (see Fig. 10-12-12a). At more positive gate voltages

$$V_{t3} \leq V_g < V_{t3} + V_{ds} \tag{10-12-12}$$

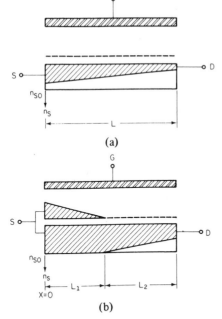

FIGURE 10-12-2. A pictorial diagram showing charge distribution in MODFETs [86]. (a) At low gate bias in the range $V_{t2} < V_g < V_{t2} + (V_{po})_{2D}$, 2DEG density (n_{s2}) along the 2D channel is always less than n_{s0}. (b) At high gate bias when $V_{t2} + (V_{po})_{2D} + V_0 > V_g > V_{t2} + (V_{po})_{2D}$ we have parallel conduction in $(Al, Ga)As$ (n_{s3}) and $n_{s2} = n_{s0}$ for $0 < x < L_1$, and $n_{s2} < n_{s0}$ $(n_{s3} = 0)$ for $L_1 < x < L$.

where V_{ds} is the drain-to-source voltage, the electron distribution in the channel looks like the distribution shown in Fig. 10-12-2b. We limit ourselves to drain-to-source voltages, $V_{ds} \leq V_{ds}^s$, where V_{ds}^s is the drain-to-source saturation voltage for the gate voltage range given by Eq. (10-10-12) (where we assume $V_{ds} = V_{ds}^s$). In this case the current $(I_{ds})_{2D}$ carried by the two-dimensional gas in region I $(0 < x < L_1)$ is given by

$$(I_{ds})_{2D} = qW\mu_2 n_{s0} V(L_1)/L_1 = I_2 V(L_1)/V_1 \qquad (10\text{-}12\text{-}13)$$

where

$$I_2 = qn_{s0}v_{s2} W \qquad (10\text{-}12\text{-}14)$$

$$V(L_1) = V_g - V_{t3} \qquad (10\text{-}12\text{-}15)$$

and

$$V_1 = v_{s2}L_1/\mu_2 \qquad (10\text{-}12\text{-}16)$$

Here v_{s2} and μ_2 are the electron saturation velocity and the low field mobility of the two-dimensional gas. On the other hand, the drain saturation current $(I_{ds}^s)_{2D}$ in Region II $(L_1 < x < L)$ can be found using our two-piece model (see Section 10-9):

$$(I_{ds}^s)_{2D} = I_2 \frac{[(V_{po})_{2D}^2 + V_2^2]^{1/2} - V_2}{(V_{po})_{2D}} \qquad (10\text{-}12\text{-}17)$$

where

$$V_2 = \frac{v_{s2}L_2}{\mu_2} \qquad (10\text{-}12\text{-}18)$$

The value of L_1 (or V_1) can be found by equating Eqs. (10-12-17) and (10-12-13) and by noting that $L = L1 + L2$, or that

$$V_1 + V_2 = V_0 \tag{10-12-19}$$

where

$$V_0 = \frac{v_{s2} L}{\mu_2} \tag{10-12-20}$$

The resulting expression for V_1 is

$$V_1 = V(L_1) \frac{V_0 + [V_0^2 + 2(V_{po})_{2D} V(L_1) + (V_{po})_{2D}^2]^{1/2}}{(V_{po})_{2D} + 2V(L_1)} \tag{10-12-21}$$

Substitution of Eq. (10-12-21) into Eq. (10-12-13) yields

$$(I_{ds}^s)_{2D} = I_2 \frac{(V_{po})_{2D} + 2V(L_1)}{V_0 + [V_0^2 + 2(V_{po})_{2D} V(L_1) + (V_{po})_{2D}^2]^{1/2}} \tag{10-12-22}$$

The saturation drain-to-source current I_{ds}^s and the device transconductance in the saturation regime g_m^s versus the gate voltage are shown in Figs. 10-12-3a and 10-3-3b. The solid curves show the dependence predicted by the two-piece charge control model which does not take into account the finite value of n_{s0} (see Section 10-9). As pointed out above for a small voltage, e.g., $V_g < (V_{po})_{2D} + V_{t2}$, these curves coincide with the dependence given by the present model. At higher biases the transconductance decreases, and finally, at $V_g = V_{t2} + (V_{po})_{2D} + V_0$ the maximum value of the current, I_2, through the two-dimensional gas is reached [see Eq. (10-12-15)]. The maximum current carried by the two-dimensional gas was analyzed in Section 10-11. Any further small increase in current must only be due to the

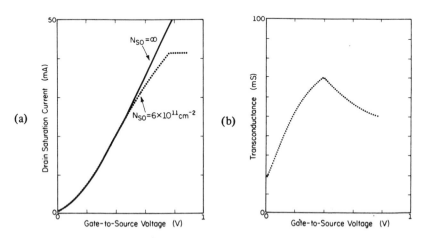

FIGURE 10-12-3. The effect of finite n_{s0} on transfer characteristics of the MODFET [86]. Maximum transconductance is obtained at $V_g = V_{t2} + (V_{po})_{2D}$. The maximum drain current carried by 2DEG (I_2) is obtained when $V_g > V_{t2} + (V_{po})_{2D} + V_0$. (a) Transfer characteristic, (b) transconductances. The values of parameters used for calculation are $V_{t2} = -0.06$ V, $\mu_2 = 4,000$ cm^2/V s, $v_{s2} = 1.5 \times 10^7$ cm/s, $L = 1$ μm and $W = 290$ μm.

conduction through the doped (Al, Ga)As layer. This current $(I_{ds}^s)_{3D}$ may be found using the model developed in Chapter 7 and used in Section 9-5:

$$(I_{ds}^s)_{3D} = \frac{2\varepsilon v_{s3} W(V_g - V_{T3})^2}{t_d[(V_{po})_{3D} + 3v_{s3}L/\mu_3]} \tag{10-12-23}$$

where μ_3 and v_{s3} are the mobility and saturation velocity in (Al, Ga)As, t_d is given by Eq. (10-12-8), and

$$V_{T3} = V_{t2} + (V_{po})_{2D} + V_0 \tag{10-12-24}$$

This theory was compared with experimental data obtained for modulation doped field effect transistors fabricated at the University of Illinois [86, 87].

Single interface (Al, Ga)As/GaAs modulation doped heterostructures were used for device fabrication and were grown by molecular beam epitaxy. Epitaxial layers were grown on (001) Cr doped substrates in the following sequence: 1 μm of undoped GaAs, a thin layer of undoped (Al, Ga)As with a thickness d_i, and 600 Å of Si doped (Al, Ga)As. After epitaxy the crystals were characterized by Van der Pauw–Hall measurements at 300 and 77 K. The Hall samples consisted of cloverleaf lamella with alloyed Sn contacts.

Transistors were fabricated by first removing the indium (used to mount the crystal for epitaxy) from the backside of the crystal and lapping it down to 400 μm. Isolation mesas were photolithographically defined in AZ 1450J positive photoresist and etched in a solution of HF:H$_2$O:H$_2$O (1:1:3). This etch provided a gradual edge slope that enhanced metal step coverages. AuGe/Ni/Au was evaporated and lifted off to form the source and drain. Ohmic contacts were obtained by alloying the metallization for one minute at 500°C in a hydrogen atmosphere. The 1 × 290-μm gate contact was made using a chlorobenzene lift-off enhancement treatment followed by a brief channel etching with NH$_4$OH:H$_2$O$_2$:H$_2$ (3:1:150) for threshold adjustment. The 3000-Å-thick Al gate metallization was then evaporated and lifted off.

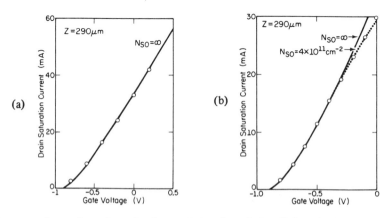

FIGURE 10-12-4. Comparison of transfer characteristics of two devices [86]. Solid dots are experimental values. The lines are best theoretical fitting. (a) This device shows no transconductance degradations; it may be modeled well assuming $n_{s0} \to \infty$ (solid line). (b) This device shows apparent transconductance degradation, $n_{s0} = 4 \times 10^{11}$ cm^{-2} (dotted line).

At this point the devices were probed on wafers. For each probing, only half of the split source transistor, corresponding to a gate width of 145 μm, was tested. The data reported in this section are the three terminal characteristics measured under the probe station. For this reason the source and drain resistance reported may be slightly overestimated owing to the series resistance of the probe.

Implications of finite n_{s0} and, as a result, the finite gate voltage swing are illustrated by Fig. 10-12-4. The transfer characteristic shown in Fig. 10-12-4a does not show any transconductance degradation up to current levels as high as 40 mA. On the other hand, the characteristic shown in Fig. 10-12-4b shows transconductance degradation when the drain current exceeds 15 mA, indicating a small value of $n_{s0} \approx 4 \times 10^{11}$ cm^{-2}, obtained from the best fit to the experimental data. Both devices are normally on and the gate current is negligible at all gate voltages. Therefore the decrease in the transconductance at large gate bias cannot be attributed to the gate leakage current.

10-13. DEVICE CHARACTERIZATION [86, 87]

In order to compare the extended charge control model with the experimental I-V characteristics the series source resistance, R_s; the series drain resistance, R_d; the device threshold voltage, V_{t2}; the parasitic MESFET threshold voltage, V_{t3}; the electron mobilities in the two-dimensional electron gas, μ_2, and in doped AlGaAs layer, μ_3; the thickness of the doped AlGaAs layer, d_d; the thickness of the undoped AlGaAs spacer layer, d_i; and the electron saturation velocities in the two-dimensional gas, v_{s2}, and in the doped AlGaAs layer, v_{s3}, must be known.

Some of these parameters may be known from the fabrication process, others have to be either determined from the device characteristics or assumed to have some "typical values."

Several characterization techniques which allow us to deduce FET parameters from the FET I-V characteristics were discussed in Section 7-9. Another approach, specifically for MODFETs, was developed in Ref. 88. However, GaAs MESFET and MODFET characterization remains one of the most difficult tasks in their modeling. As far as MODFETs are concerned there are several basic difficulties—the contribution from the "subthreshold" current at voltages close to V_{t2}, the contribution from the gate current at voltages close to the voltage at which the transconductance reaches its maximum value, effects related to the gate voltage dependence of the source series resistance, short channel effects responsible for the output conductance in the saturation region, and effects related to traps (see Section 10-14) may substantially change the current voltage characteristic. All these important effects are not taken into account by the charge control model. In particular, the role of the subthreshold conductance is clearly illustrated by Fig. 10-13-1 [89] where the charge control model results are compared with the experimental data. As can be seen from Fig. 10-13-1a, the zero field drain-to-source conductance coincides with the measured curve in a relatively small range. A subthreshold drain-to-source conductance is quite substantial even at voltages which are 0.2 V less than the deduced value of the threshold voltage.

A qualitative understanding of the MODFET operation is provided by the charge control model and is corroborated by two-dimensional simulations. However, a good characterization technique is absolutely necessary for computer data

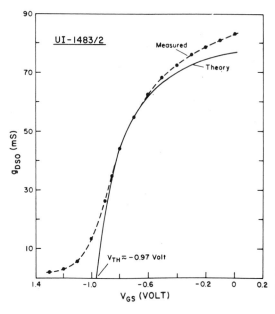

FIGURE 10-13-1. MODFET drain-to-source conductance at zero drain-to-source voltage [89].

acquisition on a chip and for computer-aided design of MODFET integrated circuits. An example of a simplified MODFET characterization procedure [87, 88] is considered in this section. We should note, however, that this technique may not be adequate for devices with relatively small values of n_{s0} where the voltage swing is limited and the difficulties mentioned above become important. Also, based on our experience in characterizing many devices from different laboratories, the deduced values of the effective electron saturation velocity may vary greatly from device to device (as much as from 1.5×10^5 to 3×10^5 m/s at 77 K and from 1×10^5 to 2×10^5 at room temperature). Similar (though perhaps somewhat smaller) variations of the deduced saturation velocity are found also for GaAs MESFETs where other factors (such as the doping profile nonuniformity) may play a role; see Chapter 7.

In the characterization example discussed below, the doping level, N_d, of the (Al, Ga)As layer and the thickness of the undoped layer, d_i, are known from the device fabrication procedure. All other parameters are determined from the experimental curves.

Our characterization starts from a crude estimate of the gate voltage range $V_{t2} < V_g < V_{t2} + (V_{po})_{2D}$. At this point the objective is simply to find gate voltages within this range without attempting an accurate determination of V_{t2} and V_{t3}. We then plot the drain-to-source saturation current, I_{ds}^s and the saturation transconductance g_m^s as functions of V_g. According to the two-piece charge control model (see Section 10-9), we find the drain-to-source saturation current at large values of V_g:

$$I_{ds}^s = I_2 \frac{V_g - V_{t2} - V_0}{(V_{po})_{2D}} \qquad (10\text{-}13\text{-}1)$$

where

$$V_0 = v_{s2} L / \mu_2$$

[see Eq. (10-12-20)]. Hence $V_{t2} + V_0$ may be found from the linear asymptote of the saturation current at large gate biases (but below $(V_g)_{max}$), corresponding to the maximum transconductance. We take $V_{t2} + V_0 = V_{g1}$ as a low bound of the gate voltage used for the characterization procedure described below. The value of

$$V_{g2} = V_{t2} + V_0 + 3/4[(V_g)_{max} - V_{t2} - V_0] \qquad (10\text{-}13\text{-}2)$$

is used as an upper bound of this gate voltage range.

As an example, plots of I_{ds}^s vs. V_g and g_m^s vs. V_g measured for two devices fabricated at the University of Illinois [86, 87] are shown in Figs. 10-13-2 and 10-13-3. The points marked by $\triangle$ in Figs. 10-13-2a and 10-13-3a are shifted owing to the use of two sweep ranges for the gate in the measurements: -1–0 V for points marked by open circles and 0–0.8 V for points marked $\triangle$. This offset voltage is due to the shift of the threshold voltage caused by traps (see Section 10-14) and is taken into account by a parallel transfer of the measured data in the gate voltage range $0 < V_g < 1$ V (points marked in Figs. 10-13-2a and 10-13-3a by solid circles). By drawing asymptotes at large gate voltages, we find from Fig. 10-13-2a that $V_{g1} \approx V_{t2} + V_0 \approx -1.6$ V, $V_{g2} \approx 0$ V and from Fig. 10-13-3a we determine $V_{g1} \approx -1.4$ V and $V_{g2} \approx 0$ V, respectively.

In the voltage range $V_{g1} < V_g < V_{g2}$ the simple charge control model for the two-dimensional gas (see Sections 10-8, 10-9) may be applied. According to this model the channel resistance, R_{ds}, at very low drain-to-source voltages is given by

$$R_{ds} = \frac{L}{\mu_2 C_{2D}(V_g - V_{t2})W} + R_{S+D} \qquad (10\text{-}13\text{-}3)$$

where $R_{S+D} = R_s + R_d$. Hence R_{S+D} and V_{t2} may be determined from the best least-squares linear fit to the measured R_{ds} vs. $1/(V_g - V_{t2})$ in the gate voltage range

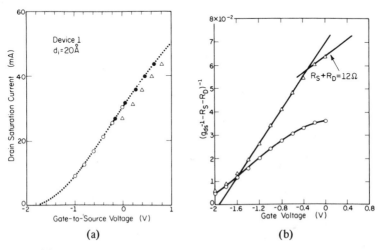

FIGURE 10-13-2. (a) Transfer characteristic of device No. 1 [86]. Open circles and triangles are experimental points. Solid circles are the triangles ($V_g > 0$) shifted by -0.16 V (see text). The dotted line is a theoretical curve using data in Table 10-13-1 assuming two-dimensional electron conduction only. (b) Drain conductance data for device No. 1 [86]. Circles are experimental points and triangles are intrinsic drain conductance data assuming $R_s + R_d = 12\ \Omega$.

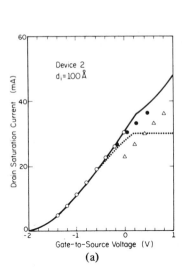
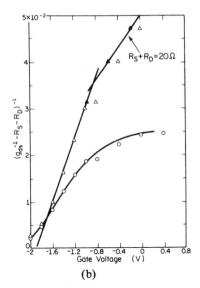

(a)   (b)

FIGURE 10-13-3. (a) Transfer characteristics of device No. 2 [86]. Open circles and triangles are
experimental points. Solid circles are the triangles ($V_g > 0$) shifted by -0.36 V (see text). The dotted line
is a theoretical curve using data in Table 10-13-1 assuming two-dimensional electron conduction only.
The solid line is obtained assuming conduction by the two-dimensional electrons and undepleted carriers
in the (Al, Ga)As MESFET section. (b) Drain conductance data for device No. 2 [86]. Circles are
experimental points and triangles are intrinsic drain conductance data assuming $R_s + R_d = 20\,\Omega$. The
data for $V_g > -0.8$ V are moved by -0.17 V to compensate for the threshold voltage shift.

$V_{g1} < V_g < V_{g2}$ (see Figs. 10-13-2b and 10-13-3b). If the device structure is symmetrical then $R_s = R_d = R_{S+D}/2$. The value of R_s determined in this manner is in
agreement with those determined independently using an "end" resistance technique
(see Chapter 7).

Now we find the thickness of the doped (Al, Ga)As based on the value of V_{t2}
deduced above and Eq. (10-8-6a). The values of the barrier height $\phi_b = 1$ V, $\Delta E_c =$
0.34 V, and $\varepsilon = 10^{-10}$ F/m were used in this calculation. For the devices discussed
above with $N_d = 10^{18}$ cm^{-3}, $V_{t2} \approx -1.9$ V and the values of $d_d = 570$ Å were deduced.
These values are in good agreement with those expected from the fabrication
procedure.

Now we find the values of C_{2D} and C_{3D} from equations given in Section 10-12
using the value of d_d determined above; the undoped layer thickness d_i is known
from the growth procedure and the value of $\Delta d = 80$ Å is deduced in Section 10-8.
For the devices discussed above, $C_{2D} = 0.22$ pF, $C_{3D} = 0.35$ pF (device of Fig.
10-13-2) and $C_{2D} = 0.2$ pF, $C_{3D} = 0.31$ pF (device of Fig. 10-13-3).

Once the values of $R_s = R_d$ and V_{t2} are measured and the values of d_d, C_{2D},
and C_{3D} are deduced as described above, the threshold voltage of the (Al, Ga)As
"MESFET" V_{T3}, the pinch-off voltage of the two-dimensional electron gas ($V_{po})_{2D}$,
the low field electron mobility of the two-dimensional electron gas (μ_2) and of the
doped (Al, Ga)As layer (μ_3), and the equilibrium electron concentration (n_{s0}) can
be determined. We use the plot of the intrinsic drain-to-source conductance at a
low drain-to-source voltage

$$g_{di} = \frac{1}{(1/g_{ds})_{measured} - R_{S+D}}$$ (10-13-4)

versus the gate voltage, V_g.

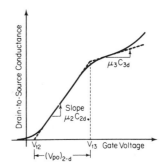

FIGURE 10-13-4. A qualitative sketch for drain conductance versus gate voltage at low drain voltage (solid line) [86]. The two dashed lines correspond to equations of the charge control model.

The qualitative sketch of g_{di} vs. V_g is shown in Fig. 10-13-4. When $V_{t2} < V_g < V_{t3}$, we have drain conductance due to only two-dimensional electron gas

$$g_{di} = \frac{\mu_2 C_{2D}(V_g - V_{t2})}{L} \qquad (10\text{-}13\text{-}5)$$

which is shown with dashed lines in Fig. 10-13-4. But the slope of this curve at large gate biases changes due to the transition from "MODFET" to (Al, Ga)As "MESFET" operation when the two-dimensional electron gas becomes totally undepleted. This dramatic change occurs because $\mu_2 \gg \mu_3$. When $V_g > V_{t3}$

$$g_{di} = \frac{\mu_2 C_{2D}(V_{po})_{2D}}{L} + \frac{\mu_3 C_{3D}(V_g - V_{t3})}{L} \qquad (10\text{-}13\text{-}6)$$

which is valid for $V_g - V_{t3} \ll (V_{po})_{3D}$ (see dashed line in Fig. 10-13-4).

The values of μ_2 and μ_3 can be determined from the slopes shown by dashed lines in Fig. 10-13-4 using Eq. (10-13-5) and Eq. (10-13-6), respectively. Furthermore, V_{t3} can be determined as shown in the same figure. The pinch-off voltage of the two-dimensional gas $(V_{po})_{2D}$ is then found as

$$(V_{po})_{2D} = V_{t3} - V_{t2} \qquad (10\text{-}13\text{-}7)$$

and the value of n_{s0} as

$$n_{s0} = \frac{C_{2D}(V_{po})_{2D}}{q} \qquad (10\text{-}13\text{-}8)$$

We can now deduce the saturation velocity, v_{s2}, from the drain-to-source current $(I_{ds})_{max}$ at $V_g = (V_g)_{max}$, i.e., at the gate voltage when the transconductance reaches its maximum value. This value of the current may be found using the charge control model of Section 10-9:

$$(I_{ds})_{max} = I_2 \frac{[(V_{gt})_{max}^2 + V_0^2]^{1/2} - V_0}{(V_{po})_{2D}} \qquad (10\text{-}13\text{-}9)$$

where

$$I_2 = q n_{s0} v_{s2} W$$

Here

$$(V_{gt})_{max} = (V_{gs})_{max} - (I_{ds})_{max} R_s - V_{t2} \qquad (10\text{-}13\text{-}10)$$

TABLE 10-13-1. Parameters Deduced from Drain-to-Source Conductance Data and Transfer Curves[a]

Device No.	R_s	From drain-to-source conductance data						From transfer characteristic
		μ_2	μ_3	V_{t2}	$(V_{po})_{2D}$	d_d	n_{s0}	v_{s2}
1	6	1,400	560	−1.93	1.58	570	1.47	1.5×10^7
2	10	1,800	610	−1.88	1.02	570	0.85	1.5×10^7

[a] Units are cm^2/Vs for μ_2 and μ_3, volts for V_{t2} and $(V_{po})_{2D}$, angstroms for d_d, 10^{12} cm^{-2} for n_{s0}, and cm/s for v_{s2} [86].

and $(V_{gs})_{max}$ is the measured value of the gate-to-source voltage at $g_m = (g_m)_{max}$. From Eq. (10-13-9) we find

$$v_{s2} = \frac{\mu_2(I^s_{ds})_{max}}{L[K^2(V_g)^2_{max} - 2K(I^s_{ds})_{max}]^{1/2}} \qquad (10\text{-}13\text{-}11)$$

where $K = \mu_{2D}C_{2D}/L$. The value of v_{s3} is not very important because the "MESFET" operation is governed by Shockley's model due to a very low mobility of AlGaAs.

The values of the parameters determined in this fashion for two of the devices fabricated at the University of Illinois are given in Table 10-13-1. The fabrication data for these devices are shown in Table 10-13-2. The I–V characteristics calculated for devices No. 1 and 2 using these parameters are shown in Figs. 10-13-2a and 10-13-3a. As can be seen from the figures, the agreement with experimental data is quite good. We should note that device No. 1 (see Fig. 10-13-2) has a thin undoped layer, $d_i = 20$ Å, and, hence, a large n_{s0} (see Tables 10-13-1 and 10-13-2). In this case the role of the parallel conduction in the (Al, Ga)As layer is quite small. Device No. 2 has $d_i = 100$ Å, which is quite large. As a result, n_{s0} is relatively low and the role of the conduction in the doped (Al, Ga)As layer is quite considerable. The values of n_{s0} determined from the drain conductance measurements (as discussed above) are in good agreement with our theoretical calculation (see Section 10-4). All other parameters are predicted with reasonable accuracy.

One exception is the low field mobility, μ_2, of the two-dimensional electron gas at 300 K. As indicated in Table 10-13-1, these values are consistently low. For devices 1 and 2 they seem to be close to the Hall mobilities we measured on the same wafers. However, as we discussed in Section 10-6, the low values of the Hall

TABLE 10-13-2. Growth and Hall Data for the Devices of Table 10-13-1[a]

Device No.	Growth data			Hall data				Device dimensions (μm)	
				300 K		77 K		Gate length	Gate width
	N_d	x	d_i	μ^H	n^H	μ^H	n^H		
1	1	0.4	20	1,320	9.7	1,810	7.7	1	145
2	1	0.33	100	1,290	9.7	3,230	2.4	1	145

[a] The units are 1×10^{18} cm^{-3} for N_d, angstroms for d_i, cm^2/Vs for μ^H, and 10^{12} cm^{-2} for n^H. The Hall data are not corrected for parallel conduction in (Al, Ga)As [86].

mobility are caused by the conduction through the doped (Al, Ga)As layer. When this contribution is taken into account, the deduced values of the Hall mobility for the two-dimensional gas at room temperature are consistently close to 8000–9000 $cm^2/V\,s$. Further studies are necessary to find out the causes for this anomalously small low field mobility in some short gate devices.

We should also notice that the values of the effective saturation velocity of the two-dimensional electron gas deduced from the I-V characteristics vary from device to device, depending, perhaps, on the interface quality.

10-14. TRAPS IN MODULATION DOPED STRUCTURES

10-14-1. *Threshold Voltage Shift with Temperature*

Current–voltage characteristics of typical modulation doped transistors are dependent on temperature, illumination, and bias conditions. They typically show an increase in the threshold voltage of about 0.2 V at 77 K with respect to room temperature. This positive voltage shift can be reduced or eliminated by illuminating the sample with white light as shown in Fig. 10-14-1 [93]. While a relatively large shift of the threshold voltage with the illumination is observed at 77 K, the change in the threshold voltage with illumination at 300 K is small. As shown in Ref. 91,

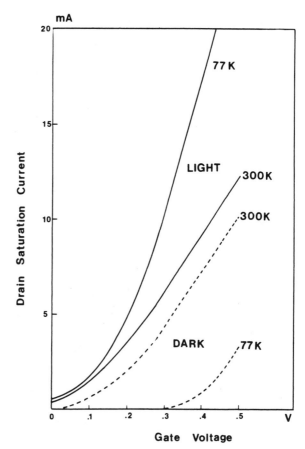

FIGURE 10-14-1. Transfer characteristics of a normally-off FET at 300 and 77 K [93]. Solid line, under light; dashed line, under dark.

this behavior may be explained by the presence of traps in the (Al, Ga)As layer which are the type of traps responsible for the persistent photoconductivity effects in the bulk (Al, Ga)As.

Figure 10-14-2 shows the measured shift in the threshold voltage with temperature for two MODFETs (normally-on and normally-off). The devices were fabricated at the University of Minnesota [91] from (Al, Ga)AlGaAs layers grown by molecular beam epitaxy (MBE). The substrates were semi-insulating (100) GaAs prepared by etching in $H_2SO_4 : H_2O_2 : H_2$, followed by a H_2O rinse. The resulting surface oxide was removed by *in situ* thermal desorption prior to film growth. Tin-doped GaAs layers grown on substrates prepared in this manner have exhibited electron concentrations of 6.3×10^{14} cm^{-3} with mobilities of 8,480 cm^2/V s at 300 K and 89,000 cm^2/Vs at 77 K. The AlGaAs/GaAs layers were grown at 630°C and consisted of 1-μm-thick undoped GaAs buffer layer, covered by 80 Å of undoped $Al_{0.3}Ga_{0.7}As$, covered by 2000 Å of Si-doped (1×10^{18} cm^{-3}) $Al_{0.3}Ga_{0.7}As$. The AlGaAs/GaAs heterojunction structures exhibited mobilities of 6,800 to 7,450 cm^2/V s (300 K) and 22,000 to 43,600 cm^2/V s (77 K) with sheet electron concentrations of $(7–13) \times 10^{11}$/cm^2.

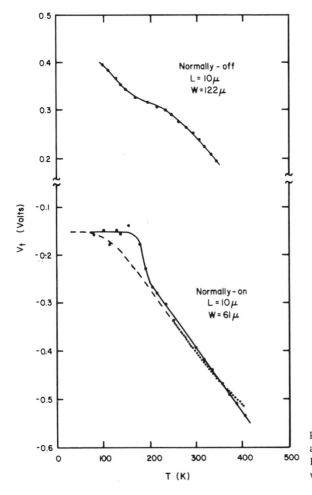

FIGURE 10-14-2. Threshold voltage as a function of temperature for MODFETs [91]. *L*, gate length; and *W*, gate width.

Recessed-gate FETs were formed by thinning the AlGaAs layer down to about 600 Å using a solution of $H_2O_2 : NH_4OH : H_2O$, mesa etching for device isolation, deposition and alloying of Ni/Au/Ge layers for source and drain ohmic contacts, and etching the AlGaAs in the gate region prior to deposition of a 2000-Å-thick film of Al for Schottky-gate metalization. A cross section of the finished device is shown in Fig. 10-14-3. Both normally-on and normally-off devices were fabricated by varying the thickness of the AlGaAs layer underneath the gate. Devices were fabricated with gate lengths of 10 μm, source-to-drain spacing of 30 μm, and gate widths of either 61 or 122 μm.

The static drain current–gate voltage characteristics of a normally-off device in saturation are shown in Fig. 10-14-4 for several different temperatures. At $V_g = 0.8$ V, the intrinsic g_m was 36 mS/mm at 300 K and 55 mS/mm at 77 K. The data of Fig. 10-14-4 were fitted to the model discussed in Sections 10-9 and the channel mobility and the source resistance per gate width were found to be 7,000 cm^2/V s, 20.7 Ω mm, at 77 K, respectively. The source resistance at 300 K is apparently dominated by the sheet resistance of the 2D electron gas in the ungated region between the gate and source electrodes. The source resistance R_s was crudely estimated from the slope of the I–V characteristics at forward gate bias and at low drain-to-source voltages assuming that $R_s \approx (1/3)(dV_{ds}/dI_{ds})$ at $V_{ds} = 0$ as is sometimes done for MESFETs.

As can be seen from Figs. 10-14-2 and 10-14-4, the threshold voltage decreases for both normally-on and normally-off devices with increasing temperature. However, the normally-on devices show little change in V_t until the temperature exceeds about 180 K. Both the rate of change and the total shift in V_t are larger for the normally-on devices.

As has been shown in Section 10-8, the threshold voltage V_t is determined by the barrier height ϕ_b, the conduction band discontinuity ΔE_c, the pinch-off voltage V_{p2} of the (Al, Ga)As layer, and by the difference ΔE_{F0} between the Fermi level and the bottom of the conduction band in GaAs at the heterointerface extrapolated to $n_s = 0$ [see Eqs. (10-8-5)–(10-8-6)]:

$$V_{\text{off}} = \phi_b - \Delta E_c - V_{p2} + \Delta E_{F0}(T) \qquad (10\text{-}14\text{-}1)$$

As can be seen from the calculation given in Section 10-8 only a small part of the threshold voltage shift with temperature (25–35 mV out of the observed 200–400-mV shift) may be explained by the shift of the Fermi level with temperature [$\Delta E_{F0}(T)$ term in Eq. (10-14-1)].

The values of ϕ_b and ΔE_c are also nearly independent of temperature and cannot contribute much to the shift in V_t.

A larger contribution to this shift may come from a subthreshold current. As can be seen from Fig. 10-8-5, owing to the subthreshold effects a noticeable charge

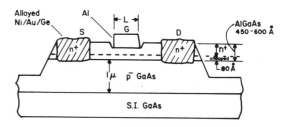

FIGURE 10-14-3. Cross section of a long channel MODFET [91].

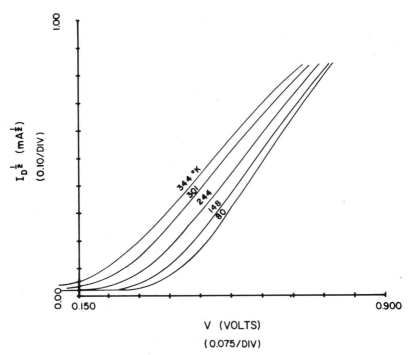

FIGURE 10-14-4. Transfer characteristics of a long channel normally-off FET [91]. The gate length was 10 μm and gate width was 122 μm.

is induced into the channel at voltages smaller than V_t by approximately 2–$3k_BT/q$. This is confirmed by a more accurate analysis [68]. However, this is still not enough to explain a much larger shift in V_t observed experimentally.

Most of the threshold voltage shift has to come from the change in

$$V_{p2} = \frac{qN_d d_d^2}{2\varepsilon} \qquad (10\text{-}14\text{-}2)$$

i.e., due to the change in the effective concentration of donors in the (Al, Ga)As layer.

According to Ref. 91 such a change is related to deep traps in the AlGaAs layer. The existence of deep traps (so called DX centers) in the MBE (Al, Ga)As layers has been revealed by studies of a persistent photoconductivity which occurs at low temperatures (see Section 10-15). It has been demonstrated [90] that deep donor traps exist at an energy (E_T) of the order of 50 meV below the bottom of the conduction band and, in addition, have an energy barrier (E_B) for electron capture of the order of 0.3 eV (see Fig. 10-14-5). The emission time from the traps is given by

$$\tau = \tau_0 \exp[(E_T + E_B)/k_BT] \qquad (10\text{-}14\text{-}3)$$

where T_0 can be estimated as

$$\tau_0 = 1/(sN_cv_T) \qquad (10\text{-}14\text{-}4)$$

where s is the capture cross section, N_c is the effective density of states in the

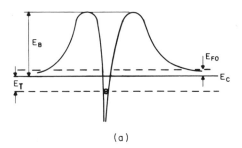

FIGURE 10-14-5. Energy band diagram of AlGaAs with deep traps characterized by capture energy E_B and emission energy $E_T + E_B$ [91]. (a) At equilibrium; (b) in a AlGaAs/GaAs MODFET near threshold. E_F is the Fermi level.

conduction band of AlGaAs, and v_T is the thermal velocity. At low temperatures τ can be much longer than the measurement time. Thus if at the flat-band thermal equilibrium condition, the deep donors are filled (see Fig. 10-14-5a), the occupation of the traps and their charge contributing to V_t will remain unchanged for times less than τ. This is true even if the AlGaAs layer is pinched off and the Fermi level is far below the bottom of the conduction band (see Fig. 10-14-5b). Hence, the value of V_t measured at low temperatures will be different from that measured at high temperatures, where τ is small compared to the measurement time. This effect should be present not only in normally-on but also in normally-off devices where the traps may be at least partially filled by the forward bias, prior to V_t measurements.

The shift in V_t due to the change in the trap occupation may be estimated as follows. If we assume that the deep donors are in thermal equilibrium with the conduction band then the initial value of the threshold voltage (after the instantaneous application of a negative gate bias) is given by

$$V_t(T) = \phi_b - \Delta Ec - \frac{qN_d d_d^2}{2\varepsilon} - \frac{qN_{dT}d_d^2}{2\varepsilon\{1 + g\exp[(E_F + E_T)/k_BT]\}} \quad (10\text{-}14\text{-}5)$$

where N_d is the density of the ionized shallow donors, N_{dT} is the density of deep traps, g is the degeneracy factor of the deep donor level (unity is assumed here), and E_F is the Fermi level for electrons in the AlGaAs layer at thermal equilibrium (measured from the bottom of the conduction band). The fourth term in Eq. (10-14-5) is simply the concentration of the ionized deep donors at thermal equilibrium. [For simplicity we have neglected $\Delta E_{F0}(T)$ in (10-14-5) because this term is small compared to the threshold voltage shift.] This model and Eq. (10-14-5) apply only when the emission time constant is large compared to the measurement time and the traps are in the thermal equilibrium prior to the measurement.

Assuming $d_d \backsimeq 500$ Å and 350 Å for the normally-on and normally-off devices, respectively, we roughly estimate $E_{F0} + E_T \approx 42$ meV from the best fit of the experimental data using Eq. (10-14-5). Here E_{F0} is the difference between the Fermi level and the bottom of the conduction band in (Al, Ga)As at the thermal equilibrium. The concentrations of shallow and deep donors determined from the experimental data were $N_d = 4 \times 10^{17}$ cm^{-3} and $N_{dT} = 7 \times 10^{17}$ cm^{-3}, respectively [91].

The characteristic measurement time was of the order of one second in the temperature range of 250–270 K. Therefore the values of E_T, N_d, and N_{dT} were found by fitting data taken below 250 K [91]. Since the model is applicable over a narrow temperature range, the values of E_T, N_d, and N_{dT} should only be considered as estimates. Nevertheless, if we neglect E_{F0} relative to E_T (since N_d is of the order of the density of states in the conduction band), the value of $E_T \approx 42$ meV deduced from the shift in V_t is in a reasonably good agreement with the values of $E_T = 65$ meV deduced from the analysis of the persistent photoconductivity measurements [90]. This model is also in agreement with the transient capacitance measurements (see below).

At the same time it does not explain a relatively large shift in the threshold voltage at high temperatures (300–400 K) in Fig. 10-14-2 even when the variation of the Fermi level with temperature is included. This may indicate that other types of traps (most likely in the GaAs layer and/or at the interface with AlGaAs) may also contribute to the threshold voltage shift. A more detailed and accurate analysis of this effect is given in Ref. 128.

10-14-2. Transient Capacitance and Current Measurements [91]

More detailed information about the traps, in particular about the value of the barrier energy E_B, may be obtained from transient capacitance [91] or drain current measurements [71].

A large-area Schottky capacitor (area $= 10^{-3}$ cm^{-2}), fabricated using the recessed gate metal, was used for transient capacitance studies in Ref. 71. An ohmic contact was provided to the 2D electron gas with source–drain metallization. This capacitor was used to make transient capacitance measurements as a function of temperature using a digital measurement system.

The qualitative shapes of the transient voltage and capacitance waveforms are shown in Fig. 10-14-6. The highly nonexponential capacitance transients may be explained by the time change of the threshold voltage due to the change in the occupation of traps.

When a large forward gate voltage V_F is applied (see Fig. 10-14-6a) the donor traps in the AlGaAs layer are filled, leading to an increase in the threshold voltage. This results in a drop of the capacitance when the bias voltage is decreased to the value V_R (the reverse bias) because the capacitance changes with the threshold voltage according to curves shown in Fig. 10-8-5. As time goes on, the electrons from the donor traps are emitted and the threshold voltage starts decreasing again, leading to nonexponential capacitance transients of the type shown in Fig. 10-14-6b.

A quantitative analysis of the effect should be based on Eq. (10-14-5) where we now take into account the time dependence of N_d due to the emission from the traps:

$$V_t = \phi_b - \Delta E_c - \frac{q N_d^+(t) d_d^2}{2\varepsilon} \qquad (10\text{-}14\text{-}6)$$

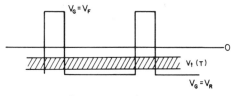

a) Voltage Waveform

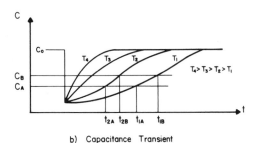

FIGURE 10-14-6. Qualitative transient (a) voltage and (b) capacitance waveform for modulation doped Schottky diode [91].

b) Capacitance Transient

Here

$$N_d^+(t) = N_{d\,\text{tot}} - n_{\text{dt}}(t) \qquad (10\text{-}14\text{-}7)$$

where $n_{\text{dt}}(t)$ is the concentration of the electrons at the deep donor traps and $N_{d\,\text{tot}}$ is the total donor density including all shallow and deep donors. Owing to electron emission $n_{\text{dt}}(t)$ varies with time as

$$n_{\text{dt}}(t) = n_{\text{dt0}}(T)\,e^{-t/\tau} \qquad (10\text{-}14\text{-}8)$$

The concentration of the trapped electrons $n_{\text{dt0}}(T)$ at $t = 0$ depends on how many traps are filled and thus is a function of V_F and T. The emission time constant is given by Eq. (10-14-3). Combining Eqs. (10-14-6)–(10-14-8), we obtain

$$V_t = V_t(t \to \infty) + [qn_{\text{dt0}}(T)/2\varepsilon]d_d^2\,e^{-t/\tau} \qquad (10\text{-}14\text{-}9)$$

where

$$V_t(t \to \infty) = \phi_b - \Delta E_c - (qN_{d\,\text{tot}}/2\varepsilon)d_d^2 \qquad (10\text{-}14\text{-}10)$$

[At low temperatures the value of $V_t(t \to \infty)$ may never be reached because of extremely long time constants.] In the framework of this model we assume that the time dependence of C is determined solely by the time dependence of $V_t(t)$. Therefore the emission time constant $\tau(T)$ may be deduced from the experimental transient curves by picking a value of C and measuring the time it takes for the capacitance to reach this value. This time interval should be a function of temperature. Indeed, for a fixed capacitance C_A the value of V_t should also be fixed. Thus for two temperatures T_1 and T_2 we find from Eq. (10-14-8)

$$n_{\text{dt0}}(T_2)\exp[-t_{2A}/\tau(T_2)] = n_{\text{dt0}}(T_1)\exp[-t_{1A}/\tau(T_1)] \qquad (10\text{-}14\text{-}11)$$

(see Fig. 10-14-6). For another fixed capacitance value C_B we find, in a similar way,

$$n_{\text{dt0}}(T_2)\exp[-t_{2B}/\tau(T_2)] = n_{\text{dt0}}(T_1)\exp[-t_{1B}/\tau(T_1)] \qquad (10\text{-}14\text{-}12)$$

From Eqs. (10-14-10) and (10-14-11) we obtain

$$\exp[(-t_{2A} + t_{2B})/\tau(T_2)] = \exp[(-t_{1A} + t_{1B})/\tau(T_1)] \qquad (10\text{-}14\text{-}13)$$

and

$$(t_{2B} - t_{2A})/\tau(T_2) = (t_{1B} - t_{1A})/\tau(T_1) \qquad (10\text{-}14\text{-}14)$$

Thus by plotting $\ln(t_B - t_A)$ vs. $1/k_B T$, we should be able to obtain the activation energy for emission. If $n_{dt0}(T)$ is a weak function of temperature, we can plot $\ln(t_B)$ vs. $1/k_B T$ to find the same parameter. This means that if we plot C as a function of $\ln(t)$, we should have the same shape of the capacitance curves for different temperatures.

In Fig. 10-14-7 the experimental capacitance vs. $\ln(t)$ curves with temperature as a parameter are presented. The curves have nearly the same shape, which is close to the shape of C_g vs. $V_g - V_t$ curves shown in Fig. 10-8-5. If $n_{dt0}(T)$ is temperature independent, then the distance between the curves at two different temperatures shown in Fig. 10-14-7 should be proportional to $(E_T + E_B)/k_B T$ [see Eq. (10-14-3)] and should be independent of time.

The activation plots $\ln(t_B)$ and $\ln(t_B - t_A)$ vs. $1/T$ are shown in Fig. 10-14-8 for the values of C_A and C_B given in Fig. 10-14-7. Both curves yield the same activation energy for emission (approximately 450 mV), which is somewhat larger than the value of ≈ 300 mV deduced from photoconductivity data [90, 92].

The pre-exponential factor τ_0 in Eq. (10-14-3) was estimated from the measured data to be approximately 10^{-12} s, leading to values of τ of the order of seconds at 200 K and several tenths of a millisecond at 300 K. The estimate for τ at 200 K is

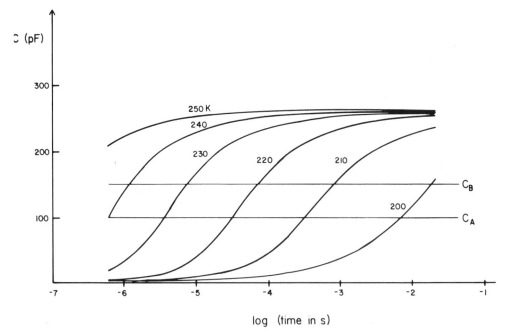

FIGURE 10-14-7. Measured capacitance transients C versus $\ln(t)$ [91]. The values C_A and C_B were used to find the data in Fig. 10-14-8.

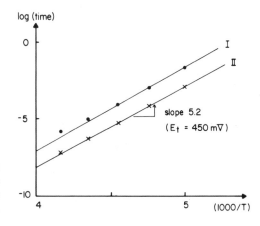

FIGURE 10-14-8. Activation energy plot based on transient capacitance measurement [91]. The slope of the straight line corresponds to the energy for emission from traps in the AlGaAs. I $\ln(t_B)$; II $\ln(t_B - t_A)$.

consistent with the rapid change in the V_t vs. T curve shown in Fig. 10-14-2 in this temperature range. This change may occur because the characteristic measurement time becomes comparable with the emission time of the traps. At 77 K the estimated value of τ is extremely large, which is in agreement with the observed stability of the current–voltage characteristics at 77 K. The emission cross section estimated from τ_0 using $N_c = 1 \times 18$ cm^{-3} and $v_T = 10^7$ cm/s is 10^{-13} cm^2.

A similar transient response should be observed in the current–voltage characteristics as well [127]. The saturation current I_{ds}^s of a long channel device is given by

$$I_{ds}^s = \beta[(V_G - V_t(T, t)]^2/2 \qquad (10\text{-}14\text{-}15)$$

where

$$\beta = \frac{\varepsilon W \mu}{(d + \Delta d)L} \qquad (10\text{-}14\text{-}16)$$

Equation (10-14-15) follows from Eq. (10-9-7) when $V_{sl} = F_s L$ tends to infinity and R_s tends to zero.

According to the model discussed above the threshold voltage should be given by Eq. (10-14-5) when the characteristic measurement time t_m is much smaller than the emission time τ and by Eq. (10-14-10) when the measurement time t_m is much larger than the emission time τ. At a temperature corresponding to $t_m \approx \tau$ there should be a peak in the transient current response. From the activation plot of the log (measurement time) versus the inverse temperature of the peak the activation energy $E_T + E_B$ may be deduced. The results of such measurement are in good agreement with the transient C-V measurements [91].

10-15. BIAS DEPENDENCE AND LIGHT SENSITIVITY OF MODFETs AT 77 K

10-15-1. Introduction

As stated in Section 10-14, by illuminating MODFETs with white light, the positive threshold voltage shift of MODFETs, observed when devices are cooled down to 77 K, can be reduced or eliminated (see Fig. 10-14-1). The data discussed in Section 10-14 suggest that the reduced threshold shift may be attributed to the change of the occupation of traps. The dominant traps in AlGaAs are donor-complex traps (DX). These traps are responsible for a persistent photoconductivity (PPC) in bulk (Al, Ga)As/GaAs modulation doped structures [94]. They also lead to a transient capacitance response with a long time constant of the order of seconds at about 200 K as was discussed in the previous section.

The background impurity level in undoped GaAs grown by MBE is typically less than or of the order of 10^{14} cm^{-3} and is assumed to make only a small contribution to the total light sensitivity of the structure. The light sensitivity of the FETs can be characterized by measuring the changes in the threshold voltage and in the saturation current at cryogenic temperatures under various illumination conditions. If all traps affecting device performance are located in the (Al, Ga)As, then for a given trap density the light sensitivity will depend on the structural parameters of the (Al, Ga)As layer: the doping level, the Al mole fraction, and the thickness of the undoped (Al, Ga)As layer.

In this section we describe the dependence of the I-V characteristics of MODFETs on illumination and the bias applied prior to the measurement. We also demonstrate that the light sensitivity and threshold voltage shift with the growth temperature and Al composition.

10-15-2. Light and Bias Sensitivity

The light sensitivity of MODFETs with Al mole fractions of 0.24 and 0.33 with different growth temperatures of the (Al, Ga)As layer was studied in Ref. 95. Hall measurements made on each structure showed a persistent increase in the transferred electron density after illumination. The $x = 0.33$ layers were prepared with the growth temperature of the (A, Ga)As layer (T_s) between 600°C and 620°C. The $x = 0.24$ transistors were grown as a series with only the growth temperature of the (Al, Ga)As layer varied from 580°C to 695°C. At 300 K the average Hall mobility was 7000 cm^2/V s and the average sheet carrier concentration was 9.5×10^{11} cm^{-2}. Upon exposure to light the mobilities increased or decreased by less than 15% of the dark value. For larger sheet carrier concentrations intersubband scattering became important and caused the mobility to decrease.

In all cases except one (growth temperature $T_s = 600$°C) the mobilities and sheet carrier concentrations in the dark (after the light was removed) were only slightly less than those with illumination. These mobilities are less than those obtained in high mobility structures (see Section 10-6) because only a 20-Å-thick undoped layer was used to separate donors from electrons.

The MODFETs were characterized by measuring the saturation current (I^s_{ds}) versus gate voltage (V_g) characteristics at 300 K, cooling the device to 77 K in the dark, and measuring I^s_{ds} vs. V_g in the dark, under illumination, and again in the dark for persistent effects. MODFETs with $x = 0.33$ typically displayed positive threshold voltage shifts (ΔV_t) of ~0.2 V at 77 K with respect to room temperature and less light sensitivity than those with $x = 0.24$. When exposed to high intensity microscope light illumination ΔV_t decreased to a value between -0.05 and 0.1 V. When the illumination was removed V_t returned to the original dark value $+0.05$ V.

Of the MODFETs with $x = 0.24$ only those with the (Al, Ga)As layer grown at 620°C or below showed good transfer characteristics. At 300 K each device had a threshold voltage of -0.2 V, which shifted to about $+0.2$ V at 77 K. At 300 K the maximum saturation currents and transconductances were greater than 100 mA and 175 mS per mm gate width, respectively. The 77 K saturation current versus gate voltage characteristics are shown in Figure 10-15-1. The measurements were made in the dark after exposure to light with a maximum drain bias of 1.2 V and a maximum gate bias of +0.8 V. The theoretical curve was generated from the known parameters of the layers and the estimated maximum saturation current and source resistances of the layer grown at 620°C. The maximum currents are nearly the same in all three devices; however, the transconductance increases slightly with T_s. This is attributed to improved interface quality and a larger electron saturation velocity [96]. Devices from the layers with $T_s = 660$°C and 695°C showed a degradation in transconductance at the midpoint of the gate voltage swing. The reverse diode characteristics of the gate-drain Schottky diode showed an abrupt breakdown at 0.2–0.3 V for devices fabricated from layers grown at 660°C and 695°C. The breakdown voltage for devices on layers grown at 620°C or less was around 3 V.

Detailed low-temperature FET characterization was performed only for devices on layers grown at $T_s = 580$–620°C. The FETs were cooled in the dark under zero bias conditions. The gate bias was set to +0.2 V/step with two steps and the drain bias was then gradually increased from 0 to 3 V. Except for the devices grown at $T_s = 580$°C, which had the largest dark sheet carrier concentration ($n_s = 9 \times 10^{11}$ cm^{-2} at 77 K), the drain I-V characteristics collapsed at low drain bias when the drain-to-source voltage, V_{ds}, exceeded 0.75 V (see Fig. 10-15-2). Increasing the drain bias further resulted in a larger distortion of the characteristics. The curves shown in the bottom of Fig. 10-15-2 are typical of a device cooled to 77 K in the dark with a maximum V_{ds} of 3 V on the right. Under illumination the original device characteristics were restored, although the drain resistance increased and the saturation current decreased slightly for drain voltages greater than 2 V. When the illumination was removed the distortion of the saturation characteristics did not appear for drain voltages less than 1.2–1.5 V.

The drain I-V characteristics shown in Fig. 10-15-2 were measured for the $T_s = 620$°C device at 77 K after exposure to light. The characteristic on the left is

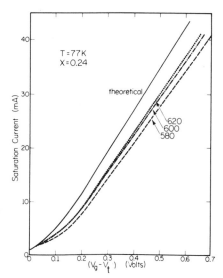

FIGURE 10-15-1. Saturation current versus gate voltage at 77 K for the series of MODFETs with Al$_{0.24}$Ga$_{0.76}$As grown at 580, 600, and 620 C [95]. The theoretical curve was generated to match the estimated source resistance and saturation current of the 620°C device.

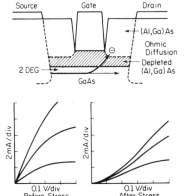

FIGURE 10-15-2. A cross section of a MODFET showing the current conduction path through the two-dimensional electron gas in the GaAs and electron injection into the AlGaAs for large drain voltages [95]. The I-V characteristics before (left) and after (right) applying a 3-V drain bias show the distortion of the I-V at low drain voltage when electrons are trapped after injection into the AlGaAs.

for maximum $V_{ds} < 0.5$ V. The characteristic changed to that on the right after the device had been stressed with a 3-V drain bias. While the I-V characteristic became compressed for $V_{ds} < 0.5$ V, the transconductance measured at $V_{ds} = 1.2$ V was similar to that measured at 1.2 V before stress. When the source and drain leads were reversed the transconductance decreased by a factor greater than 15. The device was then illuminated at zero bias and returned to the dark. The stress cycle was repeated with the source and drain reversed prior to stress with identical results. As expected from the FET geometry, device operation as well as the distortion mechanism is symmetrical. It should be noted that the distortion induced by increasing the drain bias was not a persistent effect. For $V_{ds} < 0.5$ V spontaneous recovery was observed to take place over a span of about 30 min.

When gate biases greater than 0.7 V were applied transconductances decreased and a persistent increase in the threshold shift was observed. By forward biasing the gate-drain diode it was possible to turn the device completely off. Unlike the distortion caused by large drain voltages the threshold shift induced by large positive gate biases was persistent over a 12-h period at 77 K. The threshold shift could be eliminated only by illuminating the FET or by cycling the device back to room temperature.

The current–voltage characteristics of a MODFET depend strongly upon the free electron concentration in the (Al, Ga)As. The activation energy for Si donors in (Al, Ga)As is less than 25 meV [97]. For this activation energy all of the donors in the region depleted by electron transfer should be ionized independent of temperature. The presence of traps, however, can change the effective donor concentration and make the free electron concentration substantially less than the donor concentration (see Section 10-14). The illumination empties the traps and increases the free electron concentration. This effect is responsible for the light sensitivity of MODFETs. We may use curves such as shown in Figs. 10-15-3 and 10-15-4 in order to estimate the change in the free electron concentration in (Al, Ga)As with the effective donor concentration. These curves are calculated using the theory developed in Section 10-4 assuming zero donor activation energy to simplify the calculation of the free electron density in the (Al, Ga)As. As can be seen from these figures, for a fixed trap concentration, increasing the doping level, increasing the Al mole fraction, or decreasing the undoped layer thickness will result in a decrease in light sensitivity, because of a weaker dependence of the interface carrier density on the effective donor density.

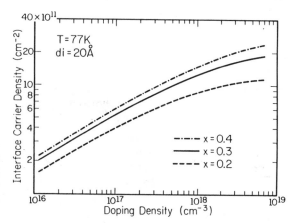

FIGURE 10-15-3. Interface carrier density versus free electron density in AlGaAs [95]. Calculations are for 77 K and an undoped layer thickness $d_i = 20$ Å. The Al mole fraction, x, is used to generate a family of curves.

The devices with $x = 0.33$ had a threshold voltage shift ΔV_t of approximately 0.2 V when cooled to 77 K. Illuminating the MODFETs reduced the threshold shift and in some cases made it negative. After the illumination was removed the threshold voltage shift returned to 0.2 ± 0.05 V. The shift of the threshold voltage with temperature and illumination may be explained by the changes in the occupation of the persistent photoconductivity (PPC) traps considered in Section 10-14. Let us consider, for example, a PPC trap with a net binding energy of 0.1 eV and with thermal barriers for capture and emission of the order of 0.2 and 0.3 eV, respectively. Such a trap has been characterized as a donor-vacancy complex in bulk (Al, Ga)As [92]. The barriers for emission and capture of electrons are due to a large lattice relaxation as the temperature is lowered. Such traps are believed to be responsible for persistent photoconductivity [101, 92–94, 106], and transient capacitance response in (Al, Ga)As/GaAs modulation doped structures as discussed in the previous section.

A reduced threshold voltage shift after illumination may be attributed to the excitation of electrons from the PPC traps in the (Al, Ga)As layer. This may be explained as follows. The threshold voltage is given by Eq. (10-14-1) where a small term $\Delta E_{FO}(T)$ has been neglected:

$$V_t = \phi_b - \Delta E_c(T) - q N_d^+ d_d^2 / (2\varepsilon) \qquad (10\text{-}15\text{-}1)$$

When a device is cooled to 77 K in the dark the threshold voltage is larger (more

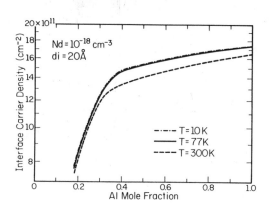

FIGURE 10-15-4. Interface carrier density versus Al mole fraction, x, at 10, 77, and 300 K [95]. The free electron concentration in the AlGaAs is 1×10^{18} cm^{-3} and the undoped layer thickness $d_i = 20$ Å.

positive) than at room temperature by an amount related to the number of electrons frozen into PPC traps [see Eq. (10-14-5)]. At low temperatures the trapped electrons cannot be reemitted from the trap even when the trap level is above the Fermi level in the (Al, Ga)As (see Figure 10-14-5b) because the barrier to emission is too large. Electrons optically excited from these traps are not recaptured (because of the barrier to capture) resulting in a persistent increase in the free electron density in the (Al, Ga)As layer and, hence, in an increase in the transferred electron density. This leads to a threshold voltage close to that at 300 K. The increase in the threshold voltage in the dark after illumination is partially due to the recombination of excess carriers generated under steady state illumination. These carriers typically recombine in about five minutes upon removal of the illumination [90]. In some cases V_t is larger than prior to the illumination, possibly indicating the presence of a second trapping center. The presence of such a center may explain the temperature dependence of V_t at temperatures larger than 300 K (see Section 10-14).

For $x = 0.24$ a series of devices with different growth temperatures for the (Al, Ga)As layer was prepared. The study of these devices shows that, while the interface quality apparently improved with increasing the growth temperature, surface effects associated with high growth temperatures result in poor device performance. Neither large gate nor drain biases cause a degradation of the drain $I\text{-}V$ characteristics of these devices under illumination, although large drain biases do increase the drain resistance and decrease the saturation current slightly. Large drain biases applied to the MODFETs in the dark resulted in the collapse of the drain $I\text{-}V$ characteristics at $V_{ds} < 0.5$ V while transconductances were unaffected for $V_{ds} > 1$ V. Similar behavior has been observed in Si MOSFETs [98, 99] and CdSe thin-film transistors [100]. In those cases such behavior was attributed to hot electrons being injected into and trapped by the oxide layer near the drain. The trapped electrons cause a thin depletion region to be formed in the channel beneath them. To obtain conduction through the channel a drain voltage large enough to punch through the depletion region must be applied. When the drain and source leads are reversed the depletion region appears as a large source resistance, resulting in greatly reduced transconductances.

The (Al, Ga)As layer in MODFETs may act similarly to the oxide layer in a MOSFET. Ideally, the current in a MODFET flows through the two-dimensional electron gas in the GaAs between the n^+ diffusions beneath the source and drain. The potential barrier to electron injection into the (Al, Ga)As, however, is approximately the discontinuity in the conduction bands at the heterojunction, which is about 300 meV. At low temperatures the presence of deep traps in (Al, Ga)As or at the surface in the gate–drain channel or the drain metal–semiconductor interface would prevent the electrons from transferring back into the channel. This is illustrated in Fig. 10-15-2 where it is assumed the electrons are being injected into the (Al, Ga)As on the drain side of the gate where the electric field in the channel is a maximum. The increase in V_{ds} required to distort the $I\text{-}V$ characteristics after cooling in the dark and after illumination at 77 K is attributed to the emptying of the PPC traps in the (Al, Ga)As increasing the free electron density. We should notice that according to the model of Fig. 10-15-2 this injection mostly occurs in the drain-to-gate spacing. This is consistent with the experimental studies of self-aligned FETs which seem to indicate that the collapse of the current–voltage characteristics in self-aligned MODFETs is much less pronounced [129]. This mechanism is similar to the "real space transfer effect" first studied by the University of Illinois group. However, more recently it was pointed out that the presence of the conduction channel under high drain bias is a necessary prerequisite for I-V collapse in MODFETs [130]. This means that the field enhanced trapping in AlGaAs plays an important role in the I-V collapse and that the hot electron transfer from the channel into AlGaAs may not provide enough electrons to fill the traps.

The characteristics of all of the devices indicated the presence of bulk traps in the (Al, Ga)As due to native defects. There are also traps related to the crystal surface. The concentration of these traps seems to be related in part to the growth temperature and Al mole fraction. On the basis of the results obtained for GaAs MESFETs it is very probable that photolithography related carbon contamination at the metal–semiconductor interfaces plays an important role in low-temperature device performance [102].

Recent studies of MODFET characteristics under different gate biases [105] in a temperature range between 77 and 300 K confirmed a dominant effect of deep traps in AlGaAs on device characteristics. These studies showed that low-temperature threshold voltage, transconductance, and the saturation current may either decrease or increase depending on the gate bias conditions (and, hence, the trap occupation during cool-down).

10-16. SELF-ALIGNED MODULATION DOPED TRANSISTORS

10-16-1. Conventional Self-Aligned MODFETs

A tight threshold voltage control and low source series resistance are required in order to fabricate large-scale MODFET integrated circuits (see Chapter 9). These requirements are very difficult to meet using a standard recessed gate etching technique. Consequently, a completely planar self-aligned process has been developed for MODFETs [107–111].

A cross-sectional view of a self-aligned MODFET is shown in Fig. 10-16-1 [107]. The modulation doped structures used in the device fabrication were grown on $\langle 100 \rangle$ oriented semi-insulating GaAs substrates by molecular beam epitaxy. Van der Pauw measurements made on as-grown material typically yielded sheet electron concentration n_{s0} between 8×10^{11} cm^{-2} and 1.5×10^{12} at both 300 and 77 K. The measured Hall electron mobility varied between 4000 and 8000 cm^2/V s at 300 K and 33,000 and 100,000 cm^2/V s at 77 K. The fabrication process [107–111] was

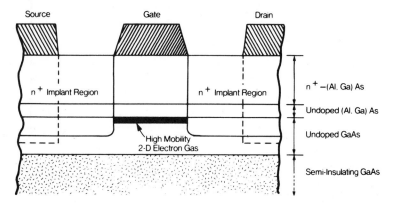

FIGURE 10-16-1. A cross-sectional view of the planar self-aligned MODFET structure [107].

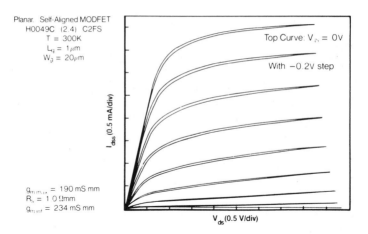

Planar. Self-Aligned MODFET
H0049C (2.4) C2FS
T = 300K
L_q = 1μm
W_g = 20μm

$g_{m\,m\,u}$ = 190 mS mm
R_s = 1 0 Ωmm
$g_{m\,int}$ = 234 mS mm

I_{ds} (0.5 mA/div)

Top Curve: V_{gs} = 0V

With −0.2V step

V_{ds} (0.5 V/div)

FIGURE 10-16-2. The drain-to-source saturation current of a normally-on planar self-aligned MODFET at 300 K [107].

entirely planar with no gate recess required. The ion implantation technique; to form the source and drain regions and to isolate devices is completely compatible with a similar self-aligned process for GaAs integrated circuits. Rapid annealing techniques were used to minimize dopant or impurity drift during the annealing process.

The obtained values of the transconductance were as high as 249 mS/mm gate at 300 K and 300 mS/mm gate at 77 K. Both normally-on and normally-off devices were fabricated. A typical current–voltage characteristic of a normally-on device is shown in Fig. 10-16-2 [107]. In Fig. 10-16-3 we show transfer characteristics compared with the theoretical curves calculated using the charge control model (see Section 10-9).

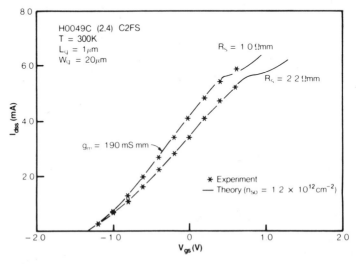

H0049C (2.4) C2FS
T = 300K
L_q = 1μm
W_q = 20μm

R_s = 1 0 Ωmm

R_s = 2 2 Ωmm

g_m = 190 mS mm

∗ Experiment
— Theory (n_{s0} = 1 2 × 10^{12} cm^{-2})

I_{dss} (mA)

V_{gs} (V)

FIGURE 10-16-3. The transfer characteristics of two self-aligned MODFETs having different gate–source spacing at 300 K [107]. Parameters used in the calculation: effective barrier height $\phi_b - \Delta E_c = 0.7$ V, electron mobility in the two-dimensional gas 3000 cm^2/V s, electron saturation velocity in the two-dimensional gas 1.5×10^7 cm/s, threshold voltage −1.42 V.

10-16-2. Multilayer Self-Aligned MODFET [112]

It has been suggested that the dopant diffusion during the annealing process depends on the electric field strength at the (Al, Ga)As/GaAs interface [113]. Novel multilayer MODFET structures showed less than 5% reduction in electron mobility at 77 K after an anneal at 800°C for 15 min. In Ref. 112 normally-off multilayered self-aligned modulation doped FETs with a low electric field at the heterointerface were described and it was demonstrated that the effective values of the electron velocity deduced from the experimental data are, indeed, larger than for the conventional self-aligned MODFETs described above. The values obtained for the electron saturation velocity were 1.9×10^5 m/s and 2.7×10^5 m/s at 300 K and 77 K, respectively, i.e., consistent with the maximum values previously reported for recessed gate devices.

The doping and composition profiles of the modulation doped layers grown by MBE and used to fabricate the multilayer self-aligned MODFETs are shown in Fig. 10-16-4. The composition of the top undoped graded $Al_xGa_{1-x}As$ layer varied linearly from 0.01 at the surface to 0.3. The reason for that was to have a surface close in composition to GaAs because MODFETs with AlGaAs surfaces are known to deteriorate over several months. Also, such a grading may help to reduce the specific contact resistance of alloyed ohmic contacts. Van der Pauw measurements made on these structures yielded typical electron mobilities of 6500 cm^2/V s at 300 K and 120,000 at 77 K. The sheet electron concentration was 5.4×10^{11} cm^{-2} at 300 K and 7.8×10^{11} cm^{-2} at 77 K. The fabrication procedure for the self-aligned gate MODFETs consisted of (1) delineation of the Schottky metal silicide gates, (2) Si ion implantation into the regions on either side of the gate, (3) an 800°C anneal for 15 min with a GaAs proximity cap, (4) ohmic contact formation, and (5) boron implantation to isolate devices.

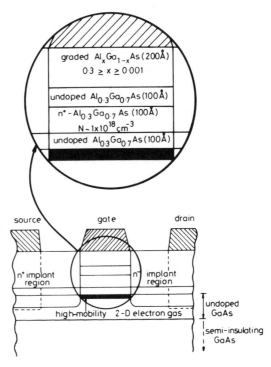

FIGURE 10-16-4. Self-aligned gate multi-layered AlGaAs/GaAs MODFET [112].

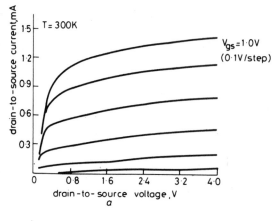

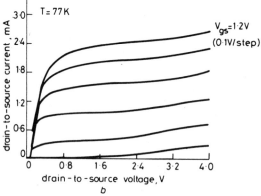

FIGURE 10-16-5. Current–voltage characteristics of self-aligned gate multilayered MODFET at (a) 300 K and (b) 77 K. Gate length 1.1 μm, gate width 20 μm [112].

Typical I–V characteristics for these self-aligned MODFETs are shown in Fig. 10-16-5. The typical measured extrinsic transconductances were 175 mS/mm at 300 K and 290 mS/mm at 77 K. Figure 10-16-6 shows the comparison between the measured drain saturation current and the calculated curves using the charge-control model [112].

10-17. GaAs GATE HETEROJUNCTION FET [114]

As was shown in Sections 10-14 and 10-15 deep traps in doped AlGaAs affect the threshold voltage of modulation doped FETs and the variation of the threshold

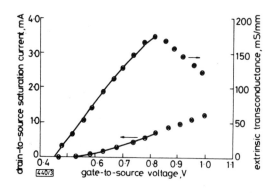

FIGURE 10-16-6. Comparison between the measured transfer curves and the charge control model [112]. Electron mobility of 3300 cm^2/V s and 10,730 cm^2/V s, saturation velocity of 1.9×10^5 and 2.7×10^5 m/s, $n_{s0} = 4.9 \times 10^{11}$ and 3.9×10^{11} cm^{-2} at 300 and 77 K, respectively, were used in the calculation.

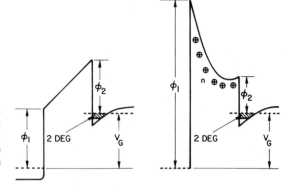

FIGURE 10-17-1. Comparison of band diagrams of GaAs gate FET (left) and conventional MODFET (right) [114]. Dashed lines indicate the position of the Fermi levels. Both devices have the same forward bias applied.

voltage with temperature and illumination. In 1980 J. Rosenberg (see Ref. 3 of 114) proposed to replace the Schottky gate by a GaAs n^+ gate and to eliminate doping in the AlGaAs layer.

The band diagram of such a device is compared with the band diagram of a conventional MODFET in Fig. 10-17.1 [114]. The threshold voltage for both devices may be approximately expressed as

$$V_T = \phi_1 - \phi_2 - qN_D d_d^2/2\varepsilon \qquad (10\text{-}17\text{-}1)$$

where ϕ_1 is the barrier height of the gate/AlGaAs interface, ϕ_2 is the barrier of the AlGaAs/channel interface, N_D and d_d are doping and thickness of the doped AlGaAs layer. For a conventional MODFET $\phi_1 \approx 1$ V, $\phi_2 \approx 0.3$ V and their difference is compensated by the last term in Eq. (10-17-1) if low or negative threshold voltage is desired. In a GaAs gate device $\phi_1 \approx \phi_2$ and $V_T \approx 0$ for $N_D = 0$, i.e., for the undoped AlGaAs layer. Hence, this device is quite similar to a silicon gate MOSFET [114].

The problems with the GaAs device include a substantial leakage current related to a relatively low barrier between the gate and the channel. Studies of GaAs gate capacitor structures [115, 116] showed, however, that the charge density of 10^{12} cm^{-2} may be achieved in the channel with a gate leakage current in a typical device of only 1 μA at 77 K for Al mole fraction in AlGaAs of 0.4. Even smaller leakage current may be achieved in devices with a larger Al mole fraction. Also, the structure should be self-aligned because the ungated regions will provide blocking contacts as AlGaAs is undoped.

A schematic cross section of a GaAs gate FET is shown in Fig. 10-17-2 [114]. The GaAs/AlGaAs/GaAs structures were grown by MBE on (100) semi-insulating GaAs substrates. A 1-μm undoped GaAs buffer layer, a 600-Å Al$_{0.4}$Ga$_{0.6}$As layer, and a 0.4-μm GaAs n^+ silicon doped layer were grown. The doping of the top

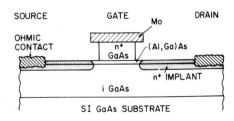

FIGURE 10-17-2. Schematic cross-section of the GaAs gate FET [114].

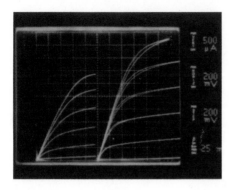

FIGURE 10-17-3. Drain characteristics of GaAs gate FET at room temperature (left) and at 77 K (right) [114]. The bottom trace at both temperatures corresponds to zero gate voltage.

GaAs layer was increased near the surface to improve the gate contact. A 0.15-μm Mo layer was deposited by e-beam evaporation. The layer was patterned using reactive ion etching and CF_4-O_2 plasma. The GaAs was then patterned using a CCl_2F_2-He plasma which stops at the AlGaAs surface [117]. By over-etching the GaAs the Mo stencil was undercut by a controlled amount. Source/drain regions were then implanted using Si at 100 keV at a dose of 5×10^{13} cm^{-2}. The devices were annealed at 700°C for 2 s and Au/Ni/Au-Ge ohmic contacts were formed.

The current–voltage characteristics at room temperature and at 77 K for devices with a gate length of 1.5 μm and a gate width of 25 μm are shown in Fig. 10-17-3 [114]. Maximum measured transconductance was about 240 mS/mm with a source-drain resistance of 2 Ω mm. Typical contact resistances were 0.1–0.3 Ω mm. Gate leakage currents at 77 K were less than 10 μA for gate voltages ranging from −4 to 1.4 V.

10-18. p-CHANNEL MODULATION DOPED FIELD EFFECT TRANSISTORS AND COMPLEMENTARY LOGIC GATES

Advantages of complementary logic include a smaller power dissipation and a larger voltage swing. GaAs JFET complementary circuits have been built [118] and demonstrated a reasonable switching speed. This speed is limited, however, by a low hole mobility in GaAs leading to a very low transconductance in p-MESFETs of only 5 mS/mm [118].

In modulation doped heterostructures, in addition to the conduction band discontinuity, there is a valence band discontinuity (see Fig. 10-2-1b). Originally the valence band discontinuity was assumed to be 15% of the difference in the energy gaps, ΔE_g (so-called Dingle rule) but more recent studies (see, for example,

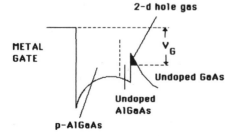

FIGURE 10-18-1. Qualitative band diagram of a p-channel MODFET.

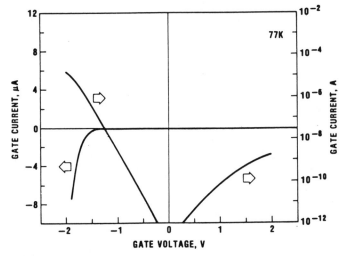

FIGURE 10-18-2. Gate current–voltage characteristic of a *p*-channel MODFET at 77 K [124]. The same turn-on voltage (≈ -1.5 V) was observed also at 300 K.

Ref. 119) seem to indicate that it is a considerably larger fraction (perhaps, 35%) of ΔE_g. As a consequence, the two-dimensional hole electron gas may be localized at the heterointerface [120, 121]. It has been demonstrated that the mobility of the two-dimensional hole gas follows quite closely the hole mobility in high purity undoped GaAs with the values as high as 5000 cm^2/V s at 77 K and 43,000 cm^2/V s at 4.2 K [121]. These values compare quite favorably with electron mobilities in GaAs MESFETs at 300 K.

The qualitative band diagram of a *p*-type modulation doped structure is shown in Fig. 10-18-1. It is similar to the inverted band diagram of an *n*-type modulation

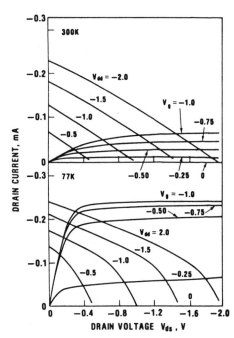

FIGURE 10-18-3. Drain current–voltage characteristics of *p*-channel driver transistor and *p*-channel load transistor [124]. V_{dd}, supply voltage; V_g, gate voltage.

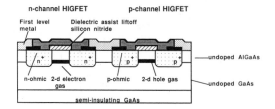

FIGURE 10-18-4. Cross-section of a complementary HIGFET inverter [134].

doped structure. Beryllium doping was used in Ref. 121 and it is interesting to notice that the *p*-type Be-doped modulation doped structures did not exhibit persistent photoconductivity [121]. Similar to *n*-channel modulation doped layers, *p*-channel modulation doped structures may be used to fabricate *p*-channel MODFETs and *p*-channel MODFET logic gates [122–126].

Modulation doped *p*-channel (Al, Ga)As/GaAs FETs with transconductances of 28 mS/mm at 77 K [122], 35 mS/mm at 77 K [123], and 44 mS/mm at 4.2 K [122] have been reported. At 300 K the peak transconductance was 5 mS/mm [123].

DCFL *p*-channel MODFET inverters described in Ref. 124 were fabricated on a four-layer Be-doped structure grown by MBE. The deposition of 1 μm of undoped GaAs was followed by the deposition of 75 Å of undoped $Al_{0.6}Ga_{0.4}As$, 450 Å of *p*-$Al_{0.6}Ga_{0.4}As$ doped at 2×10^{18} cm^{-3}, and 50 Å of *p*-type GaAs doped at 4×10^{18} cm^{-3}. A hole sheet concentration of 1.5×10^{12} cm^{-2} was obtained with a hole mobility of 190 cm^2/V s and 1800 cm^2/V s at 300 K and 77 K, respectively. Au–Be contacts were fabricated. The contact resistance of 19 Ω at 300 K was determined from TLM measurements.

Ti/Au 1.5-μm-long gates were chemically recessed into *p*-AlGaAs. The depth of recess controlled the threshold voltage of the driver FETs and the saturation current of load transistors.

The gate current–voltage characteristics of a driver *p*-MODFET are shown in Fig. 10-18-2 [124]. The drain current–voltage characteristics of the driver and load FETs at 77 and 300 K are presented in Fig. 10-18-3 [124]. These curves clearly illustrate the increase in transconductance at 77 K because of a larger hole mobility.

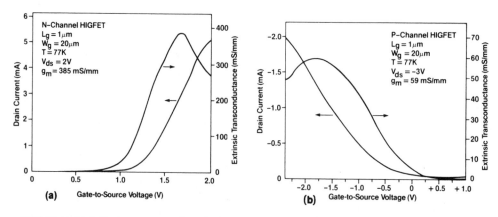

FIGURE 10-18-5. Drain current and extrinsic transconductance for *n*-channel (a) and *p*-channel (b) HIGFETs [133]. Measured (crosses) and calculated (solid lines) current-voltage characteristics of *n*- and *p*-channel HIGFETs.

Eleven-stage ring oscillators studied in Ref. 124 exhibited a propagation delay of up to 233 ps at 77 K with a power dissipation of 0.31 mW for a supply voltage -2 V. The lowest obtained power-delay product was 9.1 fJ. No oscillations were observed at 300 K [124].

Complementary p-MODFET and heterojunction n-type GaAs MESFETs were described in Ref. 126.

Recently a complementary heterostructure insulated gate FET (C-HIGFET) technology based on n- and p-channel AlGaAs/GaAs heterostructure devices was developed [131-134]. This technology offers the potential for high-speed, low-power gates for VLSI circuit applications. In addition to circumventing the trapping problems associated with DX centers in MODFET devices, C-HIGFET circuits are more tolerant to threshold voltage variations than DCFL circuits thus resulting in higher yield. Complementary AlGaAs/GaAs heterostructure insulated gate field effect transistor (C-HIGFET) test circuits (see Fig. 10-18-4) have been fabricated using MBE and a self-aligned ion-implantation process. The transconductance of n-channel devices was as high as 385 mS/mm at 77 K and 218 mS/mm at 300 K [131-132] (see Fig. 10-18-5a). The transconductance of pn-channel devices was as high as 59 mS/mm at 77 K and 28 mS/mm at 300 K [131-132] (see Fig. 10-18-5b). Ring oscillators have been demonstrated with a minimum gate propagation delay of 76 ps [134]. Minimum stand-by power as low as 23 μW has been achieved in complementary RAM cells. Charge control models of HIGFETs including effects of gate current were implemented in a UM-SPICE program to simulate the complementary circuits. Simulations show that the gate current in p- and n-channel devices plays a dominant role in determining the propagation delay and power consumption. The design (i.e. the optimum ratio of the n- and p-channel device widths) is also dependent on the gate currents as well as on the n-HIGFET threshold voltage, V_{tn}. The lowering of V_{tn} from the present value of 0.9 V to 0.3 V should further decrease the gate propagation delay. Gate propagation delay of close to 50 ps is estimated for a 1-μm C-HIGFET technology. Further improvements in speed and power dissipation can be achieved by increasing the AlAs molar fraction, decreasing the AlGaAs thickness, and going to submicron dimensions [134].

REFERENCES

1. W. Shockley, U.S. Patent 2,569,347 (1951).
2. A. I. Gubanov, *Zh. Tekh. Fiz.* **21**, 304 (1951); *Zk. Eksp. Teor. Fiz.* **21**, 721 (1951).
3. H. Kroemer, Theory of a wide-gap emitter for transistors, *Proc. IRE* **45**, 1535 (1957).
4. R. A. Anderson, *IBM J. Res. Dev.* **4**, 283 (1960).
5. R. L. Anderson, Experiments on Ge-GaAs heterojunctions, *Solid State Electron.* **5**, 341 (1962).
6. M. J. Adams and A. Nussbaum, A proposal for a new approach to heterojunction theory, *Solid-State Electron.* **22**, 783-791 (1979).
7. O. von Roos, Theory of extrinsic and intrinsic heterojunctions in thermal equilibrium, *Solid-State Electron.* **23**, 1069-1075 (1980).
8. H. Kroemer, Critique of two recent theories of heterojunction lineups, *IEEE Electron Devices Lett.* **EDL-4**(2), 25-27 (1983).
9. H. Unlu and A. Nussbaum, Band discontinuities as heterojunction device design parameters, *IEEE Trans. Electron Devices*, **ED-33**, 616-619 (1986).
10. A. G. Milnes and D. L. Feucht, *Heterojunctions and Metal-Semiconductor Junctions*, Academic, New York, 1972.
11. B. L. Sharma and R. K. Purohit, *Semiconductor Heterojunctions*, Pergamon, London, 1974.

12. H. C. Casey, Jr. and M. B. Panich, *Heterojunction Lasers*, Academic, New York, 1978.
13. T. L. Taunsley, *Heterojunction Properties in Semiconductors and Semimetals*, Vol. 7, pp. 294–366, Academic, New York, 1971.
14. H. Kressel and J. K. Butler, *Semiconductor Lasers and Heterojunction LED's*, Academic, New York, 1977.
15. H. Kressel and J. K. Butler, *Heterojunction Laser Diodes in Semiconductors and Semimetals*, Vol. 14, pp. 66–192, Academic, New York, 1979.
16. L. Esaki and R. Tsu, Internal Report RC 2418, IBM Research, March 26, 1969.
17. R. Dingle, H. L. Stormer, A. C. Gossard, and W. Wiegmann, *Appl. Phys. Lett.* **37**, 805 (1978).
18. T. Mimura, S. Hiyamizu, T. Fujii, and K. Nanbu, A new field effect transistor with selectively doped GaAs/n-Al$_x$Ga$_{1-x}$As heterojunctions, *Jpn. Appl. Phys.* **19**, L225–L227 (1980).
19. S. S. Perlman and D. L. Feucht, p-n heterojunctions, *Solid State Electron.* **7**, 911–923 (1964).
20. J. A. Van Vechten, *Phys. Rev.* **187**, 1007 (1969).
21. See the review by F. Stern, Quantum properties of surface space-charge layers, *CRC Crit. Rev. Solid State Sci.*, 499 (1974).
22. F. Stern, Self-consistent results for n-type Si inversion layers, *Phys. Rev. B* **5**, 4891 (1972).
23. F. F. Fang and W. E. Howard, Negative field-effect mobility on (100) Si surfaces, *Phys. Rev. Lett.* **16**, 797 (1966).
24. D. Delagebeaudeuf and N. T. Ling, Metal-(n)AlGaAs–GaAs two dimensional electron gas FET, *IEEE Trans. Electron Devices* **ED-29**, 955 (1982).
25. Kwyro Lee, Michael Shur, Timothy J. Drummond, S. L. Su, W. G. Lyons, R. Fisher, and Hadis Morkoc, Design and fabrication of high transconductance modulation-doped (Al, Ga)As/GaAs FETs, *J. Vac. Sci. Technol. B* **1**(2), 186–189 (1983).
26. W. B. Joyce and R. W. Dixon, Analytic approximations for the Fermi energy of an ideal Fermi gas, *Appl. Phys. Lett.* **31**, 354 (1977).
27. Kwyro Lee, Michael Shur, Tim Drummond, and Hadis Morkoc, Electron density in the two-dimensional electron gas in modulation doped layers, *J. Appl. Phys.* **54**(4), 2093–2096 (1983).
28. See for example, R. A. Smith, *Semiconductors*, Cambridge University Press, Cambridge, Second Edition, p. 83 (1978).
29. T. J. Drummond, H. Morkoc, K. Lee, and M. S. Shur, Model for modulation doped field effect transistor, *IEEE Electron Device Letters* **EDL-3**(11), 338–341 (1981).
30. T. J. Drummond, W. Kopp, M. Keever, H. Morkoc, and A. Y. Cho, Electron mobility in single and multiple period modulation-doped (Al, Ga)As/GaAs heterostructures, *J. Appl. Phys.* **53**(2), 1023–1027 (1982).
31. K. Lee, M. S. Shur, J. Klem, T. J. Drummond, and H. Morkoc, Parallel conduction correction to measured room mobility in (Al, Ga)As–GaAs modulation doped layers, *Jpn. J. Appl. Phys.* **23**(4), L230–231 (1984).
32. A. A. Grinberg and M. S. Shur, Density of two-dimensional electron gas in modulation-doped structure with graded interface, *Appl. Phys. Lett.* **45**(5), 573–574 (1984).
33. A. A. Grinberg and M. S. Shur, Modulation-doped structures with graded interfaces, *J. Appl. Phys.* **57**(4), 1242–1246 (1985).
34. F. Stern and W. E. Howard, *Phys. Rev.* **163**, 816 (1967).
35. A. A. Grinberg and M. S. Shur, Effect of image charges on impurity scattering of two-dimensional electron gas in AlGaAs/GaAs, *J. Appl. Phys.* **58**(1), 382–386 (1985).
36. Kwyro Lee, Michael Shur, Timothy J. Drummond, and Hadis Morkoc, Low field mobility of 2-d electron gas in modulation doped structures, *J. Appl. Phys.* **54**, 2093 (1983).
37. J. V. DiLorenzo, R. Dingle, M. Feuer, A. C. Gossard, R. Hendel, J. C. M. Hwang, A. A. Kastalsky, V. G. Keramidas, R. A. Keihl, and P. O'Connor, *IEDM Tech. Dig.* **25**, 578 (1982).
38. P. J. Price, *J. Vac. Sci. Technol.* **19**, 599 (1981).
39. K. Hess, *Appl. Phys. Lett.* **35**, 484 (1979).
40. S. Mori and T. Ando, *J. Phys. Soc. Jpn.* **48**, 865 (1980).
41. Y. Takeda, H. Kamei, and A. Sasaki, *Electron. Lett.* **18**(7), 309 (1982).
42. F. Stern, *Phys. Rev. Lett.* **44**(22), 1469 (1980).
43. F. Stern, *Phys. Rev. Lett.* **18**, 546 (1976).
44. N. T. Linh, in *Festkorperprobleme* (*Advances in Solid State Physics*), Ed. by P. Grosse, Vieweg, Braunschweig, 1983, Vol. XXIII, p. 227.
45. H. L. Stormer, *Surf. Sci.* **132**, 519 (1983).
46. R. Ando, *J. Phys. Soc. Jpn.* **51**, 3900 (1982).
47. G. Fishman and D. Calecki, *Physica B* **117/118**, 744 (1983).

48. F. Stern, *Appl. Phys. Lett.* **43**, 974 (1983).
49. B. Vinter, *Appl. Phys. Lett.* **44**, 307 (1984).
50. Hadis Morkoc, Modulation doped $Al_xGa_{1-x}As/GaAs$ field effect transistors (MODFETs): Analysis, fabrication, performance, in *Molecular Beam Epitaxy and Heterostructures*, Nato Adv. Study Institute, Ed. by L. L. Chang and K. Kloog, Martinus Nijhoff, The Hague, 1983.
51. P. F. Maldacue, *Surf. Sci.* **73**, 296 (1978).
52. J. V. DiLorenzo, R. Dingle, M. Feuer, A. C. Gossard, R. Hendel, J. S. M. Hwang, A. Kastalsky, V. G. Keramidas, R. A. Kiehl, and P. O'Connor, Material and device considerations for selectively doped heterojunction transistors, *IEDM Tech. Dig.* **25**(1), 578 (1982).
53. B. Vinter, Phonon-limited mobility in GaAlAs/GaAs heterostructures, *Appl. Phys. Lett.* **45**(5), 581-583 (1984).
54. D. L. Rode, in *Semiconductors and Semimetals*, Ed. R. K. Willardson and A. C. Beer, Academic, New York, 1975, Vol. 10, pp. 4-28.
55. S. Hiyamizu, J. Saito, K. Nanbu, and T. Ishikawa, *Jpn. J. Appl. Phys.* **22**, L609 (1983).
56. N. Sano, H. Kato, and S. Chika, *Solid State Commun.* **49**, 123 (1984).
57. J. C. Hwang, A. Kastalsky, H. L. Stormer, and V. G. Keramidas, *Appl. Phys. Lett.* **44**, 802 (1984).
58. M. Heiblum, E. E. Mendez, and F. Stern, *Appl. Phys. Lett.* **44**, 1064 (1984).
59. L. F. Eastman, Private Communication, 1984.
60. K. Hess, in Proc. 3rd Int. Cont. on Hot Carriers in Semicond., Montpellier, 1981, *J. Phys. (Paris)* **C7**, 3 (1981).
61. M. Inoue, S. Hiyamizu, H. Hida, H. Hashimoto, and Y. Inuishi, in Proc. 3rd Int. Cont. on Hot Carriers in Semicond. Montpellier, 1981, *J. Phys. (Paris)* **C7**, 19 (1981).
62. M. Inoue, S. Hiyamizu, M. Inayama, and Y. Unuishi, in Proc. Int. Conf. Solid State Devices, Tokyo 1982, *Jpn. J. Appl. Phys.* **22**, Suppl. 22-1, 357 (1983).
63. T. J. Drummond, M. Keever, W. Kopp, H. Morkoc, K. Hess, and B. G. Streetman, Field dependence of mobility in $Al_{0.2}Ga_{0.8}As/GaAs$ heterojunctions at very low fields, *Electron. Lett.* **17**(15), 545-546 (1981).
64. T. J. Drummond, M. Keever, and H. Morkoc, *Jpn. J. Appl. Phys.* **21**, L65 (1982).
65. M. Inoue, M. Inayama, and S. Hiyamizu, Parallel electron transport and field effects of electron distributions in selectively doped GaAs/n-AlGaAs, *Jpn. J. Appl. Phys.* **22**(4), L213-L215 (1983).
66. M. Tomizawa, K. Yokoyama, and A. Yoshii, Hot-electron velocity characteristics of AlGaAs/GaAs heterostructures, *IEEE Electron Devices Lett.* **EDL-5**(11), 464-465 (1984).
67. S. Blakemore, Semiconductor and other major properties of GaAs, *J. Appl. Phys.* **53**(10), R123-R181 (1982).
68. T. H. Chen, High speed GaAs device and integrated circuit modeling and simulation, Ph.D. thesis, University of Minnesota, 1984.
69. M. B. Das, A high ratio design approach to millimeter-wave HEMT structures, *IEEE Trans. Electron Devices* **ED-32**(1), 11-17 (1985).
70. S. Hiyamizu and T. Mimura, MBE-grown selectively doped GaAs/N-AlGaAs heterostructures and their application to high electron mobility transistors, *Semiconductor Technologies*, Ed. by J. Nishizawa, North Holland, Amsterdam, The Netherlands, 1982, p. 258-271.
71. N. C. Cirillo, Jr., M. Shur, and J. K. Abrokwah, Inverted GaAs/AlGaAs Modulation-Doped Field-Effect Transistors with extremely high transconductances, *IEEE Electron Device Lett.*, **EDL-7**(2), 71-74 (1986).
72. K. Lee, M. S. Shur, T. J. Drummond, and H. Morkoc, Current-voltage and capacitance-voltage characteristics of modulation doped field effect transistors, *IEEE Trans. Electron Devices* **ED-30**(3), 207-212 (1983).
73. P. L. Hower and G. Bechtel, Current saturation and small-signal characteristics of GaAs field-effect transistors, *IEEE Trans. Electron Devices* **ED-20**, 213 (1973).
74. C. H. Hyun, M. S. Shur, and N. C. Grillo, Jr., Design and simulation of modulation doped integrated circuits, *IEEE Trans. Computer Aided Design*, **CAD-5**(2), 284-292 (1986).
75. J. Yoshida and M. Kurata, Analysis of high electron mobility transistors based on two-dimensional numerical model, *IEEE Electron Device Lett.* **EDL-5**(12), 508-510 (1984).
76. T. J. Drummond, R. Rischer, P. Miller, H. Morkoc, and A. Y. Cho, Influence of substrate temperatures on electron mobility in normal and inverted single period modulation doped $Al_xGa_{1-x}As/GaAs$ structures, *J. Vac. Sci. Technol.* **21**, 684-688 (1982).
77. R. E. Thorne, R. Rischer, S. L. Su, W. Kopp, T. J. Drummond, and H. Morkoc, Performance of inverted structure modulation doped Schottky barrier field effect transistors, *Jpn. J. Appl. Phys. Lett.* **21**, L223-L224 (1982).

78. D. Delagebeaudeuf and N. T. Linh, Charge control of the heterojunction two-dimensional electron gas for MESFET application, *IEEE Trans. Electron Devices* **ED-28**(7), 790 (1981).

79. K. Lee, M. S. Shur, T. J. Drummond, H. Morkoc, Charge control model of an "inverted" GaAs-(AlGa)As modulation doped structure, *J. Vac. Sci. Technol. B* **2**(2), 113–116 (1984).

80. H. Morkoc, Current transport in modulation doped (Al, Ga)As/GaAs heterostructures: Applications to high speed FETs, *IEEE Electron Devices Lett.* **EDL-2**, 260 (1981).

81. T. J. Drummond, S. L. Su, W. G. Lyons, R. Fischer, W. Kopp, H. Morkoc, K. Lee, and M. S. Shur, Enhancement of electron velocity in modulation doped (Al, Ga)As/GaAs FETs at cryogenic temperatures, *Electron. Lett.* **18**(24), 1057 (1982).

82. T. Mimura, S. Hiyamizu, K. Joshin, and K. Hikosaka, Enhancement-mode high electron mobility transistors for logic applications, *Jpn. J. Appl. Phys.* **20**(5), L317 (1981).

83. T. J. Drummond, R. Fisher, S. L. Su, W. G. Lyons, H. Morkoc, K. Lee, and M. S. Shur, Characteristics of modulation doped $Al_xGa_{1-x}As$/GaAs field effect transistors: Effect of donor-electron separation, *Appl. Phys. Lett.* **42**(3), 262 (1983). See also T. J. Drummond, S. L. Su, W. Kopp, R. Fischer, R. E. Thorne, H. Morkoc, K. Lee, and M. S. Shur, High velocity *N*-on and *N*-off modulation doped GaAs/$Al_xGa_{1-x}As$ FETs, *IEEE, Proc. IEDM Tech. Dig.* **25**(1), 586 (1982).

84. M. Laviron, D. Delagebeaudeuf, P. Delescluse, J. Chaplart, and N. T. Linh, Low-noise two-dimensional electron gas FET, *Electron. Lett.* **17**, 536 (1981).

85. J. H. Baek, M. S. Shur, and N. C. Cirillo, Jr., Temperature and gate length dependence of MODFET parameters, unpublished.

86. K. Lee, M. S. Shur, T. J. Drummond, and H. Morkoc, Parasitic MESFET in (Al, Ga)As/GaAs modulation doped FETs and MODFET characterization, *IEEE Trans. Electron Devices*, **ED-31**(1), 29–35 (1984).

87. K. Lee, M. S. Shur, T. J. Drummond, and H. Morkoc, A unified method for characterizing (Al, Ga)As/GaAs MODFETs including parasitic MESFET conduction in the (Al, Ga)As, in Proceedings of Biannual IEEE Conferences on High Speed Devices, Cornell Univ., August 1973, pp. 177–186.

88. K. Lee, M. S. Shur, A. J. Valois, G. Y. Robinson, X. C. Zhu, and A. van der Ziel, A new technique for characterization of "end" resistance in modulation doped FETs, *IEEE Trans. Electron Devices*, **ED-31**, 1394–1398 (1984).

89. M. B. Das, W. Kopp, and H. Morkoc, Determination of carrier saturation velocity in short-gate-length modulation doped FETs, *IEEE Electron Device Lett.* **EDL-5**, 446–448 (1984).

90. J. F. Rochette, P. Delescluse, M. Laviron, D. Delagebeaudeuf, J. Chevrier, and N. T. Linh, Low temperature persistent photoconductivity in two-dimensional GaAs FETs, Proceedings 1982 Symposium on GaAs and Related Compounds, Albuquerque, New Mexico.

91. A. J. Valois, G. Y. Robinson, K. Lee, and M. S. Shur, Temperature dependence of the *I–V* characteristics of modulation-doped FETs, *J. Vac. Sci. Technol. B* **1**(2), 190–195 (1983).

92. R. J. Nelson, Long-lifetime photoconductivity effect in *n*-type GaAlAs, *Appl. Phys. Lett.* **31**(5), 351 (1977).

93. Kwyro Lee, Modulation doped $Al_xGa_{1-x}As$/GaAs heterojunction field effect transistors, Ph.D. thesis, University of Minnesota, 1983.

94. T. J. Drummond, W. Kopp, R. Fischer, H. Morkoc, R. E. Thorne, and A. Y. Cho, Photoconductivity effects in extremely high mobility modulation-doped (Al, Ga)As/GaAs heterostructures, *J. Appl. Phys.* **53**(2), 1238 (1982).

95. T. J. Drummond, W. G. Lyons, S. L. Su, W. Kopp, H. Morkoc, K. Lee, and M. S. Shur, Bias dependence and light sensitivity of (Al, Ga)As/GaAs MODFETs at 77 K, *IEEE Electron Devices* **ED-30**(12), 1806–1811 (1983).

96. H. Morkoc, T. J. Drummond, and R. Fischer, Interfacial properties of (Al, Ga)As/GaAs structures: Effect of substrate temperature during growth by MBE, *J. Appl. Phys.* **53**, 1030–1033 (1982).

97. R. Fischer, C. G. Hopkins, C. A. Evans, Jr., T. J. Drummond, W. G. Lyons, J. Klem, C. Colvard, and H. Morkoc, The properties of Si in $Al_xGa_{1-x}As$ grown by molecular beam epitaxy, Proceedings 1982 Symposium on GaAs and Related Compound, Albuquerque, New Mexico.

98. T. H. Ning, C. M. Osburn, and H. N. Yu, Effect of electron trapping on IGFET characteristics, *J. Elect. Matl.* **6**, 65–76 (1977).

99. L. Forbes, E. Sun, R. Alders, and J. Moll, Field induced reemission of electrons trapped in SiO_2, *IEEE Trans. Electron Devices* **ED-26**, 1816–1818 (1979).

100. J. J. Wysocki, Drain-current distortion in CdSe thin-film transistors, *IEEE Trans. Electron Devices* **ED-29**, 1798–1805 (1982).

101. D. V. Lang, R. A. Logan, and H. Jaros, Trapping characteristics and a donor-complex (DX) model for the persistent-photoconductivity trapping center in Te-doped $Al_xGa_{1-x}As$, *Phys. Rev. B* **19**, 1015–1030 (1979).

102. T. H. Miers, Schottky contact fabrication for GaAs MESFETs, *J. Electrochem. Soc.* **129**, 1795–1799 (1982).

103. M. S. Shur and N. C. Cirillo, Temperature and gate length dependency of important MODFET parameters, presented at WOCSEMMAD-85.

104. K. Hess, H. Morkoc, H. Shichijo, and B. G. Streetman, *Appl. Phys. Lett.* **35**, 469 (1979); see also K. Hess, *Physica* **117B**, 723 (1983) and references therein.

105. J. Y. Chin, R. P. Holmstrom, and J. P. Salerno, Effect of traps on low-temperature high electron mobility transistor characteristics, *IEEE Electron Device Lett.* **EDL-5**(9), 381–384 (1984).

106. J. Klem, T. Masselink, D. Arnold, R. Fischer, T. J. Drummond, H. Morkoc, K. Lee, and M. S. Shur, Persistent photoconductivity in (Al, Ga)As/GaAs modulation doped structures: Dependence on structure and growth temperature, *J. Appl. Phys.* **54**(9), 5214–5217 (1983).

107. N. C. Cirillo, J. K. Abrokwah, and M. S. Shur, Self-aligned modulation-doped (Al, Ga)As/GaAs field-effect transistors, *IEEE Electron Device Lett.* **EDL-5**(4), 129–131 (1984).

108. N. C. Cirillo, J. K. Abrokwah, and M. S. Shur, S self-aligned gate process for ICs based on modulation doped (Al, Ga)As/GaAs FETs, Proc. 42nd Dev. Res. Conf., June 1984, p. IIA-4.

109. M. S. Shur, T. H. Chen, C. H. Hyun, P. N. Jenkins, and N. C. Cirillo, Jr., Design and simulation of self-aligned modulation doped AlGaAs/GaAs ICs, ISSCC 1985 Digest of Technical papers, pp. 264–265, published by Lewis Winner, Coral Gables, Florida 33134, 1985.

110. N. C. Cirillo, J. K. Abrokwah, and S. Jamison, A self-aligned gate modulation doped AlGaAs/GaAs FET IC process, 1984, GaAs IC symposium Technical Digest, pp. 167–170, October 1984.

111. M. Shur, J. K. Abrokwah, R. R. Daniels, D. K. Arch, and N. C. Cirillo, Jr., *Extended Abstracts of the 18th (1986 International) Conference on Solid State Devices and Materials*, Tokyo, 1986, pp. 363–366.

112. N. C. Cirillo, A. Fraasch, H. Lee, L. F. Eastman, M. S. Shur, and S. Baier, Novel multilayer modulation doped (Al, Ga)As/GaAs structures for self aligned gate FETs, *Electron. Lett.* **20**(21), 854–855 (1984).

113. H. Lee, G. Wicks, and L. F. Eastman, High temperature annealing of modulation doped GaAs/AlGaAs heterostructures for FET applications, Proc. of IEEE/Cornell Conf. on high-speed semiconductor devices and circuits, August 1984, pp. 204–208.

114. P. Solomon, C. M. Knoedler, and S. L. Wright, A GaAs gate heterojunction FET, *IEEE Electron Device Lett.* **EDL-5**(9), 379–381 (1984).

115. P. M. Solomon, T. W. Hickmott, H. Morkoc, and R. Fischer, *Appl. Phys. Lett.* **42**, 821 (1983).

116. T. W. Hickott, P. M. Solomon, R. Fischer, and H. Morkoc, *Appl. Phys. Lett.* **44**, 90 (1984).

117. K. Hikosaka, T. Mimura, and K. Joshin, *Jpn. J. Appl. Phys.* **20**, L847 (1981).

118. R. Zuleg, J. K. Notthoff, and G. L. Troeger, Double-implanted GaAs complementary JFET, *IEEE Electron Device Lett.* **EDL-5**, 21, 23 (1984).

119. D. Arnold, A. Ketterson, T. Henderson, J. Klem, and H. Morkoc, Determination of the valence band discontinuity between GaAs and (Al, Ga)As by the use of p^+-GaAs-(Al, Ga)As-p^--GaAs capacitors, *Appl. Phys. Lett.* **45**(11), 1237–1239 (1984).

120. H. L. Störmer and W. T. Tang, *Appl. Phys. Lett.* **36**, 685 (1980).

121. H. L. Störmer, A. C. Gossard, W. Wiegman, R. Blondel, and K. Baldwin, Temperature dependence of the mobility of two-dimensional hole systems in modulation-doped GaAs-(Al, Ga)As, *Appl. Phys. Lett.* **44**, 139–141 (1984).

122. H. L. Störmer, K. Baldwin, A. C. Gossard, and W. Wiegman, Modulation-doped field effect transistor based on a two-dimensional hole gas, *Appl. Phys. Lett.* **44**, 1062–1064 (1984).

123. W. I. Wang and S. Tiwari, p-channel $Ga_{0.5}Al_{0.5}As$/GaAs MODFETs, presented at 42nd Annual Device Res. Conf., Santa Barbara, June 1984.

124. R. A. Kiehl and A. C. Gossard, p-channel (Al, Ga)As/GaAs modulation doped logic gates, *IEEE Electron Devices Lett.* **EDL-5**(10), 420–422 (1984).

125. R. A. Kiehl, H. L. Störmer, K. Baldwin, A. C. Gossard, and W. Wiegman, Modulation doped field effect transistors and logic gates based on two-dimensional hole gas, presented at 42nd Annual Device Res. Conf., Santa Barbara, June 1984.

126. R. A. Kiehl and A. C. Gossard, Complementary p-MODFET and n-HB MESFET (Al, Ga)As Transistors, *IEEE Electron Devices Lett.* **EDL-5**(12), 521–523 (1984).

127. A. J. Valois and G. Y. Robinson, *IEEE Electron Device Lett.*, **EDL-4**, 360 (1983).

128. S. Subramanian, *IEEE Trans. Eletctron Dev.*, **ED-32**(5), 865–870 (1985).
129. J. K. Abrokwah, M. Shur, R. R. Daniels, and D. K. Arch, Effect of traps on current-voltage characteristics of self-aligned Modulation Doped Field Effect Transistors, unpublished.
130. A. Kastalski and R. A. Kiehl, *IEEE Trans. Electron Dev.*, **ED-33**(3), 414–423, 1986.
131. N. C. Cirillo, M. Shur, P. J. Vold, J. K. Abrokwah, R. R. Daniels, and O. N. Tufte, *IDEM Tech. Digest*, 317–320, 1985.
132. N. C. Cirillo, M. Shur, P. J. Vold, J. K. Abrokwah, and O. N. Tufte, *IEEE Electron Device Lett.*, **EDL-6**, 645–647 (1985).
133. T. Mizutani, S. Fujita, Y. Yanagawa, *Electronics Lett.*, **21**, 1116–1117 (1985).
134. R. R. Daniels, R. Mactaggart, J. K. Abrokwah, O. N. Tufte, M. Shur, J. Baek, and P. Jenkins, Complementary heterostructure insulated gate FET circuits for high-speed, low-power VLSI, *IEDM Tech. Digest*, paper 17.3, 448–451 (1986).

11

Novel GaAs Devices

11-1. INTRODUCTION

As the dimensions of GaAs devices shrink, the effective electron velocity should increase, leading to a shorter transit time and to a ballistic or near-ballistic mode of operation (see Chapter 2). At the same time the power consumption drops as a consequence of smaller device dimensions.

Several novel device structures have been proposed in order to take advantage of higher electron velocity and mobility in GaAs and smaller device sizes. They include modulation doped FETs (see Chapter 10), heterojunction bipolar transistors (HBTs) with a wide band gap emitter [1-10], permeable base transistors [11-16], vertical ballistic transistors [17], planar doped barrier devices [18-24], and hot electron injection devices [25-29] based on the effect of real-space hot electron transfer [30].

In a modulation doped field effect transistor the electrons move from the source to the drain in a thin two-dimensional inversion layer formed at the boundary between wide-gap AlGaAs and undoped GaAs. A thin layer of undoped AlGaAs separates the electrons in the two-dimensional GaAs from the donors in the doped layer of AlGaAs under the gate. A substantial decrease in the impurity scattering enhances the low field mobility, especially at low temperatures, leading to low values of the source series resistance and high transconductance. Propagation gate delays of 10.2 ps at 300 K and 5.8 ps at 77 K switching time per gate have been achieved, making this device the fastest solid state device known (see Chapter 10).

In a heterojunction bipolar transistor (see Fig. 11-1-1) an energy gap discontinuity creates a barrier for holes coming from the p-type GaAs base into the n-type AlGaAs emitter (see Fig. 10-2-1). This barrier is larger than the barrier for the electrons coming from the emitter into the base, a property that greatly enhances the emitter injection efficiency. An additional advantage is that the electrons from the emitter are coming into the base with high energies with respect to the bottom of the conduction band in the base and hence with high velocities. This feature should lead to a very short transit time in a submicron base and large current gain. Cutoff frequencies between 100 and 200 GHz have been predicted for this device [31].

In a permeable base transistor an ultrafine metal grid is incorporated into an epitaxial GaAs film (see Fig. 11-1-2). The metal forms a Schottky barrier with GaAs and the built-in voltage totally depletes the openings in the grid. When a sufficient positive bias is applied to the base electrode the depletion layer shrinks and a

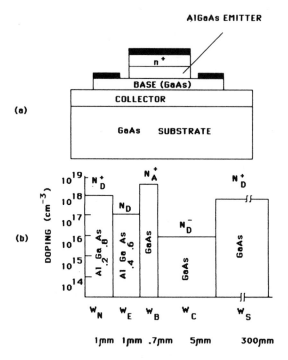

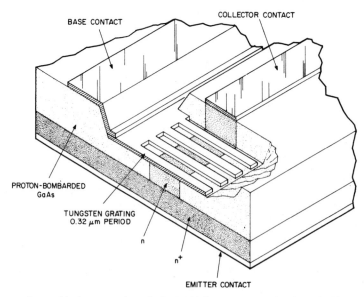

FIGURE 11-1-1. Heterojunction bipolar transistor [5]. (a) Schematic diagram; (b) possible doping profile.

conductive path forms between the collector and emitter. The cutoff frequency of such a device is a strong function of the finger size and is presently close to 40 GHz. Cutoff frequencies close to 200 GHz are possible if the finger size is decreased to about 500 Å [31].

A vertical ballistic transistor (see Fig. 11-1-3) has a vertical structure like a permeable base transistor. The gate electrodes deplete the conduction path by

FIGURE 11-1-2. Permeable base transistor [11]. A 300-Å tungsten grating is embedded into a single crystal of GaAs. The electrons flow from emitter to collector through the slits in the grating.

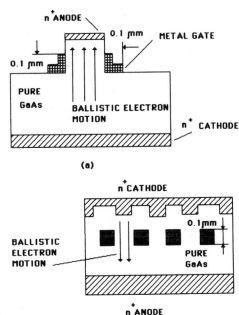

(a)

(b)

FIGURE 11-1-3. Vertical ballistic transistor (VBT)
[17]. (a) Single section VBT; (b) periodical VBT.

fringing fields. The main advantage of this structure is that the gate electrodes are on the surface of the device, not embedded in the device as in a permeable base transistor, making the structure much easier to make. The device performance should be close to the predicted performance for a permeable base transistor.

In planar doped barrier transistors (see Fig. 11-1-4) the narrow regions of p doping are used to create emitter and collector barriers. The p planes are totally depleted of carriers under all biases. Therefore the planar doped barrier transistor is a majority carrier device. As in a heterojunction wide band gap emitter bipolar junction transistor the electrons are injected into the base at high energies with respect to the bottom of the conduction band in the base, leading to a potentially higher drift velocity in the base. Preliminary estimates indicate that a cutoff frequency close to 200 GHz may be achieved [24]. A two-terminal switching device utilizing the planar doped barrier has also been proposed [22].

A device which employs a planar doped barrier as an injector of energetic high-velocity electrons into a vertical ballistic transistor is a very interesting possibility [24].

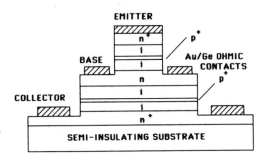

FIGURE 11-1-4. Schematic cross section of planar doped barrier transistor [21].

The basic idea of hot electron injection devices may be illustrated by comparing the structure shown in Fig. 11-1-5a to a vacuum tube diode (Fig. 11-1-5b) [30]. In a vacuum tube diode the anode current may be controlled by varying the cathode temperature. In a hot electron injection device the current I_{HOT} over the barrier separating the channel from the substrate may be controlled by varying the electron temperature in the channel. Three new device concepts—the charge injection transistor (CHINT), the negative resistance field effect transistor (NERFET), and the hot-electron erasable progammable random access memory (HE^{2}PRAM), were proposed based on this principle [25-29].

In many of the novel structures mentioned above the characteristic feature size limiting the device performance is in the vertical (and not the horizontal) dimension. An epitaxial process (typically MBE or MO CVD; see Chapter 3) is required for device fabrication. For all novel devices accurate control of the vertical device dimensions and doping profile is needed to maintain device uniformity across the wafer, which is a necessary precondition for achieving a high yield and/or large scale integration.

The necessity to achieve a low contact resistance and/or a low base spreading resistance is another practical limitation.

At the present time modulation doped field effect transistors (see Chapter 10) are more developed than other novel device structures.

Another contender for high-speed applications is the heterojunction bipolar transistor because of potential ease of maintaining a good uniformity of turn-on

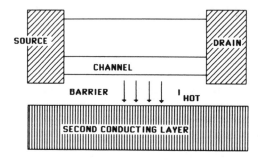

(a)

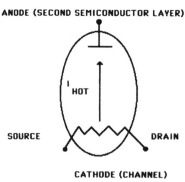

(b)

FIGURE 11-1-5. Principle of operation of hot electron injection devices [29]. (a) Schematic device structure; (b) vacuum tube analogy.

voltages. More studies are required, however, to evaluate the full potential of different novel device structures.

11-2. HETEROJUNCTION BIPOLAR TRANSISTORS

11-2-1. Principle of Operation

The idea of using a heterojunction bipolar transistor was introduced by W. Shockley [1] and was later developed by Kroemer [2, 6, 7].

The basic theory describing the operation of a wide-gap emitter device was given by Kroemer [6]. The principle of operation may be understood by analyzing the different current components flowing in the device as shown in Fig. 11-2-1. The emitter, base, and collector currents I_E, I_B, and I_C can be represented as

$$I_E = I_{nE} + I_{pE} + I_{rE} \qquad (11\text{-}2\text{-}1)$$

$$I_C = I_{nE} - I_r + I_{gC} \qquad (11\text{-}2\text{-}2)$$

$$I_B = I_{pE} + I_{rE} - I_{gC} + I_r \qquad (11\text{-}2\text{-}3)$$

where I_{nE} is the electron current injected from the emitter region into the base (this is the principal emitter and collector current component for the active forward mode, which we are now considering), I_{pE} is the current of holes injected into the emitter region, I_{rE} is the electron–hole recombination current due to carrier recombination in the emitter-base forward-biased depletion region, I_r is the recombination current due to the recombination of the carriers within the base outside of the depletion region, I_{gC} is the generation current in the collector-base reversed-biased depletion region. The common emitter current gain β is given by

$$\beta = \frac{I_C}{I_B} = \frac{I_{nE} - I_r + I_{gC}}{I_{pE} + I_{rE} + I_r - I_{gC}} < \beta_{\max} \qquad (11\text{-}2\text{-}4)$$

where

$$\beta_{\max} = \frac{I_{nE}}{I_{pE}} \qquad (11\text{-}2\text{-}5)$$

is the upper limit for the gain which is related to the emitter injection efficiency γ:

$$\gamma = \frac{I_{nE}}{I_E} \qquad (11\text{-}2\text{-}6)$$

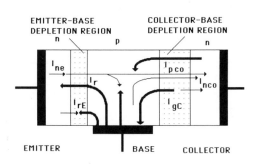

FIGURE 11-2-1. Current components in a bipolar junction transistor. Thin lines represent the electronic component of the current; thick lines correspond to the current carried by holes. A forward bias across the emitter–base junction and a reverse bias across the collector-base junction are assumed.

In the frame of the elementary diffusion theory of the uniform gap transistor we have

$$\beta_{max} = \frac{D_n N_{DE} X_E}{D_p N_{AB} X_B} \qquad (11\text{-}2\text{-}7)$$

where X_E and X_B are widths, N_{DE} and N_{AB} are the doping levels for the emitter and base regions, respectively, and D_n and D_p are diffusion constants for electrons and holes. As pointed out by Kroemer [6], a more realistic estimate of β_{max} is given by

$$\beta_{max} = \frac{N_{DE} v_{nB}}{N_{AB} v_{pE}} \qquad (11\text{-}2\text{-}8)$$

where v_{nB} and v_{pE} are effective velocities of electrons in the base and holes in the emitter region, which include contributions from the carrier diffusion and drift.

In a heterojunction bipolar junction transistor (HBT) the emitter region has a wider band gap than the base. If we assume for illustrative purposes that the discontinuity of the band gaps is entirely related to the valence band discontinuity, then the band diagram of an HBT will look as shown in Fig. 11-2-2. In this case the expression for β_{max} for a wide-gap emitter device is given by

$$\beta_{max} = \frac{N_{DE} v_{nB}}{N_{AB} v_{pE}} \exp(\Delta E_g / k_B T) \qquad (11\text{-}2\text{-}9a)$$

As a result, a very high value of β_{max} may be achieved even when N_{DE} is smaller than N_{AB}.

In fact, the band diagram of heterojunction looks like that shown in Fig. 10-2-1b, and, as a consequence, Eq. (11-2-9a) should be rewritten as

$$\beta_{max} = \frac{N_{DE} v_{nB}}{N_{AB} v_{pE}} \exp(\Delta E_v / k_B T) \qquad (11\text{-}2\text{-}9b)$$

where $\Delta E_v = \Delta E_g - \Delta E_c$. This leads to a less dramatic improvement in β_{max} but the basic principle remains intact. Moreover, the decrease in the maximum gain related to the decrease of the potential barrier for holes may be partially compensated by the increase in the electron velocity in the base caused by the spike. As pointed out by Kroemer [6], electrons in this "spike-notch" structure enter the base with a very

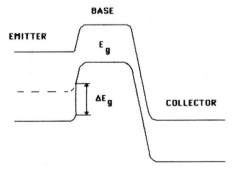

FIGURE 11-2-2. Simplified energy band diagram of a heterojunction bipolar transistor [6]. For simplicity all the difference in the energy gaps is related to the valence band (neglecting the spike in the conduction band; see Section 11-2-2). The dashed line corresponds to a homojunction transistor and is shown for comparison.

large energy (close to ΔE_c) and, as a consequence, may have very high velocities (of the order of several times 10^5 m/s). Because of the directional nature of the dominant polar optical phonon scattering they may traverse the base region maintaining a very high velocity. Kroemer described the conduction band spike as "a launching pad" for ballistic electrons. Also, the magnitude of the spike can be reduced by grading the composition of the wide band gap emitter near the heterointerface [6] as further discussed in Section 11-2-3 (see also Ref. 34).

This discussion shows that the emitter injection efficiency of an HBT may be made very high. The transistor gain in this case is limited by the recombination current

$$\beta = \frac{I_{nE}}{I_{rg} + I_r} \tag{11-2-10}$$

and could be as high as a few thousand or more if the heterojunction interface is relatively defect free so that I_{rE} is not excessively high.

The recombination current in the base can be estimated as

$$I_{rbulk} \approx qkn_p(0) W_B / \tau \tag{11-2-11}$$

where k is a numerical constant of the order of unity, $n_p(0)$ is the concentration of minority carriers (electrons) at the emitter end of the base, W_B is the base width, and τ is the lifetime. Thus, if I_{rE} can be neglected in Eq. (11-2-10) and if we assume

$$I_{nE} \approx qn_p(0)v_{nB} \tag{11-2-12}$$

we obtain

$$\beta \approx \frac{v_{nB}\tau}{kW_B} = \frac{t}{t_{TR}}$$

where t_{TR} is the electron transit time across the base. For a sufficiently short base (say, $W_B \approx 10^{-5}$ cm) $\beta > 10^3$ can be obtained even if the lifetime is only of the order of a nanosecond.

For comparison, in a conventional homojunction silicon transistor the dependence of the energy gap on the doping level leads to a narrowing of the energy gap in the emitter region. The change in the energy gap starts at 10^{17} cm^{-3} and is roughly proportional to $\ln(N_D)$:

$$\Delta E_g(\text{mV}) \approx 17 \ln[N_D(\text{cm}^{-3})/10^{17}] \tag{11-2-13}$$

As a result the ratio I_{nE}/I_{pE} decreases quite substantially compared to the estimate given by Eq. (11-2-8), up to a factor of 20 for $N_D = 10^{19}$ cm^{-3}. This energy gap narrowing represents one of the dominant performance limitations for conventional Si BJTs [43].

In addition to high injection efficiency and, as a consequence, high current gain, heterojunction bipolar junction transistors have a number of other advantages over conventional bipolar transistors. As a consequence of higher base doping the base spreading resistance is smaller. Because of the relatively low doping of the emitter region the emitter–base capacitance could be made small. All these factors lead to a higher speed of operation.

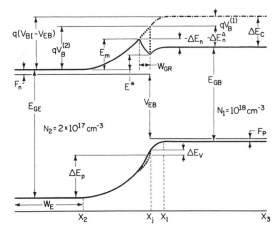

FIGURE 11-2-3. Band diagram of the graded emitter–base junction [10]. Notation is explained in text. Dotted line, an abrupt heterojunction; dashed-dotted line, $qV(x)$, where $V(x)$ is the electron potential; ΔE_n^A is the value of ΔE_n for the abrupt heterojunction.

Recent advances in MBE technology have made it possible to obtain abrupt or graded heterojunctions with a high degree of reproducibility. Common emitter gains in excess of 1600 have been achieved for AlGaAs–GaAs bipolar junction transistors [3–10].

Equations presented in this section merely illustrate the principle of operation of a HBT. More detailed and accurate analysis of HBT operation should be based on the heterojunction theory described below.

11-2-2. Thermionic-Diffusion Model of an HBT with an Abrupt Emitter–Base Junction [10]

The band structure of the emitter–base junction of an *npn* HBT is shown in Fig. 11-2-3. As discussed in Section 11-2-2 and as can be seen from Fig. 11-2-3, the discontinuity in the conduction band edge gives rise to a barrier which impedes the injection of electrons into the base region. Electron transport across this barrier can be described using a thermionic model as discussed by Anderson [32]. However, a conventional diffusion model is valid in the remainder of the structure at low injection levels.

In the depletion region where the total current is much smaller than either the drift or diffusion components (i.e., the two components nearly cancel one another),

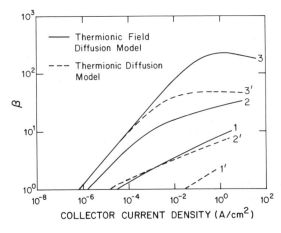

FIGURE 11-2-4. Common emitter current gain β as a function of the collector current for the abrupt heterojunction [10]. $N_2 = 2 \times 10^{17}$ cm^{-3}, $T = 300$ K. For 1 and 1′, $N_1 = 4 \times 10^{17}$ cm^{-3}; for 2 and 2′, $N_1 = 3 \times 10^{17}$ cm^{-3}; and in 3 and 3′, $N_1 = 2 \times 10^{17}$ cm^{-3}.

the electron and hole quasi-Fermi levels are nearly constant. Assuming quasi equilibrium for the n–p heterojunction, the quasi-Fermi levels are nearly constant in the depletion layers except in the immediate vicinity of the heterointerface, as shown in Fig. 11-2-3. As one departs from quasiequilibrium, the region in the n-side of the junction over which the quasi-Fermi level is non-constant will increase. Further details of this are discussed in Ref. 33. If the quasi-Fermi levels are constant across the depletion region as described above, then the electron concentrations at the boundaries of the depletion regions and in the vicinity of the graded region can be related by Boltzmann factors.

For an abrupt heterojunction the electron current density across the heterointerface is found as the difference between two opposing fluxes

$$J_n(X_j) = -q\frac{v_n}{4}[n(X_j^-) - n(X_j^+)\exp(-\Delta E_c/k_B T)] \qquad (11\text{-}2\text{-}14)$$

where ΔE_c is the conduction band discontinuity at the heterointerface,

$$V_N = \left\{\frac{8K_B T}{\pi m_n}\right\}^{1/2}$$

is the mean electron thermal velocity, m_n is the electron effective mass in the emitter region, and $n(X_j^-)$ and $n(X_j^+)$ are the electron concentrations at each side of the heterointerface. Although the electron effective mass is different on either side of the junction, its effect is only a multiplicative constant as compared to the exponential factor in the conduction band discontinuity. We therefore neglect the difference of effective mass in these expressions. As discussed above, these concentrations may be related to the electron concentrations at the boundaries of the space charge region X_1 and X_2 (see Fig. 11-2-3):

$$n(X_j^-) = n(X_2)\exp(-qV_B^{(2)}/k_B T)$$
$$n(X_j^+) = n(X_1)\exp(+qV_B^{(1)}/k_B T) \qquad (11\text{-}2\text{-}15)$$

where

$$V_B^{(1)} = (V_{bi} - V_{EB})\xi$$
$$V_B^{(2)} = (V_{bi} - V_{EB})(1 - \xi) \qquad (11\text{-}2\text{-}16)$$

$$\xi = N_2\varepsilon_2/(N_1\varepsilon_1 + N_2\varepsilon_2) \qquad (11\text{-}2\text{-}17)$$

Here ε_1 and ε_2 are the dielectric constants and N_1 and N_2 are the doping levels in the base and emitter regions, respectively, V_{bi} is the built-in potential, and V_{EB} is the applied forward bias voltage. The electron current density given by Eq. (11-2-14) should be equal to the electron current density due to diffusion $J_n(X_1)$ at the boundary of the space charge region in the base (X_1 in Fig. 11-2-3)

$$J_n(X_1) = -\frac{qD_{n1}}{L_1}\frac{[n(X_1) - n_1]\cosh(W_B/L_1) - [n(X_3) - n_1]}{\sinh(W_B/L_1)} \qquad (11\text{-}2\text{-}18)$$

Here n_1, D_n, and L_1 are the equilibrium concentration, diffusion coefficient, and diffusion length of the electrons in the base, respectively, W_B is the base width, and X_3 is the boundary coordinate of the base–collector space charge region in the base.

Since the currents given by Eqs. (11-2-14) and (11-2-18) must be equal, the two expressions may be equated in order to solve for the excess carrier concentration at the edge of the emitter depletion region, namely, $n(X_1) - n_1$.

Equating Eq. (11-2-14) and Eq. (11-2-18) and using Eq. (11-2-15) we find

$$n(X_1) - n_1 = (1/R_n)\left\{ \eta_n \exp\left(-\frac{\Delta E_n}{k_B T}\right) [n(X_3) - n_1] + n(X_2) \right.$$

$$\left. \times \exp[(-qV_{bi} + qV_{EB} + \Delta E_c)/k_B T] - n_1 \right\} \qquad (11\text{-}2\text{-}19)$$

where

$$\eta_n = 4D_{n1}/[v_n L_1 \sinh(W_B/L_1)]$$

$$R_n = 1 + \eta_n \cosh(W_B/L_1) \exp(-\Delta E_n/k_B T) \qquad (11\text{-}2\text{-}20)$$

$$\Delta E_n = qV_B^{(1)} - \Delta E_c$$

Substituting Eq. (11-2-19) into Eq. (11-2-20) and using the following relationships:

$$n_1 = n_2 \exp[(\Delta E_c - qV_{bi})/k_B T] \qquad (11\text{-}2\text{-}21\text{a})$$

$$n(X_3) = n_1 \exp(-qV_{CB}/k_B T) \qquad (11\text{-}2\text{-}21\text{b})$$

we find

$$J_n(X_1) = \frac{-J_{nE}}{R_n}\left\{ \left[\exp\left(\frac{qV_{EB}}{k_B T}\right) - 1 \right] \cosh\left(\frac{W_B}{L_1}\right) - \left[\exp\left(-\frac{qV_{CB}}{k_B T}\right) - 1 \right] \right\}$$

$$(11\text{-}2\text{-}22)$$

$$J_{nE} = qD_{n1}n_1 \left/ \left[L_1 \sinh\left(\frac{W_B}{L_1}\right) \right] \right. \qquad (11\text{-}2\text{-}23)$$

Here, n_2 is the equilibrium electron concentration in the emitter and V_{CB} is the collector base voltage. We should note that Eq. (11-2-21b) may not be accurate if the carrier velocity in the base becomes comparable to the carrier saturation velocity. This may be the case in a short base transistor ($W_B \sim D_{n1}/v_s$, where v_s is the saturation velocity). However, in the normal active mode, the values of $n(X_3)$ will still be small compared to n_1 and, therefore, should not affect the results significantly. A similar derivation yields the following equation for the hole component of the current density:

$$J_p(X_2) = \frac{J_{pE}}{R_p}\left[\exp\left(\frac{qV_{EB}}{k_B T}\right) - 1 \right] \cosh\left(\frac{W_E}{L_2}\right) \qquad (11\text{-}2\text{-}24)$$

Here

$$J_{pE} = qD_{p2}p_2 \left/ \left[L_2 \sinh\left(\frac{W_E}{L_2}\right) \right] \right. \qquad (11\text{-}2\text{-}25)$$

$$R_p = 1 + \eta_p \cosh\left(\frac{W_E}{L_2}\right) \exp(-\Delta E_p / k_B T)$$

$$\Delta E_p = q V_B^{(2)} \tag{11-2-26}$$

$$\eta_p = 4 D_{p2} \Big/ \left[v_p L_2 \sinh\left(\frac{W_E}{L_2}\right) \right]$$

L_2 and D_{p2} are the hole diffusion length and diffusion coefficient in the emitter region, respectively, p_2 is the equilibrium hole concentration in the emitter, and v_p is the mean thermal velocity of holes.

The electron component of the collector current is determined by substituting the quantity $[n(X_1) - n_1]$ from Eq. (11-2-19) for the electron current density at the edge of the collector depletion region into the following equation, which is found using the diffusion model in the base region:

$$J_n(X_3) = \frac{-J_{nE}}{n_1} \{ [n(X_1) - n_1] - [n(X_3) - n_1] \cosh(W_B / L_1) \} \tag{11-2-27}$$

This substitution yields

$$J_n^{(c)} = \frac{-J_{nE}}{R_n} \left\{ \left[\exp\left(\frac{q V_{EB}}{k_B T}\right) - 1 \right] - \left[\exp\left(-\frac{q V_{CB}}{k_B T}\right) - 1 \right] \left[\cosh\left(\frac{W_B}{L_1}\right) \right. \right.$$
$$\left. \left. + \eta_n \sinh^2\left(\frac{W_B}{L_1}\right) \exp\left(-\frac{\Delta E_n}{k_B T}\right) \right] \right\} \tag{11-2-28}$$

The expression for the hole component of the collector current density is the same as the equation given by the conventional diffusion model of a BJT:

$$J_p^{(c)} = J_{pc} \cosh\left(\frac{W_C}{L_3}\right) \left[\exp\left(-\frac{q V_{CB}}{k_B T}\right) - 1 \right] \tag{11-2-29}$$

Here W_c is the width of the neutral part of the collector region and L_3 is the hole diffusion length in the collection. Also,

$$J_{pc} = \frac{q D_{p3} p_3}{L_3 \sinh(W_C / L_3)}$$

where D_{p3} is the diffusion coefficient for holes in the collector region and p_3 is the equilibrium hole concentration in the collector.

Since one of the advantages of the HBT is that the base doping level can be made as large as practical without degrading performance, the doping density in the base region is usually either larger or at least comparable to the doping density in the emitter region. As a consequence, in a typical HBT the concentration of the injected electrons in the base region is smaller than the concentration of majority carriers (holes). Hence we may assume

$$n(X_2) \cong n_2 \quad \text{and} \quad p(X_1) \cong p_1$$

Having solved for the current components at each junction, the equations for the emitter and collector currents may be rewritten in a form similar to the Ebers–Moll equations for a homojunction BJT:

$$J_E = A_{11}[\exp(qV_{EB}/k_BT) - 1] + A_{12}[\exp(-qV_{CB}/k_BT) - 1] - J_{RG}^E$$
$$J_C = A_{21}[\exp(qV_{EB}/k_BT) - 1] + A_{22}[\exp(-qV_{CB}/k_BT) - 1]$$

$$(11\text{-}2\text{-}30)$$

where

$$A_{11} = -\left[\frac{J_{nE}}{R_n}\cosh\left(\frac{W_B}{L_1}\right) + \frac{-J_{pE}}{R_p}\cosh\left(\frac{W_E}{L_2}\right)\right]$$

$$A_{22} = \left\{\frac{J_{nE}}{R_n}\left[\cosh\left(\frac{W_B}{L_1}\right) + \eta_n\sinh^2\left(\frac{W_B}{L_1}\right)\exp\left(-\frac{\Delta E_n}{k_BT}\right)\right] + J_{pc}\cosh\left(\frac{W_C}{L_3}\right)\right\}$$

$$(11\text{-}2\text{-}31)$$

$$A_{12} = -A_{21} = J_{nE}/R_n$$

and J_{RG}^E is the recombination current density in the space-charge region of the emitter–base junction. The common emitter current gain may be found from Eqs. (11-2-30) and (11-2-31)

$$\beta = \left[\cosh\left(\frac{W_B}{L_1}\right) - 1 + \frac{R_nJ_{pE}}{R_pJ_{nE}}\cosh\left(\frac{W_E}{L_2}\right) + \frac{J_{RG}^E R_n}{J_{nE}\exp(qV_{EB}/k_BT)}\right]^{-1}$$

$$(11\text{-}2\text{-}32)$$

On the other hand, conventional diffusion theory for BJTs yields the following equation for the common emitter current gain:

$$\beta = \left[\cosh\left(\frac{W_B}{L_1}\right) - 1 + \frac{J_{pE}}{J_{nE}}\cosh\left(\frac{W_E}{L_2}\right) + \frac{J_{RG}^E}{J_{nE}}\exp\left(-\frac{qV_{EB}}{k_BT}\right)\right]^{-1}$$

$$(11\text{-}2\text{-}33)$$

Comparing these two expressions we see that the difference between these models lies in the factors R_n and R_p on the right-hand side of Eq. (11-2-32). When $\Delta E_n \gg k_BT$ and $\Delta E_p \gg k_BT$ the results predicted by this model become identical to the conventional diffusion model. When $\Delta E_n < 0$ the position of the interface spike is higher than the bottom of the conduction band, E_c, in the neutral region of the base. Recalling the expression for ΔE_n,

$$\Delta E_n = q(V_{bi} - V_{EB})\xi - \Delta E_c$$

$$(11\text{-}2\text{-}34)$$

we see that ΔE_n becomes negative at forward emitter–base bias. As a consequence, the injected electron current becomes substantially smaller than the value predicted by diffusion theory. On the other hand, the condition $\Delta E_p \gg k_BT$ is typically fulfilled in HBTs for all values of the bias voltages (see Fig. 11-2-3).

In a typical device,

$$L_1 \gg W_B, \qquad L_2 \gg W_E, \qquad \Delta E_p \gg k_BT$$

and the expression for the common emitter current gain β [i.e., Eq. (11-2-32)] can be simplified. With these inequalities, R_n can be written as

$$R_n = 1 + \frac{4D_{n1}}{v_n W_B} \exp(-\Delta E_n/k_B T) \tag{11-2-35}$$

and if a phenomenological equation for the recombination current density of the form

$$J_{RG}^E = J_{RG} \exp(qV_{EB}/mk_B T) \tag{11-2-36}$$

is used, Eq. (11-2-32) becomes for $-\Delta E_n > k_B T$,

$$\beta = \left[\frac{1}{2}\left(\frac{W_B}{L_1}\right)^2 + \frac{J_{pE}}{J_{nE}^*}\left(\frac{J_c}{J_{nE}^*}\right)^{\xi/(1-\xi)} + \left(\frac{J_{RG}}{J_{nE}^*}\right)\left(\frac{J_c}{J_{nE}^*}\right)^{[1/m(1-\xi)]-1} \right]^{-1} \tag{11-2-37}$$

where

$$J_{nE}^* = \frac{qv_n n_1}{4} \exp[(qV_{bi}\xi - \Delta E_c)/k_B T]$$

The dependences of β on the collector current calculated using Eqs. (11-2-32), (11-2-35), and (11-2-36) for different doping levels N_1 are shown in Fig. 11-2-4 (dashed lines). The parameters used in the calculation are given in Table 11-2-1. An apparent saturation of β at high collector currents for $N_1 = N_2$ (curve 3') is related to a small value of the exponent $\{1/[(1 - \xi)m] - 1\}$ in the last term in the right-hand side of Eq. (11-2-37). For nonsymmetrical doping (curves 1' and 2') β increases with increasing collector current. The change of the slope in curves 2' and 3' with the increase in the collector current is related to the change of sign of ΔE_n. For curve 3, only the region with $\Delta E_n < 0$ appears in the figure.

The results presented in Fig. 11-2-4 demonstrate that the values of the common emitter current gain in HBTs with an abrupt emitter–base junction are quite small

TABLE 11-2-1. Material Parameters Used in the HBT Calculation[a,b]

	Symbol	Value
Emitter doping	N_2	2×10^{17} cm^{-3}
Emitter width	W_E	0.5 μm
Base width	W_B	0.1 μm
Electron diffusion coefficient (base)	D_{n1}	78.0 cm^2/s
Hole diffusion coefficient (emit.)	D_{p2}	9.0 cm^2/s
Electron lifetime in the base	τ	2×10^{-9} s
Pre-exponential factor of the recombination current	J_{RG}	4.0×10^{-12} A/cm^2
Ideality factor	m	1.92

[a] Reference 10.
[b] The dependencies of the effective mass, energy gap, and electron affinity of Al$_x$Ga$_{1-x}$As on composition are given in Ref. 38.

because the electron injection is limited by the interface spike barrier. This conclusion agrees with a recent qualitative analysis by H. Kroemer [7]. However, the values of β predicted by the thermionic-diffusion model are considerably smaller than the experimental values (see Section 11-2-4). As shown in Ref. 10 the reason for the discrepancy is that current enhancement by tunneling near the peak of the interface spike leads to a higher current gain. The effect of tunneling is roughly equivalent to a decrease of the interface barrier height for electrons. As pointed out by Kroemer [7] and also discussed by Ankri and Eastman [34], grading of the composition of the emitter near the heterointerface also allows one to diminish the interface spike and to greatly enhance the emitter injection efficiency. The theory of HBTs with a graded emitter region is considered in Section 11-2-3 [10].

11-2-3. Thermionic Model for a Graded HBT

A qualitative analysis of the effects of grading the AlGaAs composition at the heterointerface (see Fig. 11-2-3) was given by Kroemer [7]. Also, a discussion of the role played by grading may be found in Ref. 37. The analysis given below is based on the results obtained in Ref. 10.

The energy diagram of a graded emitter region is shown in Fig. 11-2-3 (solid lines). Here we assume that the composition x of $Al_xGa_{1-x}As$ in the emitter region varies linearly over the grading length W_{GR}. We take $x = 0$ to be at the heterointerface. As can be seen from the figure, the effect of grading is to make ΔE_n less negative. For practical values of grading length, the spike is abrupt enough to justify the use of the thermionic emission model. Hence, the only necessary change in the theory of Section 11-2-2 is the replacement of ΔE_n by its value for the graded junction.

For simplicity, we only consider the case of uniform doping profiles in the emitter and base regions. In this case, the potential distributions in the depletion layers are given by

$$V(X) = -\frac{qN_2}{2\varepsilon_2}(X_D + X)^2 \qquad -X_D \leqslant X \leqslant 0 \qquad (11\text{-}2\text{-}38)$$

$$V(X) = \frac{qN_1}{2\varepsilon_1}(X - X_A)^2 - V_{bi} + V_{EB} \qquad 0 \leqslant X \leqslant X_A \qquad (11\text{-}2\text{-}39)$$

where

$$X_D = \left[\frac{2N_1(V_{bi} - V_{EB})\varepsilon_1\varepsilon_2}{qN_2(\varepsilon_1N_1 + \varepsilon_2N_2)}\right]^{1/2}$$

$$X_A = X_DN_2/N_1$$

$$V_{bi} = E_{GB} - F_n - F_p + \Delta E_c \qquad (11\text{-}2\text{-}40)$$

$$V_{bi} - V_{EB} = \frac{qN_2X_D}{2}(X_D/\varepsilon_2 + X_A/\varepsilon_1)$$

Here, F_n and F_p are the electron and hole quasi-Fermi levels and E_{GB} is the energy gap in the base. The shape of the conduction band edge in the emitter region is given by the following equation:

$$E_c(X) = -qV(X) - \frac{(X + W_{GR})}{W_{GR}}\Delta E_c, \qquad W_{GR} \leqslant X \leqslant 0 \qquad (11\text{-}2\text{-}41)$$

This is valid for linear grading assuming that the electron affinity and energy gap vary linearly with composition. The energy E_m (see Fig. 11-2-3) at $X = -W_{GR}$ is given by

$$E_m = \frac{q^2 N_2}{2\varepsilon_2}(X_D - W_{GR})^2 \qquad (11\text{-}2\text{-}42)$$

Hence

$$\Delta E_n = V_{bi} - V_{EB} - \Delta E_c - E_m \qquad (11\text{-}2\text{-}43)$$

These equations are valid when the grading length is smaller than the width of the depletion layer in the emitter region. If this is not the case or if

$$\left.\frac{dE_c}{dX}\right|_{X=-W_{GR}} > 0$$

then $E_m = 0$. The collector current J_c and the short circuit common emitter current gain may now be found from

$$\beta = [\tfrac{1}{2}W_B^2/L_1^2 + (J_{pE}/J_c)\exp(qV_{EB}/k_BT) + (J_{RG}/J_c)\exp(qV_{EB}/mk_BT)]^{-1} \qquad (11\text{-}2\text{-}44)$$

$$J_c = (J_{pE}/R_n)\exp(qV_{EB}/k_BT) \qquad (11\text{-}2\text{-}45)$$

The collector current versus emitter–base voltage curves calculated using Eq. (11-2-45) for different grading lengths are shown in Fig. 11-2-5 (dashed lines). The parameters used in the calculation are given in Table 11-2-1. Notice a dramatic rise in the collector current with the increase in W_{GR}. For large values of W_{GR} ($W_{GR} > 300$ Å) the resulting current approaches the values predicted by diffusion theory.

Shown in Fig. 11-2-6 are plots of current gain, β, versus collector current, J_c, for several different graded layer thicknesses. The dramatic rise in β for increasing graded layer thickness illustrates the importance of grading. Further, upon comparing Figs. 11-2-6a and 11-2-6b, the influence of the doping ratio N_2/N_1 can be seen.

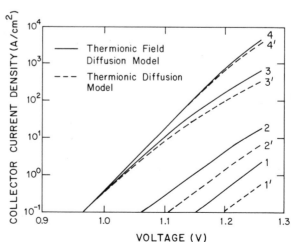

FIGURE 11-2-5. Voltage dependence of the collector current for different graded layer thicknesses [10]. $T = 300$ K. $N_1 = 10^{18}$ cm^{-3}, $N_2 = 2 \times 10^{17}$ cm^{-3}. In 1 and 1', $W_{GR} = 50$ Å; for 2 and 2', $W_{GR} = 100$ Å; for 3 and 3', $W_{GR} = 200$ Å; and in 4 and 4', $W_{GR} = 300$ Å.

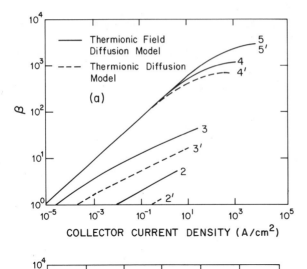

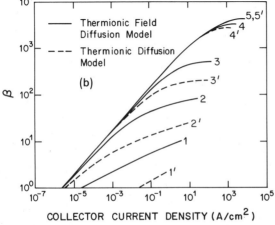

FIGURE 11-2-6. Common emitter current gain for different grading lengths W_{GR} vs. collector current [10]. (a) $N_1 = 10^{18}\ \text{cm}^{-3}$; (b) $N_1 = 4 \times 10^{17}\ \text{cm}^{-3}$. In 1 and 1', $W_{GR} = 0$; in 2 and 2', $W_{GR} = 50\ \text{Å}$; for 3 and 3', $W_{GR} = 100\ \text{Å}$; for 4 and 4', $W_{GR} = 200\ \text{Å}$; and for 5 and 5', $W_{GR} = 300\ \text{Å}$.

The curves of Fig. 11-2-6a were generated using $N_1 = 10^{18}\ \text{cm}^{-3}$ while those of Fig. 11-2-6b had $N_1 = 4 \times 10^{17}\ \text{cm}^{-3}$. All other parameters were as indicated in Table 11-2-1 and the figure caption. As the N_2/N_1 ratio is increased, the current gain is also seen to increase for the same grading length. The primary reason for this dependence is that the height of the interface spike is a function of the N_2/N_1 ratio.

The magnitude of the emitter current and the value of the short circuit current gain for junctions with abrupt interfaces is one or two orders of magnitude less than typical experimental results. As mentioned earlier, this discrepancy can be explained by electron tunneling near the peak of the interface spike. In the next section we introduce the effects of tunneling into the model.

11-2-4. Thermionic-Field-Diffusion Model

When electron tunneling is taken into account Eq. (11-2-14) should be rewritten as

$$J_n(X_j) = -q\frac{v_n}{4}\left[n(X_j^-) - n(X_j^+) \exp\left(-\frac{\Delta E_c}{k_B T} \right) \right] \gamma_n \qquad (11\text{-}2\text{-}46)$$

where

$$\gamma_n = 1 + \exp(E_m/k_BT)\frac{1}{k_BT}\int_{E^*}^{E_m} D(E_x/E_m)\exp(-E_x/k_BT)\,dE_x \quad (11\text{-}2\text{-}47)$$

Equation (11-2-47) is obtained by the integration over the range of barrier energies available for tunneling. Here $D(E_x/E_m)$ is the barrier transparency. Using an approach found in Refs. 35 and 36 we find for a triangular barrier that

$$D(X) = D_0(X) = \exp\left(-\frac{E_m}{E_{00}}\{(1-X)^{1/2} + \tfrac{1}{2}X \ln X - X \ln[1 + (1-X)^{1/2}]\}\right)$$

$$(11\text{-}2\text{-}48)$$

where $D_0(X)$ is defined as the barrier transparency for a triangular barrier and

$$E_m = qV_B^{(2)}$$

$$E^* = \begin{cases} qV_B^{(2)} - \Delta E_c & \text{for } qV_{bi} > \Delta E_c \quad (\text{see Fig. 11-2-1}) \\ 0 & \text{otherwise} \end{cases}$$

and

$$E_{00} = \frac{\hbar q}{2}\frac{N_2}{m_{n2}^*\varepsilon_2} \quad (11\text{-}2\text{-}49)$$

For the graded junctions discussed in Section 11-2-3 we derived the following expression for the barrier transparency:

$$D(X) = D_0(X)\exp\left(-\frac{E_m}{E_{00}}\left\{([1 - X + f(X)](1 - X))^{1/2}\right.\right.$$

$$\left.\left. -f(X)\ln[([(1-X)/f(X)] + 1)^{1/2} + \left[\frac{1-x}{f(x)}\right]^{1/2}\right\}\right) \quad (11\text{-}2\text{-}50)$$

where

$$f(X) = X + \frac{1}{4}\left(\frac{\Delta E_c}{E_m}\right)^2\left(\frac{X_D}{W_{GR}} - 1\right)^2 - \frac{\Delta E_c}{E_m}\left(\frac{X_D}{W_{GR}} - 1\right) \quad (11\text{-}2\text{-}51)$$

and

$$E^* = \begin{cases} qV_B^{(2)} - \Delta E_c & \text{for } qV_B^{(2)} - \Delta E_c < E_m \\ 0 & \text{for } qV_B^{(2)} - \Delta E_c \geq E_m \end{cases} \quad (11\text{-}2\text{-}52)$$

Equation (11-2-52) is valid only when

$$\left.\frac{dE_c(X)}{dx}\right|_{X=-W_{GR}} < 0 \quad (11\text{-}2\text{-}53)$$

(this condition was discussed in Section 11-2-3).

When the inequality (11-2-53) is not valid, the conventional diffusion model applies. However, Eq. (11-2-53) holds in a typical experimental situation. The comparison between Eq. (11-2-46) and Eq. (11-2-14) indicates that the effect of tunneling may be accounted for by replacing v_n by $v_n\gamma$ in the equations for the emitter and collector currents and in the expression for the common emitter current gain.

We computed the device characteristics shown by solid lines in Figs. 11-2-4, 11-2-5, and 11-2-6 using this substitution. The same parameters as in the thermionic-diffusion model discussed in Section 11-2-2 were used.

As can be seen from comparison with Fig. 11-2-4, the increase of β_{max} and $(J_c)_{max}$ due to tunneling is about a factor of 10 for the abrupt junction. Thus, tunneling effects play a very important role in HBTs.

For HBTs with graded junctions, the effect of tunneling is less pronounced and becomes negligible for $W_{GR} > 300$ Å (see Figs. 11-2-5 and Fig. 11-2-6).

In Fig. 11-2-7 we compare the results of the calculation with the experimental data for a device with $W_{GR} = 250$ Å. The recombination current density J_{RG} was taken as $J_{RG} = 8.2 \times 10^{-12}$ A/cm². For this value, the computed maximum value of β is very close to the experimental results. The more rapid decrease of β with decreasing collector current than that predicted by the model may be caused by the presence of an interfacial recombination current. A more detailed device characterization is necessary in order to more accurately determine the device parameters, especially the recombination current, for a better comparison between the theory and experiment.

The dependence of the maximum current gain on the grading length is shown in Fig. 11-2-8, where it is compared with the experimental data. The spread in the experimental data may be related to the difference in other parameters for the transistors with different grading length. According to this theory, no further increase in current gain occurs after the grading length becomes comparable to the width of the depletion region. Depending on the doping level in the emitter, this corresponds to $W_{GR} \cong 350$–400 Å. Experimentally, however, we do not see any rise in current gain with grading length after about 250 Å.

The figure demonstrates that grading should increase the maximum common emitter gain. Experimentally, however, the optimum grading length seems to be smaller than predicted by this theory. The reasons for this discrepancy are not understood at the present time.

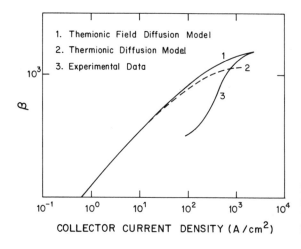

FIGURE 11-2-7. Experimental and theoretical curves β vs. J_c [10]. $N_1 = 10^{18}$ cm^{-3}, $N_2 = 2 \times 10^{17}$ cm^{-3}, $W_{GR} = 250$ Å, $T = 300$ K.

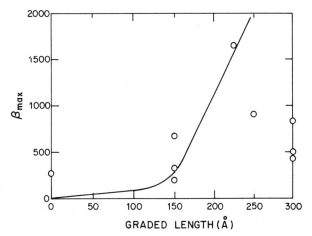

FIGURE 11-2-8. Maximum common emitter current gain vs. grading length [10]. $N_1 = 10^{18}\,\text{cm}^{-3}$, $N_2 = 2 \times 10^{17}\,\text{cm}^{-3}$, $T = 300\,\text{K}$, (O) experimental data.

11-2-5. HBT Performance

As was mentioned above, the heterojunction bipolar transistor allows us more flexibility in choosing the doping profiles in the emitter and base regions (see Fig. 11-1-1, where a possible doping profile of an HBT is depicted).

High doping in the base helps to reduce the base spreading resistance r_{bb}^{1}, which is inversely proportional to the Gummel number

$$Q_B = q \int_0^{W_B} N_{AB}\, dx \qquad (11\text{-}2\text{-}54)$$

For the circular and planar geometry, respectively, we find [39]

$$r_{bb'} = \begin{cases} \dfrac{1}{8\pi\mu_p Q_B} & (11\text{-}2\text{-}55) \\[3mm] \dfrac{1}{12(h/l)\mu_p Q_B} & (11\text{-}2\text{-}56) \end{cases}$$

(see Fig. 11-2-9). The low doping of the emitter region makes it possible to decrease the emitter base transition capacitance

$$C_{TE} \approx A_E \left[\frac{q\varepsilon N_{DE}}{2(V_{bi} - V)} \right]^{1/2} \qquad (11\text{-}2\text{-}57)$$

The cutoff frequency of a bipolar junction transistor f_T is given by

$$\frac{1}{2\pi f_T} = \tau_E + \tau_C + \tau_{CT} + \tau_{BT} \qquad (11\text{-}2\text{-}58)$$

where

$$\tau_E = r_e(C_{TE} + C_{DE}) \approx \frac{4V_{TH}}{I_E} C_{TE}(0) \qquad (11\text{-}2\text{-}59)$$

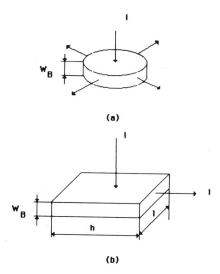

(a)

(b)

FIGURE 11-2-9. Base spreading resistance. (a) Circular geometry; (b) planar geometry.

is the emitter capacitance charging time, V_{TH} is the thermal voltage, r_e is an emitter–base junction resistance, and C_{DE} is the diffusion capacitance of the emitter–base junction [Eq. (11-2-59) may be only considered as a crude estimate]. As can be seen from Eqs. (11-2-58) and (11-2-59), the decrease in N_{DE} reduces τ_E and increases f_T.

The collector capacitance charging time

$$\tau_C = r_{sc} C_{CT} \tag{11-2-60}$$

where C_{CT} is the collector base junction transition capacitance and r_{sc} is the collector series resistance. The increase in the doping density between the collector region and collector contact helps to decrease r_{sc} and hence τ_C. The same objective is achieved by decreasing C_{CT} due to the low collector doping.

The effective base transit time is

$$\tau_{BT} = \frac{W_B}{v_{nB}} \tag{11-2-61}$$

When the drift of the minority carriers is neglected the effective velocity is

$$V_{nB} = \frac{2D_{nB}}{W_B} \tag{11-2-62}$$

where

$$D_{nB} = \frac{\mu_n k_B T}{q} \tag{11-2-63}$$

is the diffusion constant of electrons in the base. Finally

$$\tau_{CT} = \frac{X_{cd}}{v_s} \tag{11-2-64}$$

is the transit time of carriers across the collector depletion region X_{cd}, v_s is the saturation velocity of carriers.

The beneficial effect of high electron mobility in GaAs is clearly seen from Eqs. (11-2-62) and (11-2-63)—it leads to the decrease of the base transit time. A higher mobility also leads to the decrease of the collector series resistance r_{sc} [see Eq. (11-2-60)].

An extra (and important) advantage is the increase of the critical current density at which the base widening becomes important. This current density is proportional to the electron mobility [41]:

$$J_{\text{BWC}} = q\mu_{nC}N_{DC}\frac{V_{CB}}{W_C} \tag{11-2-65}$$

Here W_C is the width of the collector region.

Another factor to consider in the design is the emitter current crowding which limits the transistor power. It becomes important when the emitter current density becomes larger than [42]

$$J_{\text{EEC}} = \frac{8}{l^2} D_{pB}Q_B\beta \tag{11-2-66}$$

As can be seen from Eq. (11-2-66) it is beneficial to have a higher doping level in the base (higher Q_B).

All these factors offer a rationale for a doping profile shown in Fig. 11-1-1.

A schematic drawing of an AlGaAs–GaAs structure described in Ref. 5 is shown in Fig. 11-1-1a. The devices were grown by LPE and a good material quality resulted in lifetimes as high as 35 ns for $N_{AB} = 2 \times 10^{18}$ cm^{-3} and 3 ns for $N_{AB} = 10^{19}$ cm^{-3}. An extra highly doped N$^+$ emitter layer was used to reduce the ohmic contact resistance area. The device performance was characterized by

$$\beta \simeq 850$$
$$BV_{\text{ceo}} = 25 \text{ V} \tag{11-2-67}$$
$$f_T \simeq 1 \text{ GHz}$$

Different time constants in Eq. (11-2-28) were estimated to be

$$\tau_C = 48 \text{ ps}$$
$$\tau_{BT} \simeq 30 \text{ ps}, \qquad \tau_{CT} \simeq 4.5 \text{ ps}$$

A considerable part of r_s came from the emitter contact resistance. Better ohmic contacts and shorter base width may considerably improve the device performance.

In Fig. 11-2-10 the cutoff frequency of GaAs/GaAlAs transistors is compared with the cutoff frequency of a silicon homojunction transistor of similar geometry. Figure 11-2-10 allows us to compare Si BJTs and AlGaAs/GaAs HBTs. A more impressive performance of AlGaAs/GaAs HBTs achieved recently is illustrated by Fig. 11-2-11 [44]. As can be seen from the figure a bandwidth gain product of 25 GHz was obtained. The cutoff frequency f_T increased with the collector current (see Fig. 11-2-12) up to a very high current density of 10^4 A/cm^2. No Kirk effect or current crowding effects seemed to be important. The parameters of the epitaxial

632
11. NOVEL GaAs DEVICES

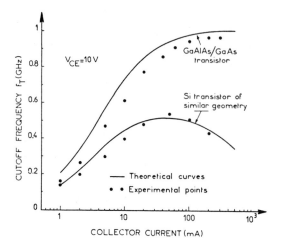

FIGURE 11-2-10. Cutoff frequency vs. collector current for the AlGaAs–GaAs heterojunction transistor and for the Si homojunction transistor of similar geometry [5].

layers used in the transistor fabrication are given in Table 11-2-2. The devices were fabricated by a conventional liftoff technique. Base and collector layers were recessed by wet chemical etch. Ohmic AuGe/Ni and Cr/Au contacts were used for n and p layers, respectively. The transistor had two emitter fingers 4.5 μm wide and 10 μm long. The base–collector junction area was 200 μm^2. Base sheet resistance was 1 Ω per square. The devices were isolated by proton implantation. The transistors exhibited current gains close to 90. Larger area devices (50 × 50 μm emitters) had a larger gain (up to 230), indicating the importance of surface recombination caused by fabrication processes.

The emitter intrinsic resistance $r_E = nk_BT/qI_E$ was estimated at 3.6 Ω; the emitter series resistance was about 5 Ω. The emitter-base transition capacitance was close to 0.6 pF. From these values τ_E was calculated to be 5.2 ps. From the cutoff frequency we deduce $\tau = 1/2\pi f_T = 6.4$ ps. This shows that τ_E (and, more specifically, r_E, the emitter series resistance) was a factor limiting the device performance [44]. By reducing r_E one may increase f_T to 50 GHz or above.

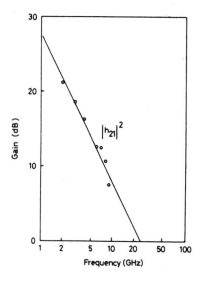

FIGURE 11-2-11. Gain vs. frequency for an HBT [44].

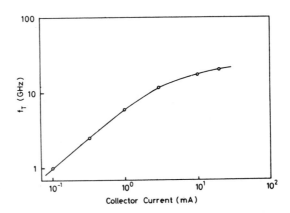

FIGURE 11-2-12. Cutoff frequency vs. collector current for an HBT [44].

It is interesting to notice that the upper bound for the transit time deduced in Ref. 44 is 6.4–5.2 = 1.2 ps. The sum of the base and collector transit times assuming the electron saturation velocity of 1.5×10^5 m/s and using a conventional diffusion model is 2.5 ps [44]. As pointed out in Ref. 44, this reduction may be caused by ballistic effects enhanced by the high energy of electrons injected into the base as suggested in Refs. 6, 45, and 46.

This is further confirmed by the Monte Carlo simulation [47], which predicts electron velocities in the base as high as 5×10^5 m/s and indicates that a cutoff frequency as high as 150 GHz may be achieved with AlGaAs/GaAs HBTs.

High-speed HBTs may find applications in GaAs integrated circuits where they may have a number of advantages over GaAs MESFET circuits (see Refs. 6, 47, 51, 52 and Section 9-2-12).

11-3. PERMEABLE BASE TRANSISTOR

The idea of improving the frequency response of a transistor by decreasing the transit time of carriers led to the development of the so-called permeable base transistor (PBT) [1–16, 53–55]. In this device a thin metal grating of tungsten which is embedded in a single crystal of GaAs forms a transistor base. A three-dimensional drawing of this structure is presented in Fig. 11-1-2. The device consists of four layers: the n^+ substrate, the n-type emitter layer, the thin film tungsten grid, and

TABLE 11-2-2. Epitaxial Layer Parameters of Fabricated Heterojunction Bipolar Transistors[a]

Layer	Type	Doping (cm^{-3})	Thickness (μm)
Cap	n^+ GaAs	3×10^{18}	0.2
Emitter	n Al$_{0.3}$Ga$_{0.7}$As	5×10^{17}	0.1
Base	p^+ GaAs	1×10^{19}	0.1
Collector	n GaAs	1×10^{17}	0.3
Buffer	n^+ GaAs	3×10^{18}	1.0
Substrate	S.I. GaAs	Cr/O$_2$	

[a] Reference 44.

the n-type collector layer. In the initial devices [10-15] the tungsten grid consists of lines and spacings of 1600 Å in a 300-Å layer of tungsten.

Tungsten forms a Schottky barrier with GaAs. The carrier concentration N_D in the n layer is such that the zero bias depletion width a_0 of the Schottky barrier is larger than the space between the tungsten strips [10]:

$$a_0 = \left(\frac{2\varepsilon V_{bi}}{qN_D}\right)^{1/2} > d > L \qquad (11\text{-}3\text{-}1)$$

where L is the film thickness. As a result the current is zero at zero emitter–base voltage. When the positive bias is applied to the base the depletion layer shrinks and the conductive path forms between the collector and emitter.

The behavior of this device has been modeled [14] based on the solution of the two-dimensional Poisson equation

$$\nabla^2 V = -\frac{q(N_D - n)}{\varepsilon} \qquad (11\text{-}3\text{-}2)$$

and the continuity equation

$$\nabla \cdot (q\mu n \nabla E_{Fn}) = 0 \qquad (11\text{-}3\text{-}3)$$

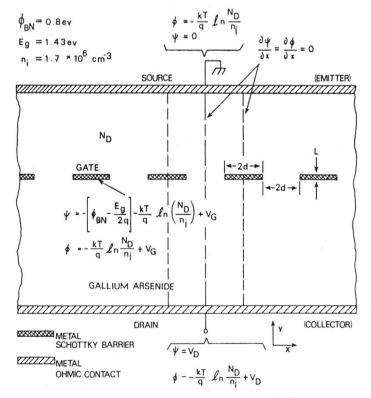

FIGURE 11-3-1. The device geometry used in the numerical simulation [14]. Simulation is done for the cell limited by the dotted lines.

Here V is the electric potential,

$$\mu = v/F \qquad (11\text{-}3\text{-}4)$$

is the field-dependent electron mobility, F is the electric field, v is the electron velocity,

$$E_{Fn} = -\frac{k_B T}{q} \ln \frac{n}{n_i} \qquad (11\text{-}3\text{-}5)$$

is the electron quasi-Fermi level. The device geometry used in the simulation is shown in Fig. 11-3-1, where the boundary conditions are also given. The results of the simulation are presented in Figs. 11-3-2–11-3-4 [14].

At small voltages V_{BE} a potential barrier exists (negative voltages in Fig. 11-3-2) impeding the electron flow and leading to an exponential dependence of the current (see Fig. 11-3-4). As V_{BE} is increased, the current shown in Fig. 11-3-4 begins to deviate from an exponential dependence on V_{BE} when the mobile charge in the opening is approximately $0.05 N_D$. As V_{BE} continues to increase, the collector current rises and the mobile charge in the opening increases until it exceeds N_D with a resulting accumulation of a net negative space charge in the base opening and surrounding regions. At higher values of V_{BE} the current through the device is limited by the space charge injection through the opening. Using the curve-fitting procedure the resulting current–voltage characteristics may be approximated by [14]

$$I_c = I_0^* e^{B V_{BE}}, \qquad V_{BE} \leqslant V_T \qquad (11\text{-}3\text{-}6)$$

$$I_c = I_0^* e^{B V_T} + K(V_{BE} - V_T)^2, \qquad V_{BE} > V_T \qquad (11\text{-}3\text{-}7)$$

The dependence of parameters I_0^*, B, K, and V_T on the tungsten film thickness L

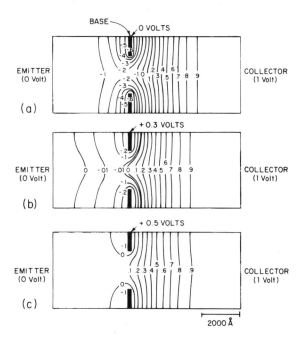

FIGURE 11-3-2. Contours of constant potential for a PBT with $N_D = 10^{16} \, \text{cm}^{-3}$, $d = 1000 \, \text{Å}$, $L = 200 \, \text{Å}$ [14].

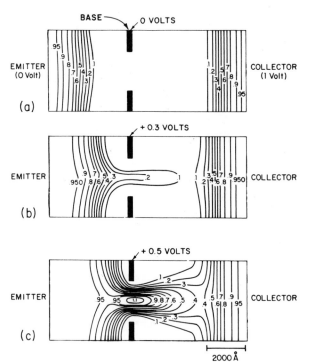

FIGURE 11-3-3. Contours of constant mobile charge normalized to N_D for the device in Fig. 11-3-1 [14].

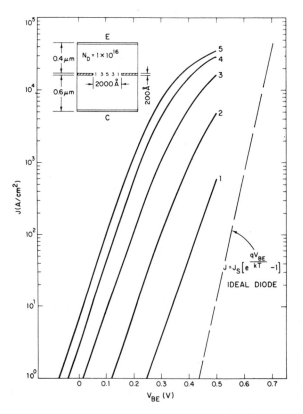

FIGURE 11-3-4. Current density in the base opening of a PBT as a function of base-to-emitter voltage for the device in Fig. 11-3-1 [14].

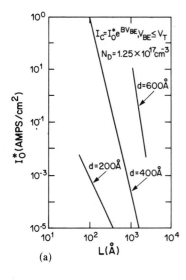

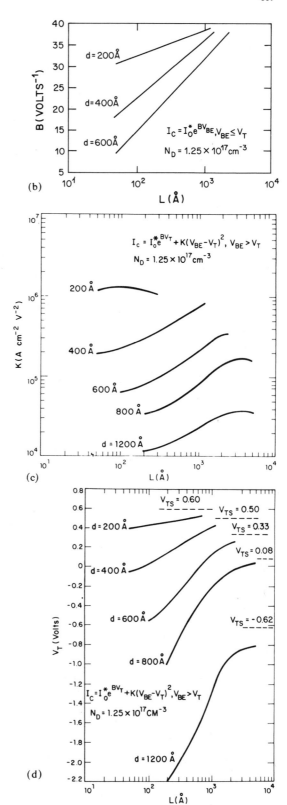

FIGURE 11-3-5. Parameters I_0^* (a), B (b), V_T (c), and K (d) in Eqs. (11-3-6) and (11-3-7) as functions of L and d [14]. $V_{CE} = 1$ V. V_{TS} is the value of the threshold voltage predicted by the Shockley model.

and the opening in the grid d are shown in Fig. 11-3-5 [14]. Parameters V_T and B have been found to be quite insensitive to a particular shape of the v vs. F curve. Both I_0^* and K are nearly proportional to the effective electron saturation velocity v_s, but rather insensitive to the low field mobility value [14]. Figure 11-3-5 could be used for a preliminary PBT design. It also allows us to estimate the transconductance g_m and the open circuit voltage gain $g_m R_0$, where R_0 is the output impedance. For the devices with $L = 200$ Å, $N_D = 1.25 \times 10^{17}$ cm^{-3}, $V_{BE} = 0.5$ V, and $V_{CE} = 1$ V, the open circuit voltage gain changes from 7.5 to 15 when d is reduced from 1200 to 400 Å. As d is reduced from 400 and 200 Å, $g_m R_0$ increases to a value above 75 [14].

The cutoff (unity gain) frequency is given by

$$f_T = \frac{g_m}{2\pi C_{\text{eff}}} \tag{11-3-8}$$

where

$$C_{\text{eff}} = \frac{\Delta Q}{\Delta V_{BE}} \tag{11-3-9}$$

Here ΔQ is the total change in the charge in the device produced by the change ΔV_{BE} in V_{BE}.

In a different form

$$f_T = \frac{1}{2\pi\tau} \tag{11-3-10}$$

where the effective time constant

$$\tau = \tau_E + \tau_C + \tau_B \tag{11-3-11}$$

includes contributions due to the charging of the effective emitter capacitance (τ_E), the effective collector capacitance (τ_C), and the effective base transit time (τ_B). At low V_{BE} the effective resistance of the device is large, leading to a large value of τ_E. Therefore f_T increases rapidly with the increase in V_{BE}.

The effect of the doping density and device dimensions on f_T is shown in Fig. 11-3-6 [14]. The device dimensions are scaled down with N_D to keep L/a_0 and d/a_0 constant (i.e., $\sim 1/\sqrt{N_D}$). The threshold voltage V_T then remains nearly constant with the change in doping. As can be seen from the figure, f_T increases rapidly with the doping owing to the higher current densities in the device and higher g_m.

The PBT fabrication sequence includes epigrowth of a GaAs layer, electron beam evaporation of the tungsten grating using x-ray lithography, and liftoff [11–15]. Tungsten shorting bars prevent the overgrowth of GaAs above them. The devices are isolated using the proton bombardment technique [56]. Standard nickel–gold–germanium contacts have been used.

The current–voltage characteristics and a small-signal equivalent circuit of a PBT are shown in Figs. 11-3-7 and 11-3-8 [14]. The parameters of the small signal equivalent circuit have been determined from dc and S-parameter measurements. For the device $f_T \simeq 37$ GHz ($f_{\text{max}} \simeq 10.4$ GHz), which is much smaller than the expected value close to 200 GHz for this design. The possible reasons for this

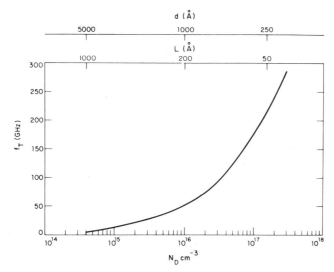

FIGURE 11-3-6. f_T vs. doping for a PBT [14]. Device dimensions are scaled down with the increase in doping.

disagreement include high resistance of the tungsten grating (up to 360 times higher than what can be expected from the bulk resistivity of tungsten) and defects in GaAs crystal structure in the vicinity of the tungsten grid. Much higher values of f_{max} (extrapolated value of $f_{max} \sim 100 \text{ GHz}$) have been reported recently [53]. The measured gain of 16 dB at 18 GHz was observed [53].

Application of PBTs to logic circuits has also been proposed [15]. The calculated switching time delay for a GaAs PBT logic gate is shown in Fig. 11-3-9 [15]. Currently fabricated devices already correspond to a minimum time delay of 4.3 ps [15].

A self-aligned dual grating GaAs PBT was reported in Ref. 55. The device structure is illustrated by Fig. 11-3-10, where the fabrication sequence is shown. The advantage of this transistor is a possiblility of four-terminal device operation and the potential for both horizontal and vertical integration.

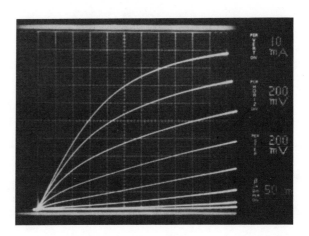

FIGURE 11-3-7. The collector characteristic of a PBT with the top curve having the most positive V_{BE} of +0.5 V [14].

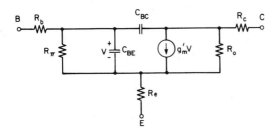

R_c = 2.88 Ω

R_b = 28 Ω

R_π = 215 Ω

R_e = 3.65 Ω

R_o = 235 Ω

g_m = 0.159 mhos

C_{BC} = 0.299 pF

C_{BE} = 0.372 pF

$g_m' = g_m e^{j\pi f\tau} \dfrac{\sin \pi f\tau}{\pi f\tau}$

τ = 4.2 × 10^{-12} sec

V_{BE} = 0.0 V

V_{CE} = 2.0 V

I_c = 37.0 mA

ACTIVE AREA = 3.2 × 10^{-6} cm^2

FINGER LENGTH = 8.0 × 10^{-4} cm

FIGURE 11-3-8. The equivalent circuit of the PBT and element values derived from measured data [14]. V_{CE} = 2.0 V; V_{BE} = 0.0 V.

11-4. VERTICAL GaAs FETs

The first discussion of a vertical FET was given by Shockley [57] and further developed by Zuleeg [58]. A vertical FET called a static injection transistor (SIT) was proposed by Nishizawa *et al.* [59]. As in many other structures such as HBTs, PBTs, and planar doped barrier transistors (PBDTs) the critical device dimension of a vertical transistor is controlled by a deposition process which allows one to achieve a very high frequency of operation. In Refs. 17 and 18 a vertical ballistic transistor (VBT) was proposed. GaAs vertical FETs were fabricated by MIT–Lincoln Labs [60], Westinghouse [61], Cornell [62, 63], Thomson CSF–Cornell [64], Texas Instruments [65], Caltech [66], and other groups.

GaAs vertical FETs have a number of advantages over Si devices such as higher electron velocity and mobility and an ability to utilize heterojunctions as "launching

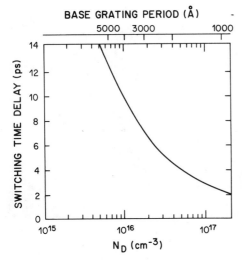

FIGURE 11-3-9. The calculated switching time delay for GaAs PBT logic gates having an enhancement mode switching transistor and an ideal current source as the load. The grating dimensions and carrier concentration are scaled so the transistor remains normally off [15].

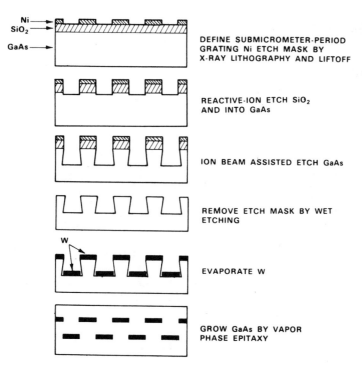

FIGURE 11-3-10. Self-aligned dual-grating GaAs permeable base transistor fabrication sequence [55].

pads" for electrons injected into the active region with high energy (this idea was discussed in Section 11-2 with respect to HBTs). Compared to horizontal channel FETs, vertical FETs, in addition to a shorter effective channel length, may have smaller series resistances, especially if we compare them to non-self-aligned MESFETs where the source resistance is higher because of the surface depletion.

Schematic diagrams of GaAs and AlGaAs/GaAs vertical FETs are shown in Fig. 11-4-1 [66]. In this design a thin p^+ region between the drain and channel is

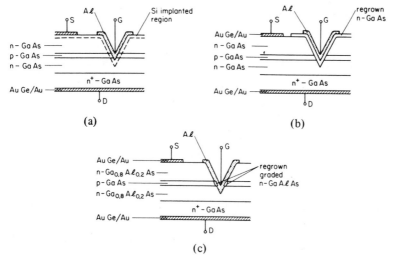

FIGURE 11-4-1. GaAs and GaAs/AlGaAs vertical FETs [66]. (a) Ion implanted vertical GaAs MESFET. (b) Vertical GaAs MESFET with a regrown n-type GaAs. (c) Heterojunction vertical FET.

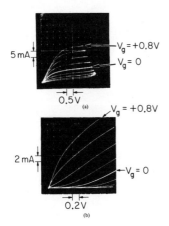

FIGURE 11-4-2. Drain current–voltage characteristics of ion-implanted vertical GaAs MESFETs [66]. (a) LPE grown layers. Gate width of 100 μm. (b) MBE grown layers. Gate width of 120 μm.

used in order to reduce the short channel effects caused by the extension of the depletion region from the drain into the channel. A lightly doped n region between the p^+ region and the n^+ drain is used to increase the device breakdown voltage.

Ion implanted GaAs vertical FETs shown in Fig. 11-4-1 [66] were fabricated using epitaxial layers grown by LPE and MBE on n^+ GaAs substrate using the following layer thicknesses and doping levels: 3 μm n (1.5×10^{17} cm^{-3}), 0.15 μm p $1 - 2 \times 10^{17}$ cm^{-3}), 1.5 μm n (3×10^{15} cm^{-3}). Grooves of 2.5 μm depth were etched with 1:8:8 H_2SO_4:H_2O_2:H_2O. The dose of 6×10^{12} cm^{-2} Si atoms was implanted at 120 keV. Following annealing AuGe/Au ohmic contacts were deposited for the drain and source and Al gates were defined. Schottky gates had a breakdown voltage of -4 V. The gate-to-source series resistance was $\approx 1 \, \Omega$ mm. The current–voltage characteristics for two ion-implanted vertical GaAs FETs are shown in Fig. 11-4-2 [66]. These characteristics show that at low gate voltages the space charge limited current is important. A typical device transconductance was 250 mS/mm gate with highest transconductance of 280 mS/mm gate achieved.

An alternative MBE grown structure shown in Fig. 11-4-1c has a larger barrier height because of the higher energy band of AlGaAs and allows one a larger flexibility in optimizing the doping profiles. A transconductance of 280 mS/mm was also achieved for this structure [66].

The results of microwave measurements of GaAs vertical transistors were reported in Ref. 65 (see Fig. 11-4-3). The frequency response was limited by parasitics. A maximum oscillation frequency f_{max} of 12 GHz was obtained.

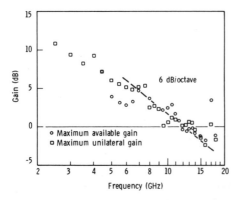

FIGURE 11-4-3. Gain as a function of frequency for GaAs vertical FETs [65].

Further study of heterostructure vertical FETs must be undertaken in order to realize the vertical ballistic FET concept [17, 18]. Such a device should be able to achieve performance similar to a PBT.

11-5. PLANAR DOPED BARRIER DEVICES

A planar doped barrier (PDB) device is another example of a structure which may make it possible to take advantage of high electron velocity in submicron layers [19–23, 67–80]. Different devices utilizing planar doped barrier structures include high-speed logic switches [68], mixers [69], fast photodiodes [70], BARITT diodes [71], and thermionic emission transistors [72].

The structure uses a plane of a p-type dopant positioned within an undoped region sandwiched between two n^+ regions (see Fig. 11-5-1). The p-region doping is such that it is fully depleted. Two narrow positively charged depletion regions in the n^+ layer also exist, maintaining the charge neutrality. As a result of this charge distribution the field is constant between the n^+ layers and the p plane with the field discontinuity induced by the positive charge of the p plane:

$$-F_1 d_1 = F_2 d_2 \qquad (11\text{-}5\text{-}1)$$

$$F_1 - F_2 = Q_p/\varepsilon \qquad (11\text{-}5\text{-}2)$$

leading to a triangular potential barrier (see Fig. 11-5-1).

The zero bias barrier height can be found from Eqs. (11-5-1) and (11-5-2):

$$\phi_{b0} = \frac{-d_1 d_2 Q_p}{(d_1 + d_2)\varepsilon} \qquad (11\text{-}5\text{-}3)$$

where Q_p is the total charge per unit area in the p plane. Thus the barrier height can be modified by changing the doping in the p plane (Q_p) or the device geometry (d_1 and d_2). In a sense, the basic idea of this device has some similarity with attempts to change the effective barrier height of Schottky barriers by implanting n and p regions near the surface [81].

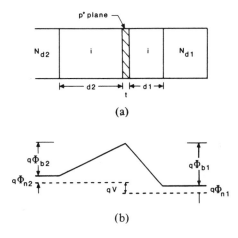

FIGURE 11-5-1. Planar doped barrier diode [73]. (a) Schematic structure. (b) Conduction band edge diagram under applied forward bias.

644   11. NOVEL GaAs DEVICES

Under the applied bias the barrier becomes asymmetrical (see Fig. 11-5-1b). The voltage drop V is related to the electric fields F_1 and F_2 as follows:

$$F_1 d_1 + F_2 d_2 = -V \tag{11-5-4}$$

with Eq. (11-5-2) still valid. From Eqs. (11-5-2) and (11-5-4) we find

$$\phi_{b1} = \phi_{b0} + \frac{d_1}{d_1 + d_2} V \tag{11-5-5}$$

$$\phi_{b2} = \phi_{b0} - \frac{d_2}{d_1 + d_2} V \tag{11-5-6}$$

If the distances d_1 and d_2 are such that the voltage drop across d_1 and d_2 is much larger than $k_B T/q$ the current over the barrier in both directions will be determined by the thermionic emission. A crude estimate of the current may be obtained using Eqs. (11-5-5) and (11-5-6) [19]:

$$J = A^* T^2 \exp(-q\phi_{b0}/k_B T)[\exp(\alpha_2 V) - \exp(\alpha_1 V)] \tag{11-5-7}$$

where J is the current density, A^* is the Richardson constant, T is the lattice temperature, and

$$\alpha_1 = \frac{-qd_1}{(d_1 + d_2)k_B T}, \qquad \alpha_2 = \frac{qd_2}{(d_1 + d_2)k_B T} \tag{11-5-8}$$

The capacitance of the PDB device should be approximately constant and given by

$$C = \frac{\varepsilon S}{d_1 + d_2} \tag{11-5-9}$$

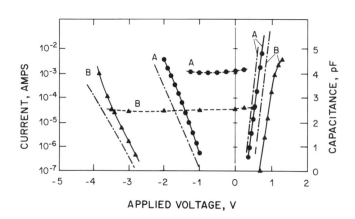

FIGURE 11-5-2. Current- and capacitance-voltage characteristics of a planar doped barrier diode [19]. Device A: $d_1 = 500$ Å; $d_2 = 2000$ Å; $N_p = 10^{12}$ cm^{-2}. Device B: $d_1 = 250$ Å; $d_2 = 2000$ Å; $N_p = 2 \times 10^{12}$ cm^{-2}. — current (measured); – · – current [Eq. (11-5-4)]; --- capacitance (measured).

where S is the device cross section. As can be seen from Fig. 11-5-2 these predictions are in agreement with the experimental results [19].

A planar doped barrier diode could be used as a mixer [69] with the advantage that the I–V characteristic could be made symmetrical. It could also be used as a switch [68] and as a photodiode [70]. An opportunity to tailor the barrier height provided by the structure may be used for launching fast ballistic or near-ballistic electrons into a submicron structure [21].

A more realistic model describing the PDB structure should include the effect of the mobile carrier distribution on the barrier height (see Fig. 11-5-3). The theory of PDB diodes which takes into account the electron injection was developed in Ref. 23. It is based on the assumption that the charge injected into the i regions is located close to the n–i interfaces where the electric field F is still small so that the electron drift velocity

$$v = \mu F \qquad (11\text{-}5\text{-}10)$$

where μ is the low field mobility. In higher electric fields (further from the interfaces with n^+ regions) the electrons move with the saturation velocity but their density is quite small and the free carrier charge in the high field region may be neglected. The current density J is given by

$$J = qn\mu F + \mu k_B T \partial n/\partial x \qquad (11\text{-}5\text{-}11)$$

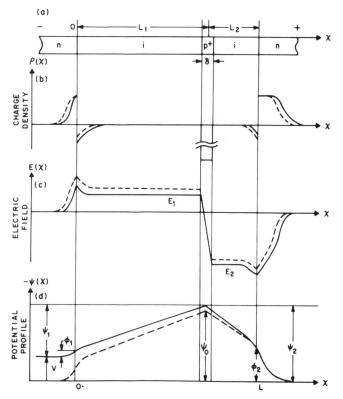

FIGURE 11-5-3. Charge density, electric field, and potential distributions in the PDB structure [23]. --- thermal equilibrium; — applied bias.

The electric field distribution may be found from the solution of the Poisson equation which (in the i-region) is given by

$$\frac{\partial F}{\partial x} = -qn/\varepsilon \qquad (11\text{-}5\text{-}12)$$

From (11-5-11) and (11-5-12) we obtain the following equation:

$$J = -\mu\varepsilon\frac{\partial}{\partial x}\left(\frac{1}{2}F^2 + \frac{k_B T}{q}\frac{\partial F}{\partial x}\right) \qquad (11\text{-}5\text{-}13)$$

When integrated over x Eq. (11-5-13) becomes a nonlinear first-order differential equation of Riccati type and its solution is

$$F(x) = F_1\left[\coth\left(\frac{qF_1 x}{k_B T} + C_1\right) - \frac{J}{J_1}\frac{qF_1 x}{k_B T}\right] \qquad (11\text{-}5\text{-}14)$$

where $J_1 = q\varepsilon\mu F_1^3/k_B T$, F_1 is the electric field at the position of the p plane in region 1 (see Fig. 11-5-4), and C_1 is the integration constant. The solution given by Eq. (11-5-14) is valid for region 1; the solution for region 2 is obtained by changing the subscripts from 1 to 2 and changing x to $x - d_1 - d_2$. Terms of the order of $(J/J_1)^2$ and $(J/J_2)^2$ have been neglected in Eq. (11-5-14) because for a typical PDB device

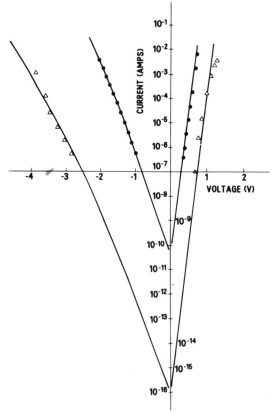

FIGURE 11-5-4. Theoretical and experimental log I vs. V characteristics for two diodes A and B (same devices as in Fig. 11-5-2) [23]. Experimental data [19] are shown by $\triangle$ for diode A and by $\bigcirc$ for B. Solid lines represent the results of the calculation.

$J \ll J_1, J_2$. Equations relating the constants of integration C_1 and C_2 and fields F_1 and F_2 are obtained from the requirement of field and concentration continuity at $x = 0$ and $x = d_1 + d_2$. For sufficiently steep slopes when

$$F_1 > (4k_B T N_d / \varepsilon)^{1/2} \tag{11-5-15}$$

and

$$F_2 > (4k_B T N_d / \varepsilon)^{1/2} \tag{11-5-16}$$

the expression for the I-V characteristic could be reduced to a simple form

$$J = A^* T^2 \exp\{q[\phi_{b0} - qV^2/(qN_d(d_1 + d_2)^2)]/k_B T\}$$
$$\times [\exp(qVL_1/k_B T) - \exp(-qVL_2/k_B T)] \tag{11-5-17}$$

where

$$L_{1,2} = \frac{d_{1,2}}{d_1 + d_2} + \frac{2Q_p}{q(d_1 + d_2)N_d}\left(1 - 2\frac{d_{1,2}}{d_1 + d_2}\right) \tag{11-5-18}$$

The first term in Eq. (11-5-18) corresponds to the simple model which has been discussed above. The second term describes the influence of the space charge layers near the n^+-i interfaces.

The calculations performed using this model are compared with the experimental results [19] in Fig. 11-5-4 [23]. As can be seen from the figure, the agreement with experiment is quite good. It should be pointed out, however, that the doping density of $N_D = 7 \times 10^{18}$ cm^{-3} was used in the calculation, which is two times smaller than assumed in Ref. 19. The calculation presented in Ref. 23 corroborates the conclusion reached by Malik *et al.* that the thermionic emission over the barrier determines the current density in the PDB structure.

The barrier height in planar doped barrier devices may be affected by the electron spill-over (see Section 2-9). As shown in Ref. 83 the spill-over may considerably increase the barrier height compared to the simple geometric model. Using two planar doped barrier devices back-to-back one can build a planar doped

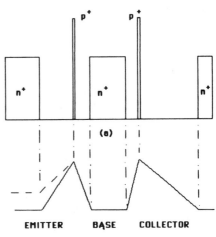

FIGURE 11-5-5. Planar doped barrier transistor [21]. (a) Doping profile; (b) band diagram. Solid line, thermal equilibrium; dashed line, forward emitter base bias.

transistor [21]. The doping profile for such a structure is shown in Fig. 11-5-5. As in a wide-gap emitter HBT the hot electrons are injected into the base with very high velocity (up to 6×10^5 m/s), which they may preserve at distances up to 0.3 μm. As a result the cutoff frequency of the structure would be limited by the RC constants of the emitter and collector circuits. Simple estimates show that cutoff frequencies as high as 100 GHz may be achieved.

The quantum mechanical reflection of electrons has to be minimized for efficient operation of a planar doped barrier transistor. This problem was analyzed in Ref. 82, where it was shown the transmission coefficient is relatively insensitive to the lengths of the barrier arms and depends primarily on the slopes of these arms near the barrier peak. With a careful design transmission coefficients as high as 99% may be achieved [82].

REFERENCES

1. W. Shockley, U.S. Patent No. 2,569,347, issued 1951.
2. H. Kroemer, Theory of a wide-gap emitter for transistors, *Proc. IRE* **45**, 1535-1537 (1957).
3. W. P. Dumke, J. M. Woodall, and V. L. Rideout, GaAs-GaAlAs heterojunction transistor for high frequency operation, *Solid State Electron.* **15**, 1339-1343 (1972).
4. M. Konagai, K. Katsukawa, and K. Takahashi, (GaAl)As/GaAs heterojunction photoresistors with high current gain, *J. Appl. Phys.* **48**, 4389-4394 (1977).
5. J. P. Bailbe, A. Marty, P. H. Hicp, and G. E. Rey, Design and fabrication of high-speed GaAlAs/GaAs heterojunction transistors, *IEEE Trans. Electron Devices* **ED-27**(6),1160-1164 (1980).
6. H. Kroemer, Heterostructure bipolar transistors and integrated circuits, *Proc. IEEE* **70**, 13-25 (1982).
7. H. Kroemer, Heterostructure bipolar transistors: What should we build? *J. Vac. Sci. Technol.* **B 1**(2), 126-130 (1983).
8. D. Ankri, A. Scavennec, C. Besombes, C. Courbet, F. Heliot, and J. Riou, Diffused epitaxial GaAlAs-GaAs heterojunction bipolar transistor for high frequency operation, *Appl. Phys. Lett.* **40**, 816 (1982).
9. S. L. Su, R. Fischer, W. G. Lyons, O. Tejayedi, D. Arnold, J. Klem, and H. Morkoc, Double heterojunction GaAs/$Al_x Ga_{1-x}$As bipolar transistors prepared by molecular beam epitaxy, *J. Appl. Phys.* **54**, 6725-6731 (1983).
10. A. A. Grinberg, M. S. Shur, R. J. Fisher, and H. Morkoc, Investigation of the effect of graded layers and tunneling on the performance of AlGaAs/GaAs heterojunction bipolar transistors, *IEEE Trans. Electron. Devices* **ED-31**(12), 1758-1765 (1984).
11. C. O. Bozler, G. D. Alley, R. A. Murphy, D. C. Flanders, and W. T. Lindley, Permeable base transistor, in Proc. 7th Bien. Cornell Conf. on Active Microwave Semiconductor Devices, Aug. 1979.
12. G. D. Alley, C. O. Bozler, R. A. Murphy, and W. T. Lindley, Two dimensional numerical simulation of the permeable base transistor, in Proc. 7th Bien. Cornell Conf. on Active Microwave Semiconductor Devices, Aug. 1979.
13. C. O. Bozler, G. D. Alley, R. A. Murphy, D. C. Flanders, and W. T. Lindley, Fabrication and microwave performance of the permeable base transistor, IEEE Int. Electron Devices Meet. Tech. Dig., pp. 384-387, Dec. 1979.
14. C. O. Bozler and G. D. Alley, Fabrication and numerical simulation of the permeable base transistor, *IEEE Trans. Electon Devices* **ED-27**(6), 1128-1141 (1980).
15. C. O. Bozler and G. D. Alley, The permeable base transistor and its application to logic circuits, *Proc. IEEE* **70**(1), 46-52 1982.
16. U. Mishra, E. Kohn, N. J. Kaweni, and L. F. Eastman, Permeable base transistor—A new technology, in Abstracts of 40th Annual Device Research Conference, June 1982, Colorado State University, Fort Collins, P: VI B-3.
17. L. F. Eastman, R. Stall, D. Woodard, N. Dandekar, C. Wood, M. S. Shur, and K. Board, Ballistic electron motion in GaAs at room temperature, *Electron Lett.* **16**, 524-525 (1980).
18. L. F. Eastman, Very high electron velocity in short gallium arsenide structures, in *Adv. Solid State Physics*, Vol. 12, Ed. by J. Treush, Vieweg, Braunschweig, 1982, p. 173.

19. R. J. Malik, T. R. AuCoin, R. L. Ross, K. Board, C. E. C. Wood, and L. F. Eastman, Planar doped barriers in GaAs by molecular beam epitaxy, *Electron Lett.* **16**, 836–837 (1980).
20. R. J. Malik, K. Board, L. F. Eastman, C. E. C. Wood, T. R. AuCoin, R. L. Ross, and R. O. Savage, GaAs planar doped barriers by molecular beam epitaxy, in Proceedings of IEDM, Washington, DC, 1980, pp. 456–459.
21. R. J. Malik, M. A. Hollis, L. F. Eastman, D. W. Woodard, C. E. C. Wood, and T. R. AuCoin, GaAs planar-doped barrier transistors grown by moleclar beam epitaxy, in Proceedings of Eight Biennial Cornell Electrical Engineering Conference, August, 1981, Ithaca, New York, pp. 87–96.
22. K. Board, K. Singer, R. Malik, C. E. C. Wood, and L. F. Eastman, A planar doped barrier switching device, in Proceedings of Eighth Biennial Cornell Electrical Engineering Conference, August 1981, Ithaca, New York, pp. 115–124.
23. R. F. Kararinov and S. Lurji, Charge injection over triangular barriers in unipolar semiconductor structures, *Appl. Phys Lett.* **38**, 810 (1981).
24. L. F. Eastman, Ballistic Transistors, presented at WOCSEMMAD 81.
25. A. Kastalsky and S. Luryi, *IEEE Electron Device Lett.* **EDL-4**, 334 (1983).
26. S. Luryi, A. Katalsky, A. C. Gossard, and R. Hendel, *IEEE Trans. Electron Devices* **ED-31**, 832 (1984).
27. A. Katalsky, S. Luryi, A. C. Gossard, and R. Hendel, *IEEE Electron Device Lett.* **EDL-5**, 57 (1984).
28. S. Luryi, A. Kastalsky, A. C. Gossard, and R. Handel, *Appl. Phys. Lett.* **45**, 1294 (1984).
29. S. Luryi and A. Kastalsky, Hot electron injection devices, presented at International Conference on Superlattices, Microstructures and Microdevices, Champaign, Illinois.
30. K. Hess, H. Morkoc, H. Shichijo, and B. G. Streetman, *Appl. Phys. Lett.* **35**, 469 (1979); also K. Hess, *Physica* **117B**, 723 (1983) and references therein.
31. R. Eden and B. M. Welch, Ultra high speed GaAs VLSI: Approaches, potential, and progress, in *VLSI Electronics: Microstructure Science*, Vol. 3, Ed. by N. Einspruch, Academic, New York, 1981.
32. R. L. Anderson, Expeiments on GeGaAs heterojunctions, *Solid-State Electron.* **5**, 341–351 (1962).
33. S. S. Perlman and D. L. Feucht, *P–N* heterojunctions, *Solid-State Electron.* **7**, 911–923 (1964).
34. D. Ankri and L. F. Eastman, GaAlAs–GaAs ballistic heterojunction bipolar transistor, *Electron Lett.* **18**, 750 (1982).
35. R. Stratton, Theory of field emission from semiconductors, *Phys. Rev.* **125**, 67–82 (1969).
36. F. A. Padovani and R. Stratton, Field and thermionic-field emission in Schottky barriers, *Solid-State Electron.* **9**, 695–707 (1966).
37. J. R. Hayes, F. Capasso, R. J. Malik, A. C. Gossard, and W. Weigman, Optimum emitter grading for heterojunction bipolar transistors, *Appl. Phys. Lett.* **43**, 949–951 (1984).
38. H. C. Casey and M. B. Panish, *Heterostructure Lasers, Part A: Fundamental Principles*, Academic, New York, 1978.
39. R. M. Warner and J. N. Fordemwalt, *Integrated Circuits*, McGraw-Hill, New York, 1965, pp. 108–109.
40. E. S. Yang, *Fundamentals of Semiconductor Devices*, McGraw-Hill, New York, 1978.
41. F. A. Lindholm, S. W. Director, and D. L. Bowler, *IEEE J. Solid State Circuits* **SC-6**, 213–222 (1971).
42. G. Ray and J. P. Bailbe, *Solid State Electron.* **17**, 1045–1057 (1974).
43. J. S. Slotboom and H. C. de Graaf, Measurement of bandgap narrowing in Si bipolar transistors, *Solid-State Electron.* **19**(10), 857–862 (1976).
44. H. Ito, T. Ishibashi, and T. Sugeta, High-frequency characteristics of AlGaAs/GaAs heterojunction bipolar transistors, *IEEE Electron Device Lett.* **EDL-5**(6), 214–216 (1984).
45. D. Ankri, W. Schaff, P. Smith, C. E. C. Wood, and L. F. Eastman, Enhancement of the electron velocity in GaAlAs–GaAs heterojunction bipolar transistor with abrupt emitter–base interface, in IEDM Technical Digest, 1982, p. 788.
46. D. Ankri, W. J. Schaff, P. Smith, and L. F. Eastman, High-speed AlGaAs–GaAs heterojunction bipolar transistors with near-ballistic operation, *Electron. Lett.* **19**, 147 (1983).
47. P. Solomon, Semiconductor devices for high speed logic, *Proc. IEEE* **70**(5), 489–509 (1982).
48. M. Asbeck, D. L. Miller, R. J. Anderson, and F. H. Eigen, Emitter-coupled logic circuits implemented with heterojunctin bipolar transistors, in Technical Digest 1983 GaAs IC Symposium, p. 170.
49. M. Asbeck, D. Miller, R. Anderson, R. Deminy, R. Chen, C. Liehti, and F. Eisen, Application of heterojunction bipolar transistors to high-speed small scale digital ICs, 1984 GaAs IC Symposium Technical Digest, pp. 133–136, October 1984.
50. M. Asbeck, D. L. Miller, R. J. Anderson, L. D. Hou, R. Deminy, and F. Eisen, Nonthreshold Logic Ring Oscillators Implemented With GaAs/GaAlAs Heterojunction Bipolar Transistor, *IEEE Electron Device Lett.* **EDL-5**(5), 181–183 (1984).
51. H. T. Yuan, GaAs bipolar gate array technology, 1982 GaAs Symposium Technical Digest, p. 100, New Orleans, published by IEEE (1982).

52. M. V. McLevige, H. T. Yuan, W. M. Duncan, W. R. Frensley, F. H. Doerbeck, H. Morkoc, and T. J. Drummond, GaAs/AlGaAs heterojunction transistor for integrated circuit applications, *IEEE Electron Device Lett.* **EDL-3**, 43 (1982).

53. G. D. Alley, C. O. Bozler, N. P. Economou, D. C. Flanders, M. W. Geis, G. A. Lincoln, W. T. Lindley, R. W. McClelland, R. A. Murphy, K. B. Nichols, W. J. Piacentini, S. Rabe, J. P. Salerno, and B. A. Vojak, Multimeter-wavelength GaAs permeable base transistors, presented at the 1982 Device Research Conference, Ft. Collins, Colorado.

54. T. W. Tang, L. Sha, and D. H. Navon, Improved high frequency performance of the permeble base transistor, in Proceedings of IEEE/Cornell Conference on High Speed Semiconductor Devices and Circuits, Cornell University, Ithaca, New York, 1983, pp. 250–259.

55. B. A. Vojak, R. W. McClelland, G. A. Lincoln, A. R. Calawa, D. C. Flanders, and M. W. Geis, A self-aligned dual-grating GaAs permeable base transistor, *IEEE Electron Device Lett.* **EDL-5**(7), 270–272 (1984).

56. J. P. Donnelly, C. O. Bozler, and R. A. Murphy, Proton bombardment for making GaAs devices, *Circuits Manuf.* **18**, 45–58 (1978).

57. W. Shockley, *Proc. IRE* **40**, 1289–1313 (1952).

58. R. Zuleeg, *Solid State Electron.* **10**, 449–460 (1967).

59. Jun-Ichi Nishizawa, Takeshi Terasaki, and Kiro Shibata, *IEEE Trans. Electron Devices*, **ED-22**(4) 185–197 (1975).

60. C. O. Bozler, G. D. Alley, R. A. Murphy, C. D. Flanders, and W. T. Lindley, in Proc. Seventh Biennial Cornell Conference on Active Microwave Devices and Circuits, 1979.

61. R. C. Clarke, H. C. Nathanson, J. G. Oakes, and G. T. Hardison, presented at Device Research Conference, June 1982.

62. U. Mishra, E. Kohn, N. J. Kawai, and L. F. Eastman, presented at Device Research Conference, June 1982.

63. U. Mishra, E. Kohn, and L. F. Eastman, Submicron GaAs vertical electron transistor, in 1982 IEDM Technical Digest, paper 25.5, pp. 594–597, 1982.

64. E. Kohn, J. Magarshack, U. Mishra, and L. F. Eastman, Novel vertical GaAs FET structure with submicrometre source-to-drain spacing, *Electron. Lett.* **19**(24), 1021–1023 (1983).

65. W. R. Frensley, B. Bayraktaroglu, S. E. Campbell, H. D. Shih, R. E. Lehmann, and R. E. Williams, Microwave operation of a gallium arsenide vertical MESFET, in Proc. IEEE/Cornell Conference on High-Speed Semiconductor Devices & Circuits, Ithaca, New York, 1983.

66. Z. Rav-Noy, U. Schreter, S. Mukai, E. Kapon, J. S. Smith, L. C. Chiu, S. Margalit, and A. Yariv, Vertical FET's in GaAs, *IEEE Electron Device Lett.* **EDL-5**(7), 228–230 (1984).

67. J. M. Woodcock and J. J. Harris, Bulk unipolar diodes in MBE GaAs, *Electron Lett.* **19**, 181–183 (1983).

68. C. E. C. Wood, L. F. Eastman, K. Board, K. Singer, and R. J. Malik, Regenerative switching device using MBE grown gallium arsenide, *Electron. Lett.* **18**, 676–677 (1982).

69. R. J. Malik and S. Dixon, A subharmonic mixer using a planar doped barrier diode with symmetric conductance, *IEEE Electron Device Lett.* **EDL-3**, 305 (1982).

70. C. Y. Chen, A. Y. Cho, P. A. Gabrinski, and C. G. Bethea, An ultrahigh speed modulated barrier photodiode made on *p*-type gallium arsenide substrates, *IEEE Electron Device Lett.* **2**, 290–291 (1981).

71. S. Luryi and R. F. Kazarinov, Optimum BARITT structure, *Solid-State Electron.* **25**, 943–945 (1982).

72. R. F. Kazarinov and S. Luryi, Majority carrier transistor based on voltage-controlled thermionic emission, *Appl. Phys. A* **28**, 151–160 (1982).

73. D. C. Streit and F. G. Allen, Silicon triangular barrier diodes by MBE using solid-phase epitaxial regrowth, *IEEE Electron Device Lett.* **EDL-5**(7), 254–255 (1984).

74. M. A. Hollis, Fabrication and performance of GaAs planar-doped barrier transistors, Ph.D. dissertation, Cornell University, Ithaca, New York, May 1983.

75. A. Chandra and L. F. Eastman, Quantum mechanical reflection at triangular "planar doped" potential barriers for transistors, *J. Appl. Phys.* **53**, 9165 (1982).

76. S. C. Palmateer, P. A. Maki, M. A. Hollis, and L. F. Eastman, A study of substrate effects on planar doped barrier structures in GaAs grown by molecular beam epitaxy, in Proc. 1982 Int. GaAs Symp. (Albuquerque, New Mexico), 1982, pp. 149–156.

77. F. Buot, J. Krumhansl, and J. Socha, The onset of diffusion-drift emission regime and the transition from exponential to linear current–voltage characteristic of triangular barrier semiconductor structures, *Appl. Phys. Lett.* **40**, 814 (1982).

78. M. A. Littlejohn, R. J. Trew, J. R. Hauser, and J. M. Golio, Electron transport in planar-doped barrier structures using an ensemble Monte Carlo method, *J. Vac. Sci. Technol. B* **1**, 449 (1983).

79. M. A. Hollis, S. C. Palmateer, L. F. Eastman, N. V. Dandekar, and P. M. Smith, Importance of electron scattering with coupled plasmon–optical phonon modes in GaAs planar-doped barrier transistors, *IEEE Electron Device Lett.* **EDL-4**(12), 439–443 (1983).

80. S. E.-D. Habib and K. Board, Theory of triangular-barrier bulk unipolar diodes including minority-carrier effects, *IEEE Trans. Electron Devices* **ED-30**, 90–96 (1983).

81. S. M. Sze, *Physics of Semiconductor Devices*, Wiley, New York, 1981, p. 294.

82. A. Chandra and L. F. Eastman, Quantum mechanical reflection at triangular "planar-doped" potential barriers for transistors, *J. Appl. Phys.* **53**(12), 9165–9169 (1982).

83. M. Shur, Spill-over effects in planar doped barrier diodes, *Appl. Phys. Lett.* **47**(8), 869–871 (1985).

New Developments and Recent References

GaAs technology is a rapidly developing area. Thousands of relevant papers are published every year, and it may be impossible to compile a complete and up-to-date list of references. Nevertheless, I would like to mention important review articles and books more recent than the references given in the book.

In addition to an excellent review paper by Blakemore [1] quoted extensively in this book, important review papers on material parameters of GaAs, AlAs, and $Al_xGa_{1-x}As$ [2] and on electronic properties of semiconductor alloy systems [3] appeared recently.

We should mention recent books on molecular beam epitaxy technology [4]; silicon MBE technology [5]; GaAs technology [6]; GaAs materials, devices, and circuits [7]; GaAs processing [8]; microwave mixers [9]; and microwave solid state devices [10], and a review paper on modulation doped field effect transistors [11] and novel devices [12].

Important recent developments in compound semiconductor technology include chemical beam epitaxy [13] and GaAs growth on silicon substrates [14].

New heterojunction cathode transferred electron devices have been developed [15, 16] and demonstrated more than two times higher efficiency and output power than comparable conventional Gunn oscillators.

Novel compound semiconductor devices, such as an induced base transistor [17] and a tunneling emitter bipolar transistor [18], have been proposed.

Resonant tunneling devices pioneered by R. Tsu and L. Esaki in 1973 [19] have recently received a lot of attention (see, for example, Ref. 20).

The ideas and device concepts developed for compound semiconductors stimulated interest in silicon molecular beam epitaxy [21] and alternative heteroepitaxial systems (see, for example, Ref. 22).

After many years of research and development GaAs integrated circuits have become commercially available. They include integrated circuits from Gigabit Logic, Inc., bit-slice ICs from Vitesse Electronics Corporation [23], and GaAs gate arrays from Honeywell, Inc. GaAs microwave monolithic integrated circuits have also become commercially available [24].

An ambitious program of developing GaAs RISC microproccessors is well underway [25] and a GaAs supercomputer project has been launched by Cray Research, Inc.

The prospects for GaAs technology seem now to be better than ever.

REFERENCES

1. J. S. Blakemore, Semiconducting and other major properties of gallium arsenide, *J. Appl. Phys.* **53**(10), R123-181, (1982).

2. S. Adachi, GaAs, AlAs and $Al_xGa_{1-x}As$: Material parameters for use in research and device applications, *J. Appl. Phys.* **58**(3), R1-R-29, (1985).

3. M. Jaros, Electronic properties of semiconductor alloy systems, *Rep. Prog. Phys.* **48**, 1091-1154 (1985).

4. *Molecular Beam Technology and Heterostructures*, Ed. By Leroy L. Chang, and Klaus Ploog, Martinus Nijoff Publishers, Dordrecht, Boston/Lancaster, 1985.

5. *Silicon Molecular Beam Epitaxy*, Ed. by E. Kasper and J. C. Bean, CRC Uniscience Series, CRC Press, to be published.

6. *Gallium Arsenide Technology*, Ed. by D. K. Ferry, Howard W. Sams & Co., Inc., 1985.

7. *Gallium Arsenide Materials, Devices and Circuits*, Ed. by M. J. Howes and D. V. Morgan, John Wiley and Sons, Chichester, 1985.

8. R. E. Williams, *Gallium Arsenide Processing and Techniques*, Artech House, Inc., Dedham, MA, 1984.

9. S. A. Maas, *Microwave Mixers*, Artech House, Inc., Dedham, MA, 1986.

10. S. Y. Liao, *Microwave Solid-State Devices*, Prentice-Hall, Inc., Englewood Cliffs, New Jersey, 1985.

11. T. J. Drummond, W. T. Masselink, and H. Morkoç, Modulation Doped GaAs/(Al/Ga)As Heterojunction Field Effect Transistors: MODFETs, *Proc. IEEE* **74**(6), 773-822 (1986).

12. K. Board, New unorthodox semiconductor devices, *Rep. Prog. Phys.* **48**, 1595-1635 (1985).

13. W. T. Tsang, Chemical beam epitaxy of III-V compounds, in Extended Abstracts of the 18th (1986 International) Conference on Solid State Material and Devices, Tokyo, 1986, pp. 611-617.

14. R. Fisher and H. Morkoç, III-V semiconductors on Si substrates: New directions for heterojunction electronics, *Solid State Electronics* **29**, 269-271 (1986).

15. M. R. Friscourt, P. A. Rolland, and M. Pernisek, Heterojunction cathode transferred electron oscillators, *IEEE Electron Dev. Lett.* **EDL-6**, 497-499 (1985).

16. Z. Greenwald, The effect of a high energy electron injection cathode on the performance of the Gunn oscillator, Ph.D. thesis, Cornell University, Ithaca, NY, 1986.

17. S. Luryi, An induced base hot-electron transistor, *IEEE Electron Dev. Lett.* **EDL-6**, 178 (1985).

18. J. Xu and M. Shur, A tunneling emitter bipolar transistor, *IEEE Electron Dev. Lett.* **EDL-7**, 416-418 (1986).

19. R. Tsu and L. Esaki, *Appl. Phys. Lett.* **22**, 562 (1973).

20. N. Yokoyama, Resonant Tunneling Hot Electron Transistors (RHET): Potential and applications, in Extended Abstracts of the 18th (1986 International) Conference on Solid State Material and Devices, Tokyo, 1986, pp. 347-350.

21. S. Luryi and S. M. Sze, Possible device applications of silicon molecular beam epitaxy, in *Silicon Molecular Beam Epitaxy*, Ed. by E. Kasper and J. C. Bean, CRC Uniscience Series, CRC Press, to be published.

22. Y. Condo, T. Takahashi, K. Ishii, Y. Hayashi, E. Sakuma, S. Misawa, H. Daimo, M. Yamanaka, and S. Yoshida, Experimental 3C-SiC MOSFET, *IEEE Electron Dev. Lett.* **EDL-7**, 404-406 (1986).

23. Bit-slice ICs kick off era of commercial GaAs LSI, *Electronics*, Sep. 18, 1986.

24. J. Arnold and D. C. Smith, Commercial availability of GaAs MMICs challenges system designers, *Microwave Systems News* **16**(10), 119 (1986).

25. GaAs microprocessor technology, Ed. by Veljko Milunovic, *Computer* (Special Issue) **19**(10), Oct. 1986.

Appendixes

A

Some Room-Temperature (300 K) Properties of GaAs*

a. Mechanical, thermal, and dielectric properties

Crystal density	$\rho_{300} = 5.317 \text{ g/cm}^3$
Bulk modulus (compressibility^{-1})	$B_s = 7.55 \times 10^{11} \text{ dyn/cm}^2$
Shear modulus	$c' = 3.26 \times 10^{11} \text{ dyn/cm}^2$
Linear expansion coefficient	$\alpha_{300} = 5.73 \times 10^{-6} \text{ K}^{-1}$
Volume expansion coefficient	$3\alpha = \gamma = 1.72 \times 10^{-5} \text{ K}^{-1}$
Specific heat	$C_p = 0.327 \text{ J/g K}$
Effective Debye temperature	$\theta_{300} = 360 \text{ K}$
Lattice thermal conductivity	$\kappa_L = 0.55 \text{ W/cm K}$
Static dielectric constant	$\kappa_0 = 12.85$
Infrared refractive index	$n_\infty = 3.299$

b. Energy band separations and derivatives

Direct (zone center) intrinsic gap	$\varepsilon_i = 1.423 \text{ eV}$
Pressure derivative	$(\partial\varepsilon_i/\partial P) = +0.0126 \text{ eV/kbar}$
Temperature coefficient	$(\partial\varepsilon_i/\partial T) = -0.000452 \text{ eV/K}$
Direct exciton transition energy	$\varepsilon_{xl} = 1.419 \text{ eV}$
Spin–orbit splitting energy	$\Delta_{a0} = 0.341 \text{ eV}$
L_6 conduction band gap	$\varepsilon_L = 1.707 \text{ eV}$
Pressure derivative	$(\partial\varepsilon_L/\partial P) = -0.0055 \text{ eV/kbar}$
Temperature coefficient	$(\partial\varepsilon_L/\partial T) = -0.000506 \text{ eV/K}$
Energy elevation $(\varepsilon_L - \varepsilon_i)$	$\Delta_{\Gamma L} = 0.284 \text{ eV}$
X_6 conduction band gap	$\varepsilon_x = 1.899 \text{ eV}$
Pressure derivative	$(\partial\varepsilon_x/\partial P) = -0.0015 \text{ eV/kbar}$
Temperature coefficient	$(\partial\varepsilon_x/\partial T) = -0.000385 \text{ eV/K}$
Energy elevation $(\varepsilon_x - \varepsilon_i)$	$\Delta_{\Gamma x} = 0.476 \text{ eV}$
X_7–X_6 band separation	$\approx 0.40 \text{ eV}$

c. Intrinsic properties

Intrinsic carrier pair density	$n_i = 2.25 \times 10^6 \text{ cm}^{-3}$
Intrinsic electrical conductivity	$\sigma_i = 3.0 \times 10^{-9} \, \Omega^{-1} \text{ cm}^{-1}$

---

* This appendix is taken from J. S. Blakemore, *J. Appl. Phys.* **53**(10), R175 (1982).

Minimum conductivity (for $p_0 = bn_0$) $\qquad \sigma_{\min} = 1.15 \times 10^{-9}\,\Omega^{-1}\,cm^{-1}$
Intrinsic Fermi energy $\qquad\qquad\qquad (\psi - \varepsilon_v) = 0.752\ eV$

d. Parameters for lowest conduction band
Band-edge effective mass $\qquad\qquad\qquad m_\infty = 0.0632 m_0$
Conduction electron rms speed $\qquad\qquad v_c(rms) = 4.4 \times 10^7\ cm/s$
(Nondegenerate) effective density of states $\qquad N_c' = 4.21 \times 10^{17}\ cm^{-3}$
Drift mobility (weak doping) $\qquad\qquad\qquad \mu_n = 8000\ cm^2/V\,s$
Hall mobility (weak doping, weak field) $\qquad \mu_{H0} = 9400\ cm^2/V\,s$
Weak-field Hall factor $\qquad\qquad r_{H0} = (\mu_{H0}/\mu_n) = 1.175$

e. Parameters for upper conduction bands
For L_6 conduction band:
Density of states effective mass $\qquad\qquad m_L = 0.55 m_0$
Conductivity mobility $\qquad\qquad\qquad\qquad \mu_L \sim 2500\ cm^2/V\,s$
Fraction of all conduction electrons $\qquad\quad \sim 0.0004$
For X_6 conduction band:
Density of states effective mass $\qquad\qquad m_x = 0.85 m_0$
Conductivity mobility $\qquad\qquad\qquad\qquad \mu_x \sim 300\ cm^2/V\,s$
Fraction of all conduction electrons $\qquad\quad < 10^{-6}$

f. Parameters for valence bands
Heavy-hole density of states effective mass $\qquad m_h = 0.50 m_0$
rms heavy-hole speed $\qquad\qquad\qquad v_h(rms) = 1.65 \times 10^7\ cm/s$
Light-hole density of states effective mass $\qquad m_l = 0.088 m_0$
rms light hole speed $\qquad\qquad\qquad v_l(rms) = 3.4 \times 10^7\ cm/s$
Light holes as fraction of total $\qquad\qquad\qquad 0.069$
Combined heavy/light effective density of states $N_v' = 9.51 \times 10^{18}\ cm^{-3}$
Drift mobility (weak doping) $\qquad\qquad\qquad \mu_p = 320\ cm^2/V\,s$
Hall mobility (weak doping, weak field) $\qquad \mu_{H0} = 400\ cm^2/V\,s$
Weak-field Hall factor $\qquad\qquad r_{H0} = (\mu_{H0}/\mu_p) = 1.25$
Splitoff band effect mass $\qquad\qquad\qquad m_\infty = 0.15 m_0$
Splitoff holes as fraction of total $\qquad\qquad < 10^{-8}$

B

Microwave Bands

The term microwave frequencies is generally used for frequencies from 1 to 300 GHz (corresponding roughly to the wavelengths from 30 cm to 1 mm). There are two widely used microwave band designations adopted by the United States Department of Defense in August 1969 (see Table B-1) and in August 1970 (see Table B-2).

TABLE B-1. U.S. Military Microwave Bands

Designation	Frequency range (GHz)
P Band	0.225–0.390
L Band	0.390–1.550
S Band	1.550–3.900
C Band	3.900–6.200
X Band	6.200–10.900
K Band	10.900–36.000
Q Band	36.000–46.000
V Band	46.000–56.000
W Band	56.000–100.000

TABLE B-2. U.S. Military Microwave Bands

Designation	Frequency range (GHz)	Designation	Frequency range (GHz)
A Band	0.100–0.250	H Band	6.000–8.000
B Band	0.250–0.500	I Band	8.000–10.000
C Band	0.500–1.000	J Band	10.000–20.000
D Band	1.000–2.000	K Band	20.000–40.000
E Band	2.000–3.000	L Band	40.000–60.000
F Band	3.000–4.000	M Band	60.000–100.000
G Band	4.000–6.000		

Index